MOLECULES AND GRAINS IN SPACE

50th INTERNATIONAL MEETING OF PHYSICAL CHEMISTRY

ACKNOWLEDGEMENTS

This meeting has received financial support from the following:
- Centre National d'Etudes Spatiales
- Centre National de la Recherche Scientifique, Secteur Chimie
- C.N.R.S. Groupement de Recherche "Physico-Chimie des Molecules et Grains Interstellaires"
- Commissariat à l'Energie Atomique, Direction des Sciences de la Matière, DRECAM
- Direction des Recherches, Etudes et Techniques
- Ministère de l'Education Nationale
- European Office of Aerospace Research and Development, U.S.D.O.D.

This support is gratefully acknowledged.

AIP CONFERENCE PROCEEDINGS 312

MOLECULES AND GRAINS IN SPACE

50th INTERNATIONAL MEETING OF PHYSICAL CHEMISTRY
MONT SAINTE-ODILE, FRANCE SEPTEMBER 1993

EDITOR:
IRÈNE NENNER
L.U.R.E. ORSAY

ASSOCIATE EDITORS:
**PHILIPPE BRÉCHIGNAC
YVES ELLINGER
SYDNEY LEACH
ALAIN LÉGER
RONALD McCARROL
EVELYNE ROUEFF
CLÉMENT TROYANOWSKY**

American Institute of Physics   New York

Authorization to photocopy items for internal or personal use, beyond the free copying permitted under the 1978 U.S. Copyright Law (see statement below), is granted by the American Institute of Physics for users registered with the Copyright Clearance Center (CCC) Transactional Reporting Service, provided that the base fee of $2.00 per copy is paid directly to CCC, 27 Congress St., Salem, MA 01970. For those organizations that have been granted a photocopy license by CCC, a separate system of payment has been arranged. The fee code for users of the Transactional Reporting Service is: 0094-243X/87 $2.00.

© 1994 American Institute of Physics.

Individual readers of this volume and nonprofit libraries, acting for them, are permitted to make fair use of the material in it, such as copying an article for use in teaching or research. Permission is granted to quote from this volume in scientific work with the customary acknowledgment of the source. To reprint a figure, table, or other excerpt requires the consent of one of the original authors and notification to AIP. Republication or systematic or multiple reproduction of any material in this volume is permitted only under license from AIP. Address inquiries to Series Editor, AIP Conference Proceedings, AIP Press, American Institute of Physics, 500 Sunnyside Boulevard, Woodbury, NY 11797-2999.

L.C. Catalog Card No. 94-72615
ISBN 1-56396-355-8
DOE CONF-9309317

Printed in the United States of America.

CONTENTS

Foreword xi

OBSERVATIONS AND MODELS

Recent Developments in Astrochemistry 3
    D. A. Williams

From Millimeter to Optical Wavelengths

Observations of Vibrationally Excited H_2 in NGC2023 through High Resolution Imaging of the S(1) Line Emission at 2.121 μm 29
    D. Field, M. Gerin, S. Leach, J. L. Lemaire, G. Pineau des Forêts,
    F. Rostas, D. Rouan, and D. Simons

Millimeter Observations of ^{12}CO and ^{13}CO in the Diffuse Region around Z OPH 33
    M. Kopp, M. Gerin, and E. Roueff

HCO^+ and HCN Observations toward Dark Clouds and Herbig Ae/Be Stars 39
    F. Scappini, G. Bruni, G. G. C. Palumbo, and P. Bergman

Molecular Abundances in Translucent Clouds 47
    R. Gredel

The 30 Micron Emission Band in Carbon-Rich Pre-Planetary Nebulae 53
    A. Omont, P. Cox, H. Moseley, W. J. Glaccum, S. Casey, and T. Forveille

A Quantitative Study of the 7.7 and 11.3 μm Emission Bands Based on IRAS/LRS Spectra 59
    A. Zavagno, P. Cox, and J. P. Baluteau

Observation of the Diffuse Galactic Emission at 6.2 μm 65
    I. Ristorcelli, M. Giard, C. Mény, G. Serra, J. M. Lamarre, C. Le Naour,
    J. Léotin, and F. Pajot

Possible Detection of Solid Formaldehyde towards the Embedded Source GL 2136 73
    W. A. Schutte, P. A. Gerakines, E. F. van Dishoeck, J. M. Greenberg,
    and T. R. Geballe

Grain Processing and IRAS Color Variations 81
    S. D. Taylor

Diffuse Interstellar Absorption and Emission Bands 87
    R. E. Hibbins, J. R. Miles, and P. J. Sarre

Optical Observations of Metals in Circumstellar Envelopes 99
    N. Mauron and Ch. Guilain

Infrared and Radio Spectroscopy of Comets: Recent Results and New Problems 107
    J. Crovisier and D. Bockelée-Morvan
Complex Molecules in the Solar System: The Giant Planets and Titan 117
    T. Encrenaz

From Diffuse Clouds to Comets

The Effect of Varying Cosmic Ray Ionization Rates on the Chemistry of Interstellar Clouds 135
    P. R. A. Farquhar, T. J. Millar, and E. Herbst
Neutral-Neutral Reactions in the Chemistry of Interstellar Clouds 141
    E. Herbst, H. H. Lee, D. A. Howe, and T. J. Millar
Radiation Fields Inside Dark Clouds with Variable Extinction Laws 149
    S. Aiello, C. Cecchi-Pestellini, and B. Barsella
Chemistry of Star-Forming Cores 155
    S. B. Charnley
Bistability in Dark Clouds 161
    J. Le Bourlot, G. Pineau des Forêts, and E. Roueff
Atomic Carbon in Interstellar Clouds 167
    J. Le Bourlot, G. Pineau des Forêts, E. Roueff, and D. R. Flower
The Evolution of Dust Particles in Dense Clouds 173
    B. van den Hoek
Iron-Aromatics Chemistry in Interstellar Clouds 183
    P. Marty, G. Serra, B. Chaudret, and I. Ristorcelli
Models of Molecular Processes in the Low Mass Star-Forming Region B335 189
    J. M. C. Rawlings, N. J. Evans, and S. Zhou
The Mid Infrared Continuum in the Planetary Nebula NGC 7027: Evidence for the Presence of Amorphous Carbon Grains 195
    A. Zavagno and J. P. Baluteau
Rotational Excitation of Interstellar Water Vapor by Near Infrared PAH Photons 199
    A. D'Heeger and M. Giard
Physical Chemistry of Comets: Models, Uncertainties, Data Needs 205
    C. Arpigny
Gas-Grain Chemistry in Protosolar Nebula and Composition of Comets 239
    Z. Moravec and V. Vanysek
On a Possible Mechanism for Production of S_2 from Comets 247
    V. Vanysek
Excitation Mechanisms of the C_2 Swan Bands in Comet Halley's Spectrum 255
    P. Rousselot, B. Goidet-Devel, J. Clairemidi, and G. Moreels

Identification of Phenanthrene in Halley's Inner Coma................... 261
  G. Moreels, J. Clairemidi, P. Rousselot, P. Hermine, and P. Brechignac

SMALL MOLECULAR SYSTEMS: STRUCTURE AND PHOTON-INDUCED DYNAMICS

Millimeter and Submillimeter-Wave Spectroscopy of Molecular Ions and Other Transient Species............................ 269
  J.-L. Destombes, M. Bogey, M. Cordonnier, C. Demuynck, and A. Walters

Millimeter Wave Spectroscopy of Stable Molecules of Astrophysical Interest.. 289
  G. Wlodarczak, J. Burie, M. Le Guennec, and J. Demaison

Gas Phase Synthesis of Hetero-sila-alkenes of Cosmochemical Importance... 293
  J.-M. Denis and J.-L. Ripoll

The Laboratory Millimeter-Wave Spectrum of MgCN ($X^2\Sigma^+$).............. 297
  M. A. Anderson, T. C. Steimle, and L. M. Ziurys

Interstellar CH_3D: Deuterated Methane in the Orion Hot Core?........... 305
  M. Womack, L. M. Ziurys, A. J. Apponi, and J. T. Yoder

Metal-Containing Molecules in the Laboratory and in Space............... 311
  L. M. Ziurys, M. A. Anderson, A. J. Apponi, and M. D. Allen

Quantum-Chemical Calculations on Molecules of Astrochemical Interest.. 321
  P. Botschwina, S. Seeger, M. Horn, J. Flügge, M. Oswald, M. Mladenović, U. Höper, R. Oswald, and E. Schick

Physical Chemistry of Silicon Containing Molecules...................... 337
  A. Spielfiedel, N. Feautrier, I. Drira, G. Chambaud, P. Rosmus, and Y. P., Viala

MCSCF-CI Study of the Isomerization Reaction HCSi→HSiC.............. 343
  H. Lavendy, J. M. Robbe, D. Duflot, and J. P. Flament

High Resolution VUV Laser Measurements of the CO A $^1\Pi$ - X $^1\Sigma^+$ (v',0) Absorption Cross Sections.................................... 349
  D. Malmasson, A. Vient, J. L. Lemaire, A. Le Floch, and F. Rostas

A 2+1 REMPI Study of the E-X Transition in CO....................... 355
  J. Baker, J. L. Lemaire, S. Couris, A. Vient, D. Malmasson, and F. Rostas

Modelling of the Rydberg-Valence Predissociating Interactions in the CO Molecule...................................... 361
  W.-Ü L. Tchang-Brillet and P. S. Julienne

The A $^2\Pi_i$–X $^2\Sigma^+$ and B $^2\Sigma^+$–A $^2\Pi_i$ Electronic Transitions of CS^+ by High Resolution Fourier Transform Spectroscopy..................... 367
  D. Cossart, M. Horani, and M. Vervloet

A Laser Vaporization Source for Spectroscopic Characterization
of Refractory Compounds... 373
    J. Chevaleyre, C. Bordas, A. M. Valente, V. Boutou, J. Maurelli,
    B. Erba, and J. d'Incan

Radiative Relaxation of Spin-Orbit States in Xe^+ and Kr^+ and of Vibrational
and Spin-Orbit States in HBr^+ and DBr^+ 381
    M. Heninger, S. Jullien, J. Lemaire, G. Mauclaire, R. Marx, and S. Fenistein

Photodissociation and Rotational Excitation of CO and its Isotopomers
in Interstellar Clouds.. 387
    S. Warin, J. J. Benayoun, and Y. P. Viala

Constants for Astrophysical Gas-Phase Molecules: Photodissociation
Rates, Microwave and Infrared Spectra................................ 393
    J. Crovisier

Photodynamics of Acetylene 395
    J. H. Fillion, N. Shafizadeh, and D. Gauyacq

COLLISIONS AND REACTIVITY

Rate Constant Formulae for Fast Reactions 405
    D. C. Clary

Reactions between Neutral Species without a Significant Barrier:
The C+NO and $C+N_2O$ Reactions 423
    M. Costes, C. Naulin, N. Ghanem, and G. Dorthe

Preliminary Experimental and Theoretical Results on the Dynamics
of the Reaction N+CH→CN+H 429
    G. Dorthe, P. Caubet, N. Daugey, M. T. Rayez, J. C. Rayez,
    P. Halvick, P. Millié, and B. Levy

Astrophysically Important Reactions Involving Excited Hydrogen 437
    J. M. C. Rawlings, J. E. Drew, and M. J. Barlow

Experimental Studies of Gas-Phase Reactions at Extremely Low
Temperatures... 445
    B. R. Rowe, I. R. Sims, P. Bocherel, and I. W. M. Smith

Ammonia in the Interstellar Medium 463
    C. M. Walmsley

The Rotational Excitation of NH_3 by Ortho- and Para-H_2: A
Status Report 477
    D. R. Flower and A. Offer

Recent Progress in Experimental Studies of Ion-Molecule Reactions
Relevant to Interstellar Chemistry.................................... 489
    D. Gerlich

Ion-Trap Experiments on $C_3H^+ + H_2$: Radiative Association vs.
Hydrogen Abstraction 505
    A. Sorgenfrei and D. Gerlich

About the Formation of Interstellar SiN................................. 515
    O. Parisel, M. Hanus, and Y. Ellinger
Reaction of C^+ Ions with Molecules at Low Temperatures.................. 519
    M. Ramillion, R. McCarroll, and M. Gargaud
Gas Phase Reaction of Atomic Hydrogen, Atomic Nitrogen Radical
with Transition Metal Carbonyl Cations 529
    M. Sablier, L. Capron, H. Mestdagh, C. Rolando, N. Billy,
    G. Gouédard, and J. Vigué
Reaction of Anthracene with He^+ and Ar^+ at Room Temperature 537
    H. Abouelaziz, J. C. Gomet, D. Pasquerault, L. Nedelec, A. Canosa,
    C. Rebrion, B. R. Rowe, and P. Lukac
Is Stripping of Polycyclic Aromatic Hydrocarbons a Route
to Molecular Hydrogen? 543
    P. Cassam-Chenaï, F. Pauzat, and Y. Ellinger
Dissociative Recombination of H_3^+ 549
    A. E. Orel, K. C. Kulander, and B. H. Lengsfield III
Photodissociation of Small Polyatomic Molecules 557
    R. Schinke

CLUSTERS AND LARGE HYDROCARBONS

Isomerization of Pure Carbon Cluster Ions: From Rings to Fullerenes........ 571
    J. M. Hunter, J. L. Fye, E. J. Roskamp, and M. F. Jarrold
Excited State Lifetime Broadening Effects on Rotational Band Contours
of C_{60} Calculated for Comparison with the Diffuse Interstellar Bands 589
    S. A. Edwards and S. Leach
Electronic Structures and Stabilities of M_pC_n Microclusters. I.
Si_pC_n ($n+p \leq 6$) 595
    J. Leclercq, G. Pascoli, M. Leleyter, and M. Comeau
Electronic Structures and Stabilities of M_pC_n Microclusters. II.
B_pC_n ($n<6$, $p=1,3$) 605
    M. Comeau, M. Leleyter, J. Leclercq, and G. Pascoli
Photophysical Studies of Jet-Cooled PAHs: Emission Spectra, Lifetimes
and van der Waals Clusters Astrophysical Implications..................... 613
    P. Brechignac and P. Hermine
IR Spectroscopy of Laboratory-Simulated Interstellar PAHs: Role of
Temperature on the Band Positions...................................... 629
    C. Joblin, L. d'Hendecourt, A. Léger, and D. Défourneau
UIR Bands: Computational Experiments on Model PAHs................... 635
    F. Pauzat, D. Talbi, and Y. Ellinger
Computational Checking of Aromatic Models for Interstellar VUV
Absorption.. 649
    O. Parisel and Y. Ellinger

Effects of Molecular Size on the Dissociation Rates of PAH Cations 659
    H. W. Jochims, E. Rühl, H. Baumgärtel, S. Tobita, and S. Leach
Photofragmentation of PAH Ions: Laboratory Experiments on Long Timescales 667
    P. Boissel, G. Lefèvre, and Ph. Thiébot
Ionized Polycyclic Aromatic Hydrocarbon Molecules and the Interstellar Extinction Curve 675
    W. Lee and T. J. Wdowiak
A Laboratory Investigation of the Diffuse Interstellar Bands and Large Linear Molecules in Dark Clouds 687
    T. J. Wdowiak, W. Lee, and L. W. Beegle
The Formation of the Hydrocarbon Component of Carbonaceous Chondrites from Interstellar Polycyclic Aromatic Hydrocarbons 693
    W. Lee and T. J. Wdowiak
Iron Aromatics Coordination: Ion Trap Experiments on Fe^+ $(C_{10}H_8)_n$ Complexation 699
    P. Boissel
Evaluation of the Role of Organometallic Species in the Chemistry of Interstellar and Circumstellar Media 705
    A. Klotz, I. Ristorcelli, D. deCaro, G. Serra, B. Chaudret, J.-P. Daudey, M. Giard, J.-C. Barthelat, and P. Marty
On the Large Organic Molecules in the Interstellar Gas 711
    P. Thaddeus

GRAINS, ICES AND CARBONACEOUS MATERIALS

Physical and Chemical Processes in Icy Grain Mantles 735
    B. Schmitt
Determination of the Optical Constants of Solids in the Mid and Far Infrared 759
    F. Trotta and B. Schmitt
Molecular Photoproduction Rate Coefficients in Icy Grain Mantles as Applied to Dust/Gas Chemical Modeling 767
    O. M. Shalabiea and J. M. Greenberg
Photodesorption from CO Ices 773
    L. Hellner, G. Dujardin, T. Hirayama, L. Philippe, M. J. Ramage, G. Comtet, and M. Rose
Can Mg/Fe Sulphides Solve the Problem of the 30 μm Band of Carbon Stars? 781
    B. Begemann, H. Mutschke, J. Dorschner, and Th. Henning
Experimental Study of Laboratory-Synthesized Carbonaceous Grains and Astrophysical Implications 789
    L. Colangeli, V. Mennella, E. Bussoletti, G. Monaco, P. Merluzzi, P. Palumbo, and A. Rotundi

Amorphous Carbon Grains and the Circumstellar Extinction around C-Rich Objects 795
    A. M. Muci, A. Blanco, S. Fonti, and V. Orofino

Vibrational Excitation of Hydrogen Desorbed from a Carbon Surface 801
    C. Schermann, S. F. Gough, F. Pichou, M. Landau, R. I. Hall, and I. Čadež

Recent Measurements on Coal in the Near and Mid-IR as a Model of Interstellar Dust 811
    K. Ellis, O. Guillois, I. Nenner, R. Papoular, and C. Reynaud

Coal Model for the UV-Visible Interstellar Extinction Curve 817
    O. Guillois, R. Papoular, C. Reynaud, and I. Nenner

The Dust Heating Mechanism in the Coal Model 823
    O. Guillois, I. Nenner, R. Papoular, and C. Reynaud

Panel Discussion on Possible Carriers of Unidentified Spectra 831
    J. Lequeux and Panel Members

CONCLUDING OVERVIEW

Molecules and Grains in Space: An Overview 841
    A. Dalgarno

List of Participants 847

Author Index 855

Foreword

Cross-fertilization is probably one of the most hackneyed ideas in research, and one of the least commonly achieved. As research becomes every day more specialized, those engaged in it have for many years tended to become more and more the members of specialists' clubs, attending specialized meetings and remaining happily, or rather sadly, ignorant of what was not their field. We know this is not true of all domains, and that for instance molecular biology has needed and brought together biologists along with chemists and physicists. When dealing with astrophysics, the name itself implies a blending of astronomy and physics. And, since chemical physicists see no clear boundary between chemistry and physics... .

Things are not so simple, and the conference committee was clearly aware of the fact that the many fields concerned are far from mixing fully. It had to build a really interdisciplinary meeting on the "Physical Chemistry of Molecules and Grains in Space." There was additional proof in the attendance: since the subject is hot, competition is active, and our friends of the Royal Society of Chemistry—Faraday Division—had started first with a closely related conference held in Birmingham in December 1992. We therefore expected at most 100 participants to our conference in September 1993—there were over 130. Many did not hesitate to cross the Atlantic for the second time in nine months—despite severe budget restrictions that spare nobody—and the last contribution to the meeting, certainly "post deadline," was submitted to the committee on the day the conference started.

Despite the closeness in subject and in time, the Faraday meeting and ours were clearly different, each being the occasion to bring a wealth of new results. It seemed to be satisfactory to astronomers, astrophysicists, physicists, and chemists alike to have been brought together and realize the extent to which their respective fields overlapped. The feeling was palpable that meeting colleagues from outside one's club was profitable to all. This book of Proceedings will bear testimony to the variety of topics dealt with, the discussion will give an idea of what it still controversial, and the occasion of lively debate.

We met in Mont Sainte-Odile, a mountaintop convent 25 miles from Strasbourg, where everybody worked hard throughout four days. There, weather permitting, one can also enjoy a gorgeous view on the plain of Alsace and, when taking a relaxation break, go sightseeing to the old part of Strasbourg or discover the Alsatian vineyard and taste its products.

We owed this sensible choice to the meeting committee, chaired by Irène Nenner and including Philippe Bréchignac, Yves Ellinger, Sydney Leach, Alain Léger, Ronald McCarrol and Evelyne Roueff. Despite a tight budget—the sorry state of the world's economy is strongly felt in the interstellar medium—they managed to attract nearly 100 contributions which covered an impressive lot of ground:

- Observation data and models: from millimeter to optical wavelengths, from diffuse clouds to comets
- Structure and photon-induced dynamics of small molecular systems
- Collisions and reactivity
- Ices and carbonaceous materials in grains and on surfaces

For small systems there was evidence of marked advances in the millimeter range, in quantum chemistry, and the gathering of an impressive number of fresh data, far from being all interpreted. Nearly 100 neutral or ionic species have already been identified.

Our understanding of reactive collisions showed also important progress, despite the difficulty of very low temperature studies. Some interpretations are still conflicting, and the gas phase synthesis of large molecules remains a matter of debate.

Research on interstellar dust is exceptionally active, with respect to its nature—fullerenes, PAH, coal—and the assignment of the many diffuse bands observed between the far infrared and the ultraviolet. Several models, some very recent, led to detailed discussions bearing on the size of these systems, their state of ionization or dehydrogenation, and the choice between the coal or PAH models.

"Ices"—water, ammonia, formaldehyde—are a research field still young, where our knowledge is still incomplete: mechanisms of the interaction between grains and surfaces, understanding of their complex photochemistry. The chemistry of these grains is of major importance, and major developments are to be expected.

In short, it was a conference rich in new results and understanding, which also "located" a number of the main problems research aims at solving.

We owe very much to the many authors who brought us their newest results and engaged in active and rewarding discussions. I believe we shall all be agreed in expressing our particular gratitude to Alex Dalgarno, who accepted the thankless task of giving a concluding overview of the conference. This implies listening to everybody and looking at all posters, an impressive—and exhausting—achievement. We gained from it a limpid and humorous survey in which nobody and nothing was neglected, where every result was acknowledged and placed in proper perspective, including some softly worded skepticism. It was a beautiful and fitting conclusion to our meeting.

Clément Troyanowsky
Honorary Secretary
Division de Chimie Physique/SFC

OBSERVATIONS AND MODELS

RECENT DEVELOPMENTS IN ASTROCHEMISTRY
David A. Williams
Mathematics Department, UMIST, Manchester M60 1QD, U.K.

ABSTRACT

I describe the astronomical regions in which chemistry is known to occur, and list the types of molecules that have been positively identified. Some of the successes and failures of chemistries that have been proposed to give rise to these molecular species are reviewed. The evidence which supports the view that processes on and in interstellar dust contribute to the overall chemistry are highlighted. Present ideas concerning the nature of interstellar dust and its response to the changing conditions in the interstellar medium are presented, so that the rôle of dust in interstellar chemistry can be considered. Current problems and directions of future work are indicated.

1. INTRODUCTION

The title of this meeting is "The Physical Chemistry of Molecules and Grains in Space". It is a remarkable fact that the last 25 years or so have revolutionized our view of the importance of chemistry and dust in the Universe. In the 1960s only a few molecular species (CH, CH^+, CN) had been identified in the very tenuous gas towards bright stars. Now we know that chemistry is occurring in many of the most interesting and active regions of the Galaxy (and other galaxies), and that observations of the molecules present allow us to study in detail the conditions within those regions. Thus, in this context, chemistry is a tool of astronomy which helps to determine local densities, temperatures, velocities, elemental abundances, particle and photon fluxes, etc. In return, astronomy provides some of the most testing challenges to physics and chemistry and the opportunity to study material under what are –by terrestrial standards – extreme conditions (see ref. 1). Our view of interstellar and circumstellar dust, too, has changed very significantly over the last 25 years; and it will change even more as new observing facilities such as ISO and FIRST become available. Although an enormous amount of work remains to be done before our understanding of dust is at all complete, nevertheless we now know, in contrast to the 1960s, that dust is not merely an irritating fog impeding the observation of distant stars and galaxies but a crucial component of the Galaxy. Without the presence of dust, this Galaxy and others would be very different. The study of cosmic dust has also led us into those interesting régimes at the interface between macro- and micro-physics and into the study of free-flying large molecules.

In this article I shall describe very briefly the main regions of space where molecules are presently observed (Section 2) and the observed properties of interstellar dust (Section 3). In Section 4 I describe the chemical schemes that are thought to provide an entry into the chemistry occurring in several of these regions, and some problems of current interest. This discussion shows that many problems in astrochemistry are associated with cosmic dust,[2] and I describe some areas of interest, including the gas-dust interaction, in Section 5.

In addition to consideration of the regions of space described in Section 2, one may also infer that a simple chemistry of hydrogen (including deuterium) and helium played a very important role in the development of the early (post-recombination) Universe. Although we shall not consider it further here, this chemistry was important because it provided molecular coolants (H_2, HD) which controlled the collapse and fragmentation of gas into proto-clusters of galaxies, proto-galaxies, and proto-globular clusters within galaxies. Thus, astrochemistry is almost as old as the Universe. Its product molecules have helped control the formation of almost all the components within it: cluster of galaxies, galaxies, stellar clusters, stars and planets.

2. CURRENTLY CHEMICALLY-ACTIVE REGIONS OF SPACE

The chemically active regions of space are generally those where the temperature is typically low ($\sim 10- \sim 100\ K$) and generally not greater than a few thousand degrees Kelvin. Number densities in these regions are significantly greater than the mean interstellar value (~ 1 H nucleus cm^{-3}). In interstellar clouds the number density is typically around one thousand times larger than that mean value, and in star forming regions and circumstellar envelopes the densities are much larger. It is worth noting that while these relatively cool neutral regions occupy only a few percent of the volume of the interstellar medium and are embedded in hot ($\sim 10^4$ K) or very hot ($\sim 10^6$ K) even more tenuous gas, they contain nearly all the mass of the interstellar medium, i.e. about one tenth of the mass of the Galaxy. Interstellar molecular clouds are important in astronomical terms because they are the sites of star formation, and the material of which they are composed is the raw material of the next generation of stars. We have the opportunity of studying the star formation process by observing molecular line emission from dense cores of gas in the process of collapse. We can observe the interaction of very young stars (of masses comparable to that of the Sun) on their environments through the impact of stellar winds and jets or gas clouds. Towards the end of their lives, stars of moderate mass develop extended cool molecular envelopes within which dust may form and be ejected to the interstellar medium, together with the gaseous envelope. Conditions in these envelopes can promote the stimulated emission of radiation; intense molecular masers involving OH, SiO, CH_3OH etc. have been observed in such regions. Transfer of matter on to neighbouring collapsed stars can trigger an outburst called a nova, which is also a location of molecules and dust. In regions where massive stars ($\sim$ ten solar masses) have formed, we can infer the nature of the pre-stellar gas from the interaction of the star and dense molecular remnants of this gas. Masers of OH and H_2O stimulated by the intense radiation fields are also observed in such environments and are regarded as tracers of massive star formation. Massive stars burn their fuel rapidly and end this phase of their lives in a supernova explosion, in which a significant fraction of the stellar mass is ejected at high speed into space. Even in this ap-

Table I Typical parameters for some regions of interstellar space (after ref. 4)

Region	No. density (cm^{-3})	Temperature (K)	Timescale (yr)	Process
Diffuse clouds				
quiescent	10^2	10^2	10^6	supernova
shocked	10^2	10^3	10	cooling
Giant molecular clouds	10^3	10	10^7	gravitational collapse
Quiescent clouds				
interclump	$10^2 - 10^3$	$10-10^2$	$10^6 - 10^7$	gravity
clump	$10^4 - 10^5$	10	10^6	gravity or ablation
Hot cores	10^7	10^2	10^4	photodissociation
T Tauri winds	10^9	$10^3 - 10^4$	1	expansion
Circumstellar Envelopes				
inner	10^9	10^3	10^4	expansion
outer	10^5	10^2	10^4	dilution
Novae ejecta	$10^{11} - 10^{13}$	$10^3 - 10^4$	1	chemistry
Supernovae ejecta	10^9	$10^3 - 10^4$	10	expansion

parently unlikely location chemistry has a role to play and dust can form. Together with novae and mass loss from cool envelopes from solar-type stars, supernovae enrich the interstellar gas with the products of thermonuclear burning, principally C, N and O, together with some refractory dust. In Table I we summarize the main parameters for many regions of astrochemical interest. (The Solar System is also a region of prime chemical interest, of course, but is outside the scope of this article). From the chemical point of view, the problem is to identify the routes by which molecules are formed and destroyed in these diverse regions, noting that the regions are almost entirely hydrogen and contain only traces of other elements

Table II Mean parameters for the interstellar medium

Mean number density: 1 H nucleus cm^{-3}

Fractional abundances of other elements relative to hydrogen
He	O	C	N	Ne	Fe	Si	Mg	S
0.1	7×10^{-4}	3×10^{-4}	10^{-4}	8×10^{-5}	4×10^{-6}	3×10^{-5}	3×10^{-5}	2×10^{-5}

Dust grains: carbons, silicates, ices; $\sim 1\%$ by mass of interstellar gas

Radiation field: mean flux in visual/UV $\simeq 10^{10}$ photons m^{-2}s^{-1}nm^{-1}
(outside the radiatively ionized regions
the radiation field is truncated for $h\nu \leq 13.6\text{eV}$).

Cosmic rays: gas is pervaded by relativistic protons and electrons,
causing ionization at rate $\sim 10^{-17}$s^{-1}.

(apart from He, nearly always $< 0.1\%$), that densities are such that gas phase three-body reactions are almost always excluded. In addition, interstellar space is in principle a harsh environment for chemistry, being swept by cosmic ray particle and stellar UV radiation fluxes (see Table II for a summary).

No references have been supplied in this Section. Further background information may be found in refs. 1, 2.

3. COSMIC DUST AND LARGE MOLECULES

Dust has an important rôle in many interstellar and circumstellar regions. A brief description of its observed properties is therefore given here.

Historically, dust was revealed to be present in the interstellar medium through optical photography: an apparent absence of stars in local regions ("lanes", "globules") against the background of an otherwise rich star field was found to be due to the absorption and scattering of starlight by dust. In fact, dust is generally well-mixed with gas, and these dark regions are also regions of denser gas, i.e. molecular clouds. The dust is distributed throughout the interstellar medium with the gas, and in fairly tenuous regions extinguishes starlight in a characteristic way, rising almost monotonically with photon energy from the near infrared to far UV. The near-linear rise through the visual region of the spectrum is conventionally interpreted as requiring grains with a size comparable with the wavelength of visual light. In fact, the entire curve requires a distribution of sizes of dust grains ranging, typically but not uniquely, from about 3nm to 300nm and with a density distribution varying with radius, a, as $a^{-3.5}$. The commonly accepted candidate dust materials are silicates and carbons of various forms, though the details are by no means secure. Obviously, the smallest particles may well be in the "molecular" régime. However, the detection of interstellar polarization (a selective extinction of one plane of polarization with respect to another) indicates that grains must be asymmetric (by a factor ~ 2) and that their rotations must be partially aligned by paramagnetic (or superparamagnetic) dissipative effects. Polarization effects therefore require the presence of macroscopic dust. Nevertheless, the contribution of molecular absorption to extinction remains a topic for discussion, with ionized or neutral PAH molecules (polycyclic aromatic hydrocarbons), and fullerenes (possibly hydrogenated, possibly ionized) being under current consideration. Significant differences exist between extinction curves along different lines, of sight, especially in the UV. Therefore, the dust – or the molecular contributors – must also vary. However, the extinction in the infrared along all lines of sight appears to obey a uniform law, characteristic of amorphous carbon of relatively low hydrogen content. The prominent absorption near 220nm is conventionally attributed to small graphite particles or inclusions. A large number of less prominent absorptions longward of 443nm, the Diffuse Interstellar Bands, have been known for many years, but their origin remains unknown. Neither grain nor molecular origin is excluded at present.

Dust grains also emit radiation. A broad emission peaking in the red region of the spectrum, the Extended Red Emission, is observed in reflection nebulae and as a reduced extinction in the general interstellar medium. Continuum emission in the infrared (the infrared "cirrus") was detected by IRAS in bands at 12, 25, 60, and 100μm. The emission at the longer wavelengths is generally accepted as thermal emission from dust. The emission at shorter wavelengths seems to require an alternative origin associated with dust and PAH material. Cirrus at 3.3μm has also been detected to be widespread in the Galaxy. This almost certainly arises in some PAH material. In the near infrared a set of emission features at 3.3, 6.2, 7.7, 8.6, 11.3μm is observed in ionized regions and reflection nebulae around bright stars. It is generally accepted that the carriers of these features are PAH material, either free-flying or condensed. The features are associated with Photodissociation Regions (PDRs) which form an interface between ionized regions around hot stars and a nearby molecular cloud. The features may also be excited by photons in the visual region of the spectrum.

Dust must be chemically active. The association of H atoms at the surface of dust is required by the observations of H_2 in diffuse clouds to be an efficient process. Hydrogenation of other species almost certainly occurs and is a contribution to the general chemical network. Dust in clouds of sufficient opacity accumulates molecular mantles containing H_2O, CO, CH_3OH and other species. Chemical processing of this ice is likely to occur, with the possibility that new species are created. Evaporation of mantles when dust is warmed by the radiation from a nearby star can create transient chemical imbalances.

No references have been supplied in this Section. Further background information may be found in ref. 5.

4. ASTROCHEMISTRY OF SELECTED REGIONS

In this section we shall discuss the chemistry of several regions in which there is current interest. We shall indicate the main chemical routes, refer to recent work, and identify areas of chemistry where further laboratory or theoretical work is required.

4.1 *Diffuse clouds* are those for which the optical depth at visual wavelengths is less than about unity, so that the ambient interstellar radiation field (with a photon energy cut-off $h\nu \leq 13.6\text{eV}$) penetrates fairly readily, and as a consequence carbon, sulphur, and silicon atoms are ionized, while oxygen and nitrogen are neutral. Molecular species that have been detected in these objects are given in Table III. Although these were the first molecule-bearing clouds to be detected

Table III Molecules detected in diffuse interstellar clouds

$H_2(J=0-J=7)$, HD, CH, CH^+, CO, C_2, NH, CN, OH, CS

---

in the interstellar medium, their astronomical significance is probably relatively minor. However, they provide severe tests for chemistry because the ambient radiation field destroys molecules on a timescale of about 100 years. Even though they clearly represent the simplest astrochemical systems, diffuse clouds are not at present fully understood, and recent developments have actually increased the severity of the observational constraints.

Gas phase chemistry in diffuse clouds proceeds where possible in reactions with H_2, the most abundant molecule, assumed to be formed on dust. The ion O^+, created in charge exchange with H^+ (arising from the ionization of hydrogen by cosmic ray particles), reacts successively with H_2

$$O^+ \xrightarrow{H_2} OH^+ \xrightarrow{H_2} OH_2^+ \xrightarrow{H_2} OH_3^+ \xrightarrow{e} H_2O$$

The final dissociative recombination provides the neutral molecule H_2O, which is photodissociated to give the detected species OH

$$H_2O \xrightarrow{h\nu} OH \xrightarrow{h\nu} O$$

Thus, if these are the major routes forming and destroying OH, then their effectiveness is determined by the cosmic ray ionization rate and the radiation field. Nitrogen atoms, ionized directly by cosmic rays, may follow a similar sequence of reactions to form NH. The reaction of carbon ions with H_2 is endothermic and proceeds only at elevated temperatures ($\stackrel{\sim}{>} 10^3$ K)

$$C^+ + H_2 \rightarrow CH^+ + H - 0.4\text{eV}$$

but at the low temperatures typical of interstellar clouds the radiative association

$$C^+ + H_2 \rightarrow CH_2^+$$

followed by

$$CH_2^+ \xrightarrow{H_2} CH_3^+ \xrightarrow{e} \begin{cases} CH \\ CH_2 \xrightarrow{h\nu} CH \xrightarrow{h\nu} C \end{cases}$$

provides CH. Exchange reactions with these primary products may provide other species detected; e.g.:

$$OH \xrightarrow{C^+} CO^+ \xrightarrow{H_2} HCO^+ \xrightarrow{e} CO$$

$$CH \xrightarrow{N} CN$$

Conventional models (e.g. ref. 6) adopt the radiation field intensity to give the observed H_2 high rotational state population, and adjust the cosmic ray ionization rate to give the observed OH abundance. Density and temperature (and other factors) are free parameters, adjusted to give best fits to the observations.

These models appear to be unsatisfactory, at least for several well-studied lines of sight. Firstly, where data exist, the distribution of high rotational levels of H_2 is incompatible with UV pumping[7]. Secondly, the production of CO seems to be inadequate, at least in the cloud towards ζ Oph[7]. Thirdly, the adoption of a faster CN photodissociation rate causes the simple route for CN production described above to fail[8]. Fourthly, the detection of NH towards two diffuse clouds implies an abundance that cannot be met by cold cloud gas phase chemistry. Fifthly, revised elemental abundances from Hubble Space Telescope[9,10] measurements are straining the models even further; in particular, oxygen toward ζ Oph is more strongly depleted than previously assumed, so all the calculated abundances of O-bearing molecules predicted by the models are correspondingly reduced.

A solution to these difficulties may lie in the contribution that grain surface chemistry can make[11]. If the populations of high rotational levels of H_2 are determined by the formation mechanism on dust grain surfaces, then the radiation field can be allowed to take a value determined by the chemistry. A plausible contribution of NH_3 (or NH) to the gas phase from surface reactions can satisfy the observational requirement of NH, and through the reaction of NH with carbon provide an additional source of CN. The reduced oxygen abundance implies that the cosmic ray ionization rate must be increased to a high value ($\sim 10^{-16}$ s^{-1}). However, if grain surface reactions are also highly effective in producing H_2O then the cosmic ray ionization rate may take its canonical value (1.3×10^{-17} s^{-1}).

Topics of chemical interest requiring further work, and arising from these diffuse cloud studies, include:

i. the efficiency of H_2 formation on dust, and the kinetic, vibrational, and rotational distribution of product molecules;

ii. the efficiency of H_2O and NH_2 formation on dust and injection into the gas in diffuse clouds; are the molecules saturated before injection?

iii. the recombination of ions and electrons at low temperatures (current determinations are for temperatures $> 10^3\ K$);

iv. the temperature dependence of ion-dipole reactions at low temperatures (present studies assume an increase according to simple theories).

Finally, the well-known problem of CH^+ remains to be resolved[13]. Chemistry in cold clouds cannot produce enough CH^+, by a large factor, to account for the observations. The line profiles are broad, indicating that CH^+ must be located in a warm region where the direct reaction of C^+ and H_2 can proceed. The relevant chemistry is probably known, but the problem appears to be an astrophysical one: understanding the nature of the warm region. Heating by shocks has been the subject of detailed study, but problems of velocity shifts and of H_2 rotational excitation have made these venues less favourable (but see ref. 107). It has been suggested that the formation of CH^+ may occur instead in warm interfaces between diffuse clouds and hot ambient gas. Studies of chemistry in astrophysical mixing layers are in progress. An alternative scenario, CH^+ in intense UV fields, has recently been revived[14].

4.2 *Dark clouds* (sometimes also called quiescent clouds) are here assumed to be those where massive star formation is not occurring. However, low mass star formation is normally detected in such regions, and – when young – these low mass stars drive winds which cause significant dynamical effects in the clouds. Such clouds have been favourite objects for molecular line observations and are rich sources of molecules. Table IV lists molecular species that have been identified in a well-studied dense clump (TMC-1) of the larger molecular cloud HCL2. A recent detailed study[15] of TMC-1 has emphasized the great complexity even with this dense clump. The maps of emission from sulphur-containing carbon chains and other molecules[15] reveal that TMC-1 has at least six cores and confirm that there is very significant chemical differentiation between them. The complexity of dark clouds has been emphasized by Falgarone[16]. As traced in the $100\mu m$ emission from dust[17] and in the line integrated CO rotational emission[18] and atomic hydrogen emission[19] the following common properties emerge: there is a large connectivity of the emission between peaks assumed to be previously isolated; unresolved structure exists in all maps; there is scale invariance in maps of integrated emission; the gas in these regions is highly turbulent, and the dissipation zones may be hot. The complex structure of all the maps can be described in terms of fractal geometry of dimension about 1.5. This complexity, and these surprising geometrical concepts, have not generally been taken into account in chemical models of such regions; the word "quiescent" has perhaps been misinterpreted. However, a series of models in which the chemistry has been modulated by the effects of low mass star formation has been explored[20–22], though even these models are probably simplistic when compared with the observations reviewed by Falgarone.

It is necessary to preface our remarks about chemistry in dark clouds with the caution given above, since any apparent failures in the chemical modelling may in fact be due to inadequacies in the physical description of the region. The following

Table IV Molecular species detected in TMC-1

CO HCO$^+$ H$_2$CO C$_2$O H$_2$C$_2$O C$_3$O C$_5$O HC$_2$CHO

OH

C$_2$ C$_3$H$_2$ H$_2$C$_4$

CH C$_2$H C$_3$H C$_4$H C$_5$H C$_6$H

CN HCN HCNH$^+$ HC$_3$N HC$_3$NH$^+$ HC$_5$N HC$_7$N HC$_9$N HC$_{11}$N

HNC HNC$_3$ HC$_2$NC

NH$_3$ N$_2$H$^+$ NO HNCO C$_3$N CH$_2$CN CH$_2$CHCN

H$_2$S SO OCS

CS HCS$^+$ C$_2$S C$_3$S H$_2$CS

CH$_3$OH CH$_3$CN CH$_3$CHO CH$_3$C$_2$H CH$_3$C$_3$N CH$_3$C$_4$H

NH$_2$ has been recently discovered[104] in conditions that may be similar to TMC-1

---

chemistry conventionally adopted for dark clouds[1] assumes that the interstellar radiation field is heavily extinguished by dust, so that the cloud is almost entirely neutral; the chemistry is driven largely by cosmic rays which generate both H$_3^+$ ions

$$H_2 \xrightarrow{c.r.} H_2^+ \xrightarrow{H_2} H_3^+$$

and an internal radiation field caused by the energetic electrons (released in the ionization) exciting electronic transitions in H$_2$, which subsequently relax radiatively, emitting UV photons. The H$_3^+$ ion has a low proton efficiency, and readily transfers its proton to many other species, for example, to carbon atoms, which after successive reaction with H$_2$ molecules,

$$C \xrightarrow{H_3^+} CH^+ \xrightarrow{H_2} CH_3^+ \xrightarrow{H_2} CH_5^+ \xrightarrow{e, CO} CH_4, CH_3$$

form methane or the methyl radical. Complex molecules form through radiative association of ions (such as CH$_3^+$) with neutral molecules (such as H$_2$O) followed by dissociative recombination leading to neutral products (methanol, for example):

$$CH_3^+ \xrightarrow{H_2O} CH_3^+ \cdot H_2O \xrightarrow{e} CH_3OH$$

Carbon monoxide readily undergoes dissociative charge transfer with He^+ formed by cosmic ray ionization

$$CO + He^+ \longrightarrow C^+ + O + He$$

to give a ready source of carbon ions which can undergo insertion reactions with hydrocarbons, e.g.

$$C_m H_n \xrightarrow{C^+} C_{m+1} H_{n-1}^+$$

to construct carbon chains. Such molecules may also arise in condensation reactions, e.g.

$$CH_3^+ + C_2H \rightarrow C_3H_3^+ + H$$

Most hydrocarbon ions are unreactive with H_2 at low temperatures (which explains the prevalence of unsaturated species in dark clouds) but it is possible to attach neutrals such as nitrogen to chains, e.g.

$$C_3H_3^+ + N \longrightarrow H_2C_3N^+ + H$$

which may contribute to the cyanopolyyne abundance, thus:

$$H_2C_3N^+ \xrightarrow{e} HC_3N$$

The reactions of neutral species, too, are important[23], and their effects may be significant for CN. Reactions with H_2D^+ incorporate deuterium into interstellar molecules. Since the deuterated species is of slightly lower energy than H_3^+, the ratio H_2D^+/H_3^+ can very greatly exceed D/H in cold clouds. This degree of fractionation is then also largely transferred to other species. Observations of normal and deuterated versions of a molecule provide useful constraints on cloud conditions and chemical routes.

Gas phase chemistry of large molecules is limited by the freeze-out of atoms and molecules on to dust grains. The timescale for freeze-out is likely to be less than one million years in molecular clouds. The detection of molecular ices as mantles on dust in dark clouds[5] confirms the significance of this process. Complex solid state chemistries may occur in the mantle, and these have been studied by extensive computational models[24,25,105]. Mantles may also be removed by intermittent processes associated with star formation, or by continuous processes driven by cosmic rays, by chemistry, by thermal evaporation, or by turbulence[12]. A great many models of objects such as TMC-1 have been constructed, with some success in explaining the relative abundances of various species. By suitable adjustment of the many free parameters reasonable agreement can be obtained between theory and observation. However, this encouraging harmony is probably fortuitous and may merely reflect the fact that the available parameter space is large. No models made to date properly reflect the complexity of structure and chemical diversity found within TMC-1. There is no doubt that new observational approaches should help to define the important parameters. The oxygen budget is being addressed by various authors[26-28], and useful upper limits for O_2 are now available for several dark clouds. It is now clear that the chemistry of the solid phase occurring in the icy mantles on dust, is linked to that of the gas phase. Models of chemistry in dark clouds must take account of freeze-out and possible return of gas phase material[29]. It has recently been suggested that interconversion between rotational spin conformers E and A of methyl cyanide may be affected by freeze-out and desorption processes[30].

The major chemical uncertainties for dark clouds reside with processes associated with dust and solid state chemistry. These are discussed in more detail in Section 5, below.

4.3 *Hot cores* provide more evidence that the solid and gas phases of chemistry in molecular clouds are intimately linked. The most intensively studied hot cores[31,32] are two in Orion, one eponymously called the Hot Core and the other the Compact Ridge; similar sources are believed to exist in all high mass star forming regions. Table V lists the molecular species that have been identified in these two regions in Orion; it is evident that significant chemical differentiation occurs between them. Both objects are thought to be remnants of the original molecular cloud that gave birth to several young massive stars, of which one (called IRc2) has a luminosity of 10^5 solar luminosities. Both regions are dense ($\sim 10^7$ H_2 cm^{-3}) and warm (~ 200 K). The Hot Core is thought to be heated directly by infrared radiation from IRc2, while the Compact Ridge is probably impacted by the wind from this source. Both processes cause the heating of these two regions, and consequently the evaporation of icy mantles established on dust grains during the lifetime of the pre-existing molecular cloud. Detailed modelling has established that the chemical differences between these two similar objects seem to require that the mantles in the Compact Ridge be methanol rich, whereas those in the Hot Core may be water rich. Further studies have attempted to relate these chemical differences in the ices to different phases of evolution in separate locations of the collapsing molecular cloud. If this view is confirmed, then the nature of the pre-existing cloud that gave rise to massive star formation will be specified in some detail.

The major chemical problems arising in the study of hot cores are to do with the chemical nature of the icy mantles, and the processes by which these may vary with position. These will be discussed further in Section 5 below.

4.4 *Circumstellar regions* provide an interesting variety of astronomical laboratories[33], ranging from the winds of very young stars to cool envelopes around highly evolved stars. Dramatic events such as novae and supernovae also have successful chemistries associated with them.

The wind of a young star can entrain ambient molecular material and create a molecular outflow which may be observed. The nature of the wind itself, and its origin, are poorly understood, but it is probably initially hot, partially ionized, dense, and strongly irradiated by stellar photons. The question has arisen as to whether chemistry may occur in such situations, and therefore whether molecular observations of the pure wind can help to define its status. The chemistry in the wind of a young star may be significantly different from that in other astrochemical regions; firstly, the densities initially may be high enough to permit efficient three-body formation of H_2

$$3H \rightarrow H_2 + H$$

and secondly, the stellar radiation field may establish a significant population of excited atomic hydrogen, $H(n=2)$, and open up a new chemical channel involving associative ionization[34]

$$H(1s) + H(2s) \rightarrow H_2^+ + e$$

Table V Molecules detected in the Orion hot cores

Molecule	Hot Core	Compact Ridge
CO	✓	✓
CS	✓	
HDO	✓	✓
H_2S	✓	
HCN	✓	✓
HNC	✓	
OCS		✓
NH_3	✓	✓
H_3O^+	✓	
C_2H_2	✓	
H_2CO	✓	✓
H_2CS		✓
HNCO	✓	
CH_4	✓	
CH_2CO		✓
HCOOH		✓
HC_3N	✓	
CH_3OH		✓
CH_3CN	✓	✓
CH_3CHO		?
CH_2CHCN	✓	
$HCOOCH_3$		✓
$(CH_3)_2O$		✓
CH_3CH_2CN	✓	

The cross section for this interaction peaks at the H_2^+ dissociation energy, and is substantial. This new channel operates in addition to the network of reactions involving H_2^+ and H^- that can form and destroy H_2 in a dust-free environment.

The associative ionization cross section in warm gas is much larger than the corresponding direct radiative association[35]

$$H(1s) + H(2s) \rightarrow H_2 + h\nu$$

A study[34] of the significance of this new route for H_2 formation (and, consequently, CO formation) predicts that it may be of importance in the denser outflows of luminous blue variables (early B-type hypergiant stars with mass loss in excess of $10^{-5} M_\odot$ yr^{-1}). Cooler, denser winds favour a richer chemistry, and the chemistry in massive young stellar outflows may be fairly extensive. Collisional dissociation is usually an effective loss route for molecules at temperatures above a few thousand degrees, so a rapid cooling aids the chemistry. Geometrical dilution of the gas suppresses the chemistry.

Late in their evolution, stars of around a few solar masses develop extended atmospheres which drift slowly outwards in the form of a "superwind". This generates circumstellar envelopes (CSEs) which are warm ($\sim 10^3$ K) and relatively dense, and are ideal chemical laboratories[36]. Dust formation also occurs, and radiation pressure on the dust accelerates the flow to a terminal velocity in the range 10–20 kms^{-1}. In addition to this dynamical rôle, dust grains in envelopes also help to shield the envelopes from the interstellar radiation field, may also act as catalysts for surface chemistry, and be sinks or sources of molecules. Chemistry in the stellar photosphere is established in local thermodynamic equilibrium. As the gas drifts outwards the relative molecular abundances remain largely fixed, until ionization initiated by the interstellar radiation field allows a burst of ion-molecule chemistry to proceed, only to be terminated through photodissociation driven by the increasing intensity of the radiation field. Thus, a chemistry exists in which "parent" molecules from the photosphere give rise to a shell of "daughter" species in the photo-chemical zone.

Oxygen-rich stars and envelopes contain oxygen-rich dust, i.e. silicates, detected by absorption in the 10μm band. In the photospheres of such stars, LTE chemistry should incorporate almost all carbon into CO, which with H_2O, N_2, and H_2 are probably the main parent species. The appearance of some carbon-bearing species in the detected daughter species (see Table VI) is, therefore, surprising. The photodissociation of CO and photoionisation of C to give C^+ ions in the photochemical region fail by several orders of magnitude to produce the amounts of HCN observed. It seems necessary that some carbon be retained by some non-equilibrium process in a more accessible form than CO (perhaps CH_4) deep in the envelope, so that it can be made available in the photochemical zone. A variety of sulphur-bearing species have also been detected in O-rich envelopes; these indicate clearly that H_2S is the parent molecule, while the daughter molecules SO and SO_2 (arising from reactions of S with OH in the photochemical region) show a shell morphology.

Carbon-rich envelopes show a particularly rich chemistry. Molecular species detected in a well-studied nearby envelope of high mass loss rate (IRC+10216)

Table VI Molecules detected in O-rich CSEs

CO	OH	SO	SiS	SiO	CN	CS
SO_2	H_2O	H_2S	OCS	HCN	HNC	HCO^+
NH_3	H_2CO					

are listed in Table VII. The parent molecules are probably H_2, CO, HCN, C_2H_2, CH_4, NH_3, N_2, and SiS. The families of carbon chain molecules $HC_{2n+1}N(n=1-5)$, $C_nH(n=2-6)$ and $C_nS(n=1-3)$ have attracted much attention. It has generally proved difficult to account theoretically for the high abundance of the larger species. Recent work[37] has proposed that the large cyanopolyynes, $HC_{2n+1}N$, are formed in radical reactions of C_3N and C_5N with acetylene. These routes are efficient and the predicted spatial distributions of HCN, CN, HC_3N and C_3N are qualitatively in agreement with observations. However, the predicted abundances of the larger cyanopolyynes is still too small. It may be that their chemistry and that of other large carbon molecules is related to the formation of carbonaceous dust[38-40]. The formation of PAHs may also be expected to occur in carbon-rich environments. However, a recent detailed study[41] suggests that although PAH formation may occur close to the photosphere of a carbon-rich star, but probably not in the envelope itself. It may be that non-steady situations promote PAH formation. It has been suggested that fullerenes may also form in such situations[42].

Eventually, a star with a circumstellar envelope exhausts its fuel, the "superwind" ceases to be replenished and becomes detached from the star which is now revealed as a hot degenerate core. The radiation from this core is intense and hard, and partially ionizes the surrounding gas creating a planetary nebula. Simple molecular detections are made in a few planetary nebulae, and some – resumably young – objects contain molecules as complicated as large cyanopolyynes. A detailed study[43] of protoplanetary nebulae shows that few molecules of the original wind can survive more than 100 years of irradiation by the stellar UV field; however, detectable amounts of small hydrocarbon and ions form behind a shock associated with the ionization of the nebula. Small dense knots of gas within the nearby planetary nebula, the Helix[44], have been shown to contain CO^{45}. The detailed chemistry of such knots has been explored[46], and it is shown that the knots are probably in a chemical steady state[47].

Novae are eruptions caused by the transfer of material within a binary system from a star with an extended envelope on to a companion white dwarf. The thermonuclear explosion in the transferred gas as it is heated and compressed at the dwarf's surface creates a transient dense hot wind, irradiated by the stellar UV. It is striking evidence of the power of chemistry that molecules are formed even in this unpromising situation. The emission near $5\mu m$ and at $2.3\mu m$ seen in many novae up to about one month after outburst have been attributed to CO,

Table VII Molecules detected in IRC + 10216

At infrared wavelengths only (6):

CH_4, SiH_4, C_2H_2, C_2H_4, C_3, C_5

At radio wavelengths (36):

CO, CN, SiC, CP, CS, SiN, SiO, SiS, NaCl, AlCl, KCl, AlF

H_2S, HCN, HNC, C_2H, SiC_2, C_2S, MgNC

NH_3, l-C_3H, C_3N, C_3S, HCCN

c-C_3H_2, H_2CCC, HC_3N, C_4H, SiC_4

H_2CCCC, C_5H, CH_3CN

HC_5N, C_6H

HC_7N, HC_9N, $HC_{11}N$

and the onset of dust formation usually occurs within a few months, causing extinction in the UV accompanied by a rise in the infrared emission. This interpretation is supported by a theoretical study[48] which suggests that some clumping or shielding is necessary for chemistry to occur. Observational work has also detected features of H_2, CN, SiO, SiO_2, SiC and PAH molecules in novae[49-52]. However, no chemical modelling is available. Since the onset of dust formation in novae generally follows closely the first detected chemistry the processes are almost certainly related[53].

The ejecta of the supernova SN1987A have been shown to contain CO (about 100 days post-outburst), SiO, and (probably) H_3^+, (cf. ref. 1). Chemistry in these ejecta is unlike that in any other astronomical location in that, in the layers of the stellar interior containing oxygen and carbon, hydrogen is almost totally absent, and therefore the direct association of C and O (or their ions) is likely to be the route by which CO forms[54,55]. The mixing of other layers containing hydrogen or helium has a pronounced effect on the chemistry, and the chemical models can place limits on the extent of the mixing. The H_3^+ must form in the outermost zone of the ejecta, from reactions similar to those in dust-free stellar winds[56]. Some clumpiness may be necessary to produce the inferred amount of H_3^+.

In general, the study of near stellar environments emphasises the intimate nature of the gas-dust interaction. The major problems for the astrochemist, and future progress in understanding these areas of astronomy, depend on understanding better the rôle of chemistry in dust nucleation and growth, and the passage of dust through hot gases.

5. COSMIC DUST AND LARGE MOLECULES – SOME PROBLEMS

Our survey of chemically active regions of space reveals that dust is present in nearly all of them and is also chemically active. It is also clear that many of the major uncertainties in astrochemistry are now associated with the dust, or with the gas-dust interaction. Purely gas-phase processes are, in general, fairly well understood. In this Section we shall therefore draw attention to some problems associated with dust and large molecules such as PAHs.

The interstellar extinction curve: most interpretations of this simple curve have introduced sufficient parameters that the fitting procedure does not provide serious constraints for any plausible model. Progress is more likely to be made from understanding the cause of variations in the extinction curve along different lines of sight. While extinction to nearby stars in the infrared is uniform[57], suggesting a commonality of dust in all regions, much larger extinction variations occur in the UV. These variations can be accounted for by varying the proportions and size ranges of components of dust; if this is the case, then there should be astrophysical interpretations to be placed on these variations. Alternatively, dust may be composed of a material that responds to its environment and changes its optical properties in different locations. Amorphous carbons have been well studied in the laboratory, and behave in such a way[58]. The importance of PAHs, fullerenes and very small grains in the interstellar medium and specifically to the extinction curve has been stressed by many authors[59-63]. Many studies of individual PAHs have been made. To quote one example[64], it is suggested that ionized naphthalene molecules may contribute to the UV – visible extinction. It also appears that ionized PAHs may not add unwanted structure to at least part of the extinction curve[65]. The extinction bump at 217nm is conventionally attributed to small graphite particles, though alternatives have been proposed[58]. The PAH molecules may also contribute[64]. The chemistry of PAHs is being investigated[113,114]. The suggestion that C_{60} may be the 217nm carrier does not appear to be supported by laboratory work[63,66]. However, the apparent lack of polarization associated with the 217nm feature in three stars observed by the Hubble Space Telescope is not inconsistent with a molecular carrier[67].

Interstellar linear polarization: the measured average interstellar magnetic field strength is sufficiently low that the required alignment of spinning grains can be achieved only if grains contain either ferromagnetic or superparamagnetic inclusions[68]. How and where these inclusions are to be established in the dust is at present unclear.

The Infrared Emission Features (to use Tielens' terminology[62]) between $3.3\mu m$ and $11.3\mu m$ are securely identified with aromatic hydrocarbon material[69-71]. It is not clear whether detailed assignments can be made. The astronomical observations have stimulated much laboratory work on PAHs; however, as Tielens remarks[62] "infrared spectroscopy is not very sensitive to the size of the emitting species; essentially, it probes the molecular bonding between neighbouring atoms". Observations show features shift and change in response to their environments[115].

This is seen in the laboratory also[116]. Tielens[62] has reviewed much of the recent work comparing laboratory results with astronomical observations. Much attention has been given to the emission expected from ionized PAHs[64]; frequency shifts in infrared features from neutrals occur, and the relative intensities are more in harmony with observations[72]. Images of several astronomical regions in 3.3 and 11.3μm have been obtained[73,74]; these data suggest that the features may be optically thick in the source known as the Red Rectangle. High resolution spectroscopy of protostars in the 3μm region has revealed new features, tentatively attributed to methanol and diamond-like carbon[75] (a component of amorphous carbon[58]). It has also been suggested that the features may be attributed to fulleranes[108-111].

The Extended Red Emission (ERE) has been studied observationally by many authors, (in reflection nebulae[76], planetary nebulae[77], HII regions[78], and the galactic cirrus[79] and the proposed origin of ERE in terms of luminescence in hydrogenated amorphous carbon[80,81] has recently received strong support from laboratory work[82]. The proposal (which is the basis of the HAC model of dust) that HAC can be graphitized (H-poor) by UV and restored by H atom irradiation is now independently justified. The possibility that the ERE arises in PAHs that are proposed as carriers for the diffuse interstellar bands in other large molecules has been recently reviewed[83]. It has also been suggested that candidates for ERE may be found among the fullerenes and fulleranes[112].

The diffuse interstellar bands have been studied for more than half a century, yet no definitive assignment has been made for any of them[84]. The current situation has been well summarized in recent reviews[5,83]. Although the correlations observed between extinction and band strengths encouraged the view that their carriers were in or on the dust, more recent proposals have focused on potential molecular carriers. Both chlorine[83] and ionized naphthalene[64,85] have recently been reported to have a number of features in common with the diffuse bands. Some problems with PAHs as the carriers have been indicated[86]. The suggestion[87] that ionized C_{60} might be the carrier of some of the bands led to a prediction of a new infrared band at 1.3μm. A search for this band on lines of sight through the Taurus dark cloud complex has proved negative[88].

The chemical activity of dust has been addressed by many authors (see, e.g., refs. 12, 61, 62, 89). The astronomical evidence from diffuse cloud observations supports the view that not only H_2 but simple hydrides (and possibly more complicated species, such as H_2CO) are formed on dust surfaces, and that the efficiency of such reactions may be high[11]. This astronomical evidence needs to be supported by detailed chemical studies[90]. The state of excitation of product molecules is of some interest; it has been suggested that for H_2 molecules formed on surfaces the vibrational line emission may be detectable[91]. A laboratory study of H_2 adsorption and formation on ices suggests that H_2 on interstellar dust may be detectable by infrared absorption[92].

The onset of molecular ice mantles of dust was first studied in detail by Whittet and collaborators[5]. A re-examination[93] of the onset of ice in the Taurus clouds has revealed that the onset of 3.4μm absorption (characteristic of sp^3 bonding in C-H) occurs at the same extinction ($A_V = 2.6$mag) as H_2O ice absorption at 3.1μm. The processes controlling onset are unclear at present. The onset of H_2O-ice and hydrocarbon mantles at a critical extinction represents a step change in the chemistry of clouds which will have significant consequences. This gas-grain interaction should be incorporated into comprehensive cloud models. A full understanding of the abrupt critical onset of mantles should be a tracer of specific conditions in the cloud[94].

The chemical processing of icy mantles occurs in dark clouds and circumstellar environments, as the astronomical evidence indicates. There has been a large laboratory programme aimed at understanding these events[95,96]. The reduction of mantle CO by hydrogenation is of particular interest, since CO is an important mantle species. Recent laboratory work[97] indicates that H_2CO may form under interstellar conditions but no evidence of CH_3OH was found. However, methanol ice is apparently detected towards a number of protostars[75,98,99]. Its presence is also indicated by studies of hot cores. Perhaps the final stage of hydrogenation to CH_3OH requires high activation. Recent studies of CO ice have shown that it arises in at least two forms, corresponding to polar and non-polar environments[100,101,106]. Modelling of the CO deposition process, taking into account various desorption mechanisms[102] has shown that both the ratio of polar to non-polar CO, and the total CO solid fraction, can be obtained if the processing in cold quiescent clouds is slow[103].

ACKNOWLEDGMENT

I am grateful to Dr. T.J. Millar for his advice during the writing of this paper.

REFERENCES

1. A. Dalgarno, J. Chem. Soc. Faraday Trans. 89, 2111 (1993).
2. *Dust and Chemistry in Astronomy*, eds. T.J. Millar and D.A. Williams, IOP Publishing, Bristol, (1993) p. 1–335.
3. T.J. Millar and D.A. Williams, Sci. Progress 75, 279 (1991).
4. D.A. Williams and T.W. Hartquist, QJRAS 31, 593 (1990)
5. D.C.B. Whittet, *Dust in the Galactic Environment*, IOP Publishing, Bristol (1992), pp. 1–295.
6. E.F. van Dishoeck and J.H. Black, ApJ Suppl. 62, 109 (1986).
7. R. Wagenblast, MNRAS 259, 155 (1992).
8. R. Wagenblast, D.A. Williams, T.J. Millar and L.A.M. Nejad, MNRAS 260, 420 (1993).
9. B.D. Savage, J.A. Cardelli and U.J. Sofia, ApJ 401, 706 (1992).
10. J.A. Cardelli, J.S. Mathis, D.C. Ebbets and B.D. Savage, ApJ 383, L23 (1991).
11. R. Wagenblast and D.A. Williams, submitted to MNRAS.
12. D.A. Williams, ref. 2, p.143.
13. D.A. Williams, Plan. Sp. Sci. 40, 1683 (1992).
14. T.P. Snow, ApJ 402, L73 (1993).

15. Y. Hirahara, H. Suzuki, S. Yamamoto, K. Kawaguchi, N. Kaifu, M. Ohishi, S. Takano, S.-I. Ishikawa and A. Masuda, ApJ 394, 539 (1992).
16. E. Falgarone, in *Astrochemistry of Cosmic Phenomena*, ed. P.D. Singh, Kluwer Academic Publishers, Dordrecht (1992), p. 159.
17. J.M. Scalo, in *Physical Processes in Fragmentation and Star Formation*, eds. R. Capuzzo-Dolcetta et al., Kluwer Academic Publishers, Dordrecht.
18. E. Falgarone, T.G. Phillips and C. Walker, ApJ 378, 186 (1991).
19. R.L. Dickman, M.A. Horvath and M. Margulis, ApJ 365, 586 (1991).
20. S.B. Charnley, J.E. Dyson, T.W. Hartquist and D.A. Williams, MNRAS 231, 269 (1988).
21. S.B. Charnley, J.E. Dyson, T.W. Hartquist and D.A. Williams, MNRAS 235, 1257 (1988).
22. L.A.M. Nejad and D.A. Williams, MNRAS 255, 441 (1992).
23. E. Herbst, H.S. Lee, D.A. Howe and T.J. Millar, MNRAS in press (1993).
24. E. Herbst, ref. 2, p. 183.
25. T.I. Hasegawa, E. Herbst and C.M. Leung, ApJ Suppl. 82, 167 (1992).
26. A. Fuente, J. Cernicharo, S. Garcia-Burillo and J. Tejero, Astron. Astrophys. 275, 558 (1993).
27. H.S. Liszt, ApJ. 386, 139 (1992).
28. T.G. Phillips, E.F. van Dishoeck and J. Keene, ApJ. 399, 533 (1992).
29. K. Willacy and D.A. Williams, MNRAS 260, 635 (1993).
30. K. Willacy, D.A. Williams and Y.C. Minh, MNRAS 263, L40 (1993).
31. C.M. Walmsley and P. Schilke, ref. 2, p. 37.
32. T.J. Millar, ref. 2, p. 249.
33. D.A. Howe, J.M.C. Rawlings and D.A. Williams, Adv.At.Mol.Opt.Phys. 30, in press.
34. J.M.C. Rawlings, J.E. Drew and M.J. Barlow, MNRAS in press, (1993).
35. W.B. Latter and J.H. Black, ApJ 372, 161 (1991).
36. A. Omont, J. Chem. Soc. Faraday Trans. 89, 2137 (1993).
37. I. Cherchneff, A.E. Glassgold and G.A. Mamon, ApJ 410, 188 (1993).
38. H.-P. Gail and E. Sedlmayr, Astron. Astrophys. 133, 320 (1988).
39. A. Goeres, Rev. Mod. Astron. 6, 165 (1993).
40. W.B. Latter, ApJ 377, 187 (1991).
41. I. Cherchneff, J.R. Barker and A.G.G.M. Tielens, ApJ 401, 269 (1992).
42. H. Kroto, Science, 242, 1139 (1988).
43. D.A. Howe, T.J. Millar and D.A. Williams, MNRAS 255, 217 (1992).
44. J. Meaburn, J.R. Walsh, R.E.S. Clegg, N.A. Walton, D. Taylor and D.S. Berry, MNRAS 255, 177 (1991).
45. P.J. Huggins, R. Bachiller, P. Cox and T. Forveille, ApJ 401, L43 (1992).
46. D.A. Howe and D.A. Williams, MNRAS submitted (1993).
47. J.E. Dyson, T.W. Hartquist, M. Pettini and L.J. Smith, MNRAS 241, 625 (1989).
48. J.M.C. Rawlings, MNRAS 232, 507 (1988).
49. A. Evans, MNRAS 251, 54P (1991).
50. R.D. Gerhz, Ann. Rev. Astron. Astrophys. 26, 377 (1988).

51. R.D. Gerhz, T.J. Jones, C.E. Woodward, M.A. Greenhouse, R.M. Wagner, T.E. Harrison, T.L. Hayward, and J. Benson, ApJ 400, 671 (1992).
52. P.F. Roche, D.K. Aitken and C.H. Smith, MNRAS 261, 522 (1993).
53. J.M.C. Rawlings and D.A. Williams, MNRAS 240, 729 (1989).
54. S. Lepp, A. Dalgarno and R. McCray, ApJ 358, 262 (1990).
55. J.M.C. Rawlings and D.A. Williams, MNRAS 246, 208 (1990).
56. S. Miller, J. Tennyson, S. Lepp and A. Dalgarno, Nature 355, 420 (1992).
57. P.G. Martin and D.C.B. Whittet, ApJ 357, 113 (1990).
58. A.P. Jones, W.W. Duley and D.A. Williams, QJRAS 31, 567 (1990).
59. L.J. Allamandola, A.G.G.M. Tielens and J.R. Barker, ApJ Suppl. 71, 733 (1989).
60. J.L. Puget and A. Léger, Ann. Rev. Astron. Astrophys. 27, 161 (1989).
61. W.W. Duley in ref. 2, p. 71.
62. A.G.G.M. Tielens in ref. 2, p. 103.
63. W. Krätschmer, J. Chem. Soc. Faraday Trans. 89, 2285, (1993).
64. F. Salama and L.J. Allamandola, ApJ 395, 301 (1992).
65. W. Lee and T.J. Wdowiak, ApJ 410, L127 (1993).
66. S. Leach, J. Chem. Soc. Faraday Trans. 89, 2305 (1993).
67. W. Somerville, J. Chem. Soc. Faraday Trans. 89, 2305 (1993).
68. M.J. Wolff, G.C. Clayton and M.R. Meade, ApJ 403, 722 (1993).
69. W.W. Duley and D.A. Williams, MNRAS 196, 269 (1981).
70. A. Léger and J.L. Puget, Astron. Astrophys. 137, L5 (1984).
71. L.J. Allamandola, A.G.G.M. Tielens and J.R. Barker, ApJ 290, L25 (1985).
72. D.J. De Frees, M.D. Miller, D. Talbi, F. Pauzat and Y. Ellinger, ApJ 408, 530 (1993).
73. J. Bregman, D. Rank, S.A. Sandford and P. Temi, ApJ 410, 668 (1993).
74. J.D. Bregman, D. Rank, P. Temi, D. Hudgins and L. Kay, ApJ 411, 134 (1992).
75. L.J. Allamandola, S.A. Sandford and A.G.G.M. Tielens, ApJ 399, 134 (1992).
76. A.N. Witt and R.E. Schild, ApJ Suppl. 62, 839 (1986).
77. D.G. Furton and A.N. Witt, ApJ 386, 587 (1992).
78. J.-M. Perrin and J-P Sivan, Astron Astrophys 255, 271 (1992).
79. P. Guhathakurta and J.A. Tyson, ApJ 346, 773 (1989).
80. W.W. Duley, MNRAS, 215, 259 (1985).
81. W.W. Duley and D.A. Williams, MNRAS 230, 1P (1988).
82. D.G. Furton and A.N. Witt preprint (1993).
83. J.R. Miles and P.J. Sarre, J. Chem. Soc. Faraday Trans. 89, 2269 (1993).
84. G.H. Herbig, ApJ 196, 129 (1975).
85. T.P. Snow, ApJ 401, 775 (1992).
86. C. Cossart-Magos and S. Leach in *Molecular Clouds*, eds. R.A. James and T.J. Millar, Cambridge University Press, Cambridge, p. 317 (1991).
87. A. Léger, L. d'Hendecourt, L. Verstraete and W. Schmidt, Aston. Astrophys. 203, 143 (1988).
88. A.J. Adamson, T.J. Kerr, D.C.B. Whittet and W.W. Duley, MNRAS, in press (1993).

89. A.G.G.M. Tielens and L.J. Allamandola in *Interstellar Processes*, eds. D.J. Hollenbach and H.A. Thronson, Jr., D. Reidel Publishing Company, Dordrecht, p. 397 (1987).
90. V. Buch and Q. Zhang, ApJ 379, 647 (1991).
91. W.W. Duley and D.A. Williams, MNRAS 260, 37 (1993).
92. S.A. Sandford and L.J. Allamandola, ApJ 409, L65 (1993).
93. R.G. Smith, K. Sellgren and T.Y. Brooke, MNRAS 263, 749 (1993).
94. D.A. Williams, T.W. Hartquist and D.C.B. Whittet, MNRAS 258, 599 (1992).
95. L.J. Allamandola and S.A. Sandford in *Dust in the Universe*, eds. M.E. Bailey and D.A. Williams, Cambridge University Press, Cambridge, p. 229 (1988).
96. J.M. Greenberg, C.X. Mendoza-Gómez, M.S. de Groot and R. Breukers, ref. 2, p. 271.
97. R.S. Bohn, L.J. Allamandola and S.A. Sandford, Astronomical IR Spectroscopy: Future Observational Directions, 41, 223.
98. W.A. Schutte, A.G.G.M. Tielens and S.A. Sandford, ApJ 382, 523 (1991).
99. C.J. Skinner, A.G.G.M. Tielens, M.J. Barlow and K. Justtanont, ApJ 399, L79 (1992).
100. T.H. Kerr, A.J. Adamson and D.C.B. Whittet, MNRAS 251, 60P (1991).
101. A.G.G.M. Tielens, A.T. Tokunaga, T.R. Geballe and F. Baas, ApJ 381, 181 (1991).
102. K. Willacy and D.A. Williams, MNRAS 260, 635 (1993).
103. K. Willacy, D.A. Williams and W.W. Duley preprint (1993).
104. E.F. van Dishoeck, D.J. Jansen, P. Schilke and T.G. Phillips, ApJ in press (1993).
105. T.I. Hasegawa and E. Herbst, MNRAS 263, 589 (1993).
106. T.H. Kerr, A.J. Adamson and D.C.B. Whittet, MNRAS 262, 1047 (1993).
107. E.L. Heck, D.R. Flower, J. Le Bourlot, G. Pineau des Forêts and E. Roueff, MNRAS 262, 795 (1993).
108. A. Webster, MNRAS 255, 41P (1992).
109. A. Webster, MNRAS 262, 831 (1993).
110. A. Webster, MNRAS 263, 385 (1993).
111. A. Webster, MNRAS 264, 121 (1993).
112. A. Webster, MNRAS 264, L1 (1993).
113. T.J. Millar, MNRAS 259, 35P (1992).
114. S. Leach, J. Chem. Soc. Faraday Trans. 89, 2312 (1993).
115. A.T. Tokunaga, K. Sellgren, R.G. Smith, T. Nagata, A. Sagata, ApJ 380, 452 (1991).
116. G.C. Flickinger, T.J. Wdowiak, P.L. Gómez, ApJ 380, L43 (1991).

DISCUSSION

BUSSOLETTI — I would like to make a comment about the fact that "*graphite*" may be the most fashionable material to account for the interstellar bump. I personally do not believe this affirmation for several reasons:
1. according to the laboratory work that my group did, for several years, first in Lecce and now in Napoli, amorphous/disordered carbon grains seem to match better the interstellar bump though some problems remain still open.
2. graphite is a highly anisotropic material; it is difficult to produce "micronic" grains; experimentally you do not obtain spherical particles with graphite (bulk) charecteristics and, finally, the UV bump which sometimes is seen in the lab falls quite far from 2200 Å. Graphite grains represent a historical bias that, I believe, should be overcome nowadays.

WILLIAMS — In my talk, I gave what I believe is the most widely held views about the carrier of the interstellar "bump" at 217 nm. I agree that the view that the carrier is graphite probably persists for historical reasons, and I personnally share your concerns about the validity of this attribution.

LEGER — Are there clear examples of a star (presently) ejecting O rich matter and where C bearing molecules are seen, with reasonable certainty they have not turned from C rich to O rich in the past ?

WILLIAMS — There are examples in the literature, but this interpretation is, I believe, still controversial.

BOTSCHWINA — You had C_5O on your list of molecules observed in dark clouds. Is the detection of that species now certain ?

WILLIAMS — The detection is claimed by B.E. Turner.

D'HENDECOURT — What about depletion of O in diffuse clouds ? formation of ice ? or what else ?

WILLIAMS — The depletion of O towards ζ Oph is now apparently larger than can be accommodated in silicates and oxides. The oxygen is, however, not in H_2O-ice. There is no published interpretation of this surprising result.

LEACH — My remark concerns matching of observed UIR bands with laboratory I.R. spectra. The UIR are emission bands whereas most laboratory work on PAHs and other carrier candidates is in absorption. Peak wavelengths therefore can be somewhat different in the two cases for the same species.

There have been some laboratory studies on I.R. emission of PAHs. For example, Brenner and Baker (Ap.J., 1992) have studied the I.R. emission of benzene and naphthalene excited by laser U.V. excitation. The peaks differ in wavelength from those in absorption by amounts of the order of those discussed in your talk. Another interesting part of their study is that they followed the profile of the I.R. emission bands as a function of time i.e. as a function of total energy content decrease.

COX — You introduced TMC-1 as being a typical cold dark cloud. Now, I was wondering why TMC-1 is so different from other dark clouds having apparently similar physical conditions. More precisely, why is TMC-1 so extremely chemically-rich ?

WILLIAMS — TMC-1 is well-studied, and is unusually rich is molecules. I agree with you that in this respect it is atypical. Your question is an extremely good one; but it is not known why it is so chemically rich. My remarks about TMC-1 were meant to emphasize that the physical nature of the region is very much more complicated than is normally assumed in chemical models. These complications (density + velocity structure) may have a bearing on the chemistry in this region.

From Millimeter to Optical Wavelengths

OBSERVATIONS OF VIBRATIONALLY EXCITED H_2 IN NGC2023 THROUGH HIGH RESOLUTION IMAGING OF THE S(1) LINE EMISSION AT 2.121μm

D.Field
School of Chemistry, University of Bristol, Bristol BS8 1TS, UK

M.Gerin, S.Leach, J.L.Lemaire, G.Pineau des Forêts, F.Rostas, D.Rouan
Observatoire de Paris-Meudon, 92195 Meudon Cedex, France

D. Simons
Canada-France-Hawaii Telescope, PO Box 1597 Kamuela, Hawaii 96743 USA

ABSTRACT

We have used the Canada-France-Hawaii Telescope to obtain an image of vibrationally excited H_2 at 2.121μm in NGC2023, at a spatial resolution of ~1". The image has structures in the form of filaments of bright emission, showing clear spatial correlations and anti-correlations with extended red emission (ERE). Our results have implications for the mechanism of the formation of ERE. Interpreting our observations in terms of the standard UV mechanism for the excitation of H_2, we deduce that the surface of the emitting zone is characterised by undulations, whose morphology traces that of the emission.

INTRODUCTION

The reflection nebula NGC2023 is well-known to emit in the H_2 S(1) line at 2.121μm[1,2,3]. Both HD37903 and star C (S108) illuminate NGC2023 [4] to form the nebula emission. The nebula is an active zone of recent star-formation. The morphology and detailed composition of the nebula remain poorly characterized, largely because of a lack of high spatial resolution data, such data being at present limited to ERE at 1" resolution[5] and 2.1μm broadband emission at similar resolution[6]. In the present work we report a spectral line image of vibrationally excited hydrogen in the S(1) line at 2.121μm, with a spatial resolution of ~1" (= 2.2x10^{-3} pc at the distance of NGC2023). A more detailed account of this work will appear elsewhere[7].

OBSERVATIONS

NGC2023 was observed with the Redeye wide field camera at the F8 Cassegrain focus of the Canada-France-Hawaii Telescope on Feb.8th 1993. The camera was equipped with a Rockwell NICMOS 3 Hg:Cd:Te array of 256x256 pixels at 40 μm pitch. The image scale is 0.5" per pixel, giving an observed field of 128"x128". Fixed frequency filters were used for wavelength selection, with a 2.121μm filter, 1% bandpass, nominally centred on the v=1-0 S(1) H_2 emission line, and a 2.18μm filter, 1% bandpass, to record the background continuum. 300 second exposures were used for both line and continuum observations. Sky flats were recorded just before each exposure and were subtracted from each image. Halos surrounding the stars arose from effects of saturation in the images. Absolute flux calibrations were performed by observing the star HR2007 (K magnitude 4.45).

In fig. 1, we show an image of NGC2023 at $2.121\mu m$ in spectral line emission alone. In fig. 2, we show an image of the background continuum at $2.18\mu m$. Fig.2 closely resembles the image recorded in ref.6. The image in Fig. 1 was obtained by subtraction of the continuum, in Fig. 2, from the total emission (less sky background) at $2.121\mu m$. We used the flux in the faint star 46" east and 11" south of HD37903 to bring the background continuum at $2.18\mu m$ and the image at $2.121\mu m$ to the same intensity scale. The process of background subtraction does not fully eliminate the saturated star images, since residuals in these regions exceed the dynamic range chosen for Fig.1. We note that within the spectral range admitted by the $2.121\mu m$ filter, other lines of H_2 may contriubte up to 15% of the emission[3]. Significant contributions from other species would appear unlikely.

DISCUSSION

The most striking feature of our image in Fig. 1 is the presence of narrow filaments or shells of emission. Because of the high spatial resolution, the emitting zone may be seen to be broken up into a number of features, which were not generally identifiable in previous observations[1,8]. For example strong discrete features 30" to 40" to the north and east of HD37903, and strong emission near star C, are now found. Our data are taken at a very similar resolution to that of the ERE[5] and we find a striking set of spatial correlations between the morphology of the H_2 emission and the ERE. Some strong features in the ERE have however no counterpart in H_2 emission and vice versa. This has implications for the mechanism of the production of ERE, since the presently favoured mechanism, involving the reaction of hot H atoms with carbonaceous material[5,9], would require that ERE should always be accompanied by H_2 vibrational fluorescence, given a UV excitation mechanism of H_2 vibrational fluorescence[10] (see below).

Our observations also allow us to deduce some features of the morphology of the nebula. Absorption of UV, in the Lyman and Werner bands, around 100 nm, with subsequent fluorescent cascade to vibrationally excited states of the ground electronic state of H_2, is well-established as the origin of H_2 vibrational fluorescence in NGC2023[3,10,11]. However, UV radiation at 100 nm penetrates into the cloud only a very small depth[11,12], of typically 3×10^{-3} pc, for a cloud of density 10^4 cm^{-3}. In order to form bright filamentary emission, we propose that the surface of the nebula is wrinkled, showing undulatory structure. H_2 emission is preferentially observed along lines-of-sight tangential to the cloud surface, since these offer the maximum optical depth of irradiated H_2. Elsewhere, the nebula surface presents more regular surface to the UV flux and smaller lines-of-sight are present and weaker emission or no emission is observed. A generally undulatory morphology is also suggested by the ERE and in the dust emission in Fig. 2. We plan to observe NGC2023 in the para-H_2 S(2) line in Nov./Dec. 1993, again using the Redeye wide field camera on the CFHT. These observations should provide a useful test of our proposed morphology.

REFERENCES

1. I. Gatley, T. Hasegawa, H. Suzuki et al., Ap.J. 318, L73 (1987)
2. T. Hasegawa, I. Gatley, R.P. Garden, P.W.J.L. Brand, M. Ohishi, M. Hayashi and N. Kaifu, Ap.J. 318, L77 (1987)
3. M.G. Burton, T.R.Geballe, P.W.J.L. Brand, and A. Moorhouse, Ap.J. 352, 625 (1990)
4. S.M. Scarrott, C.D. Rolph and M.D. Mannion, MNRAS 237, 1027 (1989)
5. A.N. Witt and D.F. Malin, Ap.J. 347, L25 (1989)
6. K. Sellgren, M.W. Werner and H.L. Dinerstein, Ap.J. 400, 238 (1992)
7. D. Field, M. Gerin, S. Leach, G. Pineau des Forêts, F.Rostas, D.Rouan and D. Simons, Astr. Astrophys., to appear.
8. M.G.Burton, A. Moorhouse, P.W.J.L.Brand, P.F. Roche and T.R.Geballe, Interstellar Dust, IAU Symposium no.135, NASA Conference Publication 3036, Ed. A.G.G.M.Tielens and L.J.Allamandola (1989)
9. A.N. Witt and R.E. Schild, Ap.J. 325, 837 (1988)
10. J.H. Black and A. Dalgarno, Ap.J. 203, 132 (1983)
11. J.H. Black and E.F.van Dishoeck, Ap.J. 322, 412 (1987)
12. H.Abgrall, J. Le Bourlot, G. Pineau des Forêts, E. Roueff, D.R. Flower and L. Heck, Astr. Astrophys. 253, 525 (1992).

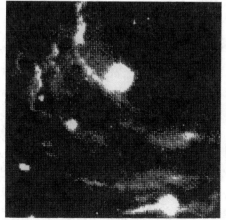

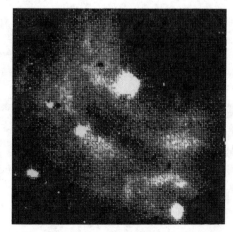

Fig. 1: An image of H_2 emission in the v=1-0 S(1) line at 2.121μm from NGC2023.

Fig. 2: An image of the continuum emission at 2.18μm from NGC2023.

MILLIMETER OBSERVATIONS OF ^{12}CO AND ^{13}CO IN THE DIFFUSE REGION AROUND Z OPH [*]

M. Kopp
ESO, Observatory La Silla, Casilla 19001, Santiago 19, Chile

M. Gerin
DEMIRM, URA 336 DU CNRS, Observatoire de Meudon, 92195 Meudon, France

E. Roueff
DAEC, URA 173 DU CNRS, Observatoire de Meudon, 92195 Meudon, France

INTRODUCTION

The line of sight towards the bright star ζ Ophiuchii presents one of the best studied diffuse cloud, which has been often used to test chemical models of the interstellar gas (Av ≈ 1). The understanding of the diffuse cloud has progressed with the detection of emission lines of carbon monoxyde CO [1,2,3] and of its isotopomer ^{13}CO [1,4]. There are indeed two velocity components (V_{LSR} = -0.7 and 0.5 kms^{-1}) along the line of sight which were not resolved in the UV absorption lines. Only recently have these two velocity components been separated in visible absorption lines of CN and CH [5]. However a detailed modelling and understanding of the abundance of interstellar molecules requires a good knowledge of the physical conditions in the cloud. The density and temperature can be derived either from absorption or from emission lines. Since all methods have their inherent uncertainties, it is better to compare several data sets. As for the ζ Oph clouds, both methods conclude to the presence of rather warm gas 30 - 100 K, but give conflicting results for the density. It is useful to obtain new informations on the clouds.in order to derive the physical conditions with more precision. This can be done by mapping the interstellar clouds responsible for the absorption lines. Indeed, Liszt [6] has shown that the same velocity components can be detected as far as 1° from the star and that the star itself is located in a minimum of emission. Two clouds located northern and southern of the star can be recognized on a large scale map [7] at the border of the Ophiuchus cloud complex. In order to derive more information about these clouds, we have observed a long cut in ^{12}CO (J=1-0) through the southern cloud starting from the star position. We have also observed ^{12}CO(J=2-1), ^{12}CO(J=3-2), ^{13}CO(J=1-0) and ^{13}CO(J=2-1) at selected positions along this cut.

---

[*] Based on observations obtained at the European Southern Observatory (ESO), La Silla (Chile).

2. OBSERVATIONS

The observations have been performed in January 1993 with the Swedish-ESO Submillimeter Telescope (SEST) at La Silla (Chile) under good weather conditions. The telescope half power beam width is 42" at 2.6 mm, 22" at 1.3 mm [8] and 22" at .8mm [9]; the pointing accuracy stood below 4" rms. At 2.6 mm, we used a Schottky mixer in SSB and the resulting system temperature was 670 K and lowered to 450 when we observed ^{13}CO(J=1-0). At 1.3 mm, we used an SIS mixer tuned in SSB and we obtained system temperatures ranging from 750 to 1800 K. CO (3-2) observations were performed at the SEST in July 1993 using a SIS mixer operated in SSB. The receiver temperature was 380K and the total system temperature ranged from 520K (at the highest elevation of the source) to 800K. Pointing and subpointing were checked every hour on the brigt source IRAS 15194 and was found accurate to better than 5"rms. We chose to perform all 2.6 mm observations in FS mode due to the large cloud size; at 1.3mm and 0.8mm we observed in PS mode with the same reference position as [3]. The spectra were calibrated with hot and cold loads and are presented in the T_A^* scale. This scale should be appropriate since the clouds are much more extended than the main beam, and the forward spillover and scattering efficiency is very close to 1; therefore, the data presented in Table II may be considered as lower limits to the true values of the brightness temperature. As a backend we used a 1000 channels A.O.S with a channel width of 43 kHz. The integration time was about 7-15 minutes for the ^{12}CO(J=1-0) spectra and was increased to 45 minutes to search for ^{13}CO lines.

Fig. 1

Top: ^{12}CO(J=1-0) (light line), ^{12}CO(J=2-1) (thick line) and ^{12}CO(J=3-2) (dotted line) at 3 selected positions: (0, 52) left, (0, 144) middle and (0, 228) right.
Bottom: ^{13}CO(J=1-0) spectra at the same positions.

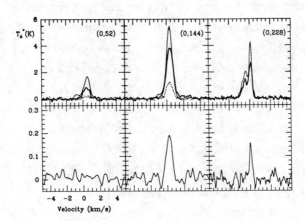

To study in details the variations of the line intensity as entering the clouds, we chose a map step of 44", and observed a five points map at 1.3 and 0.8 mm for each selected position in order to compare both lines at the same spatial resolution. As a whole, the data set comprises 65 $^{12}CO(J=1-0)$ positions, of which 6 have been observed in $^{12}CO(J=2-1)$, 2 in $^{12}CO(J=3-2)$ and 6 in $^{13}CO(J=1-0)$. Fig 1 presents the results for three typical positions in the cloud.

3. RESULTS AND CONCLUSION

We present the results of two different models of "translucent" interstellar clouds permeated by an isotropic incident UV radiation field:

a) an isochoric model (A) with a density of $n = 10^3$ cm^{-3}

b) an isobaric model (B) with pressure $p = 10^4$ K cm^{-3}

The geometry of the cloud is plane parallel and the radiation field, isotropically incident on one side of the cloud, is increased by a factor of 10 over the "standard" interstellar UV radiation field of Mathis et al.[10]. The abundances of the atomic and molecular species are calculated, as a function of the optical depth of the cloud, in a steady-state approach as described in Le Bourlot et al[11]. A total of 77 species are considered, including all isotopic substitutions of CO. The gas phase elemental abundances, relative to n_H, ($n_H = n(H) + 2n(H_2)$) are given in Table I together with the different parameters characterizing the ζ Oph environment. The turbulent velocity is taken to be 1 km/s, which does not affect the results of the models.

The atomic to molecular transition is dominated by discrete UV line photodissociation (especially the H_2 Lyman and Werner band dissociating lines) described in [14]. Fig 2a and 2b give the H, H_2 and CO column densities for the two models. The models differ essentially in the boundary of the cloud where the H/H_2 transition is dependent on self-shielding effects in the UV dissociating lines.

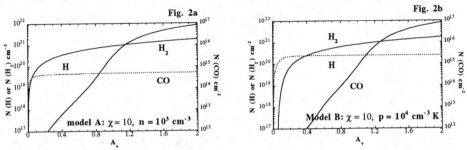

Fig. 2a: Column densities of H, H_2 and CO as a function of the visual extinction A_V for the isochoric model (A) at $n = 1000$ cm^{-3};

Fig. 2b: The same for the isobaric model (B) at $p = 10^4$ K.cm^{-3};

Table I: Physical parameters adopted in the models for the ζ Oph environment. Values in parenthesis refer to powers of ten.

Parameter	
[C] / n_H & depletion $\delta = C_{cosmic}/C_{\zeta Oph}$	7.35 (−5) [12] $\delta_C = 5$
[O] / n_H and depletion δ	1.80 (−4) [12] $\delta_O = 4.7$
$^{12}C/^{13}C$	70 [12, 13]
$^{16}O/^{18}O$	500 [12]
Grain radius	0.1 µm
M_{grain} / M_{gas}	0.01
χ (UV field strength)	10
ζ (Cosmic ray ionization rate)	1 (−17) s^{-1}
v_t (turbulent velocity)	1 km s^{-1}
R = A$_V$ / E(B−V)	3.1 [14]
N_H / E(B−V)	5.8 10^{21} cm^{-2} mag^{-1}
Extinction curve	Fitzpatrick & Massa [14]

The column density of atomic hydrogen in the isobaric model is about 2 10^{20} cm^{-2}, comparable to large scale observations [16]. The CO molecule is formed deeper inside inside the cloud. The same trend is found in Fig.3 which shows the temperature determined from the thermal balance and the derivative of the brightness temperatures of the CO(1-0) line regard to the visual extinction, which gives the location of the maximum of emissivity of the line, occuring at A$_V$ close to 1. This line is formed in a narrow region and becomes saturated deeper inside the cloud whereas the ^{13}CO line remains optically thin over a large A$_V$ range.

We give in Table II observed peak intensities and emissivity ratio at three positions, together with model results obtained at different visual extinctions. Both models predict reasonable values for the peak intensity of ^{12}CO and ^{13}CO (J=1-0) lines, but the isobaric model predictions for the 2-1/1-0 intensity ratio are closer to the observed data. Furthermore, the differences are more pronounced at the cloud boundary as shown from the temperature variations shown in Fig 3. This is a first attempt to model this transluscent cloud; the isobaric model seems to provide a better fit to the present observations than the isochoric model, since it includes a larger transition zone from atomic to molecular gas, with a low density and higher temperature (T ≈ 100 K, n ≈ 100

cm^{-3}). We shall explore the role of the radiation field in a more systematic way. Other molecules, such as CS and CN will be searched for in order to obtain better constraints on the models.

Table II
Comparison between observations and results of the two models A and B at different extinctions. P.I. refer to the peak intensities. The rms noise level of our CO(1-0), CO(2-1), CO(3-2) and ^{13}CO(1-0) spectra is respectively 65mK, 75mK, 50mK and 20mK.

	A_v	^{12}CO (1-0) P.I in K	^{13}CO (1-0) P.I. in K	^{12}CO (3-2) P.I. in K	^{12}CO/^{13}CO (1-0)	$T_{(2-1)}/T_{(1-0)}$ CO
pos 52		1.7	< 0.1	< 0.1	>20	0.5
pos 144		5.4	0.19	≈ 0.5	30	0.7
pos 228		3.8	0.15		27	0.7
Model A	0.5	1.1(-2)	2.2(-4)	4.(-3)	52	1.1
	1	1.6	7.5(-2)	4.8(-1)	22	0.88
	1.5	6.4	0.55	2.6	12	0.85
	2	8.3	1.0	3.2	8.2	0.80
Model B	0.5	1.8(-3)	3(-5)	1.6(-4)	66	0.46
	1	0.57	2.3(-2)	8.5(-2)	24	0.65
	1.5	5.0	0.42	1.7	12	0.80
	2	6.9	0.86	2.5	8.0	0.76

Fig. 3: Temperature profiles and derivative of the emissivity of the CO (1-0) line for the two models A and B

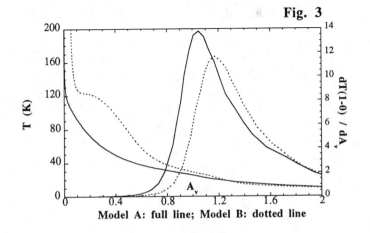

Model A: full line; Model B: dotted line

REFERENCES

1   W.D. Langer, A.E. Glassgold, R.L. Wilson : *ApJ* **322**, 450 (1987)

2   R.M.Crutcher, S.R. Federman, *ApJ* **316**, L71 (1987)

3   Le Bourlot J., Gerin M., Pérault M. , *A&A* **219**, 279 (1989)

4   T.L. Wilson, R. Mauersberger, W.D. Langer, A.E. Glassgold, R.W. Wilson, *A&A* **262**, 248 (1992)

5   D.L. Lambert , Y. Sheffer, P. Crane, *ApJ* **359**, L19 (1990)

6   H.S. Liszt, *ApJ* **390**, 226 (1992)

7   E.J. De Geus, L. Bronfman, P. Thaddeus *A&A* **231**, 137 (1990)

8   R.S.Booth, G. Delgado, M. Hagström, L.E.B. Johansson, D.C. Murphy, M: Olberg, N.D. Whyborn, A. Greve, B. Hansson, C.O. Lindström, A. Rydberg, *A&A* **216**, 315 (1989)

9   N Whyborn, L-A. Nyman, W. Wild, G. Delgado, *ESO Messenger* **45**, (1992)

10   J. S.Mathis, P. G. Mezger, N. Panagia, *A&A* **128** 212 (1983)

11   J Le Bourlot., G. Pineau des Forêts, E. Roueff & D.R. Flower, *A&A.* **267**, 233 (1993)

12   J.A. Cardelli, B.D. Savage, D.C Ebbets, *Ap J* **383** L23 (1991)
E.Anders, N. Grevesse, Geochim. Cosmochim. Acta **53** 197 (1989)

13   O. Stahl, T.L. Wilson, C Henkel, I. Appenzeller*A&A* **221** 321 (1989)

14   E.L. Fitzpatrick, D. Massa, *Astrophys. J..* **328**, 734 (1988).

15   H Abgrall, J. Le Bourlot, G. Pineau des Forets, E. Roueff, D.R.Flower, L.Heck, 1992, A&A. **253**, 525.

16   C.E. Cappa de Nicolau, W.G.L. Pöppel, *Astron. Astrophys.* **164**, 274 (1986)

HCO⁺ AND HCN OBSERVATIONS TOWARD DARK CLOUDS AND HERBIG Ae/Be STARS

F. Scappini, G. Bruni
Istituto di Spettroscopia Molecolare del CNR, Via de' Castagnoli, 1
40126 Bologna, Italy

G.G.C. Palumbo
Dipartimento di Astronomia, Università degli Studi, Via Zamboni, 33
40126 Bologna, Italy and ITESRE-CNR, Via de' Castagnoli, 1
40126 Bologna, Italy

P. Bergman
Onsala Space Observatory, 43900 Onsala, Sweden

ABSTRACT

We have studied a sample of seven dark clouds (Bok globules) and eight Herbig Ae/Be stars in the J=1→0 transition of HCO⁺, H^{13}CO⁺, HCN, and H^{13}CN. The most abundant isotopomers were found in almost all the sources (detection rate 70-90%).

From the derived physical parameters and column densities it seems that a quite similar scenario applies for the dark clouds and the gas around the Herbig stars. These similarities, together with the already known features from literature, suggest that the presently studied globules are likely to be sites of low-mass star formation.

INTRODUCTION

The direct observation of the molecular gas content in regions where star formation is supposed to take place or where stars have been recently formed is relevant to the understanding of the phenomena associated with the earliest stages of stellar evolution.

In order to contribute data to this problem we are conducting a systematic study of molecular lines in: (i) dark molecular clouds, also known as Bok globules, and (ii) regions surrounding Herbig Ae/Be stars.

The first class of objects has the characteristic of being rather isolated, in the sense that they have no associated large molecular cloud, which would affect the region of interest [1]. Moreover, they are small enough to ensure a relatively simple interpretation of the occurring physical and chemical processes. The Herbig Ae/Be stars are objects of intermediate mass in their pre-main sequence stage of evolution. Because of their youth they are still embedded in the dark clouds which have originated them [2,3].

The molecular species that we have observed so far are: HCO⁺ (J=1→0),

HCN (J=1→0), and the corresponding ^{13}C isotopomers. These species are known to exist in regions where stars are forming [4]. Thus, their detection, together with their abundance, should allow us to characterize the chemical and physical conditions of the different stages in the evolution sequence of a small molecular cloud toward star formation.

OBSERVATIONS

Observations were made in February and May 1990 with the 20m radiotelescope at Onsala. The half-power beamwidth of the telescope was 45" at 90 GHz. The spectrometer used was a 256-channel filter-bank with 250 kHz resolution (0.83 kms^{-1}). The SSB-tuned SIS receiver was operated in a dual beam switching mode and a chopper wheel calibration technique was used. The main beam efficiency was determined to be 0.43. We position switched, with one on-source per off-source measurement and 90s integration time on each position.

DESCRIPTION OF THE SAMPLE

In Table I we present a tabulation of the sources, consisting of seven Bok globules and eight Herbig Ae/Be stars.

For the choice of the globules our selection criteria are: (i) they have IRAS point sources, (ii) these point sources have infrared colours corresponding to those of Beichman[5] and Emerson[6] samples for T-Tauri stars and cores. Finally, (iii) they have quite large CO linewidths. The above selection criteria clearly privilege star forming clouds.

For the choice of the Herbig Ae/Be stars our criteria are: (i) they have optical outflows and (ii) they have P-Cygni profiles (Hα lines).

RESULTS

Figures 1 and 2 show the spectra of HCO$^+$ (J=1→0) and HCN (J=1→0) in CB34 (Bok globule) and Figures 3 and 4 show the spectra of the same species in LkHα234 (Herbig Ae star).

HCO$^+$ was detected in all seven Bok globules and around six Herbig Ae/Be stars. HCN was found in five globules and around five stars. In those sources where HCO$^+$ and HCN exhibited the strongest signals we searched also for H^{13}CO$^+$ and H^{13}CN. In fact, H^{13}CO$^+$ was found in one Bok globule and around three Herbig Ae/Be stars and H^{13}CN around one star.

The HCO$^+$/HCN column density ratio distribution among the investigated Bok globules is 0.3-1.3 and among the Herbig Ae/Be stars is 0.4-1.9. This shows that there is a large overlapping density interval between the two categories of objects. Table II compares the HCO$^+$/HCN column density ratio, excitation temperature, density and linewidth of the present sample with

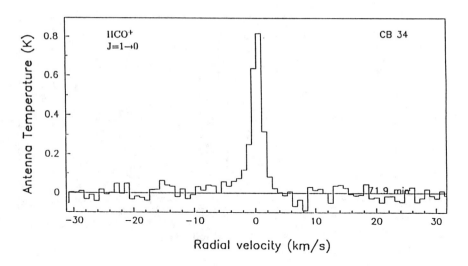

Figure 1: Line profile of the J=1→0 transition of HCO$^+$ in CB34 (Bok globule)

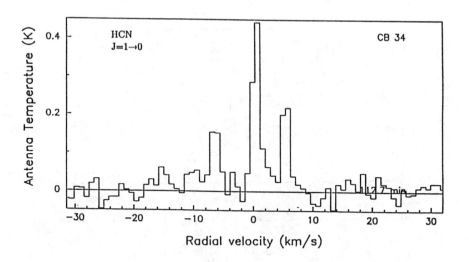

Figure 2: Same as Fig. 1, but for HCN

42 HCO$^+$ and HCN Observations

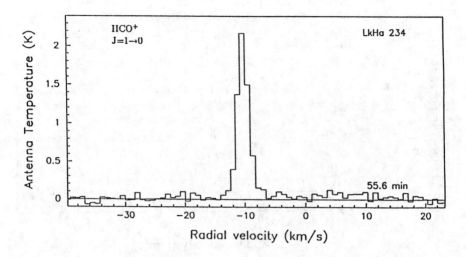

Figure 3: Line profile of the J=1→0 transition of HCO$^+$ in LkHα234 (Herbig Ae star)

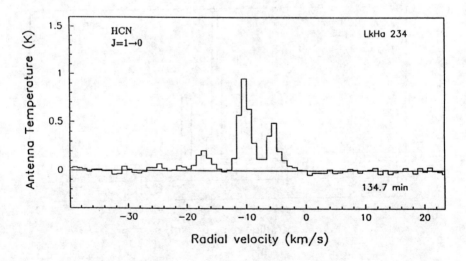

Figure 4: Same as Fig. 3, but for HCN

those of other known regions. The chemical and physical scenario of the investigated objects looks similar to those found in cold regions, but linewidths are larger than thermal broadening alone. Moreover, the observed Bok globules and the gas around the Herbig stars show similar molecular abundances and physical conditions. These similarities, together with other features, such as embedded infrared sources, water maser emission, CO outflow, and large CO linewidth, suggest that the presently studied globules are likely to be sites of low mass star formation.

Table I   List of the sources and of their physical properties

Source name	α_{1950} (hms)			δ_{1950} (° ′ ″)			v_{LSR}^a (kms^{-1})	Comments
	Bok globules							
LBN594[1]	00	25	59.0	56	25	32	-38.3	H$_2$O maser[b]
LBN613[1]	00	46	34.0	50	28	25	-12.5	
L1534[2]	04	36	31.6	25	35	56	6.1	
CB34[1]	05	44	03.0	20	59	07	0.7	
L810[3]	19	43	21.0	27	43	37	15.8	H$_2$O maser[c]
L797[1]	20	03	44.0	23	17	54	12.6	
L1262[1]	23	23	48.0	74	01	07	4.0	
	Herbig Ae/Be stars							
LkHα198[4]	00	08	44.0	58	33	06	-0.7	
RRTau[5]	05	36	23.3	26	20	56	-5.4	
HD250550[4]	05	59	06.5	16	30	58	2.1	
V645Cyg[4]	21	38	10.6	50	00	43	-44.6	H$_2$O maser[d]
LkHα234[4]	21	41	57.0	65	53	09	-7.8	H$_2$O maser[e]
BD463471[4]	21	50	38.5	46	59	34	7.0	
LkHα233[6]	22	32	28.2	40	24	33	0.1	
MWC1080[4]	23	15	14.9	60	34	19	-29.1	

References - (1) Clemens and Barvainis 1988; (2) Meyer et al. 1987; (3) Nekel et al. 1985; (4) Lada 1985; (5) Loren 1981; (6) Levreault 1988.

[a] From CO millimeter-wave measurements in the clouds (CB) and in the star environments (Cantò et al. 1984).
[b] Scappini et al., 1991.
[c] Neckel et al., 1985.
[d] Lada et al., 1981.
[e] Rodriguez et al., 1980.

Future observations will be aimed at other molecular species and also at more quiescent globules in order to eventually detect differences between active and inactive regions.

Table II The HCO$^+$/HCN column density ratio for different regions together with excitation temperature, density and linewidth information

Source	HCO$^+$/HCN	$T(K)$	$n(\text{cm}^{-3})$	$\Delta v(\text{kms}^{-1})^a$	Ref.
TMC-1	0.4	10-20	10^3-10^5	0.2-0.9	7
L134N(L183)	2.0	10-20	10^3-10^5	0.2-0.9	7
Orion ridge	0.15	50-100	10^4-10^6	2.5-4	7
Orion plateau	0.03	90-150	10^6-10^7	20	7
Bok globules	0.3-1.3	10	10^3-10^4	1.1-2.2	b
Herbig Ae/Be stars	0.4-1.9	10		1.3-3.1	b

^{a}FWHM linewidth.
^{b}Present work. The density of the Bok globules is taken from Leung, 1985 (Ref. 8)

REFERENCES

1. D. P. Clemens and R. Barvainis, Ap. J. Suppl. 68, 257 (1988).
2. G. H. Herbig, Ap. J. Suppl. 4, 337 (1960).
3. U. Finkenzeller and R. Mundt, Astron. Astrophys. Suppl. Ser. 55, 109 (1984).
4. T. W. Harquist, J. M. C. Rawlings, D. A. Williams, and A. Dalgarno, Q. J. R. Astr. Soc., 34, 213 (1993).
5. C. A. Beichman, P. C. Meyers, J. P. Emerson, S. Harris, R. Mathieu, P. J. Benson and R. E. Jennings, Ap. J. 307, 337 (1986).
6. J. P. Emerson, Star Forming Regions, Proc. IAU Symp 128, eds. M. Peimbert and J. Jugaku (Kluwer, Dordrecht, 1987), p. 19.

7. W. M. Irvine, F. P. Schloerb, Å. Hjalmarson and E. Herbst, Protostars and Planets II, eds. D.C. Black and M.S. Matthews (The University of Arizona Press, Tucson, 1985), p. 579.
   W. M. Irvine, P. F. Goldsmith and Å. Hjalmarson, Interstellar Processes, eds. D.J. Hollenbach and H.A. Thronson, Jr. (Reidal, Dordrecht, 1987), p. 561.
8. C. M. Leung, Protostars and Planets II, eds. D.C. Black and M.S. Matthews (The University of Arizona Press, Tucson, 1985), p. 104.

MOLECULAR ABUNDANCES IN TRANSLUCENT CLOUDS

Roland Gredel
European Southern Observatory, Casilla 19001, Santiago 19, Chile

ABSTRACT

Sensitive searches for ^{13}CO and C^{18}O $J=1\to 0$ emission are presented for a sample of translucent molecular clouds previously studied by optical absorption lines towards background stars. C^{18}O has been detected in only 5 clouds, most of which have extinctions well in excess of 2 mag. The inferred ^{13}CO/C^{18}O abundance ratios range from 7–25, and the lower limits from >13 to >35. These values are as much as five times larger than the overall interstellar $([^{13}C]\cdot[^{16}O])/([^{12}C]\cdot[^{18}O])$ ratio, suggesting that isotope–selective photodissociation plays a role in at least some of the clouds. Searches for other molecules at millimetre wavelengths have been made for a few of the best characterized clouds. Surprisingly, no emission was detected from the C$_2$H or C$_3$H$_2$ molecules, even though the abundances of diatomic C$_2$ and CH are quite large. On the other hand, the abundance of HCO$^+$ appears comparable to that found in denser clouds, and the abundance of HCN may be up to an order of magnitude larger than the predictions of models in two clouds.

INTRODUCTION

In a recent series of papers, we have presented optical absorption line observations of a sample of southern translucent clouds (van Dishoeck & Black 1989; Gredel, van Dishoeck & Black 1991, 1993; Gredel et al. 1992). These clouds have visual extinctions in the range A_V=1–5 mag, and are termed translucent because photons from the interstellar radiation field can influence the chemistry. The optical observations have resulted in column densities of CH, CH$^+$, CN and C$_2$ towards the stars, have allowed determination of the kinetic temperature, density and electron fraction of the foreground material, and have provided constraints on the kinematics through determinations of the Doppler parameter along the lines of sight.

We present here complementary millimetre observations of CO and other species for the same set of clouds. Millimetre observations of molecules such as CS, C$_2$H and HCO$^+$, which have not been studied optically but which are expected to be significant in translucent clouds, will yield important tests of the chemistry. The main advantage of the present work is that the reddening towards the background stars is well known, which allows in principle an accurate determination of the foreground total hydrogen column densities.

OBSERVATIONS AND RESULTS

The observations were carried out between 1989 and 1993 with the Swedish–

ESO Submillimetre Telescope (SEST) (Booth et al. 1989). Emission from ^{13}CO $J = 1 \to 0$, $C^{18}O$ $J = 1 \to 0$, and various other molecules was detected towards a few stars. In general, the emission occurs at velocities close to those of the absorption lines.

LINE RATIOS

The ratio of the ^{12}CO $J = 1 \to 0$ and ^{13}CO $J = 1 \to 0$ antenna temperatures of our sample of translucent clouds increases significantly if T_A^* (^{12}CO) < 3 K. A similar variation was found from comparison of ^{12}CO and ^{13}CO emission at various positions within a single translucent cloud, i.e. the cloud towards HD 210121 (Gredel et al. 1992; see also Stark 1993). As discussed in those papers, the variation arises primarily because the ^{12}CO millimetre lines become optically thick, whereas those of ^{13}CO remain optically thin. The average of all measurements presented here is T_A^* (^{12}CO) / T_A^* (^{13}CO) = 6.2 ± 5.5 but reaches values > 20 for the most diffuse clouds. A similar value has been determined by Polk et al. (1988) who found 6.7 ± 0.7 for a large-scale average of diffuse and dense material in the Galactic plane.

The $C^{18}O$ $J = 1 \to 0$ line was detected in 12 lines of sight. The ^{13}CO / $C^{18}O$ antenna temperature ratios range from 7 (HD 94413) and 13 (HD 62542) to values around 18 (HD 169754) and 25 (HD 169454). $C^{18}O$ was not detected towards the stars HD 80077, HD 110432, and BD $-14°5037$. The respective lower limits to the ^{13}CO / $C^{18}O$ ratios are 13, 16, and 35. The overall interstellar ($[^{13}C]\cdot[^{16}O])/([^{12}C]\cdot[^{18}O])$ isotope ratio is in the range 5.5–8 in the solar neighborhood. Thus, the ^{13}CO /$C^{18}O$ ratios are generally consistent with or larger than the overall isotope ratio, suggesting that $C^{18}O$ may be underabundant by up to a factor of five.

MOLECULAR COLUMN DENSITIES

A major uncertainty in the determination of column densities from the emission line observations is the density in the cloud. The excitation of the C_2 and CN molecules obtained from optical absorption line observations usually indicates rather low densities of typically a few hundred cm^{-3} (Black & van Dishoeck 1991). On the other hand, millimetre observations of CO $J=3\to2$ indicate in some cases higher densities of up to a few thousand cm^{-3} (van Dishoeck et al. 1991). The column densities inferred towards HD 29647, HD 62542, HD 94413, HD 154368, and HD 169454 are presented in Table 1, which includes the values for CH, C_2 and CN derived from the optical measurements. The upper entry for each molecule indicates the column density if the lowest density permitted by observations is used; the second entry the result if the highest density is taken. For HD 62542, we also list results with $n(H_2) \approx 2 \times 10^4$ cm^{-3}, as inferred by Cardelli et al. (1990). The temperatures were taken to be those derived from the C_2 analyses.

Table 1 includes the results from the translucent model of van Dishoeck & Black (1989) for the HD 29647 cloud, as well as observations of other molecules towards this cloud by Crutcher (1985). The most remarkable finding is that the C_2H molecule is not detected in any of the clouds. The gas–phase chemistry of this radical is intimately connected to that of C_2, which has quite large column densities. For the low end of the density range, the observed values are still just consistent with the models, but if the higher densities apply, they are factors of 3–10 lower. The most likely explanation would be an underestimate of the C_2H photodissociation rate, although it is unlikely to be in error by as much as an order of magnitude (van Dishoeck 1988). The complex carbon–bearing molecule C_3H_2 is also not detected in any of the clouds, even though Madden et al. (1989) find a very weak line at 18.3 GHz towards HD 29647. The column density inferred from this line is a factor of two below the upper limits listed in Table 1 from searches for the 85.3 GHz line. This lack of more complex carbon–bearing molecules is somewhat surprising, since they are found quite abundantly in some dense cores, where their formation is thought to be stimulated by large amounts of atomic C and C^+ in the gas phase (Suzuki 1983, Suzuki et al. 1988, 1992). These translucent clouds contain plenty of atomic carbon, but even C_2H is hardly seen.

On the other hand, the HCO^+ ion is detected in a number of clouds, most notably towards HD 62542. Because of rapid dissociative recombination with the abundant electrons in the cloud, this is quite surprising and the observed values are up to an order of magnitude larger than the models if the low densities apply. The large HCN column densities inferred for HD 29647 and HD 62542 are also remarkable, although the uncertainty in density precludes any conclusions about significant variations from cloud to cloud. At low densities, the observed HCN values are again significantly higher than the model results, but at higher densities they are comparable. Note also that the model HCN values are expected to increase at higher density.

The observed CS column densities are reasonably consistent with the model results of Drdla et al. (1989), especially if the variation of nearly an order of magnitude due to uncertainties in key molecular parameters and the gas–phase sulfur abundance is taken into account.

The column densities of the various molecules with respect to that of H_2 are summarized in Table 2. The H_2 column densities along the lines of sight were estimated from the measured total A_V and from the empirical relation between CH and H_2 (Danks et al. 1984; see van Dishoeck & Black 1989). Although some of the millimetre emission may come from material located behind the star, this is not expected to change the numbers by more than a factor of two for the clouds listed in Table 2. The resulting abundances $X=N(AB)/N(H_2)$ can be compared with those found in other, denser clouds. A particularly interesting case is the small reflection nebula IC 63, for which Jansen et al. (1993) have recently presented accurate abundances of a number of the same species. Compared with the translucent clouds studied in this work, IC 63 is exposed to more intense radiation ($I_{UV} \approx 900$) and is denser, $n(H_2) \approx 5 \times 10^4$ cm^{-3}. The IC 63 cloud resembles that towards HD 62542 in the sense that both of them may be rather

Table 1: Molecular Column Densities (in cm^{-2})[a]

Species	HD29647	HD62542	HD94413	HD154368	HD169454	Model[b]
H_2	3.0(21)	1.0(21)	1.3(21)	1.8(21)	1.6(21)	3.0(21)
CO	$\geq$4(16)	5.0(16)	1.2(16)	1.5(16)	1.8(16)	2.9(17)
	$\geq$1(16)	1.4(16)	7.0(15)	6.0(15)	5.5(15)	
^{13}CO	1.0(16)	1.1(15)	2.0(15)	1.0(15)	2.0(15)	
	4.0(15)	6.5(14)	1.4(15)	6.0(14)	9.0(14)	
$C^{18}O$	1.2(15)[c]	2.0(13)	1.2(14)	...	4.0(13)	
	6.0(14)[c]	1.5(13)	1.0(14)	...	2.0(13)	
CH[d]	1.5(14)	3.8(13)	4.4(13)	5.4(13)	4.4(13)	1.1(14)
C_2[d]	1.7(14)	8.0(13)	3.5(13)	5.8(13)	7.0(13)	1.6(14)
CN[d]	1.6(14)	4.2(13)	5.2(13)	3.3(13)	4.2(13)	8.0(13)
CS	2.6(14)[e]	6.5(13)	1.7(13)	...	5.7(13)[e]	
	1.5(13)[e]	8.0(12)	2.2(12)	...	8.5(12)[e]	
		1.0(12)				
C_2H	<1.5(14)[e]	<2.0(13)	...	<9.0(13)	5.4(13):	2.7(13)
	<1.0(13)[e]	<3.0(12)	...	<2.5(13)	1.0(13):	
		<7.0(11)				
HCN	5.0(14)[e]	8.0(13)	<5.0(13)	<1.1(14)	<1.4(14)	2.4(12)
	2.5(13)[e]	1.0(13)	<7.0(12)	<1.8(13)	<1.8(13)	
		1.2(12)				
HCO^+	9.0(13)[e]	2.2(13)	2.5(13):	2.3(13)	1.4(13)	3.4(12)
	5.0(12)[e]	2.5(12)	3.2(12):	4.5(12)	2.0(12)	
		4.0(11)				
C_3H_2	<3.0(13)	<1.5(13)	<9.0(12)	...	...	
	<5.0(12)	<2.5(12)	<1.3(12)	...	...	
		<2.5(11)				

[a] The upper entry for each species corresponds to the lowest density allowed by the observations, the second entry to the highest density consistent with the data. The third entry for HD 62542 corresponds to $n(H_2) = 2 \times 10^4$ cm^{-3}. A colon indicates an uncertain value. Adopted physical parameters for each star: HD 29647: T=15 K, $n(H_2)$=350 and 5000 cm^{-3}; HD 62542: T=40 K, $n(H_2)$=300, 2000 and 20000 cm^{-3}, HD 94413: T=40 K, $n(H_2)$=500 and 3000 cm^{-3}, HD 154368: T=25 K, $n(H_2)$=250 and 1000 cm^{-3}, HD 169454: T=15 K, $n(H_2)$=300 and 1500 cm^{-3}.
[b] Model for HD 29647 cloud by van Dishoeck & Black (1989).
[c] Based on observations of Crutcher (1985).
[d] From optical absorption lines.
[e] Based on observations of Drdla et al. (1989).

Table 2: Molecular Abundances relative to H_2

Species	HD29647	HD62542	HD94413	HD154368	HD169454	IC 63	TMC-1
H_2	1.0	1.0	1.0	1.0	1.0	1.0	1.0
CO	$\geq$1.3(-5)	5.0(-5)	9.2(-6)	8.3(-6)	1.1(-5)	6.0(-5)	8(-5)
	$\geq$4.0(-6)	1.4(-5)	5.4(-6)	3.3(-6)	3.4(-6)		
^{13}CO	3.3(-6)	1.1(-6)	1.5(-6)	5.6(-7)	1.3(-6)	4.4(-7)	4(-6)
	1.3(-6)	6.5(-7)	1.1(-6)	3.3(-7)	5.6(-7)		
$C^{18}O$	4.0(-7)	2.0(-8)	9.2(-8)	...	2.5(-8)	8.0(-8)	
	2.0(-7)	1.5(-8)	7.7(-8)	...	1.3(-8)		
CH^b	5.0(-8)	3.8(-8)	3.4(-8)	3.0(-8)	2.8(-8)	...	2(-8)
C_2^b	5.7(-8)	8.0(-8)	2.7(-8)	3.2(-8)	4.4(-8)	...	...
CN^b	5.3(-8)	4.2(-8)	4.0(-8)	1.8(-8)	2.6(-8)	4.8(-9)	3(-8)
CS	8.7(-8)	6.5(-8)	1.3(-8)	...	3.6(-8)	1.5(-9)	3(-8)
	5.0(-9)	8.0(-9)	1.7(-9)	...	5.3(-9)		
		1.0(-9)					
C_2H	<5.0(-8)	<2.0(-8)	...	<5.0(-8)	3.4(-8):	4.0(-9)	8(-8)
	<3.3(-9)	<3.0(-9)	...	<1.4(-8)	6.3(-9):		
		<7.0(-10)					
HCN	1.7(-7)	8.0(-8)	<3.8(-8)	<6.1(-8)	<8.8(-8)	3.6(-9)	2(-9)
	8.3(-9)	1.0(-8)	<5.4(-9)	<1.0(-8)	<1.1(-8)		
		1.2(-9)					
HCO$^+$	3.0(-8)	2.2(-8)	1.9(-8):	1.3(-8)	8.8(-9)	3.6(-9)	2(-8)
	1.7(-9)	2.5(-9)	2.5(-9):	2.5(-9)	1.3(-9)		
		4.0(-10)					
C_3H_2	<1.0(-8)	<1.5(-8)	<6.9(-9)	...	...	...	2(-8)
	<1.7(-9)	<2.5(-9)	<1.0(-9)	...	...		
		<2.5(-10)					

^{a} The upper entry for each molecule corresponds to the lowest density allowed by the observations, the second entry to the highest density consistent with the data. The third entry for HD 62542 refers to $n = 2 \times 10^4$ cm^{-3}. A colon indicates an uncertain value. See Table 1 for details.

^{b} From optical absorption line observations.

dense clumps of gas with stellar winds having blown away most of the surrounding more diffuse gas. It is seen that the abundances of most species are similar within an order of magnitude. Diatomic molecules such as CN and CS appear relatively more abundant towards HD 62542, whereas polyatomic species such as C_2H are probably more prominent in the IC 63 cloud.

Another interesting comparison is that between the cloud towards HD 29647 and the TMC–1 core, which is located only 10′ away. The abundances of most polyatomic molecules appear significantly higher in TMC–1, especially for C_2H and C_3H_2, emphasizing again the unique nature of TMC–1 regarding complex carbon–bearing molecules.

REFERENCES

Black, J.H., van Dishoeck, E.F., 1991, ApJ 369, L9
Booth, R.S., Delgado, G., Hagström, M., Johansson, L.E.B., Murphy, D.C., Olberg, M., Whyborn, N.D., Greve, A., Hansson, B., Lindström, C.O., Rydberg, A., 1989, A&A 216, 315
Cardelli, J.A., Suntzeff, N.B., Edgar, R.J., Savage, B.D., 1990, ApJ 362, 551
Crutcher, R.M., 1985, ApJ 288, 604
Danks, A.C., Federman, S.R., Lambert, D.L., 1984, A&A 130, 62
Drdla, K., Knapp, G.R., van Dishoeck, E.F., 1989, ApJ 345, 815
Gredel, R., van Dishoeck, E.F., Black, J.H., 1991, A&A 251, 625
Gredel, R., van Dishoeck, E.F., de Vries, C.P., Black, J.H., 1992, A&A 257, 245
Gredel, R., van Dishoeck, E.F., Black, J.H., 1993, A&A 269, 477
Jansen, D.J., van Dishoeck, E.F., Black, J.H., 1993, A&A in press
Madden, S.C., Irvine, W.M., Matthews, H.E., Friberg, P., Swade, D.A., 1989, AJ 97, 1403
Polk, K.S., Knapp, G.R., Stark, A.A., Wilson, R.W., 1988, ApJ 332, 432
Stark, R., 1993, Ph. D. Thesis, University of Leiden, The Netherlands
Suzuki, H., 1983, ApJ 272, 579
Suzuki, H., Ohishi, M., Kaifu, N., Kasuga, T., Ishikawa, S., Miyaji, T., 1988, Vistas Astron. 31, 459
Suzuki, H., Yamamoto, S., Ohishi, M., Kaifu, N., Ishikawa, S.-I., Hirahara, Y., Takano, S., 1992, ApJ 392, 551
van Dishoeck, E.F., 1988, in *Rate Coefficients in Astrochemistry* (eds. T.J. Millar & D.A. Williams), p. 49, Kluwer, Dordrecht
van Dishoeck, E.F. Black, J.H., 1989, ApJ 340, 273
van Dishoeck, E.F., Black, J.H., Phillips, T.G., Gredel, R., 1991, ApJ 366, 141

1

The 30 micron Emission Band in Carbon-Rich Pre-Planetary Nebulae

A. OMONT[1], P. COX[2], H. MOSELEY[3],
W.J. GLACCUM[3], S. CASEY[3], T. FORVEILLE[4] et al.

[1] *Institut d'Astrophysique de Paris*
[2] *Observatoire de Marseille and MPIfR, Bonn*
[3] *Goddard Space Flight Center, NASA, Greenbelt*
[4] *Observatoire de Grenoble*

Introduction

Circumstellar envelopes around evolved stars the main known site of dust formation. Amongst them, those in transition between the AGB red giants and planetary nebulae (often called pre-planetary nebulae, PPNe) are particularly interesting because of the rapid variations of the physical conditions and possibly of the dust composition. One of the most intriguing class of such PPNe exhibits a strong unidentified band at 21 μm (Kwok, Volk and Hrivnak, 1989). These sources, known to be carbon-rich, show strong emission features in the 6 to 8 μm range, a feature at 3.3 μm as well as strong 12 - 15 μm emission plateaux (Buss et al., 1990). Although the bands are reminiscent of PAHs the ratios between the individual bands are quite different from those observed in ordinary PAH sources. This fact, together with the need to achieve the excitation with visible light, lead Buss et al. to propose that the emitter is some hydogenated solid amorphous carbon or cluster of PAHs. In this context, we explored with the GSFC spectrophotometer on board of the Kuiper Airborne Observatory the spectral range between 23 and 50 μm in order to search for possible additional features which could help to identify the carrier of the 21 μm band. A detailed analysis of these observations are given in Omont et al. (1994) and we present here the main results of this study.

Results

Fig. 1 displays the 16 to 48 μm spectra obtained with the GFSC spectrophotometer of the four PPNe with the 21 μm band. Three facts are immediately striking: i) the wide variety of shapes and relative intensities of the 21 μm band, ii) the absence of any narrow feature similar to the 21 μm feature in

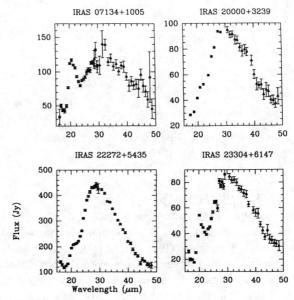

Fig. 1.1. The KAO GFSC 16 – 48 μm spectra of the four carbon-rich protoplanetary nebulae clearly displaying the 21 μm emission feature. The filled squares show the first channel (16 – 29 μm, $\Delta\lambda = 0.5\,\mu$m), and the filled triangles the 30 to 48 μm channel with $\Delta\lambda = 1.8\,\mu$m.

the 23 – 48 μm range, iii) the strong excess of emission from 24 to ~ 45 μm particularly in IRAS 22272+5435.

The composite infrared spectra of these sources are shown in Fig. 2 from 5 to 48 μm together with the 60 and 100 μm IRAS fluxes. Beside the 21 and 30 μm emission bands, the PPNe show a broad feature peaking around 11 μm and extending out to 15 μm. For IRAS 22272+5435 and IRAS 07134+1005, the KAO measurements by Buss et al. (1990) display a strong plateau of emission from 6 to 8 μm together with a 6.2 μm emission feature attributed to circumstellar PAH molecules.

In order to quantify the characteristics of the broad emission bands, we display the continuum emission expected i) from the sum of two modified black-bodies $\nu^\beta\,B_\nu(T)$ at temperatures T_1 and T_2 (with a dust emissivity index of $\beta = 1$), and ii) from a detailed radiative transfert analysis where we assumed the dust to be amorphous carbon (sample AC of Rouleau and Martin, 1991). Both methods provide reasonable approximations for the underlying continuum but clearly neither of them do account for the excess emission observed at shorter and longer wavelengths.

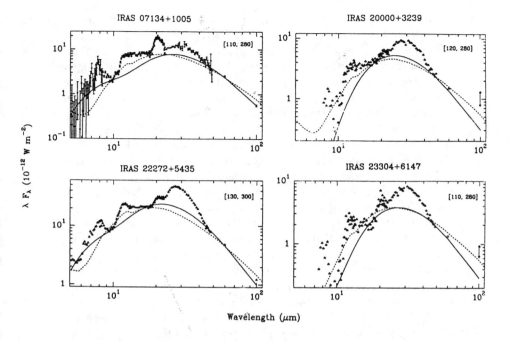

Fig. 1.2. Composite infrared spectra of the sources shown in Fig. 1 combining the present KAO GSFC data with the *IRAS* LRS spectra, and for IRAS 07134 and IRAS 22272 the KAO FOGS 5 − 8 μm spectra of Buss et al. (1990). Solid lines represent the emission expected from grains at temperatures [T_1, T_2] with emissivities proportional to λ^{-1}. Dashed lines are the results of a transfert model described in the text.

The main results of this work are:

The identification of the 21 μm band emitter remains open. The present observations have strengthened the broad variety of the shape and the strength of this feature and, hence, the likely transient nature of its carrier. Moreover, this feature appears to be only seen in the short-lived PPN phase since it has never been detected in the spectra of carbon-rich AGB stars and planetary nebulae. The relation of the 21 μm feature to carbonaceous features observed at shorter wavelengths is the most appealing explanation, but it is not fully demonstrated since there is yet no definite candidate.

The principal result of our observations is the strong evidence of the 30 μm emission excess in all the observed sources and its large strength in many of them. Our observations further prove that it corresponds to a major dust compound in a large fraction of C-rich envelopes. Previous measurements have shown that the 30 μm emission band accounts for an important part of the emissivity between 25 and 45 μm in prominent cold C-rich envelopes at various stages of evolution including the AGB stars IRC+10216, AFGL 3068, the planetary nebula IC 418, NGC 6572 and NGC 7027 (see Cox, 1993 and references therein). Its apparent absence in warmer envelopes such as V Cyg or S Cep could be due to the need of an efficient dust condensation process at low temperature implying a large mass-loss rate.

The fact that *all* the sources in which the 30μm emission band is present are C-rich, seems to exclude oxygen-rich species as its carriers (none of the O-rich AGB sources observed so far show the presence of the 30μm band). In particular, this makes species such as carbonates unlikely although they are known to have a prominent 30 μm band providing a reasonable fit to the observed band (Mc Carthy et al., 1978).

It has been suggested by Goebel and Moseley (1984) that the emitter is solid MgS. Recently, Mutschke et al. (1993) derived the optical constants of Mg - Fe sulphides in the wavelength range 10 to 500 μm (see also B. Begeman, this volume). The absorption coefficient of MgS is shown in Fig. 3 and compared to the excess emission in IRAS 22272+5435. Although the agreement is rather good, we note that the strength of the emission in IRAS 22272 and IC 418 (in both sources the band contains up to 20 % of the bolometric luminosity) puts strong abundance constraints, requiring at least most of the sulfur and magnesium unless the MgS is in a thin layer around an amorphous carbon core (Goebel and Moseley, 1985). But clearly one should also carefully consider the case of various carbonaceous material which remain an alternative explanation since they are major compound of dust, with a variety of possible structures and composition.

Conclusions

In conclusion, the present 16 to 48 μm KAO spectra of C-rich PPNe show that the 30 μm band is variable being in some instances quite weak whereas in other sources it accounts for about 20 % of the bolometric luminosity. However, its strength is not correlated with that of the 21 μm feature implying different carriers for both features. Unlike the 21 μm feature which

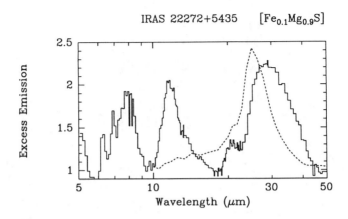

Fig. 1.3. Excess emission of IRAS 22272+5435 compared to the normalized absorption coefficient of small grains of magnesium sulphide (dashed line) using the optical constants measured by Mutschke et al. (1993).

has only be detected in PPNe, the 30 μm band is also seen in sources both in the AGB and PN stages implying that it is not a transient form of dust. Given its strength, its carrier could be an abundant dust component in C-rich circumstellar shells, and possibly an hitherto overlooked, major component of interstellar dust as well. Although solid MgS provides a good fit to the observed band, this identification should be adressed again because of the possible stringent abundance requirements. Obviously carbon-related particles are good potential candidates and the presence in the mid-infrared of other broad bands attributed to carbonaceous species further strengthens such an identification.

References

Buss Jr., R.H., Cohen, M., Tielens, A.G.G.M., Werner, M.W., Bregman, J.D., Witteborn, F.C., Rank, D., and Sandford, S.A.: 1990, *Astrophys. J.* **365**, L23
Cox, P.: 1993, in *Astronomical Infrared Spectroscopy*, ed. S. Kwok, p. 163
Goebel J.H. and Moseley S.H.: 1985 *Astrophys. J. Lett.* **290**, L35
Kwok, S., Volk, K.M., and Hrivnak, B.J.: 1989, *Astrophys. J. Letter* **345**, L51
Mc Carthy, J.F., Forrest, W.J., and Houck, J.R.: 1978, *Astrophys. J.* **224**, 109
Mutschke, H., Begemann, B., Dorschner, J. and Henning, Th.: 1993, *Astrophys. J.* submitted
Rouleau, F. and Martin, P.G.: 1991, *Astrophys. J.* **377**, 526

A QUANTITATIVE STUDY OF THE 7.7 AND 11.3 µm EMISSION BANDS BASED ON IRAS/LRS SPECTRA

A. ZAVAGNO, P. COX, J.P BALUTEAU
OBSERVATOIRE DE MARSEILLE
2 PLACE LE VERRIER
13248 MARSEILLE CEDEX 4, FRANCE

ABSTRACT

We are presented a quantitative analysis of the emission bands at 7.7 and 11.3 µm in IRAS/LRS spectra of 113 sources (HII regions, reflection nebulae, planetary nebulae and galaxies).
Relations of the intensities of the emission bands with physical conditions such as the excitation, and the relative contribution of emission bands and continuum around 10 µm to the mid-infrared and far-infrared fluxes are presented and analyzed.
A strong linear correlation is found between the (7.7+8.6) µm and the 11.3 µm band luminosities. The correlation confirms, on a large number of sources, that the carriers of these emission bands belong to the same family, most likely aromatic molecular species such as PAH molecules.
The ratio of the 7.7 and 11.3 µm intensities is equal to 8.3 (±3). The average contribution of these features to the mid-infrared luminosity is 50%, and the importance of the underlying continuum is stressed in planetary nebulae, since this contribution is weaker (about 30%). The contribution of the near-infrared emission bands to the total infrared luminosity is a few percent, implying that one to five percent of the cosmic carbon is locked in aromatic molecules.
From simple energetic considerations, it is shown that the excitation of the 7.7 and 11.3 µm emission bands is dominated by ultraviolet photons in the range 91 to 365 nm.

INTRODUCTION

The aim of this study is to look at the physical properties of emission bands seen at 7.7 and 11.3 µm. We have extracted 113 spectra from the IRAS/LRS database to obtain an homogeneous sample, allowing a good statistical study. This article has been published in 1992 (Zavagno et al., A&A 259, 241).
In Section 1, we present data reduction. Results are presented in Section 2. Conclusions are given in Section 3.

DATA REDUCTION

The intensities of PAH emission bands were measured by integrating the flux between 7.5 and 10 µm for the (7.7+8.6) µm band and from 11 to 12 µm for the 11.3 µm band and by substracting a linear continuum underlying the bands. We also define a "continuum" which flux is equal to the total bands flux substracted to the total integrated flux between 7.5 and 13.5 µm (the LRS1 band). Figure 1 shows the IRAS/LRS spectra of eight typical sources, chosen in order to illustrate the various kinds of spectra encountered in this study. Also shown for one of the source, the adopted continuum.

RESULTS

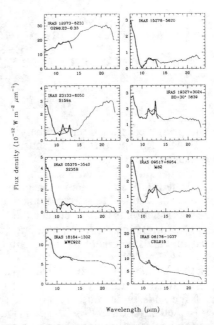

Fig. 1 Typical IRAS/LRS spectra from the selected sample for this study.

Figure 2 presents the luminosity of the 11.3 μm band versus the luminosity of the (7.7+8.6) μm band. The correlation found here indicates that the carriers of the bands belong to a same family, most likely aromatic species.

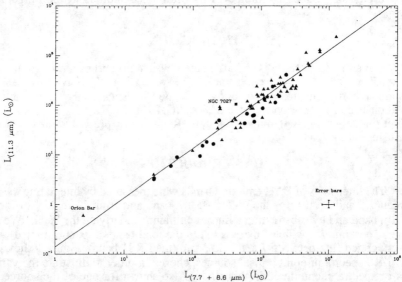

Fig. 2 Luminosity of the 11.3 μm band versus the luminosity of the (7.7+8.6) μm band. The straight line represent the least-squared fit.

Figure 3 presents the distribution of the ratio of the (7.7+8.6) μm band flux to the 11.3 μm band flux. The peak at about 8 indicates, from recent results[1], in the PAH hypothesis, that the carriers of these bands are ionized rather than neutral.

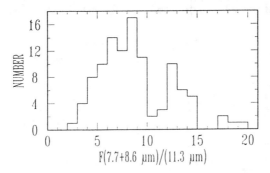

Fig. 3 Histogram of F(7.7+8.6 μm)/F(11.3 μm). Note the peak of the distribution around 8.

We have studied the contribution of the emission bands flux to the total flux emitted in the range 7.5 to 13.5 μm (the LRS1 band). Result, presents on Figure 4, shows that this contribution is about 50% for HII regions and reflection nebulae. Note that the contribution is clearly weaker in planetary nebulae, where the "continuum" seems to dominate, compared with the emission bands. This last point is studied with a fit of the mid-infrared continuum in the planetary nebula NGC7027 [2].

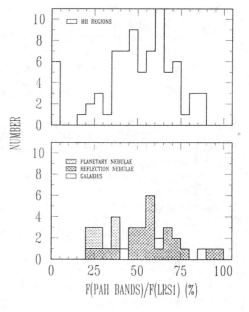

Fig. 4 Histograms of the distribution of the contribution (in percent) of the emission bands at 7.7, 8.6 and 11.3 μm to the total flux measured between 7.5 and 13.5 μm.

A simple model based on energetic considerations has been made to study the excitation mechanism of the emission bands. Figure 5 presents the luminosity contained in the bands versus the number of Lyman continuum photons. We can see from this figure that UV photons as well as visible ones can excite the emission bands. This confirms a result obtained by Sellgren et al. (1990) [3].

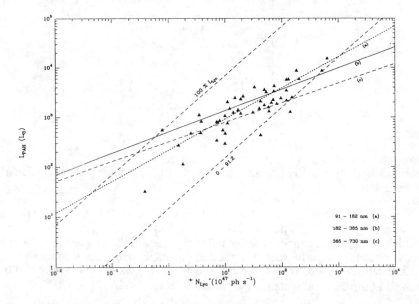

Fig. 5 Luminosity contained in the emission bands at 7.7, 8.6 and 11.3 µm versus the number of Lyman continuum photons. The curves labelled (a), (b) and (c) show the highest possible luminosity available for the corresponding spectral range indicated on the figure, whereas the dotted lined labelled "0-91.2" corresponds to the Lyman continuum domain. The dotted line labelled "100% L_{Lyc}" corresponds to a 100% conversion efficency of photons with hv > 13.6 eV into the three emission bands.

CONCLUSIONS

The main results of this study are:

a strong correlation is found between the luminosity of the (7.7+8.6) µm and the 11.3 µm emission bands. This indicates that the carriers of the bands belong to a same family, most likely large aromatic molecules. Acording to recent theoretical calculations, the fact that the (7.7+8.6) µm band is about eight times stronger than the 11.3 µm band suggests that, in the present sources, the molecules are ionized rather than neutral. A theoretical ratio of about 0.2 is expected for neutral species.

The contribution of the bands to the mid-infrared flux is about 50% for HII regions and reflection nebulae. This contribution is weaker in planetaries and lead us to point out the existence of a strong mid-infrared "continuum" emission in this kind of sources.

The excitation of the emission bands is due predominantly to photons in the range 91 to 365 nm.

REFERENCES

1. J. Szczepanski & M. Vala, Nature 363, 699 (1993)
2. A. Zavagno & J.P. Baluteau, in preparation (1993)
3. K. Sellgren, L. Luan, M.W. Werner, ApJ 359, 384 (1990)

OBSERVATION OF THE DIFFUSE GALACTIC EMISSION AT 6.2 μm

I. Ristorcelli [1], M. Giard [1], C. Mény [3], G. Serra [1], J.M.Lamarre [2], C. Le Naour [1],
J. Léotin [3], François Pajot [2].

1- Centre d'Etude Spatiale des Rayonnements du CNRS, UPR 8002,
   associé Université Paul Sabatier, Toulouse, France.
2- Institut d'Astrophysique Spatiale du CNRS, associé Université Paris XI Orsay.
3- Laboratoire de physique des Solides, Université Paul Sabatier, Toulouse.

ABSTRACT

We present the observation of the 6.2μm emission of the inner galactic disk obtained with the AROME balloon-borne experiment. The galactic coordinates of the covered region are $-5° \leq l \leq 35°$ and $|b| \leq 6°$, with an angular resolution of 0.7°. The measurements reveal the existence of an emission feature at 6.2μm all over this region. The averaged 6.2μm surface brightness ($|b| \leq 1°$, $8° \leq l \leq 35°$) is $\lambda I_\lambda = (1.2 \pm 0.3)10^{-5}$ $Wm^{-2} sr^{-1}$, with a continuum $\lambda I_\lambda = (5.9 \pm 1.2)10^{-6}$ $Wm^{-2}sr^{-1}$ and a feature's intensity $\Delta\lambda I_\lambda = (6.1 \pm 1.3)10^{-7}$ $Wm^{-2}sr^{-1}$. We can compare this value to the 3.3μm emission observed with AROME (Giard et al. 1988[1]) and to the 12μm(IRAS) surface brightness: the ratios are respectively $\Delta\lambda I_\lambda(6.2/3.3) = 9\pm2$ (extinction corrected) and $\Delta\lambda I_{\lambda 6.2}/\lambda I_{\lambda 12} = 0.09\pm0.02$; these color ratios are similar to those obtained on UV excited nebulae, despite very different radiation field intensities. This supports the idea of a common origin for the near infrared radiation from excited nebulae and the diffuse interstellar medium: emission of transiently heated very small grains or large molecules like PAHs, as it was originally proposed by Puget et al. (1985[2]).

INTRODUCTION

The family of emission bands in the near infrared (NIR) spectrum of peculiar objects (Planetary nebulae, Reflexion nebulae, HII regions, galaxies) has been tentatively attributed to the emission of Polycyclic Aromatic Hydrocarbon molecules (PAHs), (Léger and Puget 1984 [3]). This kind of molecules has been proposed to be present everywhere in the interstellar medium so as to explain the near and mid-infrared diffuse galactic spectrum. This implies that the family of NIR bands should be observed in the IR emission of the galactic disk, which is dominated by the emission of the diffuse medium (Cox and Mezger 1989 [4], Ghosh et al. 1986 [5]). In order to test this hypothesis, an observational program was started in 1985, consisting in measuring the flux radiated in the IR bands toward the galactic dust disk. A first configuration of a balloon-borne experiment called AROME was designed in 1986/87 and was devoted to the observation of the 3.3μm feature. With a beam value around 0.5°, this experiment allowed the first detection of the emission feature in the diffuse interstellar medium and demonstrated its ubiquity in the galactic disk (Giard et al. 1988[1]/1989 [6]). After these measurements at 3.3μm, it was decided to plane news observations at 6.2μm. This feature has a complementary interest because it is attributed to vibrational modes of the carbon skeleton, while the 3.3μm emission corresponds to a streching mode of the C-H bond in aromatic molecules.

INSTRUMENTATION

The AROME balloon-borne experiment was previously designed for the detection of the 3.3μm feature in the diffuse emission of the galactic plane. A technical description of the whole system has been detailed in previous papers (Giard et al 1988[1], Sales et al. 1991[7]). For a measurement of the emission at 6.2μm , the experiment had to be modified. The InSb photodiodes detectors were replaced by SiGa photoconductors needing a much cooler operating temperature. This condition has been satisfied with a ^{4}He cryostat (T≤1.5 K thanks to natural pumping at balloon flight level, P≤4mB). The electronics was changed in order to provide, for each detector, a bias optimised for the best signal to noise ratio. New interference filters have been used to select two photometric bands: a narrow band A and a wide band B, both centered at 6.2μm : their spectral response are given in Fig.1. The values of the integrated incident fluxes measured in each two bands allow us to assess the continuum brightness value and a possible emission feature at 6.2μm.

Detectors and filters are located on the cold plate of the ^{4}He cryostat. The warm optics consist of two similar Cassegrain telescopes. The main characteristics of the instrument are summarized in Table I

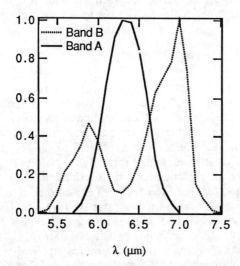

Figure 1: spectral response in the two photometric bands

TABLE I

Telescopes	2 Cassegrains, eqv. focus = 641mm
Primary mirror useable diameter	$\emptyset$ = 109mm = f/5.9
Oscillating secondary mirrors	f ~ 15 Hz , $\Delta \alpha = \pm 0.85°$
Field of-view	0.5°
Throughput	2.9 10^{-7} m^{2}Sr
Elevation	12° to 48°
Azimuthal scanning	$\Delta Az = \pm 8°$, velocity=0.8°/s
Azimuthal pointing accuracy	±15' absolute , 3' rms oscillations

OBSERVATIONAL PROCEDURES

All along the flight, the telescopes and the cryostat are jointly rotating at constant speed and elevation, producing a fast azimuthal scanning. The absolute pointing is obtained thanks to a magnetometer used as a sensor in a servo controlled system.
The electrical signal provided by each detector is amplified, filtered, demodulated and synchronously numerised on board, before beeing transmitted by telemetry to the ground station.

The observations presented in this paper were obtained during the transmediterraneen 15 hours flight of the experiment launched from Trapani (Sicily) on July the 30th in 1992 at 18^{h}30 TU, as part of the CNES (France), the ASI (Italia) and the INTA (Spain) cooperative launch campaign. In spite of a perfect running of the whole instrument, the observability time was reduced because of telemetry problems to about 7 effective hours.

The galactic plane was surveyed in the longitude range : $-5° \leq l \leq 35°$, with an azimuthal scanning direction almost perpendicular to the galactic disk.

CALIBRATION AND SIGNAL PROCESSING

The instrument's signal has first been calibrated in the laboratory from the modulated observation of a differential extended blackbody. An inflight calibration was carried out on bright IR stars: αSco, Mira Ceti and VYCMa, and allowed to correct for the effect of the atmospheric transmission on the laboratory calibrations (~ 59% for a 5m path lengh). The final relative precision on absolute flux values is estimated to ±20%. The signal has been restored by an inverse Fourier transform, as described in previous papers (Sales et al., 1991[7]). The final resolution is estimated to 0.7°.

Once the signal is restored in the two bands centered at 6.2μm, we have been able to separate the feature and the continuum fluxes, $\Delta\lambda I_{\lambda feat}$ and $\lambda I_{\lambda\, cont.}$, in the following way: we assume that the 6.2μm emission feature has a bandwidth equal to $\Delta\lambda = 0.17\mu m$ (Léger et al, 1989[8]) : $I_{\lambda\, feat} = I^0_{feat} \exp(-(\lambda-\lambda_0)^2/k)$, $(k=\Delta\lambda^2/4\ln 2)$, and considering a continuum law: $\lambda I_{\lambda\, cont.} = I^0_c (\lambda/\lambda_0)^p$, we can show that:

$\Delta\lambda_i F_i^{obs}/\lambda_0 = a_i \ I^0_{feat} + b_i \ I^0_c$, with:

- i= A, B photometric band; $a_i = \int_{\Delta\lambda_i} \tau_\lambda^i e^{-(\lambda-\lambda_0)^2/k} d\lambda \quad b_i = \int_{\Delta\lambda_i} \tau_\lambda^i (\lambda/\lambda_0)^p d\lambda$.
- ($\Delta\lambda_i$, τ_i) : the equivalent bandwidth and the spectral response,
- F_i^{obs}: the measured integrated fluxes in each band.

RESULTS

We present in Fig.2 the fluxes measured in the two photometric bands, versus the galactic latitude and averaged over the inner galactic part, $5° \leq l \leq 30°$. It shows an emission excess in the narrow band A and an emitting disk thickness of around 1.6°FWHM. We have plotted our 6.2μm measurements on a spectrum of the averaged galactic λI_λ surface brightness obtained over $8.5° \leq l \leq 35°$ and $|b| \leq 1°$, in the infrared and submillimeter range (see Fig.3). The set of data has been homogenized by subtracting linear baselines on the same latitude range: $2.5° \leq |b| \leq 5°$. The 6.2μm plots give the surface brightness directly measured in each photometric band : $\lambda I_\lambda(A) = (1.2\pm0.25)10^{-5}$ Wm^{-2}sr^{-1} and $\lambda I_\lambda(B) = (7.5\pm 1.5)10^{-6}$ Wm^{-2}sr^{-1}, from which we deduce the feature's intensity $\Delta\lambda I_\lambda = (6.1\pm1.3)10^{-7}$ Wm^{-2}sr^{-1} and the continuum $\lambda I_\lambda = (5.9\pm1.2)10^{-6}$ Wm^{-2}sr^{-1}. We just mention that the error bars allocated to the 6.2μm fluxes plotted on the spectrum come from calibration precision and are proportional in the two bands.

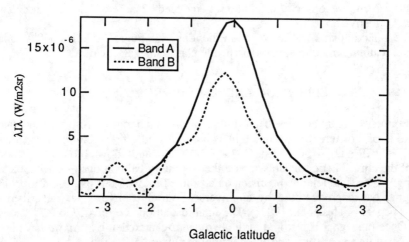

Figure 2: Latitude profile of the emission fluxes measured in each photometric band (e.g. the narrow "feature" band A and the wide "continuum" band B); the fluxes have been averaged over the longitude range: $5° \leq l \leq 30°$.

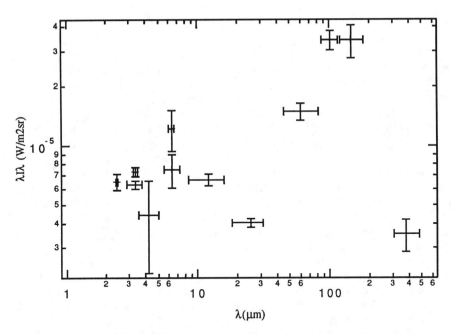

Figure 3: the averaged λI_λ surface brightness observed in the inner Galaxy ($8.5° \leq 1 \leq 35°$ and $-1° \leq b \leq 1°$), in the IR and submillimeter range. Horizontal bars show the experimental bandwidths, and the vertical bars represent error bars. The 3.3μm and 6.2μm plots are from AROME measurements (respectively 1988[1]/89[6] and 1992 this work). The other data are from Hayakawa et al. 1981[9](2.4μm), Price 1980[10](4.2μm), IRAS satellite (12, 25, 60, 100 μm) and Caux and Serra 1986[11](145 and 380μm).

The ratio of the averaged intensity emitted in the 3.3μm and 6.2μm feature is $r(6.2/3.3) \approx 13\pm3$ ($8° \leq l \leq 35°$, $|b| \leq 1°$). It is increased in the Galactic Center direction where it reaches 25. Assuming a representative column density value of $N_H = 13.2 \cdot 10^{22}$ atmHcm^{-2} in this direction, while in average, for the diffuse galactic disk : $N_H = 5.7 \cdot 10^{22}$ atmHcm^{-2}, and an absorptivity ratio $Qt(3.3\ \mu m)/Qt(6.2\mu m) = 2.3$, with $\tau_{3.3}/N_H \approx 2.5 \cdot 10^{-23}$cm^2/H (Mathis et al. 1990 [12]), the different ratio in the Galactic Center can be interpreted by an extinction effect. We adopt a typical ratio value, extinction corrected, equal to 9 ± 2 for the diffuse galactic emission. The averaged ratio deduced from the observed spectra of peculiar galactic nebulae (Cohen et al., 1986[13]) is comparable ($<r> \approx 8$). The ratio of the 6.2μm feature's intensity to the 12μm(IRAS) surface brightness $r(6.2/12)=0.09$ is similar to the one found for the planetry nebulae NGC 2023 for instance (Sellgren et al. 1985 [14]).

DISCUSSION

The dust models for the origin of IR emission bands can be gathered in two classes: carbonaceous grains (Duley and Williams 1981[15], Borghesi et al.,1987[16], and Guillois et al. 1993[17] this meeting) or large PAH molecules (Léger and Puget 1984,[3] Allamandola et al. 1985[18]). Concerning the former class, the coal model of Guillois et al. 1993[17], presented at this meeting, seems the more convincing since the peculiar optical properties of coal allow such grains to reach equilibrium temperatures of order 250K to 400K under radiation field intensities typical of UV excited nebulae, $10^{-6} \leq I \leq 10^{-4}$ W cm^{-2}. However, this kind of model will predict negligible NIR emission from the diffuse interstellar medium where $10^{-10} \leq I \leq 10^{-7}$ Wcm^{-2}, since the equilibrium temperature of the grains will not be high enough. On the contrary, the PAH model predicts about the same NIR spectrum whatever is the intensity of the exciting radiation field (see Désert et al. 1990[19]). We think that this is the major difference between the two classes of models and that it may help discriminating between them when appropriate observations are used.

It is clear, for instance, that the 12μm emission from "cirrus" clouds detected by IRAS (Boulanger et al. 1985[20]) can only be attributed to large molecules. However, the emission bands have not yet been directly observed in such clouds and in the absence of spectral information, very little can be derived on the nature of the large molecules present there.

Concerning the integrated IR emission toward the inner galaxy, $5° \leq l \leq 30°$, our measurements at 3.3μm and 6.2μm, demonstrate the presence of the NIR emission bands and of an underlying continuum at 6.2μm. Moreover, the overall IR spectrum shown in fig.3, is very similar to the spectrum of UV excited nebulae (if one excepts the peak wavelength which reflects the equilibrium temperature of standard grains).
We have shown that the ratios constructed from the 3.3μm feature/ 6.2μm feature/12μm IRAS intensities are about the same if we compare UV excited nebulae and the observed inner Galaxy. Then, the question raised is: can we account for the observed quantities in the 3.3μm and the 6.2μm features with AROME, assuming that the measured galactic flux is the additionnal contribution of discrete sources ? If we use the fact that the fractions of the 12μm IRAS band to the total IR emission for HII regions (Ghosh et al. 1986[5]) and for the inner Galaxy , are also very close (~5%), we can infer that the fractions of total IR flux emitted in the 3.3μm and the 6.2μm features are the same for the inner galactic disk and for individual nebulae. In addition, it is quite well established that the far infrared emission toward the galactic disk originates from the diffuse medium where the radiation field intensity is smaller than 10^{-7} Wcm^{-2} (see Cox and Mezger 1989[4], Sodroski et al. 1989[21], Bloemen et al. 1990[22]). Synthetic models of the FIR emission of the Galaxy actually don't take into account regions of very high excitation (e.g. $I > 10^{-6}$ Wcm^{-2}) so that 10% can be taken as a conservative upper limit of their contribution. In consequence, using the 3.3μm and 6.2μm to total IR ratios derived above, we can also set to 10% the maximum contribution of excited nebulae to our measurements.
Moreover, this is supported by the analysis of Ghosh et al. 1986 as they have shown that the 12μm Galactic emission measured by IRAS can't be explained by discrete

point-like sources and must have an interstellar origin involving transiently heated very small particles.

So, our conclusion is that most of the 3.3μm and 6.2μm features intensities measured with the AROME experiment, namely 90%, arises from the diffuse interstellar medium where only dust model including large molecules can apply. Then, the detection of the 3.3μm and 6.2μm features emission from this phase of the interstellar medium favors the PAH hypothesis.

REFERENCES

[1] M. Giard, F. Pajot, J.M. Lamarre, G. Serra, E. Caux, R. Gispert, A. Léger and D. Rouan: 1988, A&A, 201, L1.
[2] J.L. Puget, A. Léger, and F. Boulanger : 1985, A&A, 142, L19.
[3] A. Léger and J.L. Puget: 1984, A&A, 137, L5.
[4] P. Cox and Mezger: 1989, Annu. Rev. Astron. Astrophys., 27, 49-82
[5] S.K. Ghosh, S. Drapatz and U.C. Peppel:1986, A&A, 167, 341-350.
[6] M. Giard, F. Pajot, J.M. Lamarre, G. Serra and E. Caux: 1989, A&A, 215, 92-100.
[7] N. Sales et al : 1991, Exp. Astron., 1, 1-26.
[8] A. Léger, L. d'Hendecourt and D.Défourneau: 1989, A&A, 216, 148-164.
[9] S. Hayakawa, T. Matsumoto, M. Murakani, K. Uyama, J.A. Thomas and T. Yamagani: 1981, A&A, 100, 116.
[10] S.D. Price and L.P. Marcotte: 1980, " An infared survey of the diffuse emission within 5° of the galactic plane", AFGL-TR-8-0182.
[11] E. Caux and G. Serra: 1986, A&A, 165, L5.
[12] J.S. Mathis: 1990, Annu. Rev. Astron. Astrophys., 28, 37-70.
[13] M. Cohen, L. Allamandola, A.G.G.M. Tielens, J. Bregman, J.P. Simpson, F.C. Witteborn, D. Wooden and D. Rank: 1986, ApJ, 302, 737-749
[14] K. Sellgren, L.J. Allamandola, J.D. Bregman, M.Werner and D. Wooden: 1985, Ap.J , 299, 416
[15] W.W. Duley and D.A. Williams: 1981, MNRAS,196, 269
[16] A. Borghesi, E. Bussoletti and L. Colangeli: 1987, ApJ 314, 422
[17] O. Guillois, I. Nenner, R. Papoular and C. Reynaud: 1993, this meeting
[18] L.J. Allamandola, A.G.G.M. Tielens and J.R. Barker: 1985, ApJ, 290, L25
[19] F.X. Désert, F. Boulanger and J.L. Puget: 1990, A&A, 237, 215
[20] F. Boulanger, B. Baud and G.D. Van albada: 1985, A&A, 144, L9
[21] T.J. Sodroski, E. Dwek, M.G. Hauser and F.J. Ken : 1989, ApJ, 336,762
[22] J.B. Bloemen, E.R. Deul and P. Thaddeus: 1990, A&A, 233,437

POSSIBLE DETECTION OF SOLID FORMALDEHYDE TOWARDS THE EMBEDDED SOURCE GL 2136

W. A. Schutte, P. A. Gerakines, E. F. van Dishoeck and J. M. Greenberg
Leiden Observatory Laboratory, Huygens Laboratory, Leiden

T. R. Geballe
Joint Astronomy Center, Hilo, Hawaii

ABSTRACT

We present new medium resolution spectroscopy ($\lambda/\Delta\lambda \approx 1000$) between 3.37 and 3.66 μm towards the deeply embedded source GL 2136. Besides an absorption feature at 3.535 μm due to solid CH_3OH which was earlier observed towards other embedded sources, a new feature is apparent at 3.477 μm. From comparison with laboratory data of low temperature ices, we tentatively assign this feature to solid formaldehyde. Our data combined with earlier observations of CO and H_2O indicate towards GL 2136 an ice composition of $H_2O : CH_3OH : H_2CO : CO = 100 : 3.6 : 2.5 : 3.0$.

INTRODUCTION

Formaldehyde (H_2CO) is a potentially important molecule in interstellar ices, since it may be produced both by grain surface reactions of CO with H^1 or UV photolysis of CH_3OH or CO/H_2O ice mixtures[2]. Determining its abundance could therefore greatly contribute to our understanding of the ice chemistry. Formaldehyde has two strong infrared features at 5.82 and 6.68 μm, which lie unfortunately in the spectral region obscured by telluric water vapor. The strongest H_2CO features accessible from the ground fall at 3.47 and 3.54 μm, each with ~ 4 times smaller cross section than the 5.82 μm feature[3]. Earlier spectroscopic observations towards 4 embedded sources did not reveal these bands, giving upper limits to the H_2CO abundance of a few percent relative to solid H_2O[4].

In our search for promising sources to probe solid H_2CO, we investigated earlier very low resolution observations of embedded sources by the Kuyper Airborne Observatory[5] ($\lambda/\Delta\lambda \approx 50$) for signs of the strong 5.82 and 6.68 μm H_2CO features. We found that the regularly observed broad 6.0 μm (attributed to solid H_2O) and (unidentified) 6.8 μm absorption bands are slightly broadened and shifted to shorter wavelength towards the embedded source GL 2136, possibly indicating the presence of the H_2CO features. This paper reports the results of a spectroscopic search for the 3.47 and 3.54 μm features of solid formaldehyde (henceforth band 1 and band 2, respectively) with the United Kingdom Infrared Telescope (UKIRT) towards this source.

OBSERVATIONAL RESULTS

GL 2136 was observed by UKIRT with the CGS4 spectrometer at a resolution $\lambda/\Delta\lambda \approx 1000$. For full coverage of the relevant region two grating settings were used giving the 3.378 - 3.581 μm and 3.572 - 3.775 μm spectral regions. The integration times were 2 and 6 minutes at these settings, respectively. The spectra were ratioed by standard stars at approximately the same airmass for calibration and removal of telluric features.

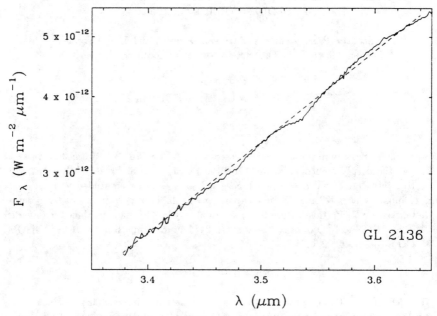

Fig. 1. Semi-logarithmic plot of the medium resolution ($\lambda/\Delta\lambda \approx 1000$) spectrum of the embedded source GL 2136 between 3.378 - 3.65 μm. The dashed line represents the "baseline" which was subtracted to make the optical depth plot shown in figure 2.

Figure 1 shows the relevant part of the observations. The general slope of the spectrum is caused by the commonly observed shoulder on the red side of the strong H_2O 3 μm ice band. To highlight the finer structure in the spectrum a straight (in the λ - $\log(F_\lambda)$ plane) "baseline" was subtracted from the data, resulting in the optical depth plot shown in Figure 2. Two sharp features at 3.477 μm and 3.535 μm can be distinguished in this figure. Furthermore a broad band is apparently present extending from beyond the short wavelength side of the spectrum down to $\sim$ 3.6 μm. The feature at 3.535 μm has been observed earlier towards the embedded sources W33 A and NGC 7538 IRS 9 and was identified with solid methanol (CH_3OH) [4,6]. The broad structure stems from the curvature present in the red shoulder on the H_2O band[7], and was recently suggested to originate from diamond like structures in interstellar carbonaceous material[4]. The 3.477 μm band, finally, has not been observed earlier towards other embedded sources.

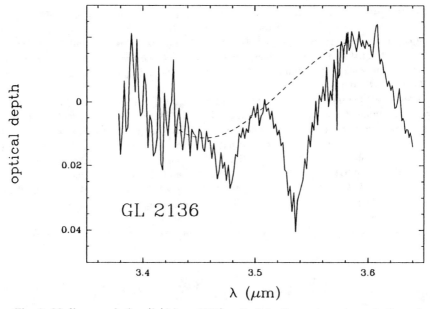

Fig. 2. Medium resolution ($\lambda/\Delta\lambda \approx 1000$) optical depth spectrum towards the embedded source GL 2136. The baseline used to construct the spectrum is shown in figure 1. The dashed line represents a 3rd order polynomial fit of the broad band absorption underlying the narrow features at 3.477 and 3.535 μm.

COMPARISON WITH LABORATORY SOLID FORMALDEHYDE DATA

Table I shows the positions, widths and relative cross sections for the formaldehyde bands 1 and 2 in a number of binary mixtures of formaldehyde with astrophysically relevant molecules. It can be seen that the position and the width of the observed 3.477 μm band fall within the range spanned by band 1 in the various mixtures.

Table I. Position, FWHM, and relative cross sections of band 1 and band 2 of solid H_2CO in pure ice and in some binary mixtures, as compared to the 3.477 μm feature towards GL2136.

Molecule	Composition Molec:H_2CO	Pos. 1 μm	Pos. 2 μm	FWHM$_1$ μm	FWHM$_2$ μm	σ_2/σ_1	ref.
H_2O	30	3.465	3.541	0.036	0.046	1.5	3
CH_3OH	2.1	3.473	3.555	0.022	0.022	1.1	8
H_2CO	pure	3.469	3.544	0.028	0.034	1.1	3
CO	250	3.494	3.578	0.006	0.006	1.7	9
CO_2	600	3.448	3.525	0.012	0.012	1.6	9
ice $\to$ GL 2136		3.477 ±0.003	—	0.019 ±0.003	—	—	

When assigning the 3.477 μm band to band 1 of H_2CO, the properties of band 2 can be inferred from the band 1 - band 2 relationships born out by Table I. Apparently, the distance between the bands is rather constant, i.e., 0.079 ± 0.004 μm, the width of band 2 is similar or slightly larger than that of band 1, while its cross section is also slightly larger. Thus we can predict that band 2 should be found at ~ 3.556 μm, be slightly deeper than band 1, and have a FWHM of 0.018 - 0.025 μm.

To investigate the possible presence of band 2 towards GL 2136, we attempted to correct for the broad "diamond" feature by fitting it with a smooth 3rd order polynomial while excluding the regions of absorption by the 3.477 and 3.535 μm bands as well as the region that could be occupied by band 2 (Figure 2). The resulting curve was subtracted, producing a spectrum which presumably shows the narrow absorption features only (Figure 3). This spectrum seems to indicate that the 3.535 μm methanol band is broadened and may have a shoulder at its long wavelength wing. The extent and depth of the broadening and the shoulder may indeed be consistent with the presence of a feature similar to the expected band 2 of H_2CO.

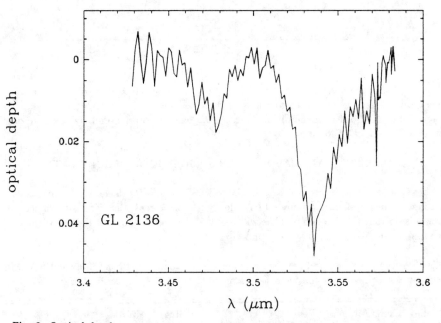

Fig. 3. Optical depth spectrum constructed by subtracting the 3rd order polynomial fit of the broad absorption band underlying the narrow features (see Figure 2).

Assigning the 3.477 μm band to H_2CO and the 3.535 μm band to CH_3OH, the column densities of these molecules can be calculated from the integrated optical depths relative to the baseline indicated in Figure 2, using the known integrated cross sections[3,10]. With the earlier determined H_2O and CO column densities[11], we find the ice abundances H_2O : CH_3OH : H_2CO : CO = 100 : 3.6 : 2.5 : 3.0 towards GL 2136.

DISCUSSION

Our tentative assignment of the 3.477 µm absorption feature indicates that formaldehyde could be an important component of interstellar grain mantles at least in some dense cloud regions. There are two mechanisms which could provide the implied amount of solid H_2CO; i.e., grain surface reactions of CO with H[1] or UV photolysis of solid methanol or (somewhat less efficient) CO/H_2O ice mixtures[2]. Direct condensation of H_2CO from the gas phase should be negligible, due to its low gas phase abundance ($< 10^{-3}$ of that of gaseous CO[12]). It has been argued that in the presence of a sufficient amount of atomic H, surface reactions would quickly convert condensed CO to CH_3OH, H_2CO being only a short-lived transition species in the process[13]. Thus, if we assume that grain surface reactions are the source of solid H_2CO, it would indicate that at least at one time during the cloud history the abundance of atomic hydrogen was sufficiently small to prevent CH_3OH formation, but large enough to convert a significant part of the condensing CO to H_2CO. Alternatively, H_2CO may be unable to react with H, making it the end-product of the CO surface hydrogenation. In that case the icy methanol should have formed in some other way than grain surface reactions, perhaps by photolysis of the formaldehyde in the water ice[14]. The final possibility, formation of the H_2CO from photolysis, would imply that UV processing in dense clouds could considerably alter the composition of the condensed grain mantles. Much more experimental, theoretical and observational work will be necessary to distinguish between these varying scenarios. In any case, the presence of H_2CO may have important implications for the solid phase chemistry. In addition, desorption of the solid formaldehyde could be the main source of the formaldehyde observed in the gas phase of dense clouds[15].

From observations of the solid CO feature at 4.67 µm it was deduced that interstellar ices consist of, at least, 2 distinct phases, one dominated by polar and one by apolar molecules[11,16]. Observations of the 9.8 µm band of solid CH_3OH towards GL 2136 pointed to the existence of a third phase rich in CH_3OH ice[17]. Assigning the 3.477 µm band to H_2CO, this feature could give important further information about the chemical differentiation of interstellar ices, since it can be seen from table I that the band position is quite sensitive to the chemical environment. Table I clearly shows that the H_2CO cannot reside in an environment dominated by H_2O ice. Since the CO feature indicates that the apolar phase is virtually absent towards GL 2136[11], this would confirm the existence of a third distinct chemical environment in the icy mantles. Additional experimental work is necessary to further constrain the chemical make-up of the environment of the solid formaldehyde from the observed width and band position[8].

Further observations are necessary to confirm the presence of solid formaldehyde towards GL 2136. First, higher S/N data should be obtained of the 3.477 and 3.535 µm features. A further conclusive test would be a space (ISO) or airborne (KAO) search for the strong 5.82 and 6.68 H_2CO bands. Indeed, such future space and airborne data towards obscured embedded and background sources in the 5 - 8 µm region would clarify whether formaldehyde is, in general, an important constituent of interstellar ices.

ACKNOWLEDGEMENTS

This work was supported by the Netherlands Organization for Scientific Research (NWO), by the Space Organization Research Netherlands (SRON), and by NASA grant no. NGR 33-018-148.

REFERENCES

[1] A. G. G. M. Tielens, and W. Hagen, A&A 114, 245 (1982).
[2] L. J. Allamandola, S. A. Sandford, and G. J. Valero, Icarus 76, 225 (1988).
[3] W. A. Schutte, L. J. Allamandola, and S. A. Sandford, Icarus 104, 118 (1993).
[4] L. J. Allamandola, S. A. Sandford, A. G. G. M. Tielens, and T. M. Herbst, ApJ 399, 134 (1992).
[5] A. G. G. M. Tielens, and L. J. Allamandola, in Physical Processes in Interstellar Clouds, eds. G. E. Morfill and M. Scholer (Reidel, Dordrecht, 1987), p. 333.
[6] R. J. A. Grim, F. Baas, T. R. Geballe, J. M. Greenberg, and W. Schutte, A&A 243, 473 (1991).
[7] R. G. Smith, K. Sellgren, and A. T. Tokunaga, ApJ 344, 413 (1989)
[8] W. A. Schutte, P. A. Gerakines, T. R. Geballe, E. F. van Dishoeck, and J. M. Greenberg, in preparation.
[9] G. P. van der Zwet, L. J. Allamandola, F. Baas, and J. M. Greenberg, A&A 145, 262 (1985).
[10] W. A. Schutte, A. G. G. M. Tielens, and S. A. Sandford, ApJ 382, 523 (1991).
[11] A. G. G. M. Tielens, A. T. Tokunaga, T. R. Geballe, and F. Baas, ApJ 381, 181 (1991).
[12] E. F. van Dishoeck, G. A. Blake, B. T. Draine, and J. I. Lunine, in Protostars and Planets III, eds. E. H. Levy, J. I. Lunine and M. S. Matthews (University of Arizona Press, Tucson, 1993), p. 163.
[13] A. G. G. M. Tielens, and L. J. Allamandola, in Interstellar Processes, eds. D. J. Hollenbach and H. A. Thronson (Reidel, Dordrecht, 1987), p. 397.
[14] N.-S. Zhao, Photochemistry of Interstellar and Cometary Ices, Ph. D. thesis, University of Leiden, the Netherlands (1990).
[15] O. M. Shalabiea, and J. M. Greenberg, this volume.
[16] S. A. Sandford, L. J. Allamandola, A. G. G. M. Tielens, and G. J. Valero, ApJ 329, 498 (1988).
[17] C. J. Skinner, A. G. G. M. Tielens, M. J. Barlow, and K. Justtanont, ApJ 399, L79 (1992).

DISCUSSION

<u>THADDEUS</u> — Has anyone been able to show theoretically that water ice should be nearly two orders of magnitude more abundant than H_2CO or CO on the grains you observe ?

<u>SCHUTTE</u> — Yes, Tielens & Hagen (1982) & d'Hendecourt et al. (~1986) showed that, especially in the early stages of cloud formation, or at low extinctions, when/where atomic O is abundant in the gas phase, surface reactions on grains of O and H should be a very efficient source of solid H_2O. Schutte & Greenberg (1991) argued that once solid H_2O has formed on the grains it can not be returned to the gas phase be any known desorption process in dense clouds.

GRAIN PROCESSING AND IRAS COLOUR VARIATIONS

S. D. Taylor
UMIST, Manchester, PO Box 88, M60 1QD, England

ABSTRACT

Maps of diffuse regions show that not only are IRAS colour ratios higher than expected from standard grain models, but also vary with total emission in a way that is difficult to explain. It is shown here that in fact the gross general features of this emission can be attributed to the degree of processing of carbonaceous grain mantles, symptomatic of the dynamical nature of such regions.

INTRODUCTION

In recent years the combination of extinction data and IR spectral features has led to a convergence of views that interstellar dust in diffuse regions is silicate and carbonaceous in nature. The structure and morphology of such dust is still a matter for debate but any consistent grain model must account for absorption and emission data, and grain evolution, and in this regard IRAS fluxes (denoted by I_{12}, I_{25}, I_{60} and I_{100} MJy sr^{-1} for bands centered at 12,25,60 and 100 microns) are particularly useful. Following an analysis[1] of observations of molecular cloud complexes in the context of the hydrogenated amorphous carbon (HAC) grain model, we try here to account more quantitatively for variations in IR colours measured in diffuse clouds using both the time and optical depth variation in the evolution of carbonaceous mantles within such clouds.

It is clear that often the ratios of the mid IR (I_{12}, I_{25}) to far IR (I_{100}) intensities cannot be explained simply by variations in the UV flux and must involve changes in the actual abundances of the emitting particles. Less attention has been paid to the I_{60}/I_{100} colour (denoted $R_{60,100}$) which also shows deviations from the behaviour expected from classical large grains. Such grains emit in thermal equilibrium - hereafter referred to as bulk thermal emission - but the observed I_{60} intensity is larger than expected and also rises or falls with opacity depending on the cloud observed. It is proposed here that the required abundance variations in the emitters merely represent epochs in the evolution of mantles accreted onto classical grains.

HAC STRUCTURE

HAC has an amorphous structure composed mostly of sp^2 and sp^3 bonded carbon. Amorphous carbon, αC, is unique amongst amorphous materials in possessing medium-range order - sp^2 sites occur in 'islands' of graphite type layers, structures that are essentially PAH's bound in a solid alkane matrix. The size of these islands depends on the H-content of the material and in the interstellar medium it is the rival processes of UV photon graphitization and H-atom addition into the matrix to form alkanes that determine the sp^2/sp^3 ratio.

In the HAC model evolutionary picture carbon deposited onto silicate cores in the presence of atomic H will form a high band-gap, sp³ bonded mantle. UV processing of this mantle leads to the loss of H and the formation of islands which increase in size when subjected to continued UV exposure. Eventually the mantle may become essentially graphitic in nature, although there will be a limit to the ratio of sp²/sp³ sites beyond which the structure may be unstable[2]. The decrease in E_g associated with this process can be related to the size of these islands and it is the continued presence of this optical gap at high sp²/sp³ ratios that indicates that the sp² structures remain as isolated islands.

In fact, these islands may be so isolated that they retain absorbed energy long enough to emit thermally as separate entities[3]. They then account for the high IRAS I_{12} and I_{25} (and even I_{60}), and also the UIR features at 3.3-11.3 microns. Since free-flying PAH's are the main alternative proposed source of the UIR's the question is only one of the environment in which the emitting PAH finds itself.

That HAC does localize heat is not in question since this is required for the structural deformation that causes graphitization to take place. The efficiency with which it converts absorbed energy to near and mid IR emission depends on the timescale for the emission of an IR photon, which may be considerably shorter than the commonly quoted value of ~1s. This is due to a combination of the high temperature to which an island is transiently heated on absorption of a photon and the unknown absorption efficiency of the islands in the IR. Since HAC is an amorphous layered structure we model the emissivity as λ^{-1} with Q=1.6πa/λ. As absorption is expected to occur in the sp² structures this value is also taken for the 'hot spot' (HS) emission from the PAH islands. With this value IR emission timescales of τ_{IR}~10^{-3}s are likely.

There are two probable mechanisms by which mantles emit non-equilibrium radiation. The first[3] presumes that islands are connected to each other by weak coupling bonds of energy quantum Δ. Absorption of a photon in one island is then followed by delocalization of energy to a further j-1 islands before IR emission takes place; j being evaluated by comparing the timescale for the sharing of energy with the timescale for emission. The temperature rise on absorption of a photon hν is given by

$$T_c = \frac{h\nu}{k(3N_c-6)} \quad (1)$$

and this collection of j islands is then an emitting cluster composed of N_c carbon atoms. In the second mechanism we consider the emission to occur from a single island. Unlike the first mechanism the emission wavelength then depends on the size of the emitting PAH, which we take to be governed by

$$\frac{dM}{dt} = \frac{\chi e^{-2.5\tau_v}}{\beta M} \quad (2)$$

where M is the number of benzene rings in an island, χ is the radiation field scaling factor, τ_v is the visual optical depth, and β is the timescale for the formation of a single ring PAH (My). The emission in this case is time and depth-dependent since further processing increases the size and decreases the temperature of the hot spot emitter.

In reality it is probable that the latter mechanism is more appropriate at low values of M since islands will be far apart from each other and are unlikely to connect.

EMISSION MODEL

It is assumed here that for every photon absorbed by a grain a fraction of the energy excites a hot spot and the rest goes into bulk thermal emission appearing in the I_{60} and I_{100} bands. The dust optical properties and size distribution are taken from Jones et al.[4]. Given that E_g has been determined from the island (or cluster) size the bulk thermal emission may be calculated, using the balance of absorption and emission to calculate the dust temperature $T_d(a)$. To model emission from a cloud an infinite slab with the full radiation field impinging from one side is used. Partitioning the cloud into a number of depth points with constant T_d, E_g and optical depth τ_v, the emitted intensity in a step Δz and a frequency band (υ_2,υ_1) can be found. Summation over all depth points gives the cloud luminosity over this band width, and this can be compared with quoted IRAS fluxes by assuming a constant υI_υ spectrum and taking the relevant IRAS response functions.

Hot spots are by definition transient emitters. Since the thrust of this paper is an explanation of colour variation rather than absolute intensity calculations we simplify the calculation by defining a grain size dependent average photon energy as a function of depth that is taken to be the energy that creates the spike temperature T_c. All grains are assumed to convert to sp^2 bonded material at the same rate, so that cluster size is merely defined by depth (and time). The intensity of hot spot emission is then given by

$$I_z = \Delta z \int_{v_1}^{v_2} \int_a B_v(T_c, a_c) Q_v \pi a^2 n(a) P \, da \, dv \qquad (3)$$

where a and a_c are the bulk grain radii and HS emitter radii. The extra factor P now is the probability that a HS is excited and therefore emitting, defined as $P=\tau_{cool}/\tau_x$, the ratio of the timescales for cooling and excitation of a HS. τ_{cool} is effectively the timescale for the loss of 1/e of the input energy, i.e. it sets the fraction of energy emerging in HS emission rather than bulk emission, and we have assumed that the HS emission is black body at temperature T_c over this period.

For large grains P may become unity at late times (high M) but the wavelength of emission is too great for such grains to make any contribution. In fact there must be limits to how small or how large an emitting HS can be - neither polymeric HAC nor αC will emit in the mid IR. We set a lower limit of M=5 (below this emission is expected to be dominated by specific side group PAH vibrations (UIR bands)).

RESULTS

We consider here that the observed colours as a function of I_{100} emission are due to cloud depth effects and the scatter in these observations at a given I_{100} intensity is a time (i.e. evolutionary) effect. We concentrate on two analyses of IRAS data; those by Boulanger et al.[5] of the molecular cloud complexes Taurus, Chamaeleon and Ursa-Major, and several isolated diffuse clouds observed at by Laureijs et al.[6]. The salient features we wish to address are:

(i) a scatter of the $R_{12,100}$ colour in the complexes of a factor of ~20 encompassing a steep decrease with increasing I_{100}. In the same observations the $R_{12,25}$ colour shows no I_{100} dependence and the scatter is only of the order of a factor of 2;

(ii) in contrast to the cloud complex observations of $R_{60,100}$ marginally decreasing the diffuse clouds G230-29N, G230-29S and G299-16 show $R_{60,100}$ to be strongly increasing with I_{100}. These particular clouds are distinctive because of their regular morphology.

At the cloud edge the full UV field impinges on the grains leading to heavy processing of the mantle and a low optical gap, but further into the cloud shielding maintains a higher sp^3/sp^2 ratio and the band gap increases. Hence the size of the emitting islands decreases into the cloud. Also the average photon energy absorbed by grains decreases, a consequence of the adopted interstellar extinction curve, so that the 'spike' temperature of a hot spot decreases into the cloud. The decrease is rather less marked in smaller grains, since they absorb much less strongly at the low photon energies near the band gap. Increasing band gap and decreasing photon energy are opposing factors when it comes to determining the wavelength of the spike emission.

The $R_{12,100}$ colour is plotted as a function of I_{100} in Fig.1 for $\beta=10^{-3}$ and $t=5.10^4 y$. Also shown are plots at times of $2.10^4 y$ and $10^5 y$. It is clear that the model roughly reproduces the observed short wavelength emission but detailed comparisons require a more intricate model. A number of authors have noted that it is not enough to alter the UV field in models in order to change colours significantly since I_{100} intensity also drops off at a comparable rate[7,8]. For our $5.10^4 y$ model the colour is then near constant for some depth, but then a stage is reached where the grain mantle has not been processed at all and localized heating will not occur. Then the I_{12} emission drops off. The decrease in $R_{12,100}$ is not as marked as might be expected since the I_{100} is also decreasing and becomes non-linear with optical depth. Since the early time model reaches this point at a lower A_v, and the late time model does not reach this point at all, it can be imagined how a number of observations taken at different values of A_v and in different regions (i.e. different evolutionary states) could produce a large scatter in $R_{12,100}$. The argument may also be made[1] that the $R_{12,100}$ ratio is controlled by shocks.

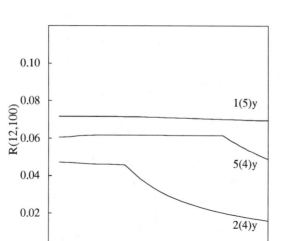

Fig. 1. Variation of $R_{12,100}$ with time and I_{100} for $\chi=1$ and $\beta=10^{-3}$

By contrast the $R_{12,25}$ colour is almost constant with depth for a given time since changes in the UV field and changes in the size of the islands affect these wavelength ranges almost equally. Also the cut-off where hot spots cease is not apparent since both I_{12} and I_{25} have the same origin. Although the scatter in this colour with time is still roughly in agreement with observations (about a factor of 2) it is of concern that the magnitude is about a factor of two too high.

The $R_{60,100}$ colour for the parameters of Fig.1 is given in Fig.2. For early times, as with the other colours, the fall with A_v is marginal until the cut-off in HS emission is reached and is less severe than for $R_{12,100}$ because there is a contribution from bulk thermal emission. This bulk thermal I_{60} emission is rather smaller than it should be since the small grain emission has been neglected. It is our contention that the $R_{60,100}$ rise with optical depth seen in the diffuse clouds is due to these clouds existing in their current state for times rather longer than 10^6y. This is supported by their highly regular structure. Then the emitting islands contain many carbon atoms, to the extent that at the cloud edge there is no HS emission at all and the mantle is either converted into αC or else has fractured and released into the gas phase. As the optical depth increases the degree of processing decreases and at some point HS emission manifests itself as I_{60} emission. Note that in such clouds the I_{12} intensity is weak. Such a scenario is capable of reproducing the observations with this single island emission model, but since it was stated in Section 2 that emission at larger wavelengths is likely to come from clusters of islands, we can use the observations to confine the value of Δ and estimate the age of the emitting clouds. In order for such a cluster to emit in I_{60} whilst not simultaneously producing shorter wavelength emission we require $\Delta \sim 0.1$eV. For the standard radiation field such a value leads to the onset of HS emission at 60 micron when I_{100} is ~1MJy sr^{-1} (roughly as observed) at $t=10^7$y.

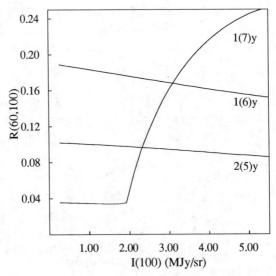

Fig. 2. $R_{60,100}$; parameters as in Fig.1.

REFERENCES

1. W.W.Duley,A.P.Jones,S.D.Taylor,D.A.Williams,MNRAS,260,415(1993).
2. J.Robertson,Adv.in Phys.,35,317(1986).
3. W.W.Duley,D.A.Williams,MNRAS,231,969(1988).
4. A.P.Jones,W.W.Duley,D.A.Williams,MNRAS,236,709(1989).
5. F.Boulanger,E.Falgarone,J.L.Puget,G.Helou,ApJ,364,136(1990).
6. R.J.Laureijs,G.Chlewicki,F.O.Clark,A&A,192,L13(1988).
7. G.Chlewicki,R.J.Laureijs,A&A,207,L11(1988).
8. J.P.Bernard,F.Boulanger,F.X.Desert,J.L.Puget,A&A,263,258(1992).

DIFFUSE INTERSTELLAR ABSORPTION AND EMISSION BANDS

R.E. Hibbins, J.R. Miles and P.J. Sarre

Department of Chemistry, University of Nottingham,
University Park, Nottingham, United Kingdom NG7 2RD

ABSTRACT

Unidentified astrophysical spectra are discussed including the diffuse interstellar absorption bands and visible emission bands from the Red Rectangle nebula and the R Coronae Borealis star V854 Cen. Assignments for the Red Rectangle optical emission and complementary diffuse absorption bands are not yet established, but comparison of the spectra allows some constraints on the nature of the carriers to be inferred.

An unidentified infrared emission feature observed towards the two Herbig Ae/Be pre-main-sequence stars HD 97048 and Elias 1, and reported by Tokunaga et. al. (Astrophys. J. **380**, 452, 1991) to be centred at 3.5325 μm in the latter, is attributed to part of the fundamental vibration-rotation spectrum of the H_3^+ molecular ion.

INTRODUCTION

Recent spectacular advances in telescope and detector technology now allow unprecedented recording of astronomical spectra in the infrared and visible spectral regions. Our interest is focused on some important problems in the identification of astronomical spectra. Apart from the obvious importance of carrier assignment, a second objective is the use of the spectral information to characterise and model the chemical and physical conditions in various astrophysical environments.

In this paper we discuss the long-standing problem of the diffuse interstellar absorption bands and particularly the relationship between some of the diffuse absorption bands and emission bands which are observed from the biconical Red Rectangle nebula and also from the R Coronae Borealis star V854 Cen. A correspondence between the strongest Red Rectangle emission bands and part of the diffuse absorption band spectrum is examined; it is of particular significance because the spectra arise from the same carriers.

On a separate topic, we discuss infrared spectra of the dust-shrouded pre-main-sequence stars HD 97048 and Elias 1 and propose assignment of an emission band near 3.5 μm to the H_3^+ molecular ion.

DIFFUSE INTERSTELLAR ABSORPTION BANDS

The last decade has seen some important advances in the long-running saga of the problem of the diffuse interstellar bands and it is to be hoped that further developments will lead, sooner rather than later, to a solution. More probably one should say solutions, as it by no means certain that solving one part will automatically lead to a solution for the rest! In recent years the observational results have generally favoured an origin in free molecules rather than grains, although the growth in knowledge of clusters illustrates that some hybrid of the two limiting forms of material should not be excluded from consideration. We concentrate here on one small subset of the bands, classified by Krełowski and Walker[1] as their third group (KW3).

The diffuse band problem dates back to the early part of this Century when the first evidence for features was discovered, although their interstellar origin was not recognised and confirmed till some years later[2]. A review of the problem is summarised in ref. 3. The diffuse bands determined by Herbig are shown in fig.1.[4,5]

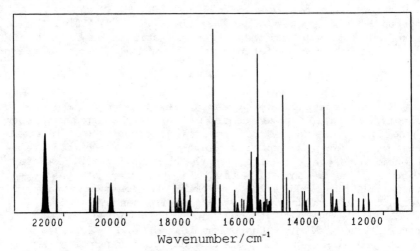

Figure 1. The diffuse band spectrum for the heavily reddened star HD 183143. The data are from refs. 4 and 5. A more recent survey[6] has revealed additional bands but the overall appearance of the spectrum is not changed substantially.

Two of the earliest findings were prominent narrow bands at 5797 and 6613 Å. In 1986 Krełowski and Walker[1] showed that these bands, together with others at 5849, 6376 and possibly 6379 Å, form a group, or 'family' (KW3), based on their common intensity behaviour along various lines-of-sight. In our view the existence of a family of bands does not imply that they all necessarily arise from the same carrier. However, the KW3 bands clearly have some other characteristics in common: the linewidths of each member are comparable in magnitude, in the region of 1 Å (FWHM) or less, which makes them amongst the very narrowest 'diffuse' bands, and the shapes in a few cases are pronounced with 5797, 6614 and possibly

6376 exhibiting a steep blue side (see ref. 3 for details).

The most striking evidence for linking the bands into a family is the observation of corresponding emission bands in the Red Rectangle nebula[7,8,9]. In Fig.2. we show a spectrum for the 5800 Å region of the Red Rectangle[10] together with a recently recorded spectrum of the heavily reddened star HD 183143[11].

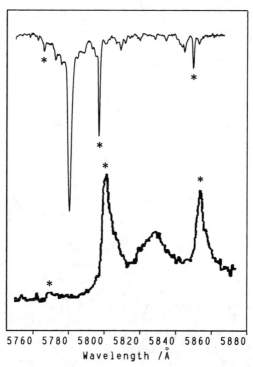

Figure 2. Absorption spectrum of HD 183143[11] (upper) compared with an emission spectrum (lower) of the Red Rectangle[10] adapted from Nature (London) **292**, 317 (1981), with permission. Note the correspondence between the 5766[12,3], 5797 and 5849 absorption bands with Red Rectangle emission features.

It is notable that there is no evidence for the famous 5780 diffuse absorption band in emission, whereas there is close correspondence between the wavelengths of the absorption features at 5766, 5797 and 5849 Å and the short-wavelength edge of the emission bands at peak wavelengths of about 5770, 5799 and 5855 Å, where the latter are shifted to longer wavelengths by a few Angstrom units[8,13]. The shift in wavelength and broadening of the band can be rationalised in terms of the rotational contour of a medium-to-large sized molecule[14,13], where the steep blue side is the rotational branch head and the band is red-degraded. In collaboration with Dr. S.M. Scarrott and Dr. T.E. Gledhill of Durham University we have recorded new spectroscopic data of the Red Rectangle to determine the evolution of the bands as a function of offset from the central star HD 44179. Figure 3. shows how the linewidth and peak wavelength for the 5799 feature change as a function of distance from the central star. In the limit of large distance from the star, the data can be extrapolated smoothly to the characteristics of the diffuse absorption band feature.

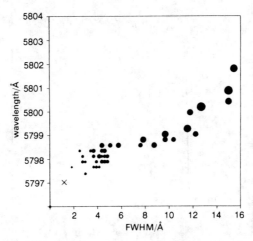

Figure 3. Peak wavelength vs. FWHM for the '5799' emission feature in the Red Rectangle, where the radius of the circle is inversely proportional to the distance from HD 44179. The values of the '5797' diffuse band wavelength and FWHM for HD 183143 are indicated by a small cross. (Reproduced with permission from J. Chem. Soc. Farad. Trans. 89, 2308 (1993).)

In an exciting recent development, Jenniskens and Désert[15] and Krełowski and Sneden[16] have made studies of the diffuse bands in the 5800 region and have discovered a number of new weak narrow bands. These include bands at 5818 ('uncertain'), 5828 and (we infer from the spectra[16]) near 5910 Å. The wavelengths of the last two are strikingly close to Red Rectangle emission features which have maxima near 5827 and 5913 Å[13]. The 5818 diffuse band may correspond to a shoulder on the short-wavelength side of the 5827 Red Rectangle emission feature (see Fig.2.). Of the remaining prominent Red Rectangle bands only the band at 5884 Å[13] appears to be without an obvious diffuse band counterpart.

We now make some tentative comments on a possible interpretation of the spectral data, to some degree in a spirit of adventure, but guided we hope by sound spectroscopic principles! Under the assumption that the spectrum is molecular it is natural to attribute the 5797 band to the origin band which we label (0,0). The fact that the band contour is red-degraded implies that there is a significant geometry change (increase in moment of inertia) on excitation. Other vibrational bands are therefore expected and the weak band at 5766 is a likely candidate for a (1,0) vibrational band from which we estimate an excited-state interval of 92.5 ± 0.5 cm^{-1}. Spacings of similar magnitude occur elsewhere in the spectrum; the spacing between the 5797 and 5828 diffuse bands is ca. 93.5 ± 0.5 cm^{-1}, and a plausible interpretation of the data is that the 5828 diffuse band corresponds to a (0,1) band. The 5849 diffuse band is the second strongest in the set and would appear to be part of the same spectrum, though this is not firmly established. In the light of the foregoing discussion, it is of interest that the weak 5818 diffuse band lies about 91.5 ± 0.5 cm^{-1} to shorter wavelength of 5849. However, if 5849 arises from the same molecule, and the assumption that the 5797 band is the (fully allowed) origin band is correct, the 5849 band would necessarily involve a vibrationally excited level of the ground state with an excitation 'energy' of 155.5 cm^{-1}. Intriguingly, the ratio of the equivalent widths of 5797 to 5849 varies along different lines of sight between 1.8 and 3.9[17]. This result, and the possible attribution of the 5828 band to '(0,1)', suggests that the possibility of excitation from vibrationally excited levels should be considered seriously, with the variability in the 5797/5849 intensity ratio reflecting different degrees of vibrational excitation depending on the local conditions.

We briefly comment on a remarkable observation of the RCB star V854 Cen at minimum light[18]. Broad emission bands are found which correspond to the most prominent unidentified emission bands seen in the Red Rectangle. The broadening in this case may be due in large part to motion of the carriers because the Na-D and other lines are also broad, in sharp contrast to the observations for sodium in the Red Rectangle. It is to be hoped that some clue as to the elemental composition of the KW3 molecular family will emerge from such studies.

INFRARED EMISSION FROM HD 97048 AND ELIAS 1

Numerous astrophysical objects display all or at least some of the so-called 'unidentified' infra-red emission bands which are widely attributed to vibrational transitions of neutral and/or ionised polycyclic aromatic hydrocarbons (PAHs) in either gas-phase or amorphous form. The discovery of infrared emission from HD 97048 by Blades and Whittet[19] in 1980, however, opened a distinct chapter because the 3 - 4 μm region is dominated by two prominent 'anomalous' features centred at ca. 3.4 and 3.5 μm, and the 3 μm feature readily detected in many objects is weak. Spectroscopic studies[20-22] have revealed that the 3.4 and 3.5 μm bands have distinctive shapes, and it has been found that a second object, Elias 1, possesses a similar spectrum[21,23,24]. A number of specific proposals for the carrier(s) have been made: initially it was thought that solid frozen formaldehyde or polyoxymethylene might be responsible[19], and this was investigated in detail in the laboratory[22,25]. However, observational studies of HD 97048 in the 8 - 13 μm range[20], and the finding that the 3.53 μm emission emanates from a relatively hot region within 0.1 arc sec of the star[26] rather than an extended region, leads to the conclusion that neither of these substances are carriers. The bands have more recently been attributed to chemical side groups on PAH molecules[25] and arguments for an origin in C-C overtones and combination bands of highly excited PAHs have been presented, together with an extensive assessment of a range of likely PAH-based carriers[27]. In a further development, the bands have been attributed to 'fullerane' molecules[28].

On reading the exchanges[29,30] in the recent Faraday Symposium on 'Chemistry in the Interstellar Medium' concerning the attribution[31] of 3.4 and 3.5 μm emission features observed in SN 1987a to H_3^+, we were led to consider whether a link might exist between the vibration-rotation spectrum of the H_3^+ molecule and the 3.43 and 3.53 μm features of HD 97048 and Elias 1. (The correspondence of the wavelengths of the 3.4 and 3.5 μm features in the supernova and those for HD 97048 had in fact been noted in 1989 in a paper on the infrared spectrum of SN 1987a[32]). Of crucial importance to our proposal of a link is a very high-resolution spectrum of Elias 1 published in 1991 by Tokunaga et. al.[33] and which is replotted in Fig. 4.

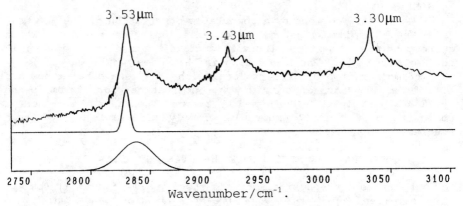

Fig. 4. Spectrum of Elias 1 first published in Astrophys. J. 380, 452 (1991) and replotted here from the original data by kind permission of A. Tokunaga. The inset shows a deconvolution of the 3.53 μm feature into a broad and narrow component.

We interpret this spectrum of Elias 1 in the following way: There are clearly two main features near 3.43 and 3.53 μm, but we find that the 3.5 μm feature can be deconvoluted into two parts. One of these is narrow which we determine to be centred at a wavelength of 3.5334 μm[33] with a FWHM of ca. 0.009 μm (6.8 cm^{-1}), the other being a broader component centred at the slightly shorter wavelength of 3.5230 μm with a FWHM of 0.04 μm (33 cm^{-1}). Although the published spectra[20,27] of HD 97048 are of lower resolution, the spectra of HD 97048 and Elias 1 are very similar in overall character and a similar deconvolution for the 3.5 μm feature of HD 97048 could also be undertaken. There may be narrow structure on top of the 3.43 μm broad feature in both HD 97048 and Elias 1 (see Fig. 4 and spectra in refs. 20 and 27), but unlike the 3.5 μm region, spectra in the 3.4 μm region are contaminated by atmospheric absorptions[33] and it is difficult to assess the reality of this possible fine structure.

Recognising that the observed narrow component has a linewidth (FWHM) of ca. 0.009 μm (7 cm^{-1}), there is close agreement between the wavelength of this component at 3.5334 μm (2830.1 cm^{-1}) and vibration-rotation lines of the fundamental vibration-rotation band of H_3^+; the $R(3,3)^-$, $R(3,1)^-$ and $R(3,2)^-$ lines have laboratory rest wavelengths (cm^{-1}) of 3.5337 μm (2829.9 cm^{-1}), 3.5319 μm (2831.3 cm^{-1}) and 3.5308 μm (2832.2 cm^{-1}) respectively[34]. It is emphasized that at this stage it is the narrow component of the 3.53 μm feature with which we concern ourselves primarily in connection with H_3^+. The broad 3.5 μm feature (and others) would appear to be rather wide to be considered as arising from vibration-rotation transitions in this molecule but this aspect warrants further investigation. In order to be assigned definitively as arising from H_3^+, a number of criteria should be satisfied, not all of which have yet been fully tested. These include: a good correspondence in wavelength between laboratory data and a high-resolution astronomical spectrum, a linewidth consistent with known or anticipated conditions of kinetic 'temperature' moderated by the resolution under which the data were recorded, an observed intensity consistent with likely abundance, presence (and/or absence) of other rovibrational lines and a plausible excitation mechanism. Parallel considerations have recently been employed in connection with the attribution of spectral features in SN 1987a to H_3^+ where the spectroscopic evidence is supported by consideration of the likely abundance of the molecule in the ejecta[30].

The wavelength criterion is well satisfied and the observed linewidth appears consistent with the resolution of the spectrum of Elias 1. Emission spectra of H_3^+ in the 3.20 - 3.66 μm (3125 - 2732 cm^{-1}) region at moderate rotational temperatures generally contain lines in addition to those mentioned above[34]. Anticipated lines under thermally excited conditions include the $R(2,2)^-$ at 3.620 μm (2762.1 cm^{-1}) as shown in figure 5 (a). It is necessary to search for an explanation as to why this line and also the $R(2,2)^+$ and $R(2,1)^+$ lines near 3.54 μm are not observed with the expected intensity. The expected spectrum at 300 K (Fig. 5 (a)) is drawn under the assumption of thermal equilibrium and the comparison of predicted and observed spectra in SN 1987a was also made in this way[31]. Attention has been drawn, however, to possible discrepancies between the observed and calculated spectra for SN 1987a[29].

We have considered a range of possible excitation mechanisms which could give rise to selective emission. The formation of vibrationally excited H_3^+ by ion-molecule reactions is not sufficiently rapid or selective to satisfy the observations, and the discrepancy for emission from thermally (collisionally) populated levels has already been discussed. An alternative mechanism, which as far as we are aware has not been considered in detail in this context is that of photo-excitation in lines of the fundamental vibration-rotation absorption band by infrared radiation. This can occur through a combination of broad-band infra-red radiation and emission lines of atomic hydrogen (e.g. Brackett-α at 4.05 µm). Using published absorption intensities[34] and Einstein A-coefficients for individual vibration-rotation lines[35], we have calculated the emission spectra of H_3^+ in the 3 - 4 µm region under conditions of broad-band 'white' infrared excitation in the ϑ_2 region (the contribution from overtone absorption was not included), and excitation by lines of atomic hydrogen, most notably the Brackett-α line at 4.05 µm. In the latter case we took the linewidth of the hydrogen line to be that observed for SN1987a as Br-α radiation is not observed from HD 97048[20,21], possibly due to radiation trapping. The results of these calculations are shown in figure 5(b) and (c) where comparison is made between (a) thermally excited, (b) broad-band photo-excited, and (c) H-line excited spectra, shown for a rotational temperature of 300 K. It is found that over a fairly wide temperature range the H-line pumping yields an emission spectrum dominated by just two vibration-rotation lines, $R(3,3)^-$ and $R(3,2)^-$ spaced by 2.28 cm^{-1} and which are of comparable intensity. On convolution with the spectral resolution for the astronomical data, the linewidth for the 3.53 µm feature is found to be in good agreement with that observed.

Fuller details of the calculations will be described elsewhere as well as comparison with the more extensive yet lower resolution spectra of SN 1987a. For the supernova case it would appear that Br-α excitation of H_3^+ would also result in an emission spectrum dominated by the 3.53 µm feature. It is of interest that an unidentified emission feature near 3.5 µm has been observed[36] in the spectrum of Nova Cygni 1975, but with little evidence for a feature near 3.4 µm; we suggest the 3.5 µm feature here may also due to H_3^+. We finally note that the $R(3,3)^-$ line of H_3^+ is found to be anomalous in intensity in the Jovian spectrum[37] and this may also be the case in Saturn[38]; whether this is related to the topics discussed here remains to be seen but is certainly worthy of investigation.

This identification represents the first evidence for H_3^+ in a star-forming region. One of the major puzzles is why the 3.53µm feature appears only in HD 97048 and Elias 1 and not in a large number of other similar objects[21]; perhaps they represent an extreme combination of hydrogen density, infrared radiation and chemical anomaly such as lack of oxygen.

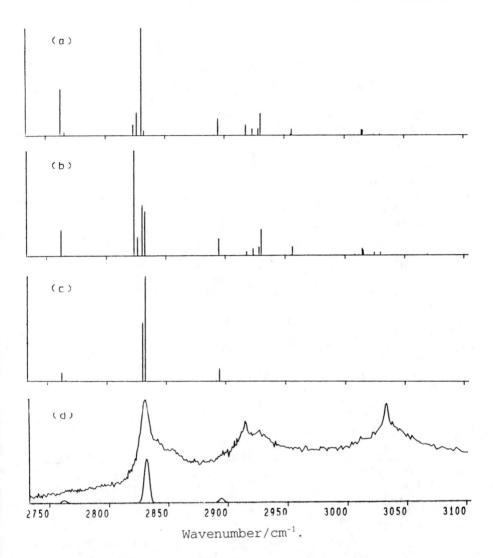

Figure 5. Calculated emission spectra of H_3^+ in the 3.20 - 3.66 μm region resulting from (a) thermally populated excited state levels at 300 K, (b) photo-excitation by a 'white' light infrared source with T_{rot} = 300 K, (c) photo-excitation by H-line radiation as observed in SN 1987a[32], (d) convoluted spectrum of (c) with resolving power of 1400, (e) astronomical data for Elias 1 (see caption to figure 4).

ACKNOWLEDGEMENTS

We are most grateful to Dr. A. Tokunaga for sending us the original data for Elias 1 and for permission to include the spectrum in this paper, Dr. G.H. Herbig for sending us the spectrum of HD 183143 and for many helpful discussions, and Professors Krelowski and Sneden for permission to quote their measurements of diffuse bands in the 5800 region. JRM and REH thank SERC for studentships.

REFERENCES

1. J. Krełowski and G.A.H. Walker, J. Roy. astr. Soc. Canada **80**, 274 (1986); Astrophys. J. **312**, 860 (1987)
2. P.W. Merrill and O.C. Wilson, Astrophys. J. **87**, 9 (1938) and references therein
3. J.R. Miles and P.J. Sarre, J. Chem. Soc. Farad. Trans. **89**, 2269 (1993)
4. G.H. Herbig, Astrophys. J. **196**, 129 (1975)
5. G.H. Herbig and K.D. Leka, Astrophys. J. **382**, 193 (1991)
6. P. Jenniskens, Ph.D. thesis, University of Leiden (1992)
7. S.J. Fossey, Ph.D. thesis, University of London (1990)
8. P.J. Sarre, Nature **351**, 356 (1991)
9. S.J. Fossey, Nature **353**, 393 (1991)
10. R.F. Warren-Smith, S.M. Scarrott and P. Murdin, Nature **292**, 317 (1981)
11. G.H. Herbig, personal communication
12. G. Chlewicki, M.S. de Groot, G.P. Van der Zwet, J.M. Greenberg, P.P. Alvarez and A. Mampaso, Astron. Astrophys. **173**, 131 (1987)
13. S.M. Scarrott, S. Watkin, J.R. Miles and P.J. Sarre, Mon. Not. Roy. astr. Soc. **255**, 11p (1992)
14. G.D. Schmidt and A.N. Witt, Astrophys. J. **383**, 698 (1991)
15. P. Jenniskens and F.-X. Désert, Astron. Astrophys. **274**, 465 (1993)
16. J. Krełowski and C. Sneden, Publ. Astron. Soc. Pacific (1993, in the press)
17. G.H. Herbig, Astrophys. J. 407, 142 (1993)
18. N.K. Rao and D.L. Lambert, Mon. Not. R. astr. Soc. **263**, L27 (1993)
19. J.C. Blades and D.C.B. Whittet, Mon. Not. Roy. astron. Soc., **191**, 701 (1980).
20. D.K. Aitken and P.F. Roche, Mon. Not. Roy. astron. Soc. **196**, 39p (1981).
21. D.A. Allen, D.W.T. Baines, J.C. Blades and D.C.B. Whittet, Mon. Not. Roy. astron. Soc. **199**, 1017 (1982).
22. F. Baas, L.J. Allamandola, T.R. Geballe, S.E. Persson and J.H. Lacy, Astrophys.J. **265**, 2990 (1983).
23. D.C.B. Whittet, P.M. Williams, M.F. Bode, J.K. Davies and W.J. Zealey, Astron. Astrophys. **123**, 301 (1983).
24. D.C.B. Whittet, A.D. McFazdean and T.R. Geballe, Mon. Not. Roy. astron. Soc. **211**, 29p (1984).

25. G.P. van der Zwet, L.J. Allamandola, F.J. Baas and J.M. Greenberg, Astron. Astrophys. 145, 262 (1985).
26. P.F. Roche, D.A. Allen and J.A. Bailey, Mon. Not. Roy. astron. Soc. 220, 7p (1986).
27. W.A. Schutte, A.G.G.M. Tielens, L.J. Allamandola, M. Cohen and D.H. Wooden, D.H., Astrophys. J. 360, 577 (1990).
28. A. Webster, Mon. Not. Roy. astron. Soc. 257, 463 (1992).
29. J.K.G. Watson, J.Chem. Soc. Farad. Trans. 89, 2170 (1993)
30. S. Miller, J.Chem. Soc. Farad. Trans. 89, 2171 (1993)
31. S. Miller, J. Tennyson, S. Lepp and A. Dalgarno, Nature 355, 420 (1992)
32. W.P.S. Meikle, J. Spyromilio, D.A. Allen and G.-F. Varani, Mon. Not. Roy. astron. Soc. 238, 193 (1989).
33. A.T. Tokunaga, K. Sellgren, R.G. Smith, T. Nagata, A. Sakata, A. and Y. Nakada, Astrophys. J. 380, 452 (1991)
34. W.A. Majewski, M.D. Marshall, A.R.W. McKellar, J.W.C. Johns and J.K.G. Watson, J. Mol. Spectrosc. 122, 341 (1987)
35. L. Kao, S. Miller, J. Tennyson, Astrophys. J. 77, 317 (1991)
36. G.L. Grasdalen and R.R. Joyce, Nature 259, 187 (1976)
37. J.-P. Maillard, P. Drossart, J.K.G. Watson, S.J. Kim and J. Caldwell, Astrophys. J. 363, L37, (1990).
38. T.R. Geballe, M.-F. Jagod and T. Oka, Astrophys. J. 408, L109, (1993).

DISCUSSION

THADDEUS — Detectable vibrational excitation of a molecule in the interstellar gas might seem paradoxical, but excitation of a level only 93 cm^{-1} above ground in a floppy molecule is possible. Vibrational excitation of the lowest frequency mode of C_4H is appreciable in the molecular shell of IRC + 10216, as shown by Guélin and co-workers, and the conditions there may not be too different from yours.

SARRE — This is an interesting comment which should be included but I have nothing to add.

LEACH — Two years ago, I considered the possibility of the 3.53 µm band in HD 97048 being due to H_3^+. I asked Steven Miller to simulate the spectrum and we compared the simulations of this and other expected features with observations. It appeared that a fit required what we thought was an unreasonably high velocity dispersion of about 3000 km.s^{-1}, so that we did not pursue this further at the time.

SCAPPINI — You mentioned possible radio observations of H_3^+ toward Herbig stars. Unfortunately, as you know, the rotational spectrum of H_3^+ is forbidden, and only a small electric dipole moment is vibrationally induced. Thus radio-observations are going to be difficult.

SARRE — I agree that radio detection might not be easy, but quite a lot of telescope time has been devoted to searches for H_3^+ in various sources, without success. Hence our plans to try towards these particular Herbig Ae/Be stars !

D'HENDECOURT — You mentioned that the 4430 and 4780 diffuse bands might arise from transitions in a pyrene cation derivative. I would comment that these bands are placed in different families by Krelowski and Walker[1].

SARRE — Although it is right that Krelowski and Walker have divided DIB's into families, this classification is by no means totally convincing and counter examples have been found. I believe that this problem of DIB's families should be reassessed.

(1)   J. Krelowski and G.A.H. Walker, J. Roy. Ast. Soc. Canada 80, 274 (1986).

OPTICAL OBSERVATIONS OF METALS IN CIRCUMSTELLAR ENVELOPES

N. Mauron and Ch. Guilain
CESR CNRS-UPS, BP 4346, 31029 Toulouse. France

ABSTRACT

Our program contributes to study refractory elements in circumstellar shells of red giants, where condensation of gas and molecules into grains is not yet fully understood. Atoms like NaI, KI, FeI, CaI, CaII, etc can be optically seen thanks to their resonance lines, informing us about the poorly known circumstellar depletion. Either absorption lines or scattering emission can be studied.
We show new detections of NaI and KI scattering around oxygen-rich red giants with very low and moderate mass-loss rates (1E-8 to <2E-6 M/yr). These shells may have no dust at all, or normal dust abundance. Our data on a limited sample of objects suggest that K and Na may or may not be depleted. This depletion is apparently not correlated with the presence of infrared-emitting silicate dust.

INTRODUCTION

Infrared observations and other techniques (polarization) showed many years ago that grains form in the gaz expelled by red giants. However the detailed physics of condensation, the nature of grains, the gaz-grain interactions are far from fully understood. Our general goal is to study circumstellar depletion of some elements in the gaz-phase and compare to interstellar one.
Alkalis like K and Na are basically in atomic form in photospheres. Condensation calculations at equilibrium predict, that Na and K remain atomic down to 1000K (Sharp and Huebner 1990)[1]. There is no prediction for colder temperatures, and/or taking into account flow effects. The molecules NaCl and KCl have been detected in IRC+10-216 only (Cernicharo et al 1987)[2], but because Cl is much less abundant than Na, NaCl is not the main Na carrier. For O-rich shells, grains like Na_2SiO_3 or K_2SiO_3 might form at about 1000K (Turner 1991)[3]. K and Na are moderatly depleted in the diffuse ISM.
The neutral atoms KI and NaI can be detected and mapped in circumstellar shells through resonance scattering in their optical lines 5890-5896Å and 7665-7699Å (e.g. Mauron and Caux 1992)[4]. The emission is usually observed at a few arcsec from the star: for example for Betelgeuse 4" at 200pc represents 1E+16cm or 200 stellar radii. This region generally has a rich chemistry. In 2 cases (Betelgeuse and Mu Cep), maps have been obtained over about 1 arcmin and resolution 2-3 arcsec. Up to now, positive detection of KI and/or NaI have been obtained for 7 oxygen-rich shells. One carbon-rich envelope (R Scl) has been recently seen in KI and NaI by Gustafsson et al.[5]

RESULTS

Figure 1 explains how the observations are carried out.

Figures 2, 3 and 4 show three new detections around red giants:
  Beta Peg (M2.5II-III; 40 pc) has a very low mass loss rate (1E-8M/yr) and does not show any trace of grains or molecules, the abundance of which is thus very low or null. Its envelope is thus mainly atomic and no depletion of elements on

grains or unseen molecules is expected. Only NaI scattering has been detected.

CE Tau is a star M2Ib or M2Iab at 400pc, whose mass loss rate is around 3E-7 M/yr. There is no detected CO or dust evidence in IRAS data; NaI has been searched for, twice independantly. KI is well detected, and NaI would have been easily detected with a standard ratio Na/K=3 (see below).

Rho Per (M4II-III;80 pc) is very similar to Beta Peg and has a circumstellar gas column density about 3 or 4 times larger. The IRAS PSC2 fluxes and the LRS spectrum do not indicate any dust infrared emission. The mass loss rate is around 2E-8 M/yr. We easily detect the NaI.

Figure 5 shows our best measured KI or NaI haloes intensities. Please note the large difference between Alpha Ori and Mu Cep: Alpha Ori is bright in KI and faint in NaI; the inverse is true for Mu Cep. The stars are however similar M supergiants.

Figure 6 displays the ratio Na/K versus the total mass-loss rate, showing large variations between envelopes. Note the similarity between Alpha Her and Mira Ceti: both have mass-loss of 1E-7 M/yr, but Mira is normally dusty whereas Alpha Her is completely dust-free. This suggests that Na and K are either not strongly depleted in the gaz-phase, or, if depleted, not fixed on the grains emitting infrared radiation.

Another interesting comparison is Alpha Ori versus Mu Cep, both at 2 to 4E-6 M/yr with both silicate emission and detected CO. In Alpha Ori, CO and dust are significantly underabundant (factor 10) by incomplete formation. In Mu Cep, normal or slightly lower abundances are favoured by observations. Note the extreme contrast on the Na/K ratio (a factor 100 difficult to ascribe to experimental errors). We believe that the best interpretation is a large depletion of Na in the gaz of Betelgeuse, but again, Na would not be locked on IR emitting silicates, otherwise Mu Cep would be also faint in Na.

Recent attempts to detect CaI 4227Å (Figure 7), FeI 3854Å and other lines scattering around Alpha Ori and Mu Cep were negative, although the absorption lines display typical circumstellar distortion. This is qualitatively consistent with some Ca and Fe being present in the gaz phase in the very inner part of the wind, followed by nearly complete condensation at some point in the wind. However, a quantitative analysis is needed and more sensitive observations are planned.

CONCLUSION

The NaI and KI scattering data on a small sample of 6 O-rich shells with moderate mass-loss and various dust abundances suggest that Na and K may be depleted, but this depletion does not seem to be correlated with the abundance of IR-emitting silicates.

REFERENCES

1. Sharp and Huebner, ApJSS **72**, 417 (1992)
2. Cernicharo & Guelin, A&A **183**, L10 (1987)
3. Turner, ApJ **376**, 573 (1991)
4. Mauron et Caux A&A **265**, 711 (1992)
5. Gustafsson et al : see Olofsson et al ApJS **87**, 267 (1993)

KI lines scattering
(7665-7699 Å), on α Ori

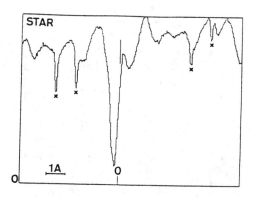

Star centered in the entrance hole

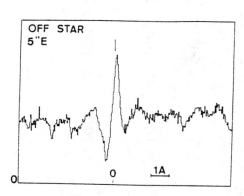

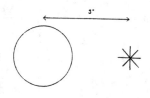

Slit centered at 5" from the star

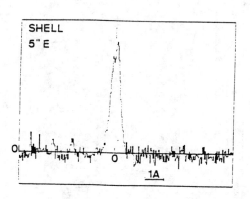

Shell spectrum with parasite scattered starlight substracted

Observation method

FIG. 1

102 Optical Observations of Metals

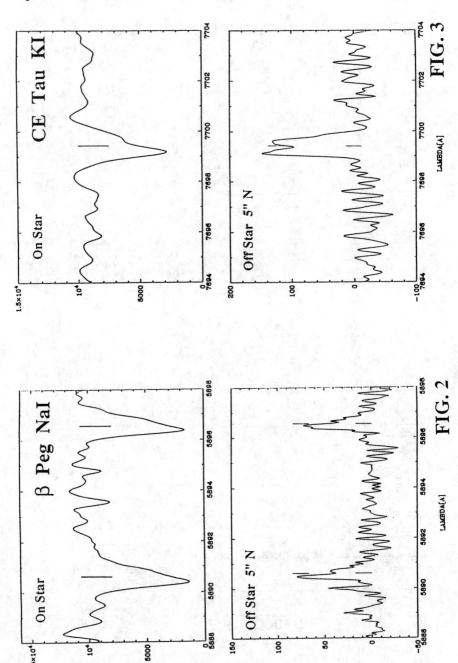

FIG. 2

FIG. 3

Figures 2,3 and 4

(Top) NaI or KI spectra lines on β Peg, CE Tau and ρ Per. The spectral resolution is about 10 km s^{-1}. The exposure time on star is 5 min. One can see on stellar spectra, typical circumstellar NaI or KI profiles.
The NaI emission cannot be confused with some sky emission thanks to a radial velocity shift of about 30 km s^{-1}.
The KI emission at 7665Å is also detected, but badly mutilated by telluric O$_2$.
(Bottom) Shell spectra were taken at typically 5 arcsec off-star. (Exposure time: 1 hour) and corrected at best for contamination by stellar light.

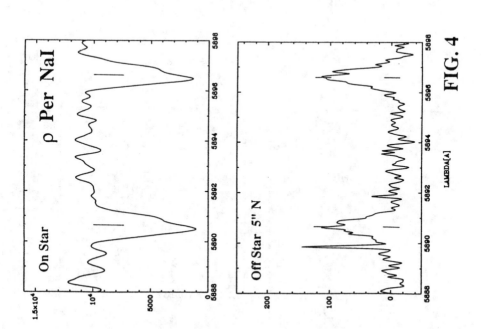

FIG. 4

104   Optical Observations of Metals

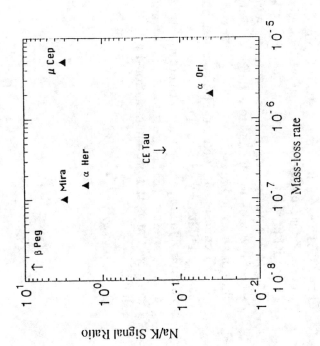

FIG. 6. Plot of the NaI over KI intensity ratio. Alpha Her, Mira and Mu Cep agree for a ratio of about 3; Betelgeuse appears strangely faint in NaI. Alternatively, if one thinks that this ratio decreases with M because of increasing NaI line opacity compared to KI, Mu Cep is exceptionally bright in NaI.

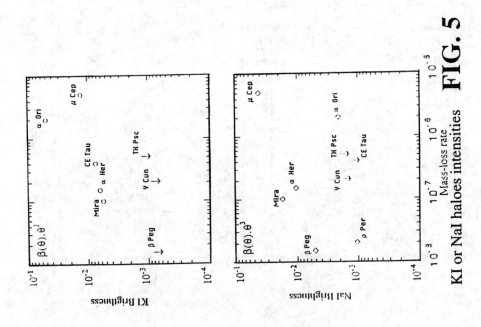

FIG. 5. KI or NaI haloes intensities

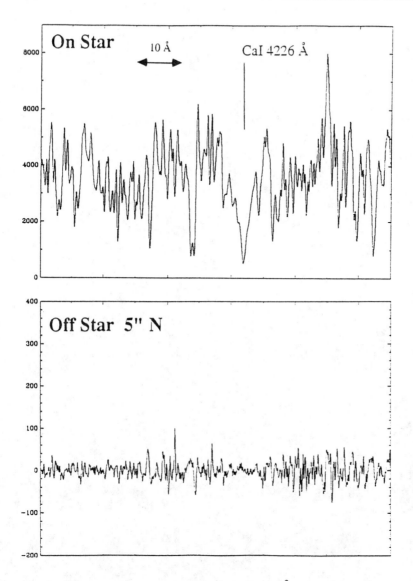

Search for scattering in CaI 4226Å line around Betelgeuse (non detection).

FIG. 7

INFRARED AND RADIO SPECTROSCOPY OF COMETS: RECENT RESULTS AND NEW PROBLEMS

J. Crovisier and D. Bockelée-Morvan
Observatoire de Paris-Meudon, F-92195 Meudon, France

ABSTRACT

The molecular composition of cometary volatiles is a basic information on the nature of comets and an important clue to their formation mechanisms. Important results were recently obtained on this topic from radio and infrared spectroscopy.

Many rotational lines of HCN, H_2S, H_2CO and CH_3OH were observed in comets by radio spectroscopy at millimetre and submillimetre wavelengths. These techniques are able to probe minor species with abundances of 10^{-3} or less relative to water.

Comets have important emissions in the 3.2-3.6 µm region. Fluorescence emission of large molecules and/or thermal emission of small organic grains were first invoked to explain this feature. We now know, from radio spectroscopy, that methanol is an abundant cometary constituent (1 to 5% relative to water) and that fluorescence of its vibrational bands accounts for a significant part of this emission. The remaining emission may be due to other CHO species. Their identification will require the confrontation of high-resolution cometary spectra with synthetic or experimental spectra of candidate species in cometary conditions (i.e. at cold temperatures).

INTRODUCTION

One of the main objectives of cometary studies is to establish the composition of cometary nuclei to get insights into the formation of these objects and into the history of our Solar System. Awaiting direct analysis of cometary nuclei by space exploration, one has to rely upon remote sensing of the sublimation products of these bodies. Since most of the parent molecules (i.e. those directly produced from the nucleus) do not have signatures in the visible or UV (most of their electronic transitions being dissociative), they must be studied from their infrared vibrational bands or their microwave rotational lines.

The goal of the present paper is to review the recent results of cometary molecular spectroscopy at radio and infrared wavelengths (for more comprehensive reviews, see[9,10,13]); to present the prospect of observations with future ground-based and space facilities; to evaluate the needs for laboratory and theoretical works in support of these investigations (see also[1]).

MICROWAVE SPECTROSCOPY

Recent results.

We undertook a program of observation of the microwave spectrum of comets with the 30-m IRAM radio telescope. It began with the detection of HCN in comet P/Halley in 1985. Since that time, successful millimetre observations were made in comets P/Brorsen-Metcalf 1989 X, Austin 1990 V, Levy 1990 XX, and recently in P/Swift-Tuttle 1992t. These studies are complemented by submillimetre observations at the CSO 10-m telescope (by F.P. Schloerb) and at the JCMT 15-m telescope[7] (in collaboration with J.K. Davies and R. Padman).

Now observed in several of its rotational transitions in several comets, hydrogen cyanide (HCN) has abundances of 0.05 to 0.2% relative to water[4,7,15]. In several cases, the CN radical is more abundant, suggesting that other CN-bearing species exist in comets.

Formaldehyde (H_2CO) was observed at 225 and 351 GHz in several comets with abundances in the range 0.03 to 1%, if coming directly from the nucleus[7,8,15]. Since there are clues that H_2CO is produced from a distributed source rather than from the nucleus[19], its true abundance may be significantly higher, but in this case, this molecule might not be in the form of simple H_2CO ice: polyoxymethylene, progressively sublimating as H_2CO monomer, was proposed by Meier et al.[19].

Hydrogen sulphide (H_2S) is observed at 169 GHz, from which abundances 0.1 to 0.5% are inferred[5,11,15]. Since this molecule condenses at a relatively low temperature (57 K), its presence testifies to the cold physical conditions of cometary formation.

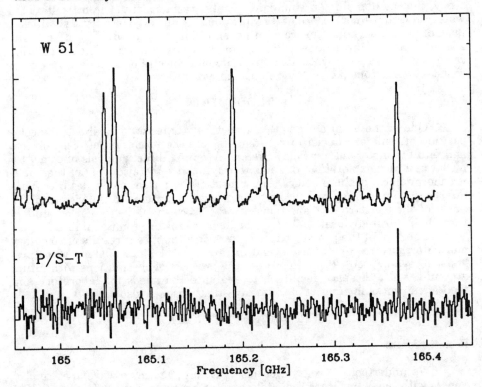

Figure 1: The methanol lines observed in P/Swift-Tuttle (from[15]) with the IRAM telescope in the 165 GHz region (*bottom*). The spectrum of the galactic HII region W 51 is also shown for comparison (*top*): more lines are present due to a larger rotational temperature; the lines are broader due to gas turbulence in the galactic source.

Methanol (CH_3OH) is observed simultaneously through several of its rotational lines (Fig. 1), which allows an evaluation of its rotational distribution and a reliable assessment of its production rate[5,6,7,15]. Abundances between 1 and 5% were evaluated, depending upon the comet. This makes methanol one of the most abundant cometary volatiles.

Upper limits on other species were also obtained, either during dedicated searches or from the systematic exploitation of the spectra secured at IRAM[12]. The most significant results were obtained in comet Levy for formic acid ([HCOOH] < 0.2%), ethanol ([C_2H_5OH] < 0.5%), acetaldehyde ([CH_3CHO] < 0.13%), isocyanic acid ([HNCO] < 0.13%), cyanoacetylene ([HC_3N] < 0.02%).

Future prospects for microwave spectroscopy.

Figure 1 shows a typical cometary spectrum together with the corresponding spectrum of a galactic HI region (W 51). One can see that cometary lines are narrower. This is due to the small expansion velocity of cometary atmospheres (about 1 km s^{-1}), compared to the larger velocity fields of galactic sources. The molecular lines are also less numerous in the comet spectrum: this reflects the low temperature of the atmosphere (rotational temperatures of the order of 30-80 K were retrieved[6,7,15]). We may thus anticipate that molecular identifications will be easier in microwave cometary spectra. We expect that, in the near future, comets will play a role as important as other astrophysical sources such as Orion, IRC+10216 or the Galactic Centre in molecular radio spectroscopic investigations.

Such investigations heavily rely upon the laboratory and theoretical spectroscopic works undertaken for interstellar studies. Comprehensive data bases exist for the microwave spectrum of simple stable species, but the data are more sparse (or even inexistent) for more complex species, radicals or ions which are likely to be present in comets. For instance, it may be noted that the spectrum of H_2O^+, one of the most important molecular ions, is still unknown.

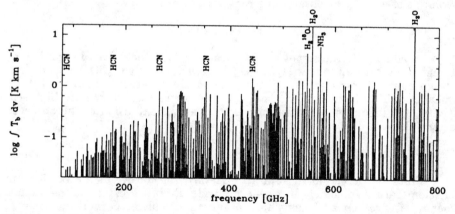

Figure 2: A synthetic cometary spectrum in the millimetre-submillimetre domain. The hypothetical comet has a water production rate of 10^{29} s^{-1} and is observed at 1 AU from the Sun and from the Earth with a 15-m telescope. A standard composition of molecular parent molecules is assumed[10], with a rotational temperature of 40 K. Most of the strong lines, besides those which are tagged, are from methanol.

The excitation of rotational lines of cometary molecules, whose modelling is a requisite for all interpretation of microwave observations, is governed by electronic and/or vibrational excitation, spontaneous decay and collisions. One of the most important problems for such modelling is the knowledge of cross sections between water (which is the main collision partner in the inner coma) and other polar molecules. Up to now, only guesstimates of these cross sections could be used (in the interstellar medium, the main collision partner is H_2, for which comprehensive cross section evaluations were possible). Cross sections for collisions between the weakly polar molecule CO and H_2O were recently evaluated[17]. An extension to more general polar-polar collisions is certainly needed.

Fig. 2 shows a synthetic cometary spectrum in the millimetre-submillimetre region, estimated for a "reasonable" cometary composition[10] and standard excitation conditions. The spectrum is dominated by a few submillimetric water lines, which have not yet been observed because telluric absorption precludes ground-based observations; these lines should be readily observable with future space facilities of even modest sizes. Next come the submillimetric transitions of ammonia and water isotopomers; the lines of these light hydrides are also difficult to observe from the ground. The remaining of the spectrum is mainly composed of HCN, CH_3OH and H_2CO lines. In real cometary spectra, other strong lines from outsiders (not considered in our rather conservative assumptions on the cometary composition) might of course be present.

Being able to probe cometary species (provided they have a significant dipole moment) with mixing ratios as low as 10^{-3}, microwave spectroscopy proved to be a very efficient tool for cometary studies. Besides Earth-based investigations, in situ observations could be still more sensitive by searching for rotational lines in absorption against the nucleus continuum. Such observations could be achieved during a rendez vous mission such as Rosetta. A heterodyne receiver dedicated to selected frequencies could thus study the distribution of trace species in the vicinity of active regions and investigate very low levels of activity by observing strong water lines.

INFRARED SPECTROSCOPY

In addition to the signatures of refractory dust, the infrared spectroscopy of comets revealed up to now H_2O, CO_2, CH_3OH, possibly H_2CO, CO and CH_4, the signature of aromatics at 3.28 μm, and puzzling features at 2.8 and 3.4 μm which are still to be identified[9,10,16]. Future progresses will require high-sensitivity high-resolution observations as well as the extension of the observed spectral domain.

The 3.2-3.6 μm emission.

Since its detection in 1986 in P/Halley, the nature of the 3.2-3.6 μm emission feature in comets has been a puzzle. Proposed explanations include resonant fluorescence of gas phase organic molecules, thermal emission of small grains with organic mantles, transient UV heating of PAH-like molecules, and even emission from biogenic material[16]. The recent identification of cometary methanol from microwave spectroscopy[5] gives new insight on this problem, since this molecule has strong bands in this spectral region. The ν_3 band of CH_3OH was subsequently identified to an isolated cometary feature at 3.52 μm[18].

Good infrared spectra around 3 µm are now available for P/Halley 1986 III, Wilson 1987 VII, Bradfield 1987 XXIX, P/Brorsen-Metcalf 1989 X, Okasaki-Levy-Rudenko 1989 XIX, Austin 1990 V, Levy 1990 XX, and recently P/Swift-Tuttle 1992t. Using an infrared fluorescence model, we investigated the contribution of the ν_2, ν_3 and ν_9 bands of methanol to the 3.2-3.6 µm emission feature[2]. The methanol production rates can be estimated from the ν_3 band at 3.52 µm. The methanol abundances vary from comet to comet between 1 and 5% relative to water. The methanol production rates are in agreement with those derived from microwave observations in comets Austin, Levy and P/Swift-Tuttle. In several cases, the flux observed from 3.3 to 3.4 µm is almost completely explained by the ν_2 and ν_9 bands of methanol. After subtracting the modelled methanol contribution, all comets present residual spectra with a distinct emission extending from 3.38 to 3.5 µm and centred on 3.43 µm (Fig. 3). Its flux is strongly correlated with the water production rate, suggesting a gaseous origin from organics present in comparable abundances from comets to comets. However, no definite identification could be found from an inspection of the laboratory spectra of small organic molecules.

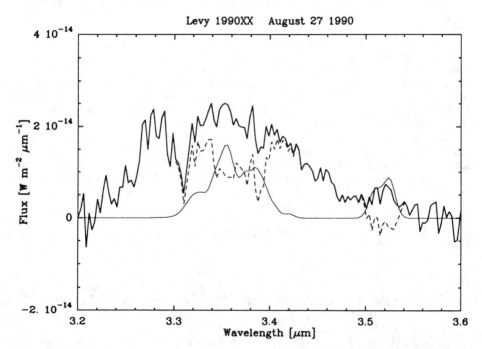

Figure 3: The spectrum of comet Levy 1990 XX in the 3 µm region. *Full line*: the observed spectrum (from[14]). *Thin line*: the modelled methanol emission[2]. *Dashed line*: the difference spectrum, which shows excess emission at 3.28 µm (attributed to aromatics) and at 3.32 and 3.43 µm (yet unidentified).

Four comets (P/Halley, Wilson, Bradfield and Levy) show a large emission excess in the 3.2-3.3 µm spectral region corresponding to the signature of aromatic compounds (PAH). Assuming UV excitation followed by internal conversion to vibrational energy, we estimate an emission rate $g = 0.2$ s^{-1} at

3.2-3.3 µm (at 1 AU from the Sun). Such a rate is much stronger than the rates of resonant fluorescence of the fundamental vibrational bands of most molecular species (typically a few 10^{-4} s^{-1}). We derive PAH abundances from 1 to 4 × 10^{-5} (in number) relative to water. This is much less than the abundance of phenanthrene deduced from the TKS visible spectra of P/Halley[20].

Pending problems.

The interpretation of the infrared spectra of comets requires the modelling of fluorescence at the cold rotational temperatures which pertain to cometary conditions. This can be readily done for species with well known vibrational spectra such as H_2O or H_2CO. This is not the case for methanol, whose infrared spectrum observed in the laboratory is too complex to allow comprehensive identifications; it is therefore difficult to infer the expected cometary methanol spectrum from the spectra observed near room temperature in the laboratory. Similar problems will also arise for other medium-size molecules likely to be present in comets (ethanol, dimethylether, acetaldehyde...). A solution may come from the development of jet-cooled spectroscopic techniques, since physical conditions in laboratory jets are similar to those in cometary atmospheres.

There are still other species to identify in the infrared spectrum of comets. Besides the excess emission at 3.3-3.4 µm, emission at 2.8 µm is also present. Part of it may be due to water hot bands[3], but presumably other OH-bearing molecules are contributing. High-resolution space observations (e.g. with ISO) are needed to elucidate their nature. There may be a common origin (CHO species) to 2.8 and 3.3-3.4 µm emision. Complementary observations at other wavelengths could help; however, emission around 10 µm is dominated by thermal emission from grains, and spectroscopy around 6 µm is difficult from the ground. The answer could also come from microwave spectroscopy. It may be anticipated that some of these species may be fairly complex, as suggested by the existence of cometary C_2 and C_3 radicals, for which the parents are still unknown.

REFERENCES

1. C. Arpigny, this conference (1993)
2. D. Bockelée-Morvan, T.Y. Brooke, J. Crovisier, *Planet. Space Sci.* submitted (1993)
3. D. Bockelée-Morvan, J. Crovisier, *Astron. Astrophys.* **216**, 278 (1989)
4. D. Bockelée-Morvan, J. Crovisier, D. Despois et al., *Astron. Astrophys.* **180**, 253 (1987)
5. D. Bockelée-Morvan, P. Colom, J. Crovisier, D. Despois, G. Paubert, *Nature* **350**, 318 (1991)
6. D. Bockelée-Morvan, J. Crovisier, P. Colom, D. Despois, *Astron. Astrophys.* in press (1993)
7. D. Bockelée-Morvan, R. Padman, D.K. Davies, J. Crovisier, *Planet. Space Sci.* submitted (1993)
8. P. Colom, J. Crovisier, D. Bockelée-Morvan, D. Despois, G. Paubert, *Astron. Astrophys.* **264**, 270 (1992)
9. J. Crovisier, In *Infrared Astronomy with ISO*, Edts T. Encrenaz and M.F. Kessler (Nova Science Publishers, 1992), p. 221.

10. J. Crovisier, in *IAU Symp. No 160 Asteroids, Comets, Meteors 1993* (Kluwer, 1993), in press
11. J. Crovisier, D. Despois, D. Bockelée-Morvan, P. Colom, G. Paubert, *Icarus* **93**, 246 (1991)
12. J. Crovisier, D. Bockelée-Morvan, P. Colom, D. Despois, G. Paubert, *Astron. Astrophys.* **269**, 527 (1993).
13. J. Crovisier, F.P. Schloerb, in *Comets in the Post-Halley Era*, Edts R.J. Newburn et al. (Kluwer, 1991), p. 149
14. J.K. Davies, S.F. Green, T.R. Geballe, *Mon. Not. R, Astron. Soc.* **251**, 148 (1991)
15. D. Despois, G. Paubert, P. Colom, D. Bockelée-Morvan, J. Crovisier, *Planet. Space Sci.* submitted (1993)
16. T. Encrenaz, R.F. Knacke, in *Comets in the Post-Halley Era*, Edts R.J. Newburn et al. (Kluwer, 1991), p. 107
17. S. Green, *Astrophys. J.* **412**, 436 (1993)
18. S. Hoban, M. Mumma, D. Reuter et al., *Icarus* **93**, 122 (1991)
19. R. Meier, P. Eberhardt, D. Krankowsky, R.R. Hodges, *Astron. Astrophys.* in press (1993)
20. G. Moreels, J. Clairemidi, P. Hermine, P. Brechignac, P. Rousselot, *Astron. Astrophys.* in press (1993)

DISCUSSION

LEE — You did not mention the identification of isotopes. Would you add some comments about the existence of some isotopes such as Deuterium and C^{13} since comets are extraterrestrial ?

CROVISIER — Microwave and IR spectroscopy of comets are still state-of-the-art techniques which are not sensitive enough to allow the observation of isotopic species. Their observation by other means are reviewed in the talk of C. Arpigny at this meeting.

BOTSCHWINA — You were stressing the need for forthcoming laboratory work on H_2O^+. The spectroscopic constants of this species are rather well-known through high-resolution IR spectroscopy and LMR spectroscopy. Is that information not sufficient for your purpose ?

CROVISIER — No, I do not think that the rotational line frequency predictions for H_2O^+ are accurate enough to allow their detection in comets.

ZIURYS — In response to the comment that infrared spectra of H_2O^+ have resulted in rotational constants sufficiently accurate to warrant a search for this species in comets, I don't believe this actually is the case. The infrared spectrum of H_2O^+ was measured using LMR, but the resulting frequencies at millimeter-wavelengths are probably known only to about $\pm$ 20 MHz, at best. Another point. On your searches for cometary molecules at mm wavelengths at IRAM, have you ever looked for the mm transitions of CO ?

CROVISIER — Because of limited alloted telescope time, we did not give priority to the observation of CO, which could be fairly well studied in the UV.

VANYSEK — The comet P/Swift-Tuttle was a "dusty" comet. Is it possible from your radio observation to draw any conclusions about relation of the parent molecules and distribution of dust in the coma ?

CROVISIER — We did not map the comet radio emissions. However, important information on the coma kinematics can be deduced from the radio line shapes. All radio lines were found to be highly asymmetric, showing that the outgassing preferentially occurred in jets towards the observer. These jets are probably related to the strong dust jets seen in the visible images of this comet.

THADDEUS — The search for "primitive" objects in the solar system has proven extremely elusive. Are any of the comets in which the larger organic molecules are observed thought to be juvenile, i.e. not cooked by previous close passages by the sun ?

CROVISIER — P/Halley and P/Swift-Tuttle are short-period comets whereas Austin and Levy were "new" comets. I think that the data on the abundances of cometary parent molecules are too sparse at the present time to allow to make definite distinctions between these two classes of comets.

D'HENDECOURT — Most of the molecules observed in comets are present in interstellar grain mantles.
Are O_2 and N_2 present in comets, because they are suspected to be abundant in interstellar grains ? Can comets be made of IS grains ?

CROVISIER — O_2 and N_2 cannot be observed in comets by present remote-sensing techniques. The abundance of N_2 is suggested to be very low ($\sim 2 \times 10^{-4}$) from the observation of the N_2^+ ion.
The similarity of composition of interstellar grains and cometary ices is consistent with comets being accreted from unprocessed interstellar grains. Other formation scenarios, however, cannot be excluded.

WALMSLEY — You cited a "rotational temperature" of 50-80 K based upon your methanol measurements of several comets. Does this imply that you can estimate the kinetic temperature in the coma on the basis of your measurements ?

116 Infrared and Radio Spectroscopy of Comets

CROVISIER — The "rotational temperature" of methanol (as well as of other cometary species) is not equal to the kinetic temperature because this species is out of equilibrium. The processes involved are collisions, IR excitation, and spontaneous decay. We have developed a model (Bockelée-Morvan et al., to appear in Astron. Astrophys.) which takes these processes into account and which allows one to constrain the kinetic temperature from the microwave observations.

MENZEL — Is it known whether the mechanism of evaporation of volatile material from comets near the sun is thermal evaporation or is induced by electronic excitations ?

CROVISIER — The sublimation of cometary volatiles is thermal and governed by heating by the sun. It must be noted that fractionation processes occur, and that the abundances of volatiles observed in the coma may not directly reflect the abundances of ices in the nucleus.

FIELD — It was suggested from the floor that it migt be possible to observe H_2O at 22 GHz, as a maser. The general consensus is that such masers in fact require temperature of ~400K (or more) and number densities of $>10^6$ cm^{-3}. These would not seem to be appropriate conditions in a comet.

CROVISIER — I agree. Tentative detections of the 22 GHz water line in comets in the past could not be confirmed.

COMPLEX MOLECULES IN THE SOLAR SYSTEM: THE GIANT PLANETS AND TITAN

Thérèse ENCRENAZ
DESPA, Observatoire de Paris, F-92195 MEUDON

ABSTRACT

The most favorable places in the Solar System for the formation of complex molecules are objects of large heliocentric distances, such as the Giant Planets and their satellites, and the comets. This paper reviews some aspects of the chemisry taking place in the atmospheres of the Giant Planets and Titan. In the case of Jupiter, auroral regions are priviledged places to observe the hydrocarbons resulting from the methane photochemistry. In the case of Titan, a very active chemistry takes place from the photodissociation of methane and the dissociation of nitrogen by high energy particles. Future progress is expected from the use of ground-based millimeter interferometry, as well as future space missions: ISO, FIRST, GALILEO and CASSINI.

INTRODUCTION

Whereas more than a hundred molecules have been detected in the interstellar medium (Table 1), there are only a few tens of different species detected altogether in the various objects of the Solar System. In addition, there are only 3 different detected Solar-System species of 8 atoms or more (Table 2), to be compared to almost 20 in the interstellar medium. The easier detection of ISM molecules can first be explained by the very long lines of sight observed in the ISM, which makes possible the detection of molecules with very low densities. Apart from this first reason, it also appears that the ISM is a more favorable medium for the development of a complex chemistry. The main reason is, in the case of the Solar System, the presence of the solar UV flux which tends to dissociate all heavy compounds. This effect is illustrated in Table 2, where it can be seen that large molecules are preferentially found at large heliocentric distances.

The most favorable cases for the formation of complex molecules in the Solar System are thus of 3 types: (1) the Giant Planets, (2) their satellites surrounded by an atmosphere, and especially Titan, and (3) the comets. The study of comets is presented in two papers by J.Crovisier and D.Bockelée-Morvan and C.Arpigny (this Conference). The present paper reviews the study of complex molecules in the Giant Planets and in Titan.

THE GIANT PLANETS

Because they are cold and massive, Giant Planets have been able to retain most of their primordial atmosphere, accreted around their cores from the surrounding primordial nebula at the time of the planets' formation. Their atmospheres thus reflect, in first approximation, the cosmic abundances,

TABLE 1
GASEOUS MOLECULES AND IONS DETECTED IN THE ISM

# of Atoms	2		3	4	5	6	7	8	≥9
	CO	PN	C_2H	NH_3	C_4H	CH_3CN	CH_3C_2H	CH_3C_3N	CH_3C_4N
	C_2	NS	HCN	C_3N	HC_3N	CH_3OH	HC_5N	CH_3OCHO	$(CH_3)_2O$
	OH	CN	HNC	HNCO	H_2C_2O	HC_2CHO	CH_2CHCN	CH_2CHCHO	HC_7N
	CH	CP	H_2O	H_2CO	HCOOH	CH_3NC	CH_3CHO	CH_3COOH	HC_9N
	CS	SiC	HDO	HCNS	NH_2CN	NH_2CHO	CH_3NH_2	$(NH_2)_2CO$	$HC_{11}N$
	SO	NH	H_2S	H_2CS	SiH_4	CH_3SH	CH_2CHOH		CH_3CH_2OH
	SiO	NaOH	C_2S	C_3H	CH_2NH	C_2H_4	CH_2CCH_2		CH_3CH_2CN
	SiS	H_2	OCS	HC_2N	CH_4	HC_2COH	C_6H		NH_2CH_2COOH (?)
	HCl	HD	HCO	C_2H_2	CH_2CN	H_2C_4			$C_4H_4O(?)$
	NaCl	NO	HNO	C_3N	H_2C_2S	C_5H			$C_4H_5N(?)$
	AlCl	KCl	SO_2	C_3O	C_3H_2				$C_3N_2H_4(?)$
			C_3	C_3S	SiC_4				CH_3COCH_3
			N_2H^+	H_3O^+	C_5				CH_3C_4H
			HCS^+	$HOCO^+$					
			HOC^+	$HCNH^+$					
			SiC_2						

Adapted from: Irvine et al, in "Protostars & Planets II, Un. of Arizona, 1985 and Crovisier, 1993 (this conference)

TABLE 2
GASEOUS MOLECULES AND IONS DETECTED IN THE SOLAR SYSTEM

OBJECT \ # of Atoms	2	3	4	≥5
VENUS	O_2 CO HCl HF	CO_2 H_2O SO_2 H_2S OCS HDO		
MARS	O_2 CO	CO_2 H_2O HDO O_3		
GIANT PLANETS	H_2O CO HD	H_2O HCN H_3^+	NH_3 PH_3 AsH_3 C_2H_2 $CH_3(?)$	CH_4 GeH_4 C_2H_4 C_2H_6 C_3H_4 C_3H_8 CH_3D C_6H_6
TITAN	H_2 N_2 CO	CO_2 HCN	C_2H_2 C_2N_2	CH_4 CH_3D C_2H_4 C_2H_6 C_3H_4 C_3H_8 C_4H_2 HC_3N CH_3CN
COMETS	CO S_2 C_2 NH OH CN CS CH OH^+ CO^+ CH^+ CN^+ N_2^+	H_2O CO_2 H_2S NH_2 OCS(?) H_2O^+ CO_2^+	H_2CO	CH_3OH $CH_4(?)$ $C_3O_2(?)$ PAH (phenanthrene)

TABLE 3
GASEOUS MOLECULAR SPECIES IN THE GIANT PLANETS

Species	Mixing ratios (versus H_2)			
	Jupiter	Saturn	Uranus	Neptune
H_2	1	1	1	1
HD	6×10^{-5}	1×10^{-4}	...	...
He	0.11	0.03	0.18	0.23
CH_4	2×10^{-3}	4×10^{-3}	2×10^{-2}	4×10^{-2}
$^{13}CH_4$	2×10^{-5}	4×10^{-5}		
CH_3D	3×10^{-7}	4×10^{-7}	10^{-5}	3×10^{-5}
NH_3	3×10^{-4}	2×10^{-4}		
$^{15}NH_3$	1×10^{-6}			
C_2H_2	$3 \times 10^{-8}(*)$	$7 \times 10^{-8}(*)$	10^{-7}	$1-9 \times 10^{-7}$
$^{12}C^{13}CH_2$	$3 \times 10^{-9}(*)$			
C_2H_6	2×10^{-6}	3×10^{-6}		3×10^{-6}
C_3H_4	$3 \times 10^{-9}(**)$			
C_3H_8	$\leq 6 \times 10^{-7}(**)$			
C_2H_4	$7 \times 10^{-9}(**)$			
C_6H_6	$2 \times 10^{-9}(**)$			
H_2O	$1 \times 10^{-6}(*)$			
CO	1.5×10^{-9}	2×10^{-9}		6×10^{-7}
PH_3	5×10^{-7}	1.5×10^{-6}		
GeH_4	7×10^{-10}	4×10^{-10}		
AsH_3	3×10^{-10}	2.4×10^{-9}		
HCN	2×10^{-9}			3×10^{-10}

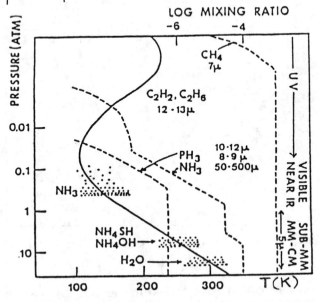

Fig.1. The Jovian atmosphere: thermal profile, vertical distributions of atmospheric consituents, and cloud structure

120 Complex Molecules in the Solar System

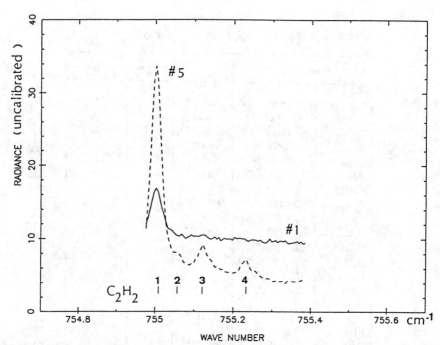

Fig.2. Spectra of acetylene in an auroral region (#5) and at the equator (#1). The figure is taken from Drossart et al (1986)

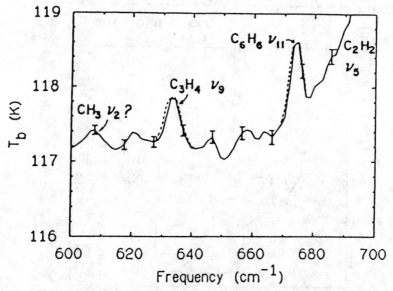

Fig.3. Hydrocarbons emissions observed in the VOYAGER-IRIS spectra in an auroral region. The figure is taken from Kim et al (1985)

with molecular hydrogen being predominant, followed by helium (6 to 20% per volume) and methane (0.1 to a few percent). Ammonia (about 0.01%) is present in Jupiter's atmosphere, but is mostly condensed in the other planets. Nitrogen could be present in Neptune, as will be discussed below, but its abundance is presently unknown. A complete list of the detected molecules and their abundances can be found in Table 3.

Most of the Giant Planets' photochemistry is dominated by the methane photodissociation, which takes place in the stratosphere, in a region where the temperature is relatively high. Some of the hydrocarbons formed in this process condense at the tropopause level, which acts as a cold trap (Fig.1). In the troposphere, several clouds are found: NH_3, NH_4OH and NH_4OH, H_2O in the case of Jupiter and probably also Saturn; in addition, CH_4 in the case of Uranus and Neptune. Stratospheric hydrocarbons are observed in emission through their strong ro-vibrational transitions in the middle infrared range (7-15 microns). In the case of Jupiter, the deep troposphere (down to a pressure level of a few bars) can be probed in the 4-5 micron spectroscopic window, free from methane and ammonia absorption.

Complex hydrocarbons in auroral regions

Aurorae were first discovered on Jupiter about twelve years ago. In the UV range, aurorae were detected both from the Voyager mission and the IUE satellite, from the hydrogen Ly-alpha emission at 1216 A (Broadfoot et al,1981). Simultaneously, infrared emissions were detected, also in the polar regions, in the methane emission band at 8 microns (Caldwell et al,1980).

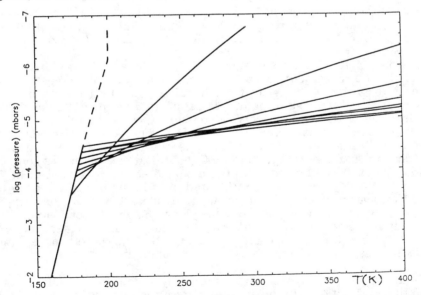

Fig.4. Thermal profiles retrieved in the Jovian upper stratosphere, inside and outside the auroral regions. The figure is taken from Drossart et al (1993)

122 Complex Molecules in the Solar System

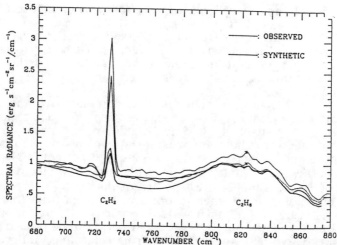

Fig.5. Observed and synthetic spectra of acetylene and ethane, inside the aurora (top) and outside (bottom). The figure is taken from Drossart et al (1993).

Localized emissions of acetylene were later detected, also in the auroral regions (Drossart et al,1986; Fig.2). The surprise then came from subsequent heterodyne spectroscopy measurements of ethane, which show no C_2H_6 enhancement in the aurorae (Kostiuk et al,1987). In the meantime, more hydrocarbons (C_2H_4, C_3H_4, C_6H_6) were detected from the Voyager spectra in the auroral spots (Kim et al,1985; Fig.3). Finally, the H_3^+ ion was also detected in the aurorae, through its ro-vibrational lines in the (1-0) and (2-0) bands, around 4 and 2 microns respectively (Drossart et al,1989).

What is the origin of these emissions? They can be explained either by an abundance enhancement of the hydrocarbons, or an increase of the temperature profile, or both, in the auroral regions. In a recent analysis, Drossart et al (1993) have been able to define an atmospheric model which provides a good fit for all observed emission spectra. In this model, the molecular vertical distributions are taken from photochemical models and are constant inside and outside the aurorae; the temperature profile is the only variable parameter, with most of the variations in the region 0.1-00.1mb (Fig.4). It is adjusted from the methane emissions observed at 7.7 microns in the Voyager spectra, inside and outside the hot spots. Then the derived thermal profiles are used to fit the hydrocarbon emissions (Fig.5). The lack of correlation between the acetylene and ethane emissions come from the fact that the corresponding weighting functions peak at very different levels (40 microbars and 2 mb respectively). The ethane emission is thus insensitive to a change of T(P) at very high altitude.

In summary, complex hydrocarbons are likely to be present all over the Jovian disk and not only in the aurorae, but, in the infrared range, it can be detected only in regions where the temperature contrast is high enough. Their study is thus a powerful means for probing the temperature profile at very high altitude levels.

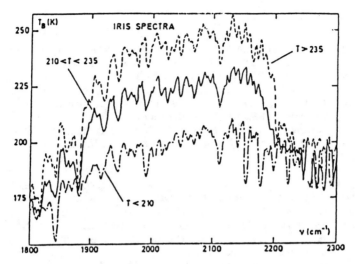

Fig.6. The 5-micron spectroscopic window in Jupiter, observed with VOYAGER-IRIS. The figure is taken from Drossart et al (1982)

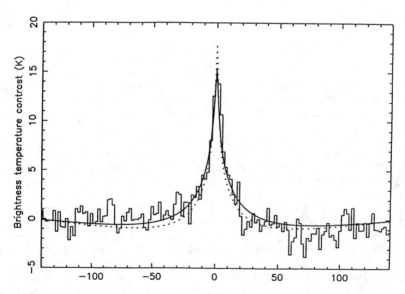

Fig.7. The CO (2-1) emission on Neptune, observed at IRAM. The figure is taken from Rosenqvist et al (1992)

Disequilibrium species in the lower troposphere

In the case of Jupiter and Saturn, several unexpected species have been detected at a pressure level of a few bars: phosphine, germane (Fig.6) and arsine. These species are not expected to be present, on the basis of thermochemical equilibrium calculations. Their origin is probably due to quenching through rapid vertical mixing (Lewis and Fegley, 1984): vertical currents carry the species from the deep levels up to the levels where they become observable, in a timescale smaller than their destruction time. Two other molecules, CO and HCN (the latter detected on Jupiter only) could also originate from the same process.

Lewis and Fegley (1984) have been able to reproduce the observed abundances of PH_3, CO and GeH_4 in a model using quenching temperatures of 1300, 800 and 1100K respectively. In the case of HCN, the model fails by several orders of magnitude, which probably indicates that HCN comes directly from the photolysis of ammonia and acetelyne (Tokunaga et al,1981). Apart from HCN, disequilibrium species are thus powerful tracers of the planets' dynamics, and their spatio-temporal monitoring, including the high latitude regions, will be highly valuable for understanding the global circulation of the planets. This will be one of the major objectives of the GALILEO and later CASSINI missions.

CO and HCN on Neptune

A major result of the few past years has been the recent discovery of significant amounts of CO (Fig.7) and HCN in the atmosphere of Neptune, from their millimeter and submillimeter transitions (Marten et al,1991,1993; Rosenqvist et al,1992). The surprise came from the abundance of CO, larger than the predicted value by a factor about 10^3. No evidence of CO and HCN was found in the stratospheres of the other Giant Planets.

The vertical distribution of CO and HCN could be derived from the shape of the spectral profiles. CO seems to be of tropospheric origin, and the discrepancy between the observations and the theoretical models is presently unexplained. HCN, in contrast, is present in the stratosphere only, and condenses at the tropopause level. HCN is believed to be produced by chemical reactions between CH_3 and nitrogen atoms. Nitrogen could come either from tropospheric molecular nitrogen, destroyed by galactic cosmic rays, or from the high atmosphere of Neptune's satellite Triton. Present models suggest that both mechanisms could be involved (Romani et al,1992). This result has an important possible implication on the presence of molecular nitrogen in the deep atmosphere of Neptune. The evidence for a strong nitrogen enhancement on Neptune, with respect to the other Giant Planets, would be a new constraint to include in the formation models of these planets.

Finally, the photochemistry of methane and the dissociation of nitrogen might lead to the formation of several hydrocarbons and nitriles, as observed in the upper atmosphere of Titan. In spite of negative results recorded up to now, the search for these species should be pursued in the future, as the available instrumentation becomes more and more sensitive.

TABLE 4
GASEOUS MOLECULAR SPECIES IN TITAN
(from Coustenis et al, 1993)

Gas		Mole fraction		
Major components				
Nitrogen	N_2	0.98		
Argon	Ar	0		
Methane	CH_4	0.018		
Hydrogen	H_2	0.002		
		Equator	*North pole*	
		~ 6 mbar	~ 0.1 mbar	~ 1.5 mbar
Hydrocarbons				
Acetylene	C_2H_2	2.2×10^{-6}	4.7×10^{-6}	2.3×10^{-6}
Ethylene	C_2H_4	9.0×10^{-8}		3×10^{-6}
Ethane	C_2H_6	1.3×10^{-5}	1.5×10^{-5}	1.0×10^{-5}
Methylacetylene	C_3H_4	4.4×10^{-9}	6.2×10^{-8}	2.0×10^{-8}
Propane	C_3H_8	7.0×10^{-7}		5.0×10^{-7}
Diacetylene	C_4H_2	1.4×10^{-9}	4.2×10^{-8}	2.7×10^{-8}
Monodeuterated methane	CH_3D	1.1×10^{-5}		
Nitriles				
Hydrogen cyanide	HCN	1.6×10^{-7}	2.3×10^{-6}	4×10^{-7}
Cyanoacetylene	HC_3N	$\leq 1.5 \times 10^{-9}$	2.5×10^{-7}	8.4×10^{-8}
Cyanogen	C_2N_2	$\leq 1.5 \times 10^{-9}$	1.6×10^{-8}	5.5×10^{-9}
Oxygen compounds				
Carbon dioxide	CO_2	1.4×10^{-8}	$\leq 7 \times 10^{-9}$	
Carbon monoxide	CO		6×10^{-5}	
			$\leq 4 \times 10^{-6}$	
Water	H_2O	1×10^{-9}	1×10^{-9}	

TITAN

Titan's atmosphere presents some interesting similarities with the terrestrial one. First, it is dominated with molecular nitrogen; second, the surface pressure is 1.5 bar; third, its surface might be covered, at least partly, by an ocean. The major difference with the Earth is the very low surface temperature (93K), and the atmospheric composition. After nitrogen, and possibly argon (the abundance of which is very uncertain), methane is the most abundant species, with a mixing ratio of about 2%. The composition of Titan's atmosphere is summarized in Table 4.

Titan's thermal structure, as the Giant Planets' one, is characterized by a convective troposphere, and, above the tropopause, a stratosphere where the temperature gradient increases with height. In this region, an active chemistry is induced by the methane photochemistry and the dissociation of nitrogen by energetic particles. From the dissociation of CH_4, ethane and acetylene are the first products to be formed; ethylene is rapidly dissociated to acetylene. From C_2H_2, heavier hydrocarbons are formed, as C_4H_2, C_3H_4, C_3H_8. Ethane is expected to condense into particles which fall on the surface. At Titan's surface, ethane should be in a liquid phase and could contribute, with methane, and nitrogen, to a global or partial ocean. HCN, on the other side, is also dissociated to form heavier species suh as HC_3N or possibly HCN polymers. These polymers could be present in the upper atmosphere of Titan in the form of a reddish and brownish aerosol layer.

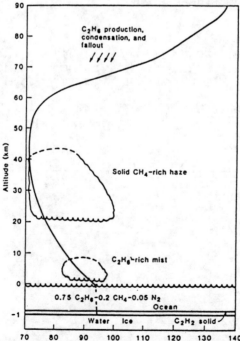

Fig.8. The atmosphere of Titan: thermal profile and cloud structure. The figure is taken from Hunten et al (1984)

Because it is continuously photodissociated, methane has to be refilled from the surface. Its reservoir could be the possible C_2H_6-CH_4-N_2 ocean (Fig.8).

Complex hydrocarbons in Titan were first detected from ground-based IR spectroscopy (Gillett et al,1975), then by the VOYAGER infrared experiment (Fig.8; see Table 4). In addition, heterodyne millimeter spectroscopy provided detections for CO (Muhleman et al, 1983), HCN (Tanguy et al,1990), HC_3N and CH_3CN (Bézard et al,1992,1993). Vertical distributions were retrieved from both the Voyager and the millimeter data. All molecules show enhanced abundances at high altitude, in the region 0.01-0.001 mb (Coustenis et al,1991).

The latitudinal variations of these species have been studied from the VOYAGER infrared spectra. A significant increase was noticed from South Pole to North Pole for C_4H_2, HC_3N, HCN and C_2H_4, associated to a slight decrease of the CO_2 abundance (Fig.10; Coustenis,1991). These variations were interpreted as a seasonal effect, as the North Pole of Titan had been in the shadow for 8 months at the time of the VOYAGER 1 encounter. As suggested by Yung et al (1984), this geometry was likely to induce an accumulation of nitriles, protected from the UV solar flux, and an inhibition of the CO_2 formation from water photolysis (Yung et al,1984; Coustenis, 1991).

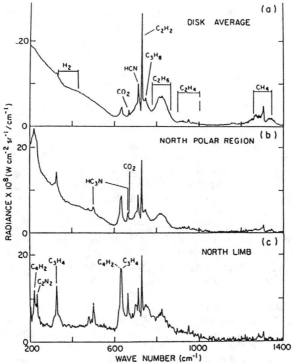

Fig.9. The infrared spectrum of Titan, recorded by VOYAGER-IRIS at 3 different locations. The figure is taken from Hunten et al (1984)

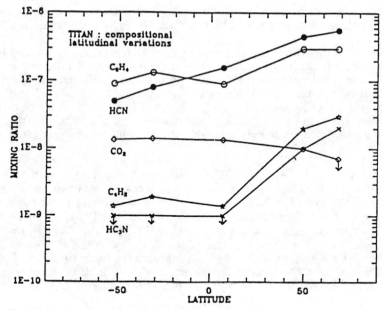

Fig.10. Latitudinal variations of various atmospheric species on Titan. The figure is taken from Coustenis (1992)

FUTURE STUDIES

Several new developments can be expected in the future, from both ground-based and space programs.

In the case of ground-based astronomy, a significant progress should be expected from the use of millimeter interferometry. Heterodyne spectroscopy has shown to be a very valuable tool, especially for the studies of Neptune and Titan. On the other hand, it has been seen that the emissions of hydrocarbons and nitriles were often highly localized on the planet or the satellite. Improving the spatial resolution of the millimeter measurement should thus allow a significant increase in sensitivity. As an example, an interferometer with a 1 km baseline would have, at 230 GHz, a diffraction limit of about 0.4 arcsec, i.e. half the diameter of Titan's disk, 0.2 times Neptune's diameter, and 0.1% times Jupiter's diameter; this should, in particular, make possible the study of the Jovian aurorae.

There are two kinds of future space missions: the Earth-orbit observatories (IUE, HST...) and the planetary "in situ" space missions. The ISO (Infrared Space Observatory) mission, to be launched by ESA in 1995, should be especially suited for the study of complex molecules in Giant Planets and Titan. The spectral resolving power of the spectrometers, in the Fabry-Pérot mode ($2 \cdot 10^4$) will allow the detection of tropospheric species like HCN and CO in Jupiter and Saturn (Bézard et al,1986), as well as an improved study of

Titan's hydrocarbons and nitriles (Coustenis et al,1993; Fig.11). In particular, a list of more complex molecules, to be searched for in Titan's atmosphere, has been derived on the basis of laboratory simulation experiments (Cerceau et al,1985). Their detectability limits with ISO (Table 5) is in the range 10^{-9}-10^{-10}, which is, in several cases, lower than the predicted abundances (Coustenis et al,1993). Later, the FIRST (Far-InfraRed and Submillimeter Telescope) mission, to be launched by ESA, will allow an improvement in sensitivity by a factor of about 25; in addition, the vertical distribution of stratospheric species such as H_2O will be retrieved with high precision from the FIRST heterodyne instrument.

Finally, the operation of descent probes in the atmospheres of Jupiter and Titan, in the context of the GALILEO and CASSINI-HUYGENS missions, is going to open a new field in the study of the atmospheric composition of planetary and satellite atmospheres. In the case of the GALILEO probe, which will enter the Jovian atmosphere in December 1995, we expect, from the mass spectrometer, the detection of all molecules in the range 2-152 AMU, with mixing ratios > 10^{-10}. Comparable performances should be reached with the HUYGENS mass spectrometer, around 2004. In addition the CIRS infrared spectrometer, aboard the CASSINI orbiter, should be able to monitor Saturn and Titan's atmospheres between 2004 and 2008 with increased detectability limits in the limb observation mode.

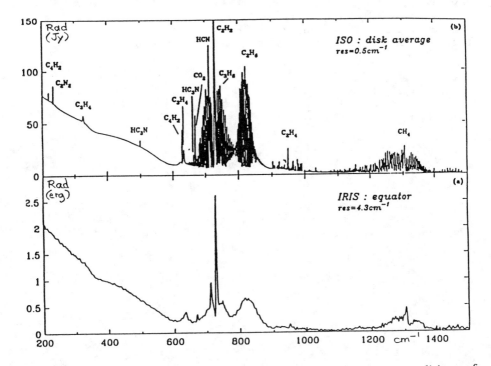

Fig.11. The synthetic spectrum of Titan, calculated under the conditions of the ISO spectrometers (grating mode). The figure is taken from Coustenis (1992)

REFERENCES

Bézard B, Gautier D and Marten A, Astron. Astrophys.,161, 387 (1986)
Bézard B, Marten A and Paubert G, BAAS,24,953 (1992)
Bézard B, Marten A and Paubert G, BAAS, in press (1993)
Broadfoot A L et al, J.Geophys.Res., 86, 8259 (1981)
Caldwell J, Tokunaga A T and Gillett F C, Icarus, 44, 667 (1980)
Coustenis A, in "Infrared Astronomy with ISO", Th.Encrenaz and M.Kessler edts, Nova Science (1992)
Coustenis A et al, Icarus, 89, 152 (1991)
Coustenis A et al, Icarus, in press (1993)
Drossart P et al, Icarus, 49, 416 (1982)
Drossart P et al, Icarus, 66, 610 (1986)
Drossart P et al, Nature, 340, 539 (1989)
Drossart et al, Icarus, in press (1993)
Gillett F C, Astrophys.J., 201, L41 (1975)
Kim S J, Caldwell J, Rivolo A R and Wagener R, Icarus, 64, 233 (1985)
Kostiuk T et al, Icarus, 72,394 (1987)
Lewis J S and Fegley B, Space Sci. Rev., 39, 163 (1984)
Marten A et al, IAUC 5331 (1991)
Marten A et al, Astrophys.J., in press (1993)
Muhleman D O, Berge G L and Clancy R T, Science, 223, 393 (1984)
Romani P et al, BAAS, 24, 972 (1992)
Rosenqvist J et al, Astrophys.J., 392, L99 (1992)
Tanguy L et al, Icarus, 85, 43 (1990)
Tokunaga A T et al, Icarus, 48, 283 (1981)
Yung Y L, Allen M and Pinto J P, Icarus, 72,468 (1984)

DISCUSSION

BUSSOLETTI — I have appreciated very much your talk and, specially, the synthetic spectra you presented for giant planets and satellites.
I would like to know if and how dust influence has been taken into account in these computations as we know that, see for instance Ulysses observations, plenty of dust is expected to sit at their locations.

ENCRENAZ — The role of aerosols is very important in the modelling of infrared spectra. Although their chemical composition is not precisely determined, their effect is empirically taken into account by fitting the existing spectra. For instance, the synthetic spectra of Titan presented here in view of ISO observations (Coustenis et al, 1993) use the aerosol continuum derived from IRIS-Voyager observations.

SCAPPINI — You have proposed to observe AsF_3 in the atmosphere of Jupiter. It looks quite a strange molecule to me: where is all this fluorine coming from ?

ENCRENAZ — AsF_3 has been suggested in the giant Planets atmospheres on the basis of thermochemical calculations starting from cosmic abundances. These models have been developed by Lewis, Pruin, Fegley, etc. Their results have been used in the calculation of synthetic IR spectra performed by Bézard et al (1986).

BOTSCHWINA — The presence of dicyano acetylene (C_4N_2) on Titan was discussed by Khanna and coworkers but this molecule was not included in your tables. Can you please comment on this ?

ENCRENAZ — My tables only included gaseous species. It is true that a spectral feature of Titan's IRIS spectrum has been interpreted as C_4N_2. This IRIS spectrum is presently reanalysed on the basis of new laboratory spectra of solids, and the interpetation of other observed features is in progress (Coustenis et al, 1993, in preparation).

SCHUTTE — To maintain the observed atmosphere of large condensible carbon-rich molecules at Titan you need a source of CH_4 (or perhaps other small & volatile C molecules ?). Is it possible that dissociation of the larger molecules by energetic particles could provide such a source ? (i.e., one would obtain a kind of equilibrium $CH_4 \leftrightarrow$ Larger species).

ENCRENAZ — As far as I know, present photochemical models of Titan do not seem to predict a possible recycling of CH_4 from its photodissociation products (see e.g. Hunten et al, 1984, in "Saturn", Un. of Arizona Press).

ROUEFF — Can you comment on the spectroscopic data used in the synthetic spectra shown in relation to future ISO observations?

ENCRENAZ — In the case of Titan, the recent calculation of synthetic spectra for ISO (Coustenis et al, 1993) has been done in close collaboration with the Laboratoire Infrarouge at Orsay who provided many new spectroscopic data. It is highly desirable that this collaboration be developed in the future.

From Diffuse Clouds
to Comets

THE EFFECT OF VARYING COSMIC RAY IONISATION RATES ON THE CHEMISTRY OF INTERSTELLAR CLOUDS

P R A Farquhar and T J Millar
Department of Mathematics, UMIST, Manchester M60lQD, UK

E Herbst
Department of Physics, The Ohio State University,
Columbus, Oh. 43210, USA

ABSTRACT

We have investigated the effects of varying the cosmic ray ionisation rate on the chemistry of interstellar clouds, including the ultraviolet photons generated by the cosmic rays. For very low cosmic ray fluxes, chemi-ionisation is able to drive a limited chemistry. Extremely large fluxes, which may be appropriate to starburst galaxies, can drive the chemistry rapidly to a steady-state. An upper limit has been derived for the cosmic ray ionisation rate in the extended molecular gas in M82 and for the dark cloud TMC-1.

INTRODUCTION

Molecular synthesis in interstellar clouds is known to be driven by ion-neutral reactions which, when exothermic, are generally rapid at the low temperatures of interstellar clouds[1]. In diffuse clouds the source of ionisation is the general interstellar ultraviolet radiation field, while in dark and dense clouds, it is low energy cosmic rays, with a generally accepted rate of around 10^{-17} ionisations s^{-1} per H-nucleus, derived from modelling the observations of ions and deuterated species[2]. It is not clear, however, that this is a universal number. Skibo and Ramaty[3] used observations of the diffuse low energy gamma ray continuum to derive an ionisation rate of 10^{-15} s^{-1}. In addition, since cosmic rays are charged particles, they can be scattered by magnetic fields in a unpredictable manner since the morphology of the field lines is unknown in interstellar clouds. Finally, cosmic rays are thought to originate in supernova explosions and hence their ionisation rate may be enhanced significantly in starburst galaxies. It is known that the ultraviolet radiation fields in such galaxies may be up to 500 times larger than that in our Galaxy[4]. For these reasons, we decided to investigate the influence of varying the cosmic ray ionisation rate by several orders of magnitude from its standard value, seeking thereby to determine its influence on the cores of dark clouds shielded from their ionising properties, as well as in starburst galaxies, which are known to contain a significant molecular component[5].

We have used the dark cloud model of Millar and Herbst[6], which contains around 1800 reactions to describe the chemistry of H, C, N, O and S, with a token metal, Si, included as a source of electrons. The basic reaction network has been augmented by the inclusion of ultraviolet photons generated by the collisional excitation and decay of H_2 by the secondary electrons produced in the cosmic ray ionisation of H_2, with photorates taken from Gredel et al.[7], and a grain albedo of 0.5. In some models, the cloud temperature was calculated in a self-consistent manner. Pseudo-time-dependent models were calculated over an evolutionary time of 4 10^8 yrs and a density of $n(H_2) = 10^4$ cm^{-3}. The cosmic ray ionisation rate was varied from 10^{-3} to 10^4 times the standard rate, as well as being set to zero. The initial chemical composition of the models was taken to be neutral and atomic, except for hydrogen which was taken to be in molecular form.

RESULTS

We begin by briefly describing the results of a model in which the cosmic ray ionisation rate was set to zero. In this case, since the initial state is neutral, only simple molecules formed in activationless, exothermic reactions can be synthesised. These molecules include CO, CN, CS, N_2, O_2 and OCS, formed, for the most part, in slow radiative association reactions. Because there is no photon field present, these molecules, once formed, are extremely long-lived and a steady-state is not reached until a time longer than 10^9 yrs. If carbon is initially taken to be ionised, rather than neutral, then the radiative association

$$C^+ + H_2 \longrightarrow CH_2^+$$

followed by the usual hydrogen abstraction reactions can drive the formation of hydrocarbon molecules at quite large fractional abundances, up to 10^{-6}, until around 3000 yrs, after which the C^+ is exhausted. On a longer time-scale, the chemi-ionisation reaction

$$CH + O \longrightarrow HCO^+ + e$$

drives the chemistry which reaches a steady-state in about 10^6 yrs. The abundances of some species can be quite large, particularly the hydrocarbons, although nitrogen-containing molecules have abundances less than those observed in dark clouds, as do proton-bearing molecular ions such as HCO^+ since there is no formation route to H_3^+ in the absence of cosmic ray ionisation.

A number of models were calculated with the ionisation rates varying by a factor S in the range 10^{-3} to 10^4 times the standard value. Since heating by cosmic rays is known to play an important role in the thermal balance of dark clouds, we computed the gas kinetic temperature in a self-consistent manner in these models. For larger S this heating mechanism dominates and the temperature is a function of time. At early times, cooling by the fine-structure lines of OI and CI dominate; at late times CO is the major coolant, while for very large values of S the cloud is highly ionised and the C^+ line at 157.7 microns dominates.

In this contribution we shall present only the steady-state results, which are much more amenable to interpretation. A full discussion is given in a separate paper[8]. Figures 1 and 2 show the fractional abundances of several species as a function of S. Figure 1 shows ion abundances. The fractional abundance of H_3^+ is proportional to S for $S < 1000$, since it is formed by cosmic ray ionisation of H_2 and destroyed by reaction with CO, whose fractional abundance remains independent of S. For larger ionisation rates, dissociative recombination with electrons dominates the loss of H_3^+ and its abundance, and those of other proton-bearing molecular ions, is reduced. For $S < 10$, HCO^+ is the dominant ion with a fractional abundance approximately proportional to $S^{0.5}$. At the largest values of S considered here, the ionisation rate is so rapid that it essentially converts hydrogen from molecular to atomic form and H^+ becomes the major ion.

The fractional abundances of several neutral molecules, for example, CH, H_2O and CO_2, are proportional to S except when S is greater than a few hundred. In this case, loss of stable neutral molecules by cosmic-ray-induced ultraviolet photons dominates and abundances generally decrease as S increases. The abundances of the sulfur-bearing molecules, SO, CS and SO_2 show a different behaviour with S. This is because the reactions between sulfur hydride ions and H_2 are endothermic[9] so that sulfur chemistry is driven by neutral-neutral reactions. For example, SO is formed by

$$S + O_2 \longrightarrow SO + O.$$

At small values of S, SO is destroyed by O atoms, to form SO_2, and by C atoms, to form CS. At large values of S, destruction of SO by H^+ causes a decrease in the SO abundance, and resulting decreases in the fractional abundances of CS and SO_2.

The time to reach a steady-state solution depends on the adopted value of S. This is shown in Figure 3 where the time at which $n(C) = n(CO)$ is plotted as a function of S. The curve levels off for smaller S due to the importance of chemi-ionisation which dominates in this regime. For S greater than 10, the chemistry reaches a steady state in 10^5 yrs.

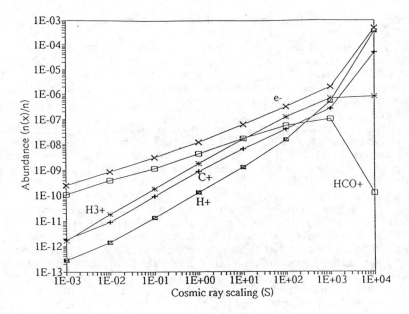

Figure 1. Plot of the steady state fractional abundances of several ions at a function of S.

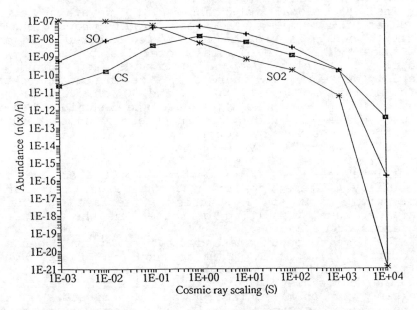

Figure 2 Plot of the steady state fractional abundances of several neutrals as a function of S.

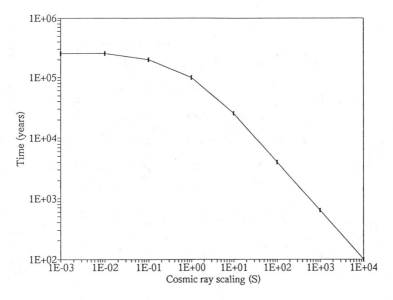

Figure 3. Plot of the time when the atomic carbon and CO abundances are equal.

DISCUSSION

We can use the chemical models together with observations of Galatic and extragalactic clouds to put limits on the cosmic ray ionisation rates. Using the observations of Gusten et al.[10] of CO toward M82, we derive an ionisation rate of about 10^{-14} s^{-1} in the warm (50-70 K), dense (n = 10^{4-5} cm^{-3}) gas in this galaxy, assuming that cosmic rays provide the heating – thus the derived rate is an upper limit. In the lower density and more extended gas in M82, the rate derived is about 5 10^{-16} s^{-1}, a factor of 40 times the standard Galactic rate. It is of course much more likely that the high temperature phase is heated by ultraviolet radiation associated with the starburst phenomena. It has been estimated that the radiation field is 500 times larger than the standard Galactic field[4].

Finally, we also made a comparison between our calculated abundances and those observed in the Galactic dark cloud, TMC-1. Correspondence is best at early-times and with an ionisation rate of (3-8) 10^{-17} s^{-1}, slightly larger than the standard value.

ACKNOWLEDGMENTS

The work of PRAF and TJM has been supported by a grant from the SERC. EH acknowledges the support of the National Science Foundation for his research programme in astrochemistry.

REFERENCES

1. E Herbst, T J Millar, in *Molecular Clouds*, eds R A James & T J Millar (Cambridge University Press, 1991),p. 209.
2. S Lepp, in *Astrochemistry of Cosmic Phenomena*, ed P D Singh, (Kluwer Academic Press, 1992), p. 471.
3. J G Skibo, R Ramaty, A&AS, , 141 (1992).
4. M Rowan-Robinson, MNRAS, 258,787 (1992).
5. R Mauersberger, C Henkel, Rev. Mod. Astron., 6, 69 (1993).
6. T J Millar, E Herbst, A&A, 231, 466 (1990).
7. R Gredel, S Lepp, A Dalgarno, E Herbst, ApJ, 347, 289 (1989).
8. P R A Farquhar, T J Millar, E Herbst, MNRAS, submitted.
9. T J Millar, N G Adams, D Smith, W Lindinger, H Villinger, MNRAS, 221, 673 (1986).
10. R Gusten, E Serabyn, C Kasemann, A Schinkel, G Schneider, A Schultz, K Young, ApJ, 402, 537 (1992).

NEUTRAL-NEUTRAL REACTIONS IN THE CHEMISTRY OF INTERSTELLAR CLOUDS

E Herbst and H H Lee
Department of Physics, The Ohio State University,
Columbus, Oh. 43210, USA

D A Howe and T J Millar
Department of Mathematics, UMIST, Manchester M601QD, UK

ABSTRACT

Recent experimental data on neutral-neutral reactions have been included in two large-scale, pseudo-time-dependent models of a cold, dense interstellar cloud. The inclusion of reactions involving CN and C atoms has the effect of reducing the calculated abundances of many complex organic molecules from those calculated in previous models, such that it is difficult to reconcile predicted and observed abundances. Specific reactions are identified for further laboratory and theoretical study.

INTRODUCTION

For twenty years, models involving ion-molecule reactions have been used to describe molecular synthesis in interstellar clouds[1]. Most interstellar molecules appear to be formed in schemes in which a series of ion-molecule reactions are terminated by a dissociative recombination reaction to produce observed neutral species. Although particular neutral-neutral reactions are recognised to be important in the formation of certain molecules, in particular those involving nitrogen, the lack of low-temperature laboratory data on such reactions has led to their relative neglect in model calculations. Since these calculations show that large abundances of complex molecules occur at so-called 'early-times', when the abundances of atomic species are high, the neglect of reactions involving the abundant O and C atoms is particularly worrying.

Recently, several laboratory studies have determined the rate coefficients for certain neutral reactions involving the CN radical at temperatures as low as 13 K[2,3,4]. These reactions can be extremely rapid – the CN + NH_3 reaction has a temperature-dependence of $T^{-1.1}$ down to 25K[3] with a rate coefficient of about $1.0 \ 10^{-9}$ cm^3 s^{-1} when extrapolated to 10 K. A reaction of potential importance to the formation of cyanoacetylene, HC_3N, is that between CN and C_2H_2 which has a T^{-1} dependence[4], and which may dominate the formation of HC_3N in both cold interstellar clouds[5] and carbon-rich circumstellar envelopes[6], although the products of this reaction have not, as yet, been determined by the laboratory studies. In addition, several reactions of C atoms with hydrocarbon molecules have been studied at room temperature[7,8] and possess large rate coefficients.

The availability of this data has encouraged us to re-examine the chemistry of cold (10 K), dense ($n = 2 \; 10^4 \; cm^3$) interstellar clouds using two large gas-phase models: the network of Herbst and Leung[5,9] and the UMIST network RATE92[10], an update of that described by Millar et al.[11]. There are certain differences between these networks. That of Herbst and Leung was developed to study the formation of large, organic molecules and uses ion-dipole enhanced rate coefficients; the UMIST network contains a more extensive description of the chemistries of smaller species, in particular those involving the elements S and Si, contains the chemistries of two elements, P and Cl, not in the Herbst-Leung network, includes cosmic-ray-induced photoreactions, and does not use ion-enhanced rate coefficients. Bettens and Brown[12] have considered the addition of rapid neutral reactions involving atomic oxygen to dense cloud models and found that complex organic molecules become difficult to form in the required quantities due to the destructive effects of O. It was our hope that the inclusion of atomic carbon-hydrocarbon reactions would alleviate this difficulty, since it is generally believed that C atom reactions aid organic synthesis. We therefore added a wide range of atomic carbon reactions to the two networks described above, taking the room temperature rate coefficients to be appropriate for 10 K, and choosing products which were exothermic and conserved spin. As an aid to building molecular complexity, we assumed that C atoms reacted only with unsaturated hydrocarbons and only via insertion reactions. There has been much controversy over the rates of C atom reactions at low temperature. Graff[13] has argued that, since in interstellar clouds C lies in its lowest fine-structure state, which is spherically symmetric and possesses no quadrupolar moment, there will be no long-range attraction between C atoms and other neutrals. However, Clary et al.[14] have pointed out that long-range dispersion forces probably dominate the interaction and lead to a non-zero rate coefficient. In addition, we up-dated the rate coefficients of several radiative association reactions, included reactions of CCH with the cyanopolyynes to form larger cyanopolyynes and with the C_nH_2 molecules to form $C_{n+2}H_2$, ($n = 3-7$).

RESULTS

We shall briefly describe our results, concentrating on the Herbst-Leung network and using results obtained by the UMIST network to identify important reactions and processes. A fuller discussion is to appear in a separate publication[15]. The addition of the neutral reactions had a dramatic effect on molecular abundances, but not in the expected manner. It was found that the abundances of complex molecules were, in fact, greatly reduced, particularly at the so-called 'early times', from previous models. Table 1 contains a comparision of observed molecular abundances in the dark cloud TMC-1 with both the 'old' and 'new' Herbst-Leung network. Most of the C_nH_m species decrease by at least an order of magnitude; some, such as C_4H, by over two orders of magnitude. The abundances of the cyanoployyne species, $HC_{2n+1}N$ (n =1-4) also decrease dramatically, HC_3N by a factor of about 1000 and HC_5N by about 30. Finally, the abundance of CN is reduced by 100.

Figure 1 shows results from the UMIST network using two models, a control calculation using RATE92 and a second model in which the neutral-neutral reactions mentioned above are added. Once again, it can be seen that the effect of the new reactions is to decrease the abundances of the complex hydrocarbon and cyanopolyyne molecules.

DISCUSSION

It was thought that the inclusion of the C atom reactions would offset the destructive effects of the O atom reactions since at early time, the abundances of C and O are comparable, while C is the more reactive species. Moreover, a carbon insertion reaction was assumed for all of the C atom reactions, thereby leading to the production of more complex, carbon-containing molecules. The reason why the C atom reactions fail to build up larger molecules in abundance is as follows.

Molecules in dark clouds are destroyed in two major ways. Firstly, by reaction with ions such as C^+ and He^+, and secondly by neutral reactions with atoms and radicals. At early times, in the standard models, the first route dominates for stable, unreactive species – C_2H_2 is an example – and the second route for reactive radicals – OH for example. The inclusion of the reaction of atomic carbon with acetylene

$$C + C_2H_2 \longrightarrow C_3H + H$$

reduces the acetylene abundance by over a factor of 100. Although this results in a formation rate for C_3H correspondingly larger, this does not feed through to its abundance since, in the new Herbst-Leung network, it is also destroyed rapidly by C atoms. In the old Herbst-Leung network, it was assumed that C_3H was destroyed slowly by O atoms. There is no similar increase in the formation rate of C_2H_2 since the reaction of C with the abundant CH_4 does not proceed as CH_4 is a stable, saturated molecule.

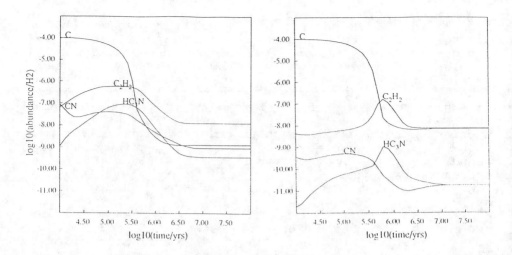

Figure 1. Fractional abundances of selected species are shown for the UMIST network using RATE92 (left) and with neutral reactions added (right)

Table 1 Observed vs early time fractional abundances in TMC-1 model. The notation a(-b) means a $\times 10^{-b}$

Species	Observed	Old	New
CN	3(-8)	7(-8)	7(-10)
HCN	2(-8)	1(-7)	1(-9)
HNC	2(-8)	7(-8)	1(-9)
HC_3N	6(-9)	8(-8)	6(-11)
HC_5N	3(-9)	8(-9)	3(-10)
HC_7N	1(-9)	2(-9)	1(-11)
HC_9N	3(-10)	2(-10)	9(-13)
C_2H	8(-8)	7(-8)	1(-9)
C_3H	5(-10)	5(-8)	2(-9)
C_4H	2(-8)	2(-7)	7(-10)
C_5H	4(-10)	6(-9)	5(-11)
C_6H	1(-9)	3(-9)	2(-11)
C_3H_2	2(-8)	3(-8)	8(-10)
CH_2CN	5(-9)	4(-8)	6(-9)
CH_3CN	1(-9)	7(-9)	3(-11)
C_3N	1(-9)	2(-8)	3(-10)

There are two major reasons for the decrease in the HC_3N and other cyanopolyyne abundances when the new reactions are included. As discussed above, the main formation route to HC_3N in the previous models is the reaction
$$CN + C_2H_2 \longrightarrow HC_3N + H$$
which has been measured to be fast down to 25 K[5], although we reiterate that the products, though exothermic, are not yet certain. The large decrease in the acetylene abundance reduces the importance of this reaction as does the inclusion of the reaction
$$O + CN \longrightarrow CO + N.$$
This reaction, in analagy with the $CN + O_2$ reaction, is assumed to have a $T^{-0.6}$ dependence giving a rate coefficient of about 10^{-10} cm^3 s^{-1} at 10 K. This reaction was included in the RATE92 network but with a temperature dependence of $T^{0.5}$. The adoption of the new dependence in the UMIST network reduces the CN abundance by a factor of 40 (see Figure 1) since it is the dominant loss for CN at early times. These reductions in the CN and C_2H_2 abundances when the new data are included affect the abundances of more complex species. For example, the abundance of the observed interstellar molecule C_4H is reduced by several orders of magnitude since its dominant formation is in the reaction of C_2H with C_2H_2, while the abundance of HC_3N is also reduced by the inclusion of the reaction
$$C + HC_3N \longrightarrow C_4N + H.$$
This reaction has not yet been studied in detail and we have not included C atom reactions with the higher cyanopolyynes in order to gauge its effect.

CONCLUSIONS

The inclusion of new laboratory data on neutral-neutral reactions in chemical models of cold, dense interstellar models is found to reduce the early time abundance of several complex organic molecules to such an extent that it is difficult to reconcile them with those observed in TMC-1. This may be due, in part, to an overestimate of the rate coefficients of the C atom reactions at 10 K, since we have assumed that the room temperature rate coefficients pertain. It is possible that small activation energy barriers exist in these reactions. The fast low temperature reactions of CN with hydrocarbons can be an important source of the cyanopolyyne molecules only if the reaction of CN with O is not fast at low temperatures. If fast, the CN abundance is reduced far below that observed and the cyanopolyynes must be made in ion-molecule reactions.

Despite the advances made in the measurement of low-temperature neutral-neutral reactions, much more work remains to be done. This includes the important task of identifying the products of such reactions, the $CN + C_2H_2$

reaction is an obvious example, and the determination of the rate coefficients of both atomic oxygen and carbon reactions at temperatures more representative of interstellar clouds. As well as the reactions discussed in this paper, the $C + O_2$ reaction could be an important loss mechanism for O_2 in interstellar clouds. Despite several sensitive searches, molecular oxygen has not yet been detected in the interstellar medium.

ACKNOWLEDGMENTS

EH acknowledges the support of the National Science Foundation for his research programme in astrochemistry and thanks the Ohio Supercomputer Center for time on their Cray Y-MP8. The work of DAH and TJM has been supported by a grant from the SERC.

REFERENCES

1. E Herbst & T J Millar, in *Molecular Clouds*, eds. R A James & T J Millar (Cambridge University Press, 1991), p. 209.
2. B R Rowe, A Canosa & I Sims, J Chem Soc Faraday Trans, 89, 2193 (1993).
3. I Sims, J-L Queffelec, A Defrance, C Rebrion-Rowe, D Travers, B R Rowe & I W M Smith, J Chem Phys, submitted.
4. I Sims, J-L Queffelec, D Travers, B R Rowe, L Herbert, J Karthauser & I W M Smith, Chem Phys Letts, submitted.
5. E Herbst & C M Leung, A&A, 233, 177 (1990).
6. D A Howe & T J Millar, MNRAS, 244, 444 (1990).
7. D Husain, J Chem Soc Faraday Trans, 89, 2175 (1993).
8. N Haider & D Husain, J Photochem Photobiol, A70, 119 (1993).
9. E Herbst & C M Leung, ApJS, 69, 271 (1989).
10. P R A Faquhar & T J Millar, CCP7 Newsletter No. 18, 6 (1993).
11. T J Millar, J M C Rawlings, A Bennett, P D Brown & S B Charnley, A&AS, 87, 585 (1991).
12. R P A Bettens & R D Brown, MNRAS, 258, 347 (1992).
13. M M Graff, ApJ, 339, 239 (1989).
14. D C Clary, T S Stoecklin & A G Wickham, J Chem Soc Faraday Trans, 89, 2185 (1993).
15. E Herbst, H H Lee, D A Howe & T J Millar, MNRAS, submitted.

DISCUSSION

SCAPPINI — There are a few mis-matches between observations and your calculated column densities. I want to know to what extent the initial conditions, like temperature, density of the medium, density of the species, etc., affect the final results of model calculations.

MILLAR — For these particular calculations which relate to cold, dark clouds, the results are fairly robust. In particular, changes in the initial conditions consistent with the observed distributions of temperature and density in dark clouds have little effect on these results.

LEACH — The sublevels of the carbon ^3P state could have very different reaction rates. The effects would be smoothed out in the measurements of Husain but could be significantly different and specific at low interstellar temperature.

MILLAR — We are aware that there may well be different reactivities among the sublevels. However, the question is still open and needs to be answered by detailed theoretical calculations.

ROUEFF — What happens when the carbon is initially on C^+ rather than on C ?

MILLAR — The results for $t \geq 10^4$ yr are insensitive to the initial form of atomic carbon. In fact, the results presented here all adopt C^+ as the initial form of carbon. The fact that the C atom abundance is large in Figure 1 at 10^4 yr is because the chemistry quickly converts C^+ to C.

ZIURYS — Do early time calculations still reproduce observed abundances in dark clouds the best, as opposed to those derived at steady state ?

MILLAR — If it turns out that the neutral-neutral reactions included in these model calculations are indeed rapid at low temperatures then early time calculations do not reproduce at all the abundances observed in dark clouds such as TMC-1.

CLARY — Ab initio calculations on the reaction of $C(^3P)$ with C_2H_4 show a path with no activation energy for a cyclic intermediate. This suggests that it might be necessary to include cyclic organic molecules in your reaction models.

MILLAR — There is a limited number of isomeric species included in the model but no attempt at including cyclic species in parallel with linear forms, for example the cyclic and linear form of C_3H_2, both observed interstellar molecules. My feeling is that there is not yet enough laboratory information available to allow one to include this type of chemistry with any degree of certainty.

SHALABIEA — Are these results obtained from pure gas phase models or dust/gas chemical models ?

MILLAR — These are, with the exception of H_2 formation, purely gas phase models.

RADIATION FIELDS INSIDE DARK CLOUDS WITH VARIABLE EXTINCTION LAWS

S. Aiello*
C. Cecchi-Pestellini**
B. Barsella***

*Dipartimento di Astronomia e Scienza dello Spazio, Universita' di Firenze, Italy
**Harvard-Smithsonian Center for Astrophysics, Cambridge (MA), USA
***Dipartimento di Fisica, Universita' di Pisa, Italy

ABSTRACT

The intensity of far-ultraviolet radiation inside interstellar clouds is a crucial factor in determining their physical and chemical evolution. The intensity of UV radiation depends on the structure of the clouds, on the optical properties of interstellar grains and on the presence of embedded sources, if any. Scattering on dust grains can greatly reduce shielding effects.

In the present work we compute the radiation intensity inside isolated dark clouds, making different assumptions on the structure of the clouds and on the extinction and scattering properties of dust grains. In particular, we consider the case in which extinction laws are different in different parts of the cloud.

1. INTRODUCTION

Evaluation of the radiation field within a dark cloud is a crucial step in the theoretical modelling of the interstellar medium. In fact, the intensity of radiation fields strongly affects the physical and chemical conditions of the matter inside clouds. Of particular importance are the UV photons which affect the lifetime of molecules against photo destruction. Even small variations in the amount of UV radiation can have significant effects on the chemical and ionization balance of dark clouds.

The degree of penetration of interstellar radiation inside dark clouds depends mainly on the absorption and scattering properties of the dust grains. Actually, when radiation penetrates inside clouds, energy is subtracted from the beam by means of physical absorption, by both dust and molecules, and through outscattering by grains. Molecular opacity, however, affects the radiation intensity only in a very narrow range (mainly corresponding to the photo destruction of CO and H_2). In the following, therefore, we will limit our analysis to the effects of dust on radiation transfer.

2. TRANSFER EQUATION

We consider the following form of the radiative transfer equation:

$$\mu\frac{\partial I(\lambda,r,\mu)}{\partial r} + \frac{1-\mu^2}{r}\frac{\partial I(\lambda,r,\mu)}{\partial \mu} = -\alpha(r)I(\lambda,r,\mu) + \frac{\alpha_s(\lambda,r)}{2}\int_{-1}^{1}I(\lambda,r,\mu')P(\mu,\mu')d\mu' \quad [1]$$

where $\alpha(\lambda,r)$ and $\alpha_s(\lambda,r)$ are, respectively, the extinction and scattering coefficients for unit length and $P(\mu,\mu')$ is the phase function. Here as follows, we adopt the Henyey-Greenstein approximation[1]. The values of the extinction and scattering coefficients depend on the dust model as well as on the cloud morphology.

To solve equation (1), we used the extension of the Spherical Harmonic Method worked out by Tine' et al.[2] in order to handle variable-density models.

3. THE CLOUD MODEL

As we have stated in previous sections, the factors which affect the penetration of radiation inside interstellar clouds are cloud morphology and the extinction and scattering properties of dust.

As regards the morphology, spectral-line observations indicate that, inside dark clouds, gas is centrally condensed, with a density gradient which approximates a power law. In general, a comparison of models with observations shows that most isolated dark globules have a core-halo density structure. Gas densities are observed to be in the range of 10^3-10^5 cm^{-3} in the central regions of these clouds[3], with a power-law radial variation as r^{-1} to r^{-2} [4-6]

The extinction law has been studied in the direction of a large number of stars located in different regions and astrophysical environments[7-8] . Said law is well established, at least up to 125 nm (beyond 125 to 100 nm is still uncertain). Although the gross features (bump, rise) are ubiquitous, there is strong variability in the shape of the UV extinction. The Mean Interstellar Extinction Curve (MIEC), commonly used to account for the presence of dust, can be considered to be representative of the general properties of dust in the diffuse medium and in old associations[9]. However, extinction and polarisation data show that, in dense clouds and in regions of recent star formation, dust grains are likely to have different properties: in particular, their size distribution appears to be biased toward larger radii, and this means a more or less marked flattening of the extinction curve in the far-UV spectral region. Moreover, inside the clouds, there may be a significant difference in properties between the dust located in the relatively low-density outer halo and that located in the high-density core. Mantle accretion and/or grain coagulation should feature in the latter, while shattering will dominate in the halo (for densities < 10^3 cm^{-3}), leading to an increase in the population of small grains[10] Actually, there is some evidence that the values of the ratio of total-to-selective extinction, R_V, inside dark clouds, increase towards the cloud center[11]. Large R_V values ($4 \leq R \leq R_V \leq 6$) are associated with areas with anomalous extinction and are generally taken as evidence of grain growth.

An straightforward use of MIEC in radiation-transfer computations may therefore be inappropriate in the case of dense clouds, at least inside their deeper parts, as it could lead to large errors (12-13)

On the basis of the previous considerations we adopted the following cloud model: core-halo clouds with envelopes surrounding cores of uniform density ρ_0 and radius r_0, with the core and the envelope density-distribution merging continuously at r_0:

$$\rho(r) = \begin{cases} \rho_0 & r \leq r_0 \\ \rho_0 \left(\frac{r_0}{r}\right)^k & r \geq r_0 \end{cases}$$

Both gas and dust are assumed to obey this equation, although the dust density distribution could have a steeper gradient than gas does, as well as a different core size.

Furthermore, envelope and core dust grains are assumed to have different extinction properties.

As shown by Cardelli et al[14], it is possible to derive a mean extinction curve for a given R_V. In our computations for the halo dust we adopted the extinction law corresponding to $R_V = 3.2$, typical of the diffuse medium, while for the core dust we adopted the extinction law corresponding to $R_V = 4.4$, a value appropriate for dense media.

The values of the albedo ω and the phase parameter g for the halo dust are derived from Draine and Lee[15]. To compute their values for the core dust, we adopted the following procedure: first, making use of the analytical expression found by Cardelli et al, we derived the mean extinction curve for $R_V = 4.4$; then, changing the size limits (a_-, a_+) and/or the exponent of a distribution of bare graphite and silicate particles with $n(a) \propto a^{-q}$ (16-17), we searched for the best fit to the curve. The best fit has been obtained with the following values: $a_-(gra) = 5$ nm, $a_+(gra) = 250$ nm; $a_-(sil) = 40$ nm, $a_+(sil) = 400$ nm; $q(gra) = 3.5$, $q(sil) = 3.0$. Finally, we computed the albedo and g for the resulting distribution. The values of the optical constants are taken from Draine [18].

4. RESULTS

We computed the transmissivities inside a cloud for various wavelengths, in the range 91-180 nm, for different values of the exponent in the density distribution and for different values of τ_V, the optical visual depth measured from edge to center.

The values of transmissivities at $\lambda = 91.16$ and 180 nm, as a function of the position inside the cloud, are shown in Fig. 1 and 2 for a density distribution exponent = 2 and an optical visual depth equal to 10.

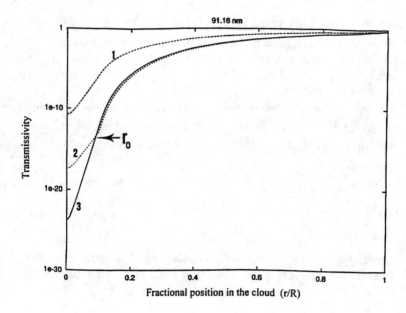

Fig.1.- The transmissivity at $\lambda = 91.16$ nm as a function of the fractional position in the cloud [2]. For comparison purposes, the transmissivity in two cases of a uniform extinction law is shown: $R_V = 4.4$ [1] and $R_V = 3.2$ [3]

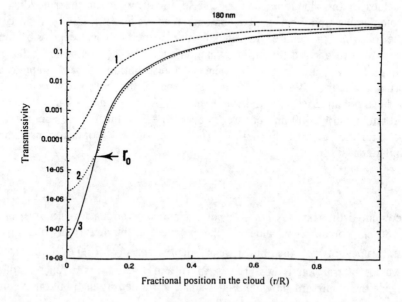

Fig. 2 - The same as in Fig. 1 for $\lambda = 180$ nm.

Fig. 3 shows the values of the *Scattering Amplification Factor*. i.e. the ratio $\frac{I_c(\lambda)}{I_0(\lambda)e^{-\tau_\lambda}}$, a measure of the shielding reduction by scattering, at the center of the cloud. Here, $I_0(\lambda)$ is the intensity of the radiation of wavelength λ outside the cloud; I_c is the intensity at the center of the cloud, and τ is the optical depth at λ.

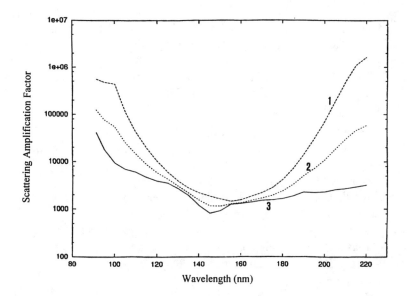

Fig. 3.- The Scattering Amplification Factor as a function of λ at the center of the cloud. Labels as in Fig. 1.

5. REFERENCES

1. L.G. Henyey and G.L: Greenstein, Astrophys,J. **93**, 70 (1941).
2. S. Tiné, S. Aiello, A. Belleni, C. Cecchi-Pestellini, J.Quan. Spectrosc. Radiat. Transfer, **47**, 95 (1992).
3. P.J. Benson and P.C. Myers, Astrophys. J. Suppl. **71**, 89 (1989)
4. S.T. Fulkerson and F.O. Clark, Astrophys. J. **287**, 723 (1894).
5. R. Arquilla and P.F. Goldsmith, Astrophys. J. **297**, 436 (1985).
6. J.A. Stuwe, Astron. Astrophys. **237**, 178 (1990).
7. D. Massa and B. Savage, in "Interstellar Dust", L.J. Allamandola and A.G.G.M. Tielens eds., pagg. 3-21 (Kluwer, 1989).
8. I S. Mathis, Ann. Rev. Astron. Ap. **28**, 37 (1990).
9. S. Aiello, B. Barsella, G. Chlewicki, J. M. Greenberg, P. Patriarchi, M. Perinotto, Astron. Astrophys. Suppl. Ser. **73**, 195 (1988).
10. A.G.G.M. Tielens, in "Interstellar Dust", pagg. 239-261
11. F.J. Vrba and A.E. Rydgren, Astrophys. J. **283**, 123 (1984).
12. S. Aiello, A. Rosolia, B. Barsella, F. Ferrini, D. Iorio, Nuovo Cimento **7**, 840 (1984).
13. I.S. Mathis, Ann. Rev. Astron. Ap. **28**, 37 (1990).
14. J.A. Cardelli, G.C. Clayton, J.S. Mathis, Astrophys. J. **345**, (1989)
15. B.T. Draine and H.M. Lee, Astrophys. J. **285**, 89 (1984).
16 J.S. Mathis , W. Rumpl, K.H. Nordsieck, Astrophys. J. **217**, 425 (1977).
17. J.S. Mathis and S.G. Wallenhorst, Astrophys. J. **244**, 483 (1981).
18. B.T. Draine, Private Communication , (1993).

CHEMISTRY OF STAR-FORMING CORES

S.B. CHARNLEY
Space Science Division, NASA Ames Research Center, CA 94035
and
Department of Astronomy, University of California, Berkeley, CA 94720

ABSTRACT

Chemical effects arising from the exchange of molecules between dust and gas in dense clumps are described. Selective desorption of CO and N_2 from grains in cool cores can account for the presence of ammonia in several cores in the NGC 2024 cloud. Evaporation of ices containing methanol and ethanol can lead to detectable abundances of $(C_2H_5)_2O$ and $CH_3OC_2H_5$. Results are presented for the hot core chemistries of sulphur and phosphorus which are initiated by evaporated hydrogen sulphide and phosphine. The implications of these studies for understanding the nature of molecular mantles, the evolution of molecular complexity in the gas phase, and the presence of small-scale abundance gradients in star-forming regions, are briefly discussed.

INTRODUCTION

In dense molecular clouds, atoms and molecules collide and stick to the surfaces of cold dust grains and form molecular ice mantles. In the earliest phases of star formation, during formation of a dense core and during isothermal collapse to form a protostar, molecular accretion is accelerated and ultimately may lead to complete removal of all species containing heavy elements[1]. The mantles formed thus contain the products of both gas and solid phase chemistries. Once the protostar is formed, it will heat surrounding dense clumps of gas and dust and evaporate the volatile ices which probably have a similar chemical composition to those initially incorporated into the protostellar system. This mantle outgassing can drive an active chemistry in the hot clump gas that is unlike any encountered in cold dark clouds[2].

So-called *hot molecular cores* - small, dense, warm regions ($n_H > 10^6 cm^{-3}$, $T_{kin} > 100K$, $T_{dust} > 40K$) associated with young protostars - are thought to represent such regions where evaporation of icy grain mantles has been important in determining the observed gas phase composition[3]. Observations of several star-forming regions demonstrate that hot core chemistry is extremely rich and that strong chemical differentiation can exist between individual cores[4,5]. The molecular inventory of hot cores is quite different from that of the surrounding molecular cloud. It is typified by large abundances of fully-saturated molecules such as water and ammonia, as well as more complex organic molecules including methanol, ethanol, dimethyl ether, methyl formate, ketene, formaldehyde, acetaldehyde, formic acid and several nitriles. Theoretical models of these sources have shown that gas phase chemistry and accretion in the cold gas, during collapse, and simple grain surface reactions can indeed qualitatively account for some aspects of their chemistry[1].

Recent studies have been able to account for the presence of many of the observed complex molecules and the chemical differentiation: evaporated mantles drive gas phase complexity and slight differences in mantle composition, particularly the presence or absence of methanol and ammonia, lead to large differences in the abundances of various complex molecules[2]. Many of the problems in dark cloud chemistry, such as the nature of the cycling of molecules between gas and dust, become manifest in cold, high-density cores ($n_H > 10^6 cm^{-3}$, $T_{kin} < 30K$). Hot cores, on the other hand, provide a natural laboratory in which to study the products of grain surface reactions and the chemistry of warm, dense gas following mantle evaporation. In this contribution some selected preliminary results are presented from recent studies of the chemistry in both cool and hot cores.

Conference Proceedings No. 312: *Molecules and Grains in Space*, edited by Irène Nenner

© 1994 American Institute of Physics

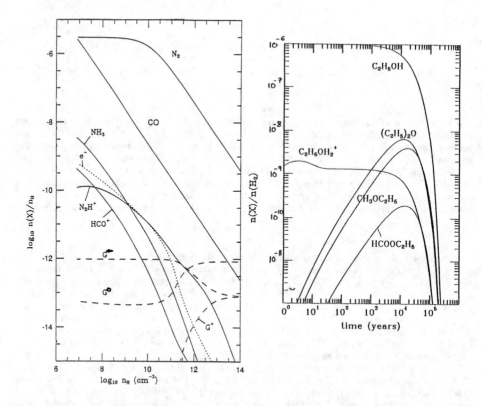

Figure 1. (left panel) Abundances as a function of density in clumps with gas and dust temperatures of 19K. **Figure 2.** (right panel) Complex molecules formed from evaporated ethanol in a hot core similar to the Orion Compact Ridge.

COLD CORES

Simple theories predict that, prior to warming by nearby protostars, there could exist ultradense cool regions ($n_H > 10^6 cm^{-3}$, $T_{kin}, T_{dust} < 30K$) where much of the heavy gas phase has condensed on to dust grains[1]. If they exist, such objects, where the gas-grain collision time-scale is so short that these processes are of paramount importance, would effectively represent the 'end' of conventional dark cloud chemistry. Seven ultradense dust emission cores (FIR1-7) have been identified in NGC 2024 ($n_H \sim 10^8 cm^{-3}$, $T_{kin} \sim 18 - 40K$, $T_{dust} \sim 19K$) which appear to show depletions of CO ($C^{18}O$ is not detected)[6]. It has been argued that these cores are cool protostellar condensations and are regions where accretion is dominating the gas phase evolution[7]. Although this interpretation has been the subject of debate[8], these sources currently represent the best candidates for studying molecular accretion at extremely high densities. Associated, and in some cases coincident, with the FIR sources are several clumps detected in ammonia[9] with fractional abundances, NH_3/H_2, in the range 10^{-10} to a few$\times 10^{-9}$. If the NGC 2024 cores are indeed regions of ongoing accretion, it is puzzling that a molecule with a large binding energy like NH_3 is present at all in gas in which material of much higher volatility is being removed! The reason for this may be connected to the observed dust temperature. Figure 1 shows results of a model calculation[10] of the chemistry in cool ultradense cores sustained by thermal desorption from grains at 19K. The theory is similar to that of Umebayashi & Nakano[11] except that account is taken of the interchange of CO, N_2 and O_2 between gas and dust; all other heavy neutrals have zero desorption rates. The cores are in a chemical steady-state and the difference in volatility of CO and N_2 means that, at 19K and for $n_H < 10^{10} cm^{-3}$, the latter molecule can remain abundant (but undetectable) in the gas where reactions with He^+ initiate ammonia formation. The theory produces NH_3 abundances similar to those observed for the NGC 2024 cores[9] and predicts that N_2H^+ should also be detectable towards these sources. The abundances of both NH_3 and N_2H^+ depend upon the total abundance (gas + dust) of N_2 available, taken here to be 3×10^{-6}, and the dust temperature: higher values yield more CO in the gas and less of these molecules.

If the above explanation is correct, then two assumptions made in the theory may have wider significance. First, there are no heavy metal atoms present in the gas phase: their thermal binding energies are too large. Second, the thermal binding energies of N_2 and CO are taken to be those for the pure substances[12] (890K & 1030K), and not those for binding to water ice[13] ($\sim$ 1740K): for slightly larger values the theory fails and both molecules are strongly depleted. However, if the lower values are applicable this may indicate the presence of a thick 'volatile crust' enclosing the water ice and containing mostly molecules with low or zero dipole moments. One naïve explanation of layered structure in the mantle is that, during core formation, H_2O ice formation occurs at a faster rate than does accretion of either CO or N_2.

HOT CORES

When dust grains are warmed to 100K and above, surface residence times are so short that no gas-grain steady-state is possible within 10^5 years and a nonequilibrium chemistry takes place that is characterised by the composition of the mantles released. Methanol-rich ices lead to high abundances of dimethyl ether (CH_3OCH_3) and methyl formate ($HCOOCH_3$), whereas ammonia-rich ices lead to several molecules containing C-N bonds[2]. This introduces the possibility that other mantle molecules could act as precursors of very large interstellar molecules, and that hot cores could contain some of the most complex molecules in the interstellar medium attainable by gas phase chemistry.

Ethanol cannot be formed efficiently in hot gas and a solid state origin is implied[2]. Self-protonation of alcohols leads to pure ethers and ion-molecule reactions between different alcohols produces mixed ethers[14,15]. Figure 2 shows the chemistry in a hot core, with physical and chemical characteristics similar to those of the Orion Compact Ridge, in which alcohol-rich ices containing formaldehyde have been evaporated[16]. The ion-molecule chemistry driven by ethanol produces diethyl ether, methyl ethyl ether and ethyl formate ($(C_2H_5)_2O$, $CH_3OC_2H_5$, $HCOOC_2H_5$). The most optimistic assumption concerning the initial abundance of ethanol[17] has been made (10^{-6}). However, other observed ethanol abundances[18] are $\sim 10^{-8}$ and are reached in the models at a time when CH_3OCH_3 and $HCOOCH_3$ come close to their observed values[2]. Thus, the lower observed abundances of ethanol do not exclude a high abundance in the original mantles and both diethyl ether and methyl ethyl ether remain detectable with abundances of $\sim 10^{-10}$. One may speculate that one facet of grain surface chemistry is to form simple alcohols: propanol and butanol similarly form large ethers, the abundances of which depend on those of the prescursor alcohols.

Hydrogen sulphide is observed to be enhanced in star-forming regions with estimated abundances that are about 100-1000 times that observed in dark clouds[19,20]. It is likely that in some sources this enhancement is due to removal of hydrogenated mantles. At hot core temperatures (less than about 300K), unlike most other abundant mantle molecules, H_2S can be attacked by atomic hydrogen to form SH, which subsequently reacts with H to release S atoms that then drive the sulphur chemistry. The reverse sequence reforming H_2S by reactions with H_2 is inhibited at these temperatures. Figure 3 shows the evolution in such a core[21]; most of the H_2S is converted to SO and SO_2. The result that SO/SO_2 ratios of less than unity are never attained within 10^5 years may allow hot core sulphur chemistry to be distinguished from other processes which form sulphur-bearing molecules, such as shock waves.

The only phosphorus-bearing molecule detected in molecular clouds to date is PN, and this only in warm regions[22,23]. As hydrogenation is clearly occurring on grain surfaces one might expect a major repository of solid phosphorus to be phosphine (PH_3). However, searches for this molecule have not been fruitful[24]. The lack of PH_3 in dark clouds is in agreement with ion-molecule chemical theory but, since it has a rather low condensation temperature[24] ($\sim$ 30K) its apparent absence in dense, warm ($T_{dust} > 40K$), dusty regions is puzzling. It has been postulated that either phosphine does not form on grains, or that it has an anomalously high binding energy, or that it is converted to another phosphorus compound when evaporated[24]. In fact, like S from H_2S, P atoms can also be broken out of phosphine by H atoms in warm gas[25]. Figure 4 shows the chemistry following PH_3 evaporation in a model of the W51 IRS2 source[26]. It appears that the tendency of the phosphorus chemistry is ultimately to convert PH_3 to atomic P and to PN. The PN abundance for $t > 1000$ years is about ten times that observed and implies that the original phosphine abundance assumed, PH_3/H_2 of 6×10^{-9}, is too large.

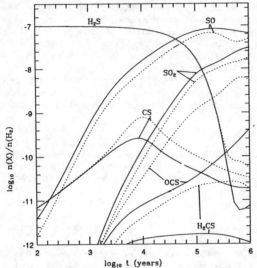

Figure 3. Hot core sulphur chemistry driven solely by hydrogen sulphide for the cases of no O_2 initially present (dotted curves) and present at an abundance of 1×10^{-6} (solid curves). ($n_H = 2 \times 10^7 cm^{-3}$, $T_{kin}=200K$).

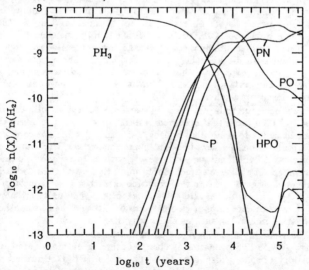

Figure 4. A model of phosphorus chemistry for W51 IRS2 following evaporation of phosphine from warm dust ($n_H = 2 \times 10^7 cm^{-3}$, $T_{kin}=150K$).

CONCLUSIONS

Slight warming of cold dust grains can induce selective desorption of mantle volatiles that produce specific molecules. Under certain assumptions concerning the appropriate binding energies and the dust temperature, it is theoretically possible for dense cores to become depleted in CO but to remain relatively rich in nitrogen-bearing molecules; this process appears to be occurring to some extent in NGC 2024. This theory predicts that heavy metal atoms should be almost entirely absent from the gas phase of high-density cores, and that a mantle of weakly bound molecules should cover water ice mantles. The presence of such a 'volatile crust' may also lower the NH_3 binding energy but probably not to a value typical of a nonpolar molecule. The interaction of Lyman photons with mantles have also been invoked to explain the presence of ammonia in the NGC 2024 cores[8]. Constructing a desorption rate involves estimating several highly uncertain quantities and, furthermore, this process is not molecule-specific: if it can desorb NH_3, with a surface binding energy[13] of about 3000K, then other abundant, less refractory, mantle species should also be present.

When alcohol-rich ices containing H_2CO are released into warm gas they form the building blocks of an ion-molecule chemistry that leads to much larger organic molecules. Specifically, large ethers can be formed from methanol, ethanol, propanol and butanol. The observed abundances of methanol and ethanol are such that the ethers formed from them, diethyl ether and methyl ethyl ether, may be detectable in warm cores. By contrast, the hot core chemistries of sulphur and phosphorus rely upon the destruction of the most abundant molecules, H_2S and PH_3, by H atoms, and the resulting neutral chemistry in each case is strongly temperature-dependent.

As it accretes material, a forming protostar will heat a surrounding (idealised) dense spherical envelope of gas and dust. The existence of radial gradients in grain temperature over spatial scales determined by the mass of the protostar[27] will mean that, at a given radius, the gas phase chemistry that obtains (see above) will largely be determined by the material that is evaporated from the dust, which in turn is influenced by the mantle composition and the distribution of binding energies, as well as by solid state reactions induced by heating. Modelling these processes[28] provides insight into the nature of molecular differentiation on small scales within regions of star-formation.

REFERENCES

1. Brown, P.D., Charnley, S.B. and Millar, T.J. 1988, *M.N.R.A.S.*, **231**, 409..
2. Charnley, S.B., Tielens, A.G.G.M. and Millar, T.J. 1992, *Ap.J.(Letters)*, **399**, L71.
3. Plambeck, R.L. and Wright, M.C.H. 1987, *Ap.J.(Letters)*, **317**, L101.
4. Blake, G.A., Sutton, E.C., Masson, C.R. and Phillips, T.G., 1987, *Ap.J.*, **315**, 621.
5. Sutton, E.C., Jaminet, P.A., Danchi, W.C. and Blake, G.A. 1991, *Ap.J.Suppl.*, **77**, 255.
6. Mezger, P.G. et al. 1988, *Astr.Ap.*, **191**, 44.
7. Mezger, P.G. et al. 1992, *Astr.Ap.*, **256**, 631.
8. Walmsley, C.M. and Schilke, P. 1992. in *The Astrochemistry of Cosmic Phenomena*, IAU Symposium No. 150, Ed. P.D. Singh, Kluwer Academic Publishers, p251.
9. Maursberger, R. et al. 1992, *Astr.Ap.*, **256**, 640.
10. Charnley, S.B. 1993, , in preparation.
11. Umebayashi, U. and Nakano, T. 1990, *M.N.R.A.S.*, **243**, 103.
12. Leger, A., Jura, M. and Omont, A. 1985, *Astr.Ap.*, **144**, 147.
13. Sandford, S.A. and Allamandola, L.J, 1993, *Ap.J.*, , submitted.
14. Meot-Ner (Mautner) M. and Karpas, Z. 1986, *J. Phys. Chem.* **90**, 2206.
15. Karpas, Z. and Meot-Ner (Mautner) M. 1989, *J. Phys. Chem.* **93**, 1859.
16. Charnley, S.B., Kress, M.E., Tielens, A.G.G.M. and Millar, T.J. 1993, *Ap.J.*, , submitted.
17. Turner, B.E. 1991, *Ap.J.Suppl.*, **76**, 617.
18. Millar, T.J. et al. 1988, *Astr.Ap.*, **205**, L5.
19. Minh, Y.C., Irvine, W.M. and McGonagle, D. 1990, *Ap.J.*, **360**, 136.
20. Minh, Y.C., Irvine, W.M. and McGonagle, D. 1991, *Ap.J.*, **366**, 192.
21. Charnley, S.B. 1993, *Ap.J.*, , submitted.
22. Turner, B.E. and Bally, J. 1987, *Ap.J.*, **321**, L75.
23. Ziurys, L.M. 1987, *Ap.J.*, **321**, L81.
24. Turner, B.E. et al. 1990, *Ap.J.*, **365**, 569.
25. Kaye, J.A. and Strobel, D.F. 1983, *Geophys. Res. Lett.* **10**, 957.
26. Charnley, S.B. and Millar, T.J. 1993, , in preparation.
27. Wolfire, M.G. and Cassinelli, J.P. 1986, *Ap.J.*, **310**, 207.
28. Charnley, S.B. and Wolfire, M.G. 1993, , in preparation.

BISTABILITY IN DARK CLOUDS

J. Le Bourlot, G. Pineau des Forêts & E. Roueff
DAEC, URA 173 du CNRS, Observatoire de Paris, F- 92195 Meudon Cedex, France

INTRODUCTION

We report the possibility of multiple solutions of the chemical reaction network describing dark interstellar molecular clouds chemistry. The physical conditions in dark clouds favour binary reactions, among which ion-molecule reactions play a predominant role [1]. These ion-molecule reactions are essentially driven by cosmic rays in dark clouds. In this paper, we show that two different stable steady-state solutions may coexist for a given set of astrophysical conditions. This *bistability* corresponds to two different types of ion-molecule reactions: proton transfer reactions prevail in a Low Ionization Phase (LIP) whereas charge exchange reactions occur in a High Ionization Phase (HIP), this leads to very different chemical behaviours [2] and has also a substantial influence on the thermal balance and thus on the temperature of the cloud. We exhibit also the unstable equilibrium solutions connecting the stable chemical states. These properties, typical of dynamical systems governed by non-linear equations, have already been demonstrated in the laboratory for other chemical systems.

THE MODEL

(i) Thermal equilibrium

The temperature is determined assuming thermal balance between heating and cooling processes in the absence of external UV radiation field (we are considering the dark part of interstellar clouds). The gas is mainly heated by formation of H_2 on grain surfaces (vibrationally hot: v = 6, and rotationally cold: T = 65K) [3], the cosmic ray ionization processes (with a mean energy of 6 eV per event) and the energy released by exothermic chemical reactions. The dominant cooling processes included in the model are radiative cooling through collisional excitation of the fine structure transitions of atomic carbon, and through excitation of rotational transitions of OH, H_2O, and ^{13}CO (we assumed that the ^{12}CO lines are optically thick, deep inside the cloud). The fine structure transitions of C^+ and O, together with the rotational transitions of H_2 are also included in the calculations but play a minor role at the temperatures of dark clouds we are considering. We take also into account the heat exchange between gas and dust

© 1994 American Institute of Physics

grains. A quantitative description of all these processes is given in Flower & Pineau des Forêts [4].

(ii) The chemical network

The chemistry includes a total of 95 species. The gas-phase elemental abundances, relative to n_H, ($n_H = n(H) + 2n(H_2)$), were taken to be: [He] = 1.0 10^{-1}, [C] = 3.62 10^{-5}, [N] = 1.12 10^{-5}, [O] = 8.53 10^{-5}, [S] = 1.85 10^{-6}, and [Fe] = 1.5 10^{-8}, although the sulphur abundance was varied. For C, N, O and S, the values given above correspond to their abundances in the solar system [5], depleted by a factor of 10. The chemical library, which characterizes dark interstellar clouds is an extended version of that published by Pineau des Forêts et al. [6, 7] and Le Bourlot et al. [8] and contains over 650 reactions. The dissociative recombination rate of H_3^+ is taken to be 1.5 10^{-7} $(T/300)^{-1/2}$ cm^3 s^{-1}, as suggested by the measurements of Canosa et al. [9]. The ionization and dissociation of the molecules due to the secondary UV photon induced by cosmic rays [10] are also included in the chemical scheme.

RESULTS AND DISCUSSION

The various formation and destruction processes of interstellar molecules lead to a set of coupled non-linear first order differential equations which describe the time dependent chemistry. It has been however a consensus in the astrochemical community to discuss *one* solution for the chemical abundances at equilibrium [11], which is obtained either in steady state calculations or in the integration of the time evolution equations. A first ambiguity about the "standard" solution has been reported by Graedel, Langer & Frerking [12], who found different solutions of the chemical equations for slight variations in the elemental depletions of the heavy elements but no further study has been made by these authors. The triggering step towards complexity in interstellar chemistry has been achieved when a phase transition has been found [13, 14] with two very different types of solutions (determined at a constant temperature: in this first approach, the thermal balance was ignored). These two chemical states can be described as a Low Ionization Phase (LIP) where proton transfer reactions are important and lead to large saturated molecules which reflects the "standard" solution of interstellar dark cloud models [15], and a High Ionization Phase (HIP) where charge exchange processes govern the chemistry.

Fig. 1 shows the degree of ionization for a series of stationnary models with densities ranging between 3 10^2 and 3 10^4 cm^{-3} and with a cosmic ray ionization rate ζ = 2 10^{-17} s^{-1} for two different values of the sulphur depletion δ_S. An unstable branch connects the two stable solutions. On the physico-chemical point of view, the dissociative recombination reaction of the H_3^+ molecular ion plays a crucial role and the transition occurs when the recombination process is of the same order than the destruction of H_3^+ by neutral species such as O, CO ... leading to proton transfer. Fig. 2 shows the steady-state gas temperature. Large temperature variations are due to the cooling through the fine structure transition of neutral carbon whose abundance can vary by two orders of magnitude between the two chemical phases (see Fig. 3).

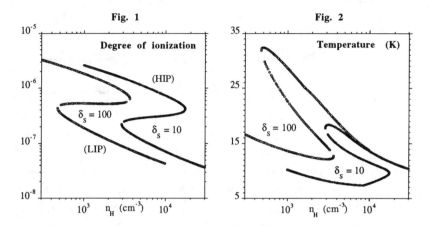

Fig. 1   Fig. 2

The behaviour of the C/CO ratio (Fig. 3) is typical of the abundances of unsaturated species or of the ionization degree for exemple. A high value (C/CO ≈ 0.1) is obtained for the HIP which is close to the preliminary intriguing observations of neutral carbon in dark clouds [17], whereas low values (C/CO ≈ 10^{-3}) are characteristic of the LIP, the standard solution found previously. In the HIP, H^+ is more abundant than H_3^+ by two orders of magnitude, and charge transfer processes are preponderant. Hence, the species undergoing charge transfer such as O_2, OH, H_2O, H_2S are much less abundant in the high ionisation phase. Conversely, the carbon bearing molecules such as CN, CS or the C_nH_m molecules are abundant in the HIP since they are mostly formed by reactions with neutral carbon or carbon ions. As another example, the behaviour of the CS/N_2H^+ ratio is shown on Fig. 4 where the variation reaches *seven orders of magnitude*. These two (detected) molecules could be good candidates to determine the chemical phase of dark interstellar clouds, in spite of the observatinal uncertainties: CS is directly proportional to the neutral carbon abondance while N_2H^+ is only present in the LIP where it is supposed to be a tracer of the (undetectable) N_2 molecule [18, 19].

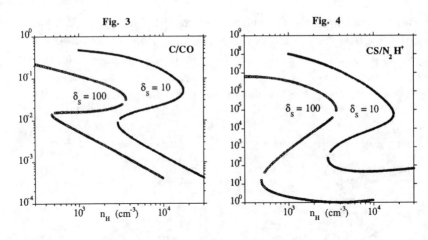

CONCLUSION

Such transitions are characteristics of dissipative systems as are the "bifurcations" studied in a variety of experiments such as the Belousov-Zhabotinsky reaction [16]. The steady-state solutions are far from LTE owing to unbalanced ionizations and dissociations by cosmic rays.

The astrophysical consequences have still to be explored but preliminary conjectures may be driven now from the extrapolation from laboratory knowledge: in particular, oscillatory or chaotic behaviour could make the traditional analysis of interstellar chemistry in terms of "early times" or "equilibrium" states obsolete. Transition zones between chemical regimes could lead to "thermochemical" turbulence. Chemical waves could propagate in the medium, leading to large abundances contrast on a yet undetermined spatial scale. We feel that this evidence of bistability in interstellar chemistry will open a fascinating new field and, if fully understood could be used as a tool to infer the history of the cloud.

REFERENCES

1. A. Dalgarno, "*The chemistry of astronomical environments*", Royal Society of Chemistry: Faraday transactions, in press (1993).

2. J. Le Bourlot, G. Pineau des Forêts, E. Roueff and P. Schilke, *Ap. J. Letters,* in press (1993).

3. W.W. Duley, D.A. Williams *MNRAS* **223**, 177 (1986).

4. D.R. Flower & G. Pineau des Forêts, *MNRAS* **247**, 500 (1990), erratum, *MNRAS* **249**, 191 (1991).

5. E. Anders & N. Grevesse, *Geochim. Cosmochim. Acta* **53**, 197 (1989).

6. G. Pineau des Forêts, E. Roueff, D.R. Flower, *MNRAS* **244**, 668 (1990).

7. G. Pineau des Forêts, E. Roueff, P. Schilke, D.R. Flower, *MNRAS* **262**, 915 (1993).

8. J. Le Bourlot, G. Pineau des Forêts, E. Roueff, D.R. Flower, *A&A* **267**, 233 (1993).

9. A. Canosa, B. Rowe, J.B.A. Mitchell, J.C. Gomet, C. Rebrion, *A&A* **248**, L19 (1991).

10. R. Gredel, in *Molecular astrophysics*, ed Hartquist T.W., Cambridge University Press (1990), p. 305.

11. W.W. Duley, D.A. Williams, "*Interstellar Chemistry*", Academic press (1984).

12. T.E. Graedel, W.D. Langer & M.A. Frerking, *ApJS* **48**, 321 (1982).

13. G. Pineau des Forêts, E. Roueff & D.R. Flower, *MNRAS* **258**, 45p (1992).

14. E. Roueff, G. Pineau des Forêts, in "*Dissociative Recombination: Theory, Experiment and Applications*", Plenum press, B.R. Rowe, B. Mitchell eds. (1993), p 257.

15. E. Herbst & C.M. Leung, *ApJS* **69**, 271 (1989).

16. P. de Kepper, A. Rossi and A. Pacault, *Comptes Rendus de l'Académie des Sciences* **283C**, 371 (1976).

17. T.G. Phillips, . in *Interstellar Processes*, eds. Hollenbach D. J. & Thronson H. A., Reidel, Dordrecht (1987), p. 707.

18. M. Womack, L.M. Ziurys & S. Wyckoff, *ApJ* **387**, 417 (1992).

19. M. Womack, L.M. Ziurys & S. Wyckoff, *ApJ* **393**, 188 (1992).

ATOMIC CARBON IN INTERSTELLAR CLOUDS

J. Le Bourlot, G. Pineau des Forêts and E. Roueff
DAEC, URA 173 du CNRS, Observatoire de Paris, F-92195 Meudon Principal Cedex, France

D. R. Flower
Physics Department, The University, Durham DH1 3LE

ABSTRACT

We present models of interstellar clouds in order to determine the transition from the warm, photon-dominated region at the edge to the cold, dark interior. We find that the equilibrium value of the C/CO ratio in the interior is crucially dependent on the fractional ionization of the gas, which varies with the cosmic ray flux, the metallicity, and the kinetic temperature. When the degree of ionization in the cloud centre exceeds a critical value, the ratio n(C)/n(CO) increases by about two orders of magnitude, to n(C)/n(CO) ≈ 0.1. However, this increase is not reflected in the intensities of the C I fine structure emission lines, when they arise predominantly in the photon-dominated region. Instead, the rotational transitions of carbon-bearing species, such as CS, may be better indicators of the atomic carbon content of the cloud interior.

1. INTRODUCTION

Many models of dark interstellar clouds have predicted that, if steady-state is attained, the fractional abundance of atomic carbon should be very small. Most of the gas-phase carbon is in the form of CO, and the ratio n(C)/n(CO) ≈ 10^{-3}. As a consequence, the computed fractional abundances of complex carbon-bearing species are much lower than observed [1,2]. Many of the observations of the C/CO ratio refer to the so-called 'photodissociation' or 'photon-dominated' regions (PDR's), where interstellar matter is believed to be exposed to a locally enhanced radiation field. The ultraviolet photons dissociate and ionize the gas at the edge of the cloud, whereas the interior is screened by the dust. Indeed, all dark interstellar clouds have edges on which impinges an ultraviolet radiation field whose intensity depends on the proximity of hot stars. The transition from the 'diffuse' medium at the edge of the cloud to the 'dark' interior has been the subject of several detailed studies [3,4], all of which have predicted a low value of the C/CO ratio in the cloud interior.

In the present paper, we return to the problem of determining the transition from the diffuse edge of the cloud to the dark interior. Following previous studies of the chemistry in dark clouds [1,5], we show that the equilibrium value of the C/CO ratio towards the cloud centre depends crucially on the degree of ionization of the gas. When the fractional ionization exceeds a critical value, the ion-molecule chemistry in the gas-phase is dominated by charge transfer reactions with H^+ (the High Ionization Phase, HIP), rather than proton-transfer reactions with H_3^+ (the Low Ionization Phase, LIP). The degree of ionization of the medium is determined by a number of factors, including the cosmic ray flux, the metallicity, the kinetic temperature of the gas, or even the proximity of a source of X-rays. Cosmic rays ionize the gas (principally H_2) directly, and the free electrons which are produced induce ionization of other species through the collisional excitation of the Lyman and Werner ultraviolet transitions of H_2 [6,7]. A change in any of the parameters which determine the degree of ionization can trigger the transition from one chemical regime to the other.

In Section 2, we describe the model that has been used to obtain the results discussed in Section 3.

2. THE MODEL

We have used the physico-chemical model of the envelopes of dark clouds which was described in a recent publication [4]. For a given incident radiation field, the model determines the chemical and thermal structure of the cloud, in steady-state, from its edge to its centre (i.e. to a point beyond which the structure no longer varies significantly with depth). The transfer of radiation in the ultraviolet absorption lines of H_2 and CO is solved numerically, using the method of Federman et al.[8].

In addition to He, C, O, and the representative metal (Fe), N and S have been incorporated in the model. The gas-phase elemental abundances (relative to H) were taken to be: $He/H = 1.00\ 10^{-1}$, $C/H = 3.62\ 10^{-5}$, $N/H = 1.12\ 10^{-5}$, $O/H = 8.53\ 10^{-5}$, $S/H = 1.85\ 10^{-6}$, and $Fe/H = 1.50\ 10^{-8}$, although the oxygen abundance was varied. For C, N, O and S, the values given above correspond to their abundances in the solar system [9], depleted by a factor of 10. Only the principal isotopes of C and of O were included. The chemical network comprised 95 species and 742 reactions (Pineau des Forêts et al.[10] and references therein) and included cosmic ray-induced ionization and photodissociation reactions [6].

The inclusion of sulphur has important consequences for the structure of the model: S^+ ions are major contributors to the degree of ionization of the medium, as will be seen below. The S/H abundance ratio adopted here is about an order of magnitude below the 'high metals' and an order of magnitude above the 'low metals' values considered by Graedel et al.[1] to be appropriate to diffuse and dense clouds, respectively. Millar & Herbst [11] have deduced $S/H = 1\ 10^{-7}$ from a time-dependent chemical model of sulphur-bearing molecules in the dense cloud TMC-1. In practice, the gas-phase abundance of sulphur depends on its degree of depletion on to grains at any given point in the history of the cloud.

3. RESULTS

3.1 DARK CLOUD MODEL

We present first results for a constant density model of a dark cloud in which $n_H = 10^4$ cm^{-3}, the elemental abundances of C, N, O and S are depleted by a factor of 10, relative to their solar system values, and T = 10 K; ζ is varied. We see from Fig. 1 that, under the above conditions, the transition between the two chemical regimes occurs for $2\ 10^{-17} \leq \zeta \leq 7\ 10^{-17}$ s^{-1}. In fact, as shown in Fig. 1 and discussed in more detail by Le Bourlot et al.[12], *two* stable steady-state solutions of the chemical rate equations coexist within this range of ζ. The solution which is actually obtained depends on the initial conditions which are adopted in the model. We note that this range of values of ζ is within that deduced from observations of diffuse interstellar clouds [13].

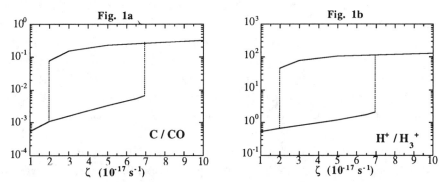

Fig. 1. The steady-state composition of a dark cloud in which $n_H = 10^4$ cm^{-3}, T = 10 K and δ = 10, as a function of the cosmic ray ionization rate, ζ (in units of 10^{-17} s^{-1}). The steady-state solution obtained in the range $2\ 10^{-17} \leq \zeta \leq 7\ 10^{-17}$ s^{-1} depends on the initial values which are adopted

The effects on the C/CO ratio and on the relative abundance of H_3^+ and H^+, are dramatic, as is shown by the Figure. Similar behaviour is found for all the chemical species and the degree of ionization. The two solutions which are possible within a range of values of ζ correspond to the Low Ionization Phase (LIP) and High Ionization Phase (HIP), to which reference was made above. These results suggest that even modest variations in the cosmic ray ionization rate, with position in the Galaxy, for example, might have profound consequences for the chemical composition of dark clouds.

3.2 PHOTON DOMINATED REGION MODELS

We discuss now the variations of the number densities of C^+, C and CO with the visual optical depth, τ_v, for a constant density model in which $n_H = 10^3$ cm^{-3}, $\zeta = 10^{-17}$ s^{-1}, and $\delta_C = 10$. The kinetic temperature was computed point by point using the thermal balance equation. The ultraviolet radiation is incident on a semi-infinite, plane-parallel cloud and its energy density as a function of wavelength is taken from Mathis et al.[14].

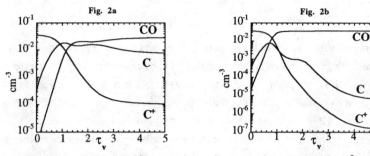

Fig. 2. Variation with visual optical depth, τ_V, of the densities of C^+, C and CO (cm^{-3}) for a model in which $n_H = 10^3$ cm^{-3}, $\chi = 1$ and (a) $\delta_O = 10$, (b) $\delta_O = 1$.

The results shown in Figs. 2a and 2b exhibit very different values of the ratio $n(C)/n(CO)$ at the cloud centre. As may be seen from Fig. 2a, this ratio is approximately 0.2 for the HIP, compared with $3 \ 10^{-3}$ for the LIP; the ratio $n(C^+)/n(CO)$ is also higher for the HIP, by about an order of magnitude. In the HIP, sulphur is by far the most abundant 'metal' and a major contributor to the degree of ionization of the medium, owing to the importance of the cosmic ray-induced ionization processes, mentioned above, and to the reaction $C^+(S, C)S^+$. Its presence ensures that, for the gas density $n_H = 10^3$ cm^{-3} and cosmic ray ionization rate $\zeta = 10^{-17}$ s^{-1} adopted in model 2a, the abundance of H_3^+ in the cloud interior is low and that of atomic carbon is high. If the gas-phase abundance of sulphur is further reduced, a higher value of ζ is necessary to ensure a degree of ionization sufficient to yield a similar result.

The atomic carbon content of the medium is critical to the formation of carbon-bearing molecules, other than CO. This point is illustrated in Tables Ia and Ib, where we compare the integral intensities of the fine structure transitions of C and C^+ and the first few rotational lines of the CS molecule, as computed in the two cases $\delta_O = 10$ and $\delta_O = 1$. When $\delta_O = 1$ and $\chi = 1$, for example, practically all the intensity of the line arises in the outer layers, $\tau_V < 1$, owing to the low abundance of C for $\tau_V > 1$. On the other hand, if $\delta_O = 10$, the greater atomic carbon content of the gas ensures that the integral line intensity continues to rise towards the cloud centre, saturating for $\tau_V > 100$, where the optical depth in the line becomes significant. However, for $\tau_V \leq 10$, the differences between the two models ($\delta_O = 10$ and $\delta_O = 1$) are not large [15]. Indeed, it is clear that the line intensity does not reflect the amount of atomic carbon in the cloud interior: a higher atomic carbon content is partially offset by a lower kinetic temperature in the region where C I fine structure transitions are important contributors to the cooling of the medium. The intensities of the CS lines, which are formed deeper in the cloud, prove to be much better indicators of the atomic carbon content at moderate densities.

Table I: Emissivities of fine structure lines of C and C⁺ together with rotational lines of CS for various enhancement factors of the radiation field χ at different optical depths τ_v for an isochoric model of density $n_H = 10^3$ cm^{-3}

a: $\delta_o=1$

χ	τ_v	C⁺ (158 µm)	C (1->0) (609 µm)	C (2->1) (370 µm)	CS (1->0) (6.12 mm)	CS (2->1) (3.06 mm)	CS (3->2) (2.04 mm)
1	1	8.2 (-7)	1.6 (-8)	2.0 (-8)	4.0 (-12)	1.1 (-11)	1.0 (-11)
	3	8.2 (-7)	1.8 (-8)	2.0 (-8)	4.8 (-12)	1.4 (-11)	1.2 (-11)
	10	8.2 (-7)	1.8 (-8)	2.0 (-8)	4.9 (-12)	1.4 (-11)	1.2 (-11)
10	1	4.5 (-6)	7.9 (-9)	1.5 (-8)	2.9 (-13)	8.6 (-13)	8.8 (-13)
	3	4.6 (-6)	2.0 (-8)	2.3 (-8)	1.5 (-12)	4.1 (-12)	3.3 (-12)
	10	4.6 (-6)	2.0 (-8)	2.3 (-8)	1.6 (-12)	4.4 (-12)	3.7 (-12)
100	1	1.1 (-5)	6.7 (-10)	2.0 (-9)	3.7 (-15)	1.2 (-14)	1.3 (-14)
	3	1.3 (-5)	2.2 (-8)	2.5 (-8)	3.6 (-13)	9.8 (-13)	7.8 (-13)
	10	1.3 (-5)	3.0 (-8)	2.6 (-8)	8.3 (-13)	2.2 (-12)	1.6 (-12)
1 000	1	1.6 (-5)	5.2 (-11)	1.9 (-10)	3.4 (-17)	1.1 (-16)	1.3 (-16)
	3	2.3 (-5)	1.9 (-8)	3.0 (-8)	3.4 (-14)	9.8 (-14)	9.2 (-14)
	10	2.3 (-5)	5.2 (-8)	3.6 (-8)	5.4 (-13)	1.4 (-12)	9.8 (-13)

b: $\delta_o=10$

χ	τ_v	C⁺ (158 µm)	C (1->0) (609 µm)	C (2->1) (370 µm)	CS (1->0) (6.12 mm)	CS (2->1) (3.06 mm)	CS (3->2) (2.04 mm)
1	1	9.9 (-7)	3.7 (-8)	3.9 (-8)	1.3 (-11)	3.6 (-11)	3.2 (-11)
	3	9.9 (-7)	7.0 (-8)	4.1 (-8)	1.1 (-10)	2.9 (-10)	2.5 (-10)
	10	9.9 (-7)	1.2 (-7)	6.5 (-8)	2.6 (-10)	1.1 (-9)	1.8 (-9)
10	1	5.9 (-6)	8.4 (-9)	1.7 (-8)	3.3 (-13)	9.9 (-13)	1.0 (-12)
	3	6.1 (-6)	7.9 (-8)	5.3 (-8)	1.9 (-11)	5.0 (-11)	3.3 (-11)
	10	6.1 (-6)	1.3 (-7)	7.5 (-8)	2.3 (-10)	9.8 (-10)	1.5 (-9)
100	1	1.4 (-5)	5.8 (-10)	2.0 (-9)	4.0 (-15)	1.3 (-14)	1.5 (-14)
	3	1.6 (-5)	8.9 (-8)	7.2 (-8)	1.9 (-12)	5.1 (-12)	3.7 (-12)
	10	1.6 (-5)	1.5 (-7)	9.2 (-8)	2.1 (-10)	8.53 (-10)	1.2 (-9)
1 000	1	2.1 (-5)	4.5 (-11)	1.8 (-10)	3.5 (-17)	1.2 (-16)	1.4 (-16)
	3	3.0 (-5)	6.7 (-8)	8.7 (-8)	1.0 (-13)	2.9 (-13)	2.6 (-13)
	10	3.0 (-5)	1.3 (-7)	1.2 (-7)	1.8 (-10)	6.8 (-10)	8.9 (-10)

REFERENCES

1   T.E. Graedel, W.D. Langer, M.A. Frerking, ApJS **48** 321 (1982)

2   E. Herbst, C.M. Leung, ApJ Suppl. **69** 271 (1989)

3   M. Meixner, A.G.G.M. Tielens, ApJ **405** 216 (1993)

    D.J. Hollenbach, T. Takahashi, A. G. G. M.Tielens. ApJ **377** 192 (1991)

    A. Sternberg, A. Dalgarno, ApJ **338** 197 (1989)

    A.G.G.M. Tielens, D.J. Hollenbach, ApJ **291** 722 (1985)

    A. G. G. M. Tielens, D. J. Hollenbach, ApJ **291** 747 (1985)

4   J. Le Bourlot, G. Pineau des Forêts, E. Roueff, D.R. Flower, A&A **267** 233 (1993)

5   G. Pineau des Forêts, E. Roueff, D.R. Flower, MNRAS **258** 45P (1992)

6   R. Gredel, in *Molecular astrophysics*, p. 305, ed Hartquist T.W., Cambridge University Press (1990)

7   S.S. Prasad, S.P. Tarafdar, ApJ **267** 603 (1983)

8   S.R. Federman, A.E. Glassgold, J. Kwan, ApJ **227** 466 (1979)

9   E. Anders, N. Grevesse, Geochim. Cosmochim. Acta **53** 197 (1989)

10 G. Pineau des Forêts, E. Roueff, P. Schilke, D.R. Flower D. R., MNRAS **262**, 915 (1993)

11 T.J. Millar, E. Herbst, A&A **231** 466 (1990)

12 J. Le Bourlot, G. Pineau des Forêts, E. Roueff, P. Schilke , ApJ Let, in press (1993)

13 E.F. van Dishoeck, in *Molecular astrophysics*, p. 55, ed Hartquist T.W., Cambridge University Press (1990)

14 J. S. Mathis, P.G. Mezger, N. Panagia, A&A **128** 212 (1983)

15. D.R. Flower, J. Le Bourlot, G. Pineau des Forêts, E. Roueff, *A&A* , in press (1993).

THE EVOLUTION OF DUST PARTICLES IN DENSE CLOUDS

Bob van den Hoek
Astronomical Institute 'Anton Pannekoek', University of Amsterdam
Kruislaan 403, NL 1098 SJ Amsterdam, The Netherlands

ABSTRACT. The formation and destruction of (refractory organic mantles on) dust particles in molecular clouds is discussed. We examine the principle processes that determine the nature of the dust recycled to the diffuse interstellar medium by molecular clouds. The impact of grain processing in dense clouds on the dust content of galaxies is considered with respect to the compound structure and composition of the dust present in the Galactic disk and the Magellanic Clouds.

1 Introduction

Interstellar dust particles are thought to consist of (elongated) heavy element cores with a surrounding mantle layer of organic refractory material. Well known examples include solid particles such as SiC, SiO_2, Al_2O_3 and Mg_2SiO_4. Dust grains are observed in the warm diffuse Galactic interstellar medium [**ISM**] where temperatures of $T \gtrsim 100$ K and densities of $n \lesssim 10^2$ cm^{-3} are common. In addition, grains coated with newly accreted ice-mantles of solid molecules such as H_2O, NH_3, CO, H_2CO, CH_3OH are detected inside cold dense molecular clouds [**MC**]. Within these clouds ($T \lesssim 30$ K, $n \gtrsim 10^2$ cm^{-3}) dust accretion on (and processing of) pre-existing core-mantle grains is found to occur. To date, grains covered with icy mantle layers have *not* been detected in the warm diffuse ISM.

As matter is continuously exchanged between MCs and the diffuse ISM component, sequential organic grain mantle formation and processing is expected to occur. Survival of the grain mantles in the diffuse ISM as well as in protostellar environments critically depends on the efficiency of appropriate desorption mechanisms such as photodesorption (e.g. Barlow 1978) and interstellar shocks (e.g. Seab 1987). In this manner, the problem of interstellar mantling of dust particles is closely connected to the stability of the accreted layers in the ISM. We here will restrict ourselves to the dense MC phase and address the fundamental question of the nature of grains formed and/or accreted onto pre-existing dust inside molecular clouds. The importance of grain processing in dense clouds for maintaining the dust content of the diffuse ISM is briefly examined.

2 Signatures of dust in dense clouds

Diffuse cloud grains represent a steady-state average of various solid particles with different histories and ages (e.g. recently ejected by evolved stars; processed by the interstellar radiation field [**ISRF**] or supernova shocks). However, detailed observations indicate that the dust grains associated with dense clouds are generally larger and different in nature compared to the grains present in the diffuse ISM:

- many infrared [IR] absorption features in the range 3–8 μm, due to stretching modes of solid molecules such as H_2O (3.08 and 6.0 μm) and CO (4.67 and 6.85 μm), have been observed towards objects embedded in or located behind dense clouds (see e.g. review Tielens 1989; Whittet & Duley 1991, hereafter WD91). Other identifications probably include solid OCS (4.9 μm), CH_3OH (3.54 μm) and H_2CO (3.48, 5.8, and 6.69 μm; Schutte et al. 1993). These extinction features provide direct physical information about the solid molecules present on grain surfaces. The solid molecules are presumably contained within newly accreted ice mantles as efficient condensation and freezing takes place only at the surfaces of cold grains (Seki & Hasegawa 1983). The strength, position and shape of the molecular absorption bands have been shown to be good diagnostics of the physical state, temperature, mantle composition and compound structure of the absorbing grains (e.g. Van de Bult, Greenberg & Whittet 1985; Tielens et al. 1991; WD91).

- the ratio of total visual to selective extinction $R = A_V/E_{B-V}$ towards a number of dense clouds such as the Ophiuchi region is observed to be as large as $R \approx 4-7$ (e.g. Vrba & Rydgren 1984). These values are considerably higher than those found for most lines of sight in the diffuse ISM where typically $R \approx 3.1$.

- the wavelength of maximum linear polarisation λ_{max}, which is primarily sensitive to the peak of the grain size distribution, is observed to increase towards many objects associated with MCs (e.g. Vrba, Coyne & Tapia 1993, hereafter VCT). Both λ_{max} and R are principle indicators of grain growth. It is interesting to note that the rate at which λ_{max} scales with R in MCs is very similar to that in the diffuse ISM (VCT).

- dense clouds ($n \sim 100$ cm^{-3}) generally exhibit greater heavy element depletions than do low density regions ($n \sim 0.5$ cm^{-3}). For instance, gas-phase abundances of elements such as Mg, P, Mn, and Fe are observed to be much lower in denser environments (e.g. Jenkins 1989; Joseph 1993). Whether or not the depletion factor is even increasing to higher cloud densities is unknown.

- the 10 μm silicate absorption feature appears to be broader in dense clouds than in diffuse clouds (e.g. Roche & Aitken 1985). This may be due to the inclusion of silicates in the strong polar matrices of icy grain mantles as is probably the case for H_2O and CO (e.g. WD91).

- the bulk of the far-IR emission from the Galactic disk as traced by the IRAS $I_\nu(100\ \mu m)$ intensity is radiated by dust grains embedded in MCs (e.g. Boulanger 1989). This long wavelength radiation is thought to originate from large cold ($T \lesssim 15$ K) grains absorbing in the optical & UV and re-emitting in the far-IR.

The grain temperature (or internal energy) generally determines whether or not the dust mantle can absorb the adsorption (and formation) energy of colliding atoms and molecules. For instance, the critical grain mantle temperature for CO frosting ($T \sim 16$ K at $n \sim 10^5$ cm^{-3}; Nakagawa 1980) is significantly lower than the temperature for H_2O ice-formation ($T \sim 90$ K at the same n). Impurities contained within the ice-mantle (e.g. a CO frost in a H_2O dominated matrix) may change this desorption temperature significantly (Schmitt et al. 1989). During the formation of the icy mantles it is possible that other grains, e.g. small amorphous carbon grains stick to the surface and become incorporated into the mantle.

For stars in the Taurus dark cloud region, visual threshold extinctions A_V^{thr} of ~ 3.3 mag for H_2O and $A_V^{thr} \sim 5.9$ mag for CO have been estimated above which molecular frosting onto dust particles presumably occurs (e.g. Whittet et al. 1988; WD91). The thresholds of the ice-deposition are argued to be directed by the intensity of the ambient radiation field at wavelengths grains absorb or molecules can be desorbed, i.e. at red or

IR wavelengths (Williams, Hartquist & Whittet 1992, hereafter WHW). Alternatively, molecular desorption may also be related to the chemical structure of the grain mantle rather than being associated with the local radiation field (Tielens et al. 1991).

The collisional deposition rate of molecules (i.e. set by the gas density and temperature) in relation with the (IR-) photon desorption rate is suggested to be the threshold mechanism (WHW). These may be simply related to a critical temperature below which molecular frosting onto grains becomes efficient. When the molecule deposition-rate is high enough to form a monolayer coverage of H_2O with a residence time longer than the photo-desorption time, a substantial ice-mantle can be formed. High-energy photons are thereafter needed to desorb the icy grain mantle as the desorption energy for solid molecules frozen within an H_2O ice-mantle is substantially higher (~ 0.5 eV) than for molecules frozen directly on grain surfaces without H-bonding (~ 0.1 eV), see further WHW. However, ice coatings on amorhous carbon-grains are not likely to occur since they are so small ($a \sim 0.007 - 0.03$ μm) that temperature spikes produced by single photon absorption or cosmic rays may prevent the ice-mantle formation (Yorke 1985). In addition, the sticking coefficient for molecules on silicate grain surfaces may be considerably higher than for graphite grain surfaces.

The relative intensity of high compared to low energy (i.e. visual to near-IR) photons decreases when moving inwards MCs according to the average interstellar extinction curve [**IEC**] at these wavelengths which in dense MCs may be somewhat different compared to the diffuse ISM (e.g. Massa & Savage 1989). Therefore, ice mantle growth is no longer restricted by the ambient stellar radiation field once an ice-monolayer has been formed although other desorption processes may be effective. For instance, deep inside dark clouds the UV field produced by cosmic ray excitation of H_2 may be a significant source of energy contributing to the desorption of molecules from grain surfaces (Duley et al. 1989). Furthermore, 4.67 μm radiation penetrates more deeply than the 3.1 μm photons according to the IEC. Consequently, the threshold for solid CO is expected to be higher than for H_2 as indeed is observed. Solid CO may freeze on top of (and within) the H_2O ice-mantles and/or directly condense onto the surfaces of silicate grains depending on e.g. the relative molecular abundances.

When moving towards the center of a dark MC from the mantle-free dust layer at the cloud edge, grains freeze probably more and more complex molecules as molecular densities increase and radiation field intensities decrease. Therefore, the dust-to-gas mass-ratio as well as the average grain size is expected to increase rapidly moving inwards a dark cloud. In this manner, the threshold extinction also determines the distance to the edge of a MC at which gas-phase molecules start to be depleted by grains. The optical depths at the CO and H_2O bands both show a linear relation with A_V above their threshold extinctions A_V^{thr} up to $A_V \sim 20$. This optical depth effect is probably caused by both larger grains *and an increasing number of grains covered with ice mantles* as more and more IR-photons are absorbed at higher visual extinction. The silicate 9.7 μm absorption depth, however, does not show a similar correlation with A_V, suggesting that the dust mantles contribute increasingly to A_V as the mantle grows (WD91). It is unclear whether the absorbing silicates are also included efficiently in the grain mantles. It appears that in dense clouds associated with star formation only a small fraction ($\lesssim 5\%$) of CO is in the solid phase (cf. WD91). However, in the Taurus *dark* cloud the depletion of CO is essentially constant at $\sim 30\%$ for $A_V > 5$ mag (as both solid and gas- phase CO column densities tend to increase linearly with A_V). In contrast, the abundance of solid H_2O may be considerably larger than gas-phase H_2O (e.g. Knacke & Larson 1991).

Grains in MCs may grow either through the accretion of molecules or by coagulation of (smaller) grains with distinct effects on the grain size distribution, the shape and chemical structure of grains. For instance, accretion should lead to an increase in the value of A_V/N_H while coagulation should have the opposite effect. Recently, evidence has mounted that coagulation is the dominant process of grain growth in ρ Oph (VCT). Since Bohlin et al. (1978) found the ratio A_V/N_H to be about a factor of 2 smaller for ρ Oph than for the diffuse ISM, grain growth via accretion seems to be largely ruled out as a significant process, at least up to $A_V \approx 5 - 7$. However, thin ice-mantles may be present and even may be required for grain-grain collisions to result in coagulation (cf. Tielens 1989; Chokshi et al. 1993).

3 Dust associated with star forming regions

Observed spectra of protostars (e.g. T Tauri stars, Herbig-Haro objects, FU Orionis stars) are generally observed to peak at wavelengths of $60-100\mu$m and often can be explained by a dust material with an emissivity in the far-IR proportional to λ^{-1} (e.g. Reipurth et al. 1993). Overall dust temperatures (determined at $\lambda \gtrsim 50\mu$m) are typically in the range $T_d \sim 10 - 40$ K. Reliable dust mass estimates, however, are difficult to obtain. This is because of possible clumping of dust, unknown optical properties of the grains present in protostellar regions, and the generally complex dust temperature distribution which also may result in uncertain contributions of cold grains ($T \lesssim 15$ K). Nothwithstanding, protostars may be important contributors to the dust content of the diffuse ISM, either by forming new grains or by returning amounts of pre-existing grains originating from protostellar material, or both.

The formation of a circumstellar disk from the inside-out collapse of a cold dense rotating cloud core is central to the scenario of star formation (Reipurth et al. 1993). The compact disk is likely to contain large amounts of dust-rich circumstellar material covering a wide range in temperature below ($T \sim 2000$ K). After this infall phase all stars pass through an outflow stage as a fundamental part of their formation process (e.g. Shu et al. 1987). Outflows are often highly energetic, variable (eruptive), non-spherically symmetric and exhibit a wide variety of observable phenomena (including e.g. Herbig-Haro objects, water masers, and "cometary" reflection nebulae). Although the explanation for the outflow phase is rather controversial, it may be partly driven by radiation pressure on dust particles. The cycling of material driven by the winds of low-mass stars is believed to be a widespread phenomenon: many dark clouds are thought to be shocked, heated and cooled by the winds of low mass stars (cf. Williams & Hartquist 1991). The cycling is generally assumed to (selectively) remove grain mantles which are frozen on dust grains in the dense phase (Nejad & Williams 1992). Also simultaneous infall and outflow of matter is observed to occur in several young stellar objects.

The energetic outflows of protostellar objects are known to contain large amounts of dust (e.g. Henning & Gürtler 1986; Churchwell et al. 1990; Kenyon & Hartmann 1991). This dust is likely to be formed in the stellar wind as the ionized wind cools down because of expansion. Alternatively, pre-existing dust-particles may be reprocessed within the parent cloud material of the protostar. Since the types of silicates present in the dust shells around very young objects is rather different from that around O-rich giants and supergiants (e.g. Henning & Gürtler 1986), significant dust processing within star forming regions is likely to occur. Intense UV fields associated with star forming regions (up to $\sim 10^5$ times the average ISRF) photo-process grain mantles accreted in the (dense) ISM. For instance, MCs surrounding embedded IR sources generally provide a less favourable environment for the condensation and survival of CO-rich mantles than dark clouds (WD91). As a result of local heating of the grain mantles by the ambient radiation field at the edges of MCs, small solid particles are expected to form (e.g. Duley 1989). Furthermore, metal containing molecules are not detected in MCs but they are in star forming regions, probably due to grain depletion (Ziurys 1993). In the hot cores of MCs like the Orion IRc2 and IRc7 sources ($T \sim$ 150-300 K), evidence is found for molecules such as C_2H_2, OCS and HCN, to be recently evaporated from grain-mantles (Blake et al. 1987; Evans, Lacy & Carr 1991). Moreover, abundances of molecules like NH_3 and H_2CO decrease orders of magnitude from cold dense MCs to the hot cores of star forming clouds. These molecules in turn provide information on the mantle composition of the grains from which they were desorbed.

Narrow IR emission features at 3.3, 3.4, 6.2, 7.7, 8.7, and 11.3 μm (3–13μm) are observed in regions irradiated with substantial UV fluxes, such as HII regions, PNe and reflection nebulae around stars of A0 or earlier (e.g. review Roche 1988). These bands are seen in emission and generally appear together. They probably originate from thermal spiking of small grains and have also been observed both in the diffuse ISM and around C-rich giant stars where dust has condensed very recently. The 7.7 μm feature is found to be correlated with the C/O ratio in PNe for C/O $\gtrsim 1.5$ suggesting that the carriers are carbon-rich. Furthermore, the principle emission bands lie close to wavelengths of C-H and C-C transitions (Duley & Williams 1981). The carbon-rich particles are detected just outside the hot ionized zone as has been shown by the spatial behaviour across the ionization front towards the Orion bar (cf. Roche 1988). This indicates that carbonaceous particles are destructed in high-excitation ionized regions. However, they may form carbon-chains in the cooling plasma at the outer edge of the ionization front, e.g. in the manner suggested by Thaddeus (1993). To distinguish between large molecules and small grains can be difficult at these wavelengths.

The 2175 Å bump of the interstellar extinction curve [**IEC**], indicating non-amorphous graphite grains, is observed to be much weaker in dense quiescent regions or even to be absent (Sorrell 1990). The bump has been observed in the circumstellar dust surrounding recently formed stars (see further Massa & Savage 1989) while being absent in newly condensing dust within the ejecta of evolved stars. This suggests that the graphite grains responsible for the 2175 Å bump are predominantly formed within the MC material around protostars. Consequently, the formation of graphite grains responsible for the 2175 Å bump may be related to the carbon-rich dust accounting for the narrow IR emission features. The formation of large organic molecules (containing heterocyclic ring compounds) from carbon chains also may be associated with the small amorphous grain component observed in the dense ISM although the abundance of these small grains in MCs is unknown (cf. Thaddeus 1993).

It remains unclear whether the small particles responsible for the narrow IR emission features are related to those emitting at 12 and 25 μm. These latter particles seem to be destroyed by strong UV fields, as a lack of 12 μm emission in regions with strong UV has been found (e.g. Cox et al. 1990; Fuente et al. 1992).

4 Dust formation and processing within dense clouds

Grain silicate cores and refractory organic core-mantle grains inside MCs are (partly) evaporated or destructed when irradiated by the intense UV fields from young hot massive stars or processed by multiple shocks caused by supernovae or high-velocity protostellar outflows. On the other hand, mantle accretion and grain formation is also known to occur in MCs. The relative importance of these processes depends on many quantities related to the interactions between dense gas, dust and the ambient radiation field: e.g. the cloud structure and chemical composition, the grain properties, the star formation efficiency and the variation of the radiation field throughout cloud. We briefly point here to shock-processing of grains and dust coagulation processes.

The SiO abundances in the shocked (high-velocity, ~ 50 km s^{-1}) gas associated with molecular outflows in L1148 are larger than those in the quiescent gas by two orders of magnitude (Martin-Pintado, Bachiller & Fuente 1992). Also an enhancement of the SiO/CO ratio at high velocities is observed in molecular outflows which may result from shock destruction of grains. The high temperatures and densities associated with shocks resulting from the interaction of a supernova remnant [**SNR**] with a dense MC may also affect the gas-chemistry and processing of pre-existing dust within the shocked matter. However, from abundance studies of several molecules in the high density component of the embedded SNR IC443 only the abundance of SiO has been found to be much larger than in quiescent clouds like TMC-1 and L134N (van Dishoeck, Jansen & Phillips 1993; hereafter DJP). The IC 443 abundances differ significantly from the only other well studied shock, Orion-KL (cf. DJP), where grain mantle destruction probably has occured. The grains in the preshocked matter of IC443 may have been pre-processed by an intense radiation field or perhaps have been recently shocked by a preceding supernova. An important question is what happens with the observed metal containing molecules when they are recycled to the diffuse ISM (Ziurys 1993).

Grain-grain collisions are thought to play an important role in dust destruction processes as well as for dust coagulation. In absence of significant radiation fields, ice accretion always proceeds fast compared to grain coagulation according to the higher velocity of the ice-molecules (Ossenkopf 1993). Therefore, it may be justified to assume that the main ice-mantle accretion is completed before coagulation becomes effective (depending on the ambient radiation field preventing ice-mantle formation; cf. VCT). The accretion of volatile ice mantles on grains has a promoting influence on the coagulation process (Tielens 1989). In fact, a thin ice-layer may be required to obtain sufficiently high sticking coefficients in grain-grain collisions (Ossenkopf 1993). For small grains ($a \sim 0.005$ μm) coated with a thin layer of ice, coagulation is efficient for grain velocities below $v_{coag} \sim 100$ m s^{-1} (Chokshi et al. 1993).

Small (carbon-rich) particles will be homogeneously embedded within the ice mantles assuming the same sticking efficiency as for the condensing molecules (Preibisch et al. 1993). This results to the accretion of a "dirty ice" mantle containing small inclusions of carbonaceous particles and probably also small silicate grains. Heated carbon-grains in the ISM tend to form aromatic structures and tend to crystallize (increasing the number of graphitic bonds) while cooler grains in denser regions stay amorphous and prefer aliphatic bondings (Sorrell 1990). However, the carbon grains within MCs and the cold parts of protostellar disks are suggested to be highly amorphous (Sorrell 1990). In this manner, small amorphous carbon grains may be reprocessed to larger, stronger bounded, polycyclic carbonaceous grains by strong UV radiation fields.

5 The dust content of the diffuse ISM

Important stellar sources of interstellar grains include AGB stars, OH/IR stars, Novae and Supernovae [SN] and Wolf-Rayet stars. Estimated total dust injection rates from evolved stars into the diffuse ISM of the Galactic disk (hereafter MW) are in the range $\dot{M}^{\text{form}} \sim 0.01-0.08$ $M_\odot$ yr^{-1} (e.g. Gehrz 1989; Mc Kee 1989; van den Hoek & de Jong 1992, hereafter VD92). However, strong observational evidence exists for interstellar shock waves to be the dominant destruction mechanism of interstellar core-mantle grains (e.g. Seab 1987; Mc Kee 1989; Dwek 1989; Jones et al. 1993). Assuming very modest average values for the *net* dust destruction efficiency of SN-shocks ϵ ($\sim$0.1), the current SN-rate ($\sim$0.02 yr^{-1}), the average dust-to-gas mass-ratio in the diffuse ISM D ($\sim$6 10^{-3}) and the shocked ISM mass per SN ($\sim$10^4 $M_\odot$), a minimum estimate for the current dust destruction rate by SNe is $\dot{M}^{\text{destr}} \gtrsim 0.12$ $M_\odot$ yr^{-1}. Therefore, the dust destruction rate almost certainly dominates over the stellar dust injection rate (at least at present in the Galactic diffuse ISM) and may be larger by more than one order of magnitude (see references above). As a consequence, MCs and star forming regions do substantially contribute to the interstellar dust mass unless the dust content of the diffuse ISM decreases at the present epoch (which is unlikely, see e.g. van den Hoek & de Jong 1993). This argument also holds to be true for the Large and Small Magellanic Cloud (hereafter LMC and SMC, respectively).

The empirical IEC ($\xi(\lambda) \equiv A_\lambda/A_B$) appears to be very similar for the MW, LMC, and SMC at IR and visual bands but differs substantially at UV wavelengths (i.e. $\lambda \lesssim 0.25$ μm). The most prominent feature of the Galactic IEC is the 2175 Å bump which is generally accepted to be caused by graphite dust particles or a slightly less well ordered form of carbon (Massa & Savage 1989; Fitzpatrick 1989; Mathis 1990). The 2175 Å bump is weak in the LMC (Koornneef & Code 1981) and almost non-evident in the SMC (Bouchet et al. 1985; Lequeux 1989). The important result that the average IEC of the SMC can be accounted for by incorporating only silicate dust grains while the required contribution of the silicate dust number fraction further decreases from 80% to 50% for the LMC and the MW (Pei 1992), respectively, suggests that *the formation of graphite dust requires specific physical conditions related to the galactic chemical evolution and star formation* [SF] *history*. We note that this result has been obtained while assuming the same grain-size distribution in the MW, LMC and SMC.

The disappearance of the 2175 Å bump for the Clouds seems to be coherent with the under-abundance of C present in gas in the Clouds, however, the abundance of solid C (in a form different from graphite grains) may be relatively high (e.g. Dufour 1984). In addition, small silicate grain cores which are probable candidates for causing the far-UV rise of the IEC (e.g. Pei 1992), are more abundant in the LMC (and even more abundant in the SMC) than in the MW. In case graphite grains are predominantly formed from small aliphatic carbon-grains within star forming MCs, as suggested in Sect. 3, grains present in the diffuse ISM of the LMC and SMC may not yet have been processed significantly by their molecular cloud content. A possible explanation may be that the recycling time between dense MCs and the diffuse ISM for the Magellanic Clouds is much longer than in the Galaxy as is expected because both the LMC and SMC have lower molecular gas mass-fractions (relative to their total ISM mass) compared to the MW, and furthermore, because the Magellanic Clouds are irregular galaxies not experiencing gas compression in spiral arms. Alternatively, the graphite dust destruction rates may be higher in the Magellanic Clouds compared to that in the MW (cf. van den Hoek & de Jong 1993).

While it is clear that the newly accreted (icy) dust layers inside dense clouds are not stable against photo-processing (cf. Greenberg 1983) and SN-shocks (e.g. Seab 1987), the dust content of the diffuse ISM may be primarily directed by the amount of available pre-existing small grain cores (ejected by evolved stars and/or formed within MCs) and their ages (e.g. Lifmann & Clayton 1989). The time that dust grains are embedded in MCs is both determined by the onset of SF after MC formation and the MC-lifetime after the onset of SF. Also the timescales for effective dust processing by SN-shocks and photo-processing by the ISRF are important. Molecular clouds may contribute to the dust in the diffuse ISM *without* altering the grain size distribution by just adding more grains with the same size distribution as initially present within the diffuse ISM from which the MCs accumulated.

6 Summary and Conclusion

The contribution of young stellar objects and their parent MCs to the dust content of the diffuse ISM may take place by: 1) the accretion and processing of mantle layers which partly survive in the diffuse ISM, and/or 2) the formation of new grains (also from pre-existing grain material). Their contribution depends strongly on the average matter exchange-rate between MCs and the diffuse ISM. Since the total dust content of the diffuse ISM comprises grains having distinct evolution histories and ages, part of the dust may have been recycled (and processed) many times between the warm diffuse and the cold dense ISM while the remaining part is more recently ejected by evolved stars.

To which extend protostars and their parent MCs contribute to the dust content of the diffuse ISM is not clear yet. We have suggested that both the grain size distribution in the diffuse ISM and the amount of non-amorphous graphite grains responsible for the 2175 Å bump may provide important clues on the molecular cloud dust contribution. More detailed observations of the regions where MC matter is recycled to the diffuse ISM, i.e. of the protostellar outflow regions and shocked cloud edges, are urgently needed. Furthermore, the grain size distribution at time of ejection by evolved stars is a very important key for a quantitative understanding (timescales and efficiencies) of the grain processing that occurs after stellar ejection into the diffuse ISM.

ACKNOWLEDGEMENTS. VDH acknowledges support from The Netherlands Foundation of Astronomical Research (ASTRON) with financial aid from the Netherlands Organization for Scientific Research (NWO) under grant 782-372-028.

References

Barlow M.J., 1978, MNRAS **183**, p. 417
Blake G.A., Sutton E.C., Masson C.R., Philips T.G., 1987, ApJ **315**, p. 621
Bohlin R.C., Savage B.D., Drake J.F., 1978, ApJ **224**, p. 132
Bouchet P., Lequeux J., Maurice E., Prévot L., Prévot-Burnichon M.L., 1985, ApJ **149**, p. 330
Boulanger F., 1989, in *'The Physics and Chemistry of Interstellar Molecular Clouds'*, G. Winnewisser and J.T. Armstrong (Eds.), Springer Verlag, p. 30
Chokshi A., Tielens A.G.G.M., Hollenbach D., 1993, accepted for publication in ApJ
Churchwell E., Wolfire M.G., Wood D.O.S., 1990, ApJ **354**, p. 247
Cox P., Deharveng L., Leene A., 1990, A&A **230**, p. 181
Dufour R.J., 1984, in IAU Symp. **108** on *'Structure and Evolution of the Magellanic Clouds'*, S. van den Bergh and K.S. de Boer (Eds.), Dordrecht: Kluwer, p. 353
Duley W.W., Williams D.A., 1981, MNRAS **196**, p. 269

Duley W.W., Jones A.P., Whittet D.C.B., Williams D.A., 1989, MNRAS **241**, p. 697
Dwek E., 1988, ApJ **329**, p. 814
Evans N.J., Lacy J.H., Carr J.S., 1991, ApJ **383**, p. 674
Fitzpatrick E.L., Massa D.M., 1986, ApJ **307**, p. 286
Fitzpatrick E., 1989, in IAU Symp. **135** on *'Interstellar Dust'*, L.J. Allemandola and A.G.G.M. Tielens (Eds.), p. 37
Fuente A., Martin-Pintado J., Cernicahro J., Brouillet N, Duvert G., 1992, ApJ **260**, p. 341
Gehrz R., 1989, in IAU Symp. **135** on *'Interstellar Dust'*, L.J. Allemandola and A.G.G.M. Tielens (Eds.), p. 445
Greenberg J.M., 1983, in *'Birth and Infancy of Stars'*, R. Lucas, A. Omont, R. Stora (Eds.), Les Houches (Session XLI), North-Holland Physics Publishing, p. 139
Henning T., Gürtler, 1986, Ap&SS **128**, p. 199
Jenkins E.B., 1989, in IAU Symp. **135** on *'Interstellar Dust'*, Allemandola L.J. and A.G.G.M. Tielens (Eds.), Kluwer Academic, p. 23
Jones A.P., Tielens A.G.G.M., Hollenbach D.J., McKee C.F., 1993, this proceedings
Joseph C.L., 1993, to be published in A&A (Letters)
Kenyon S.J., Hartmann L.W., 1991, ApJ **383**, p. 664
Knacke R.F., Larson H.P., 1991, ApJ **367**, p. 162
Koornneef J., Code A.D., 1981, ApJ **247**, p. 860
Lequeux J., 1989, in *'Recent Developments of Magellanic Cloud Research'*, K.S. de Boer, F. Spite, and G. Stasinska (Eds.), Meudon (Obs. de Paris), p. 113
Lifmann K., Clayton D.D., 1989, ApJ **340**, p. 853
Martin-Pintado J., Bachiller R., Fuente A., 1992, A&A
Massa D., Savage B.D., 1989, in IAU Symp. 135 on *'Interstellar Dust'*, Allemandola L.J. and A.G.G.M. Tielens (Eds.), Kluwer Academic, p. 3
Mathis J.S., 1990, Ann. Rev. A&A **28**, p. 37
McKee C., 1989, in IAU Symp. **135** on *'Interstellar Dust'*, L.J. Allemandola and A.G.G.M. Tielens (Eds.), p. 431
Nakagawa A., 1980, in IAU Symp. **87** on *'Interstellar Molecules'*, B.H. Andrew (Ed.), Dordrecht Reidel, p. 363
Nejad L.A., Williams D.A., 1992, MNRAS **255**, p. 441
Ossenkopf V., 1993, submitted to A&A, main journal
Pei Y.C., 1992, ApJ **395**, p. 130
Preibisch Th., Ossenkopf V., Yorke H.W., Henning Th., 1993, accepted, A&A, main journal
Reipurth B., Chini R., Krügel E., Kreysa E., Sievers A., 1993, accepted, A&A, main journal
Roche P.F., Aitken D.K., 1985, MNRAS **215**, p. 425
Roche P.F., 1988, in 22nd ESLAB Symposium on *'Infrared Spectroscopy in Astronomy'*, ESA SP-290, B.H. Kaldeich (Ed.), p. 79
Schmitt R., Greenberg J.M., Grim R.J.A., 1989, ApJ **340**, L33
Schutte W.A., Geballe T.R., van Dishoeck E.F., Greenberg J.M., 1993, this proceedings
Seab C.G., 1987, in *'Interstellar Processes'*, D.J. Hollenbach & H.A. Thronson, Jr. (Eds.), Kluwer Academic, p. 491
Seki J., Hasegawa H., 1983, Ap&SS **94**, p. 177
Shu F.H., Adams F.C., Lizano S., 1987, Ann. Rev. A&A **25**, p. 23
Sorrell W.H., 1990, MNRAS **243**, p. 570
Thaddeus P., 1993, this proceedings
Tielens A.G.G.M., 1989, in IAU Symp. **135** on *'Interstellar Dust'*, Allemandola L.J. and A.G.G.M. Tielens (Eds.), Kluwer Academic, p. 239
Tielens A.G.G.M., Tokunaga A.T., Geballe T.R., Baas F., 1991, ApJ **381**, p. 181
van de Bult C.E.P.M., Greenberg J.M., Whittet D.C.B., 1985, MNRAS **214**, p. 289
van den Hoek L.B., de Jong T., 1992, in *'The Feedback of Chemical Evolution on the Stellar Content of Galaxies'*, D. Alloin, G. Stasinska (Eds.), Meudon (Observatiore de Paris), p. 289
van den Hoek L.B., de Jong T., 1993, submitted to A&A, main journal
van Dishoeck E.F., Jansen D.J., Philips T.G., 1993, accepted, A&A, main journal (**DJP**)
Vrba F.J., Rydgren A.E., 1984, ApJ **283**, p. 123
Vrba F.J., Coyne G.V., Tapia S., 1993, AstrJ **105**, p. 1011 (**VCT**)
Whittet D.C.B., Duley W.W., 1991, A&AR **2**, p. 167 (**WD91**)
Whittet D.C.B., Bode M.F., Longmore A.J., Adamson A.J., McFadzean A.D., Aitken D.K., Roche P.F., 1988, MNRAS **233**, p. 321
Williams D.A., Hartquist T.W., Whittet D.C.B., 1992, MNRAS **258**, p. 599 (**WHW**)
Williams D.A., Hartquist T.W., 1991, MNRAS **251**, p. 351
Yorke H.W., 1985, in *'Birth and Infancy of Stars'*, R. Lucas, A. Omont, R. Stora (Eds.), Les Houches (Session XLI), North-Holland Physics Publishing, p. 645
Ziurys L.M., 1993, this proceedings

IRON-AROMATICS CHEMISTRY IN INTERSTELLAR CLOUDS

P. Marty[1], G. Serra[1], B. Chaudret[2] and I. Ristorcelli[1]

[1] Centre d'Etude Spatiale des Rayonnements
UPR-CNRS 8002, associée Université Paul Sabatier TOULOUSE
9 Avenue du colonel Roche BP 4346, 31029 Toulouse cedex (FRANCE)

[2] Laboratoire de Chimie de Coordination
UPR-CNRS 8241, associée Université Paul Sabatier TOULOUSE
205 Route de Narbonne, 31077 Toulouse cedex (FRANCE)

ABSTRACT

Serra et al. 1992 [1] proposed that efficient organometallic reactions should occur in the ISM between the PAHs and the iron atoms or ions, leading to the formation of organometallic complexes, the organometallic bonding being of the same order of stability as an aromatic C-H bond. We present in this paper a model taking into account the formation and the photodissociation reactions inside typical cloud configurations. The results show that organometallic complexes are stable inside interstellar clumps protected by an extinction Av *ca.* 4 to 5 magnitudes against the local star light or the interstellar radiation field (ISRF). A consequence of organometallic complexation which would occur, in a very short time scale (a few 10^3 years), is that Fe would be depleted in the dense cores of molecular clouds (R7 in the ρ-Oph molecular complex for example) and could participate to the dust grains formation by PAH accretion.

INTRODUCTION

Aromatics are abundant in the ISM as shown by AROME measurements (Giard et al. 1988 [2]). Following Puget and Léger 1989 [3], we suppose that free PAHs do exist as a fundamental component of the ISM, including 10 to 20% of the cosmic carbon.

Fe, Mg, Si are the most abundant metals ($\sim 4 \cdot 10^{-5}$ n_H) but Fe is the most depleted ($\delta \sim 0.01$). Assuming that the dust grains consist of carbon and silicates, it is difficult to understand why Fe is 15 times more underabundant than Si in the gas phase. Adopting the hypothesis that the grains are in olivine (fayalite & forsterite), wich are the silicates containing the largest proportion of iron, one can deduce:

n_{Fe}(in grains) = n_{Mg}(in grains) = 0.7 n_{Fe}(cosmic)

Regarding simultaneously the problem of the Fe depletion excess and the large abundance of PAHs, the possibility of the coordination of Fe on PAHs has been investigated. Previous works on the subject showed:
i) the Fe/PAHs collisions should be frequent in the ISM (Serra et al. 1992),
ii) the energy of the Fe-PAH bond is of the same order of magnitude than that of a PAH C-H bond (Chaudret et al. 1991 [4]), iii) the activation barrier is very low (see poster Klotz et al [5] in this conference).

HYPOTHESIS AND MODELLING INPUTS

The adopted hypothesis for the organometallic reactions involved in the model are described below.
i) All the free PAHs are coronene $C_{24}H_{12}$ with an abundance relative to H equal to $2.5 \cdot 10^{-6}$. ii) One PAH cannot accept more than three Fe atoms or ions. iii) The activation barrier of the reaction is zero and the resulting bond has an energy in the range 1.5-5 eV. iv) The interacting cross-section of the reaction is always larger than the "Geometrical size" of the atom and the molecule. So, a geometrical cross-section and a probability of reaction equal to 1 has been adopted for each Fe/PAH encounter.

In order to compete with the organometallic reactions, several processes have been considered.
i) A competing phenomenon depleting Fe has been introduced representing the most unfavourable conditions for the coordination. ii) The organometallic complex is supposed to be broken when receiving a photon of an energy higher than 5 eV with a probability equal to 1. Following RRKM computations, the stability of the complex increases quickly with the number of degrees of freedom (i.e. the number of atoms included in the molecule). So, this value of 5 eV is probably a lower limit.

MODELING PROCEDURE

The model takes into account the reactions of organometallic bonds creation and destruction and the competing process involving Fe.

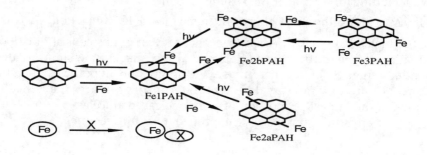

In order to simulate the evolution of the PAHs coordination state, it is necessary to solve numerically the non linear system in which the abundance variation of the species i can be described as follows:

$$\frac{dn\ i}{dt} = \sum_j (-\frac{dL\ i,\ j}{dt}) + \frac{dPi}{dt}$$

where $dL_{i,j}/dt$ is the loss rate caused by the reaction of the species i and j leading to the desappearance of i, and dPi/dt is the production rate caused by the reaction involving the species j and k leading to the formation of the species i.

Despite the time dependance of the system, the convergence time scale is small enough in the molecular clouds ($\tau < 10^5$ years) to only focuse on the stationary solution.

APPLICATION TO TYPICAL CLOUDS

Two clouds, supposed to be constituted by parallel layers, have been considered as representative samples of the ISM; the photon flux is coming from outside (Av=0) and a position inside the cloud corresponds to a visual extinction value Av.

The medium # 1 is a barely dense (100 H cm^{-3}) cloud excited by the ISRF. It is representative of a cloud halo.

The medium # 2, excited by a B8 type star situated at 1pc from the cloud, has a density of 1500 H cm^{-3}. This configuration is comparable to the molecular cloud associated to NGC 1999.

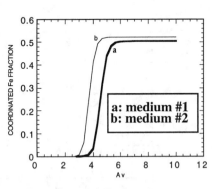
Fig. 1: Coordinated Fe fraction.

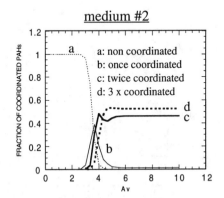
Fig. 2: Fraction of the different species relative to the initial PAHs abundance.

The coordination reactions become very efficient (see figure1) when Av is larger than a critical value (Av_{crit}). This Av_{crit} value corresponds to the maximum of the first metallisation state appearance (see figure 2). When Av is large enough, the PAHs are almost fully metallised. In consequence, the coordination reactions affect most of the PAHs from $Av=Av_{crit}$ with $Av_{crit} \sim 4$ to 5 magnitudes. This result implies many consequences for the physico-chemistry in the ISM.

APPLICATION TO A REAL CLOUD: R7 IN THE ρ-OPH MOLECULAR COMPLEX

The code was applied to R7, a 30 M_{sol} clump in the ρ-Oph molecular complex. This choice was supported by the good knowledge of the physical parameters of the cloud. It is a nearby cloud (125 pc) with a diameter of about 1.1 pc. Its spherical geometry allows the density profile deduction (Bernard et al. [6]). ρ-Oph A, a B2 type star located at 4 pc of the cloud and the ISRF are both contributing to the cloud excitation.

In order to estimate the coordination observability, the density of the excited molecules has been calculated everywhere inside the cloud for each species i. The column density of the excited non coordinated PAHs and of the coordinated PAHs have been computed along the projected star-to-cloud center direction graduated in degrees (0° corresponds to the cloud center).

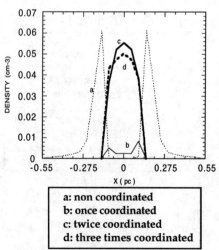

a: non coordinated
b: once coordinated
c: twice coordinated
d: three times coordinated

Fig. 3: density profiles

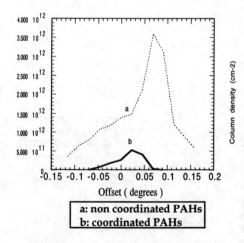

a: non coordinated PAHs
b: coordinated PAHs

Fig. 4: Column densities along the apparent star-to-cloud axis

The figure 3 shows the density profiles of the different molecules along the direction star-to-cloud center x. Organometallic complexation is an efficient process in the cloud dense core where there are few destructive photons. In this region of the cloud, most if not all the PAHs are coordinated. The figure 4 shows that the column density has its maximum at an offset of 4'12" for the non coordinated PAHs and 1'12" for the coordinated molecules. The 3' shift between the two brightness peaks lead to the existence of lines of sight where 30 % of the molecules seen (i.e. excited) are coordinated. This configuration allows to plane future observations, by scanning the cloud in a spectral feature characteristic of the organometallic bonding.

CONCLUSION

Between Fe and aromatic molecules, the processes leading to the formation of organometallic complexes are favoured by three effects:
i) the relatively high value of the Fe-arene bonding energy; ii) the increased colliding cross-section value per C atom of carbonaceous material when the structure is a plane aromatic molecule (free PAH or not) with respect to other forms of carbon; iii) the estimated low value for the barrier of activation of the Fe/PAH reaction, making each encounter an efficient way to produce an organometallic complex.

But such complexes must be dissociated by interactions with sufficient high energy UV photons (see posters Boissel et al. [7] in this conference). The presented results prove that organometallic complexes can exist and survive inside dense clouds, when an extinction of about Av = 4 to 5 magnitudes protects the medium against the UV radiation originated either by a young massive star or by the ISRF.

The coordination of transition metals and PAHs should have many consequences on the physico-chemistry of the ISM. The coordination of Fe on PAHs could be at the origin at least of a part of: i) the strong depletion of Fe, ii) the observed IR color ratio variations. A possible contribution to the 220 nm bump (the d-π^* transition of Fe-arene complexes, located around $\lambda \approx 220$ nm) has not to be excluded.

Furthermore, the organometallic bond properties imply the existence of new chemical processes on the complexes or grain surfaces. Schröder et al. [8] have produced benzene from three acetylene molecules in the coordination sphere of an Fe ion in the gas phase. So the organometallic chemistry could contribute to the PAH formation (Chaudret et al. 1991) by catalysing reactions involving such simple molecules as acetylene. Another consequence of this property is the possibility to participate to the carbonated matter accretion leading to the formation inside the clouds of interstellar grains (or aggregates), depleting Fe and the PAHs inside the dense cores of the molecular clouds. Works on these items are now in progress.

REFERENCES

[1] Serra G., Chaudret B., Saillard Y., Le Beuze A., Rabaa H. Ristorcelli I., Klotz A., 1992, A & A, 260, 489

[2] Giard M., Pajot F., Lamarre J.M., Serra G., Caux E., Gispert R., Léger A., Rouan D., 1988, A & A, 201, L1

[3] Puget J.L., Léger A., 1989, Annu. Rev.Astron.Astrophys., 27, 161

[4] Chaudret B., Le Beuze A., Rabaâ H., Saillard J.Y, Serra G., 1991, New.J.Chem., 1991, 15, 791

[5] Klotz A. et al., poster 47 in this conference

[6] Bernard J.Ph., Boulanger F., Désert X., Puget J.L., 1992, A & A, to be published

[7] Boissel P. et al., posters 14 and 35 [f] in this conference

[8] Schröder D., Sülzle D., Hrusák J., Böhme D.K., Schwarz H., Int. J. Mass Spectrom. Ion Processes, 110 (1991) 145

MODELS OF MOLECULAR PROCESSES IN THE LOW MASS STAR-FORMING REGION B335

J.M.C. Rawlings
Department of Mathematics, UMIST, Manchester, UK

N.J. Evans
Department of Astronomy, University of Texas at Austin, USA

S. Zhou
University of Illinois, USA

ABSTRACT

We have modelled chemical abundances and line profiles in the Bok globule B335. The chemical characteristics of this star-forming core are largely determined by gas-grain interaction in the inflowing material. As with L1498, we aim to identify the molecular species which have broad velocity distributions. The resulting line profile broadening for these species can be used to directly diagnose the characteristics of star-formation. By comparing high resolution observational data with our model it should be possible to determine the evolutionary status of B335 and to establish the role that surface chemistry plays in protostellar clouds.

INTRODUCTION AND BACKGROUND

Nearby, low-mass dense cores have simple geometry and kinematics with little turbulence. They are thus very good laboratories for studying the effects of gas-grain interaction in star-forming regions. There is already a wealth of observational information on these objects, although mainly restricted to common tracers such as NH_3. Particularly important observational studies include Myers & Benson [1] (L1498, and other cores) and Menten et al.[2] (B335). More recently, Zhou et al. [3] observed H_2CO and CS in the Bok globule, B335, and found strong evidence for the inside-out (Shu[4]) type of collapse. Indeed, H_2CO and CS line profiles agree very well with the observed lines both in terms of strength and profile shape – providing strong support for the "inside-out" collapse model.

To model B335 we have extended earlier models that were specific to the dark core L1498 [5]. These have the advantage of predicting both column densities and line profiles for the species and transitions of interest. The essential chemical characteristics of infall regions arise from differential depletion effects (on to the surface of grains). Thus, while some species (such as NH_3) freeze out, others (eg. N_2H^+, HCO^+, H_2S) demonstrate enhancements as collapse and depletion ensues. As an example of how this behaviour might arise, consider the primary loss channels for HCO^+. In a water-rich environment the reaction:

$$HCO^+ + H_2O \longrightarrow H_3O^+ + CO \qquad (1)$$

dominates over the (slower) dielectronic recombination of HCO^+. As depletion ensues, the H_2O abundance falls and the loss channel is suppressed, resulting in an enhancement of the HCO^+. Because the abundances of these species peak at some radius where the infall velocity is non-zero, the line profiles are expected to exhibit noticeable broadening as compared to the situation when there is no chemical variation in the flow.

THE MODEL OF B335

B335 is a low mass core that shows strong evidence for inflow associated with protostellar collapse. It is apparently at a more advanced stage of evolution than L1498 in that it has an embedded protostar and is almost certainly undergoing collapse. B335 is probably the best example to date of the inside-out type of collapse first described by Shu [4] and is an ideal test object for the chemical models of Rawlings et al. [5]. Parameters for the B335 core are given in table 1. The density and temperature distributions, as well as the velocity field, are known for B335 so that theoretical abundances and line profiles can be accurately related to the observations.

Table 1: Basic parameters for the B335 core

Temperature (Isothermal region)	13K
Turbulent velocity	0.12 kms^{-1}
Protostellar mass	0.42$M_\odot$
Effective sound speed	0.23 kms^{-1}
Collapse age	1.5×10^5 years
Infall radius	0.036 pc (30")

In the determination of the infall radius, it has been assumed that that the H_2CO abundance is constant throughout the core. There is therefore considerable uncertainty in the collapse age and infall radius.

In order to make it applicable to B335, the model of L1498 has been modified and updated with a full revision of the sulphur chemistry. A one point global chemistry is modelled in order to establish the chemical initial conditions prior to protostellar collapse. Once collapse has started a 1.5D multi-point time-dependent model is used.

The models of the molecular processes in L1498 and B335 differ from previous studies in that they are true multi-point time-dependent models which take full account of the changing physical conditions at all points in the inflow and at all times. The evolution of the (outer parts of the) core is described by three phases:

(i) The initial collapse of the cloud into a hydrostatic pressure-balanced dense core. This is assumed to follow a homologous free-fall collapse with a magnetic retardation factor included as a test for chemical sensitivity.

(ii) The hydrostatic core (with $n \propto r^{-2}$) remains in quiescent equilibrium until protostellar collapse is initiated.

(iii) Collapse starts spontaneously at the centre of the cloud and propagates outwards at the local sound speed according to the "inside-out" self-similar solutions of Shu[4]. Inside the "Collapse Expansion Wave" (CEW), matter approaches free-fall (with $n \propto r^{-3/2}$). Outside the CEW the material remains static.

There is continual chemical evolution throughout each of these phases. An important aspect of the model is the chemical composition of the pre-collapse cloud. Observed line profiles (of NH$_3$ in particular) in L1498 and other low mass cores indicate that, unlike the cloud complexes from which they came, the cores are regions of exceptional quiescence. However, the chemical abundances are somewhat larger than can be explained by simple, static, gas-phase chemistries. Our physical model assumes that low mass stars form in periodic shock-regulated cloud complexes that are both dynamically active and chemically rich[6]. A single cycle in this scenario consists of a period of cold chemical evolution (lasting about a million years) and accretion of gas phase species on to the surface of grains. This is followed by clump formation and then erosion into the wind from the newly formed star. The mass-loaded wind is then arrested in a weak reverse shock and a period of "hot" post-shock chemistry ensues. This cycle is likely to be repeated many times in the lifetime of a low-mass star-forming cloud and results in a very rich chemistry. By comparison, the early stages of the protostellar collapse are exceptionally quiescent.

CONSTRAINTS AND FREE PARAMETERS

In the case of B335, the main chemical constraints are (from [2,3,7]):

- $X(CS) \approx 3.6 \times 10^{-9} (\pm 30\%)$
- $X(H_2CO) \approx 2.8 \times 10^{-9}$
- $X(NH_3) \lesssim 2 \times 10^{-9}$ in the high density centrally condensed region of the core (from observations of the (2,2) line)

There are of course, many free parameters and uncertainties in the model. The most important free parameters relating to the B335 core are:

- the sulphur depletion from the gas phase
- the length of time that the core spends in the hydrostatic pressure-balanced state
- the age of the collapse
- the extinction (A_v) of the region in which the core is embedded

There are also free parameters relating to the accretion characteristics, such as the grain size distribution and the sticking coefficients for the various species. These, however, have a greater effect on the line profiles in the collapse phase of the evolution and are not significant in constraining the pre-collapse abundances. The observed CS and H$_2$CO abundances are somewhat lower than are predicted by the dynamical model of the inter-core medium. Moreover, there are significant uncertainties in the sulphur chemistry. As a result there are several combinations of the free parameters which satisfy the chemical and physical constraints.

LIMITATIONS OF THE MODEL

There are several limitations of the model which will be addressed in the near future. Most importantly, as can be seen from the discussion above, the model is strongly tied to a specific dynamical scenario. As we have shown, there is strong evidence for "inside-out" collapse, but this is by no means definite. However, the Shu[4] collapse model provides the clearest description available of the dynamics of the early, isothermal, phase of the collapse, and since the observational evidence supports isothermality in most collapse cores, it is probably a very reasonable approximation to the true situation (for early times at least).

The cyclic model used to establish the chemical initial conditions is not fully consistent with the model used to describe the protostellar collapse. The primary cause of this (as pointed out in Rawlings[8]) is the omission of the external and cosmic-ray induced radiation fields in the earlier cyclic models. Alternative pre-collapse dynamical and chemical scenarios are also being investigated.

As yet, the model does not account for any of the desorption mechanisms that have been proposed for releasing material from the surface of grains. Since many of the chemical characteristics of the regions are linked to the behaviour of the gas-phase abundances of important volatile species such as CO and H_2O, this needs to be given careful attention.

Surface chemistry is treated in a very approximate fashion in the pre-collapse cyclic model. At present it is assumed that total saturation takes place on the surface of the grains: All C, N and O atoms and unsaturated radicals are converted to CH_4, NH_3 and H_2O respectively.

RESULTS AND CONCLUSIONS

In B335 the chemical constraints are somewhat less stringent than for the case of L1498. In particular, the CS and H_2CO abundances do not strongly constrain the length of time that the core spends in the hydrostatic pressure-balanced state. In L1498 the high CH and NH_3 abundances imply that the protostellar collapse must have started very soon after the formation of the singular isothermal sphere. In B335 it is found that there are several combinations of the free parameters which satisfy the chemical and physical constraints. Our investigations are still at an early stage so that we have so far only considered two such combinations; one where $A_v=5.5$ and the collapse starts immediately after the formation of the hydrostatic core, another where $A_v=7.0$ and there is a pause of some 1/2 million years before collapse starts. Both models predict the same abundances for CS, H_2CO and NH_3 (to be consistent with the observations), but the larger depletion in the second model results in major species such as N_2H^+, HCO^+, CN and H_2CO having enhanced abundances at high infall velocities. Certain (optically thin) transitions of these molecules will therefore show anomolously wide line profiles, indicative of protostellar inflow.

By observing the appropriate lines (eg. $H^{13}CO^+$ J=1-0, HCO^+ J=3-2,1-0 and N_2H^+ J=1-0) it should thus be possible to constrain physical parameters such as the evolutionary status and the mass of the protostellar core, at the same time yielding information on the efficiency of the depletion and desorption processes.

Clearly, as more accurate abundances and line profiles are obtained, for a greater variety of species, the free parameters will be more rigorously constrained. At the same time the model is being further generalised so as to include in greater detail previously simplified or neglected processes (such as the role of the external radiation field, surface reactions, and grain mantle desorption mechanisms). Our aim is to develop fully flexible and observationally/theoretically self-consistent chemical models of B335 and other low mass star-forming regions.

REFERENCES

1. Myers, P.C. and Benson, P.J. Astrophys. J., **266**, 309 (1983)
2. Menten, K.M., Walmsley, C.M., Krugel, E. and Ungerechts, H., Astron. & Astrophys. **137**, 108 (1984)
3. Zhou, S., Evans, N.J., Kompe, C., and Walmsley, C.M., Astrophys. J. **404**, 232 (1993)
4. Shu, F.H., Astrophys. J. **214**, 488 (1977)
5. Rawlings, J.M.C., Hartquist, T.W., Menten, K.M. and Williams, D.A., Mon. Not. of the Roy. Ast. Soc. **255**, 471 (1992)
6. Charnley, S.B., Dyson, J.E., Hartquist, T.W. and Williams, D.A., Mon. Not. of the Roy. Ast. Soc. **231**, 269 (1988)
7. Zhou, S., Evans, N.J., Butner, H.M., Kutner, M.L., Leung, C.M., and Mundy, L.G., Astrophys. J. **363**, 168 (1990)
8. Rawlings, J.M.C., Dust and Chemistry in Astronomy (IOP publishing Ltd., 1990)

THE MID INFRARED CONTINUUM IN THE PLANETARY NEBULA NGC 7027: EVIDENCE FOR THE PRESENCE OF AMORPHOUS CARBON GRAINS.

A. ZAVAGNO, J.P BALUTEAU
OBSERVATOIRE DE MARSEILLE
2 PLACE LE VERRIER
13248 MARSEILLE CEDEX 4, FRANCE.

ABSTRACT

The importance of the "continuum" emission has already been pointed out in several sources of high excitation (such as planetary nebulae) with the quantitative study of the emission bands at 7.7 and 11.3 µm with the IRAS/LRS spectra (Zavagno et al. [1]).
We present here a model of the observed continuum emission between 2 µm and 1 mm in the source NGC7027. We find evidences for the existence of a large emission band in the range 5-15 µm. This band is characteristic of amorphous carbon grains. Using the data given by Rouleau and Martin [2], we have been able to derive the extinction coefficient for three samples of such grains. With our model, we are able to reproduce the observed emission bump using amorphous carbon grains. It is then possible, for the first time, to know the exact contribution of the emission bands with respect to the continuum and, using the derived optical depth, to estimate the mass of carbon contained in amorphous carbon grains.

INTRODUCTION

The aim of this study is to fit the mid-infrared observed continuum in the planetary nebula NGC7027. The article relative to this work is in preparation.
In Section 1, we present the data analysis and the model. Section 2 presents the results of this study. Conclusions are given in Section 3.

DATA ANALYSIS

The spectrum of NGC7027 from 2 um to 1 mm is presented on Figure 1.

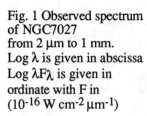
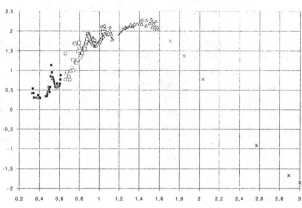

Fig. 1 Observed spectrum of NGC7027 from 2 µm to 1 mm. Log λ is given in abscissa Log λFλ is given in ordinate with F in (10^{-16} W cm^{-2} µm^{-1})

The spectrum has been arbitrarily divided in two parts: the long wavelength range - from 14 μm to 1 mm - and the short wavelength range - from 2 μm to 14 μm. The model is composed of three components: free-free emission which dominates the spectrum at long (near 1 mm) and short (near 2 μm) wavelength, grey body emission for the continuum emission of warm and cold dust grains which dominates respectively the short and the long wavelength range. Free-free emission has been extincted using a standard extinction law in the infrared (Mathis [3]). Results of this model are given hereafter.

RESULTS

Figure 2 presents the result of the fit using a black body emission at 115K and an emissivity law proportionnal to $\lambda^{-0.7}$ characteristic of amorphous carbon grains in this part of the spectrum (Hoare et al. [4] - Bussoletti et al. [5]). Recently, Midlemass [6] and Hoare et al. [4] have shown that amorphous carbon grains rather than graphite can fit the long wavelength part of the spectrum in planetary nebulae.

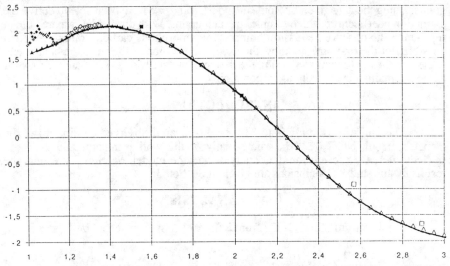

Fig. 2 The long wavelength part of NGC7027 spectrum. The fit is superimposed (full line). Black body emission at 115K, an emissivity law proportionnal to $\lambda^{-0.7}$ and free-free emission are considered here.

We have tried to fit the observed spectrum in the short wavelength range using a continuum type emission (i.e. black body emission + emissivity law + free-free emission). We have not been able to reproduce the observed spectrum by this way. A large band appears in the range 5-15 μm, under the emission bands seen at 6.2, 7.7, 8.6 and 11.3 μm. This band is characteristic of amorphous carbon grains (Koike et al. [7]). Using n and k values obtained by Rouleau and Martin [2], we have calculated extinction coefficient for three samples of amorphous carbon grains. Figure 3 presents the spectral dependance of extinction coefficients for three samples of amorphous carbon grains, calculated using n and k values with the Mie approximation. Note the bump in the 5-15 μm range.

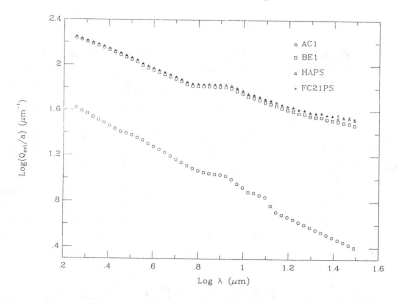

Fig. 3 Extinction coefficients for three sample of amorphous carbon grains.

We have fitted the observed spectrum in the short wavelength range using the AC1 sample given in Rouleau & Martin [2]. Figure 4 presents the result of this fit for T=300K. The two other samples are also correct for the fitting.

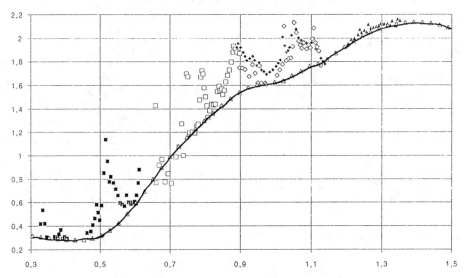

Fig. 4 The short wavelength part of NGC7027 spectrum. The fit is superimposed (full line). Black body emission at 300K, an emissivity law for amorphous carbon grains and free-free emission are considered here.

CONCLUSIONS

The study of the mid-infrared continuum in the planetary nebula NGC7027 gives new evidence for the presence of amorphous carbon grains.

The study of the mid-infrared continuum in NGC7027:
reveals the existence of a large band in the 5-15 µm range, a band which is characteristic of amorphous carbon grains.
The long wavelength part of the spectrum can be fitted using amorphous carbon grains' emissivity law.
The short wavelength part of the spectrum cannot be fitted by a continuum type emission.
Amorphous carbon grains emissivity is succesfully used to fit the short wavelength part of the spectrum.
Using optical properties of these grains, we will be able to extimate an abundance of carbon locked in amorphous carbon grains.

REFERENCES

1. A. Zavagno, P. Cox, J.P. Baluteau, J.P., A&A 259, 241 (1992)
2. F. Rouleau & P.G. Martin, ApJ 377, 526 (1991)
3. J.S. Mathis, ARA&A 28, 57 (1990)
4. M.G. Hoare, P.F. Roche, E.S.Clegg, MNRAS 258, 257 (1992)
5. E. Bussoletti, L. Colangeli, A. Borghesi, V. Orofino, A&ASS 70, 257 (1987)
6. D. Middlemass, MNRAS 244, 294 (1990)
7. C. Koike, H. Hasegawa and A. Manabe, As. Spac. Scien. 67, 495 (1980)

ROTATIONAL EXCITATION OF INTERSTELLAR WATER VAPOR BY NEAR INFRARED PAH PHOTONS

A. D'Heeger, M. Giard
CESR-CNRS, BP 4349, F 31029 Toulouse-cedex, France

ABSTRACT

Near infrared photons emitted by interstellar PAHs are able to pump the vibrational transitions of interstellar water vapor, and populate the rotational levels of the molecule by subsequent ro-vibrational de-excitation. We have solved the equations of statistical equilibrium and radiative transfer for water vapor in warm molecular clouds photodissociation regions. Our results show that the near infrared pumping of water vapor is very efficient and allows the excitation of many infrared and submillimeter transitions. If water vapor is an abundant constituent in such regions, those transitions should be detectable with the future airborne and satellite telescopes.

INTRODUCTION

On the simple basis of cosmic abundances, water vapor is expected to be an abundant molecular species in the interstellar medium. This is confirmed by the observation of ices in cold molecular clouds where about 30% of the cosmic oxygen can be in the form of solid H_2O on grains (Knacke et al. 1982 [1]). Basically, the knowledge of the gas phase abundance of H_2O is needed in order to test and constrain interstellar chemical models. But the interest in observing water vapor is increased by the role that it could play in the cooling of dense interstellar clouds (Goldsmith and Langer 1978 [2]). However, direct observations of non masing lines of interstellar water vapor are highly damped by atmospheric lines and only tentative detections have been reported up to now (Knacke et al. 1991 [3], Encrenaz et al. 1993 [4]). This is why the measurement of some low-lying rotational transitions of water vapor remains a primary objective in some future submillimeter balloon borne and space missions: PRONAOS, SWAS and ODIN.

Modelling H_2O masers, de Jong (1973) [5] has demonstrated that radiative transfer is a key point in the computation of the rotational excitation of interstellar water vapor. This is because the critical density of the molecule, $n_{H2} = 10^9$ cm^{-3}, is too high to allow thermodynamic equilibrium of rotational levels in the usual conditions of the interstellar medium. Takahashi et al. (1983 [6] and 1985 [7]) have shown how infrared dust photons are able to pump the H_2O rotational levels in warm molecular clouds. We known now that the far infrared emission of dust is always associated to large amounts of near infrared photons emitted by very small grains and large molecules coexistent with interstellar dust (Désert at al. 1990 [8]). We investigate here how the near infrared photons are able to pump the lowest vibrational states of H_2O and populate the rotational levels by subsequent ro-vibrational deexcitation. In particular, the lowest vibration rotation band of ortho H_2O (ν_2-ground) overlaps with the dust emission band at 6.2 μm attributed to large Polycyclic Aromatic Molecules (PAHs, Léger and Puget 1984 [9]), favoring the pumping mechanism. We have chosen to model conditions typical of photodissociation regions in star forming complexes for two reasons. 1/ The radiation field in such regions is high enough to allow significant vibrational pumping and 2/ Evaporation of grain mantle ices should lead to a significant water vapor abundance, making these regions favorable places for future observations.

THE MODEL

We have calculated the rotational population of ortho H_2O in the fundamental and the first excited, ν_2, vibrational states. The fraction of molecule in each energy level, x_i, is found by assuming statistical equilibrium among all levels:

$$\frac{dx_i}{dt} = \sum_{j \neq i}\left(A_{ji} + B_{ji}J_{ij} + C_{ji}n_{H_2}\right)x_j - x_i\sum_{j \neq i}\left(A_{ij} + B_{ij}J_{ij} + C_{ij}n_{H_2}\right) = 0 \quad (1)$$

where A_{ij} and B_{ij} are respectively the spontaneous and induced radiative transition rates, and C_{ij} is the collision rate. J_{ij}, the mean line intensity, is a function of the x_i's through the equation of line transfer. We have generalized the result of Takahashi et al. (1983) [6] to non-blackbody-like dust spectra, and we obtain in the escape probability approximation:

$$J_{ij} = (1-\beta)S + \beta F_d \quad (2)$$

where, β is the escape probability, S the source function and F_d the dust flux.
The source function is defined as:

$$S = \frac{2h\nu^3}{c^2}\frac{1}{g_i x_j / g_j x_i - 1} \quad (3)$$

Under the assumption of large velocity gradient we have:

$$\beta = \frac{1-\exp-\tau}{\tau} \quad (4)$$

where the line optical depth reads:

$$\tau = (x_i B_{ij} - x_j B_{ji})\frac{hc}{4\pi}\frac{N_{H_2}}{\Delta v} \quad (5)$$

and $\frac{N_{H_2}}{\Delta v}$ is the H₂ column density per unit velocity.

Using Eqs. (2), (3), (4) and (5), Eq. (1) becomes a nonlinear system of unknown x_i's. It is solved using standard numerical techniques.

Concerning the dust flux, we neglect geometrical effects and assume that it is proportional to the hydrogen column density:

$$F_d = 2N_{H_2} FIR_H \qquad (6)$$

where FIR_H is the dust emissivity per hydrogen atom. It is taken from the model of Désert et al. (1990) [8] in the case where the dust is heated by the radiation of an O5 star at a distance of 1 pc.

Finally, we take into account the 16 lowest rotational levels in each vibrational state. The collisional rates in the ground state are from Green (1980) [10] and are extended to the excited state as in Deguchi and Q-Rieu (1990) [11]. The transitions frequencies and spontaneous emission rates are from the GEISA data bank (Chedin et al. 1984 [12]).

Table I: H₂O line intensities of strongest lines predicted for photodissociation regions. Powers of ten are within brackets.

| Upper level | | Lower level | | λ | Line intensity | |
| v2 | JK-K+ | v2 | JK-K+ | (μm) | Total flux (Wm⁻²sr⁻¹) | |
					standard model	big grains only
1	110	0	101	6.18	-4.2 [-6]	-9.3 [-9]
1	110	0	221	6.64	2.2 [-6]	4.9 [-9]
0	432	0	303	40.69	-7.7 [-8]	-9.4 [-9]
0	432	0	321	58.69	4.6 [-8]	4.2 [-9]
0	330	0	221	66.43	-7.8 [-8]	-2.0 [-9]
0	321	0	212	75.38	-2.7 [-7]	-1.6 [-7]
0	221	0	110	108.1	1.3 [-7]	-8.5 [-10]
0	414	0	303	113.5	1.7 [-8]	2.2 [-9]
0	330	0	321	136.5	2.9 [-8]	1.6 [-9]
0	303	0	212	174.6	1.2 [-7]	6.1 [-8]
0	212	0	101	179.5	1.6 [-7]	1.4 [-8]
0	221	0	212	180.5	1.3 [-7]	1.6 [-8]
0	321	0	312	257.8	7.8 [-8]	2.8 [-8]
0	312	0	221	260.0	2.6 [-8]	6.0 [-9]
0	312	0	303	273.2	6.2 [-8]	2.2 [-8]
0	110	0	101	538.3	1.0 [-8]	6.8 [-10]
0	414	0	321	788.5	7.0 [-12]	8.3 [-13]

RESULTS

Fig. 1a displays the full infrared and submillimeter spectrum, H_2O lines plus dust continuum, obtained in the case where the dust mixture includes PAH molecules, very small grains and big grains as in the standard model of Désert et al. (1990) [8]. The hydrogen column density used is $N_{H2} = 10^{22}$ cm^{-2}, the hydrogen density $n_{H2} = 10^4$ cm^{-3}, the H_2O abundance $Y = 10^{-5}$ and the line width $\Delta v = 10$ km/s. For comparison, we show in Fig. 1b the spectrum obtained in the case of dust containing only big grains (no near infrared photons). For a quantitative comparison the intensities of the strongest lines

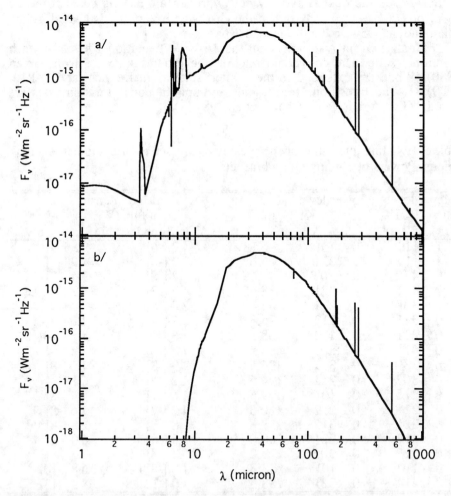

Figure 1: a/ Predicted H_2O lines and dust continuum emitted from a photodissociation region with PAH molecules and very small grains. b/ same model but the dust continuum includes only emission from big grains.

are gathered in Table I. The results show that the intensities of many of the H_2O far infrared and submillimeter lines are increased by a factor of the order of one decade if PAH and very small grains are present. Exploration of the parameter space shows that below 10^5 cm^{-3} the results depend very little on the gas density which is consistent with an excitation dominated by radiative transitions.

For a better understanding of the mechanism we show in Fig. 2 the main transition routes in the H_2O energy diagram. It demonstrates that the population of the rotational ladder in the ground state occurs mainly via the de-excitation of the v_2 110 level to the ground 221 level and subsequent far infrared and submillimeter transitions within the ground state. The v_2 110 level itself is populated by pumping from the ground 101 level with 6.2 μm PAH photons.

CONCLUSIONS

The evaporation of grain mantle ices in photodissociation regions is likely to provide large amounts of water vapor. Our model show that the near infrared photons emitted by large PAHs in the same regions are able to pump H_2O in its first vibrational excited state. This allows significant rotational excitation of the molecule in the ground state by subsequent de-excitation. As a consequence, we predict strong far infrared and submillimeter transitions of water vapor to be emitted from such regions. These lines should be observable with the future balloon and space borne instruments: PRONAOS (380 GHz), ISO (far infrared), SWAS (557 GHz) and ODIN (557 GHz). We also predict that the v_2-ground near infrared ro-vibrational lines should be observed in absorption or emission on the dust continuum. They could be valuable objectives for the near infrared spectrometers of the ISO telescope.

REFERENCES

[1] R.F. Knacke, S. McCorkle, R.C. Puetter, E.F. Erickson, and W. Krätschmer, Ap.J., 260, 141 (1982).
[2] P.F. Goldsmith and W.D. Langer, AP.J., 222, 881 '1978).
[3] R.F. Knacke and H.P. Larson, AP.J., 367, 162 (1991).
[4] P.J. Encrenaz, F. Combes, F. Casoli, M. Gerin, L. Pagani, C. Horellou and C. Gac, A&A, 273, L19 (1993).
[5] T. de Jong, A&A, 26, 297 (1973).
[6] T. Takahashi, D.J. Hollenbach and J. Silk, AP.J., 275, 145 (1983).
[7] T. Takahashi, D.J. Hollenbach and J. Silk, AP.J., 292, 192 (1985).
[8] F.X. Désert, F. Boulanger, and J.L. Puget, A&A, 237, 215 (1990).
[9] A. Léger and J.L. Puget, A&A, 137, L5 (1984).
[10] S. Green, Ap.J. Suppl., 42, 103 (1980).
[11] S. Deguchi, and Nguyen-Q-Rieu, Ap.J., 360, L27 (1990).
[12] A. Chedin, N. Husson, N.A. Scott, I. Gohen-Hallaleh and A. Berroir, *The "GEISA" Data Bank* (1984).

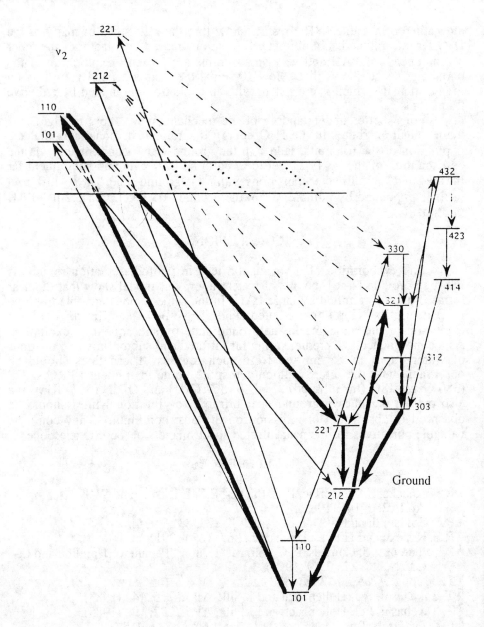

Figure 2: Transition routes in the H$_2$O energy diagram when exposed to PAH near infrared photons. The arrows and their thickness show respectively the sign and magnitude of the total transition rates (radiative and collisional).

PHYSICAL CHEMISTRY OF COMETS : MODELS, UNCERTAINTIES, DATA NEEDS

C. Arpigny
Institut d'Astrophysique, Université de Liège
B-4000 Liège, Belgium

ABSTRACT

Some recent results or discussions pertaining to the physical chemistry of comets are reviewed, with a special interest in those aspects which illustrate the close interaction between laboratory work and astronomical observations or theories. An overview is given of the present state of knowledge on the chemical composition of comets, both from the point of view of molecular abundances and as regards the relative abundances of the elements and of some isotopic species. Error bars are, unfortunately, uncomfortably large in many instances and we point out the various difficulties encountered in the determination and interpretation of production rates or abundances.

Assessing the true meaning of the concentrations measured in the atmosphere, or coma, involves the construction of elaborate cometary physico-chemical models. These models and their uncertainties are only sketched. Rather, the emphasis is laid upon the needs which appear as the most important as far as the relevant fundamental physico-chemical data are concerned: spectroscopic identifications, radiative transition probabilities, photolysis data (rate coefficients, branching ratios, energetics), collisional cross-sections, or rates of various types of chemical reactions which play a rôle in the interpretation of the structure and composition of the coma or in the models constructed to infer the primordial composition of comets. Specific examples are given in each of these areas.

INTRODUCTION

Comets are "minor" bodies of the solar system only on account of the very modest sizes of their nuclei (several 0.1 to several 10 km) and of their small masses (10^{-13} to 10^{-7} times the earth's mass), certainly not by their dimensions when fully developed (a bright comet's atmosphere can be larger than the sun, the length of its tail can reach a substantial fraction of an astronomical unit, the mean sun-earth distance). A widely held opinion considers, at any rate, that they are among the most interesting members of the solar system for different reasons. They offer, of course, a great intrinsic interest, for it would be very gratifying to learn more about the nature and origin of these fascinating objects and to have a better interpretation of their sometimes puzzling and unique behaviour. However, they are even more important owing to the numerous connections that exist between their study and a number of fundamental areas of physics, chemistry and astrophysics, such as low temperature physics and chemistry, spectroscopy, photochemistry, the physical chemistry of surfaces, plasma physics, interplanetary space physics, the study of the past and present of the solar system, of the interstellar medium... Comets have been rightly described as space or natural laboratories where research work very often takes on an interdisciplinary character.

Still, the prime reason for studying comets stems from the cosmogonic significance attached to them. According to current views, comets formed at about the same time as the sun and the planetary system, after the collapse of a dense interstellar molecular cloud (IMC) which gave rise to the solar nebula. It is generally believed that they were born and spent the major part of their lifetime at large distances from the sun, in extremely cold regions. Besides, owing to their small sizes, they did not undergo any major transformations due to internal heat and pressure as the planets did. Thus, comets can be regarded as surviving, particularly well preserved condensations of that cosmic matter from which the solar system originated. A number of different models have been proposed, from those which envisage that at least part of the matter contained in comets was heated during accretion, sublimated and recondensed under physico-chemical conditions pertaining to the primitive solar nebula (PSN)[39,43,83], to more gentle scenarios according to which comet nuclei formed by agglomeration of interstellar grains essentially without any chemical processing[56,140]. Indeed, the key questions we ask today about comets are: just how "pristine" are they, how much alteration did their matter undergo during their life history, and what is their exact relationship with the interstellar matter? There are pieces of evidence suggesting that some, but not much, processing did occur; our knowledge is still too scanty, however, to enable us to choose among the various existing theories, except to say that the truth no doubt lies in between the extremes!

Comets hold some record of their early chemical histories anyhow and one way or the other, they will ultimately provide information on the conditions that prevailed when and where they were born. To take a simple (perhaps oversimplified) example: if a highly volatile substance can be identified as issued from a comet's nucleus, then it can be inferred that the matter present in this nucleus was never, at the comet's birth or later, warmed above the sublimation temperature of that compound at the relevant pressure. The abundances of methane and ammonia relative to water, as well as ratios like CH_4/CO, NH_3/N_2, CO_2/CO, and others, should also yield significant diagnostic tests regarding cometary formation environments when we have learned how to interpret them properly. The methane vs carbon monoxide and ammonia vs molecular nitrogen mixing ratios estimated in Halley's comet would appear to lead to contradictory conclusions in the sense that CH_4/CO is < 1 and $NH_3/N_2 > 1$, whereas models usually predict that the reduced or oxidized forms of carbon and nitrogen dominate together. However, these measurements refer to a single comet and are rather uncertain. Clearly, more accurate data on the chemical composition of many more comets than available at present will be required before we can really constrain the models to get a better understanding of the physical and chemical conditions and processes in the PSN and/or in the parent IMC.

More generally, comets can be expected to provide some clues concerning the conditions under which stars, and planetary systems, are formed from interstellar clouds. They may even have played a rôle in the chemical evolution of our Galaxy and it has also been speculated that the volatile elements H, C, N, O, essential to the advent of life may have been brought onto the earth (in the form of H_2O, of prebiotic molecules,...?) by collisions with comets at a late stage in the formation of the planets.

Everyone will agree, then, that these "relics of the past" have a great deal to say about a variety of important problems. It simply remains for us to teach ourselves how to decipher their esoteric language!

In the following sections we shall first summarize what is known today about the composition of comets and then indicate how we go about determining the various mixing ratios. Owing to the very complex nature of these objects, very elaborate modeling is necessary, if we are to interpret correctly the derived abundances. Models will be mentioned briefly, but we shall be more concerned with the important needs they imply for fundamental physico-chemical data of different kinds. Much more should be said about the physical chemistry of comets than is possible here and I therefore refer the reader to some excellent reviews dealing directly or indirectly with this subject and appearing in recent books[57,64,88,103].

OBSERVED SPECIES. ABUNDANCES

In order to set the stage, it may be useful to first recall a few basic facts about the structure of comets. As a comet approaches the sun, its ices sublimate into an expanding gas cloud that gradually forms the coma around the nucleus. The latter also releases solid particles containing volatile and non volatile ("dusty") substances. The smallest of these grains are swept along by the gas, in such a way that the "bulk fluid" reaches an outflow velocity typically of the order of 1 km s^{-1}. Measured production rates show that a bright (Halley type) comet at one astronomical unit (AU) from the sun sheds $Q \sim 3 \times 10^{29}$ molecules s^{-1} and a comparable mass, 10^4 kg, of dust per sec. In this case, the total gas density, n, near the surface of the nucleus (radius R_n), is $\sim 3 \times 10^{12}$ cm^{-3}, while it has dropped down to 10^6 cm^{-3} at a distance $R_c = 5 \times 10^3$ km from the nucleus. This is the approximate radius of the "collision sphere": within this inner zone, the densities are still high enough for particle-particle interactions to be significant and to influence the flow, in particular; there too, some chemical reactions may compete to some extent with the photolytic processes, which otherwise predominate. Although R_c is an ill-defined quantity which, in fact, differs for different collisional processes according to their cross-sections, this notion can still be useful for qualitative discussions or comparison purposes. Beyond R_c a transition occurs to free molecular flow, which characterizes the outer coma or "exosphere". The n's and R_c vary in direct proportion to the outgassing rate Q, which can be

50 times lower for a fainter comet like Encke's comet e.g. Due to the interplay of cooling and heating processes, the temperature drops from about 200 K at the nucleus to a minimum of $\sim$ 20 K near 150 km ($\sim$ 30 R_n) in the Halley type comet considered above, and then rises outward to reach $\sim$ 200 K again at $R \sim R_c$.

According to our standard concept, stable molecules called "parent molecules" are issued directly from the nucleus. They may be excited to their lowest rotational levels and take part in some reactions while in the inner coma, depending upon their lifetime and particular properties, but they are eventually photodissociated or photoionized by the solar UV radiation, to give rise to "daughter products", such as the radicals, atoms or ions observed mainly in the ultraviolet and visual spectra. The parent species themselves, essentially tri- or polyatomic molecules (CO may be an exception), have no visible spectra and no sharp UV features: they can only be detected in the Radio or IR wavelength regions.

Table I. Atoms and molecules observed in cometary spectra

Species	Neutrals	Spectral region	Ions	Spectral region
Atoms	H, O	UV, Opt.		
	C, S	UV	C^+	UV
	Na, K, Ca,	Opt.	Ca^+	Opt.
	V, Cr, Mn,	Opt.		
	Fe, Co, Ni, Cu	Opt.		
Diatomics	OH	UV, Opt., Rad.	OH^+	Opt.
	CH, NH, CN	Opt.	CH^+	Opt.
	C_2	UV, Opt.		
	CO	UV	CO^+	UV, Opt.
	CS	UV	N_2^+	Opt.
	S_2	UV, Opt.		
Polyatomics	NH_2, C_3	Opt.		
	H_2O	IR	H_2O^+	Opt.
	CO_2	IR	CO_2^+	UV, Opt.
	HCN	Rad.		
	H_2S	Rad.		
	H_2CO	IR, Rad.		
	CH_3OH	IR, Rad.		
	Organics	IR		
	Silicates	IR		

The atomic and molecular species identified in cometary spectra so far[9] appear in Table I. Neutrals and ions occur in the coma. The ions, which are dragged out in an anti-solar direction by interaction with the solar wind, form the plasma tail. Except for sodium, the metallic elements appear only in the spectra of "sun grazing" comets, very close to the sun (within 0.15 AU or so).

- A few species have been reported, generally at the trace level, but have not been confirmed : NH_3 ([3]), CH_4 ([76,81]), CH_3CN ([125]), OCS ([18]), HCO ([23]), CN^+ ([120]), H_2S^+ ([23]).

- It has not been possible so far to specify the carriers responsible for the various emission "bands" observed in the IR (notably near 3.4, 10 and 18 μm). Carbonaceous compounds (among them methanol, which can explain up to $\sim$ 40 % of the 3.4 μm feature) and silicates are indicated, but their detailed identification is still lacking (for recent reviews see e.g. [27,37,59,124]).

- The presence of several species (mostly "parent molecules"), in comet P/Halley has been established or, in most cases, inferred on the basis of the measurements made by the mass spectrometers on board the Giotto and Vega space probes : H_2O, CO, CO_2, CH_4, NH_3, H_2CO, CH_3OH, HCN, H_2S ([49,79]). A number of important ions have also been detected by these instruments, in particular: He^{++}, C^+, O^+, S^+, OH^+, H_2O^+, H_3O^+, CO^+ ([11]); H_3S^+, C_3H^+, $C_3H_3^+$ ([36,87]); H_2CN^+, H_3CO^+, $CH_3OH_2^+$ ([36,49,71,90]); NH_3^+, NH_4^+ ([91]).

- On the other hand, grain destruction on impact has indicated grains composed of C, H, O, N, leading to the concept of an uncharacterized richly organic component of cometary grains, the so-called "CHON particles". These tiny (micron- and submicron-sized) particles have been described as containing "refractory organics which are highly unsaturated polycondensates rich in C = C and C - O compounds"([74]). The poor characterization of these organic compounds is reminiscent of the organic matter present in the carbonaceous meteorites, but the C, H, O, N elements are more abundant in comet Halley's grains than in CI chondrites, the most primitive meteorites. The Positive Ion Cluster Composition Analyzer (PICCA) on board Giotto detected a series of peaks from 30 to 105 amu spaced by about 15 amu. There is no satisfactory identification of the presumably complex organic material producing them, although it has been suggested that polyoxymethylene ("POM" : $-CH_2O-CH_2O-...$) may contribute to these peaks [66,95].

- Very recently, the presence of a polycyclic aromatic hydrocarbon ("PAH"), phenanthrene ($C_{14}H_{10}$), has been proposed as responsible for a few near-UV emissions (340-375 nm) detected in comet Halley by the Vega 2 spacecraft[99]. However, this identification may lead to an inconsistency, as some IR emission near 3.25 μm, which should follow the production of these UV features, would be about 100 times stronger than the maximum signal observed at that wavelength in several comets[30], including P/Halley.

Molecular abundances.

An overview of the chemical composition of the volatile component of the cometary matter is presented in Figure 1. The dominant compound in most, if not all, comets is water (about 80 % of all volatiles by number), so that it is customary to express the other abundances relative to H_2O. In fact, it is the precise knowledge of the concentrations of the less abundant species that will be crucial to get some insight into the origin of comets and of the solar system. It is important to note that the available data, which were obtained only recently, are still incomplete and affected by rather large uncertainties, even if the situation has been constantly improving during the last few years. On the other hand, the range of values shown does also reflect variations among comets in some cases. Most of the molecules listed are believed to be parent molecules. However, it has been realized that some species, certainly CO and H_2CO, are released to a large extent from an extended source (probably CHON grains) rather than from the nucleus itself[35,90]. This has introduced some doubts as to the origin of some of the other species, as well as some uncertainty regarding the derived production rates.

As for the radicals, or daughter products, not quoted in Figure 1, CH, C_2, CN, NH, NH_2, their relative abundances are typically of the order of one to a few 0.1 %, C_3 being about a factor 10 lower still. Unexplained anomalies concerning the production rates of these secondary species have been found[113]. Finally, let us point out that virtually nothing quantitative is known regarding the chemical composition of the refractory component of comets. Estimates of the dust/gas production ratio have been attempted, but they are based on indirect and often incomplete evidence and hence are not very reliable. The d/g mass ratio appears to be highly variable among comets. The "best" value for P/Halley is probably of the order of 2, with an uncertainty factor of about 2.

More details on the chemical composition of the volatile fraction of comets will be found in some recent general discussions[29,34,38,44,48,79,102,121,131].

Elemental abundances.

The determination of the relative abundances of the four light elements H, C, N, O and of 13 heavy elements, from Na to Ni, in comet P/Halley has indicated that those are closer to "solar system" values than the corresponding abundances in CI chondrites[74]. More specifically, the comet's values are solar within a factor 2, with two exceptions: an expected one, H widely underabundant (factor $\sim$ 700) because of the extreme volatility of molecular hydrogen, and a perhaps less expected one, that of N, with a milder depletion factor ($\sim$ 3-5). One may indeed wonder, at first, why N_2 should have behaved so differently from CO, which has similar volatility and yet is rather abundant in the comet. The answer may be twofold: CO, unlike N_2, can combine even if weakly with various kinds of substances and the CO measured in the coma of Halley's comet is to a large

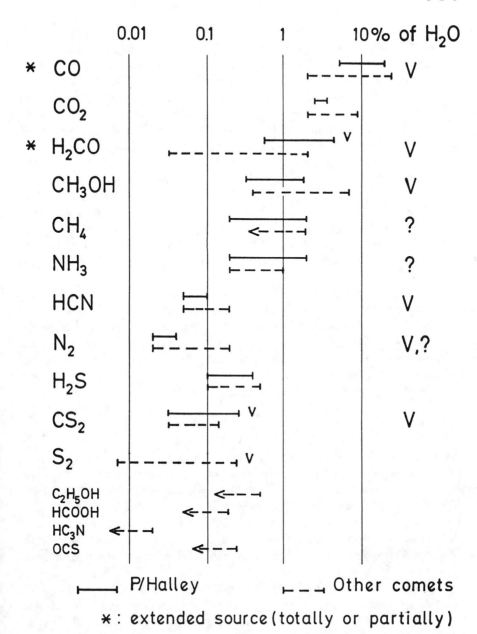

Fig.1. Relative abundances of the cometary volatiles expressed in percentage of the water content, by number. Small "v" indicates that the corresponding substance shows definite temporal variability, while the mixing ratios of the molecules marked with "V" in the last column have been observed to vary from comet to comet. Molecules with particularly uncertain abundances are labeled with a question mark. A few upper limits are also given at the bottom.

extent (if not totally) a daughter, not a parent species[50]. It should be noted that a number of assumptions enter into the derivation of the relative abundances, particularly in relation to the dust/gas ratio, which is not well known and whose value influences the derived abundances of the light elements. On the other hand, Fe and Si appear to be anomalous (just beyond the quoted overall uncertainty of a factor 2) in the sense that Fe/Si is about 4 times lower than the solar system value. If correct, this result would suggest that the refractory part of comet Halley's material went through some chemical processing somewhere along its own thermal history[4].

In comet Austin (1990 V) an upper limit has been determined[122] for He, showing this element to be down by a factor of at least 10^4, which is not particularly informative or useful to really constrain scenarios for the thermal evolution of the comet. Even less significant are some upper limits derived for Ne and Ar in the same comet and in Levy (1990 XX)[45,122]. However, these far-ultraviolet observations have given definite hope that these species should be detectable under more favourable circumstances.

The only comet other than P/Halley for which some information could be obtained concerning the metallic elements is the "sun-grazer" Ikeya-Seki (1965 VIII). The analysis of spectra taken when this comet was only 14 solar radii from the sun's centre has indicated that even so close to the furnace, the temperature was not high enough to completely vaporize the dust grains, leading to some apparent fractionation; nevertheless, the observations were consistent with the conclusion that the relative elemental abundances were essentially solar or meteoritic[8].

Isotopic ratios.

Again, almost everything we know about cometary isotopic relative abundances refers to comet Halley. D/H in the water was estimated[79] to lie in the range $(0.6 - 4.8) \times 10^{-4}$ (refined analysis will hopefully yield a better-defined value). This falls in the range found for the earth's oceans and for other low-H solar system objects (like Titan e.g.), but it is higher than in Jupiter or Saturn and hence presumably than in the PSN hydrogen. It may not be inconsistent with cometary H_2O having originated in the parent IMC where it would have been deuterated to a limited degree[50].

The $^{12}C/^{13}C$ ratio measured *in situ* ranges[74] from ~ 90 (the terrestrial or solar system value) up to as high as ~ 5000. These very high values found in some grains seem to be real and the wide range covered would imply that the dust in comet Halley comes from different sites of nucleosynthesis. The estimates for this isotopic ratio made in a few comets from C_2 Swan emission bands have large error bars but are otherwise in agreement with the solar system value[127]. So are the Halley *in situ* values for $^{18}O/^{16}O$ and $^{32}S/^{34}S$([79]).

MODELS AND THEIR UNCERTAINTIES

The determination of abundances naturally relies upon the knowledge of oscillator strengths for the observed transitions, but it also almost always requires some modeling, if only because of the limitations of our observations, and hence the introduction of a number of parameters, including different kinds of atomic and molecular data, which are unfortunately often poorly known. The various needs will become obvious as we proceed in this and the next sections.

How, then, are the relative abundances or production rates obtained? Let us start with a very simple case and consider a spectral transition $i \to k$ of a given species, for which a flux, or irradiance (Wm^{-2}), F_{ik}, has been measured from a comet located at a distance Δ from the earth. Since the main excitation mechanism of almost all electronic and vibrational cometary features is resonance-fluorescence in the solar radiation field, the photonic luminosity or emitted power, L_{ik} ($= 4\pi\Delta^2 F_{ik}/ h\nu$) is readily converted into the total number of atoms or molecules in the volume of the coma seen in the field of view:

$$M = \frac{L_{ik}}{g_{ik}}, \qquad (1)$$

g_{ik} being the fluorescence efficiency or excitation rate (s^{-1}):

$$g_{ik} = \frac{\pi e^2}{m_e c^2} \lambda_{ik}^2 \, (fH_\lambda^\odot)_{ki} \, b_{ik}, \qquad (2)$$

where e, m_e, c and λ have obvious meanings, f_{ki} is the f-value, b_{ik} the radiative branching ratio (= $A_{ik} / \sum_l A_{il}$, in terms of Einstein A coefficients), and $H_\lambda^\odot$ the solar irradiance at the comet's heliocentric distance r. Note that the latter is proportional to $1/r^2$ and must be known accurately at the excitation wavelength of the transition, account being taken of the Doppler shift associated with the comet's radial heliocentric velocity ("Swings effect"), as well as of the variation of $H_\lambda^\odot$ with the solar activity cycle (important only in the ultraviolet). Note also that we have assumed the comet's atmosphere to be optically thin in the observed emission, which is the most frequent case; important exceptions are the strong atomic resonance lines, such as H Lyα, for example. If we assume further that the instrument used has a large aperture or beam, so that it sees the whole coma in the $i \to k$ radiation, and that the comet is in a steady state, then, equating production and destruction rates immediately yields the production rate as

$$Q = \frac{M}{\tau}, \qquad (3)$$

with τ = lifetime of the observed species against photodestruction, i.e. photodissociation + photoionization (the former being by far the more important for most cometary molecules).

Thus, we have already met with three important required quantities: f, b, and τ, but we need more! For it must be realized that the idealized situation we have considered is, in fact, rarely encountered. First, the upper level of a transition can usually be reached from more than one lower level, so that the excitation rate involves a weighted sum, with the weights equal to the appropriate lower level populations. The latter are obtained by solving steady-state equations which usually include both radiative and collisional transitions, whose rates are therefore needed. Since the relative importance of these two types of excitation varies considerably with the distance from the nucleus (note that the collision terms relate only to lower levels, i.e. rotational, fine or hyperfine structure levels or sublevels), the statistical equilibrium equations have to be solved at a number of locations within the coma and the resulting populations then appropriately integrated along several lines of sight. The situation may be further complicated when it comes to interpreting optically thick transitions, e.g. the H_2O rotational transitions[13], which requires proper treatment of line transfer problems.

Another unfortunate aspect is that, for practical reasons, our instruments most often have rather limited fields of view, so that an important correction has to be brought about if we are to get the total number of particles or the correct production rate. This is achieved by adopting a model for the spatial distribution of the observed species, the most common one being Haser's two-component model[60]. This model assumes radial outflow of two decaying species: a parent produced by sublimation from the nucleus, moving with velocity v_p and giving rise, by photolysis with lifetime τ_p, to a daughter radical or atom, itself characterized by v_d and τ_d. It is still widely used in spite of its weak, unrealistic features (even enhanced by the usual assumption that $v_p = v_d =$ expansion velocity $= 1$ km s^{-1}), because it gives a simple analytic formula for the column density of a daughter species (or for a parent, since this can be regarded as a daughter whose parent has $\tau_p = 0$) as a function of the projected distance (ρ) from the centre of the comet's image, essentially in terms of the ratio of two scale lengths: $R_i = v_i \tau_i$ $(i = p, d)$, and because it can provide a fairly reasonable means of intercomparing comets or of conducting statistical studies about them. Examples of density distributions (or brightness profiles, for optically thin comas) are given in Figure 2. More elaborate models have been proposed for the density distribution: the ramdom-walk[21], vectorial[47], or Monte Carlo particle-trajectory models[20,22]. All of these take into consideration the energy acquired upon photolysis by the daughter and its consequent non-radial motion. All of them necessitate accurate knowledge not only of photodestruction rates, but also of outflow velocities and photodissociation excess energies.

In addition to the immediate needs just outlined, there are, of course, other aspects of the study of comets that require modeling, but the really basic demand is undoubtedly this: since we have no direct evidence concerning the physical and chemical properties of the nucleus, the holder of the comet's secrets, our only way to try and unveil those is to investigate all the characteristics of the coma, the main source of our information, in all possible details with a view to understanding the intricacies of the coma-nucleus connection. A global model, in which physics and chemistry are intimately linked, should cover many aspects or processes indeed, starting in fact from the outgassing by the nucleus (itself modeled, for different assumed compositions and at successive heliocentric distances) and including the gas-dust interaction (multi-fluid dynamics), the interaction of the cometary gas

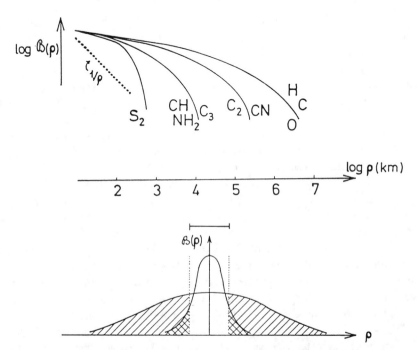

Fig. 2. Radial profiles of some cometary atoms or molecules (schematic). These cover lifetimes from less than 10^3 s to two or three days (at 1 AU from the sun). The $1/\rho$ distribution shown corresponds to a radial expansion with no decay and results from the integration of a $1/R^2$ distribution along a line of sight through the comet. The inner part of a parent molecule's profile follows this law: for example, the column density of H_2O (scale length $\sim 10^5$ km) would decrease as $1/\rho$ out to about 10^4 km (where it would be $\sim 10^{14}$ cm^{-2}, in a bright comet at 1 AU) and then more and more rapidly as it is progressively dissociated.

On the lower drawing we see that observations made with a small field of view can miss a substantial fraction of the emissions and thus lead to seriously underestimated production rates if they are not complemented by a proper model. The slit of spectrometers or spectrographs (or the beams of radiotelescopes) generally used have angular dimensions of the order of a few to a few tens of arcsecs, where we would need at least a few arcminutes in many cases.

with the solar radiation (photodissociation, photoionization, radiation pressure,...), heating and cooling processes, chemical reactions, interactions with the solar wind and associated magnetic fields,... No fully comprehensive model encompassing all these effects and in particular, dealing in detail with the important transition zone between the inner and outer comas exists as yet, although some rather sophisticated treatments have been published, each emphasizing such and such aspects or processes more than the others. These are described in a number of review papers[25,48,51,67,92,114].

Of special interest to our topic is the chemistry going on in the inner coma among radicals and ions produced mainly by photodestruction of parent molecules released by the nucleus. Besides the photochemical processes, electron impact ionizations and dissociations, ion-neutral reactions, as well as dissociative recombinations are particularly important. Although they do not have so profound an effect as to totally reshuffle the composition of the gas given off by the nucleus, chemical processes do play a role in the densest part of the coma where they produce some interesting minor (and less minor) species, in particular some ions, notably of protonated substances: H_3O^+ (the dominant ion inside 20000 km in comet Halley's coma, as recorded by the Giotto Ion Mass Spectrometer[11]), H_3CO^+, H_2CN^+, NH_4^+...([49,90,91]). Impressive networks comprising more than one thousand reactions involving over one hundred species have been constructed: a good many rate coefficients are still just "estimated" or "guessed"; a critical survey to pin down the most important lacking ones would be welcome.

On the other hand, a valuable outcome of realistic models will be the runs of density, temperature and velocity with cometocentric distance, which are needed to interpret many observed spectra. As a matter of fact, in a really complete treatment, the atomic and molecular level populations themselves would be calculated, ready to be used for predicting line or band intensities.

Before going over to a survey of the various types of physical and chemical data needs in cometary research, it may be appropriate to draw attention to a number of pitfalls we may find in our way towards ascertaining and interpreting the abundances in cometary atmospheres. Beyond the usual errors of measurements, the model deficiencies or ambiguities, and the lack of laboratory data, there are other fundamental, more subtle sources of uncertainties, some of which will now be mentioned briefly.

- First, we should be aware of the possible alteration effects that may have taken place at different epochs in the life history of comets, from the time of their formation down to their passage through the inner solar system. Thermal, collisional, or irradiation processes may have physically and/or chemically affected the cometary matter by causing e.g. phase changes in ices (amorphous → crystalline), synthesis of organic compounds, redistribution or loss of volatiles, formation of a non volatile surface deposit[89],....

• A particularly disturbing uncertainty concerns the relationship between the concentrations measured in the coma and the composition of the "primordial" matter in the nucleus. Indeed, much work has been devoted to this question in recent years[41,42,61,106,107], by studying the thermal evolution of the nucleus along the comet's orbit. Owing to the very porous structure, gas diffusion occurs inside the nucleus, both outward and inward (the latter being followed by recondensation), and a stratigraphic chemical differentiation results. Then, the consequent sublimation fractionation means that the coma abundance ratios will vary with the heliocentric distance of the comet, the more so the higher the volatility of the parent substance compared to that of water. It is therefore strongly recommended to monitor the composition of comets over the largest possible portion of their orbits and to make use, in the analysis of the observations, of thermal behaviour models built for different plausible assumed nuclear ice mixtures and porosities. Such studies will hopefully also help to explain the rather puzzling, often "bursting" activity of comets at large distance from the sun[69], of which P/Halley itself gave an example recently, while it was ~ 14 AU away.

• Temporal variability on a considerably shorter timescale (a few days or even less than a day) is also observed in some cases. This is attributed to the rotation of the nucleus, coupled with the inhomogeneity of its surface and upper layers. Outbursts also occur occasionally. It is well to remember these possibilities when comparing different objects. Comets can rarely be observed continuously and quite often different species are measured, by different techniques, at different times. All production rates are usually referred to $Q(H_2O)$, which is not always observed concomitantly! Yet another variable factor should be taken into consideration, namely the solar flux of ultraviolet radiation, which varies according to the sun's cycle of activity (the more so, the shorter the wavelength), with consequent changes in photolytic rates, excess energies, excitation rates, etc..., for a given heliocentric distance.

• Since, as already mentioned, some molecules like CO and H_2CO which were believed to be parent molecules are actually not issued from the nucleus itself but produced by a distributed source (probably only partially in the case of CO[111]), the question often arises as to whether such or such observed species is or is not a parent and this may lead to large uncertainties, if indeed we have no information on the spatial distribution of that species. For instance, if the emission is recorded through a fairly small aperture, the production rate derived assuming a point (nuclear) source may be a factor 3 or even an order of magnitude (depending on the molecular scale lengths involved) smaller than the Q that would be obtained if an extended source were adopted.

The above remarks should not be taken negatively. They merely mean that we should be aware of these limitations and exercise caution when trying to draw general cosmogonic conclusions. They also point to the need to have a comprehensive, global view of all the aspects involved in the study of the composition of comets.

PHYSICO-CHEMICAL DATA NEEDS

1. Spectroscopy

Great progress was accomplished in cometary spectroscopy during the last three decades, first through the use of high-resolution coudé spectrographs, then by the considerable widening of the wavelength coverage, and more recently with the advent of more sensitive, digital electronic detectors. Numerous illustrations of the achievements will be found in a forthcoming Atlas of Cometary Spectra[9]. Important discoveries were made in the UV, IR, mm and sub-mm ranges. However, much remains to be done as regards the identification of new transitions or of new trace species. In addition to wavelength matches appropriate to the spectral resolution used, such identifications must naturally consider other stringent criteria related to intensity ratios, oscillator strengths, level populations, for all the associated transitions, in agreement with a plausible excitation mechanism. Since the latter, in a cometary atmosphere, is most often resonance-fluorescence in the diluted solar radiation field, this may produce relative intensities differing totally from those typical of laboratory conditions; a few specific cases where chemiluminescence (or "prompt" emission) comes into play are known too. It is also essential to take account of the spatial extent of the emissions and of their distribution on each side of the centre, as this gives information concerning the nature of the carriers (lifetimes involved; neutral or ionized species, since the latter have distinct strongly asymmetric profiles, much longer in the tailward direction). In an effort to help future identification work, a survey of the literature and re-examination of some rather rich photographic spectra have been started in Liège with a view to taking stock of the situation and preparing a catalogue of cometary emissions. This will contain over two thousand lines, perhaps some 15 % of which have received no assignment.

More than four times as many lines have been recorded, in the range 380-900 nm, in CCD spectra of recent comets obtained at Lick Observatory[14]! The majority of the unidentified features are weak, but some are rather strong; most belong to *neutral* species. A most vexing example is illustrated in the upper part of Figure 3: the emissions at 483.8 and 485.1 nm have been observed for decades, in many cometary spectra[9], and yet have not been assigned so far. They resemble NH_2 features of moderate strength in their spatial distribution, being short and nearly symmetrical. The stronger of the two is not due to HCO, as suggested for one comet[23] (missing R-branch which should appear at 483.3), and these emissions seem, in fact, to form a pair as their intensity ratio remains constant from one cometary spectrum to the next. As shown in Fig. 3, the use of very high resolution has revealed that each of them has some structure. It may be noted here that the near-ultraviolet region ($\sim$ 300 - 380 nm), which has been too little studied so far, also contains unidentified emission features. Some have been reported as observed in one or two recent comets at McDonald Observatory [126].

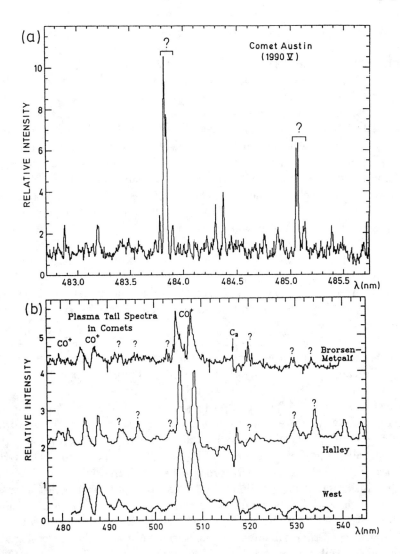

Fig. 3. Spectra of recent comets showing a number of unidentified features.
(a) Comet Austin at high resolution (0.012 nm) revealing doublet structure in the 483.8 and 485.1 nm emissions. It is known that these belong to a rather short-lived neutral species, but no assignments have been found for them so far (Lick Observatory, 3 m telescope; courtesy of M. Brown and H. Spinrad).
(b) Ion emissions at moderate resolution (0.9, 0.7, and 0.3 nm for West, P/Halley, and P/Brorsen-Metcalf, respectively). The strongest unidentified emissions observed in Halley and Brorsen-Metcalf are not detected in comet West. The peak near 520 nm may be contaminated by night sky [NI] emission in the case of Brorsen-Metcalf. (Wise Observatory 1 m telescope, Cerro Tololo Inter-American Obs. 4 m tel., and NOAO Kitt Peak 2.1 m tel., from bottom to top, resp.; courtesy of S. Wyckoff).

A number of cometary emissions due to *ions*, characteristic of the plasma tail, still remain to be identified as well, in particular the moderately strong bands near 492.3, 495.6, 529.0, 533.2, 600.2 nm, and quite a few weaker features, first detected in Halley's comet and confirmed in other comets, at medium and low resolution [138]. Some of these appear in Figure 3(b). The following ions, which have resonance transitions in the optical region, have been considered as possible candidates: NH^+, C_2^+, CS^+, SH^+, H_2S^+, CS_2^+, OCS^+. The first three of these do have bands coinciding in wavelengths with the unidentified cometary band features, but the comparison of the observed intensities with band intensities computed assuming fluorescence excitation have given no fit[137]. Could ions of some of the other neutral species observed in comets, C_3^+, NH_2^+, H_2CO^+, HCN^+, e.g., be responsible for some of these emissions? Clearly laboratory work as well as higher-resolution cometary spectra are needed in this connection.

The case of the spectroscopy of *hydrocarbons* is also quite unsatisfactory: many spectra are not available in the gas phase and the few that do exist have been obtained at low resolution, too low to be really useful for identification purposes, and usually in absorption at room temperature, whereas we need spectra at temperatures below ~ 100 K.

The rather recent pioneering investigations of comets in the *infrared* and *microwave* spectral ranges have already been remarkably productive, leading to the discovery of new transitions and of important new species [27,32,133]. Yet we are very far indeed from having fully exploited these fields and there are pressing needs, here too, in the way of wavelengths and radiative lifetimes or f-values. To quote but a few examples of lacking or poorly known spectral characteristics: pure rotation spectrum (and dipole moment) of H_2O^+, infrared spectrum of CH_3OH and CH_3OD at sufficiently low temperature and very high resolution ($\lambda/\Delta\lambda \sim 10^5$), individual line strengths in rotation-vibration spectra of known or possible parent molecules, such as H_2CO, CH_3OH, or aromatic molecules; fluorescence yields of large molecules (e.g. PAH's) excited by solar radiation, at very low rotational and vibrational temperatures to simulate cometary conditions. The general information currently available can be found in spectroscopic data banks (like GEISA, HITRAN (formerly AFGL), ATMOS) or atlases (e.g. Aldrich Library of FT-IR spectra). References and data are given in a very useful recent compilation[26], while the latest results and new problems in the field of IR and Radio spectroscopy of comets are reviewed by Crovisier and Bockelée-Morvan[30].

2. Photolysis

The importance of the photolytic processes concerning both the primary and the secondary cometary species has already been stressed: such processes determine the spatial distribution of these species and, more generally, they play crucial roles in governing the physical and chemical properties of the larger part of the neutral coma as well as in providing major sources of ionization; furthermore, knowledge of photodestruction lifetimes is required for the assessment of production rates. Data are needed in different aspects of photochemistry :

- *Photodissociation cross sections*, $\sigma(\lambda)$, are not well known, if known at all, for a number of important molecules, mainly for radicals (e.g. C_2H_2, CH_3OH; C_2, C_3, NH_2, C_2H,...). We have seen that the radial profiles of the radicals depend, in the two-component Haser formalism, upon the ratio of two scale lengths: if the scale length of the daughter radical were known with sufficient accuracy from laboratory measurements, then we could derive the parent's lifetime from the observed intensity distribution and thus obtain insight into the nature of the parent. Also very scanty is the information on photo rates for ions in general, as well as for the larger organic molecules, such as PAH's, POM's or other polymers or oligomers. A further point to note is that, except for a few cases[17,136], the temperature effect on photoabsorption cross sections has been very little investigated (laboratory measurements are usually made at room temperature, whereas the cometary molecules are considerably cooler).

- *Branching ratios* for the various possible channels of the dissociation process are needed for most, if not all, parent molecules. For instance, even for as fundamental a process as the photodissociation of water, the respective quantum yields of the available channels are still poorly known! Yet, if we knew the exact relative importance of the $H_2 + O(^1D)$ branch (after dissociation by solar UV), we would have a way to independently estimate the H_2O production rate, from the measurement of the intensity of the forbidden red doublet of oxygen, and to compare this with the Q's determined from UV and radio observations of OH (which do not completely agree with one another).

- *Photodissociation dynamics* is concerned with the partitioning of the energy difference between the photon's $h\nu$ and the threshold, the "excess energy", into kinetic energy of the fragments and internal (rotational, vibrational, and possibly electronic) energy of these. This important chapter of photodissociation studies is only now receiving the attention and efforts it deserves. Not only is this type of information providing deep insight into how the photolytic processes work, but we also need it in modeling the structure of a comet's atmosphere (the energy carried by the dissociation products contributes to the heating in the collision-dominated zone; the velocity acquired by a secondary species affects its spatial distribution), as well as in the interpretation of some observations (of prompt emissions, in particular). An excellent review of photodissociation dynamics and its influence on atmospheric chemistry has been published very recently [130]. Some results of cometary interest (regarding CO_2, CH_4, NH_3, C_2H_2) are given there, but in general only very limited data are available in this area today. For instance, uncertainties in the velocity distributions of H and OH produced by solar photodissociation of water have been pointed out [135].

To be sure sure, the wavelength dependence of all the photodestruction properties is required. For example, the photolytic rate will be the result of an integration:

$$k = \int \sigma(\lambda)\phi^\odot(\lambda)d\lambda \ , \qquad (4)$$

where $\phi^\odot(\lambda)$ is the solar photonic flux appropriate to the cometary observations (e.g. quiet or active sun). In some cases, a significant Swings effect (sensitivity to the comet's radial heliocentric velocity $\dot{r}$) may act on the dissociation rate. Thus, if a molecule is destroyed by predissociation from several lower levels j via a number of discrete transitions $j \to l$, with dissociation efficiencies η_l, then:

$$k = \sum_{j,l} x_j C_{jl} \, \eta_l \ , \qquad (5)$$

where the fractional populations x_j and the absorption rates C_{jl} depend upon $\dot{r}$ because of the very irregular spectral distribution of the solar radiation (especially in the UV). Examples of estimates that ought to be revised are those concerning CH([118]) and NH([119]), for which use should be made of more realistic x's, η's and/or solar fluxes; it is gratifying to know that work on this is in progress[33]. Extremely useful, comprehensive recent surveys and compilations of solar photo rates, with detailed lists of references exist [26,28,68]. A quality scale given in one of them [68], decreasing from A to F, indicates that about 60% of the molecules of relevance to cometary studies are qualified C or worse (C meaning branching ratios and excess energies not accurately known). The situation is particularly bad with regard to the photoionization cross sections for radicals.

Finally, as an illustration of the importance of the variability of the solar UV flux for cometary photochemistry, let me draw attention to the latest review on this topic[16], where the uncertainties are discussed in detail and formulas given for estimating the lifetimes and branching ratios of H_2O and OH for any observation date, as functions of appropriate solar activity indices.

3. Gas phase collisional processes

Even if the physico-chemistry of comets is dominated by photolytic and radiative mechanisms, it is true that a number of processes involving collisions are also of importance in determining the structure and composition of a cometary atmosphere, not only in the inner coma, but also in the outer regions (where plasma interactions are concerned), and that urgent data needs exist in this sector too. As for the currently available information relevant to cometary research, one can refer to a few extensive sources, e.g. [5,6,94]. One of them (UMIST ratefile [94]) gives a label describing the estimated accuracy of the rate coefficients, at least for ion-neutral reactions (and for photo reactions); let us note in passing that it also contains a table of electric dipole moments for various species, many of which are of cometary interest. The following list of data shortages, on the other hand, is by no means exhaustive.

- *Reaction rate coefficients applicable to the cometary environment (low T!).*

  - Ion-neutral reactions. Laboratory work is still needed on reactions of ions with atoms (H, C, N, O, S) and with radicals (OH, CN, CH_2, CH_3,...).

  - Proton transfer ("protonation") reactions: a particular case of the preceding category which deserves special mention. Indeed, retrieving fundamental molecular abundances (H_2CO, CH_3OH, NH_3, HCN, H_2S,...) from *in situ* measurements of ion density runs in comet Halley's inner coma, which is a delicate and complex task, depends heavily on the knowledge of some such reactions (in addition to relying on the solution of problems arising from difficult instrumental calibrations and from ambiguities in certain channels of the mass spectrometers). Very detailed and careful analyses based upon ion-chemical models have permitted considerable progress in this area recently [36,49,90,91]. These indicate that the uncertainties associated with the limited accuracy of reaction rates are, in general, not dramatic (not larger than $\sim 30\%$), although in some cases the absence of experimental information concerning the temperature behaviour imposes use to be made of a constant rate measured at room temperature (e.g. for $H_3CO^+ + H_2O$), and more important, reaction rates are poorly known for significant production or loss processes (such as $H_3O^+ + NH_2 \rightarrow NH_3^+ + H_2O$ and $NH_3^+ + H_2O \rightarrow NH_4^+ + OH$, which are among the principal reactions to be considered in the derivation of the relative abundance of ammonia[91]).

  - In the same context, the situation regarding rate constants for processes involving C_xH_y molecules or radicals, or their ions, appears to leave a lot to be desired and there are no data about the dissociative recombination of some ions (H_3CO^+, $C_2H_4^+$, e.g.), let alone data on the corresponding branching ratios.

  - Information on proton transfer reactions involving heavier H-C-N-O compounds would also be useful for the interpretation of the higher-mass peaks that were observed in comet P/Halley by the ion analyzers on board Giotto and Vega 2. It is believed that heavy ions up to about 120 amu may be protonated species[65].

  - Electronic recombination coefficients for such important ions as CO^+ and N_2^+ are still lacking.

- The cross sections and the temperature dependence of charge-exchange reactions are largely uncertain. Data on such processes with OH, in particular, would be welcome.

- Determinations of the abundance ratios of different nuclear spin species of a given molecule, which have hardly begun and have only been possible for the ortho-para ratio (OPR) of H_2O so far, promise to yield information related to the origin and evolution of comets[101,102]. However, here as elsewhere, the observations must be interpreted with great care and one has to ask whether the measured "nuclear spin temperature" is actually primordial or not. Thus, we need laboratory measurements on processes that might have induced spin conversions (in H_2O, H_2CO, CH_3OH, H_2S,...), such as collisions with paramagnetic substances (O_2 e.g.), or H transfer between molecules on ionization by cosmic-ray irradiation[75], and hence modified the OPR after the formation of the comets.

• *Collisional cross sections for rotational excitation* of known and potential parent molecules, or more generally of stable molecules observable in the IR or Radio wavelength regions. These parameters are of the utmost importance to properly analyse such observations, to determine production rates, to check the validity of the models used and thus obtain some insight into the physical structure of the coma (inferring neutral gas temperatures and electron temperatures as functions of cometocentric distance, for instance). On the other hand, the instrumental capabilities of infrared and microwave spectroscopy are at present in a fast-growing phase and therefore, provided appropriate fundamental data become available, it should be possible to derive full benefits from the high-quality, high spectral and spatial resolution observations which can be expected in the near future, be it with earth-bound facilities or in space. The main collision partners to be considered are H_2O, electrons and possibly ions [27,139], the neutral molecules being predominant in the innermost part of the coma. Experimental values and/or theoretical calculations for such collision rate constants are badly needed. These, again, should necessarily refer to low temperatures (T $\sim$ 10 - 100 K); besides, information concerning the variation of σ with the quantum jump, $\sigma(\Delta J)$, should be sought. Interesting suggestions have been made recently[53] with regard to the possibility of using the relationship between collision rates and spectral line shape parameters to obtain some estimates for the former.

4. Processes involving matter in the solid phase (grains, nucleus)

Some areas of cometary research which are booming nowadays deal with the thermal, structural, and compositional evolution of the nucleus along a comet's orbit (including activity at large distance from the sun), the behaviour of cometary analogues in the laboratory, or the connection between the gas- and solid-phase constituents of the cometary atmosphere. It is natural enough that these new or reopened fields should have their own fundamental data requirements. In the following, some of the most important categories of needs will be indicated.

- *Spectra of icy mixtures in the mid- and far- infrared regions, measured at different temperatures (phase changes).* The existing data and numerous references to earlier work can be found in two recent papers giving spectra of a variety of pure or mixed ices comprising, among others, H_2O, CO, CO_2, CH_3OH, CH_4, NH_3, at various temperatures from ~ 10 K to ~ 150 K, in the intervals 2.5 - 20 μm (with extension to 200 μm in some cases)[62] or 20 - 100 μm[97]. The former contains extensive tables of the real and imaginary parts of the index of refraction (n and k), needed to study radiation transfer problems, as well as integrated absorption coefficients. The optical constants n and k for low-T condensates of CO, CO_2, H_2CO, H_2CO polymers, HCN, NH_3, in the *far-IR* have not been measured until now and more studies of the spectral effects of including such compounds as H_2CO, HCN, NH_3, in particular, into cometary ice analogues are also needed[96]. These various data will be required for the interpretation of the spectra that will be secured with the forthcoming earth-orbiting IR telescopes, ISO and SIRTF. Another type of substances for which it would be useful to obtain IR spectra, vibrational intensities, dissociation rates..., is the class of hetero-dimers like H_2O - H_2CO, H_2O - CH_3OH, H_2O - $HCOOH$,..., as such hydrogen-bonded species might be present in the distributed sources found in cometary comas[100].

- *Low-temperature properties of ices and icy mixtures.* Our knowledge of a number of characteristics of icy compounds such as vapour pressure, thermal conductivity and diffusivity, density, porosity, tensile strength,... is still fragmentary or unsatisfactory. For example, the sublimation rate of water ice at very low temperature is still somewhat controversial and it has been shown to depend on the formation and growth conditions of the ice[109]. Many of the physical parameters, notably the porosity, are certainly of great importance for our understanding (a) of the structure and evolution of the nucleus[41,42,105,107] (heat and gas transport, rôle and timescale of crystallisation, chemical differentiation,...) and (b) of the still rather ill-defined rôle of the cometary grains and their relationship to the interstellar grains [55,56]. The influence on the low-T properties of the presence of "dust" (minerals and organics in varying proportions) and of the ice thickness should be examined as well.

- *Laboratory studies and cometary simulations.* The real meaning and usefulness of the elaborate comet nucleus modeling that is being carried out depend upon the support by appropriate experiments conducted on solid materials relevant to comets, under conditions pertaining to the natural cometary environment. These experiments, described or discussed in several reviews[58,70,78,140], investigate the sublimation process in detail, the effect of variable insolation, the gas and dust release, the refractory mantle formation (mineral and carbonaceous compounds), the physical and chemical modifications produced by ion and electron irradiation[98,123], the formation of clathrate hydrates[12,63,112], the production of organic molecules at very low temperature [54,66,117]. It is to be hoped that intensive work will be pursued on these and other questions related to the trapping properties and release of different components of ice mixtures; to the formation of amorphous vs crystalline ices as influenced by temperature, pressure, nature of substrate, non equilibrium ("kinetic") effects,...; to the production and destruction of oligomers and polymers, of clusters,...

• *Size effect on submicron particles properties.* The microscopic dimensions of the grains which represent an important component of the cometary matter may have a profound influence on such characteristics as their infrared spectra, specific heats, vaporization, phase transitions, reaction to sputtering,... How these may differ from the macroscopic or "bulk" properties of the corresponding solids has just started to be studied[96].

• *Yields of molecules and free radicals from grains.* Following the discovery of the CHON particles, the detection of gaseous "jets" of CN, C_2,... in comet Halley[2,19,24,116], and the evidence for extended sources of CO ([35]) and H_2CO ([90]), it has been suggested strongly that refractory grains, presumably the tiny CHON grains themselves, contribute at least a significant fraction of these gaseous components. Logical and quite plausible as it is, this interpretation will nevertheless become really convincing only when it can be based on more solid and more precise grounds, and for this we should endeavour to gain experimental information on a number of grain properties; for instance, to determine how such small particles react when exposed to solar electromagnetic and corpuscular radiation. Thus it would be interesting to measure the yields of cometary radicals (or their potential parents) through photodissociation of small organic grains, such as tholins[110] for example, by solar UV, as well as through ion or electron sputtering of such grains. Another possibility that should be looked at is the production of cometary species by sublimation off submicron grains, which can be warmed up to temperatures well above 500 K by absorption of solar light at 1 AU from the sun.

5. A few further specific examples

This last section will mention some additional pending problems whose solution is impeded by the lack of accurate physico-chemical data, or in some cases, by the absence of laboratory or theoretical data.

CO, CO_2

The discovery of emissions due to the Cameron band system, $a^3\Pi - X\ ^1\Sigma^+$, of CO in two recent comets observed with the Hubble Space Telescope[132] has, in fact, provided a unique way to estimate the cometary abundance of CO_2 (a direct determination of which has to await the availability of IR observations above the earth's atmosphere). Indeed, it can be shown that, among the various possible excitation mechanisms for the Cameron bands, the photodissociation (PD) of CO_2 producing CO in its metastable a $^3\Pi$ state (cf. Figure 4) has the largest contribution ($\gtrsim 60\%$), the other two competing processes being electron impact excitation of CO (EI, $\lesssim 30\%$) and electron impact dissociation excitation of CO_2 (ED, $\sim 10\%$). It has thus also been possible to conclude that CO_2/CO, a key abundance ratio from a cosmogonic point of view, is relatively high ($\gtrsim 1$) in a number of comets. On the other hand, we have found, in the course of this work, that some relevant information is still inaccurately known. Although our main conclusions are essentially independent of the uncertainties, improved data will undoubtedly be very useful for the interpretation of future observations, especially when we have obtained higher resolution, high signal-to-noise spectra of the CO emissions and determined their spatial distribution. Thus, further study of the following

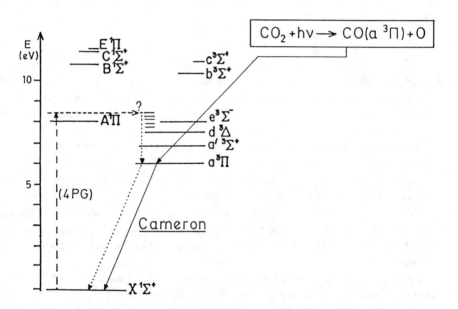

Fig. 4. Energy level diagramme of CO. Cameron band emission in comets occurs primarily (as "prompt" radiation) after photodissociative excitation of CO_2 (at λ below about 109 nm) into the a $^3\Pi$ electronic state. However, an alternative route exists: absorption in the 4th Positive Group (4PG) toward perturbed levels comprising a mixture of singlet ($^1\Pi$) states and triplet states, for instance d $^3\Delta$ states, from which cascading in the Triplet System will lead to a $^3\Pi$ states. The resulting slight effects may have to be taken into account when accurate, simultaneous abundance determinations of CO and CO_2 become possible. Should the relative abundances of the two carbon oxides vary from comet to comet, these effects might, in fact, become significant in low CO_2/CO comets.

points is warranted: branching ratios and photodissociation dynamics, as functions of wavelength, for PD itself; mixing between singlet and triplet states of CO (see Figure 4), influencing the f-values of some transitions in the 4th Positive Group[134] and producing a small contribution to the Cameron emission, after cascading in one of the triplet transitions, e.g. in the Triplet System, d-a, for which oscillator strengths are also needed (the recent determination, at high resolution, of 4PG absorption cross sections which may be of interest in this connection has been reported[86]); cross sections for ED, widely different values of which have been published[40].

C_2, C_3

Despite the fact that the principal C_2 emissions (Swan system, d $^3\Pi_g$ - a $^3\Pi_u$) are, with those due to OH and CN, the strongest features in the optical spectra of comets and as such have been extensively observed and analysed, a really satisfactory interpretation of these emissions is still lacking. Moreover, the parent, and the grandparent of the C_2 radical, which is indeed most likely the product of (a) two-step process(es), have yet to be identified. Recent detailed studies[52,104], where references to earlier work can be found, have demonstrated that some aspects of the vibrational and rotational intensity distributions of the Swan bands can be reproduced theoretically. These distributions, somewhat loosely characterized by "excitation temperatures", T_{vib} and T_{rot}, are "hot" (several 10^3 K) as expected in view of the homonuclear nature of the C_2 molecule, which cannot easily get rid of the vibrational and rotational energy acquired under fluorescence. An important point that has also been established is the central part played by the intercombination transitions, notably a $^3\Pi_u$ - X $^1\Sigma_g^+$, in controlling more than one aspect of the relative fluorescent equilibrium populations, in particular the ratio between triplets and singlets, much in favour of the former.

However, a number of as yet totally unexplained anomalies have been detected. Both T_{vib} and T_{rot} have been found to vary with the projected distance ρ, within a few 10^3 km from the nucleus, while their variations are opposite, with T_{vib} decreasing[104,128] and T_{rot} increasing[52,80] as ρ increases. Somewhat similar changes in excitation temperature with position in the coma, albeit with no clear separation between T_{vib} and T_{rot}, have just been reported[108]. All these results refer to comet P/Halley, but we have observed the increase of T_{rot} with ρ on high-resolution spectra of the (0,0) Swan band in several other comets recently. An example[10] is shown in Figure 5. On the other hand, in Halley's comet again, it was found[80] that the populations of the upper rotational levels could be described by two rotational temperatures: $T_{rot} \sim 600$ K for J' ≤ 15, ~ 3200 K for the higher J' levels; furthermore, the relative populations of the different spin substates were in sharp disagreement with the corresponding predicted populations, for the lower J' values.

A natural qualitative interpretation of the variation of the excitation conditions with distance from the nucleus in the inner coma consists in relating this to the

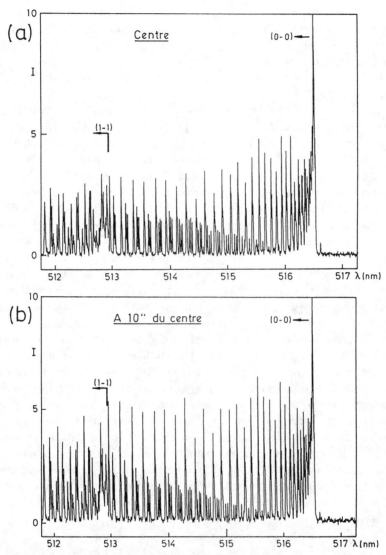

Fig. 5. The (0-0) and (1-1) C_2 Swan bands at high resolution ($\lambda/\Delta\lambda \simeq 60000$) in the spectrum of comet Aarseth-Brewington (1989 XXII) observed through a 3" circular aperture (a) centred on the comet and (b) offset by about 8000 km, when the comet was 0.4 AU from the sun and 1.1 AU from the earth.
Notice the difference in rotational intensity distribution between these two spectra. In particular, the individual lines are stronger relative to the band heads in the spectrum taken away from the comet's centre: this distribution corresponds to a higher excitation temperature and is closer to the fluorescence equilibrium distribution.
(Observatoire de Haute-Provence, 1.5 m telescope.)

production mechanism of C_2 and arguing that rather close to the centre we see a large proportion of "freshly" formed radicals which have not had time yet to reach fluorescence equilibrium, whereas further away this steady state has been attained by all the observed C_2. An important constraint that seems to emerge from the Halley results is that the C_2 radicals would be born vibrationally hot and rotationally cold.

At any rate, it is clear that significant progress toward the solution of the long-standing problem of the origin of C_2 will require (a) the spatial mapping in a few comets of the relative intensities of the singlet (e.g. the Mulliken system, $D\ ^1\Sigma_u^+ - X\ ^1\Sigma_g^+$, near 231 nm) and triplet emissions, as well as of the excitation temperatures, with a resolution of a few hundred km or better and (b) a parallel time-dependent treatment of the fluorescence of C_2 including a sufficient number of electronic, vibrational and rotational levels, and starting from different initial population distributions. To carry out the latter ambitious task, reliable experimental and/or theoretical values for a number of characteristics of the molecule are necessary. Most urgently needed are the a-X and X-a transition probabilities, to improve upon the existing set[82], obtaining those up to high J's (say ~ 60) and for the three Ω's of the Π state. Note that the theoretical evaluation of such Einstein A's itself requires accurate interatomic potentials, hence accurate energy levels, many of which are currently not available[82]. Similar data for the other intercombination transitions, c-X, as well as accurate transition moments for the Phillips, $A\ ^1\Pi_u - X\ ^1\Sigma_g^+$ system, and for the Mulliken system, in particular, would also be welcome. In addition, the photochemistry of prospective C_2 progenitors should be studied, determining the state (singlet or triplet) of the nascent C_2, as well as the initial T_{vib} and T_{rot}. Interestingly, the 193 nm photolysis of a likely grandparent, acetylene C_2H_2, producing C_2H, itself photodissociated, finally yields C_2 radicals with bimodal rotational distributions in some electronic states (singlets)[73]. Hopefully, such experiments will be continued and extended, to investigate the dependency of the results upon the intensity and wavelength of the dissociating radiation, and to explore the C_2 triplet manifold in more detail. Finally, the lifetime of cometary C_2 radicals in the solar radiation field should be determined.

The C_3 radical has at least this in common with C_2: it too comes from as yet unknown parents and grandparents and it cannot be produced by a single-step photolytic process either. On the contrary, the cometary emissions of these two radicals have different behaviours; in particular, their variation with heliocentric distance, C_3 appearing earlier and the ratio C_3/C_2 in general decreasing as r decreases. The allene molecule C_3H_4 ($CH_2=C=CH_2$) has been proposed as a likely supplier of C_3 radicals; its photochemistry (also studied at a single λ, 193.3 nm, so far) indeed shows possible channels leading to C_3 after photodecomposition of a C_3H_2 intermediate[72]. The investigation has only begun, however; it should be broadened and other potential C_3 forbears, among molecules like propyne C_3H_4

($CH_3C\equiv CH$), propylene C_3H_6, propane C_3H_8, or propynal $HC\equiv CCHO$, carbonaceous polymers, or grains should be considered as well. The spectroscopy of C_3 in comets has been little studied. A puzzling result was nevertheless pointed out[7]: T_{rot} is low ($\sim$ 300 K or less) even though C_3 has no permanent dipole moment i.e. no pure rotational emission. Could this anomaly have anything to do with the "floppiness" of the bending vibration of C_3([115,129]), allowing coupling between this motion and the rotation of the molecule, and internal transfer of energy?

CH, NH, NH_2

What fraction of the important cometary methylidyne radicals derive from methane is totally unknown. While there are several concordant values for the overall solar photodestruction rate of the latter, considerable doubt has been cast recently on earlier estimates of the quantum yields for the various dissociation channels[130]. On the other hand, other hydrocarbons such as C_2H_6, C_3H_4, C_3H_6,..., may produce CH, via the secondary decomposition of CH_2 or CH_3([141]); detailed studies of the photochemical processes concerning these molecules and radicals are therefore desirable. (Besides, more complex organic molecules, or even grains, containing the CH, CH_2, or CH_3 groups, may contribute to the production of CH). Also necessary is a re-evaluation of the photodestruction lifetime of CH itself, as already mentioned. Although the latter indeed needs to be improved, we do know that it is short, typically hundreds of seconds at 1 AU, that is of the same order as or sometimes shorter than, the time it takes the radicals to reach fluorescence equilibrium. Thus, initial conditions may have to be taken into account in the interpretation of high-resolution spectra of the A-X and B-X (0,0) bands of CH and we have here an example of the usefulness of information regarding the photodissociation dynamics (particularly T_{rot} of CH at formation in this case). Were these various kinds of data available, we might be able to constrain the elusive cometary CH_4 abundance.

Arguments can be given to suggest that NH and NH_2 have a common, unique source, namely ammonia. If so, they should lead to the same production rate $Q(NH_3)$. This is not the case if we consider, for example, the most extensive set of results found in the literature for a given comet, that is for P/Halley (see [46,91] e.g.). Expressed relative to water (which means that the risk of not being able to use the appropriate $Q(H_2O)$ should be kept in mind), remote observations give the following ranges of values for the ammonia abundance: (a) 0.1-0.4% from NH_2; (b) 0.44-0.94% from NH, whereas the *in situ* measurements indicate 1-2%. Now, the abundances under (a) should be multiplied by a factor of $\sim$ 2 because the determinations concerned either use a single NH_2 band or take an overall average derived from different bands. The reason for this correction is that (1) due to the structure of the NH_2 transitions and relevant selection rules, an even-v'_2 band involves only even-K_a lower rotational levels, an odd-v'_2 band only odd-K_a levels and (2) even and odd lower levels being about equally populated, a single band samples only approximately half the total populations. Even if after this correction, the (a) and (b) ranges come closer together and indeed overlap, there remains some discrepancy with the *in situ* results. The photolysis of NH_3 appears to be well documented (see [68,130] e.g.). A synthesis of all the available data should be performed to fulfil our exact requirements: the yields of NH and NH_2 in the photodissociation of cold NH_3 by solar UV, with information on the internal state and kinetic energy of the newborn radicals. Then, the most pressing needs are the solar photorates of NH and NH_2 themselves, with proper account of the "Swings effect". New absolute f-values for the NH_2 bands would also be welcome. As for the interpretation of the Giotto Neutral Mass Spectrometer data, further improvement may be achieved with the availability of reaction rates for some of the main NH_3^+ formation and destruction mechanisms (see above) and of more accurate rates for the NH_3^+ and NH_4^+ electron recombinations. It is not inconceivable that when this comes about, all the different estimates of $Q(NH_3)/Q(H_2O)$, from *in situ* and remote observations might point to an acceptable common value (0.8 - 1% ?) for P/Halley. On the observational side, data for both NH and NH_2, recorded at a given time, should be obtained for a number of different comets, as this would allow a more direct comparison between the results derived from the two radicals and hence the actual verification of the hypothesis that they both come from NH_3.

Ions

Key processes which contribute to the build-up of the structure and composition of cometary atmospheres by the production or destruction of ionized species are legion (in the inner as well as in the outer regions) and many are affected by our deficient knowledge of fundamental data : ion-molecule reactions, recombinations, proton or charge transfers, phtoionizations, not to mention mechanisms which have been little explored up to now, like reactions involving high energy electrons or ions, or reactions with heavy organic ions, the Alfvén critical velocity effect of collective ionization, the interaction between ions and small solid particles. A number of specific examples of data needs concerning ions have been pointed out in the preceding sections. As a further illustration of the problems we are confronted with, let us just quote here the results of recent measurements of ion abundances in cometary plasma tails[85], which show some large discrepancies between observed and calculated abundance ratios : with respect to H_2O^+, N_2^+ is predicted $\sim$ 10 times too high, $CH^+ \sim$ 100 times too low. While uncertainties in some of the parameters (reaction rates, lifetimes) which enter into the quite complex model computations are probably responsible for the major part of the discordance, it would be worthwhile revising the transition probabilities or exciting rates for a few of the observed ionic emission bands. Thus, the Swings effect on the fluorescence efficiency of the N_2^+ B-X (0,0) band (similar to the corresponding transition in the isoelectronic CN) near 390 nm should be examined, and new laboratory measurements of radiative lifetimes and relative band strengths in the $\tilde{A} - \tilde{X}$ system of H_2O^+ should be made[84,93]. It should be noted that since this system has the same structure as its NH_2 analogue, a similar correction factor of ~ 2 as pointed out above for NH_2 should be applied to single-band estimates of Q (H_2O^+), the correct recipe consisting in measuring at least two bands, of different v_2' parities, and then adding the two production rates obtained separately for each parity. Another important ratio is N_2^+/CO^+, because it can, in principle, be used to determine the corresponding ratio for the neutral molecules, of cosmogonic significance. The reasoning followed to go over to the N_2/CO ratio usually invokes arguments of similarity between the two molecules which are not necessarily valid and ought to be checked on the basis of experimental data anyway; reliable rate coefficients concerning the various formation and loss processes (collisional and radiative) for the N_2^+ and CO^+ ions are required. Ionic abundance ratios like CO^+/H_2O^+, N_2^+/CO^+, CH^+/CO^+, CO^+/CO_2^+, ... tend to be rather variable among comets. Only when improved data and models are available will it be possible, for instance, to discern how much of this kind of variability is due to differences in physical conditions, how much can be assigned to differences in original compositions or in evolutionary paths.

Sulfur-containing species
The sulfur chemistry in comets is also of great interest, since its study is expected to provide yet other useful constraints for cometary modeling in general or with regard to the temperatures and other physical parameters characterizing comet formation and evolution. Unfortunately, this area is itself marked with a number of dark patches. As recalled in several recent papers[1,31,36,77], four S-bearing species have been identified in cometary spectra until now: S, CS, H_2S, and S_2 (the latter observed in only one comet so far, and found to be highly variable[15]). Among the puzzling or unresolved questions, the following may be quoted: atomic S abundance (appreciably different values derived from different triplets; optical depth effect in the strongest?) and overall elemental S budget (are there any other significant, as yet unknown, sulfur compounds in comets?); origin of CS (is CS_2 its sole parent?); rôle, if any, of OCS; S_2 formation mechanism. In order to throw light upon these problems, it would be desirable to obtain improved, or original data regarding, in particular: the atomic S oscillator strengths; the excitation rate and lifetime of CS; the collisional cross sections for rotational excitation of H_2S by H_2O; the solar photorates for OCS and H_2CS; the predissociation probability of SO, as well as the Einstein coefficients for vibrational transitions within the ground state of this molecule; the incompletely known spectroscopy of SO_2; the cross sections for producing S_2 by proton irradiation of S-bearing compounds in the gas phase[77] (In the hope, perhaps, to resolve the "S_2 - SO puzzle" and to find processes effectively capable of creating the former without forming the latter).

CONCLUDING REMARKS

While we must continue to hope and actively prepare for the "top class" cometary space missions, rendez-vous or, later, sample-return missions, which will no doubt induce quantum leaps in cometary science, it is also true that such outstanding projects are still indeed remote in time and that they will necessarily be concerned with a mere handful of comets. It is therefore essential that sustained efforts be devoted to the study of comets with the facilities available on or near the earth and that full advantage be taken of the forthcoming technical progress in these facilities, in particular to explore the far UV and far IR and to expand the investigation of the microwave spectral region.

We want to know the detailed composition of the atmospheres of comets and we would like to make sure that we have sufficient knowledge of all the atmospheric processes which may influence the observed abundances and hence to reconstruct the true composition of the matter in the nuclei of comets, volatiles, organic grains, and minerals. We would like, then, to ascertain the evolutionary effects that the comets may have undergone during their life history, with regard to their structure and composition, so that we can, at last, reach our ultimate objective and obtain a faithful description of the original characteristics.

Clearly, much remains to be done to achieve these ambitious goals. We could not stress too much the need to secure uniform, systematic data on many comets observed over a wide range of heliocentric distances (over whole orbits whenever feasible). Two-dimensional spectra as well as spatial mapping are particularly desirable and, for the brighter objects, (very) high-resolution spectra in all wavelength regions. Comets exhibit a striking diversity in their behaviour and properties. It is crucial to determine how much of the observed variations can be explained on account of geometrical or dynamical (age) factors, or of differences in excitation conditions. We have hints that some intrinsic, primeval differences may exist. The so-called "CO-rich" comets (like Morehouse, 1908 III, and Humason, 1962 VIII) provide a famous example, strongly suggesting that the content in CO or CO-bearing substances may vary appreciably from one comet to the next. Are there any more subtle, yet diagnostic variations of the detailed relative abundances in the cometary mix ?

The realization of our programme also depends very heavily upon the construction of complex models and hence upon the availability of reliable fundamental data, experimental and theoretical. Cometary scientists and physico-chemists should work together to ensure the success of the entreprise. The task is tremendous, but also very exciting in view of the most deeply significant information that is sought, probably related to our origins.

While preparing this review, I contacted quite a few colleagues who kindly clarified such or such point or gave advice on urgent data needs in cometary science. It is a pleasure to thank them all here, collectively but warmly. I owe special gratitude to M.E. Brown, M.R. Combi, J. Crovisier, B. Donn, P. Eberhardt, P.D. Feldman, W.F. Huebner, M.H. Moore, G. Moreels, M.J. Mumma, C.Y.R. Wu, and S. Wyckoff, for their important contributions and/or for providing me with valuable unpublished material.

REFERENCES

1. A'Hearn, M.F., in *Astrochemistry of Cosmic Phenomena*, IAU Symp. 150, ed. P.D. Singh (Kluwer, Dordrecht, 1992), p. 415.
2. A'Hearn, M.F., Hoban, S., Birch, P.V., Bowers, C., Martin, R., and Kinglesmith III, D.A., *Nature* **324**, 649 (1986).
3. Altenhoff, W.J. et al., *Astron. Astrophys.* **125**, L19 (1983).
4. Anders, E. and Grevesse, N., *Geochim. Cosmochim. Acta* **53**, 197 (1989).
5. Anicich, V.G., *Astrophys. J. Suppl. Ser.* **84**, 215 (1993).
6. Anicich, V.G. and Huntress, W.T., Jr., *Astrophys.J.Suppl.Ser.* **62**, 553 (1986).
7. Arpigny, C., in *The Study of Comets*, IAU Coll. 25, NASA SP-393, eds. B. Donn et al. (1976), p. 797.
8. Arpigny, C., in *Les éléments et leurs isotopes dans l'Univers*, 22nd Liège International Astrophys. Coll., p. 189 (1979).
9. Arpigny, C., Dossin F., Woszczyk, A., Donn, B., Rahe, J., and Wyckoff, S., *Atlas of Cometary Spectra* , to be published (1994).
10. Arpigny, C. and Manfroid, J., *Ciel et Terre* **106**, 5 (1990).
11. Balsiger, H., in *Comet Halley. Investigations, results, and interpretations*, ed. J. Mason (Ellis Horwood Ltd., Chichester, England, 1990), Vol. 1, p. 129.
12. Blake, D., Allamandola, L., Sandford, S., Hudgins, D. and Freund, F., *Science* **254**, 548 (1991).
13. Bockelée-Morvan, D., *Astron. Astrophys.* **181**, 169 (1987).
14. Brown, M.E. and Spinrad, H., private communication (1993).
15. Budzien, S.A. and Feldman, P.D., *Icarus* **99**, 143 (1992).
16. Budzien, S.A., Festou, M.C. and Feldman, P.D., *Icarus* **107**, 164 (1994).
17. Caldwell, J. et al., Poster presented at the *5th International Confer. on Lab. Res. for Planetary Atmospheres*, Boulder, Colorado (October 1993).
18. Combes, M. et al., *Icarus* **76**, 404 (1988).
19. Combi, M.R., *Icarus* **71**, 178 (1987).
20. Combi, M.R., Bos, B.J. and Smyth, W.H., *Astrophys. J.* **408**, 668 (1993).
21. Combi, M.R. and Delsemme, A.H., *Astrophys. J.* **237**, 633 (1980).
22. Combi, M.R. and Smyth, W.H., *Astrophys. J.* **327**, 1026 (1988).
23. Cosmovici, C.B. and Ortolani, S., *Nature* **310**, 122 (1984).
24. Cosmovici, C.B., Schwarz, W., Ip, W.-H. and Mack, P., *Nature* **332**, 705 (1988).
25. Crifo, J.F. in *Comets in the Post-Halley Era*, eds. R.L. Newburn et al. (Kluwer, Dordrecht, 1991), p. 937.
26. Crovisier, J., in *Infrared Astronomy with ISO*, eds. T. Encrenaz & M.F. Kessler (Nova Science Publishers, 1992), p. 159; see also poster presented at this conference.
27. Crovisier, J., *ibid.*, p. 221.
28. Crovisier, J., *J. Geophys. Res. (Planets)*, in press (1994).
29. Crovisier, J., in *Asteroids, Comets, Meteors 1993*, IAU Symp. 160 (Kluwer, in press).
30. Crovisier, J. and Bockelée-Morvan, D., this conference.
31. Crovisier, J., Despois, D., Bockelée-Morvan, D., Colom, P., and Paubert, G., *Icarus* **93**, 246 (1991).
32. Crovisier, J. and Schloerb, F.P. in *Comets in the Post-Halley Era*, eds. R.L. Newburn et al. (Kluwer, Dordrecht, 1991), p. 149.
33. Dalgarno, A., private communication (1993).

34. Delsemme, A.H., in *Comets in the Post-Halley Era*, eds. R.L. Newburn et al. (Kluwer, Dordrecht, 1991), p. 377.
35. Eberhardt, P. et al., *Astron. Astrophys.* **187**, 481 (1987).
36. Eberhardt, P., Meier, R., Krankowsky, D., and Hodges, R.R., *Astron. Astrophys.*, in press (1994).
37. Encrenaz, T. and Knacke, R., in *Comets in the Post-Halley Era*, eds. R.L. Newburn et al. (Kluwer, Dordrecht, 1991), p. 107.
38. Encrenaz, Th., Puget, J.L., d'Hendecourt, L., *Space Sc. Rev.* **56**, 83 (1991).
39. Engel, S., Lunine, J.I., and Lewis, J.S., *Icarus* **85**, 380 (1990).
40. Erdman, P.W. and Zipf, E.C., *Planet. Space Sci.* **31**, 317 (1983).
41. Espinasse, S. et al., *Planet. Space Sci.* **41**, 409 (1993).
42. Espinasse, S., Klinger, J., Ritz, C., Schmitt, B., *Icarus* **92** ,350 (1991).
43. Fegley, B. and Prinn, R.G., in *The Formation and Evolution of Planetary Systems*, eds. H.A. Weaver & L. Danly (Cambridge Univ. Press, 1989), p. 171.
44. Feldman, P.D., in *Chemistry in Space*, eds. J.M. Greenberg and V. Pirronello (Kluwer, Dordrecht, 1991), p. 339.
45. Feldman, P.D. et al., *Astrophys. J.* **379**, L37 (1991).
46. Feldman, P.D., Fournier, K.B., Grinin, V.P., and Zvereva, A.M., *Astrophys. J.* **404**, 348 (1993).
47. Festou, M.C., *Astron. Astrophys.* **95**, 69 (1981).
48. Festou, M.C., Rickman, H., and West, R.M., *Astron. Astrophys. Rev.* **4**, 363; **5**, 37 (1993).
49. Geiss, J. et al., *Astron. Astrophys.* **247**, 226 (1991).
50. Geiss, J., *Rev. Modern Astron.* **1**, 1 (1988).
51. Gombosi, T.I., Nagy, A.F., and Cravens, T.E., *Rev. Geophys.* **24** , 667 (1986).
52. Gredel, R., van Dishoeck, E.F., and Black, J.H., *Astrophys. J.* **338**, 1047 (1989).
53. Green, S., *Astrophys. J.* **412**, 436 (1993).
54. Greenberg, J.M., in *Comets*, ed. L.L. Wilkening (University of Arizona Press, Tucson, 1982), p. 131.
55. Greenberg, J.M., in *Asteroids, Comets, Meteors 1993*, IAU Symp. 160, (Kluwer, Dordrecht, in press).
56. Greenberg, J.M. and Hage, J.I., *Astrophys.J.* **361** , 260 (1990).
57. Grewing, M., Praderie, F. and Reinhard, R. (eds.), *Exploration of Halley's Comet*, Springer-Verlag Berlin (1988). Also published as *Astron. Astrophysics* **187** (1987).
58. Grün, E., Benkhoff, J., and Gebhard, J., *Ann. Geophysicae* **10**, 190 (1992).
59. Hanner, M.S., Lynch, D.K., and Russell, R.W., *Astrophys. J.*, in press (1994).
60. Haser, L., *Bull. Acad. Roy. Belgique, Cl. Sc.* **43**, 740 (1957).
61. Houpis, HLF, Ip, W.H., and Mendis, D.A., *Astrophys. J.* **295**, 654 (1985).
62. Hudgins, D.M., Sandford, S.A., Allamandola, L.J., Tielens, A.G.G.M., *Astrophys. J. Suppl. Ser.* **86**, 713 (1993).
63. Hudson, R.L., and Moore, M.H., *Astrophys. J.* **404**, L29 (1993).
64. Huebner, W.F. (ed.), *The Physics and Chemistry of Comets*, Springer Verlag, Berlin (1990).
65. Huebner, W.F., private communication (1993).
66. Huebner, W.F., and Boice, D.C., *Origins of Life* **21**, 299 (1992).
67. Huebner, W.F., Boice, D.C., Schmidt, H.U. and Wegmann, R., in *Comets in the Post-Halley Era*, eds. R.L. Newburn et al. (Kluwer, Dordrecht, 1991), p. 907.
68. Huebner, W-F., Keady, J.J. & Lyon, S.P., *Astrophys. Space Sci.* **195** , 1 (1992).
69. Huebner, W-F. et al. (eds.), *Workshop on the Activity of Distant Comets* , (Southwest Research Institute, San Antonio, Texas, 1993).
70. Ibadinov, Kh.I., Rahmonov, A.A., and A. Sh. Bjasso, in *Comets in the Post-Halley Era*, eds. R.L. Newburn et al. (Kluwer, Dordrecht, 1991), p. 299.
71. Ip, W.-H. et al., *Ann. Geophysicae* **8**, 319 (1990).

72. Jackson, W.M., in *Comets in the Post-Halley Era*, eds. R.L. Newburn et al. (Kluwer, Dordrecht, 1991), p. 313.
73. Jackson, W.M., Bao, Y. and Urdahl, R.S., *J. Geophys. Res.* **96**, 17569 (1991).
74. Jessberger, E.K. and Kissel, J., in *Comets in the Post-Halley Era*, eds. R.L. Newburn et al. (Kluwer, Dordrecht, 1991), p. 1075.
75. Johnson, R.E., *J. Geophys. Res.* **96**, 17553 (1991).
76. Kawara, K., Gregory, B., Yamamoto, T., and Shibai, H., *Astron. Astrophys.* **207**, 174 (1988).
77. Kim, S.J. and A'Hearn, M.F., *Icarus* **90**, 79 (1991).
78. Klinger, J., in *Comets in the Post-Halley Era*, eds. R.L. Newburn et al. (Kluwer, Dordrecht, 1991), p. 227.
79. Krankowsky, D., in *Comets in the Post-Halley Era*, eds. R.L. Newburn et al. (Kluwer, Dordrecht, 1991) p. 855.
80. Lambert, D.L., Sheffer, Y, Danks, A.C., Arpigny, C., and Magain, P., *Astrophys. J.* **353**, 640 (1990).
81. Larson, H.P., Weaver, H.A., Mumma, M.J., and Drapatz, S., *Astrophys. J.* **338**, 1106 (1989).
82. Le Burlot, J. and Roueff, E., *J. Mol. Spectrosc.* **120**, 157 (1986).
83. Lunine, J.I., Engel, S., Rizk, B., and Horanyi, M., *Icarus* **94**, 333 (1991).
84. Lutz, B.L., *Astrophys. J.* **315**, L147 (1987).
85. Lutz, B.L., Womack, M and Wagner, R.M., *Astrophys.J.*, **407**, 402 (1993).
86. Malmasson, D., Vient, A., Lemaire, J.-L., Le Floch, A., Rostas, F., this conference.
87. Marconi, M.L., Mendis, D.A., Korth, A., Lin, R.P., Mitchell, D.L., and Reme, H., *Astrophys. J.* **352**, L17 (1990).
88. Mason, J. (ed.), *Comet Halley. Investigations, results, and interpretations*, Ellis Horwood Ltd, Chichester, England (1990).
89. McSween H.Y., Jr and Weissman P.R., *Geochim. Cosmochim. Acta* **53** , 3263 (1989).
90. Meier, R., Eberhardt, P., Krankowsky, D., and Hodges, R.R., *Astron. Astrophys.* **277**, 677 (1993).
91. Meier, R., Eberhardt, P., Krankowsky, D., and Hodges, R.R., *Astron. Astrophys.*, in press (1994).
92. Mendis, D.A., Houpis, H.L.F., and Marconi, M.L., *Fund. Cosmic Phys.* **10**, 1 (1985).
93. Meyer-Vernet, N., Strauss, M.A., Steinberg, J.L., Spinrad, H., and McCarthy, P.J., *Astron. J.* **92**, 474 (1987).
94. Millar, T.J., Rawlings, J.M.C., Bennett, A., Brown, P.D., and Charnley, S.B., *Astron. Astrophys. Suppl. Ser.* **87**, 585 (1991).
95. Mitchell, D.L. et al., *Icarus* **98**, 125 (1992).
96. Moore, M.H., private communication (1993).
97. Moore, M.H. and Hudson R.L., *Astron. Astrophys. Suppl. Ser.* **103**, 45 (1994).
98. Moore, M.H., Khana, R., and B. Donn, *J. Geophys. Res.* **96**, 17541 (1991).
99. Moreels, G., Clairemidi, J., Hermine, P., Brechignac, P., Rousselot, P., *Astron. Astrophys.* **282**, 643 (1994).
100. Mumma, M.J., private communication (1993).
101. Mumma, M.J., Blass, W.E., Weaver, H.A., and Larson, H.P., in *Proc. STScI Conf. on Origins and Evolution of Planetary Systems*(STScI, Baltimore, 1988).
102. Mumma, M.J., Weissman, P.R., and Stern S.A., in *Protostars and Planets III*, eds. E.H. Levy et al., (University of Arizona Press, Tucson, 1993), p. 1177.
103. Newburn, R.L. et al. (eds.), *Comets in the Post-Halley Era*, Kluwer, Dordrecht (1991).
104. O'Dell, C.R., Robinson, R.R., Krishna-Swamy, K.S., McCarthy, P.J., and Spinrad, H., *Astrophys. J.* **334**, 476 (1988).
105. Prialnik, D., *Astrophys. J.* **418**, L49 (1993).
106. Prialnik, D., in *Asteroids, Comets, Meteors 1993* , (Kluwer, Dordrecht, in press).
107. Rickman, H., in *Asteroids, Comets, Meteors 1993*, (Kluwer, Dordrecht, in press).

108. Rousselot, P., Clairemidi, J., and Moreels, G., *Astron. Astrophys.*, in press (1994).
109. Sack, N.J. and Baragiola, R.A., *Phys. Rev. B*, **48**, 9973 (1993).
110. Sagan, C. and Khare, B.N., *Nature* **277**, 102 (1979).
111. Samarasinha, N.H. and Belton, M.J.S., *Icarus*, in press (1994).
112. Sandford, S.A. and Allamandola, L.J., *Astrophys. J.* **417**, 815 (1993).
113. Schleicher, D.G., in *Asteroids, Comets, Meteors 1993*, (Kluwer, Dordrecht, in press).
114. Schmidt, H.U., Wegmann, R., Huebner, W.F., and Boice, D.C., *Comp. Phys. Comm.* **49**, 17 (1988).
115. Schmuttenmaer, C.A. et al., *Science* **249**, 897 (1990).
116. Schulz, R., A'Hearn, M.F., and Samarasinha, N.H., *Icarus* **103**, 319 (1993).
117. Schutte, W.A., Allamandola, L.J., and Sandford, S.A., *Icarus* **104**, 118 (1993).
118. Singh, P.D. and Dalgarno, A., in *Symp. on the Diversity and Similarity of Comets*, ESA SP-278, 177 (1987).
119. Singh, P.D. and Gruenwald, R.B., *Astron. Astrophys.* **178**, 277 (1987).
120. Smith, A.M., Stecher, T.P., and Casswell, L., *Astrophys. J.* **242**, 402 (1980).
121. Spinrad H., *Ann. Rev. Astron. Astrophys.* **25**, 231 (1987).
122. Stern, S.A., Green, J.C., Cash, W., and Cook, T.A., *Icarus* **95**, 157 (1992).
123. Strazzulla G. and Johnson, R.E., in *Comets in the Post-Halley Era*, eds. R.L. Newburn et al. (Kluwer, Dordrecht, 1991), p. 243.
124. Tokunaga, A.T. and Brooke, T.Y., *Icarus* **86**, 208 (1990).
125. Ulich, R.L. and Conklin, E.K., *Nature* **248**, 121 (1974).
126. Valk, J.H., O'Dell, C.R., Cochran, A.L., Cochran, W.D., Opal, C.B., and Barker, E.S., *Astrophys. J.* **388**, 621 (1992).
127. Vanysek, V., in *Comets in the Post-Halley Era* , eds. R.L. Newburn et al. (Kluwer, Dordrecht, 1991), p. 879.
128. Vanysek, V., Valnicek, B., and Sudova, J., *Nature* **333**, 435 (1988).
129. Vervloet, M., personal communication (1993).
130. Wayne, R.P., *J. Geophys. Res.* **98**, 13119 (1993).
131. Weaver, H.A., in *Highlights of Astronomy* **8**, ed. D. McNally, (Kluwer, Dordrecht 1989), p. 387.
132. Weaver, H.A., Feldman, P.D., McPhate, J.B., A'Hearn, M.F., Arpigny, C., and Smith, T.E., *Astrophys. J.* **422**, 374 (1994).
133. Weaver, H.A., Mumma, M.J., and Larson, H.P., in *Comets in the Post-Halley Era*, eds. R.L. Newburn et al. (Kluwer, Dordrecht, 1991), p. 93.
134. Wu, C.Y.R., private communication (1993).
135. Wu, C.Y.R. and Chen, F.Z., *J. Geophys. Res.* **98**, 7415 (1993).
136. Wu, C.Y.R., Chen, F.Z., and Judge, D.L., Poster presented at the *4th International Confer. on Lab. Res. for Planetary Atmospheres*, Munich (Oct. 1992).
137. Wyckoff, S., personal communication; publication in preparation (1993).
138. Wyckoff, S. and Theobald, J., *Adv. Space Res.* **9**, No.3, 157, (1989).
139. Xie, X. and Mumma, M.J., *Astrophys. J.* **386**, 720 (1991).
140. Yamamoto, T., in *Comets in the Post-Halley Era*, eds. R.L. Newburn et al. (Kluwer, Dordrecht, 1991), p. 361.
141. Yung, Y.L., Allen, M., and Pinto, J.P., *Astrophys. J. Suppl. Ser.* **55**, 465 (1984).

GAS-GRAIN CHEMISTRY IN PROTOSOLAR NEBULA AND COMPOSITION OF COMETS

Zdeněk Moravec and Vladimír Vanysek
Astronomical Institute, Charles University
Svedska 8, 15000 Prague, Czech Republic

ABSTRACT

The time dependent chemistry of dense interstellar clouds is applied in order to estimate the possible composition of the volatile component of a comet nucleus originated in the protosolar nebula. Both gas-phase and grain-surface chemistry are considered and initial gas-phase atomic abundances are assumed to be protosolar. Physical conditions in the protosolar nebula in the phase of dust grains formation are assumed to be similar to those in dense, quiescent interstellar clouds. The gas-phase and grain-surface chemistry scheme is based on similar models discussed other authors, namely by Hasegawa et al. [1,2]. It is assumed that resulting surface abundances of molecular species in grain mantles at times 10^6 years, when the gas phase is substantially depleted of its heavy molecules, remain preserved in cometary dust and volatile constituents. Therefore the comparison of the modeled molecular composition of cometary material with abundances derived from infrared observations and data provided by measurements "*in situ*" on the VEGA spacecraft may throw a new light on the processes preceding the formation of the solar system. The results indicate that water, as expected from generally accepted comets composition, is a dominant volatile constituents. However other relatively highly abundant species according here discussed model in the grain mantles are CH_4 and CO_2. This is in contradiction to observational data of comets in which these molecules tend to be only marginal even as parent compounds. More consistent with observed abundances seems to be models in which the desorption induced by the cosmic rays and with so called "normal" initial gas phase abundance. The preliminary results combined with another evidences concerning the cometary dust composition suggest that the dust grains coming from the interstellar environments with relatively stable mantles and are preserved in their pristine form in those primitive solar system bodies as are represented by comets.

INTRODUCTION

Comets seem to be composed of matter, which is supposed to have the same molecular composition as protosolar nebula. Although there are no unbiased evidence that cometary nuclei retain the molecular composition inherited from the protosolar cloud, the observed properties of comets indicate that there is at least a relation between cometary composition and the material properties of dense interstellar clouds. Therefore the origin of comets could be searched in the cold stages of the protosolar nebula and molecular abundances of grain mantles in this nebula may be similar to those in the cometary dust. as molecular abundances in comet nucleus. It is suggested that comets contain pristine, virtually unaltered protosolar material and their study might be very relevant way to more information about processes in early stages of the solar nebula. Our knowledge about composition of the cometary nucleus is still relatively scarce, but we can partly deduce it from data obtained either by ground-based spectroscopy or by "*in situ*" mass spectrometry from space experiments. Most important were the discovery of fluffy CHON particles composed partly or even completely from compounds containing light elements.

The attempt of this study is to simulate conditions in protosolar cloud, where were formed mantles of volatile species on the cometary dust grains. Molecules are formed both in gas phase and on grain surfaces. Gas-phase species could stick on grains and form another molecules, on the contrary species on grain mantle could evaporate into gas phase (it concerns mainly molecule

H_2). The combined grain-surface and gas phase chemistry in the interstellar quiescent clouds has been discussed by several authors (Pickles and Williams [3,4]; Allen and Robinson [5]; Thielens and Hagen [6]; d'Hendecourt, Allamandola and Greenberg [7]; Brown and Charnley [8]) grain surfaces. This work is inspired by models of Hasegawa and colleagues, who studied gas-grain chemistry in dense, quiescent interstellar clouds (Hasegawa et al. [1,2]). They considered formation of more complex molecules, evaporation from grain mantle and in new models also desorption induced by cosmic rays. In most of these models the physical conditions are held fixed at some standard values. The organic molecules can grow on the grain surface as well as in the gas phase as the molecular hydrogen is dominant (Brown [9]). On the other hand atomic hydrogen may more easily converted reactive radicals, which may be already adsorbed on the grain surface, into more into more stable hydrogen containing molecules (Hasegawa et al. [1]). If the initial gaseous abundance is dominated by an almost purely neutral atomic species, the final molecular composition on the grains consist to little if any complex molecular compounds. Since the molecular content in cometary dust is at least partly organic, there are principal constrains in the initial chemical composition: both atomic as well as the molecular hydrogen must be present in sufficient amount.

THE MODEL

In the model discussed in this study we use gas-phase reaction network of Prasad and Huntress [10] corrected and updated by Duley and Williams [11]. Grain-surface reaction network is adopted from Hasegewa et al.[1]. We assume that solar abundances of elements are the same as protosolar ones and values by Anders and Grevese [12] were used. The initial gas-phase composition is atomic and neutral. Exception is molecule H_2, already partially formed on grains. Conditions within cloud and initial elemental abundances are shown in Table 1.

Table 1. Summary of the parameters used in our model.

Temperature of gas	T_g	$10K$
Temperature of dust grains	T_d	$10K$
Total hydrogen density	n_H	$10^4 cm^{-3}$
Density of grains	n_d	$1.3 \times 10^{-8} cm^{-3}$
Grain radius	a_d	$1.0 \times 10^{-5} cm^{-3}$
Visual extinction	A_V	$5\ mag$
Cosmic ray ionisation	$\zeta_{C.R.}$	$1.3 \times 10^{-17} s^{-1}$
Initial fractional abundances	H	0.4
(relative to total hydrogen)	H_2	0.3
	He	0.0975
	C	3.6×10^{-4}
	N	1.1×10^{-4}
	O	8.5×10^{-4}
	Na	2.1×10^{-6}
	Mg	3.9×10^{-5}
	Si	3.6×10^{-5}
	S	1.8×10^{-5}
	Fe	3.2×10^{-5}

The model is generated by solution of a coupled system of differential equations :

$$\frac{df_i}{dt} = G_i - \lambda_i f_i + \xi_i^{evap} f_i^S \tag{1}$$

$$\frac{df_i^S}{dt} = \lambda_i f_i + S_i - \xi_i^{evap} f_i^S \tag{2}$$

In this equations $f_i = n_i/n_H$ is the gas phase fractional abundance of ith chemical species (where $n_H = n(H) + 2n(H_2)$ is the total hydrogen density) and $f_i^S = n_i^S/n_H$ is its grain surface fractional abundance. The formation rate of gas phase reactions G_i include net production of i by gas, formation rate of surface reactions S_i include net production of i by reactions on grain surfaces. Accretion rate is given by

$$\lambda_i = 2\pi a_d^2 n_d \sqrt{\frac{2RT_g}{\pi M_i}} \tag{3}$$

where a_g is grain radius, n_g is density of grains by number, R is the gas constant, T is gas kinetic temperature and M_i is the mass of ith species in atomic mass units. It is assumed that sticking probability is 1 for all neutral species, and 0 for all species with charge, so ions do not stick on grain surface. Evaporation rate of the ith species is given as

$$\xi_i^{evap} = (t_i^{evap})^{-1} \tag{4}$$

where t_i^{evap} is the thermal evaporation time scale, which is meaningful only for H, H_2 and He. For other species is very large so they remain on grains. In here discussed models only thermal evaporation is considered and other desorption mechanisms are neglected.

There are 1425 gas-phase reactions in the model involving 138 gas-phase species and 156 surface reaction involving 118 species on the grain surface. We developed program for solving system of differential equations (1) and (2), which is based on Rosenbrock's numerical method of solving stiff differential systems. This method is linearized implicit Runge-Kutta method. There is system of N linear equations (N is sum of species in gas phase and on grains - 254 in our model), which are solved by using Gauss' method. The model is followed to a time of 10^7 years, at which most of species are lost from the gas phase.

DISCUSSION

The models results and comparison with assumed cometary molecular abundances are summarized in Tables 1. 2. and 3. as well as in graph. It must be noted that all models in which the influence of grains is included suffer from uncertainties associated with grain chemistry. Comments to this problem can be found for instance in above cited study by Hasegawa and Herbst [2]. Since in here presented results considered only classical "sticking" process is considered, the results must be rated as some kind of tests whether or not the gas-grain chemistry in principle may produce the comparable molecular compositions as supposed in the cometary material. In the Table 2. are compared the estimated abundance of selected species in P/Halley comets compiled by Encrenaz an Knacke [13] (noted as [E91]) and Yamamoto [14] [Y91] from the infrared spectroscopy and from the mass spectroscopy provided by the VEGA fly-by experiments in 1986. with data in this study (MV) as well as with models denoted as (HH,cr) obtained by Hasegawa and Herbst [2] including the cosmic ray induced desorption. The data [E91] and [Y91] represent only the bulk chemical composition, mostly of so called parent molecules in the volatile material. The specific chemical species contained in the organic mantles of the CHON particles is, however, unknown. Nevertheless, the assumed cometary abundances represent, in respect to the total cometary mass, substantial percentage of typical species of the volatile material.

The comparison in Table 3 is concerned on the abundances in the grain mantles. From this comparison is evident that the good agreement with cometary data is in the high content of water in all models, which is not surprising results at all. Relatively good agreement is between cometary formaldehyde H_2CO and this molecule abundance in all models and between OCS and NH_3 and (MV) models. On the other hand CH_4 exhibits the best agreement with (HH,cr) models, whereas (MV) models produce extremely high abundance of methane. However, the initial high content of CH_4 may led to further production of more complex stable compounds on the grain surface. The discrepancy between assumed cometary abundance and models is evident in (MV) results where this molecule is strongly underabundant, while the (HH,cr) models for early stage of the clouds chemical history fit fairly good the cometary CO. However, the CO is very likely not the parent molecule and may have its origin in formaldehyde and to less extend from CO_2. Only less than half of the observed CO coming from volatile ices in the nucleus, while the rest from the dust in coma (see comments Rickman and Huebner [15]). The production rates of CO and H_2CO from the distributed sources (probably dust) in the coma run indicate a close relationship.

The serious discrepancy was found for abundance of the CO_2, which is supposed to be a parent compound. Both models shown high abundance of this molecule which, according to the available data, seems to be one of relatively minor species in comets. However, the low abundance of observed CO_2 in comets may be observational artifact, because only observable emission band is at $4.3\mu m$, a wavelength at which are the strong atmospheric CO_2 absorption bands and above the atmosphere for observation of comets. In this spectral range until now are available only the five individual spectra of P/Halley taken by IKS-VEGA infrared spectrometer (see Encrenaz and Knacke [13]). The CO_2^+ has been observed in many comets, but the lack of knowledge of ionization processes has inhibited any attempts to deduce the abundance of CO_2 in those objects

Table 2. Comparison between molecular abundances in cometary material and models of the grain mantles normalized to H_2O

Molecule	Observations		Model (MV)			Model (HH,cr)		
	[91E]	[91Y]	3×10^5 yr	10^6 yr	10^7 yr	3.2×10^5 yr	10^6 yr	10^7 yr
H_2O	1.000	1.000	1.000	1.000	1.000	1.000	1.000	1.000
CO_2	0.03 ± 0.01	0.02 - 0.04	0.120	0.24	0.25	0.12	0.12	0.10
H_2CO	0.04 ± 0.02	(0.04)	0.006	0.005	0.005	0.04	0.19	0.20
CO	0.05 ± 0.02	0.15 - 0.20	< 0.001	< 0.001	< 0.001	0.09	0.04	< 0.001
OCS	0.007 ± 0.003	-	0.005	0.01	0.010	< 0.001	< 0.001	< 0.001
HCN	0.001 ± 0.0005	0.001	0.011	0.016	0.016	0.06	0.04	0.03
CH_4	0.02 ± 0.01	0.005-0.020	0.44	0.32	0.31	0.02	0.02	0.02
NH_3	-	0.01 - 0.02	0.09	0.06	0.06	< 0.001	< 0.001	0.014
N_2	-	< 0.02	< 0.001	0.008	< 0.001	0.03	0.03	0.006

Molecular abundances in volatile cometary material:
[91E] ... Encrenaz and Knacke [13]
[91Y] ... Yamamoto [14]

Computed abundances in grain mantle :
Model (MV)   ... this study
Model (HH,Cr) ... model of Hasegawa and Herbst [2] with presence of cosmic ray induced desorption

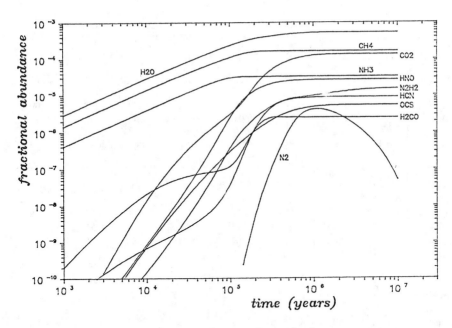

Figure 1: Time evolution of some surface molecules

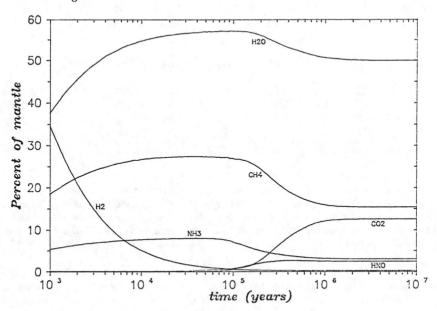

Figure 2: Time evolution of grain mantle composition

Table 3 : Molecular composition of grain mantle (in percent) and total fraction of species on grains relative to n_H

Time(yr)	H_2	N_2	HNO	C_3O	N_2H_2	H_2O_2	SiH_4	C_3H_4
1.0E+05	0.57	0.00	0.66	0.00	0.10	0.00	1.97	0.00
3.0E+05	0.29	0.01	2.53	0.01	0.61	0.01	2.42	0.12
1.0E+06	0.22	0.41	2.49	0.03	0.93	0.03	2.98	0.13
1.0E+07	0.21	0.00	2.42	0.02	1.39	0.01	3.11	0.13

Time(yr)	SiS	SiO	NS	H_2O	HCN	CO_2	H_2S	OCS
1.0E+05	0.00	0.00	0.00	57.20	0.01	0.77	0.83	0.08
3.0E+05	0.03	0.01	0.01	54.28	0.55	5.93	0.77	0.26
1.0E+06	0.11	0.08	0.02	50.67	0.83	12.14	0.62	0.47
1.0E+07	0.11	0.11	0.03	49.96	0.83	12.58	0.61	0.50

Time(yr)	NH_3	CH_4	HC_3N	CH_3CN	CH_3OH	NaH	FeH	O_3
1.0E+05	7.13	26.93	0.00	0.00	0.03	0.09	1.18	0.00
3.0E+05	4.16	21.12	0.00	0.09	0.38	0.12	1.60	0.00
1.0E+06	3.14	15.89	0.11	0.30	0.32	0.17	2.44	0.02
1.0E+07	3.04	15.38	0.00	0.30	0.31	0.19	2.90	0.00

Time(yr)	H_3C_3N	C_4H_4	C_2H_6	SO_2	H_2CO	MgH_2	HNC	Grain total
1.0E+05	0.00	0.00	0.02	0.00	0.03	2.11	0.30	4.20E-04
3.0E+05	0.02	0.01	0.70	0.04	0.31	2.64	0.95	7.81E-04
1.0E+06	0.20	0.02	0.58	0.29	0.24	3.39	0.74	1.04E-03
1.0E+07	0.31	0.02	0.56	0.41	0.23	3.59	0.72	1.07E-03

(A'Hearn and Festou [16]). Nevertheless, the abundance of CO and CO_2 is very likely not so low as seems to be only from scarce spectroscopic data. The highly volatile molecules, as are both CO and CO_2, may cause frequently observed brightness outbursts of comets at large heliocentric distances, connected with the dust ejection from the nucleus surface. On the other hand, the disproportion in the observed and modeled abundance of the CO_2 may indicate that comets were formed at temperature from 30K to 60K. The same conclusion holds for the CH_4 and NH_3.

CONCLUSIONS

The cometary matter is certainly not identical with the typical material of interstellar cool dense clouds, but it is closer to it than any other type of matter in solar system so far accessible to us. The data from comets combined with models of chemical evolution of matter in environment similar as prevailed the early stage of presolar nebula may at least impose constrains on the condition for comet formation. Here presented study is a preliminary contribution to such studies.

References

1. T.I.Hasegawa, E.Herbst, C.M.Leung: *Astrophys. J. Suppl.* **82**, 167 (1992).
2. T.I.Hasegawa, E.Herbst: *Monthly Not. R.A.S.* **261**, 83 (1993).
3. J.B.Pickles, D.A.Williams: *Astrophys & Space Sci.* **52**, 443 (1977).
4. J.B.Pickles, D.A.Williams: *Astrophys & Space Sci.* **52**, 453 (1977).
5. M.Allen, G.W.Robinson: *Astrophys. J.* **212**, 396 (1977).
6. A.G.G.M.Thielens, W.Hagen: *Astron. & Astrophys.* **114**, 245 (1982).
7. L.B.d'Hedecourt, L.J.Allamandola, J.M.Greenberg: *Astro.& Astrophys.* **152**, 130 (1985).
8. P.D.Brown, S.B.Charnley: *Monthly Not. R.A.S.* **244**, 432 (1990).
9. P.D.Brown: *Monthly Not. R.A.S.* **244**, 65 (1990).
10. S.S.Prasad, W.T.Huntress: *Astrophys. J. Suppl.* **43**, 1 (1980).
11. W.W.Duley, D.A.Williams: *Interstellar chemistry*, Academic Press (London) p.253 (1984).
12. E.Anders, N.Grevesse: *Geochimica et Cosmochimica* **53**, 197 (1989).
13. T.Encrenaz, R.Knacke: in *Comets in the Post-Halley Era.* (R.L.Newburn, M. Neugebauer, J. Rahe, eds.) Kluwer Academic Publ. Dordrecht, The Netherlands, p.107 (1991).
14. T.Yamamoto: 1991, in *Comets in the Post-Halley Era.* (R.L. Newburn, M. Neugebauer, J. Rahe, eds.) Kluwer Academic Publ. Dordrecht, The Netherlands, p.361 (1991).
15. H.Rickman, W.F.Huebner: in *Physics and Chemistry of Comets* (W.F. Huebner, ed.) Springer Verlag, Heidelberg, p.254 (1990).
16. M.F.A'Hearn, M.C.Festou: in *Physics and Chemistry of Comets* (W.F. Huebner, ed.) Springer Verlag, Heidelberg, p. 254 (1990).

DISCUSSION

WDOWIAK — I would like to point out that elemental sulfur is quite abundant in meteorites to the degree that it was discovered almost 150 years ago by Wohler. Also sulfur occurs in the sulphate form which may account for not observing SO and SO_2 as indicated in your abstract. It would be useful to search for sulfur in meteors at ultraviolet wavelengths from the perspective afforded by low earth orbit.

VANYSEK — Although it is now well accepted that interstellar material in a form of grains is present in meteorites, this may be not valid for the refractory material in comets. Especially, the iron meteorites may have different history as the nuclei of comets. From the analysis of meteoric streams associated with comets it is evident, that dust particles released from cometary nuclei have low bulk density and "fluffy" structure.

LEACH — Was the mass resolution in the PUMA experiments sufficient to identify sulfur with certainty ? What interfering species are possible at the observed m/z value?

VANYSEK — The resolution power of the PUMA experiment was about 0.5 amu, and identification of atomic S is definitively positive. No interfering species are expected at 32 amu signal. From the inspection of the reduced spectra a correlation between the occurrence of 32 amu and 34 amu (H_2S ?) has been found, which may be, however, some kind of instrumental effect. Very few signals were found at 64 amu and around 94-98 amu. Thus, the PUMA spectra provide informations only about atomic S.

ON A POSSIBLE MECHANISM FOR PRODUCTION OF S_2 FROM COMETS

Vladimír Vanysek
Astronomical Institute, Charles University
Švédská 8, 150 00 Prague, Czech Republic

ABSTRACT

Although sulfur is relatively abundant element in comets and sulfur-bearing molecules are detected in numerous cometary spectra, the S_2 molecule has been observed until now only in comet 1983 VII. From the available data is evident that the abundance of this sulfur dimmer strongly varies and very rarely is produced in observable amount. Most realistic value of the abundance relative to water seems to be less than 10^{-4}. The absence of SO and SO_2 in comets rules out the possibility that S_2 is a residue of sulfur compounds processed in the mantles of interstellar grains. An alternative process for producing of S_2 would be a product of sputtering of dust grains containing H_2S, FeS or even solid S_8 by high energy solar wind protons. It is shown, that such a process could be efficient only if the dust mass distribution reached its maximum in sub-micron size range of the grains, because of the required effective cross section and quantum yield. Therefore the an additional size redistribution of the grains is necessary. As possible mechanism is proposed the bursting of electrically charged grains under the specific conditions, namely that the interplanetary magnetic field is radially oriented with the solar wind. It is, however, rare and transient phenomenon. Therefore so rare and transient is also the occurrence of the S_2 in comets.

INTRODUCTION

The presence of sulfur-bearing molecules, as CS, SH, OCS, H_3S^+ as well as atomic S in comets has been confirmed in many cases. The sulfur dimmer S_2 has been unambiguously detected only in comet IRAS-Araki-Alcock 1983 VII (A'Hearn et al [1]). The emission spectral features of the band S_2 $B^3\Sigma_u^- - X^3\Sigma_g^-$ were observed at the wavelength interval 280-310 nm in the spectra obtained with the *International Ultraviolet Explorer* (*IUA*) satellite during the comet's closest approach to the Earth (with minimal geocentric distance $\Delta = 0.032$ AU). Since emission of sulfur dimmer was observed only in the most inner part of the coma, limited by distance about 200 km from the nucleus, it is suggested that this molecule is a compound with short lifetime in the solar radiation field and could be observed only in a case of very high spatial resolution. The fluorescence calculation have been made by Kim et al. [2] and the production rate of this sulfur dimmer has been discussed by Budzien and Feldman [3]. They estimated the upper possible limit of S_2 production rate in several comets relative to water less than 10^{-3}. However, most realistic value seem to be less than 10^{-4}, as can be inferred from the observed S_2 band in the spectra of above mentioned comet 1983 VII. Also. a tentative identification of S_2 emission in *IUE* spectra of several comets in relatively large geocentric distance (up to 2.6 AU) have been reported by Krishna Swamy and Wallis [4], but the validity of this finding is questionable. From available data is evident that the abundance of this sulfur dimmer varies strongly in short time-scale and very rarely is produced in substantial observable amount. The question how the S_2 might have been produced in comets remains still open.

Possible Mechanism for Production of S_2

The source of this molecule in cometary material has been subject of considerable debate. The probability that S_2 has been directly formed on the interstellar dust particles during the presolar nebula early stage is very low. The absence of SO and SO_2 in comets rules out the possibility that S_2 is a residue of sulfur compounds processed in the mantles of the dust grains. On the other hand, the sulfur dimmer may be produced (as a secondary product) by irradiation of sulfur compounds in icy mantles of interstellar grains (Grim and Greenberg [5]). An alternative process for producing of S_2 is by sputtering of dust grains containing H_2S, FeS, or even solid S_8 by high energy solar wind protons. Recently A'Hearn [6] has pointed out that the dust particles might be sources of this molecule. In this study we examine some relevant processes concerning the dust grains as parent particles of S_2.

SPUTTERING OF DUST GRAINS BY HIGH ENERGY PROTONS

The cometary dust seems to be a most promising candidate as a source of the sulfur dimmer in comets. Boring et al. [7] and Chrisey et al. [8] showed that S_2 was produced by sputtering solid sulfur or solid H_2S. The diagrams in Fig 1. constructed from the data obtained by the mass spectroscopy provided by the PUMA 2 experiment during the fly-by of VEGA 2 near the coma of P/Halley indicate, that S is preferably associated with C. Since the H_2S is too volatile constituents, the existence of solid H_2S grains in cometary dust is unlikely. But this molecule in presence of H_2O and Fe can be converted to FeS.

Therefore it can be suggested that S_2 is produced by sputtering of either grains containg FeS by the impact of protons or electrons. In such a case the production rate $Q_{v,j}$ of a species j per unit volume is given by

$$Q_{v,j} = F q Z_i P_j \int_0^\infty \sigma(a) n(a) da \qquad (1)$$

where σ is the cross section of the grain, n number density of grains with radius a, F is the flux of high energy photons, q is the sputtering efficiency, i.e. the yield of atoms or molecules per impact, Z_i the relative abundance of the i-th element, and P_j is the probability that the given element will be bound in the observed species j. Thus, the effective yield of this species is

$$Y_j = q Z_i P_j \qquad (2)$$

Since the production rate is derived by observation, then the reliability of the assumed production mechanism could be examined under assumption of the other observable parameters. The flux of the energetic protons F is also known from observations. In the case of the comet 1983 VII, the solar wind flux with proton energy 0.5 to 7 MeV was about in range of $8.8\ 10^7$ to $2\ 10^8 cm^{-2} sec^{-1}$ (Russell et al. [9]). The number of grains per unit volume can be substituted by the total number N_d of dust particles in the projected observed volume V of the coma (defined by the projected instrumental aperture), thus

$$N_d = V \int_0^\infty n(a) da \qquad (3)$$

Substituting the cross section $\sigma(a)$ by the mean effective cross section related to the mean radius a, with mean mass m_d of the particle, then the minimal total mass $M_{d,j} < m_d N_d$ of dust grains required for production of Q_j is

$$M_{d,j} = 4/3(a\rho Q_j/FY_j) \qquad (4)$$

where ρ is mean density of the particles. The observable column density of the short lived S_2 is virtually confined in the innermost region of the coma and the total dust mass projected into this region M_d might be

$$M_d/M_{d,j} > 1 \qquad (5)$$

Assuming that the ratio of the gas to dust production rates in comet 1983 VII was $Q_g/Q_d = 1$, and H_2O production 10^{28} molecules sec^{-1} as equivalent to Q_g, then the dust production would be $Q_d = 300$ kg sec^{-1}. With the dust outflow velocity 0.2 km sec^{-1}, the total mass of the dust in the projected area of the coma, i.e in the optically thin column with base 240 x 110 km (corresponding to the slit dimension 9.3" x 4.3" and to the minimal geocentric distance 0.035 AU) is $M_o > 3\ 10^5$ kg. The dust mass of about 10^6 kg produced in time interval $3.6\ 10^3$ sec could be assumed as upper limit of M_d in the projected area of the coma where the observable abundance of the sulfur dimmer was produced. This upper limit value of the mass must be, at last, equal to the lower limit of the mass $M_{d,j} = M_{d,s}$ required for the S_2 production.

In the Fig 2. are shown lower limits of $M_{d,s}$ for given $Y_j = Y_{j,s}$ and for different radii of the dust grains with density 500 kg m^{-3}. Other parameter fit the observed production of S_2 molecules in comet 1983 VII, i.e $Q_s = 2\ 10^{24}$sec^{-1}, and observed proton flux $F = 2\ 10^8$ cm^{-2} sec^{-1}. The dust production rates of some comets are marked for comparison. As can be expected, the observed (or indirectly determined) dust production in comets require relatively high values of Y_j if the mean radius of dust grains is $a > 0.5$ micron. Thus, there are clearly defined constraints on the dust production rates and size of dust particles. The upper limit of Y_j is essentially determined by the product of qP_j, because the third parameter, the relative abundance $Z_i = Z_s$ of sulfur in the grains varies only by a factor of 4. The sulfur abundance in cometary dust seems to somewhat enhanced relative to the average abundance in the solar system. We have found highest mean value of $Z_s = 0.08$ from a set of 517 mass spectra provided by the PUMA 2 (i.e. VEGA 2). In contrast to this high value was found only 0.013 as the average value of Z_s from another set of 1296 mass spectra classified as "good" and "normal" provided by PUMA 1 experiment for the same particles mass range (5 10^{-12} to 3 10^{-18} kg). This discrepancy can be partly explained by the substantially large number of CHON grains in the PUMA 1 samples with high H, C, O, and N content in which heavier elements appear as relatively less abundant. From data published by Jessberger and Kissel [10] can be inferred, that in the dust of comet Halley the relative abundance of S is about 0.02 and about 0.03 in CI-chondrites. Therefore we have assumed as upper value for the sulfur abundance in the cometary dust $Z_s = 0.04$. Since the probability that from all sulfur species (atomic or molecular) released from grains will be produced S_2 is $P_j = P_s < 1$, then the value of the sputtering efficiency q must be in the range of 100, because the effective yield should be $Y_s > 0.1$, if the dust production rates in majority of comets $Q_d < 10^4$ kg sec^{-1}.

Possible Mechanism for Production of S_2

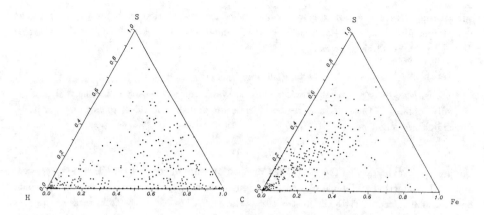

Fig. 1. Diagrams constructed from the data obtained by the mass spectroscopy provided by the PUMA 2 experiment during the fly-by of VEGA 2 near the coma of P/Halley indicate, that S is preferably associated with C.

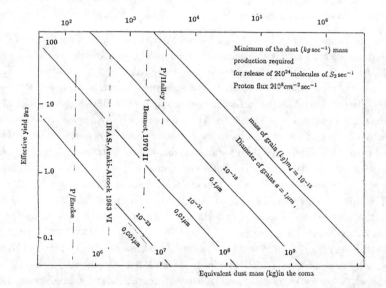

Fig. 2. Minimum of total dust mass and dust mass production required for production of $2\ 10^{24}$ S_2 molecules sec^{-1} by the solar wind proton flux $2\ 10^8$ cm^{-2}sec^{-1} as function of the effective yield Y_s and average size of the grains. Dust production rates for comets are labelled by their designation. If the upper limit for dust production is 10^4 kg sec^{-1} and upper limit of Y_s is about 4, then the required total mass of the dust must be confined in grains with diameters of about 0.01 microns.

Experimental data which provide values for parameters equivalent to above defined sputtering efficiency are in order 10 to 100 per 1.5 MeV He ion, depending on the chemical composition of the target (Strazzulla and Johnson [11]). In the case of sulfur target was found $q = 10$ and most abundant released neutral species were molecules of S_2, thus the probability $P_s = 1$. Low-energy protons (0.1 to 10 KeV) produce only 0.1 to 1 atom or molecule per proton impact (Behrisch [12,13]). In the range of MeV proton energy, the value $qP_s = 100$ seems to be the highest reasonable upper limit, and corresponding value for the effective yield Y_s could be about 4. From Fig. 2. is evident, that as far as the upper limit for $Y_s = 4$ can be assumed, then the average radius of grains which total mass corresponds to that dust mass in observed area in comet 1983 VII, ranges from $5\ 10^{-11}$ to 10^{-9}m or $a < 0.01$ microns and the smallest grains would be aggregates of only several hundredths of atoms.

SPUTTERING OF CHARGED GRAINS

Above discussed mechanism which might be responsible for the sulfur dimmer production could be efficient only if considerable amount of the dust is in form of dust grains in sub-micron size range. However, the dust size distribution obtained from in situ measurements in comet P/Halley (cf. McDonnell et al. [14]) substantial amount of total dust mass in the coma is confined in particles larger than 0.01 microns although new revision of relevant data shown, that the distribution function evidently has no cut beyond this size range and proceed almost smoothly toward very small grains. Nevertheless, the mass distribution at the nucleus of P/Halley derived by Divine [15] from the total dust fluence measured by Giotto PIA and DIDSY experiments (McDonnell et al. [14]) shows, that in the grains mass interval 10^{-15} to 10^{-18} kg the total dust mass M is at least 10^2 larger that those of the whole population of grains with mass $m_d < 10^{-18}$ kg. If the dust mass distribution - and consequently - also the size distribution of dust grains in comet Halley is typical for other comets, then mass production of grains in the size required for efficient sputtering by high energy protons is less than 10^{-5} of the total dust production. In such a case the total mass of dust grains in the size range most efficient for S_2 production could be so small that it will be not sufficient to produce any detectable amount amount of this molecule. It is the explanation of the fact that the sulfur dimmer was observed until now only in one comet. A considerable enhanced production of the sulfur dimmer could be expected in the innermost region of coma only if the maximum of the grains mass distribution will be strongly shifted toward sub-micron particles. Such a mass and size redistribution, if take place in the coma, must be caused by fragmentation of larger dust grains. Several fragmentation and evaporation processes have been proposed for cometary dust (cf. Boehnhardt et al. [16]). Rotational bursting due to to the radiation pressure on non-spherical particles can be ruled out because of its rather long time scale. For the same reason the sublimation of the particles would be also not quite efficient, although it remains as a possible mechanism which may lead to additional production of some radicals, for instance CN. The very fast fragmentation process is dust grain bursting by the electrostatic charge (Boehnhardt and Fechtig [17]). However, the electrostatic bursting of dust grains by the plasma flux in the solar wind can not be effective in the innermost cometary coma, unless one assumes following specific conditions:

a) The tensile strength of grains is not larger than 10^3 N m^{-2}, which corresponds to the strength of electric field 10^8V m^{-1}.

b) The orientation of the interplanetary magnetic field in the neighborhood of the coma is radially aligned in respect to the solar wind. In such a case, the magnetic barrier will be not created and the solar wind flow will be not braked and diverted by the bow-shock.

The first condition may be fulfilled for spherical fluffy grains of silicate-like composition and with size of about 1 micron. The equivalent of the critical surface potential ϕ is given by relation (Boehnhardt [18]):

$$\phi = a(2\mu_t e_d/e_v)^{1/2} \tag{6}$$

where μ_t is the tensile strength, e_d and e_v are dielectric constants of dust grains material and of vacuum. For silicate grains with $\mu_t = 10^2$ N m^{-2} the critical surface potential is about of 6.5 V and the fragmentation (charging) time in solar wind with average typical characteristics is about of 10 sec. The graphite particles are somewhat more stable and their charge bursting is effective for particles smaller than 0.5 microns. Small grains with radii $a < 0.3$ microns and tensile strengths up to 10^4N m^{-2} might be fragmented by charge bursting very efficiently and their critical charge time is of about 100 sec. Thus, if the solar wind could penetrate in innermost coma, then the charge bursting of dust particles may change the mass distribution maximum toward small size particles in time scale of 100 sec. However, under normal conditions, when the interplanetary magnetic field is oriented perpendicular to the solar wind, the dust grains in the comet are virtually shielded by the bow-shock and the critical surface charge will be not reached. In the case of the comet 1983 VII was the situation obviously unusual (cf. Russell et al. [9]) and the interplanetary magnetic field in the time of maximum intensity of the S_2 radially was aligned with the solar wind.

CONCLUSION

The production of S_2 in comets by the high energetic solar wind proton bombardments of dust grains will be efficient only if the maximum of the mass distribution is substantially shifted toward population of sub micron particles. Since sulfur dimmer is in solar radiation field short lived molecules (about 500 sec) its observable amount must be produced preferably in the most dense dust environment i.e. in the innermost coma. Under usual physical structure of the coma which prevails in most comets, in this particular region are not fulfilled two conditions necessary for the here discussed process. The maximum of the dust mass lies in the range of larger particles, and the solar wind is "cooled down" by the bow-shock. If the bombardments of dust particle by protons is indeed the main channel of S_2 production, then the probability that this molecule will be detected is almost nil. However, most favorable circumstance may occur if the magnetic field in the neighborhood of the comet is disturbed and the solar wind may penetrate into the dense coma. Then two independent processes take place: The electric potential of the larger particles is changed and enhanced to the critical potential which cause the bursts of individual particles and redistribution of mass with maximum among sub-micron grains in a relatively short time-scale.

This first very rarely occurred process is necessary for the substantial enhancement of the S_2 molecules. It is obviously transient and highly variable phenomenon which may explain strong variability of the S_2 production in the comet 1983 VII and absence of this sulfur dimmer in spectra of other comets observed in near UV spectral region.

REFERENCES

1. M.F. A'Hearn, P.D. Feldman, D.G. Schleicher, *Astroph. J.* **274**,*L*99 (1983).
2. S.J. Kim, M.F. A'Hearn, S.M. Larson, *Icarus* **87**, 440 (1990).
3. S.A. Budzien, P.D. Feldman, *Icarus* **99**, 143 (1992).
4. K.S. Krishna Swamy, M.K. Wallis, *Mon. Not. R.A.S.* **228**, 305 (1987).
5. R.J.A. Grim, J. M. Greenberg, *Astron. Astrophys.* **181**, 155 (1987).
6. M.F. A'Hearn, in *Astrochemistry of Cosmic Phenomena.* (P.D. Singh, ed.) Kluwer Academic Publisher, Dordrecht, The Netherlands, p. 415 (1992).
7. J.W.Boring, Z. Nansheng, D.B. Chrisey, D.J. O'Shaugnessy, J.A. Phipps, R.E. Johnson, in *Asteroids, Comets, Meteors II.* (C-I. Lagerkvist, B.A. Lindblad, H. Lundstedt, H. Rickman, eds.) Uppsala p.229 (1985).
8. D.B. Chrisey, R.E. Johnson, J.W. Boring, J.A. Phipps, *Icarus* **75**, 233 (1988).
9. C.T. Russell, J.G. Luhmann, D.J. Baker, *Geophysical Research Letters* **14**, 991 (1987).
10. E.K. Jessberger, J. Kissel, in *Comets, in the Post-Halley Era.* (R.L. Newburn, M. Neugebauer, J. Rahe eds.) Kluwer Acad. Publ. Dordrecht p. 1075 (1991).
11. G. Strazzulla, R.E. Johnson, in *Comets in the Post-Halley Era.* (R.L. Newburn, M. Neugebauer, J. Rahe, eds.) Kluwer Acad. Publ. Dordrecht p. 243 (1991).
12. R. Behrisch, *Sputtering by Particle Bombardment* Springer Verlag, Berlin Part I. (1981).
13. R. Behrisch, *Sputtering by Particle Bombardment* Springer Verlag, Berlin Part II. (1983).
14. J.A.M. McDonnell, P.L. Lamy, G.S. Pankiewicz, in *Comets in the Post-Halley Era.* (R.L. Newburn, M. Neugebauer, J. Rahe, eds.) Kluwer Acad. Publ. Dordrecht p.1043 (1991).
15. N. Divine, *ESA SP*-174 p. 25. (1987).
16. H. Boehnhardt, H. Fechtig, V. Vanysek, *Astron. Astrophys.* **231**, 543 (1990).
17. H. Boehnhardt, H. Fechtig, *Astron. Astrophys.* **187**, 824 (1987).
18. H. Boehnhardt, *Gleichgewichtpotential, Aufladenzeit und elektromagnetische Fragmentation kugelfoermiger Partikles.* PhD Thesis, University Erlangen-Nuernberg (1985).

EXCITATION MECHANISMS OF THE C_2 SWAN BANDS IN COMET HALLEY'S SPECTRUM

P. Rousselot, B. Goidet-Devel, J. Clairemidi, G. Moreels
Observatoire de Besançon, BP.1615, 25010 Besançon cedex, France

ABSTRACT

A new model suitable for calculating the populations of the triplet $d^3\Pi$, $c^3\Sigma$, $b^3\Sigma$, $a^3\Pi$ and the singlet $A^1\Pi$ and $X^1\Sigma$ states of C_2 is developed in using a Monte-Carlo method. The model shows that the time to reach fluorescence equilibrium is as long as 3000 s. It is used to calculate synthetic spectra of the C_2 $\Delta v = 0$ and $\Delta v = 1$ Swan bands. The band intensity ratio $I(1,1) / I(0,0)$ appears as an efficient parameter to compare the synthetic spectra with observational data obtained with the three-channel spectrometer of the Vega 2 mission. Two different initial situations are considered when the C_2 molecules are produced.

In the first one, the newly created C_2 molecules are not excited. The resulting spectrum provides an adequate analysis of the cometary spectrum assuming that the source is located at the nucleus. In the second one, the C_2 molecules are created with a higher rotational excitation. The synthetic spectrum in this case shows that the C_2 measured in the jets originate from a secondary diffuse source. Further refinement of the new method presented here would require a better knowledge of the singlet-triplet intercombination transitions.

INTRODUCTION

The presence of the C_2 Swan bands in the visible wavelength range is one of the most evident molecular features in cometary spectra. However, it is also one of the most puzzling emissions since the nature of the parent of C_2 is still unknown. The progenitor of this molecule is probably a grand-parent species, because the C_2 radial profile is very flat at cometocentric distances less than 10000 km. Although many cometary observations covering the range of the Swan bands are available, the C_2 fluorescence spectrum features are not, up to now, completely understood for two main reasons. First, pure vibration-rotation transitions cannot occur because C_2 is an homonuclear molecule. Intercombination transitions between singlet and triplet states also are, in principle, forbidden. As a consequence, in order to calculate the Swan band fluorescence transitions between the $a^3\Pi_u$ and $d^3\Pi_g$ electronic state, it is necessary to calculate also the populations of other levels, mainly $c^3\Sigma_u^+$, $b^3\Sigma_g^-$, $X^1\Sigma_g^+$ and $a^1\Pi_u$. Second, the production mechanisms of the C_2 molecules and their initial excitation states are not known, since the parent molecule is unidentified.

Among the first model studies of C_2, Krishna Swamy and O'Dell [1,2] included the $b^3\Sigma_g^-$ and $X^1\Sigma_g^+$ states, introducing an intercombination transition between the $a^3\Pi_u$ and $X^1\Sigma_g^+$ state. They used, as an experimental test, the intensity ratio of the $\Delta v = 1$ and $\Delta v = 0$ Swan band sequences. However, as explained by Le Bourlot and Roueff[3], additional data concerning the triplet-singlet transition moment is needed to improve the calculation of the different populations. More recently, Gredel et al.[4] developed a model including six electronic states, each of them having six vibrational levels with 60 rotational levels. From a comparison with spectra of Halley's comet ob tained with the Multiple-Mirror Telescope on Dec. 26-27, 1985 and May 4-5, 1986, they deduced that a possible value for the a-X transition moment was| Re |$^2_{a-X}$ = 3.5 10^{-6} a.u. In the following, a new approach to the C_2 problem is presented. It consists to use a Monte-Carlo algorithm to calculate the level populations. Then, a new parameter to test the validity of the model is defined as the intensity ratio of the (1,1) and (0,0) bands. The resulting synthetic spectra are compared with Halley's spectra obtained with the Vega 2 three-channel spectrometer.

A MONTE-CARLO MODEL FOR CALCULATING THE C_2 SWAN BANDS

Usual methods employed to compute fluorescence molecular spectra[1,5] are based on the inversion of matrices in order to solve a system that may include as many as several thousand linear equations. They provide the level populations once the fluorescence equilibrium is reached. In the case of C_2, this time is considerably longer than for heteronuclear molecules ; it is of the order of several thousand seconds.
As a consequence, it was felt necessary to construct a new model that would allow to study the evolution of the $d^3\Pi_g$ vibrational levels as a function of time. A Monte-Carlo method was developed to simulate the transition events of the C_2 molecules that were supposed to be a priori created in the lowest vibrational levels of the $X^1\Sigma_g^+$ or $a^3\Pi_u$ states. The method consists to follow the evolution with time of each molecule of the given sample. Let us suppose that the studied molecule is in the i state. A random number R_y, $0 < R_y < 1$, is drawn for each possible transition originating from the i level. This number is converted into a time using the formula $t_i = -\tau_{ij} \ln R_{ij}$ where τ_{ij} is the natural lifetime of the transition between the i and j levels. Then, the smallest time t_i is chosen ; the molecule suffers a quantum jump and the algorithm is repeated 10000 times.

Four allowed electronic transitions are taken into account : $a^1\Pi_u \rightarrow X^1\Sigma_g^+$ (Phillips), $b^3\Sigma_g^- \rightarrow a^3\Pi_u$ (Ballik - Ramsay), $d^3\Pi_g \rightarrow a^3\Pi_u$ (Swan) and $d^3\Pi_g \rightarrow c^3\Sigma_u^+$. Two intercombination transitions are also considered : $c^3\Sigma_u^+ \rightarrow X^1\Sigma_g^+$ and $a^3\Pi_u \rightarrow X^1\Sigma_g^+$. The transition probabilities are taken in Gredel et al.[4] for the bands taking their origin at the $^3\Pi_g$ level. They are taken in Chabalowski et al.[6] for the Ballik-Ramsay bands and in Van Dishoeck[7] for the Phillips system. In the case of the intercombination transitions, the transition moments | D_{a-X} |^2 and | D_{c-X} |^2 were both taken as equal to 2. 10^{-5} a.u.

MODEL CALCULATION OF THE $d^3\Pi$ STATE POPULATION

The first step in the calculation of the Swan band synthetic spectrum consisted in computing the $d^3\Pi_g$ state population. Two cases were considered depending upon the excitation level of the newly formed C_2 molecules. They were called distributions 1 and 2. In each case, the number of molecules was 10^4.
• Distribution 1 consisted of 75% of molecules initially created in the $X^1\Sigma_g^+$ state v = 0, J = 0-30, and 25% of molecules in the $a^3\Pi_u$ state, v = 0, J = 2-30.
• Distribution 2 consisted of 100% of molecules created in the $a^3\Pi_u$ state, but this time in the v = 1, J = 40-50 level.

The most significant results are presented in figure 1. The time required to reach the fluorescence equilibrium is long : 3000 s at least. The distributions as a function of the rotational number J are shown after a period of 100 s and after 1000s. In the case of the initial distribution 1 (boxes A and B), the final temperature obtained in fitting a Boltzmann-type distribution is 4800 K. In the case of the initial distribution 2 (boxes C and D), the final temperature is higher : 5300 K.

The following step consists in calculating the emission rate in the Swan bands. A synthetic spectrum is obtained in convoluting the calculated spectrum with the instrumental response function $\Delta\lambda \approx 6$ nm of the spectra used for comparison.

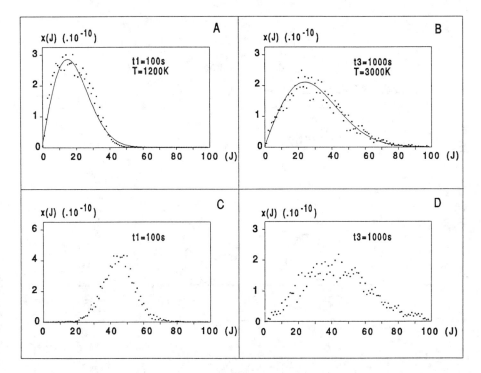

Fig. 1. Distributions of the C_2 populations in the $d^3\Pi_g$, v =0 state as a function of rotational number J. Boxes A and B refer to the initial distribution 1 defined in text : C_2 created with no excitation. Curves A and B show the rotational distribution after 100s and 1000 s. Boxes C and D refer to the initial distribution 2 defined in text : The C_2 molecules are created with higher rotational excitation. Curves C and D show the rotational distribution after 100 s and 1000 s.

COMPARISON WITH COMET HALLEY'S SPECTRA

The spectra of Halley's coma provided by the three-channel spectrometer[8] give a detailed description of an area of the coma limited by an angular sector having a radius of 40000 km, and its apex located 420 km apart from the nucleus. In this part of the inner coma, two well-differentiated jets are apparent in the cometary molecular emissions[9].

In an attempt to compare in figure 2 the synthetic spectra (A, B) with the observational data (C), a distinction was made between spectra originating from the most contrasted jet (J) and spectra originating from a region located between the two jets

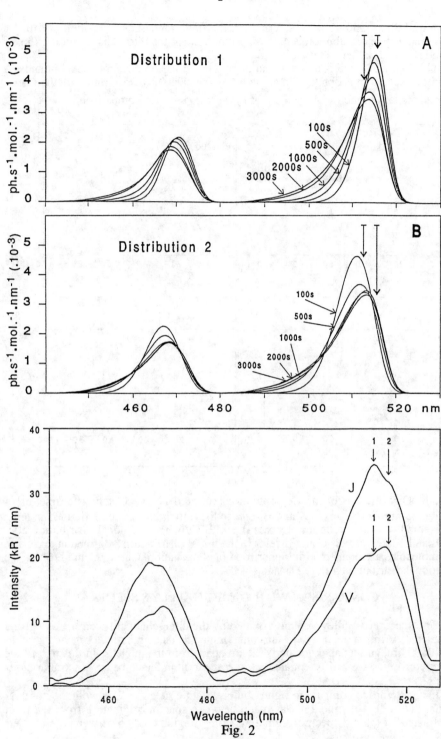

Fig. 2

Fig 2 : Comparison of the synthetic spectra obtained using the Monte-Carlo model with the cometary spectra measured by the Vega 2 three-channel spectrometer. A and B : synthetic spectra resulting of initial distributions 1 and 2. C : cometary spectra in the "valley" (V) and in the main jet (J).

called the "valley" (V). The wavelengths of the (1,1) and (0,0) bands at 512 and 516 nm are indicated by two arrows. The intensity ratio of these two bands R = I(1,1) / I(0,0) appears as an efficient function to test the different free parameters of the model.

In figure 2, box A, the C_2 molecules are supposed to be initially produced in a low excitation level previously defined as distribution 1. The $\Delta v = 0$ and $\Delta v = 1$ sequence shapes show an important evolution with time until the fluorescent equilibrium is reached. The ratio R initially is less than 1. It increases with time. In figure 2, box B, the C_2 initial distribution is supposed to be of type 2. The ratio R initially is higher than 1 and decreases with time.

The two cases considered when running the model appear suitable to propose an explanation for the spectral features of comet Halley's spectra presented in box C. The spectrum labelled V (valley) is emitted by molecules having their source at the nucleus as parent or grand-parent molecules. Given the effusion velocity ≈ 1 km.s^{-1}, the fluorescence equilibrium is not reached until the molecules have travelled over 5000 km. The spectrum labelled J (jet), on the contrary, is emitted by "freshly" produced C_2 molecules originating from the jet. The parent molecule(s) is (are) probably different from case 1. It is suggested that this second type of molecules could be produced by dust particles carried in dust jets to distances of 10000 - 30000 km. A large proportion of these particles could be made of CHON grains[10]. After having been heated by solar radiation during several hours, the particles would suffer fragmentation and release gas producing C_2 molecules in the observed high rotational excitation level.

CONCLUSION

The new model presented here, based on a Monte-Carlo algorithm appears well adapted to an analysis of the C_2 production mechanisms in comets. A comparison between synthetic spectra obtained with the model and cometary spectra provided by the Vega 2 three-channel spectrometer show that two different sources probably produce the C_2 molecules detected by their emission in the Swan bands. The intensity ratio of the (1,1) and (0,0) bands is an efficient parameter to trace a second diffuse source having its origin in a jet. The basic limitation of the model is mainly due to the lack of data concerning the intercombination transitions, principally between the $a^3\Pi_u$ and $X^1\Sigma_g^+$ states.

REFERENCES

1. K.S. Krishna Swamy, C.R. O'Dell, Ap. J. 216, p. 158 (1977)
2. K.S. Krishna Swamy, C.R. O'Dell, Ap. J. 231, p. 624 (1979)
3. J. Le Bourlot, E. Roueff, J. Molec Spectr. 120, p. 157 (1986)
4. R. Gredel, E.F. Van Dishoeck, J.H. Black, Ap. J. 338, p. 1047 (1989)
5. D.G. Schleicher, M.F. A'Hearn, Ap. J. 258, p. 864 (1982)

6. C.F. Chabalowski, S.D. Peyerimhoff, Chem. Phys. 81, p. 57 (1983)
7. E. Van Dishoeck, Chem. Phys. 77, 277 (1983)
8. G. Moreels, J. Clairemidi, J.P. Parisot et al., Astron. Astrophys. 187, 551, (1987)
9. J. Clairemidi, G. Moreels, V.A. Krasnopolsky, Icarus 86, p. 115 (1990)
10. E.K. Jessberger, A. Christoforidis, J. Kissel, Nature 332, p. 691 (1988)

IDENTIFICATION OF PHENANTHRENE IN HALLEY'S INNER COMA

G. Moreels, J. Clairemidi, P. Rousselot
Observatoire de Besançon, BP.1615, 25010 Besançon cedex, France

P. Hermine, P. Brechignac
Laboratoire de Photophysique Moléculaire, CNRS, Bât 213, France
Université de Paris Sud, 91405 Orsay cedex, France

ABSTRACT

Spectra of the inner coma of P/Halley at short distances from the nucleus were obtained in the near UV from the Vega 2 spaceprobe during the encounter session with the comet on March 9, 1986. The intensity is the sum of two components : solar radiation scattered by dust and molecular emissions mainly due to the fluorescence of the molecules in the coma. The residual spectrum obtained after subtraction of the intense dust-scattered continuum shows that a broad band emission feature between 342 and 375 nm progressively arises when the projected distance p becomes smaller than 1000 km. Three main bands at 347, 356 and 364 nm can be identified in this feature. A search for a molecule emitting fluorescence bands between 342 and 375 nm showed that phenanthrene was a possible candidate. A laser induced fluorescence experiment was conducted in the laboratory to measure the dispersed emission spectrum of phenanthrene under jet-cooled conditions. The comparison between the laboratory and the comet spectra revealed to be excellent. The agreement holds for both the wavelength positions of the peaks and their intensities. Our observations are coherent with the detection of an emission band at 3.2-3.5 micrometers in Halley's IR spectrum by the IKS spectrometer. This band was assigned to an X-CH organic compoud. A specific feature at 3.28 micrometers was suggested to be due to a polycyclic aromatic molecule. The identification of a PAH in Halley's UV spectrum emphasizes the similarities between cometary material and interstellar matter. It adds a strong argument in favor of the fact that comets would be constituted of interstellar material. It has important implications in the scenario scheme of the earlier solar system formation period.

INTRODUCTION

Since the early identifications of C_2, C_3, CN and CH in cometary spectra[1], the search for organic molecules in cometary material has been very active. In the mm and cm range, radio-astronomy has proven to be particularly efficient in identifying new molecules : HCN[1], H_2CO[2], H_2S and CH_3OH[3]. In the infrared, the spectrum revealed the presence of H_2O, CO_2[4] and of molecules of the type X-CH detected through the CH characteristic wavenumbers[5]. During their encounter with comet Halley, the Giotto and Vega spaceprobes provided unexpected information about cometary material. The Puma experiment showed that a large fraction of dust particles, called CHON, was made of light elements[6]. Another proof of the presence of organics was provided by the detection of the infrared stretch-band of CH at 3.2-3.5 nanometers by the IKS spectrometer[7].

In the following, new spectrometric data provided by the three-channel spectrometer during the Vega 2 closest approach on March 9, 1986, are presented. After subtracting the intense dust-scattered solar continuum, a broad-band emission gradually emerges in the 342-375 nm spectral range. A comparison with laboratory spectra of

© 1994 American Institute of Physics

phenanthrene shows that this polycyclic aromatic molecule is the carrier of the cometary emission.

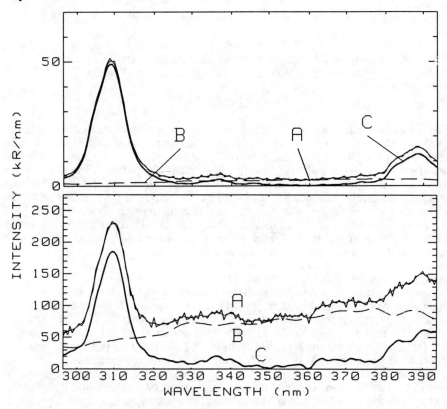

Fig 1 : Spectra of Halley's coma provided by the Vega mission three-channel spectrometer. A is the measured spectrum. B is the solar spectrum adjusted below A. C is the residual A-B spectrum, showing the molecular emissions. Upper panel : the projected distance parameter is p = 37000 km. Lower panel : The projected distance parameter is p = 1900 km. A broad-band emission progressively appears in the 342-375 nm range. Its intensity rapidly increases when the projected distance decreases.

NEAR-UV SPECTRA OF COMET HALLEY

The main molecular emissive feature in near-UV cometary spectra[8] are the bands of OH at 309 nm ($A^2\Sigma^+ \rightarrow A^2\Pi_i$ system), of NH at 336 nm ($A^3\Pi_i \rightarrow X^3\Sigma^-$ system) and of CN at 388 nm ($B^2\Sigma^+ \rightarrow X^2\Sigma^+$ system). The region between the NH and CN bands appears free of emission except if a high sensitive spectrometer is employed[9]. In this case, additional weak bands are detected : OH (0,1) at 346-348 nm, CN (1,0) at 357-359 nm, CO_2^+ ($\Delta v=2$) at 325-328 nm, CO_2^+ ($\Delta v = 0$) at 350-353 nm, CO_2^+ ($\Delta v = -1$) at 367-369 nm and C_3 at 372-376 nm. These molecules are produced in photolytic processes of parent molecules released by the nucleus or by dust grains in the coma. Their intensity radial distribution therefore does not decrease sharply with increasing cometocentric distance in the first thousand km.

The spectra of comet Halley presented here (fig. 1) were obtained with the three-channel spectrometer of the Vega 2 mission[10] during the encounter day, March 9, 1986. They were taken at short distance from the nucleus. The projected distance p between the nucleus and the line of sight varied from 40000 to 420 km. The spectra are the sum of the two components : dust-scattered continuum and molecular emissions. The continuum is subtracted in adjusting a solar spectrum below the cometary calibrated spectrum. The residual spectrum contains the molecular emissions of the fluorescent species present in the line of sight.

IDENTIFICATION OF PHENANTHRENE IN HALLEY'S INNER COMA

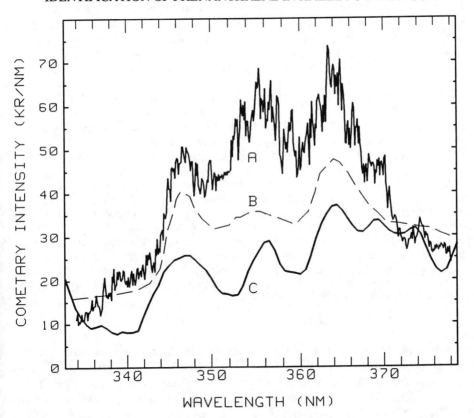

Fig 2 : Comparison of the laboratory fluorescence spectrum of the jet-cooled phenanthrene (A) with the cometary spectrum (C). The cometary NH (0,0) band at 336 nm has been subtracted. Curve (B) is the fluorescence spectrum of phenanthrene in cyclohexane excited by a radiation of wavelength 252 nm.

At small cometocentric distances (p< 1000 km), a broad-band emission progressively arises in the 342-375 nm range. This emission presents three main peaks at 347,

356 and 364 nm. Its average intensity is 26 kR nm^{-1} in a 30 nm bandpass, which corresponds to a column emission rate of 7.8 X 10^{11} photons cm^{-2} s^{-1} in 4π steradians at a projected distance p = 489 km. This intensity varies as a function of distance to the nucleus approximately as 1/p. More precisely, the intensities of the three peaks, once the dust-scattered continuum is subtracted, vary in following the same 1/p law. Consequently, it is suspected that the broad band emission is due to a unique carrier which is a parent molecule. A model calculation of the intensity distribution was performed assuming that the parent molecule has a density n(r) = Q exp (-r/l$_0$)/4πr^{2}v where Q is the production rate, r the distance to the nucleus, v the expansion velocity and l$_0$ the scalelength of the parent molecule. The result was that the parameter l$_0 \approx$ 3x10^{4}s.

Among the possible parent molecules, CH$_3$OH and H$_2$CO were considered. In the case of methanol, there is no emissive transition between 280 and 370nm. The molecule presents an absorption continuum at λ < 200 nm where it is photodissociated in producing CH$_2$0, CH$_3$O or CH$_3$. In the case of formaldehyde, H$_2$CO, this molecule presents regularly spaced bands[12] at 370, 378, 387, 397, 405 and outside our spectral range. It is not possible to conclude that it is present in our spectra, because there is no coincidence for the two bands.

Our search was oriented toward polycyclic aromatic molecules because they have a high fluorescence efficiency and because their eventual presence was suggested by the infra-red band at 3.2 -3.5 µm, mainly by the feature at 3.26 -3.28 µm. Given the wavelength range in the near UV where the fluorescence emission was observed, it could be deduced that the carrier had 2, 3 or 4 benzenic cycles. A comparison with the fluorescence spectrum of phenanthrene in solution in cyclohexame[13] showed a very agreement with the cometary spectrum. Since no vapor-phase spectrum was found in the litterature, we decided to conduct an experiment that would provide an adequate simulation of the astrophysical situation. The fluorescence of super-cooled gas phase molecules in a supersonic jet was excited in using a narrow band frequency-doubled pulsed dye laser [14]. The laboratory spectrum obtained in exciting the fluorescence at 282.6 nm is shown in Fig. 2 (curve A). It may be compared with the cometary spectrum (curve C) and with the fluorescence spectrum in cyclohexane (curve B). The comparison shows that the agreement between the spectra is excellent for both the position of the peaks and their relative intensities.

DISCUSSION AND CONCLUSION

A quantitative evaluation of the production rate of phenanthrene based on the present identification shows it is 1.5 x 10^{27} molecules s^{-1} and is a fraction of 1.5 x 10^{-3} the production rate of H$_2$O. This figure may be compared with the production rate of C$_2$ 1.6 10^{27}s^{-1} . Recent studies[15] of the action of visible photons on C$_{14}$H$_{10}$ and C$_{14}$H$_{10}^+$ show that a C$_2$H$_2$ molecule is released. Under cometary conditions, the production of C$_2$ and CH appears as a plausible photochemical scheme.

The astrophysical interest of the detection of a specific PAH in comet P/Halley may be measured in reference to the large amont of research work initiated following the suggestion[16]. that PAHs might be the carriers of the unidentified IR bands between 2 and 12 µm. This detection suppports the idea of an analogy in the composition of cometary and interstellar material. It adds a strong argument in favor of the model according to which comet nuclei were originally made by the aggregation of interstellar dust particles with refractory organic mantles. These particles would be a constituent of the interstellar matter existing well before the solar system formation.

REFERENCES

1. P. Swings, L. Haser, Atlas of representative cometary spectra, University of Liège, Contract 61 (514)-628
2. L.E. Snyder, P. Palmer, I. De Pater, The Astron. J.97, 246 (1989)
3. D. Bockelée-Morvan, P. Colom, J. Crovisier, D. Despois, G. Paubert, Nature 350, 318 (1991)
4. M.J. Mumma, H.A. Weaver, H.P. Larson, D.S. Davis, M. Williams, Science 232, 1523 (1986)
5. F. Baas, T.R. Geballe, D.M. Walther, Ap. J. 311, L97 (1986)
6. E.K. Jessberger, A. Christoforidis, J. Kissel, Nature 332, 691 (1988)
7. M. Combes, V.I. Moroz, J. Crovisier et al., Icarus 76, 404 (1988)
8. M.F. A'Hearn, in Comets, ed. L.L. Wilkening (Univ. Ariz. Press, 1982), p. 433
9. J.H. Valk, C.R. O'Dell, A.L. Cochran, W.D. Cochran, C.B. Opal, E.S. Barker, Ap. J. 388, 621 (1992)
10. G. Moreels, J. Clairemidi, J.P. Parisot et al., Astron. Astrophys. 187, 551 (1987)
1. G. Moreels, J. Clairemidi, P. Hermine, P. Bréchignac, P. Rousselot, Astron Astrophys., in press (1993)
12. K.Y. Tang, E.K.C. Lee, Chem. Phys. Lett., 43,2 (1976)
13. W. Karcher, A.J. Fordham, J.J. Dubois, P.G.J.M. Glaude, J.A.M. Lightart, Spectral Atlas of Polycyclic Aromatic Compounds, Reidel, Dordrecht (1985)
14. P. Hermine, P. Bréchignac, in Physico-Chimie des molécules et grains dans l'espace, Mont Saint-Odile 6-10 sept. 1993, p. 40
15. P. Boissel, G. Lefevre, P. Thiébot, in Autour de la Physique Moléculaire, Paris 3-5 mai 1993, p. 38
16. A. Léger, J.L. Puget, Astron. Astrophys. 137, L5 (1984)

SMALL MOLECULAR SYSTEMS:
STRUCTURE AND PHOTON-INDUCED DYNAMICS

MILLIMETER AND SUBMILLIMETER-WAVE SPECTROSCOPY OF MOLECULAR IONS AND OTHER TRANSIENT SPECIES

Jean-Luc Destombes, Marcel Bogey, Michel Cordonnier, Claire Demuynck and Adam Walters.

Université de Lille I
Laboratoire de Spectroscopie Hertzienne associé au CNRS
F-59655 Villeneuve d'Ascq, France

ABSTRACT

From the very beginning of the great adventure of the study of Interstellar Molecules, millimeter-wave spectroscopy has been a crucial step in the identification of new interstellar species, either by providing astronomers with the characteristic frequencies of a new molecule first observed in the laboratory, or in confirming unequivocally the identification of a new spectrum discovered by radioastronomy. The recent discoveries of interstellar, or potentially interstellar molecules, show clearly that *ab initio* calculations are also a key step in the identification process.

With the development of sensitive radiotelescopes, spectroscopic needs are directed towards more and more exotic molecules. It follows that laboratory spectrometers need to parallel this evolution by improvements in sensitivity, and also in efficiency of production methods.

These two different aspects of laboratory millimeter-wave spectroscopy will be reviewed and illustrated by recent results obtained in our laboratory. Special emphasis will be given to methods of production, which have been used to study protonated acetylene $C_2H_3^+$, and to synthesize several new silicon containing molecules.

1 - INTRODUCTION

Interstellar chemistry is known to produce a number of molecules (about 100 species are presently known), some of them being highly reactive species like free radicals and molecular ions. They appear as transient molecules in the chemical networks which lead to the complex organic and inorganic interstellar molecules. The discovery of these exotic molecules has been a very strong motivation for radioastronomers, theoreticians and laboratory spectroscopists, and many of the most beautiful discoveries (HCO^{+1}, $C_3H_2{}^2$, ...) are the result of a tight interplay between these three fields.

Among the various spectroscopic techniques now available, millimeter-wave spectroscopy which deals with pure rotational spectra, is of special interest because it allows a direct observation of the same rotational transitions as those observed by radio-astronomy. This is the method generally used to definitely confirm the attribution of unidentified lines first detected in the interstellar medium.

When only a restricted set of measured lines (in the laboratory, or in the interstellar medium, or both) are avalaible, good theoretical models can be used to predict the unobserved lines of astrophysical interest, with an accuracy sufficient for a radioastronomical search. When no experimental data are avalaible, accurate *ab initio* calculations are invaluable to predict the geometrical and electronic structures, and then the rotational spectrum of a new species.

In view of the interdisciplinarity of this meeting, we will first describe the basic principles of millimeter-wave spectroscopy. Since other aspects of microwave spectroscopy of reactive molecules are discussed elsewhere in this issue, we will present recent specific examples obtained in our laboratory to give an overview of the methods used to produce and identify new exotic molecules and to illustrate the potentialities of this high resolution technique.

2 - MILLIMETER-WAVE SPECTROSCOPY: A HIGH RESOLUTION, HIGH SENSITIVITY METHOD TO STUDY HIGHLY REACTIVE SPECIES

Conventional millimeter wave spectroscopy is basically an absorption method, involving a *source*, an *absorption cell* and a *detector*. In spite of the low energy of the microwave photons, high sensitivity can be obtained by using monochromatic sources whose frequency can be swept accross the absorption feature to be studied. A sensitive incoherent detector, generally a broadband InSb bolometer, is used to detect the power absorption resulting from the interaction between the molecules and the electromagnetic field.

Owing to the reactivity of the free radicals and molecular ions, their stationary concentration is generally very low, and the main quality required of a spectrometer is a detection threshold as low as possible. This is generally obtained by using a frequency modulation technique to improve the (Signal)/(Noise) ratio as well as the (Signal)/(Spurious Background) ratio. Using well known relations[3] and using typical numerical values for the various parameters, it is possible to estimate the lowest detectable concentration N_{min} (per cubic cm) of an absorbing species. Some examples relevant to astrochemistry are given in Table I, assuming a spectrometer sensitivity $\Delta P/P = 10^{-7}$ per cm of absorbing path, which is a conservative value in the 50-450 GHz frequency range provided suitable sources are used.

Table I. Lowest detectable concentration N_{min} (cm^{-3}) assuming a spectrometer sensitivity of 10^{-7} cm^{-1}.

μ(D)	ν(GHz)	T(K)	N_{min}(cm^{-3})	example
3.5	360	300	10^8	C_3H_2
4.	360	100	$8 \cdot 10^6$	HCO^+
0.5	450	100	10^8	Si_2H_2, $HCNH^+$

It is clear from this Table that millimeter-wave spectroscopy is a competitive method to study reactive species, especially closed-shell molecules which are difficult to study by conventional visible or UV spectroscopy.

2-a. A typical millimeter-wave spectrometer.

In Lille, we have recently developed a new version of our computer-controlled spectrometer[4] (Fig. 1).

Below 340 GHz, we use harmonic generation from klystrons or Gunn oscillators emitting in the 50-75 GHz frequency range. The sources are phase-locked on the emission of a 12-18 GHz synthesizer (Giga Instrument) by comparing the IF frequency beat, near 320 MHz with the emission of a 3325B Hewlett-Packard synthesizer. Small frequency scans are obtained by sweeping the emission of the HP synthesizer. For large frequency scans, the same technique is used together with step incrementation of the Giga synthesizer.

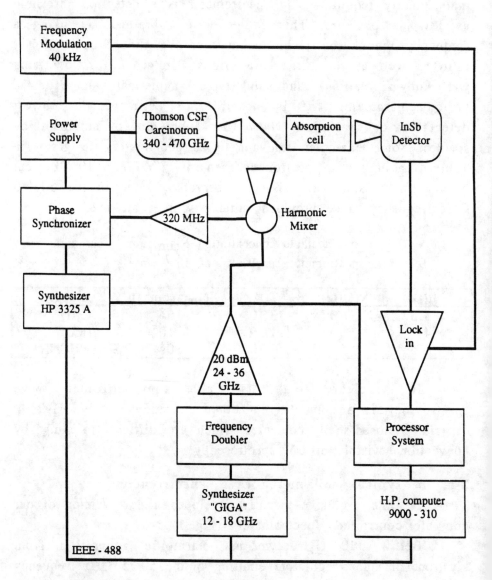

Fig. 1. Block diagram of the spectrometer.

Above 340 GHz we use two carcinotrons from Thomson CSF emitting in the 340-470 GHz frequency range. They are phase-locked on the second harmonic of the Giga synthesizer (Fig. 1). The high power delivered by these Backward Wave Oscillators allows a very good sensitivity to be reached in their frequency range (typically 10^{-8} cm^{-1} for a 10 Hz bandwidth detection).

The detection is achieved by a helium-cooled InSb detector from QMC Instruments. The emission of all the millimeter and submillimeter sources are frequency modulated at 40 kHz, the signal being demodulated at twice this frequency, which provides a second derivative lineshape.

A HP 9000-310 microcomputer is interfaced through a microprocessor to a A/D converter, and drives also the synthesizers. It automatically sweeps the frequency of the microwave source and records simultaneously the spectrum. It is then possible to use computer processing to measure the frequency of the lines, or to improve the S/N ratio.

2-b. Computer processing.

In a microwave spectroscopy experiment, the sensitivity is usually limited by the presence of standing waves which have an amplitude much larger than the weak absorption signal which is searched for.

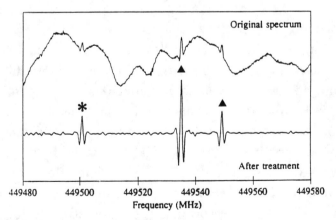

Fig. 2. A spectrum example before and after data processing showing lines of Si(H$_2$)Si (noted by *) and of Si(H)SiH (noted by ▲)

In order to solve this problem, a computer processing technique has been developed. The basic idea of this treatment is to extract from the experimental spectrum any signal that "looks like" a line. It is based on a cross-correlation routine between a spectrum sample and a reference lineshape[4,5]. Fig. 2 shows a typical scan before and after processing.

3 - METHODS OF PRODUCTION

The spectrum of very short-lived species can be detected only if these species are produced directly in the absorption cell. Various methods can be used, depending on the type of molecule to be studied. Relatively simple molecules and all molecular ions are produced and observed within a plasma excited by a DC glow discharge. More fragile molecules need softer chemical methods such as vacuum flash thermolysis or vacuum gas-solid phase reactions.

3-a. Plasma production.

The cell consists of a 2.5 m long, 5 cm internal diameter Pyrex tube. Both ends are closed by two Teflon windows at Brewster incidence to improve the transmission and to minimize the standing waves. The temperature of the gas can be reduced down to 77K by flowing liquid nitrogen through a jacket surrounding the cell. Condensable gases are introduced through a 2 m long axial glass tube drilled all along its length.

Generally, the *positive column* (I= 50-200 mA) is used when radicals or reactive molecules are searched for. The *negative glow* is more suitable for the search for molecular ions, especially protonated stable molecules. We lengthen the negative glow by applying a 200 G axial magnetic field which limits the ambipolar diffusion[6]. This magnetic field is also used to discriminate ion lines from neutral ones. The intensities of the former are strongly correlated to the strength of the field[6,7]. Furthermore, in the case of a positive column, it is used to check the Zeeman effect of the observed lines in order to discriminate paramagnetic species from others.

3-b. Chemical methods.

These techniques are well illustrated by the method used to produce the 2H-Azirine molecule, a cyclic isomer of methyl cyanide which has been produced in two successive steps: a gas-solid-phase reaction followed by a vacuum flash thermolysis[8].
More generally either of these two steps, or both, can be used, depending on the molecules to be synthesized. The general scheme of the experiment is given in Fig. 3: the gas-phase dehydrochlorination of a convenient precursor is achieved over solid t-butoxide in a 2.5 cm reactor tube (length 30 cm) which is warmed up to 50-80°C by oil circulation. The t-butanol by-product is trapped at -75°C.

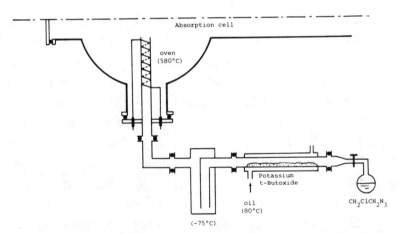

Fig. 3. Experimental device used to produce 2H-Azirine[8].

Thermolysis is achieved inside the absorption cell itself, in an oven made of a quartz tube wrapped with a tungsten wire. Temperatures up to 1000°C are easily obtained. The cell consists of a Pyrex sphere (i.d. 35 cm to limit wall recombination) which contains the oven, followed by a tube (i.d. 10 cm, length 140 cm). To prevent overheating of the sphere during high temperature thermolysis an annular shower has been used.

Several new molecules of astrophysical interest have been produced by using these soft methods : the $2H$-azirine NCH_2CH[8], the formyl cyanide $HCOCN$[9], the thioformyl cyanide $HCSCN$[10].

4 - SOME SPECIFIC EXAMPLES

Once we have a sensitive spectrometer and efficient methods of production, routine experiments in any gas mixture discharge will lead to the discovery of hundreds of lines. The main problem for the spectroscopist is then to give a name to a molecule which manifests itself through only some anonymous rotational lines! To illustrate this key step in the identification process, we chose the pedagogical example of the fortuitous discovery of the disilyne molecule Si_2H_2[11]. Some other recent experiments will be then described.

4-a. The exotic disilyne molecules: Si(H₂)Si and Si(H)SiH.

In the course of a search for the molecular ion SiH^+, unidentified lines were observed in a silane plasma around 450 GHz. The lines had a typical behaviour, being maximum in intensity in the abnormal regime of the discharge and disappearing almost completely in the positive column regime. However, they did not show the intensity evolution characteristic of ionic lines when the confinement magnetic field was varied[6,7]. As they did not exhibit any Zeeman splitting, and after some chemical tests, we concluded they were due to *a closed-shell neutral molecule containing only silicon and hydrogen atoms*.

As shown on the stick diagram of Fig. 4, the submillimeter-wave spectrum of this unknown molecule is completely dominated by a very characteristic c-type Q branch pattern. The assignment of the lines by first and second frequency difference methods was straightforward[12] and showed that the observed lines belonged to the $^RQ_1$ branch ($K_{-1} = 1 \rightarrow 2$). Extension of the initial search to a wider frequency range allowed some P and R lines to be observed. Finally, a total of 87 lines was measured in the 347-472 GHz range. The spectrum was fitted using Watson's A-reduced Hamiltonian, with a standard deviation of 20 kHz.

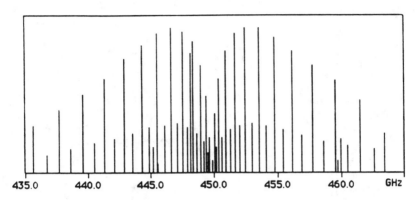

Fig. 4. Stick diagram of the ^cQ branch observed in a SiH$_4$/Ar discharge. Note the 1:3 intensity alternation.

More has been learned about the nature of this unknown molecule by carefully examining the values of the rotational constants A, B, C:

(i) Comparison of B and C ($\approx$7250 MHz) with those of molecules having similar masses (B(S$_2$)$\approx$8.8 GHz; B(SiCl)$\approx$7.6 GHz; B(Si$_2$)$\approx$7.2 GHz) clearly indicates that the molecule contains *two heavy atoms*.

(ii) The relatively high value of A ($\approx$157 GHz) shows that A is mainly determined by a few H atoms out of the heavy atom axis. Comparison with known molecules suggests that most likely $n(H)\leq 3$.

(iii) 1:3 alternation in the intensity of adjacent lines indicates that the molecule contains *one pair of exchangeable hydrogen atoms*. This result is consistent with (ii) and then suggests that *the unknown molecule is the disilyne Si$_2$H$_2$*, a fundamental molecule never observed spectroscopically.

The identification of the molecule was consistent with the available theoretical results[13]:

(j) From *ab initio* calculations, the ground state of Si$_2$H$_2$ has a singlet non classical bridged structure (Fig. 5). The C$_{2v}$ symmetry implies the existence of a permanent dipole moment.

Moreover, all *ab initio* structures show that this dipole moment lies along the *c*-axis, in agreement with the experiment.

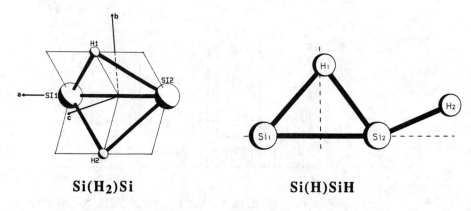

Fig. 5. Structure of the two isomers of Si_2H_2.

(jj) The A, B, C values calculated from the various *ab initio* structures available are very close to the experimental ones.

(jjj) The silasilene polar isomer H_2SiSi can be excluded because it would give rise to an *a*-type spectrum. Other potential molecules can be excluded by the same type of argument. For example, Si_2H_4, which could marginally fulfill (ii) and (iii), has a trans-bent geometry in its ground state with a center of symmetry[14,15], and then no permanent dipole moment and no pure rotational spectrum.

Owing to the symmetry of Si_2H_2, only 3 geometrical parameters (Si-H, Si-Si distances, dihedral angle η, see Fig. 5) are needed to describe its structure. Relations between the principal moments of inertia and these parameters were easily derived and were used to calculate the r_0 structure of the molecule.

Table II. The experimental r_0 structure of $Si(H_2)Si$ compared with recent *ab initio* calculations.

	r_0 structure	*ab initio*[13]
Si - Si (Å)	2.2154	2.216
Si - H (Å)	1.6680	1.668
η(deg.)	104.22	104.0

Table II shows the striking agreement between experimental and theoretical structures.

In order to experimentally confirm the identification and the geometry of the molecule, this r_0 structure was used to predict the A, B, C values of the ^{29}Si and ^{30}Si mono-isotopomers. Owing to the low natural abundance of these isotopes (4.7% and 3% respectively), the signals expected were at the limit of the spectrometer sensitivity. Nevertheless, a total of 31 isotopic lines were observed leading to the determination of the rotational constants. The agreement was excellent and definitively confirmed the identification. The observation of the deuterated species finally lead to the determination of the substitution structure, generally a better approximation of the true equilibrium structure[4].

About one year later, nearly the same story occured during the search for the SiH$_3$ rotation-inversion spectrum. Using the same line of reasoning, we were able to identify the unknown molecule as an isomer of the disilyne molecule Si$_2$H$_2$[16]. Once again, *ab initio* results were invaluable in identifying the molecule and showed unambiguously that this isomer was the astonishing monobridged form, 8.7 kcal/mol higher in energy than the dibridged one[13,17]. It has been definitely confirmed by the observation of the deuterated isotopomer Si(D)SiD. The structures of the two isomers are shown in Fig. 5.

From the astrophysical point of view, Si$_2$H$_2$ is potentially interesting, since the carbon analogue C$_2$H$_2$ is known to be a very abundant interstellar and circumstellar species. Unfortunately, the dipole moment of the most stable form Si(H$_2$)Si is relatively small (0.4 D). The chemistry of silicium in the interstellar medium is also not well understood and for the moment none of the two isomers has been detected.

4-b. A fascinating molecular ion : C$_2$H$_3^+$.

The observation of protonated species has been suggested for a long time as an indirect mean of detecting non polar molecules like C_2H_2 in the interstellar medium[18]. Moreover, protonated acetylene $C_2H_3^+$ is thought to play a key role in the hydrocarbon chemistry of the interstellar medium[19]. The chemistry of this ion has been recently reexamined by Glassgold, Omont and Guélin[20] who concluded that some millimeter-wave transitions of $C_2H_3^+$ could be detected with the existing large radiotelescopes in some interstellar clouds as well as in C-rich circumstellar shells.

$C_2H_3^+$ has also received considerable attention from the quantum chemists fascinated by the possible existence of classical and non-classical (bridged) structures in this small carbo-cation. From the numerous theoretical papers reporting *ab initio* calculations, it seems to emerge that the classical structure of protonated acetylene is a transition state between two equivalent forms of the non-classical isomer[21,22]. The height of the barrier is presently estimated to be 3.7±1.3 kcal/mol. (about 1300±400 cm^{-1})[21], which is low enough to allow tunneling on a measurable timescale.

In contrast to these numerous theoretical results, very little was known experimentally on the spectroscopic properties of this ion, until the pioneering work of Oka's group: they first observed I.R. lines of $C_2H_3^+$ in 1985, but it took them four years to get conclusive results. In a recent paper[22], they gave their main conclusion: the very rich I.R. spectrum in the 3.2 μm region is due to the non-classical (bridged) structure of $C_2H_3^+$. Moreover, tunneling splittings expected from the finite barrier height were experimentally resolved in the excited vibrational state. However, in spite of extensive theoretical developments[23-26], there is presently no theoretical model to quantitatively interpret this tunneling within the experimental accuracy. Consequently, Crofton *et al.*[22] used I.R. line combination differences to determine the B and C rotational constants of $C_2H_3^+$ in the vibrational ground state, but were unable to determine A, preventing any accurate prediction of the millimeter-wave *b*-type rotational spectrum to be made. However this pure rotational spectrum was eagerly demanded by the radioastronomers, since, in view of a reasonably large dipole moment ($\approx$1.25 D according to Lee and Schaefer[27]), it could be observed at cm and mm wavelengths[20].

The ions were produced in a negative glow discharge extended by an axial magnetic field[6] in a mixture of Ar/H$_2$/C$_2$H$_2$, as described above. A 5 GHz range was scanned around 432.3 GHz, the frequency of the intense 9_{18}-9_{09} line predicted from the constants of Crofton et al.[22], leading to the discovery of several lines. After eliminating the lines corresponding to neutral molecules by using the characteristic evolution of the ionic line intensity vs magnetic field[6,7], we were left with only one ionic line of moderate intensity, about two order of magnitude smaller than that observed for HN$_2^+$ and HCO$^+$ in the same apparatus. Several chemical tests were performed in order to ascertain that the ion contains only C and H atoms. Assuming this line was due to C$_2$H$_3^+$, its frequency was included in a least square fit together with the combination differences of Crofton et al.[22], permitting a preliminary determination of A and a prediction of the frequencies of the other lines, which were readily observed.

17 b-type lines were finally accurately measured and fitted with a standard S-reduced Watson Hamiltonian leading to the determination of the three rotational constants A, B, C and four centrifugal distortion constants.

Table III. Molecular constants of C$_2$H$_3^+$

	mmw data a)	mmw+IR data b)	Crofton et al. c)
A (MHz)	399955.983(18)	399956.00(17)	396925.0 (fixed)
B (MHz)	34237.5395(66)	34237.532(58)	34234.8(20)
C (MHz)	31371.760(14)	31371.74(12)	31371.8(20)

a) Fit including only the millimeter-wave transitions measured in this work.
b) Fit including the millimeter-wave transitions and the combination differences deduced from the I.R. spectrum of Crofton et al.[22].
c) Results of Crofton et al.[22].

The molecular constants are presented in Table III where they are compared with : (i) the values obtained in a global fit including our millimeter-wave lines and the combination differences of Crofton et al.[22]; (ii) the values deduced by fitting only the combination differences. The consistency of all these results is remarkable and confirms that the molecular ion observed by Oka's group and by us is the same carbo-cation, i.e. $C_2H_3^+$, in its non-classical (bridged) form. *Ab initio* rotational constants are all consistent with these results.

One of the most exciting problems concerning this molecule is related to the internal motion of the protons. In spite of an extensive search, no doublet resulting from this tunneling was observed in the millimeter-wave spectrum, indicating that the barrier may be higher than the 1400 cm^{-1} *ab initio* value[21].

Finally, the frequencies of some non-observed lines of potential astrophysical interest[20] have also been predicted by using our set of molecular constants, with an accuracy small enough for all astrophysical purposes[28].

4-c. A new method for a new molecule.

In the examples given above, we were in a favourable situation because : (i) for Si_2H_2, we started from the non polar precursor SiH_4 and the spectra were not too dense and had very characteristic patterns; (ii) in the case of $C_2H_3^+$, we had a method to distinguish the ionic lines from the neutral ones and we had reasonably good frequency predictions from the I.R. spectrum, restricting the search to a 5 GHz scan. In the general case however, an electric discharge in gas mixtures gives rise to numerous anonymous lines.

We were recently confronted by such a situation in the study of a SiH_4/O_2 plasma, searching for the H_2SiO millimeter-wave spectrum which was predicted using the results of *ab initio* calculations[29-31]. After eliminating all the ground state and vibrationally excited lines of SiO, we were left with about 150 lines in the 165-220 GHz frequency range, with no characteristic pattern. From chemical tests, it was concluded that the molecules responsible for the lines contained H, Si and O atoms.

As the lifetimes of the unknown carriers of all these lines seemed to be very short, we decided to measure them with a good accuracy by modulating the discharge current and using a transient digitizer, the frequency of the microwave source being stabilized on the peak absorption of a rotational line. A check of the technique on the SiO lines clearly showed that the measured lifetime is a characteristic of a given molecule in a given vibrational state (Fig. 6).

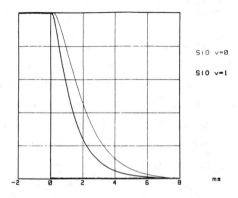

Fig. 6. Lifetime measurements of SiO in v=0, 1 states, using the J=5-4 rotational transition.

We then applied the method to the study of the strongest unidentified lines and showed that they can be classified in two groups characterized by the lifetimes 1.24 and 1.42 ms. It was then easy to recognize that the lines with the shortest lifetime were arranged in a typical *a*-type R branch pattern consistent with spin statistics characteristic of two exchangeable H atoms. The identification of the lines was then straightforward and allowed the prediction and the measurement of the frequencies of other lines, confirming the rotational quantum number attribution.

The preliminary A, B, C values are given in Table IV and compared with the *ab initio* values of various isomers of the silanone molecule H_2SiO[29-31]. This comparison suggests strongly that we have observed the C_{2v} isomer of H_2SiO, although a definite confirmation will be obtained only after the study of other isotopomers[32].

Table IV Molecular constants of silanone (MHz).

	exper.	ab initio results		
		H_2SiO (C_{2v})[30]	HSiOH cis[31]	HSiOH trans[31]
A	166661.4	168774	176506	180097
B	18679.40	18890	16485	16461
C	16742.75	16938	15077	15083

From *ab initio* calculations, it appears that the isomer of C_{2v} symmetry is not the most stable one, but lies 5.3 kcal/mol above the HSiOH cis and trans isomers, with a barrier of about 70 kcal/mol between them[31]. These results suggest that the lines with the longest lifetime are due to the low lying HSiOH isomers. Identification of their spectra will be more difficult because the strongest lines are predicted to be of *b*-type and also because the two isomers are nearly iso-energetic and separated by a relatively low barrier, according to the *ab initio* results[31].

In addition to the probable identification of the elusive molecule H_2SiO, the main conclusion to be retained from this experiment is that the lifetime of a transient species can be accurately measured by millimeter-wave spectroscopy, giving very consistent results. This method can be used to sort out the lines of a given molecule from a crowded rotational spectrum and is then a very useful aid in the identification of a new molecule.

5 - CONCLUSION

Major improvements in molecule-production methods as well as in spectrometer sensitivity and frequency coverage have made possible recent developments in the field of high-resolution rotational spectroscopy of transient species. Clearly, as shown all along by the examples given above, *ab initio* calculations also play a key role in the identification of a new molecule

The necessary interplay between theory, laboratory spectroscopy and radioastronomy is marked out for still new exciting discoveries, which are exemplified by the recent astronomical and laboratory detection of the first magnesium-containing interstellar molecule[33].

This research was partially supported by the European Communities (project 892 001 59/OP1: "Structure and Dynamics of Molecular Ions") and by the C.N.R.S. (G.D.R. "Physico-Chimie des Molécules et Grains Interstellaires").

REFERENCES

1. D. Buhl and L.E. Snyder, Nature **228**, 267 (1970); W. Klemperer, Nature **227**, 267 (1970); U. Wahlgreen, B. Liu, P.K. Pearson and H.F. Schaefer III, Nature **246**, 4 (1973); W.P. Kraemers and G.H.F. Diercksen, Ap. J. **205**, L97 (1976); R. C. Woods, T.A. Dixon, R.J. Saykally and P.G. Szanto, Phys. Rev. Lett. **35**, 1269, (1975).
2. P. Thaddeus, C.A. Gottlieb, R. Mollaaghababa and J.M. Vrtilek, J. Chem. Soc. Faraday Trans., in press (1993) and references therein.
3. W. Gordy and R.L. Cook, "Microwave Molecular Spectra", *in* Techniques of Chemistry, vol.18, edited by A. Weissberger (J. Wiley & sons, New-York, 1984).
4. M. Bogey, H. Bolvin, C. Demuynck, J.L. Destombes and A. Csaszar, in preparation.
5. W.H. Press, B.P. Flannery, S.A. Teukolsky, W.T. Vetterliing, Numerical recipes. The art of scientific computing (Cambridge University Press, 1986).
6. F.C. De Lucia, E. Herbst, G.M. Plummer and G.A. Blake, J. Chem. Phys. **78**, 2312 (1983).
7. J.L. Destombes, C. Demuynck and M. Bogey, Phil. Trans. R. Soc. Lond. A**324**, 147 (1988).
8. M. Bogey, J.L. Destombes, J. M. Denis and J.C. Guillemin, J. Mol. Spectrosc. **115**, 1 (1986).
9. M. Bogey, J.L. Destombes, Y. Vallee and J.L. Ripoll, Chem. Phys. Lett. **146**, 227 (1988).
10. M. Bogey, C. Demuynck, J.L. Destombes, A. Gaumont, J.M. Denis, Y. Vallee and J.L. Ripoll, J. Am. Chem. Soc. **111**, 7399 (1989).
11. M. Bogey, H. Bolvin, C. Demuynck and J.L. Destombes, Phys. Rev. Lett. **66**, 413 (1991).

12. G. Winnewisser, M. Winnewisser and W. Gordy, J. Chem. Phys. **49**, 3465 (1968).
13. B.T. Colegrove and H.F. Schaefer III, J. Phys. Chem. **94**, 5593 (1990).
14. B.T. Luke, J.A. Pople, M.B. Krogh-Jespersen, Y. Apeloig, M. Karni, J. Chandrasekhar and P. von Rague Schleyer, J. Am. Chem. Soc. **108**, 270 (1986).
15. H. Lischka and H.J. Köhler, Chem. Phys. Lett. **85**, 467 (1982); G. Olbrich, Chem. Phys. Lett. **130**, 115 (1986); H.J. Köhler and H. Lischka, J. Am. Chem. Soc. **104**, 5884 (1982); K. Krogh-Jespersen, J. Phys. Chem. **86**, 1492 (1982); J. Am. Chem. Soc. **107**, 537 (1985); H. Teramae, J. Am. Chem. Soc. **109**, 4140 (1987).
16. M. Cordonnier, M. Bogey, C. Demuynck and J.L. Destombes, J. Chem. Phys. **97**, 7984 (1992).
17. R.S. Grev and H.F. Schaefer, J. Chem. Phys. **97**, 7990 (1992).
18. E. Herbst, S. Green, P. Thaddeus and W. Klemperer, Ap. J. **215**, 503 (1977).
19. E. Herbst and C.M. Leung, Ap. J. S. **69**, 271 (1989).
20. A.E. Glassgold, A. Omont and M. Guélin, Ap. J. **396**, 115 (1992).
21. R. Lindh, J.E. Rice and T.J. Lee, J. Chem. Phys. **94**, 8008 (1991).
22. M.W. Crofton, M.F. Jagod, B.D. Rehfuss and T. Oka, J. Chem. Phys. **91**, 5139 (1989).
23. J.T. Hougen, J. Mol. Spectrosc. **123**, 197 (1987).
24. R. Escribano and P.R. Bunker, J. Mol. Spectrosc. **122**, 325 (1987).
25. R. Escribano, P.R. Bunker and P.C. Gomez, Chem. Phys. Lett. **150**, 60 (1988).
26. P.C. Gomez and P.R. Bunker, Chem. Phys. Lett. **165**, 351 (1990).
27. T.J. Lee and H.F. Schaefer III, J. Chem. Phys. **85**, 3437 (1986).
28. M. Bogey, M. Cordonnier, C. Demuynck and J.L. Destombes, Ap. J. **399**, L103 (1992).
29. A. Tachibana, H. Fueno, M. Koizumi, T. Yamabe and K. Fukui, J. Phys. Chem. **92**, 935 (1988); S. Sakai, M. Gordon and K.D. Jordan, J. Phys. Chem. **92**, 7053 (1988).
30. A. Csaszar, private communication (1993).
31. H.F. Schaefer III, private communication (1993).
32. A. Walters, S. Bailleux, M. Bogey, C. Demuynck and J.L. Destombes, in preparation.
33. K. Kawaguchi, E. Kagi, T. Hirano, S. Takano and S. Saito, Ap. J. **406**, L39 (1993).

DISCUSSION

SCAPPINI —
1. What kind of MW sources do you use ?
2. How did you proceed to assign the MMW spectrum of $C_2H_3^+$?

DESTOMBES —
1. Up to about 300-350 GHz, we use harmonic generation from phase locked klystrons or Gunn diodes. In the 340-470 GHz frequency range, we use B.W.O. from Thomson CSF, which are powerful, allowing a very good sensitivity to be obtained.
2. By using the molecular constants derived from the IR work of Oka and coworkers, we predicted the pure rotational spectrum. We then scanned ±5 GHz around the frequency of the most intense line and were finally able to recognize only one "ionic" line. By assuming that this line was the $C_2H_3^+$ line, we were able to derive a preliminary value for the unknown molecular constant A. We then made new predictions. The lines were searched for and detected. From the fit of the 17 pure rotational lines we finally measured, we deduced a set of molecular parameters. Note that the B and C rotational constants are in good agreement with the value of Oka. However, A differs by more than 3 GHz from the Oka's estimated value, which explains why it was impossible to make an accurate prediction of the rotational spectrum from the IR lines only.

LEGER — Are the spectra of amino acids and DNA bases known and characteristic ?

DESTOMBES — The rotational spectra of glycine and some other molecules of biological interest like adenine,... are known: I don't have the references here, but R.D. Brown in Australia made extensive studies of such molecules. These molecules are relatively heavy, they often have several conformers; so their rotational spectra are certainly dense, weak and not very characteristic. At low temperature, they will appear in the cm region. However, if some of these molecules have very low vibrational modes, submm, or even mm-wave, radiation could be used to probe the corresponding vibrational transitions (see the example of C_3O_2 : the ν_7 band is observed by millimeter-wave spectroscopy around 500 GHz).

LEACH — The use of a magnetic field enables one to strike the electric discharge at much lower gas pressures. Under these conditions, the species produced and their relative concentrations could differ from those under higher pressures and under non magnetic field conditions. Did you do any experiments at the lower presures mode possible by use of the magnetic field ?

DESTOMBES — It's clear that some species can be produced only in discharges of special type. The negative glow extended by a magnetic field is generally used to produce molecular ions. However, in some cases, we found it very useful to produce some neutral reactive species, as for example the two isomers of Si_2H_2 (see the written version of the talk). These two molecules are efficiently produced in the negative glow but disappear completely if you change the discharge from the negative to the positive column regime.

FIELD — In view of the chemical soup which is created using discharges, would it not be possible to use a more controlled method of exciting reactive species and ions? For example photodissociation and photoionization using synchrotron radiation might be a means of tuning in the specific species. Do you have any plans to attempt this, by attaching your system to a beam-line at Super ACO for example ?

DESTOMBES — We dind'nt make calculations to see the feasibility of such an experiment at super ACO. However, we are setting up an experiment using a (borrowed) 193 nm excimer laser to explore the possibilities of using photodissociation in our experiment.

ROUEFF — During your search of $C_2H_3^+$, you mentioned a series of 19 unidentified remaining lines in the 430 GHz range. Have you propositions for their identification ?

DESTOMBES — These 19 lines are still unidentified, but their are only a few lines among hundred of unidentified lines you get when you strike a discharge in acetylene. It doesn't mean however that they are all of the greatest interest, because many of them (most of them ?) are certainly coming from isotopic forms of known species, or from vibrationally excited states of such molecules.

MILLIMETERWAVE SPECTROSCOPY OF STABLE MOLECULES OF ASTROPHYSICAL INTEREST

G. Wlodarczak, J. Burie, M. Le Guennec and J. Demaison
Laboratoire de Spectroscopie Hertzienne, UA CNRS 249, Université de Lille 1,
F-59655 Villeneuve d'Ascq

ABSTRACT

We report the analysis of the laboratory spectra of some stable molecules of astrophysical interest in the millimeter wave range: deuterated isotopic species of methyl cyanide CH_3CN and methylacetylene CH_3CCH which are abundant species in interstellar clouds and complex molecules like methylamine CH_3NH_2 and propene $CH_3CH=CH_2$. These two molecules present large amplitude motions and have complex spectra which necessitate specific treatment for their analysis.

INTRODUCTION

Methyl cyanide CH_3CN (see 1 for a list of references) and propyne CH_3CCH (see for instance 2 for a list of references) have been detected in several interstellar clouds as well as in planetary atmospheres [3, 4] but very little was known about the single deuterated isotopic species. The same situation occured for propene (old measurements up to 40 GHz were only available [5]) and for methylamine (previous studies limited to 90 GHz [6]), due to the complexity of the spectra. Methylamine has still been detected in Orion A and SgrB2. The recent interstellar detection of other deuterated compounds like D_2CO induced partly this work and interstellar search of CH_2DCN, CH_2DCCH and CH_3CCD.

EXPERIMENTAL DETAILS

Rotational spectra between 100 and 300 GHz were measured with a computer-controlled millimeter wave spectrometer using superheterodyne detection [7]. The transitions between 340 and 470 GHz were measured with a source modulated spectrometer using phase-stabilized submillimeter BWOs (Thomson-CSF) as sources and an He-cooled bolometer as detector. The accuracy of the measurements is better than 50 kHz.
Some lower frequency spectra were recorded in Kiel with microwave Fourier transform spectrometers
The following reactions were used for the synthesis of the deuterated compounds :

$CH_2DI + KCN \rightarrow CH_2DCN + KI$ (solvent: dimethylsulfoxyde) [8]

$CH_2DI + NaCCH \rightarrow CH_2DCCH + NaI$ (solvent: liquid NH_3) [9]

$CH_3CCLi + D_2O \rightarrow CH_3CCD + LiOD$ [10]

DEUTERATED SPECIES

The identification of the spectra was straightforward because the lines are strong and sparse. The main species together with ^{13}C and ^{15}N isotopic species were analyzed. For CH_3CCD the usual expression for the frequency of rotational transitions for symmetric tops was used while the asymmetric top (A) reduction in the I^r

representation was used for CH_2DCCH and CH_2DCN [11]. From the rotational and centrifugal distortion constants determined by this analysis very accurate predictions are available between 0 and 600 GHz for radioastronomers.

In collaboration with the radioastronomy group from Meudon we detected two of these species in interstellar medium. CH_2DCN was detected in Orion-IRc2 and G34.3 at millimeter wavelengths mostly with the IRAM 30m telescope at Pico Veleta, Spain [12]. Some lines were also identified in a previous spectral line survey of Orion. In these two clouds HDO was previously found [13] and towards IRc2 CH_2DCN peaks like HDO. The isotopic ratio [CH_2DCN] / [CH_3CN] is about 0.01 in IRc2 and equal column densities were found for ortho and para species.

We also detected two lines of CH_2DCCH at 97 GHz in TMC1 while CH_3CCD was not detected [14]. We also searched for these two molecules in other clouds (OMC1, G34.3, DR21(OH)) without success. In TMC1 the isotopic ratio [CH_2DCCH] / [CH_3CCH] was found between 0.05 and 0.06, the highest known in TMC1. Some reaction schemes to explain this high isotopic ratio are discussed in Ref. 14.

COMPLEX MOLECULES

Some interstellar molecules present complex spectra due to their non-rigidity. When large amplitude motions occur (internal rotation of a methyl group, inversion), the description of rotational spectra necessitates more complicated models than those used in the previous section. We have developed such models for molecules with internal rotation (one-top and two-top models) in order to analyze the spectra of molecules like acetaldehyde CH_3CHO, methyl formiate $HCOOCH_3$, dimethylether $(CH_3)_2O$, acetone $(CH_3)_2CO$, acetic acid CH_3COOH. Our predictions lead to the detection of acetone in SgrB2 [15] but acetic acid was not detected despite an intensive search in various clouds [16]. The other cited species were previously known to be present in the interstellar medium.

In collaboration with the Universty of Kiel we measured new rotational transitions of propene in the centimeter wave range and we extended the measurements to the millimeter wave domain. The rotational, centrifugal distortion and internal rotation analysis were carried using the IAM (Internal Axis Method) [17]. Predictions are available in the range 0-300 GHz and despite its low dipole moment (0.36 D) propene can now be searched in interstellar clouds or cometary atmospheres, as suggested in Ref. 18.

Methylamine is another floppy molecule with two large amplitude motions: internal rotation of the methyl group and inversion of the amino group. This molecule was detected in Orion and SgrB2 but the rotational spectra was known up to 90 GHz only and accurate predictions were not available in the millimeter range. We measured the spectra up to 470 GHz in the ground state and the first excited torsional state. In collaboration with M. Kreglewski these spectra were analyzed and new molecular parameters determined which allow accurate predictions for a large number of rotational frequencies up to 500 GHz [19]. These new predictions should be helpful for the analysis of interstellar spectra recorded with the recent and sensitive radiotelescopes working in the millimeter range.

AKNOWLEDGEMENTS

The authors thank Dr. J. M. Denis and Dr. J. C. Guillemin for their help during the synthesis of the samples of deuterated compounds. This work was supported by the CNRS (GDR Physicochimie des molécules et grains interstellaires).

REFERENCES

1. R. Bocquet, G. Wlodarczak, A. Bauer and J. Demaison, J. Mol. Spectrosc. 127,382 (1988).
2. G. Wlodarczak, R. Bocquet, A. Bauer and J. Demaison, J. Mol. Spectrosc. 129, 371 (1988).
3. W. C. Maguire, R. A. Hanel, D. E. Jennings, V. G. Kunde and R. E. Samuelson, Nature 292, 683 (1981).
4. G. Graner and G. Wagner, J. Mol. Spectrosc. 144, 389 (1990) and references therein.
5. D. R. Lide and D. E. Mann, J. Chem. Phys. 27, 868 (1957).
6. N. Ohashi, K. Takagi, J. T. Hougen, W. B. Olson and W. J. Lafferty, J. Mol. Spectrosc. 126, 443 (1987).
7. J. Burie, D. Boucher, J. Demaison and A. Dubrulle, J. Phys. (Paris) 43, 1319 (1982).
8. R. A. Smiley and C. Arnold, J. Org. Chem. 25, 257 (1960).
9. G. K. Speirs and J. L. Duncan, J. Mol. Spectrosc. 51, 277 (1974).
10. C. McDade and J. E. Bercow, J. Organomet. Chem. 279, 281 (1985).
11. J. K. G. Watson, in Vibrational Spectra and Structure (J. R. Durig, Ed.,Elsevier, Amsterdam, 1977), Vol. 6, p. 1.
12. M. Gerin, F. Combes, G. Wlodarczak, T. Jacq, M. Guelin, P. Encrenaz and C. Laurent., Astron. Astrophys. 259, L35 (1992).
13. T. Jacq, C. M. Walmsley, C. Henkel, A. Baudry, R. Mauersberger and P. R. Jewell, Astron. Astrophys. 228, 447 (1990).
14. M. Gerin, F. Combes, G. Wlodarczak, P. Encrenaz and C. Laurent., Astron. Astrophys. 253, L29 (1992).
15. F. Combes, M. Gerin, A. Wootten, G. Wlodarczak, F. Clausset and P.J. Encrenaz, Astron. Astrophys.180, L13 (1987).
16. A. Wootten, G. Wlodarczak, J. G. Mangum, F. Combes, P. J. Encrenaz and M. Gerin.,Astron. Astrophys. 257, 740 (1992).
17. G. Wlodarczak, J. Demaison, N. Heineking and A. Cszaszar, in preparation.
18. J. Crovisier, D. Bockelée-Morvan, P. Colom, D. Despois and G. Paubert, Astron. Astrophys., in press (1993).
19. M. Kreglewski and G. Wlodarczak., J. Mol. Spectrosc. **156**, 383 (1992)

GAS PHASE SYNTHESIS OF HETERO-SILA-ALKENES OF COSMOCHEMICAL IMPORTANCE

Jean-Marc Denis
URA CNRS 704, Université de Rennes I, F 35042 Rennes, France

Jean-Louis Ripoll
URA CNRS 480, ISMRA, Université de Caen, F 14050 Caen, France

ABSTRACT

The syntheses of non-substituted silanone (**1**), silanimine (**2**) and silanethione (**3**) were undertaken using monomolecular decompositions under flash vacuum thermolysis conditions. The results already obtained from several precursors, and, particularly, the NMR identification of the trimer of **3**, cyclotrisilathiane **10**, are coherent with the formation of these species. However, whereas the presence of silicon monoxide and silicon monosulfide, dehydrogenated derivatives of **1** and **3**, was confirmed, the absorption lines corresponding to compounds **1** - **3** have not been identified until now in the millimeter wave spectra of the thermolysis products. Investigations are presently continued towards their determination.

INTRODUCTION

The unsubstituted hetero-sila-alkenes such as silanone (**1**), silanimine (**2**), and silanethione (**3**) can be reasonably postulated as cosmic molecules, owing to the existence of their carbon counterparts $H_2C=O$, $H_2C=NH$ and $H_2C=S$, as well as to the recent discovery of several silicon species, in the interstellar medium [1].

$$H_2Si=O \qquad H_2Si=NH \qquad H_2Si=S$$
$$\mathbf{1} \qquad\qquad \mathbf{2} \qquad\qquad \mathbf{3}$$

Compounds **1-3** are however expected to be extremely unstable and their laboratory synthesis and identification in the gas phase constitute an actual chemical challenge [2].

Under the auspices of the CNRS "GDR PCMGI", we have endeavoured, in collaboration with the Laboratoire de Spectroscopie Hertzienne (Lille, URA 249), to set up the synthesis of compounds **1-3** by flash vacuum thermolysis (FVT) [3] and the determination of their millimeter wave spectra.

METHODS, FIRST EXPERIMENTS

The allyloxysilane **4** and allylsilanamine **5** have been prepared by reacting allyl alcohol or allylamine with di-t-butylchlorosilane.

4 **5**

These compounds, when submitted to FVT, can undergo both a retro-ene reaction and double β-elimination leading, by loss of propene and isobutene, to the expected species **1** and **2**, respectively (see ref. 4 for previous paper on related topics).

$$\xrightarrow[\beta\text{-elimination}~(-\text{isobutene})]{\text{FVT 850°C, retro-ene reaction}~(-\text{propene})} H_2Si=X \xrightarrow{(-H_2)} SiX$$

4	X = O	**1**
5	X = NH	**2**

However, the millimeter wave spectrum of the compounds resulting from the FVT of **4** showed only, besides propene and isobutene, the presence of SiO ($J_{6\leftarrow 5}$ at 217104.610 MHz) resulting from the thermal dehydrogenation of **1**. A similar result is likely in the FVT of **5**, although the too short life-time of SiNH did not allow its characterization.

FVT of **4** or **5** at lower temperatures resulting in incomplete decomposition, other precursors allowing easier thermolyses were to be found.

LOWERING THE TEMPERATURE OF FVT

Several synthetic strategies have been applied, aimed at reducing the temperature required for complete thermolysis:

- introduction of a group conjugating upon FVT, thus yielding an appreciable energetic gain.

$$\text{[allyl alcohol + Si(Cl)(H)(tBu)}_2\text{]} \longrightarrow \mathbf{6} \xrightarrow[\beta\text{-elimination}\ (-\text{ isobutene})]{\text{FVT retro-ene reaction }(-\ 1,3\text{-pentadiene})} H_2Si{=}O \quad \mathbf{1}$$

- presence of a hetero-atom such as sulfur, inducing the retro-ene reaction at a lower temperature, when compared with oxygen or nitrogen.

$$\text{[allyl thiol + Si(Cl)(H)(tBu)}_2\text{]} \longrightarrow \mathbf{7} \xrightarrow[\beta\text{-elimination}\ (-\text{ isobutene})]{\text{FVT retro-ene reaction }(-\text{ propene})} H_2Si{=}S \quad \mathbf{3}$$

- replacement of the allylic group by a propargylic one, resulting in a more favored transition state in the retro-ene reaction, or replacement of the retro-ene reaction itself by a more favored $2\pi+2\pi$ thermal cycloreversion.

$$\text{[propargyl amine + Si(Cl)(H)(tBu)}_2\text{]} \longrightarrow \mathbf{8} \xrightarrow[\beta\text{-elimination}\ (-\text{ isobutene})]{\text{FVT retro-ene reaction }(-\text{ allene})} H_2Si{=}NH \quad \mathbf{2}$$

$$\mathbf{5} \xrightarrow{[Pt]} \mathbf{9} \xrightarrow[\beta\text{-elimination}\ (-\text{ isobutene})]{\text{FVT }2+2\text{ cycloreversion }(-\text{ propene})} H_2Si{=}NH \quad \mathbf{2}$$

RECENT RESULTS

The FVT of compounds **6, 7** appeared to be complete at ca. 600°C (i.e., 150°C lower than in the case of precursors **4** and **5**). Although the retro-ene reaction took still place at somewhat lower temperatures, the t-butyl groups were no more, or only partially, eliminated (H_3Si-containing precursors have not been retained until now owing to their rather unworkable reactivity).

A preliminary NMR investigation of the thermolyses of **6, 7** confirmed in each case the expected formation of 1,3-pentadiene (from **6**) or propene, and isobutene. Furthermore, in the FVT of sulfide **7**, a resonance signal at 5.29 (^1H) ppm should belong to the trimer of **3**, cyclotrisilathiane **10** (lit.: 5.23 ppm [5]).

The first results concerning the direct gas phase investigation of the thermolysis of **7** by millimeter wave spectroscopy showed the formation, increasing with the FVT temperature, of silicon monosulfide ($J_{16 \leftarrow 15}$ at 290380.770 MHz). Signals belonging to **3** were however not yet identified, perhaps due to its too rapid trimerisation. This investigation is presently pursued, as well as those concerning the precursors **6, 8** and **9**.

$$\mathbf{7} \xrightarrow[\text{(- propene)}]{\text{FVT 600°C}} H_2Si=S \xrightarrow{\times 3} \mathbf{10}$$
$$\text{(- isobutene)} \quad \mathbf{3} \quad \xrightarrow{\text{(- }H_2\text{)}} SiS$$

REFERENCES

1. D. Smith, Chem. Rev. <u>92</u>, 1473 (1992); A. Dalgarno, J. Chem. Soc. Faraday Trans. 2111 (1993).
2. G. Raabe, J. Michl, Chem. Rev. <u>85</u>, 419 (1985).
3. R. F. C. Brown, Pyrolytic Methods in Organic Chemistry (Academic Press, N. Y., 1980).
4. J. M. Denis, P. Guenot, M. Letulle, B. Pellerin, J. L. Ripoll, Chem. Ber. <u>125</u>, 1397 (1992).
5. A. Haas, M. Vongehr, Z. Anorg. Allg. Chem. <u>447</u>, 119 (1978).

THE LABORATORY MILLIMETER-WAVE SPECTRUM OF MgCN ($X^2\Sigma^+$)

M. A. Anderson, T. C. Steimle, and L. M. Ziurys
Department of Chemistry
Arizona State University, Tempe, Arizona 85287-1604

ABSTRACT

The pure rotational spectrum of the linear MgCN radical ($X^2\Sigma^+$) has been recorded in the laboratory using millimeter/sub-mm direct absorption spectroscopy. Twenty seven rotational transitions were observed in the frequency range 101-376 GHz. Spin rotation interactions were resolved in the spectra, but no hyperfine splittings were observed. The rotational and spin-rotation constants were determined from a non-linear least squares fit to the data using a $^2\Sigma$ Hamiltonian. MgCN is the metastable isomer of MgNC, and thus is of astrophysical interest.

INTRODUCTION

The only astronomical source were metal-bearing compounds have been securely detected thus far is in the late-type carbon star IRC+10216. In this object, several metal halides species have been observed, including AlCl[1], NaCl[1], KCl[1], and AlF[1,2]. More recently the magnesium-containing free radical MgNC has been detected as well.[3] Curiously, while the metal halides are present in the inner envelope of the star[1,2], MgNC appears to arise from the outer shell. Thus, metal-bearing species exist in regions where temperatures and densities are high enough such that chemical thermodynamic equilibrium prevails, but also are present in gas where photodissociation may be occurring. Therefore, the chemistry of metals must be rather complex. Additional observational input certainly appears to be necessary to understand the various chemical processes that are occurring in IRC+10216.

One species of possible interest for circumstellar chemistry is MgCN. This radical is the metastable isomer of MgNC. Theoretical calculations suggest that MgCN lies ~500 cm^{-1} higher in energy than MgNC, with a bent transition state[4]. Given conditions of chemical equilibrium, MgNC would be the favored species. However, at the outer edges of the circumstellar shell of IRC+10216, physical conditions approach those found in cool interstellar clouds, where the chemistry is kinetically rather than thermodynamically controlled. In fact, HNC, the metastable isomer of HCN, is relatively abundant in the outer shell of the carbon star[5], although it lies ~4000 cm^{-1} higher in energy than its more stable form[6]. HNC and HCN are thought to be produced from the same parent molecule, HCNH$^+$, and are almost always found in the same gas, although their relative abundance ratio may range from ~1-1000[7]. Considering the presence of both HCN and HNC in

IRC+10216, it is likely that MgCN exists in the outer shell where MgNC is observed. In fact, measuring the MgCN/MgNC ratio may give valuable insights into the formation mechanisms of magnesium-bearing molecules.

Because of our general interest in the interstellar chemistry of metals, we have measured the pure rotational mm/sub-mm spectrum of MgCN. Twenty-seven transitions of the free radical were recorded in the frequency range 100-400 GHz. In this paper our results are presented.

EXPERIMENTAL

The spectra of MgCN were measured using the Arizona State University mm/sub-mm spectrometer, which is described in detail elsewhere[8]. Briefly, this instrument consists of tunable mm radiation source, in this case phase-locked Gunn oscillators, a gas absorption cell, and a helium-cooled InSb detector. Higher frequencies are obtained by using Schottky diode multipliers. The cell is double pass system to which a Broida-type oven is attached. The radiation is quasi-optically propagated through the cell by a scalar feedhorn, a polarizing grid, and a series of teflon lenses. Phase-sensitive detection is accomplished through FM modulation of the Gunn oscillators, and the spectrometer is operated under computer control.

The MgCN radical was created in a D.C. glow discharge with a mixture of magnesium vapor, argon, and cyanogen gas. The discharge current used was ~200 mA. The metal vapor was produced in a Broida-type oven. The partial pressures of argon and cyanogen gas were 30 and 10 m torr, respectively. When the gas mixture was discharged, a bright green glow was observed, likely arising from atomic magnesium. Also, very strong lines of MgNC were observed as well, including several of its magnesium isotopomers and excited vibrational modes. MgCN could also be synthesized by a D.C. discharge of a mixture of magnesium vapor, argon, and methyl cyanide (CH_3CN). The observed intensities of the spectral lines of MgCN produced via this method were less than half of those formed when cyanogen gas was used. The linewidths of MgCN recorded were typically 200-800 GHz over the range 101-376 GHz.

RESULTS

The rotational transition frequencies measured for MgCN are listed in Table 1. As the table shows, twenty seven rotational transitions of this radical were recorded. The absolute accuracy of the measurements is estimated to be ±150 kHz. MgCN is predicted to be linear with a $^2\Sigma^+$ ground electronic state.[4,9] The unpaired electron in this species has net angular momentum which results in spin-rotation interactions, which were observed in every transition measured. The average spin-rotation splitting is ~15.0 MHz, slightly smaller than that observed for MgNC ($\gamma = 15.219$ MHz)[3]. Hyperfine interactions, which will arise from presence of the nitrogen nucleus in the molecule (I = 1), however, were not observed in any transitions measured. This result is not unexpected because the transitions observed

Table 1: Observed transition frequencies of MgCN: $X^2\Sigma^+$ (v=0)

$N \to N'$	$J \to J'$	ν_{obs} (MHz)	$\nu_{obs}-\nu_{calc}$ (MHz)
9→10	17/2→19/2	101,877.556	0.084
	19/2→21/2	101,892.557	0.081
10→11	19/2→21/2	112,063.443	0.036
	21/2→23/2	112,078.440	0.031
12→13	23/2→25/2	132,433.041	0.028
	25/2→27/2	132,448.060	0.050
13→14	25/2→27/2	142,616.595	0.044
	27/2→29/2	142,631.584	0.038
14→15	27/2→29/2	152,799.206	0.048
	29/2→31/2	152,814.172	0.022
15→16	29/2→31/2	162,980.808	0.043
	31/2→33/2	162,995.839	0.084
16→17	31/2→33/2	173,161.337	0.029
	33/2→35/2	173,176.328	0.034
17→18	33/2→35/2	183,340.745	0.026
	35/2→37/2	183,355.678	-0.023
18→19	35/2→37/2	193,518.946	0.015
	37/2→39/2	193,533.904	-0.006

Table 1: (continued -)

N→N'	J→J'	ν_{obs} (MHz)	$\nu_{obs}-\nu_{calc}$ (MHz)
19→20	37/2→39/2	203,695.952	0.073
	39/2→41/2	203,710.802	-0.052
20→21	39/2→41/2	213,871.491	0.003
	41/2→43/2	213,886.491	0.025
21→22	41/2→43/2	224,045.684	-0.028
	43/2→45/2	224,060.674	-0.005
22→23	43/2→45/2	234,218.443	-0.022
	45/2→47/2	234,233.459	0.031
23→24	45/2→47/2	244,389.675	-0.012
	47/2→49/2	244,404.674	0.028
24→25	47/2→49/2	254,559.227	-0.084
	49/2→51/2	254,574.229	-0.036
25→26	49/2→51/2	264,727.276	0.005
	51/2→53/2	264,742.177	-0.043
26→27	51/2→53/2	274,893.432	-0.067
	53/2→55/2	274,908.361	-0.082
27→28	53/2→55/2	285,057.868	-0.062
	55/2→57/2	285,072.811	-0.058
28→29	55/2→57/2	295,220.418	-0.079
	57/2→59/2	295,235.384	-0.046

Table 1: (continued -)

N→N'	J→J'	ν_{obs} (MHz)	$\nu_{obs} - \nu_{calc}$ (MHz)
29→30	57/2→59/2	305,381.094	-0.039
	59/2→61/2	305,395.999	-0.061
30→31	59/2→61/2	315,539.740	-0.032
	61/2→63/2	315,554.633	-0.060
31→32	61/2→63/2	325,696.292	-0.055
	63/2→65/2	325,711.326	0.064
32→33	63/2→65/2	335,850.803	0.012
	65/2→67/2	335,865.659	-0.041
33→34	65/2→67/2	346,002.999	-0.039
	67/2→69/2	346,018.038	0.097
34→35	67/2→69/2	356,153.082	0.060
	69/2→71/2	356,167.920	0.002
35→36	69/2→71/2	366,300.734	0.059
	71/2→73/2	366,315.576	0.011
36→37	71/2→73/2	376,446.022	0.090
	73/2→75/2	376,460.849	0.055

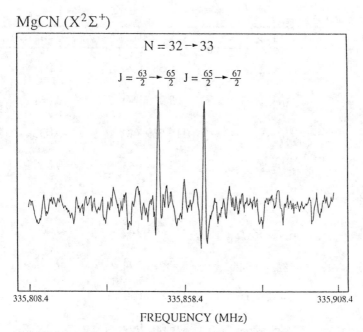

Fig. 1 Spectra of the N = 32→33 transition of the MgCN radical ($X^2\Sigma^+$) observed in this work near 335.8 GHz. The two components present arise from spin-rotation interactions.

here concern N = 9→10 and higher, and hyperfine splittings were not resolved in MgNC for comparable transitions. Moreover, the unpaired electron in MgCN probably resides primarily on the magnesium nucleus, far away from the nitrogen atom which has the nuclear spin. A sample spectrum is shown in Fig. 1.

Table 2: Molecular Constants for MgCN: $X^2\Sigma^+$(v=0)

	Constant	Millimeter-Wave (MHz)	Theoretical (MHz) c)
MgCN	B_0	5094.80351(62)	5,083.69[c]; 5,077.51[d]
	D_0	0.00277421(33)	
	γ_0	15.014(24)	
	γ_D	-0.000032[b]	

a) Errors quoted are 3σ statistical uncertainties and apply to last quoted digits.
b) Held fixed.
c) B_e value from Ref. 4.
c) B_e value from Ref. 9.

The data listed in Table 1 were fit using a non-linear least-squares routine employing a $^2\Sigma$ Hamiltonian. The rotational constants derived from this analysis are listed in Table 2. These constants reproduce the observed frequencies to better than $\nu_{obs}-\nu_{calc} \leq 90$ kHz. Also given in the table are the theoretical equilibrium rotational constants B_e, calculated by Ishii et al.[4] and Bauschlicher, Langhoff, and Partridge[9]. As the table shows, the B_0 constant derived here and the theoretical B_e values are in close agreement.

DISCUSSION

The close agreement between theoretical and observed rotational constants is strong evidence that spectra observed here arise from MgCN. Furthermore, the spin rotation constant of MgCN is smaller than that of MgNC, as is expected. Also, the spectra were produced when discharging a mixture of magnesium vapor and cyanogen gas-a technique that additionally created large concentrations of MgNC. Because MgCN is the less stable isomer and lies higher in energy than MgNC, spectra of this species should be weaker than those of its more stable form, which also was found.

These data confirm the linear structure of MgCN, as predicted by theory[4,9]. Also, failure to observe hyperfine interactions suggests the ionic configuration Mg^+CN^- dominants the bonding in this species. The bonding in this radical thus is similar to that in MgOH[10].

Measurement of the rest frequencies for MgCN now enables a viable astronomical search to be carried out for this species. Detection of MgCN in IRC+10216, or even obtaining a meaningful upper limit to its abundance, should provide useful insight into the formation of MgNC.

REFERENCES

1. C. Cernicharo and M. Guélin, *A&A* **183**, L10(1987).
2. L. M. Ziurys, A. J. Apponi, and T. G. Phillips, *Ap. J. (Letters)* submitted.
3. K. Kawaguchi, E. Kagi, T. Hirano, S. Takano, and S. Saito, *Ap. J. (Letters)* **406**, L39 (1993).
4. K. Ishii, T. Hirano, U. Nagashima, B. Weis, and K. Yamashita, *Ap. J. (Letters)* **410**, L43 (1993).
5. J. Bieging and N. -Q. Rieu, *Ap. J. (Letters)* **329**, L107 (1988).
6. R. D. Brown, F. R. Burden, and A. Cuno, *Ap. J.* **347**, 855 (1989).
7. P. F. Goldsmith, W. D. Langer, J. Ellder, W. Irvine, and E. Kollberg, *Ap. J.* **249**, 524 (1981).
8. L. M. Ziurys, W. L. Barclay, Jr., M. A. Anderson, D. A. Fletcher, and J. W. Lamb, *Rev. Sci. Instr.*, submitted.
9. C. W. Bauschlicher, Jr., S. R. Langhoff, and H. Partridge, *Chem. Phys. Lett.* **115**, 124 (1985).
10. W. L. Barclay, Jr., M. A. Anderson, and L. M. Ziurys, *Chem. Phys. Lett.* **196**, 225 (1992).

INTERSTELLAR CH$_3$D: DEUTERATED METHANE IN THE ORION HOT CORE?

M. WOMACK[*], L.M. ZIURYS[**], A.J. APPONI[**], J.T. YODER[**]
[*]Physics & Astronomy, Northern Arizona Univ., Flagstaff, AZ,
[**]Chemistry, Arizona State Univ. Tempe, AZ

ABSTRACT

The $J = 1_0 - 0_0$, $2_0 - 1_0$, and $2_1 - 1_1$ transitions of CH$_3$D have been tentatively identified toward Orion using the CSO 10-m and NRAO 12-m telescopes. The radial velocities and linewidths of these features correspond to the Orion hot core. Using millimeter- and submillimeter-wavelength direct absorption spectroscopy, the frequency of the $J = 1 - 0$ transition was remeasured and the frequencies of the two K components of the $J = 2 - 1$ transition were measured for the first time. The observational and laboratory spectra are presented and implications for the chemistry of methane in the Orion hot core are discussed.

INTRODUCTION

Chemical models predict CH$_4$, the simplest organic molecule, to be one of the most abundant polyatomic species in interstellar dense clouds, since it only involves one carbon bonded to four hydrogen atoms (e.g., Mitchell 1977, Brown & Rice 1986). In addition, methane is thought to have an increased abundance in hot, dense gas, such as the Orion *hot core*, due to evaporation of grain mantles (e.g., Brown, Charnley, & Millar 1988). Given the unusually high D/H ratios observed in many dense clouds, one might therefore expect the deuterated form of methane, CH$_3$D, to be also present in high numbers. Although CH$_3$D has a ro-vibrational spectrum, molecular clouds are probably too cool to excite these transitions. The molecule does possess a pure rotational spectrum with transitions that are likely to be excited in dense clouds in the millimeter/ submillimeter - wavelength regime. Although the fundamental transition has already been measured in the laboratory (Pickett, Cohen, & Phillips 1980), the two K components of $J = 2 - 1$ transition have never been observed. Here we present measured frequencies for all three transitions and the results of a search for all three lines toward the Orion-KL star-forming region.

RESULTS AND DISCUSSION

Experimental

Spectra of CH3D were measured using a millimeter/ submillimeter wavelength direct absorption spectrometer, which is described in detail elsewhere (Ziurys et al. 1993). In brief, the experiment consists of a tunable source of millimeter/sub-millimeter radiation, a gas absorption cell and a detector. The source for these measurements is a phase-locked Gunn oscillator, which is quadrupled when used in conjunction with a Schottky diode multiplier to operate in the frequency range of 200 to 500 GHz. The radiation is quasi-optically propagated through the absorption cell, which is a double pass system. The detector is a helium cooled InSb "hot electron" bolometer. The spectra of CH3D at 232 GHz and 465 GHz were recorded at a pressure of 50 mtorr using a sample from a commercial source. The cell is constructed from a Quartz tube 6.3 cm in diameter and 400 cm in length. The gas was introduced to the desired pressure and then sealed to enable long duration signal averaging of each spectral feature.

TABLE 1
OBSERVATIONS OF CH$_3$D TOWARD ORION-KL

Transition	Frequency[a] (MHz)	V_{LSR} (km s^{-1})	$\Delta v_{1/2}$ (km s^{-1})	T_R^* (K)	N(CH$_3$D) (cm^{-2})
$J_K = 1_0 - 0_0$	232,644.301	5.2 ± 1.3	3.8 ± 0.6	0.03±0.01	(3.9±2.0) × 10^{18}
$J_K = 2_0 - 1_0$	465,250.691	5.3 ± 1.0	3.7 ± 1.0	0.35±0.07	(2.9±1.8) × 10^{18}
$J_K = 2_1 - 1_1$	465,235.540	5.4 ± 1.0	3.7 ± 1.0	0.35±0.07	(2.9±1.8) × 10^{18}

[a]Measured frequencies are accurate to ± 75 kHz.

TABLE 2
ROTATIONAL CONSTANTS OF CH$_3$D

Constant	This Work (MHz)	Tarrago et al. (1976) (MHz)
B_o	116325.30975	116325.308 ± 0.009
D_j	1.579625	1.57229 ± 0.00749
D_{jk}	3.78775	3.78038 ± 0.02938

Figure 1 shows the spectrum of the $J_K = 1_0 - 0_0$ transition of CH$_3$D and Figure 2 shows the spectrum of the $J_K = 2_1 - 1_1$ and $J_K = 2_0 - 1_0$ transitions obtained with direct absorption spectroscopy techniques. Measured frequencies of the lines are given in Table 1 and are estimated to have an accuracy of ± 75 kHz. The rotational constants B_o, D_j and D_{jk} were determined from the measured frequencies and are given in Table 2. An accurate evaluation of the uncertainties in these constants cannot be determined due to the lack of observable transitions; however, all of the constants presented here are well within the error limits reported by Tarrago et al. (1976) on their work on the molecule's infrared vibrational bands.

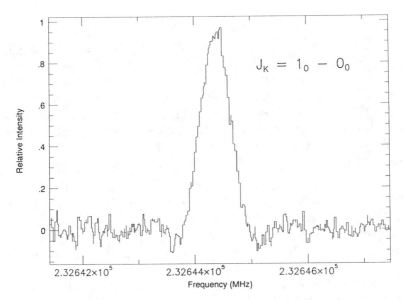

FIG. 1.— Millimeter-wavelength spectrum of CH_3D $J_K = 1_0 - 0_0$ transition obtained with a direct absorption spectrometer.

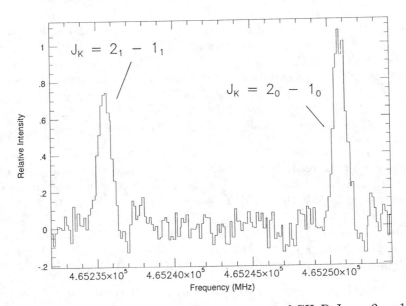

FIG. 2.— Submillimeter-wavelength spectrum of CH_3D $J_K = 2_1 - 1_1$ and $J_K = 2_0 - 1_0$ transitions obtained with a a direct absorption spectrometer.

Observational

Spectra near the frequency of the CH_3D J_K $1_0 - 0_0$ transition (232,644.3 MHz) were obtained toward Orion-KL (α = 05:32:46.8; δ = -05:24:23) using the National Radio Astronomy Observatory (NRAO) 12-m telescope on Kitt Peak, Arizona. The spectrometer utilized two filterbanks with resolutions of 500 kHz and 1000 kHz and was in the double sideband mode. The half-power beam width at this frequency was $\theta_b = 25''$, the system temperature was $T_{sys} \sim 1000$ K, and the beam efficiency was $\eta_b \sim 0.55$. The pointing accuracy is estimated to be $\sim 5''$.

Spectra were also obtained toward Orion-KL at the central frequency of 465,236.7 MHz using the Caltech Submillimeter Observatory (CSO) 10.4-m telescope on Mauna Kea. Both frequencies (see Table 1) corresponding to the $J_K = 2_1 - 1_1$ and $J_K = 2_0 - 1_0$ transitions of CH_3D were observed simultaneously in the 500 MHz bandpass in double sideband mode using with an acousto-optical spectrometer (AOS). The beam width was $\theta_b = 15''$, the system temperature was $T_{sys} \sim 2800$ K, and the beam efficiency was $\eta_b \sim 0.36$. The pointing accuracy is estimated to be $\sim 5''$.

Using the newly measured frequencies of CH_3D, a search was performed toward Orion-KL at 232 GHz and 465 GHz. Figure 3 shows the composite spectra obtained at 232,644 MHz using the NRAO 12-m telescope, and Figure 4 shows the spectrum at $\sim 465,236$ MHz with the CSO 10.4-m telescope. In each spectrum the expected position of the CH_3D lines are indicated. Unfortunately, the $J_K = 2_1 - 1_1$ transition is blended with a nearby feature in the same sideband. The radial velocities of the lines were computed using the newly measured frequencies, and are listed in Table 1, along with the measured linewidths. As the table shows, the three lines each have a radial velocity of $V_{LSR} \sim 5.3$ km s^{-1}, and a linewidth of $\Delta v_{1/2} \sim 3.7$ km s^{-1}. These line profiles correspond to the *hot core* component of the Orion-KL star-forming region.

Total column densities were calculated for the CH_3D molecule from each of the three rotational lines, and are listed in Table 1. As the table shows, the column densities calculated from the different transitions are in excellent agreement with eachother, with an average value of $N(CH_3D) \sim 3.5 \times 10^{18}$ cm^{-2}. The fact that the lines all predict the same total column density is consistent with all three of the transitions arising from CH_3D. Assuming that $N(H_2) = 10^{24}$ cm^{-2} for the Orion hot core, from which these lines arise, a fractional abundance (with respect to H_2) of $f(CH_3D) \sim 4 \times 10^{-6}$ is derived.

Although ro-vibrational spectra of methane have been obtained toward Orion (Lacy et al. 1991), no abundances have been measured for the hot core, the region toward which our candidate CH_3D lines are observed. Chemical models predict methane to have enhanced abundances in the Orion hot core due to evaporation from grain mantles, with $f(CH_4) \sim 10^{-4}$ (e.g., Brown, Charnley,

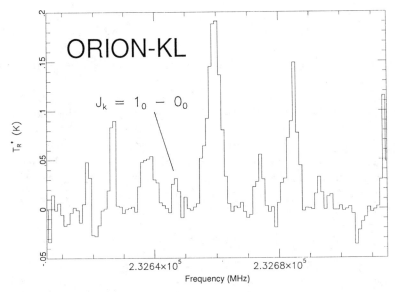

FIG. 3.— Spectrum obtained toward Orion-KL with the NRAO 12-m telescope. The location of the CH_3D $J_K = 1_0 - 0_0$ transition is indicated.

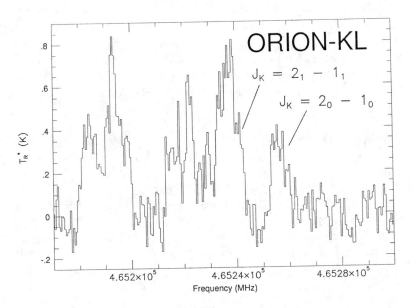

FIG. 4.— Spectrum obtained toward Orion-KL with the CSO 10.4-m telescope. The positions of the $J_K = 2_1 - 1_1$ and $J_K = 2_0 - 1_0$ transitions are identified.

& Millar 1988). If CH_4 suffers no significant deuterium fractionation, and a cosmic D/H ratio of 10^{-5} applies, then an extremely high fractional abundance of $f(CH_4) \sim 0.4$ is derived toward the Orion hot core. If, however, methane experiences significant amounts of deuterium fraction, as has been observed in several molecules in the hot core (e.g., Henkel et al. 1987; Walmsley et al. 1987), then a D/H ratio of 0.05 – 0.001 may be more appropriate. In this case, a methane abundance of $f(CH_4) \sim 8 \times 10^{-5}$ to 4×10^{-3} is derived from our CH_3D measurements. Conversely, if a methane abundance of 10^{-4} is assumed, as predicted by Brown, Charnley, & Millar (1988), then a D/H ratio of 0.04 is derived for methane. Therefore, if the three lines shown in Figures 1 and 2 are indeed attributable to CH_3D, we conclude that methane has enhanced abundances in the Orion hot core, and experiences deuterium fractionation, which is in good agreement with chemical models.

CONCLUSIONS

Millimeter- and submillimeter- wavelength spectra of the fundamental transition, and the two K components of the J = 2 – 1 transition of CH_3D were obtained using a direct absorption spectroscopy. Measured frequencies from these spectra were used to perform an interstellar search of CH_3D. Three lines were observed toward Orion-KL at the correct frequencies, and all the lines possess the radial velocities and linewidths typical of molecules found in the hot core component of Orion. Therefore, we tentatively identify these lines as the three lowest-lying rotational transitions of CH_3D. Assuming that these lines are CH_3D, then an enhanced methane abundance is found in the Orion hot core, which is predicted by chemical models. In addition, a high D/H ratio is derived for methane, as is observed in several other molecules in hot cores.

REFERENCES

Brown, R.D., Charnley, S.B., & Millar, T.J. 1988, MNRAS, 231, 409
Brown, R.D., & Rice, E.H.N. 1986, MNRAS, 223, 405
Henkel, C., Mauersberger, R., Wilson, T.L., Snyder, L.E., Menten, K., & Wouterloot, J.G.A. 1987, A&A, 182, 299
Lacy, J.H., Carr, J.S., Evans, N.J., Baas, F., Achtermann, J.M., & Arens, J.F. 1991, ApJ, 376, 556
Mitchell, G.F. 1977, A&A, 55, 303
Pickett, H.M., Cohen, E.A., & Phillips, T.G. 1980, ApJ, 236, L43
Tarrago, G., Poussigue, G., Dang-Nhu, M., Valentin, A., & Cardinet, P., 1976, J Mol Spec, 60, 429
Walmsley, C.M., Hermsen, W., Henkel, C., Mauersberger, R., & Wilson, T.L. 1987, A&A 172, 311
Ziurys, L.M., Apponi, A.J., & Anderson, M.A. 1993, in preparation

METAL-CONTAINING MOLECULES IN THE LABORATORY AND IN SPACE

L. M. Ziurys, M. A. Anderson, A. J. Apponi, and M. D. Allen
Department of Chemistry
Arizona State University, Tempe, Arizona 85287-1604

ABSTRACT

Laboratory and observational studies of various metal-containing compounds are presented. The millimeter and sub-mm spectra of several hydroxide, hydride, fluoride and cyanide metal free radicals have been measured at high resolution using direct absorption spectroscopy. The species have subsequently been searched for in the interstellar medium.

INTRODUCTION

The chemistry of metal-bearing species in the interstellar medium is something of an enigma. To date, only a few metal-containing molecules have been detected in interstellar gas, and only towards the envelope of the late-type carbon star IRC+10216. This group includes the metal halide species AlCl[1], NaCl[1], KCl[1], and tentatively AlF[1], and most recently an isocyanide radical, MgNC[2]. However, many additional metal compounds have been searched for unsuccessfully towards IRC+10216 and other interstellar sources, among them MgO[3], NaH[4], AlO[5], MgS[5] and FeO[5]. Why simple hydrides, oxides, and sulfides are not observed in IRC+10216, while chlorides and isocyanides are, is quite puzzling. Also, it is curious that metal-bearing species have not yet been detected in molecular clouds.

The negative results for metal-bearing species could arise because the majority of metals, being refractory elements, are condensed out onto dust grain surfaces. Yet, silicon-bearing species such as SiO and SiS are commonly found in interstellar gas[6]. Therefore, some refractory elements do enter the gas phase, and it is difficult to imagine a process that would release silicon from grains and not iron or magnesium.

Another possibility is that metals exist in some molecular form, but the particular carriers have yet to be searched for at sufficiently sensitive levels. Indeed, there are many simple metal compounds whose rotational rest frequencies are completely unknown. Laboratory data is not readily available for such species because they are usually highly reactive and difficult to create in the laboratory. Because interstellar chemistry is unpredictable, these other species must be searched for before any general conclusions can be made about the carriers of metals in astronomical sources.

In an effort to better understand the interstellar chemistry of metals, we have measured the pure rotational spectra of several metal-containing species of astrophysical interest, including MgOH, CaOH, AlOH, CaH, MgH, MgF, CaF, and MgCN. We have subsequently searched for these compounds in the interstellar medium. In this paper we summarize our results.

Conference Proceedings No. 312: *Molecules and Grains in Space*, edited by Irène Nenner
© 1994 American Institute of Physics

LABORATORY EXPERIMENTS

A block diagram of the Arizona State University millimeter/sub-mm spectrometer[7] is shown in Figure 1. The spectrometer operates in the frequency range of 65-550 GHz with a resolution of 200-1000 kHz. The instrument utilizes phase-locked Gunn oscillators as the tunable, coherent source of radiation from 65-140 GHz. Higher frequencies are obtained with Schottky diode multipliers. The gas cell and optics path are designed utilizing Gaussian beam optics to achieve maximum interaction between molecules and the mm-wave radiation in the reaction region. Scalar feedhorns and a series of PTFE lenses are used to propagate the source signal, as shown in the figure. The gas cell is a double-pass system comprised of a cylindrical tube 0.5 m in length with a rooftop reflector at one end and a detachable Broida-type oven. The detector is a helium-cooled InSb bolometer. Phase-sensitive detection is achieved by FM modulation of the Gunn oscillators and use of a lock-in amplifier. Spectra are recorded by electrical tuning of the Gunn oscillator, which is done under computer control.

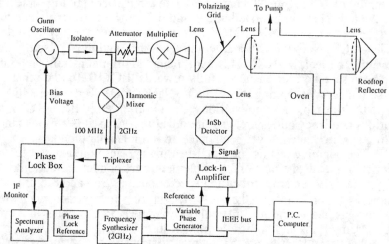

Fig. 1. Block diagram of the Arizona State University mm/sub-mm spectrometer.

The molecules studied with this spectrometer were generated by mixing metal vapor, produced in the Broida-type oven, with a reactant gas. To form the hydroxide species, approximately 15-50 mtorr of H_2O_2 was added to the metal vapor. The hydride and flouride species, on the other hand, were synthesized by reacting hydrogen or flouride atoms with the metal. The atoms were generated in a microwave discharge of H_2 or F_2 (~10-20 mtorr). To create the cyanide compounds, the metal vapor was reacted with cyanide gas in a D.C. glow discharge located over the top of the oven.

A summary of the molecules studied thus far with the spectrometer is given in Table 1. As the table shows, these species include the

Table I: Summary of Laboratory Measurements

Molecule	Ground State	B_0 (GHz)	Isotopomers Observed	Vibrational Analysis	Frequency Range	Comments	Reference
CaOH	$^2\Sigma^+$	10.0231	--	yes	80.1-340.6	--	8, 16
MgOH	$^2\Sigma^+$	14.8225	^{25}MgOH ^{26}MgOH MgOD	yes	88.9-385.2	^{25}Mg hfs resolved; quasilinear species	9, 17
MgH	$^2\Sigma^+$	171.9762	^{26}MgH MgD	yes	179.5-360.0	hfs resolved	11
CaH	$^2\Sigma^+$	126.7729	CaD	no	130.9-391.8	hfs resolved	12
AlOH	$^1\Sigma^+$	15.7403	AlOD	yes	157.4-377.6	quadrupole hfs resolved; quasilinear species	10, 17
MgF	$^2\Sigma^+$	15.4968	^{25}MgF ^{26}MgF	yes	93.0-368.4	^{25}Mg and ^{19}F hfs resolved	13
CaF	$^2\Sigma^+$	10.2675	--	yes	82.1-369.3	hfs resolved	14
MgCN	$^2\Sigma^+$	5.0948	--	no	101.8-376.5	--	15

monohydroxide compounds CaOH[8], MgOH[9], and AlOH[10], the hydrides MgH[11] and CaH[12], the flourides MgF[13] and CaF[14], and the cyanide radical MgCN[15]. The pure rotational transitions of these molecules were measured in the frequency range ~80-400 GHz with an accuracy of ±50-100 kHz. The number of transitions observed for a given species depended on the B_0 value, which is also listed in Table 1. With the exception of AlOH, which is closed-shell with a $^1\Sigma^+$ ground state, all species observed have $^2\Sigma^+$ ground electronic states. For these radicals, spin-rotation interactions were resolved in every case, but hyperfine structure for the main isotopic species was only observed in the hydride and flouride compounds. In the ^{25}Mg isotopomers of MgOH and MgF,

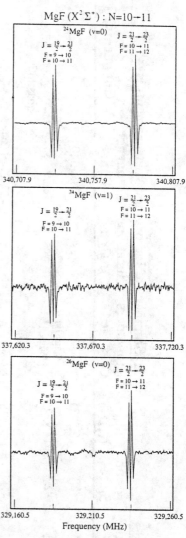

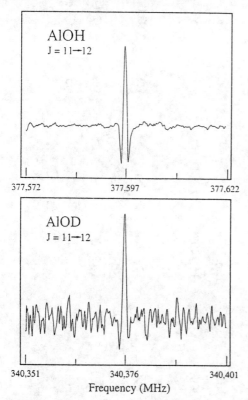

Fig. 2 Spectra of the J = 11→12 rotational transitions of AlOH and AlOD from Apponi, Barclay and Ziurys.[10]

Fig. 3 Spectra of the N = 10→11 transitions of MgF (v=0 and v=1) and the isotopomer ^{26}MgF (v=0) from Anderson, Allen, and Ziurys[13].

however, hyperfine splittings arising from the magnesium nucleus were additionally measured, and quadrupole interactions were partially resolved in AlOH. Sample spectra are shown in Figures 2 and 3.

For many of the species examined, excited vibrational states were observed as well. For the hydroxide species, rotational transitions arising from several quanta of the (0, 1, 0) bending and the (1, 0, 0) heavy-atom stretching modes were recorded. A detailed analysis of these excited modes strongly suggests that MgOH and AlOH are quasi-linear in their ground electronic states[17]. In contrast, CaOH is a fairly well-behaved linear molecule[16].

OBSERVATIONAL STUDIES

The species listed in Table I were all searched for in the interstellar medium, and the results are summarized in Table II. As the table shows, the astronomical searches were carried out using the National Radio Astronomy Observatory's 12 m telescope (Kitt Peak, AZ), the Caltech Submillimeter Observatory (CSO), and at IRAM. The measurements were conducted in the 3, 2, 1.2 and 0.8 mm windows, depending on the species involved (see Table II). Several transitions were usually searched for a given species. The molecules were sought towards molecular clouds near star forming regions (Orion-KL, SgrB2, G34.3, W51M), carbon-rich stars (IRC+10216), and oxygen rich stellar envelopes (W Hydra, VY Canis Majoris, NML Tau, etc.). With the exception of CaH and MgCN, the results of the astronomical work have been negative. The hydroxide species MgOH, CaOH, and AlOH do not appear to be present in detectable quantities in molecular clouds or late-type stars. Upper limits to their column densities are typically $N_{tot} < 10^{12}$-10^{13} cm^{-2}. MgF and CaF were not detected in IRC+10216 to limits of $N_{tot} \lesssim 10^{13}$-10^{14}cm^{-2}, as well. In the case of MgH, believable spectral features were observed at the frequencies of several hyperfine components of the $N = 1 \rightarrow 0$ transition towards Orion-KL, but other components were not detected.[18] Outside of possible anomalous excitation affects, MgH does not appear to be present in this source. Searches in other clouds and towards IRC+10216 have also proved negative for this species.

In the case of CaH, the $N = 1 \rightarrow 0$ transition of this radical was not observed in IRC+10216 down to upper limits of $N_{tot} < 5 \times 10^{10}$ cm^{-2}, using the NRAO 12 m telescope at 254 GHz. However, we have a tentative identification of this radical towards a few molecular clouds, which we are currently pursuing. We are also presently attempting to confirm our tentative detection of MgCN towards IRC+10216.

An additional result has been the confirmation of AlF in IRC+10216 by Ziurys, Apponi, and Phillips[19]. Using the CSO, three rotational transitions of this species were detected at 0.8 and 1.2 mm. The $J = 10 \rightarrow 9$, $8 \rightarrow 7$ and $7 \rightarrow 6$ lines of AlF at 230, 263, and 329 GHz, respectively, were observed towards IRC+10216, and are shown in Figure 4. Combined with the past tentative data obtained for this species at IRAM at 2 and 3 mm, these

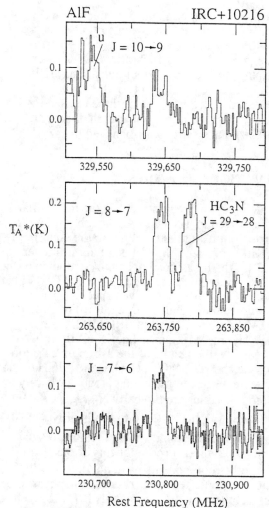

Fig. 4 Spectra of the J = 7→6, 8→7 and 10→9 transitions of AlF measured towards IRC+10216.

measurements confirm the presence of the metal halide species in this carbon-rich circumstellar shell. Analysis of the CSO and IRAM data suggests that AlF arises from a source $\theta_s \sim 5" - 10"$ in extent and is chiefly present in the inner envelope of IRC+10216. Its estimated column density is $N_{tot} \sim 0.3 - 1.1 \times 10^{15} cm^{-2}$.

The chemistry of the flouride species in IRC+10216 is perhaps most easily understood. The large abundance of AlF relative to CaF and MgF can be interpreted in terms of chemical thermodynamics. The small source size derived for AlF suggests it arises in the inner envelope of IRC+10216. In this part of the shell, the gas is thought to be relatively hot and dense such that chemical thermodynamic equilibrium prevails. Hence, it is very likely that a closed-shell, $^1\Sigma$ species such as AlF is more abundant in this region than MgF or CaF, which are both free radicals. The lack of MgH and CaH is more difficult to understand, however, since MgNC is present in this source. This radical comes from the outer part of the shell where chemical equilibrium does not exist. It would seem likely that the metal hydrides would be present here as well. The failure to detect any metal hydroxide compounds in IRC+10216 may be because most of the oxygen is tied up in CO.

Table II: Summary of Astronomical Searches

Species	Dipole Moment (D)	Telescope	Frequency Region	Column Density (cm^{-2})[a] Molecular Clouds	IRC+10216	O-rich Stars
CaOH	1.47	NRAO 12m, IRAM	2, 3 mm	<1.3 × 10^{13}[b]	<2.9 × 10^{13}[b]	--
MgOH	1.20	NRAO 12m, IRAM, CSO	1.2, 2, 3 mm	<6.9 × 10^{12}[c]	<8.3 × 10^{12}[c]	<4.2 × 10^{13}[c]
MgH	?	CSO	0.8 mm	<5.7 × 10^{11}[d]	<6.7 × 10^{12}[d]	--
CaH	?	NRAO 12m, CSO	1.2 mm	≲1.4 × 10^{11}[d]	<5.2 × 10^{10}[d]	<5.2 × 10^{11}[d]
AlOH	1.04	NRAO 12m, CSO	1.2, 2, 3 mm	<1.6 × 10^{12}[e]	<0.7-2.7×10^{14}[e]	<5.0 × 10^{12}[d]
AlF	1.53	CSO	0.8, 1.2 mm	--	0.3-1.1×10^{15}[f]	--
MgF	3.64	NRAO 12m, CSO	1.2, 2, 3 mm	<6.9 × 10^{11}[g]	<1.4-5.4×10^{13}[g]	--
CaF	3.07	CSO	1.2 mm	--	<0.8-3.6×10^{14}[f]	--
MgCN	?	NRAO 12m	3 mm	--	≲6.0 × 10^{11}[h]	--

a) Not corrected for beam dilution unless noted; assumes $T_{rot} \approx$ 50K for molecular clouds; $T_{rot} \approx$ 30K for stars.
b) Based on N = 5→4 transition near 100.2 GHz observed at IRAM ($\theta_b \approx$ 30").
c) Based on N = 5→4 transition near 148.2 GHz observed at NRAO 12m ($\theta_b \approx$ 40").
d) Calculations done assuming μ_0 = 6 D, based on NaH[4].
e) Based on N = 7→6 transition near 220.3 GHz observed with the NRAO 12m ($\theta_b \approx$ 30"); assumes $\theta_s \approx$ 5"-10" for IRC+10216.
f) From Ziurys, Apponi, and Phillips 1993 [Ref. 19]; assumes θ_s = 5"-10".
g) Based on the N = 5→4 transition near 154.9 GHz, observed with NRAO 12m; assumes $\theta_s \approx$ 5"-10" for IRC+10216.
h) Based on N = 11→10 transition near 112.1 GHz, observed with NRAO 12m; assumes μ_0=5 D.

The metal hydroxide species are predicted to be common constituents of oxygen-rich stars. Yet, searches for these species in such objects have proved negative. The column densities in O-rich stars may not be large enough to allow observation of these trace species.

The most puzzling case thus far for understanding metal chemistry is the molecular clouds. Although there are tentative detections, a secure identification of a metal-bearing species has yet to be accomplished in these objects. However, the refractory silicon molecules SiO and SiS are relatively abundant in these clouds, especially near star-forming regions[6]. If grain destruction is producing the silicon compounds, where are the metal species? Perhaps one explanation is that gas phase atomic metals quickly undergo charge exchange with other non-metal ions. Such charge exchange processes are expected to be fast[20]. The metal ions may stay exclusively in atomic form, or some may react to form metal-bearing molecular ions. Perhaps more likely molecular carriers of metals in interstellar clouds are ions rather than neutral species. Certainly additional studies of metal-bearing compounds, both in the laboratory and in space, are necessary before the chemistry of these intriguing species can be thoroughly understood.

REFERENCES

1. C. Cernicharo and M. Guélin, *A&A* **183**, L10 (1987).
2. K. Kawaguhi, E. Kagi, T. Hirano, S. Takano, and S. Saito, *Ap J.* **406**, L39 (1993).
3. B. E. Turner and T. C. Steimle, *Ap. J.* **299**, 956 (1985).
4. R. L. Plambeck and N. R. Erickson, *Ap. J.* **262**, 606 (1982).
5. L. M. Ziurys, to be published.
6. L. M. Ziurys, P. Friberg, and W. M. Irvine, *Ap. J.* **343**, 201 (1989).
7. L. M. Ziurys, W. L. Barclay, Jr., M. A. Anderson, D. A. Fletcher, and J. W. Lamb, *Rev. Sci. Instr.*, submitted.
8. L. M. Ziurys, W. L. Barclay, Jr., and M. A. Anderson, *Ap. J. (Letters)* **384**, L63 (1992).
9. W. L. Barclay, Jr., M. A. Anderson, and L. M. Ziurys, *Chem. Phys. Lett.* **196**, 225 (1992).
10. A. J. Apponi, W. L. Barclay, Jr., and L. M. Ziurys, *Ap. J. (Letters)*, in press.
11. L. M. Ziurys, W. L. Barclay, Jr., and M. A. Anderson, *Ap. J. (Letters)* **402**, L21 (1993).
12. W. L. Barclay, Jr., M. A. Anderson, and L. M. Ziurys, *Ap. J. (Letters)*, **408**, L65 (1993).
13. M. A. Anderson, M. D. Allen, and L. M. Ziurys, *J. Chem Phys.*, in press.
14. M. A. Anderson, M. D. Allen, and L. M. Ziurys, *Ap. J.*, in press.
15. M. A. Anderson, T. C. Steimle, and L. M. Ziurys, *Ap. J. (Letters)*, submitted.
16. D. A. Fletcher, W. L. Barclay, Jr., M. A. Anderson, and L. M. Ziurys, in preparation.
17. L. M. Ziurys, D. A. Fletcher, W. L. Barclay, Jr., M. A. Anderson, and A. J. Apponi, in preparation.
18. L. M. Ziurys and T. G. Phillips, in preparation.
19. L. M. Ziurys, A. J. Apponi, and T. G. Phillips, *Ap. J. (Letters)*, submitted.
20. T. E. Graedel, W. D. Langer, and M. A. Frerking, *Ap. J. Supp.* **48**, 321 (1982).

DISCUSSION

THADDEUS — Have you considered metal-carbon chains, analogous to Si CCCC ?

ZIURYS — No, we haven't studied such species yet. This is an interesting idea and should be pursued. Metal chains such as Mg CC might be present in detectable concentrations in interstellar gas.

SCAPPINI — The energy difference between the two isomers Mg NC and Mg CN may be the explanation for your non-detection of Mg CN. How do you comment on this ?

ZIURYS — First of all, we have a tentative detection of Mg CN, not a non-detection. We have to observe additional transitions, however, before we can claim an identification. The tentative Mg NC/Mg CN ratio is ~ 50-100, which is probably not unreasonbale, considering their energy difference of ~ 400 cm^{-1}. secondly, HNC, the metastable isomer of HCN, is commonly observed in interstellar gas, including in IRC + 10216. The energy difference between these two species is over 1000 cm^{-1}.

ROUEFF — Are you planning to study alkali-hydride species such as NaH, LiH... ?

ZIURYS — The spectrum of NaH at mm-wavelengths has already been recorded by Delucia's group. (The $J = 1 > 0$ line is near 289 GHz). We have no plans at the moment to study the alkali hydrides, but KH might be interesting astrophysically. The interstellar lithium abundance is thought to be quite low, so LiH is not likely to be important astrochemically.

SERRA — Iron is one of the most abundant metal and could play an important role in the organometallic astrochemistry (organometallic complexes formation, catalytic reactions...). Have you studied or planned to study molecular species containing Fe ?

320 Metal-Containing Molecules

ZIURYS — Yes, we started to study some iron-bearing compounds. Iron, however, cannot be vaporized in our current Broida oven. (It doesn't run hot enough). We have been attempting to synthesize iron compounds using organometallic precursors or iron salts, but these methods haven't been as sucessful. We are currently attempting to make FeCl and FeF. We were able to produce FeO, so we may be on the right track. Iron is more difficult to work with than magnesium or calcium.

SARRE — There is a problem in atmospheric chemistry in that the 'sink' for meteoric metals is not clear. One suggestion is that molecules such as $NaCO_3$ are involved. Is there any chance you could study such molecules in the laboratory and in astrophysical objects ?

ZIURYS — Yes, studying CO_3-containing metal compounds could be studied in the lab probably fairly easily. We haven't yet attempted such species , because di- and triatomic species are perhaps more likely interstellar candidates. However, these CO_3 molecules might be worth trying at some point, both in the lab and at the telescope.

QUANTUM–CHEMICAL CALCULATIONS ON MOLECULES OF ASTROCHEMICAL INTEREST

P. Botschwina, S. Seeger, M. Horn, J. Flügge, M. Oswald, M. Mladenović, U. Höper, R. Oswald and E. Schick
Institut für Physikalische Chemie der Universität Göttingen, Tammannstraße 6, D– 37077 Göttingen, Germany

ABSTRACT

Making use of large–scale ab initio calculations carried out mainly at the CCSD(T) level we report on spectroscopic properties of known interstellar and circumstellar molecules and potential candidates. Rotational constants in different vibrational states are investigated for HOC^+ and HC_5N. The B_0 prediction for CH_3C_2NC of 2195 ± 3 MHz is in excellent agreement with a recent experimental value. Accurate equilibrium geometries, which were obtained by combination of experimental and theoretical data, are reported for nine different interstellar molecules. Reliable predictions for the electric dipole moments of C_5S, CH_3C_3N, CH_3C_2NC, CH_3C_5N and CH_3C_4NC are made. Absolute IR intensities for stretching vibrational transitions of HCN up to high overtones as calculated from a new CCSD(T) electric dipole moment function are in excellent agreement with recent experimental values; the previous discrepancy in the intensity of the $5\nu_1+\nu_3$ band has been clarified in favour of theory. Large–scale MRCI calculations yield a value of 2417 cm^{-1} for the barrier height to dissociation in the $\tilde{B}$ state of the methyl radical, in very good agreement with an approximate value of 2200 cm^{-1} deduced from lifetime measurements.

INTRODUCTION

For a period of about 20 years the study of interstellar molecules has been characterized by a fruitful interplay between radio or infrared astronomy, laboratory spectroscopy and ab initio quantum chemistry. Ab initio calculations are of particular value for reactive species (e.g., radicals, carbenes, cations and less stable isomers of stable molecules) which are difficult to observe in the laboratory and already so big that the corresponding interstellar lines cannot be assigned without any guidance. Our previous work on molecules of current interest to interstellar cloud chemistry has been reported recently.[1] In this work we will concentrate on the following seven topics: 1) performance of state–of–the–art ab initio calculations for the rotational constants in different vibrational states of a triatomic molecule with large–amplitude bending motion (HOC^+); 2) prediction of rotational constants for CH_3C_2NC and CH_3C_4NC; 3) dependence of rotational constants on the vibrational state for HC_5N and DC_5N; 4) accurate equilibrium geometries through suitable combination of experimental and theoretical data; 5) electric dipole moments for C_nS (n = 3, 5) and $CH_3(C_2)_nCN$ and $CH_3(C_2)_nNC$ with n = 0, 1, 2; 6) absolute IR intensities for stretching vibrational transitions of HCN up to high overtones and 7) photodissociation of the methyl radical.

ROTATIONAL CONSTANTS

The most important molecular quantities for radioastronomy are ground–state rotational constants and electric dipole moments. The former are usually calculated as the sum of equilibrium values (A_e, B_e and C_e) and the corresponding ground–state vibrational contributions.

Table I Calculated and experimental rotational constants for HOC^+ and DOC^+ (in MHz)

		Ref.[11]	Ref.[12]	Ref.[7]	this work	exp.
HOC^+	B_0	44759	44789	44388	44690	44744
	$B_e - B_0$	91	77	-128	55	
	$B_0 - B_{010}$	-189	-150	-464	-198	-196
	$B_0 - B_{001}$	357	336	366	359	
	$B_0 - B_{100}$	333	153	306	291	287
DOC^+	B_0	38212	38241	38000	38134	38193
	$B_e - B_0$	-81	-102	-368	-115	
	$B_0 - B_{010}$	-309	-285	-668	-324	
	$B_0 - B_{001}$	255	243	259	259	
	$B_0 - B_{100}$	339	300	342	302	

[a]Refs. 13 – 17

As a part of a larger project aimed at the construction of a global potential energy surface for the isomerization process $HOC^+ \longrightarrow HCO^+$ we have recently set up a preliminary analytical potential for HOC^+. HOC^+ has not yet been detected in the interstellar medium (ISM) with certainty, but its more stable isomer HCO^+ (the carrier of the famous "X-ogen" line) is rather abundant. The potential for HOC^+ is based on 124 energy points which cover the range $1.37\ a_0 \leq r(OH) \leq 2.87\ a_0$, $1.784\ a_0 \leq R(OC) \leq 2.684\ a_0$ and $0 \leq \theta \leq 110°$ where θ measures the deviation from linearity. The points were obtained by approximate solution of the electronic Schroedinger equation by means of the CCSD(T) method using a large basis set of 133 contracted Gaussian–type orbitals (see ref. 1, table 1 for details). This acronym stands for a Coupled Cluster method involving single and double substitutions (CCSD)[2] plus a quasi–perturbative treatment of connected triples.[3] During the past three years this method has been widely applied (the first code is due to T. J. Lee) and proved to yield excellent results provided that large basis sets are employed. Recent applications to spectroscopic properties of triatomic and tetra–atomic molecules may be found in refs. 4 – 7. The present CCSD(T) calculations as well as the other electronic structure calculations of this work have been carried out with the MOLPRO92 suite of programmes.[8] The efficient implementation of the CCSD method is described in detail in ref. 9; the addition of the contributions arising from connected triple substitutions is due to Deegan and Knowles.[10]

Calculated and experimental rotational constants for HOC$^+$ and DOC$^+$ in different vibrational states are given in table I. The present ground–state rotational constants (B_0) differ from experiment[13-15] by only 0.12 and 0.15 %. The calculated differences in rotational constants $B_0 - B_v$ from this work are in excellent agreement with the experimental values of Nakanaga and Amano[16,17]. The deviations amount to only 1%. The previous calculations [7,11,12] have considerably larger errors. We wish to emphasize that HOC$^+$ is a non–trivial molecule since its bending potential is extremely flat (origin of bending vibration: 260 cm^{-1}) and the effect of connected triple substitutions on its shape is substantial.

Table II Equilibrium geometries (in Å and degree) and equilibrium rotational constants of CH$_3$C$_5$N und CH$_3$C$_4$NC

	CH$_3$C$_5$N	CH$_3$C$_4$NC
r_e(CH)	1.0892	1.0898
α_e(∠HCH)	108.69	108.49
R_{1e}(C$_{(1)}$C$_{(2)}$)	1.4667	1.4675
R_{2e}(C$_{(2)}$C$_{(3)}$)	1.2133	1.2128
R_{3e}(C$_{(3)}$C$_{(4)}$)	1.3727	1.3753
R_{4e}(C$_{(4)}$C$_{(5)}$)	1.2159	1.2122
R_{5e}(C$_{(5)}$C$_{(6)}$/C$_{(5)}$N)	1.3785	1.3178
R_{6e}(C$_{(6)}$N/NC$_{(6)}$)	1.1642	1.1860
A_e (GHz)	160.06	160.28
B_e (MHz)	770.64	799.71

Another potential candidate for radioastronomy is methylisocyanoacetylene (CH$_3$C$_2$NC). The unsubstituted species (HC$_2$NC) has been recently detected in TMC-1.[18] The equilibrium geometry of CH$_3$C$_2$NC has been calculated by CEPA-1[19] (coupled electron pair approximation, version 1) using a basis set of 123 cG-TOs. All valence electrons were correlated and canonical molecular orbitals were employed. The equilibrium geometrical parameters are: r_e(CH) = 1.0893 Å, α_e (HCH) = 108.57°, R_{1e}(C$_{(1)}$C$_{(2)}$) = 1.4691 Å, R_{2e}(C$_{(2)}$C$_{(3)}$) = 1.2072 Å, R_{3e}(C$_{(3)}$N) = 1.3245 Å and R_{4e} (NC$_{(4)}$) = 1.1841 Å. The corresponding equilibrium rotational constants are A_e = 160.28 GHz and B_e = 2173.4 MHz. An estimate of the ground-state value B_0 was obtained by scaling the calculated B_e value with a factor of 1.010. The latter was determined from comparison of the results of analogous calculations for CH$_3$CN, CH$_3$NC, CH$_3$C$_3$N, HC$_2$NC, HC$_3$N and HC$_5$N with experiment. We thus arrived at B_0 = 2195 MHz and estimated the uncertainty of this value to be ±3 MHz. Our prediction turned out to be of substantial value in the search of pure rotational transitions of CH_3C_2NC. In June of this year, Gripp and Guarnieri managed to detect several lines and determined a B_0 value of 2196.3 MHz.[20]

Using analogous basis sets for CH_3C_5N and CH_3C_4NC (165 cGTOs) the equilibrium geometries and the corresponding equilibrium rotational constants have been calculated and are listed in table II. The ratio $B_0(exp.)/B_e(CEPA-1)$ for CH_3C_5N is 1.0096. Our B_0 prediction for CH_3C_4NC amounts to 808 ± 1 MHz.

Table III Vibration-rotation coupling constants (in MHz) for HC_5N and DC_5N

	HC$_5$N exp.[a]	HC$_5$N theor.[b]	DC$_5$N theor.[b]
α_1		0.886	1.508
α_2	4.366	4.387	4.052
α_3		3.293	3.062
α_4		2.492	2.207
α_5		3.225	2.970
α_6		1.198	1.102
α_7	-0.268	-0.801	-1.155
α_8	-1.750	-1.182	-1.084
α_9	-1.593	-1.800	-1.584
α_{10}	-2.457	-2.400	-2.144
α_{11}	-2.782	-2.705	-2.530

[a] Calculated as $\alpha_i \approx B_0 - B_i$ from ref. 21, 22
[b] Calculated by 2nd order perturbation theory in normal coordinate space from CEPA-1 cubic force field.

In hotter local regions of the interstellar medium molecules may be vibrationally excited and the knowledge of rotational constants in excited vibrational states is of interest. The linear dependence of rotational constants on the vibrational quantum numbers is described by the vibration–rotation coupling constants α_i which are obtainable by conventional second–order perturbation theory from the cubic force field. Unless the excited vibrational states are perturbed, the calculated α_i values are usually in good agreement with the corresponding differences of experimental rotational constants: $\alpha_i \approx \Delta B_i = B_0 - B_i$.

Results for HC_5N and DC_5N are given in table III. Very good agreement between theory and experiment is found for modes i = 2, 9, 10 and 11. The v_8 = 1 state appears to be perturbed through Fermi interaction with v_{10} = 2. Our CEPA-1 calculations yield ω_8 = 506 cm^{-1} and $2\omega_{10}$ = 514 cm^{-1}. The v_7 = 1 state is certainly perturbed as well but the nature of the perturbation is probably complex, i. e. produced by more than one perturbing state. The harmonic wavenumber of the lowest bending vibration is calculated to be 107 cm^{-1}. This is in rather poor agreement with an estimate of 75 cm^{-1} made on the basis of experimental data by Hutchinson, Kroto and Walton[23] but agrees nicely with a very recent semirigid bender prediction of 102 cm^{-1} made by Ross.[24]

ACCURATE EQUILIBRIUM GEOMETRIES

For small molecules with elements of the first two periods (up to Ne) state-of-the-art ab initio calculations are approaching an accuracy of 0.001 Å in equilibrium bond lengths and 0.2° in equilibrium bond angles. The excellent performance of large-scale CCSD(T) calculations has recently been demonstrated for ten different linear molecules with 2 – 4 atoms.[25]

Table IV Equilibrium bond lengths (in Å) for linear triatomic interstellar molecules

Molecule	geometrical parameter	pert.	var.	CCSD(T) 133 cGTO
HCO$^+$	r_e(CH)	1.0919	1.0919	1.0910
	R_e(CO)	1.1058	1.1058	1.1064
HCN	r_e(CH)	1.0650	1.0651	1.0643
	R_e(CN)	1.1534	1.1534	1.1538
HNC	r_e(NH)	0.9954	0.9956	0.9945
	R_e(NC)	1.1685	1.1686	1.1692

A simple alternative which is easily applicable to larger molecules and molecules with heavier atoms and which is usually still more accurate consists in the combination of experimental and theoretical data. Ground-state rotational constants (B_0, etc.) for a sufficiently large number of different isotopic species are taken from experiment while the differences between equilibrium and ground-state values (ΔB_0, etc.) are calculated from ab initio potential energy functions either variationally (var.) or by means of conventional second-order perturbation theory in normal coordinate space (pert.). The equilibrium rotational constants are obtained as $B_e = B_0(\text{exp.}) + \Delta B_0(\text{theor.})$. Equilibrium geometries are then determined by least-squares fit to the corresponding equilibrium moments of inertia. Unless the molecule under investigation undergoes large-amplitude vibrational motions the differences between approaches "var." and "pert." are usually very small. Table IV reports the results for three triatomic interstellar molecules with linear equilibrium structure. Here, the differences amount to 0.0002 Å or less. The excellent performance of CCSD(T)/133 cGTO is also obvious.

Approach "pert." has also been applied to a number of interstellar molecules with 4 and 5 atoms such as HCNH$^+$,[26] C$_3$O,[27] C$_3$S,[28] HC$_3$N,[29] HC$_2$NC,[30] and H$_2$C$_3$.[31] The results (bond lengths in Å, angles in °) are given in fig. 1.

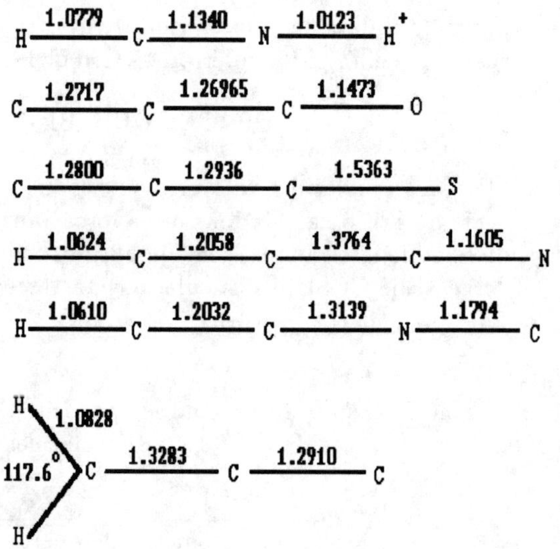

Fig. 1
Accurate equilibrium geometries for interstellar molecules

H —1.0779— C —1.1340— N —1.0123— H⁺

C —1.2717— C —1.26965— C —1.1473— O

C —1.2800— C —1.2936— C —1.5363— S

H —1.0624— C —1.2058— C —1.3764— C —1.1605— N

H —1.0610— C —1.2032— C —1.3139— N —1.1794— C

H
  \ 1.0828
117.6° C —1.3283— C —1.2910— C
  /
H

ELECTRIC DIPOLE MOMENTS

The intensities of radio lines corresponding to pure rotational transitions are proportional to the square of the permanent electric dipole moment. Accurate ab initio predictions of the latter are thus of considerable interest. As has been demonstrated recently for CO[32] and HCN[33] the CCSD(T) method is capable of obtaining equilibrium dipole moments μ_e accurate to a few thousands of a Debye unit.

In our recent investigation of C_3S,[28] a molecule detected in TMC-1[34] and IRC+10216,[35] we obtained a μ_e value of 3.886 D from CCSD(T) calculations with a basis set of 160 cGTOs. The negative end of the dipole is situated at the terminal carbon site. The ground–state value μ_0 was estimated to be 3.8 ± 0.1 D. It turned out to be in poor agreement with the previous experimental value of 2.81(7) D.[36] As noted by the experimentalists this value – due to the limited frequency shift and possible ionic shielding – was likely a lower limit. Quite recently, the μ_0 value of C_3S has been remeasured by Suenram and Lovas.[37] Their new value of 3.704(9) D compares nicely with our prediction.

Following laboratory microwave work[38] a line found at 23990.2 ± 0.2 MHz in IRC+10216 was recently tentatively assigned to the J = 13 – 12 transition of C_5S.[40] Using a large basis set of 208 cGTOs and the CCSD(T) method we predict a μ_e value of 5.38 D. The corresponding equilibrium geometry is $R_{1e} = 1.2786$ Å, $R_{2e} = 1.2924$ Å, $R_{3e} = 1.2707$ Å, $R_{4e} = 1.2800$ Å and $R_{5e} = 1.5459$ Å where the bonds are numbered according to the chemical formula. Our μ_0 estimate is 5.2 ± 0.1 D. Adopting the above tentative assignment this yields a column density of $2.7 \cdot 10^{13}$ cm^{-2}. Earlier ab initio values of the equilibrium dipole moment[38] from calculations with the relatively small 6–31G** basis set are 3.78 D (SCF), 5.25 D (MP2) and 3.93 D (MP3). Apparently, the authors gave most credit to the MP3 value.

Equilibrium dipole moments for various molecules with cyanide and isocyanide groups have been calculated at the CEPA–1 level (basis: 10s, 5p, 2d/5s, 2p contracted to [7,3,1/3,1]), with the valence electrons being correlated. As far as comparison with experiment is possible, the ratio μ_e(CEPA–1)/μ_0(exp.) is close to unity. For CH$_3$CN, CH$_3$NC, HC$_3$N and HC$_2$NC it amounts to 0.981, 1.003, 0.989 and 1.000, respectively. We are thus confident that our calculated μ_e values for CH$_3$C$_3$N (-4.91 D), CH$_3$C$_5$N (-5.75 D), CH$_3$C$_2$NC (-4.12 D) and CH$_3$C$_4$NC (-4.62 D) are reliable predictions of the ground–state dipole moments of these species. Previous estimates of the electric dipole moments of CH$_3$C$_3$N and CH$_3$C$_5$N have been made by Arnau et al.[39] on the basis of SCF/6–31G calculations plus empirical corrections. Their values of -4.85 D and -5.60 D are in good agreement with the present ones.

A graphical comparison of the dipole moments in the cyanide and isocyanide series is made in fig. 2. With the exception of n = 0 (HCN/HNC and CH$_3$CN/CH$_3$NC) the more stable cyanides have larger magnitudes of the dipole moments and are thus the better candidates for radioastronomy. Methyl substitution increases the magnitude of the dipole moment by roughly 0.9 D + 0.3 D * n for the nitriles and 0.8 D + 0.3 D * n for the isonitriles.

ABSOLUTE INTENSITIES OF STRETCHING VIBRATIONAL TRANSITIONS FOR HCN

The absolute IR intensities of rovibrational transitions for HCN up to high overtones have attracted much interest during the past ten years both theoretically and experimentally.[43] The connection with astronomical problems like the opacities of high–temperature late carbon stars is obvious. Until quite recently there was a substantial discrepancy between the calculated intensity for the $5\nu_1+\nu_3$ band (ν_1: ~ CH stretch, ν_3: ~ CN stretch) of 4.53 cm mol^{-1} and the experimental values of 13.9 and 16.7 cm mol^{-1} determined by Smith et al. [45,46] The former measurement made use of high–resolution photoacoustic detection whereas the latter was based on absorption within a traditional long–path white cell. The conflict has now been resolved in favour of our theoretical value by Romanini and Lehmann[47] who performed ring–down cavity absorption spectroscopy and obtained an intensity of 3.5 cm mol^{-1}. Calculated and experimental values for the absolute IR intensities of stretching vibrational transitions are listed in table V.

328 Quantum-Chemical Calculations

Fig.2

Dipole moments of $R(CC)_nCN$ and $R(CC)_nNC$ (R=Me,H)

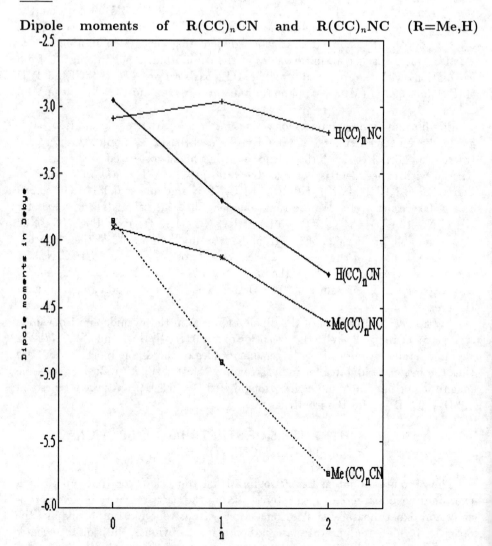

In addition to the results obtained with a CEPA-1 electric dipole moment function (EDMF)[44] this table contains the results obtained with a newly calculated CCSD(T) EDMF (basis set: 110 cGTOs). Two-dimensional vibrational wavefunctions calculated earlier[44] are used in both cases. As is obvious from the results of table V remarkable agreement between theory and experiment is observed even for the intensities of transitions to stretching vibrational states involving seven and eight vibrational quanta. Note that these values are several orders of magnitude smaller than that for the ν_1 fundamental. In particular the performance of the CCSD(T) EDMF is excellent. As is apparent from ref. 47 extension of the calculations to full (3D) dimensionality is expected to lead only to minor changes in the intensities of transitions to highly excited vibrational states.

PHOTODISSOCIATION OF THE METHYL RADICAL

The CH_3 radical has not yet been detected in the ISM (no permanent dipole moment). However, it is an ingredient of reaction schemes of dense interstellar cloud chemistry.[48] In addition, the methyl radical appears to play an important role in the photochemistry of planetary atmospheres.

Fig.3

Saddle point region for the dissociation of the $\tilde{B}\ ^2A_1'$ state of CH_3

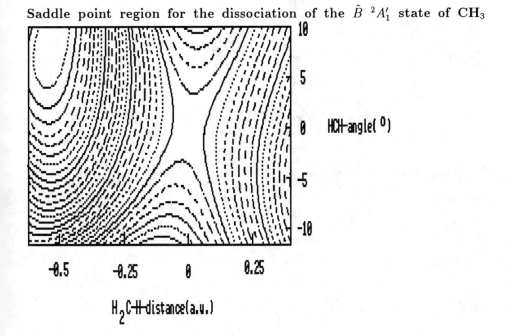

Table V Absolute IR intensities (in cm mol^{-1}) for stretching vibrational transitions of HCN arising from the vibrational ground–state

v_1	v_2	v_3^a	CEPA–1	CCSD(T)	Exp.^{b}
0	0	1	10871.	11918.	<2520.0
1	0	0	57.83e5^c	61.24e5	60.83e5
0	0	2	20.83	17.57	
1	0	1	2868.	2901.	2589.
0	0	3	16.15	21.29	23.21
2	0	0	1.343e5	1.263e5	0.8499e5
1	0	2	16.74	17.77	20.92
0	0	4	0.12	0.01	
2	0	1	1536.	1327.	1010.0
1	0	3	16.96	14.16	10.53
3	0	0	3798.	3024.	2547.0
2	0	2	16.87	10.06	6.79
3	0	1	184.8	119.6	100.0
400+203			285.0	176.2	166.11
3	0	2	5.79	3.81	2.89
401+204			28.92	18.17	17.75
5	0	0	28.20	17.58	17.5
3	0	3	0.21	0.21	0.05
205+402			1.57	1.40	2.1
5	0	1	4.53	3.62	3.51
6	0	0	2.88	2.39	2.61
5	0	2	0.39	0.52	0.58
6	0	1	0.69	0.85	0.89
7	0	0	0.28	0.37	0.55
5	0	3	0.026	0.073	0.092
6	0	2	0.092	0.20	0.22
7	0	1	0.098	0.20	0.24

^{a} Approximate quantum numbers of final state. ^{b} Ref. 43 and 47. ^{c} Reads $57.83 \cdot 10^5$

Early absorption experiments in the far–ultraviolet and vacuum ultraviolet regions were carried out by Herzberg.[49] Westre et al.[50] employed Resonance Raman Spectroscopy to study the predissociation dynamics of the $\tilde{B}$ state of CH_3. Analysis of the predissociation lifetimes of CH_3 by means of a simple one–dimensional model yielded an estimate of 2200 cm^{-1} for the height of the barrier to dissociation into CH_2 ($\tilde{a}^1A_1$) and H. This is in poor agreement with an earlier ab initio value of 4115 cm^{-1}, which was obtained from CISD (Configuration Interaction with single and double substitutions) calculations.[51] Making use of internally contracted MRCI [52, 53] and a large basis set of 213 cGTOs the barrier height to dissociation was obtained to be 2417 cm^{-1},[54] in good agreement with the above experimental value. The geometrical parameters of the planar saddle point, calculated at MRCI/130 cGTOs, are: r_S (methylenic CH) = 1.114 Å, α_S (methylenic CH_2 angle) = 111.8° and R_S (C — H) = 1.384 Å. For comparison, the equilibrium bond lengths in the $\tilde{X}^2A_1$" and $\tilde{B}^2A'_1$ states are calculated to be 1.076 Å and 1.109 Å , respectively. A contour plot showing the saddle point region of the $\tilde{B}$ state of CH_3 (at constant r_S) is shown in fig. 3. Contour lines are given in intervals of 100 cm^{-1}. Coordinate values are given relative to the saddle point geometry quoted above.

ACKNOWLEDGEMENT

Thanks are due to the Höchstleistungsrechenzentrum Jülich and the Gesellschaft für wissenschaftliche Datenverarbeitung Göttingen for providing computation time. We thank Prof. H.-J. Werner (University of Bielefeld) for supplying us with a copy of MOLPRO92. Financial support from the Deutsche Forschungsgemeinschaft and the Fonds der Chemischen Industrie is gratefully acknowledged.

REFERENCES

1. P. Botschwina, M. Horn, J. Flügge and S. Seeger, J. Chem. Soc. Faraday Trans. 89, 2219 (1993).

2. G. D. Purvis and R. J. Bartlett, J. Chem. Phys. 76, 1910 (1982)

3. K. Raghavachari, G. W. Trucks, J. A. Pople and M. Head–Gordon, Chem. Phys. Lett. 157, 479 (1989).

4. T. J. Lee and G. E. Scuseria, J. Chem. Phys. 93, 489 (1990).

5. T. J. Lee, Chem. Phys. Lett. 188, 154 (1992).

6. J. M. L. Martin and T. J. Lee, Chem. Phys. Lett. 200, 502 (1992).

7. J. M. L. Martin, P. R. Taylor and T. J. Lee, J. Chem. Phys. 99, 286 (1993).

8. MOLPRO92 is a package of ab initio programs written by H.-J. Werner and P. J. Knowles, with contributions of J. Almlöf, R. Amos, M. Deegan, S. Elbert, C. Hampel, W. Meyer, K. A. Peterson, R. Pitzer, E. A. Reinsch and A. Stone.

9. C. Hampel, K. A. Peterson and H.-J. Werner, Chem. Phys. Lett. 190, 1 (1992).

10. M. Deegan and P. J. Knowles, unpublished (1992).

11. P. R. Bunker, P. Jensen, W. P. Kraemer and R. Beardsworth, J. Mol. Spectrosc. 121, 450 (1987).

12. P. Jensen and W. P. Kraemer, J. Mol. Spectrosc. 129, 172 (1988).

13. C. S. Gudeman and R. C. Woods, Phys. Rev. Lett. 48, 1344 (1982).

14. G. A. Blake, P. Helminger, E. Herbst and F. C. de Lucia, Astrophys. J. 264, L69 (1983).

15. M. Bogey, C. Demuynck and J. L. Destombes, J. Mol. Spectrosc. 115, 229 (1986).

16. T. Nakanaga and T. Amano, J. Mol. Spectrosc. 121, 502 (1987).

17. T. Amano, J. Mol. Spectrosc. 139, 457 (1990).

18. K. Kawaguchi, M. Ohishi, S. Ishikawa and N. Kaifu, Astrophys. J. 386, L51 (1992).

19. W. Meyer, J. Chem. Phys. 58, 1017 (1973).

20. J. Gripp and A. Guarnieri, private communication from 30 June 1993.

21. K. Yamada and G. Winnewisser, Z. Naturforsch. 36a, 1052 (1981).

22. K. Yamada, private communication.

23. M. Hutchinson, H. W. Kroto and D. R. M. Walton, J. Mol. Spectrosc. 82, 394 (1980).

24. S. C. Ross, J. Mol. Spectrosc. 161, 102 (1993).

25. P. Botschwina, M. Oswald, J. Flügge, Ä. Heyl and R. Oswald, Chem. Phys. Lett. 209, 117 (1993).

26. P. Botschwina, Ä. Heyl, M. Horn and J. Flügge, J. Mol. Spectrosc., in press.

27. P. Botschwina and H. P. Reisenauer, Chem. Phys. Lett. 183, 217 (1991).

28. S. Seeger, P. Botschwina, J. Flügge, H. P. Reisenauer and G. Maier, J. Mol. Struct. (Theochem), in press.

29. P. Botschwina, M. Horn, S. Seeger and J. Flügge, Mol. Phys. 78, 191 (1993).

30. P. Botschwina, M. Horn, S. Seeger and J. Flügge, Chem. Phys. Lett. 295, 427 (1992); 200, 651 (1992).

31. C. A. Gottlieb, T. C. Killian, P. Thaddeus, P. Botschwina, J. Flügge and M. Oswald, J. Chem. Phys. 98, 4478 (1993).

32. G. E. Scuseria, M. D. Miller, F. Jensen and J. Geertsen, J. Chem. Phys. 94, 6660 (1991).

33. P. Botschwina, J. Chem. Soc. Faraday Trans. 89, 2255 (1993).

34. N. Kaifu, H. Suzuki, M. Ohishi, T. Miyawi, S. Ishikawa, T. Kasuga, M. Morimoto and S. Saito, Astrophys. J. Lett. 317, L111 (1987)

35. J. Cernicharo, M. Guélin, H. Hein and C. Kahane, Astron. Astrophys. 181, L9 (1987).

36. F. J. Lovas, R. D. Suenram, T. Ogata and S. Yamamoto, Astrophys. J. 399, 325 (1992).

37. R. D. Suenram and F. J. Lovas, J. Mol. Spectrosc., submitted for publication.

38. Y. Kasai, K. Obi, Y. Ohshima, Y. Hirahara, Y. Endo, K. Kawaguchi and A. Murakami, Astrophys. J. 410, L45 (1993).

39. A. Arnau, I. Tuñón, J. Andrés and E. Silla, Chem. Phys. Lett. 166, 54 (1990).

40. M. B. Bell, L. W. Avery and P. A. Feldman, Astrophys. J., in press.

41. Y. Hirahara, Y. Ohshima and Y. Endo, Astrophys. J. 408, L113 (1993).

42. R. A. Kendall, T. H. Dunning and R. J. Harrison, J. Chem. Phys. 98, 1358 (1993).

43. A. M. Smith, W. Klemperer and K. K. Lehmann, J. Chem. Phys. 94, 5040 (1991).

44. P. Botschwina, J. Chem. Soc. Faraday Trans. 2 84, 1263 (1988).

45. A. M. Smith, U. G. Jørgensen and K. K. Lehmann, J. Chem. Phys. 87, 5649 (1987).

46. A. M. Smith, W. Klemperer and K. K. Lehmann, J. Chem. Phys. 90, 4633 (1989).

47. D. Romanini and K. K. Lehmann, J. Chem. Phys., in press.

48. D. Smith, Chem. Rev. 92, 1473 (1992).

49. G. Herzberg, Proc. R. Soc. London Ser. A 262, 291 (1961).

50. S. G. Westre, P. B. Kelly, Y. P. Zang and L. D. Ziegler, J. Chem. Phys. 90, 6977 (1991).

51. H. T. Yu, A. Sevin, E. Kassab and E. M. Eveleth, J. Chem. Phys. 80, 2049 (1984).

52. H.-J. Werner and P. J. Knowles, J. Chem. Phys. 89, 5803 (1988).

53. P. J. Knowles and H.-J. Werner, Chem. Phys. Lett. 145, 514 (1988).

54. P. Botschwina, E. Schick and M. Horn, J. Chem. Phys. 98, 9215 (1993).

DISCUSSION

LEACH — CD_3 has a much smaller predissociation rate than CH_3, as is known from the spectroscopic studies of Herzberg. It would therefore be interesting to do calculations on the $CD^*_3 \rightarrow CH_2 + H$ dissociation reaction using the same methods as for CH_3 dissociation via de Rydbergisation.

BOTSCHWINA — Our work devoted to a theoretical investigation of the photodissociation of the methyl radical in the $\tilde{B}$ state is still in its beginning. So far, we have been engaged in the construction of the potential energy hypersurface and have not performed any dynamical calculations. However, new experimental data are available which clearly show the difference in predissociation rates between CH_3 and CD_3. Westre et al.[1] measured lifetimes of different rotational states in the vibrational ground state of CH_3 and CD_3 using Resonance Raman Spectroscopy. The CH_3 lifetimes vary from 90 fs for J' = 4 up to 70 fs for J' = 11. The corresponding lifetimes of CD_3 decreases from 760 to 340 fs for J' = 2 to J' = 15.

(1) S.G. Mestre, P.B. Kelly, Y.P. Zang and, L.D. Ziegler, J. Chem. Phys. **94**, 270 (1991).

SCAPPINI — How does the equilibrium structure theoretically calculated compare with the experimental one, when this is available ?

BOTSCHWINA — There are not too many accurate equilibrium structures determined solely on the basis of experimental data. In the cases compared so far our mixed experimental/theoretical method produces results in excellent agreement with those. E. g., our equilibrium structure for HCN (r_e = 1.0651 Å and R_e = 1.1534 Å) agrees nicely with that of Carter, Mills and Handy[1] (r_e = 1.06501(8) Å and R_e = 1.15324(2) Å). Likewise, the present equilibrium geometry of HNC (r_e = 0.9956 Å and R_e = 1.1686 Å) is in very good agreement with the very recent experimental structure (r_e = 0.9960 Å and R_e = 1.1684 Å) of Okabayashi and Tanimoto[2]. A very accurate experimental structure is available for carbon dioxide[3]: R_e = 1.15996 Å. Making use of approach "pert.", the experimental B_o values for $^{12}C^{16}O_2$ and $^{12}C^{18}O_2$ and the CCSD(T) cc-pVTZ cubic force field of Martin, Taylor and Lee[4] we arrive at R_e = 1.16011Å, in excellent agreement with the above value. Analogous work on N_2O makes use of experimental Bo values for 5 different isotopic species[5] and the ab initio force constants of the above authors[4] We arrive at R_{1e} (NN) = 1.12702 Å and R_{2e} (NO) = 1.18545 Å to be compared with the experimental values of Teffo and Chédin[6] (R_{1e} = 1.12729 Å and R_{2e} = 1.18509 Å).

(1) S. Carter, I.M. Mills and N.C. Handy, *J. Chem. Phys.*, **97**, 1606 (1992).
(2) T. Okabayashi and M. Tanimoto, *J. Chem. Phys.* **99**, 3268 (1993).
(3) G. Graner, C. Rossetti and D. Bailly, *Mol. Phys.* **58**, 627 (1986).
(4) J. M. L. Martin, P. R. Taylor and T. J. Lee, *Chem. Phys. Lett.* **205**, 535 (1993).
(5) B. A. Andreev, A. V. Burenin, E. N. Karyakin, A. F. Krupnov, and S. M. Shapin, *J. Mol. Spectrosc.* **62**, 125 (1976).
(6) J. L. Teffo and A. Chédin, *J. Mol. Spectrosc.* **135**, 389 (1989).

WDOWIAK — Can you say anything about the electron affinity of molecules such as H_2C_3 and H_2C_5 ?

BOTSCHWINA — To our knowledge there are no reliable experimental or theoretical data available for the electron affinities of H_2C_3 and H_2C_5. A quick CEPA-1 calculation (basis set: 122 cGTOs) for the former species yielded an adiabatic electron affinity of 1.6 eV; this value may have an uncertainty of ca. 0.3 eV.

PHYSICAL CHEMISTRY OF SILICON CONTAINING MOLECULES

A. Spielfiedel, N. Feautrier, I. Drira
DAMAP, Observatoire de Paris, F-92195 Meudon, France

G. Chambaud
Université de Marne la Vallée, France

P. Rosmus
Fachbereich Chemie der Universität, D-6000 Frankfurt, Germany

Y. Viala
DEMIRM, Observatoire de Paris, F-92195 Meudon, France

ABSTRACT

Much attention has been paid recently to silicon containing molecules whose chemistry may play an important role in outer space.

Using large scale ab initio calculations, we have first determined the spectrum of the Si_2C molecule in its electronic ground state X^1A_1. This molecule has not yet been detected even though it can be expected to be a constituent of carbon star atmospheres or molecular clouds. The search for the spectra of this species will be facilitated by such accurate predictions.

Secondly, we have calculated the energies and the spectroscopic constants of the valence states of SiO. They compare very well with the existing experimental data. The next step will be the determination of higher excited states which are important for photodissociation studies.

INTRODUCTION

Much work has been devoted in the recent years to the silicon containing molecules present in the gas phase of molecular clouds and circumstellar envelopes. Observations suggest that these molecules exist in star-forming regions and one might expect some of them to coexist on the surface of the grains. Hence, their chemistry is particularly rich and our purpose is to investigate two aspects of it by providing :
- new ab-initio data for unknown species in their ground electronic state in order to facilitate both their laboratory spectroscopic study and their search in outer space by radio-astronomy.
- reaction rates, and particularly photodissociation cross sections.

We report here ab-initio results relative first to the Si_2C molecule in its ground state and secondly to some excited states of SiO which are important for future studies of the spectroscopy and the photodissociation of this latter molecule.

THEORETICAL VIBRATION-ROTATION SPECTRUM OF Si_2C

The three-dimensional potential energy and electric dipole moment functions of the electronic ground state of Si_2C have been generated in large scale ab-initio multireference configuration interaction calculations (MRCI). The atomic orbital basis set comprised 142 contracted functions including f functions on the three atoms. In the subsequent MRCI calculations all valence excitations from the reference wavefunction were included. The calculated potential energies and electric dipole moments have been used in the evaluation of the vibration-rotation energy levels and eigenfunctions by the approach of Carter-Handy[1]. The Carter-Handy Hamiltonian includes the full rotation-vibration coupling and is particularly well suited for triatomic molecules with large amplitude bending modes. The resulting eigenfunctions and the three-dimensional functions of the electric dipole moments were used in calculations of the dipole matrix elements and the line intensities[2].

In table 1 the harmonic frequencies, the rotational constants and the centrifugal distorsion constants τ calculated by the perturbation theory are given. The equilibrium geometry agrees well with recent calculations [3,4] within the estimated accuracy of our results (±0.008Å for the equilibrium Si-C distance R_e, ±0.5° for the angle α_e)

Table 1 MRCI values for Rotational and Vibrational Constants for the X^1A_1 State of $^{28}Si_2^{12}C$.

Constant	Value
ω_1/cm^{-1}	840.77
ω_2/cm^{-1}	135.25
ω_3/cm^{-1}	1207.29
A/MHz	68892.0
B/MHz	4217.2
C/MHz	3973.9
τ_{AAAA}/MHz	-116.52
τ_{BBBB}/MHz	-0.048
τ_{ABAB}/MHz	-0.014
τ_{AABB}/MHz	2.238
R_e(SiC)/Å	1.698
α_e/deg	119.02

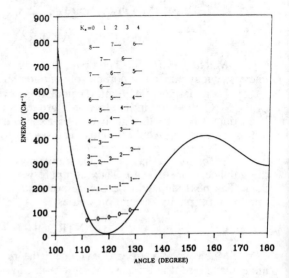

Figure 1
One-dimensional cut of the MRCI ground state potential energy function of Si_2C along the bending coordinate with the calculated energy positions of the bending levels for $v_2 = 0$ to 8 and $K_a = 0$ to 4.

The height of the barrier to linearity (figure 1) has been calculated to be 310 cm^{-1}. The shallow potential energy function along the bending coordinate results in large amplitude bending modes (figure 2). For the low bending modes and the fundamental stretching modes, vibrational band intensities and rotational line strengths have been evaluated from the dipole moment functions. In agreement with the experimental data determined in Ar matrix[5], we have calculated the antisymmetric stretching mode to be the most intense fundamental transition, followed by the symmetric stretching mode and the (v_2+v_3) combination mode[6]. The dipole moment in the vibrational ground state has been calculated to be 0.90 D .

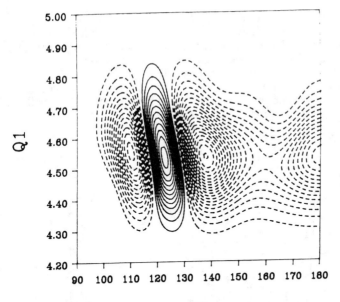

Figure 2
Contour plot of the $v_2=2$ (J=0) vibrational wavefunction of Si_2C in the coordinates $Q_1 = (1/2)^{1/2}(R_1 + R_2)$, and $Q_2 = \alpha$

ELECTRONIC STRUCTURE OF THE SiO EXCITED SINGLET STATES

We have performed ab-initio calculations of the energies of the first excited singlet states. The atomic basis sets comprised 129 contracted functions including f functions on both atoms.

In the molecular region, the equilibrium distances R_e, the harmonic vibrational frequencies ω_e and the excitation energies T_e have been calculated at the MRCI level for the ground state and the first valence excited states ($C^1\Sigma^-$, $D^1\Delta$, $A^1\Pi$) (Table 2).

States		R_e (Å)	ω_e (cm^{-1})	T_e (cm^{-1})	μ_e (Debye)
$X^1\Sigma^+$	MRCI	1.521	1238	0	-3.013
	MCSCF[10]	1.514	1262	0	-3.006
	MCSCF-CI[11]	1.515	1241	0	-3.010
	CEPA-1[11]	1.519	1240	0	-3.067
	exp[7]	1.510	1241	0	-3.098
$C^1\Sigma^-$	MRCI	1.743	744	38865	0.364
	exp[7]	1.727	740	38624	
$D^1\Delta$	MRCI	1.743	740	38974	0.320
	exp[7]	1.729	730	38823	
$A^1\Pi$	MRCI	1.639	850	44016	-0.946
	exp[7]	1.620	853	42835	

Table 2
Spectroscopic constants for valence electronic states of $^{28}Si^{16}O$.

These ab initio results compare very well with the experimental data[7].

Figure 3 shows the potential energy functions as a function of the Si-O distance for the first valence states together with the $E^1\Sigma^+$ state which presents both Rydberg and valence characters. All these states dissociate in $Si(^3P) + O(^3P)$. The MRCI dissociation energy D_e of the ground state is 8.010 eV, which compares reasonnably well with the experimental value[8] (8.33 eV). As first pointed out by Robbe et al.[9], the latter $E^1\Sigma^+$ state presents a barrier to the dissociation around 2.25 Å, resulting from interactions with higher $^1\Sigma^+$ states dissociating in $Si(^1D) + O(^1D)$.

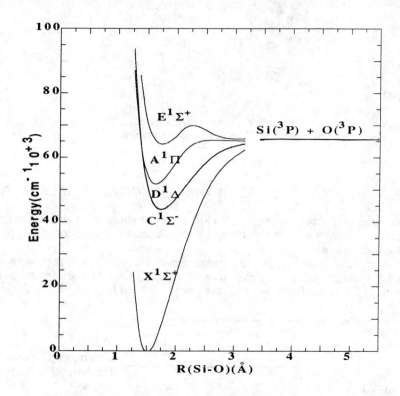

Figure 3
Potential energy functions of the first electronic states of SiO

REFERENCES

1. S. Carter, N. C. Handy, Molec. Phys. 52, 1367 (1984).
2. S. Carter, J. Senekowitsch, N. C. Handy, P. Rosmus, Molec. Phys. 65, 143 (1988).
3. C. M. L. Rittby, J. Chem. Phys. 95, 5609 (1991).
4. E. E. Bolton, B. J. DeLeeuw, J. E. Fowler, R. S. Grev, H. F. Schaefer III, J. Chem. Phys. 97, 5586 (1992).
5. J. D. Presilla-Marquez, W. R. Graham, J. Chem. Phys. 95, 5612 (1991).

6. W. Gabriel, G. Chambaud, P. Rosmus, A. Spielfiedel, N. Feautrier, Astrophys. J. 398 706 (1992).
7. J. Hormes, M. Sauer, R. Scullman, J. Mol. Spectrosc. 98, 1 (1983).
8. G. Herzberg, K. P. Huber, Molecular Spectra and Molecular structure IV, Constants of diatomic molecules; Van Nostrand; Princeton N.J (1979)
9. J. M. Robbe, J. Schamps, H. Lefebvre-Brion, G. Raseev, J. Mol. Spectrosc. 74, 375 (1979).
10. H. J. Werner, P. Rosmus, M. Grimm, Chem. Phys. 73, 169 (1982).
11. P. Botschwina, P. Rosmus, J. Chem. Phys. 82, 1420 (1985)

MCSCF-CI STUDY OF THE ISOMERIZATION REACTION HCSi --> HSiC

H. LAVENDY, J.M. ROBBE, D.DUFLOT

Laboratoire de Dynamique Moléculaire et Photonique, URA No.779 du CNRS,
Université des Sciences et Technologie de Lille, UFR de Physique, F-59655. Villeneuve d'Ascq Cedex, France

J.P. FLAMENT

Laboratoire des Mécanismes Réactionnels,DCMR,URA No.1307 du CNRS,
Ecole Polytechnique, 91128 Palaiseau Cedex.

Abstract

Ab-initio electronic structure theory (MCSCF+CI) has been used to determine the more important features of the potential energy surface for the simple isomerization reaction HCSi --> HSiC.

Our preliminary results indicate that H-Si-C has a bent structure while H-C-Si is linear. More, the H-Si-C ground state is $^2A'$ and it is $^2\Pi$ for H-C-Si.

Potential barriers for isomerization are calculated.

Geometries, energies and dipole moments are given for ground, excited states and transition states.

Introduction

The family of HAB molecules is the simplest class of systems for which the isomerization reaction can be studied on a chemically meaningful basis.

In this study, we perform ab initio MCSCF+multireference CI calculations about the H-C-Si and H-Si-C radicals in order to characterize their low lying states and their equilibrium conformation and also to determine the depths of potential wells along the reaction path. The theoretical investigations concerning these species are guided by the astrophysical interest and also, by the fact that the H-C-Si radical can probably be observed in laboratory.

Theoretical approach

Ab initio molecular orbital calculations have been carried out using the Hondo8 program package (1). The MCSCF geometry optimisation calculations were done using analytic gradient techniques(2). We used the triple zeta atomic orbitals gaussian basis sets given by Huzinaga and Dunning (3,4) consisting of a 5s1p contracted 3s1p for hydrogen, 11s6p2d contracted 5s3p2d for carbon and 13s9p2d1f contracted 6s5p2d1f for silicon. The additional d type functions for carbon and f type functions for silicon are necessary to take into account the polarization effects. We included the electron correlation effects in calculating the CI energies with the CIPSI algorithm(5) in which a variational zeroth-order wavefunction is built in an iterative selection of the most important determinants, the other ones being taken into account through a second-order Moller-Plesset perturbation. The determinants having a coefficient larger than 0.05 in the first order wavefunction have been included in the zeroth-order wavefunction at the final step. Single and double substitutions from these determinants were treated in a second-order Moller-Plesset perturbation using a diagrammatic version of CIPSI (6,7).

Results and discussion

The MCSCF optimized geometries for the two lowest $^2\Pi$ and $^2\Sigma^+$ electronic states of the linear HCSi are given in Table 1, just as the energies obtained at the CI level, compared with MRD-CI calculations done by Buenker et al (8).

Table 1: Equilibrium geometries of the two lowest electronic states of linear HCSi, and corresponding energies.

	C-H (Å)	C-Si (Å)	E (hartree)	ΔE kcal.mol^{-1}
our results* H-C-Si $^2\Pi$	1.091	1.710	-327.3690	0.0
$^2\Sigma^+$	1.084	1.624	-327.3248	27.7
our results** H-C-Si $^2\Pi$			-327.6567	0.0
$^2\Sigma^+$			-327.6080	29.3
ref.1 H-C-Si $^2\Pi$	1.074	1.704	-327.4599	0.0
$^2\Sigma^+$	1.074	1.619	-327.4138	28.9

*MCSCF
**IC

In the H-Si-C geometry, the molecule is not linear and the HSiC angle are 122° and 148°5 for $^2A'$ and $^2A''$ respectively. We also determined the saddle points for $^2A'$ and $^2A''$ states. The nature of the stationary points was characterized through computation of second derivatives and vibrational frequencies.

Table 2 : Energies, bond lengths, dipole moments and HCSi angles for the two low lying states.

	$^2\Pi$	$^2\Sigma^+$	$^2A'$	$^2A''$	$^2A'$	$^2A''$
	H-C-Si		Transition state		H-Si-C	
E(MCSCF)	-.369036	-.324820	-.284161	-.266431	-.292395	-.270648
E(CI)	-.656726	-.607992	-.573884	-.557077	-.576180	-.560089
R(C-H)	1.091	1.084				
R(C-Si)	1.710	1.624	1.726	1.836	1.716	1.780
R(H-Si)			1.542	1.534	1.502	1.513
μ	0.41	1.52	1.21	1.85	1.75	3.15
θ (HCSi)	180	180	82.6	92	122	148.5

Table 2 : All energies values are shifted of -327 a.u; R is given in Å; μ in Debye; θ in degrees

In Table 2, we give the MCSCF and CI energies, the bond length between C-H, C-Si and Si-H atoms. We give also the dipole moment and the angles between C-Si and Si-H vectors for the linear $^2\Pi$ and $^2\Sigma^+$ geometries, for the transition states and for the bended geometries of $^2A'$ and $^2A''$. We observed on one hand, that the variation in the C-Si bond length is very small and does not play a crucial role in the reaction, (a greater value of the C-Si distance is obtained for the transitions states) and on the other hand, the dipole moment is more important in $^2\Sigma^+$ and $^2A''$ states than in $^2\Pi$ and $^2A'$ states. Figure 1 shows the CI energy diagram for the H-C-Si --> H-Si-C isomerization reaction. We observe that the transition state for the $^2A'$ state is located at 1.4 kcal.mol^{-1} above the $^2A'$ state in the H-Si-C equilibrium geometry, and the transition state of $^2A''$ is located at about 2 kcal.mol^{-1} above the corresponding $^2A''$ in the H-Si-C configuration. These small differences of energies show that the molecule H-Si-C is probably unstable and that only the linear configuration H-C-Si could be observed in laboratory.

Fig. 1 CI energy diagram for the HCSi --> HSiC reaction

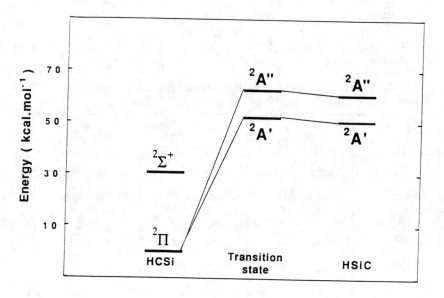

Table 3 : MCSCF vibrational frequencies for the stable states, and for the transition state geometries (in cm^{-1})

H-C-Si		H-Si-C		Saddle point geometries	
$^2\Pi$	$^2\Sigma^+$	$^2A'$	$^2A''$	$^2A'$	$^2A''$
bend 504.0	848.7	bend 505.9	279.5	zH 644.1i	xH 197.5i
C-Si 1000.4	1150.9	C-Si 939.0	884.6	C-Si 946.4	791.2
C-H 3196.0	3249.6	H-Si 2126.5	2037.6	xH 1945.0	zH 1970.6

In Table 3 we give the MCSCF harmonic vibrational frequencies, using analytical second derivatives (9), for the stable $^2\Pi$ and $^2\Sigma^+$ states, for the $^2A'$ and $^2A''$ in the H-Si-C configuration, and for the transition states. For these transition states we observed imaginary bending frequencies given by the components x and z. The z axis contains the C-Si atoms and the hydrogen atom lies in zx plane.

Conclusion

This preliminary study of the isomerization reaction H-C-Si --> H-Si-C has revealed that :
- the H-C-Si is linear with a ground $^2\Pi$ state and its first excited state is a linear $^2\Sigma^+$.
- the potential barrier for the isomerization is very small (about 2 kcal.mol-1), which indicates that it would probably be very difficult to observe H-Si-C .

Detailed calculations are in progress to determine the geometries of other excited states of H-C-Si as well as rovibrational study of the ground $^2\Pi$ state.

References

1- M.DUPUIS, A.FARAZDEL, S.P KARNA and S.A.MALUENDES, Motecc, modern techniques in computational chemistry, ed. E.CLEMENTI (Escom, Leiden, 1990).

2- D.SPANGLER, I.H.WILLIAMS and G.MAGGIORA, J.Comp.Chem. 4, 524 (1983).

3- S.HUZINAGA, J.Chem.Phys. 42, 1293 (1965).

4- T.H.DUNNING, J.Chem.Phys. 53, 2823 (1970).

5- B.HURON, J.P.MALRIEU and P.RANCUREL, J.Chem.Phys. 58, 5745 (1973).

6- R.CIMIRAGLIA, J.Chem.Phys. 83,1746 (1985)

7- R.CIMIRAGLIA and M.PERSICO, J.Comput.Chem. 8, 39 (1987)

8- R.J.BUENKER and P.J.BRUNA and S.D.PEYERIMHOFF, Israel Journal of Chemistry 19, 309-316 (1980).

9- J.A.POPLE, R. KRISHNAN, H.B. SCHLEGEL and J.S. BINKLEY, intern.J.Quantum Chem.Symp. 13, 225 (1979).

HIGH RESOLUTION VUV LASER MEASUREMENTS OF THE CO $A^1\Pi$ - $X^1\Sigma^+$ (v',0) ABSORPTION CROSS SECTIONS

D.Malmasson [1], A.Vient [1], J.L.Lemaire [1,2], A.Le Floch [1,3], F.Rostas [1]

1: Observatoire de Paris-Meudon, DAMAp & URA 812 du CNRS, 92195 MEUDON CEDEX, France
2: Université de Cergy-Pontoise, 95806 CERGY CEDEX, France
3: Département de Physique, Université de Tours, 37200 TOURS, France

ABSTRACT

Using the resonant difference frequency mixing technique in Krypton, VUV radiation continuously covering the range 117 to 143 nm is generated. It is used to measure the rotationally resolved absorption cross-section of the $A^1\Pi$ - $X^1\Sigma^+$ (v',0) transition of CO (0<v'<17). Preliminary results show the importance of the saturation effects caused by the VUV laser linewidth.

INTRODUCTION

Accurate absorption cross section measurements are needed in order to interpret absorption spectra of CO in the A-X band obtained in interstellar clouds by spaceborn telescopes such as Copernicus, I.U.E., H.S.T.[1]. The CO abundances and isotopic ratios derived from these observations depend directly on the experimental oscillator strengths.

Experimental results obtained by optical methods [2,3] tend to give cross sections which are significantly larger for v' values above 5 than those obtained recently by the electron energy loss technique [3]. The latest optical absorption results published [2] were obtained at low resolution using monochromatized synchrotron radiation as the background continuum.

The recent availability in this laboratory of the resonant difference frequency four wave mixing technique in Krypton for the generation of VUV laser radiation (Fig.1) between 117 and 143 nm allows direct rotationally resolved measurements of these cross sections up to v'=12 with the existing setup and v'=17 in the near future. Such measurements provide new values which are completely independent of the previous ones and constitute a much needed verification of the available experiments.

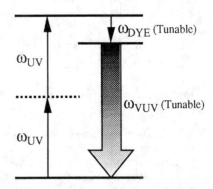

Fig.1: Resonant difference frequency four wave mixing:
$$\omega_{VUV} = 2\omega_{UV} - \omega_{DYE}$$

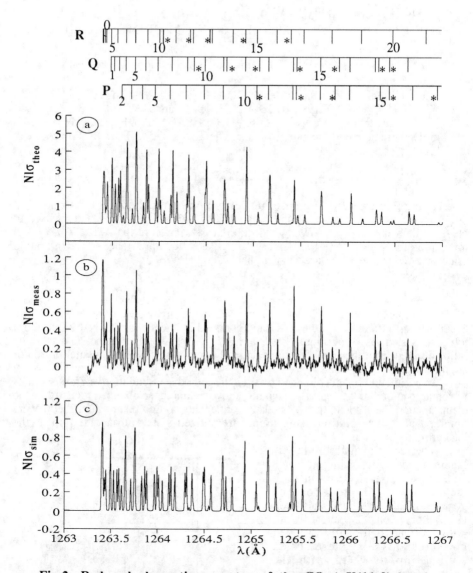

Fig.2: Reduced absorption spectra of the CO A-X(11-0) band.
a) Expected spectrum using spectroscopic data [6] of R.W.Field, a band integrated absorption cross section [2] $\sigma_B = 4\ 10^{-18}$ cm^{2}Å and a Doppler linewidth of 1.8 mÅ.
* indicates extra lines due to perturbations of the A state.
b) Experimental spectrum. $Nl = 3.04\ 10^{17}$ cm^{-2}, Doppler linewidth: 1.8 mÅ, laser linewidth: 11 mÅ.
c) Simulation of the experimental spectrum including saturation effects resulting from convolution with a gaussian apparatus function of 11 mÅ FWHM and a band integrated cross section $\sigma_B = 4\ 10^{-18}$ cm^{2}Å.

Note the drastic reduction in peak optical depth and in dynamic range in spectra b & c.

EXPERIMENTS

A preliminary study of the (11–0) and (9–0) bands has been conducted in a 41.7 cm long cell at pressures ranging from 10^{-3} to 10^{-4} bar. Well resolved spectra (Fig.2b) have been obtained. However, the band integrated absorption cross sections derived directly from these spectra are very much lower than expected. Both the relative and absolute line intensities differ from the expected ones even when the spectroscopic perturbations are taken into account.

Figure 2a displays the expected reduced absorption spectrum of the (11-0) band calculated from available spectroscopic [6] and oscillator strength [2] data, and figure 2b the corresponding experimental spectrum. In both cases the proper intensity unit is used i.e. the dimensionless optical depth $\tau(\lambda)=Nl\sigma(\lambda)$ where N is the volume density of CO, l is the cell length and $\sigma(\lambda)$ is the absorption cross section. It is seen that the observed line intensities are considerably smaller than the predicted ones and also that the range of intensities is severely compressed in the experimental spectrum.

DISCUSSION

The large observed discrepancies are due to saturation effects which can be important even at high resolution. These effects stem from the fact that the laser linewidth (11 mÅ FWHM) is about six times larger than the absorption linewidth (1.8 mÅ FWHM) which is Doppler limited. In such a case, the peak and line integrated absorptions do not vary linearly with pressure and the measured cross sections are severely in error.

The Beer-Lambert law gives the transmitted intensity:

$$I(\lambda) = I_0(\lambda) e^{-Nl\sigma(\lambda)} \tag{1}$$

so that in the ideal case the optical depth τ at wavelength λ is given by

$$\tau(\lambda) = Nl\sigma(\lambda) = \ln \frac{I_0(\lambda)}{I(\lambda)} \tag{2}$$

where $\sigma(\lambda)$ is the true absorption cross section at λ. If $P(\lambda)$ is the apparatus function, the measured intensity at λ_0 is the resultant of the convolution between the apparatus function and the transmitted intensity:

$$I_{meas}(\lambda_0) = \int P(\lambda - \lambda_0) I(\lambda) d\lambda \tag{3}$$

In the absence of correction, the apparent cross section at λ_0 is obtained through the expression:

$$I_{meas}(\lambda_0) = I_0(\lambda_0) e^{-Nl\sigma_{app}(\lambda_0)} \tag{4}$$

If $I_0(\lambda)$ can be assumed constant, the relation between true and apparent cross sections is then:

$$Nl\sigma_{app}(\lambda_0) = -\ln\left[\int P(\lambda - \lambda_0) e^{-Nl\sigma(\lambda)} d\lambda\right] \tag{5}$$

This expression shows that the apparent cross section is always less than the true one. In conventional emission spectra also, the peak line intensity is reduced by convolution between the apparatus profile and the line profile. But in such a case the apparatus profile is directly convoluted with the line profile $\sigma(\lambda)$, not through an exponential function, and the area under the emission line, $\int\sigma(\lambda)d\lambda$, is conserved, so that the integrated line intensity is not affected by instrumental resolution.

In absorption, expression (4) shows that the area under the line is not conserved, except for very small values of $Nl\sigma$ where $\exp(-Nl\sigma) \approx 1-Nl\sigma$. If the above instrumental width effect is not corrected for, the value of $\int\sigma(\lambda)d\lambda$ will be underestimated by a factor which increases as either the optical depth or the apparatus width increase.

The apparent optical depth $\tau_{app}=Nl\sigma_{app}$ at line peak, as defined in (5), has been calculated as a function of the true one for different ratios of instrument to line width when both are assumed gaussian (Figure 3). The width ratio of 6.1 corresponds to the present experimental conditions. It is seen that, even when the instrument width equals the line width, a saturation effect sets in for apparent optical depths above 1 (63% peak absorption). In the present experimental conditions, deviations from linearity will appear for apparent optical depths as small as 0.2, i.e. measured peak absorptions larger than 20%.

In figure 4, the deviation of the measured line integrated cross section with respect to the true one is plotted, versus the measured peak absorption, for different line width ratios. It is seen that, for the present experimental conditions, the measured peak absorptions have to be kept below 10% in order to be able to measure directly the line absorption cross sections. This clearly requires unrealistic signal/noise ratios in view of the difficulty in compensating the shot to shot variation in VUV laser intensity, which can reach up to 30%. Thus, indirect methods have to be used to obtain integrated band cross sections from measured spectra which suffer sizable saturation effects.

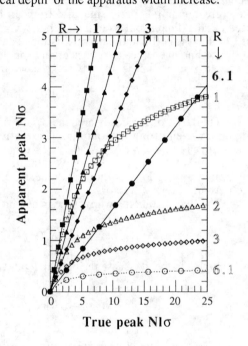

Fig.3: Saturation effects on absorption line intensities
■▲◆● emission lines
□△◇○ absorption lines
The ratio R between the instrument and the absorption linewidths is indicated on each curve. R=6.1 corresponds to the present experimental situation where the VUV laser width is 11 mÅ and the Doppler linewidth is 1.8 mÅ.

We have experimented a technique where the spectrum is simulated, including the saturation effects, and compared to the experimental spectrum. Such a simulation is shown on figure 2c. It is seen to reproduce quite accurately the experimental spectrum, when the published band integrated cross section is used. It is interesting to note that the lines fall in three categories:

a) $Nl\sigma_{peak} \approx 0.4$. From fig.3 it is seen that this corresponds to the saturation limit for strong isolated lines.

b) $Nl\sigma_{peak} < 0.4$. These are weak lines which fall in the first part of the curve and thus are not completely saturated.

c) $Nl\sigma_{peak} > 0.4$. These correspond to partially overlapped lines which act as if they were broader than the individual Doppler width and, in effect, reach a higher saturation limit, corresponding to a smaller width ratio in figure 3.

All these effects are well reproduced in the synthetic spectrum. This indicates that the saturation model we have used is adequate. The quality of the simulation rests largely on the high accuracy of the spectroscopic data

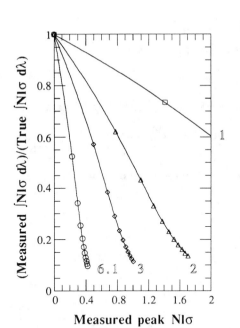

Fig.4: Saturation effects on absorption line integrated intensities
The ratio R between the instrument and the absorption linewidths is indicated on each curve. R=6.1 corresponds to the present experimental situation where the VUV laser width is 11 mÅ and the Doppler linewidth is 1.8 mÅ.

used. This includes the analysis of the perturbations affecting the $A^1\Pi$ state which allows an accurate prediction of the positions and intensities of the perturbed lines and of the associated extra-lines which borrow their intensity from the $A^1\Pi$-$X^1\Sigma^+$ transition [6].

On the basis of this successfull simulation we plan to use a least squares fitting procedure to determine the band integrated cross section which can be set as a free parameter in the fit. The instrumental width and the baseline level could also be introduced as fitting parameters in order to improve the precision of the measured cross section.

CONCLUSION

Provided the methods outlined above are carefully applied, the VUV laser source will allow rotationally resolved measurements of small band absorption cross sections free of saturation problems. Using rotationally resolved spectra permits measuring much smaller integrated cross sections, and thus should allow extending the range of measurements to v'=17 and resolving the contradictions between available experimental values.

REFERENCES

1 Y. Sheffer, S.R. Federman, D.L. Lambert and J.A. Cardelli,
    Ap.J. **397**,482 (1992)
2 M. Eidelsberg, F. Rostas, J. Breton, B. Thieblemont,
    J. Chem. Phys. **96**, 5585 (1992)
3 R.W. Field, O. Benoist d'Azy, M. Lavollée, R. Lopez-Delgado and A. Tramer,
    J. Chem. Phys. **78**, 2838 (1983)
4 W.F. Chan, G. Cooper, C.E. Brion, Chem. Phys. **170**, 123 (1993)
5 A.C. Le Floch, F. Launay, J. Rostas, R.W. Field, C.M. Brown and K. Yoshino,
    J. Mol. Spectr. **121**, 337 (1987)
6 R.W.Field, Thesis, Harvard University (1971) and private communication

A 2+1 REMPI STUDY OF THE E-X TRANSITION IN CO

J. Baker, J.L. Lemaire,* S. Couris,** A. Vient, D. Malmasson, and F. Rostas
DAMAP et URA 812 du CNRS, Observatoire de Paris-Meudon,
92195 Meudon Cedex, France.

ABSTRACT

Two-photon $E^1\Pi(v=1,0) - X^1\Sigma^+(v=0)$ resonant, three-photon ionisation spectra of the four isotopomers $^{12}C^{16}O$, $^{13}C^{16}O$, $^{12}C^{18}O$, and $^{13}C^{18}O$ have been recorded over the laser wavelength range, 210-216 nm, using a time of flight mass spectrometer, and rotational analysed. Both e and f parity levels of the $E^1\Pi$ state have been observed and localised perturbations have been identified in the v=1 vibrational level. These perturbations, which occur in both parity levels, have the characteristics of an indirect predissociation similar to the well-known accidental predissociation in the J=31 level of $^{12}C^{16}O(E^1\Pi, v=0)$. The present results favour a $^3\Pi$ valence perturbing state which is itself predissociated by a $^3\Pi$ repulsive state correlating to the lowest discocion limit; $C(^3P)+O(^3P)$.

INTRODUCTION

This work complements recent one-photon studies of the E-X transition of CO mainly carried out in absorption on the 10m spectrograph at Meudon.[1,2] In those studies only the e parity levels in general were resolved for the $E^1\Pi$ state since for a one photon $^1\Pi - {}^1\Sigma^+$ transition only the Q branch carries information on the f levels and this branch was not in general resolved in the spectrograph studies due to a very small $\Delta B = B' - B''$, for the E-X transition. In the two photon transition five rotational branches are expected O, P, Q, R, and S. In this case the O, Q, and S branches carry information on the e parity levels of the $^1\Pi$ state whilst the easily resolvable P and R branches carry information on the f parity levels of the $^1\Pi$ state.

A local perturbation in the E state of $^{12}C^{16}O$ was first observed in the one-photon E-X(0-0) absorption study of Tilford et al.[3] They photographed this band and observed a doubling in both the R(30) and P(32) lines and noted that these perturbed lines appeared more intense than the neighbouring lines, implying some type of unusual perturbation in the $J_e=31$ level of $^{12}C^{16}O(E^1\Pi, v=0)$, where the subscript on J is used here to denote the parity. Simmons and Tilford[4] photographed this same transition in emission and absorption. They considered that the apparent intensity enhancement of the perturbed lines in absorption was caused by an increased linewidth resulting from a predissociation, and confirmed this by noting the absence of the R(30) line in emission. The overall effect was then explained by an accidental predissociation, where the $J_e=31$ level is resonantly perturbed by another bound state which is itself strongly predissociated by a repulsive state. Recently, a similar effect has been observed in the $J_e=7$ level of the

---

* Also: Université de Cergy-Pontoise, 95806 Cergy Cedex, France.
** FORTH-IESL, P.O. Box 1527, Heraklion, Crete, Greece.

$E^1\Pi$ (v=1) states of $^{12}C^{16}O$, $^{13}C^{16}O$, and $^{12}C^{18}O$, respectively.[2] However, due to lack of information, especially on the effects on the f levels, the perturbing state has not yet been identified.

In this study we have been able to resolve both the e and f levels by recording 2+1 REMPI spectra of the E-X(1,0-0) transitions.

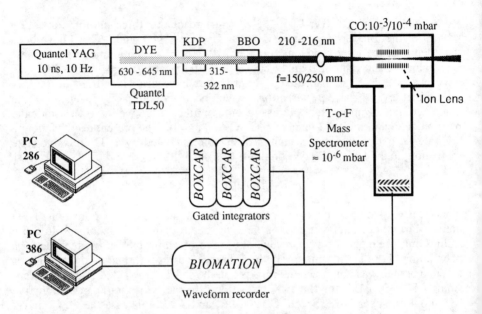

Fig.1. A schematic diagram of the experimental apparatus

EXPERIMENTAL

A schematic diagram of the experimental apparatus used in this study is given in Fig. 1. UV radiation in the wavelength range 210-216 nm was generated by first frequency doubling the fundamental output of a pulsed Quantel Nd/YAG pumped dye laser in a KDP crystal followed by frequency mixing of the doubled and fundamental frequency in a BBO crystal (Quantel/UVX-3). This radiation was then focused into a vacuum chamber. Three types of gases were used: (i) 99% natural isotopic purity $^{12}C^{16}O$, (ii) 99.8% isotopic purity $^{12}C^{18}O$ (Euriso-Top, CEA), and (iii) a ca. 60:40 isotopic mixture of $^{13}C^{16}O$ and $^{13}C^{18}O$ (Euriso-Top, CEA). The photoions generated in the 2+1 REMPI process were detected using a home built time of flight mass spectrometer enabling the separation in time of the various isotopomers. The signal was then fed either into a three boxcar integrator unit interfaced to a PC for data storage, enabling the simultaneous recordings of up to three different isotopomers, or into a transient digitizer (Biomation 6500, 2ns time resolution) interfaced to a PC with software boxcar integration capabilities for simultaneous recordings in up to eight time channels. Wavelength calibration was achieved by deriving the O, Q, and S, E-X two photon line positions from the known unperturbed $E^1\Pi$ e parity level termvalues in the v=0 and v=1 vibrational levels and the known $X^1\Sigma^+$ ground state constants, and using a weighted cubic spline fit.

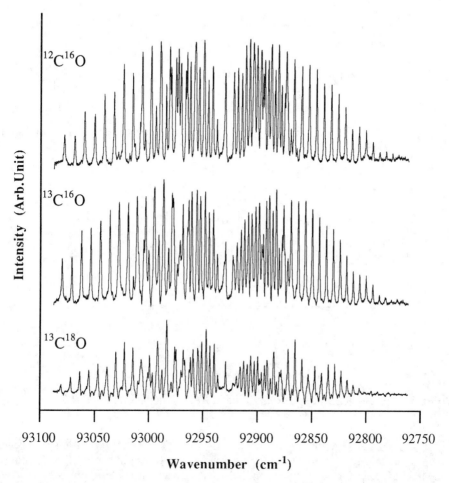

Fig.2. Simultaneous recordings of the 2+1 REMPI spectra of the E-X(0-0) transitions of $^{12}C^{16}O$, $^{13}C^{16}O$, and $^{13}C^{18}O$ respectively.

RESULTS AND DISCUSSION

Fig. 2 shows a simultaneous recording of the 2+1 REMPI spectra of the E-X(0-0) transitions in $^{12}C^{16}O$, $^{13}C^{16}O$, and $^{13}C^{18}O$ in the two-photon energy range 93100 - 92750 cm^{-1}, using the Biomation transient digitizer mentioned above. In the spectra the rotational levels of the upper state were observed up to J ≤ 27. For the $E^1\Pi(v=0)$ state, for all four isotopomers of CO studied, no perturbations were observed.

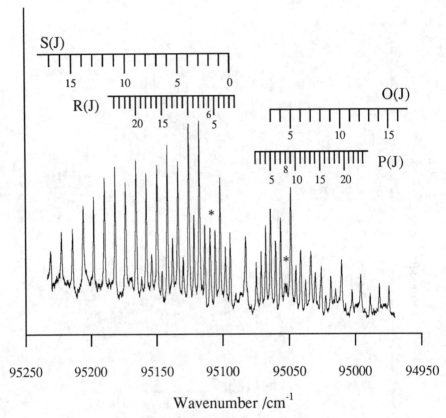

Fig. 3. The 2+1 REMPI spectrum of the E-X(1-0) transition in $^{12}C^{16}O$.

The 2+1 REMPI spectra of the $E^1\Pi(v=1)$ - $X^1\Sigma^+(v=0)$ transitions were approximately twenty times weaker than the E-X(0-0) origin bands. Fig. 3 shows a 2+1 REMPI spectrum of the E-X(1-0) transition in the 95250 - 94950 cm^{-1} two-photon energy range for $^{12}C^{16}O$. Although the spectrum is poor with an uncorrected decrease in laser power from left to right, which explains the overall intensity asymmetry of the band, a perturbation is observable in the P(8) and R(6) rotational lines which corresponds to a perturbation in the J_f =7 level of the $E^1\Pi(v=1)$ state. The P(8) line appears as a doublet and has a reduced intensity compared to the neighbouring lines. The R(6) line is overlapped by the S(2) line, however its reduced intensity is clearly evident in the spectrum. The 'extra line' associated with the J_e=7 level, although observed in a previous one-photon study,[2] was not observed here. Clearly, for $^{12}C^{16}O$, the localised perturbation in the J_f =7 level seems much stronger than that observed in the J_e=7 level.

New perturbations in the $E^1\Pi(v=1)$ state have also been observed at $J_f = 7$ for $^{13}C^{16}O$ and $^{12}C^{18}O$ in addition to the known perturbations at $J_e=7$. Furthermore, it was found that the relative intensities of the perturbed lines in the 2+1 REMPI spectra decreased by as much as 90% compared to the neighbouring unperturbed lines as the laser power was reduced which is indicative of predissociation in these levels. In this way a new localised perturbation/predissociation was found at $J=6$ for $^{13}C^{18}O$ in both e and f parity levels.

The two main results of this study have been the measurement of Λ-doubling constants for all four isotopomers in both the $v = 0$ and 1 vibrational levels of the E state and the characterisation of an indirect predissociation. The large Λ-doubling constants obtained for the E state, between 0.0095 and 0.02 cm^{-1}, is due to the strong pure-precession type interaction with the neighbouring $C^1\Sigma^+$ state, which, together with the E state, forms the 3p Rydberg complex. We have also observed perturbation/predissociation effects in all four isotopomers in the $E^1\Pi(v=1)$ level. Strong localised perturbations observed at low J in both parity levels are inconsistent with previously suggested $^1\Sigma$ or $^3\Sigma$ perturbing states.[5]

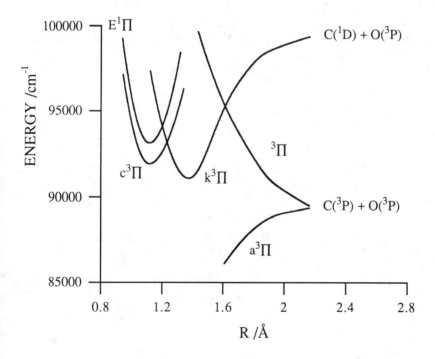

Fig. 4. A schematic diagram showing the potential energy curves of CO involved in the indirect predissociation of the E state. The perturbing state is the $k^3\Pi$ state. The $c^3\Pi$ and $a^3\Pi$ states are also indicated on the diagram.

Analysis of the data suggests that the perturbing state may be assigned to a $^3\Pi$ valence state. A detailed report of this work shall be published elsewhere.[6] A recent study of triplet-singlet absorption transitions in the 98000-90000 cm^{-1} energy range in $^{12}C^{16}O$, has confirmed the $^3\Pi$ character of the perturbing state.[7] Fig. 4 gives a schematic representation of the potential energy curves of the states of CO involved in these perturbations/predissociations.

JB wishes to thank the Ministère de l'Education Nationale for a postdoctoral research grant and S.C. thanks the Programme International de Corporation Scientifique (PICS No. 152) supported by the CNRS and the Foundation for Research and Technology, Greece.

REFERENCES

1. M. Eidelsberg and F. Rostas, Astron. Astrophys. 235 , 472 (1990).
2. M. Eidelsberg, A. Le Floch, F. Launay, J. Rostas, J.-Y. Roncin, and M. Vervloet, to be published.
3. S.G. Tilford, J.T.Vanderslice and P.G. Wilkinson, Can. J. Phys. 43, 450 (1965).
4. J.D. Simmons and S.G. Tilford, J. Mol. Spectry. 49 , 167 (1974).
5. P. Klopotek and C.R. Vidal, J. Opt. Soc. Am. B 2 , 869 (1985).
6. J. Baker, J.L. Lemaire, S. Couris, A. Vient, D. Malmasson, and F. Rostas, Chem. Phys., 178, (1993) in press.
7. J. Baker and F. Launay, J. Mol. Spectry, in press.

MODELLING OF THE RYDBERG-VALENCE PREDISSOCIATING INTERACTIONS IN THE CO MOLECULE

W.-Ü L. Tchang-Brillet

Département Atomes et Molécules en Astrophysique et URA812 du CNRS, Observatoire de Paris-Meudon, 92195 Meudon, France

P.S. Julienne

Molecular Physics Division, National Institute of Standards and Technology, Gaithersburg, Maryland 20899, U.S.A.

ABSTRACT

In a previous work, a close coupling model of the strong interaction between $3s\sigma$ $B^1\Sigma^+$ Rydberg state and the D' valence state has been built. It satisfactorily reproduces most of the experimental results on level shifts and widths in B state, as well as the relative intensities of the B-X transitions. The model is now used as a starting point for investigations of the higher Rydberg states, i.e. $3p\sigma$ $C^1\Sigma^+$ state and the $ns\sigma$ series. Preliminary results show the influence of the B-D' strong coupling through the presence of resonances in the photodissociation cross section as a function of energy. Differences in predissociation widths of two isotopes are qualitatively demonsrated although quantitative agreement is not yet achieved.

INTRODUCTION

The photodissociation of the CO molecule has been shown to proceed through predissociation of bound states corresponding to a line absorption spectrum[1,2]. This fact has consequences on the models of astrophysical media as it causes strong shielding and isotopic fractionation effects[2,3]. The VUV absorption spectrum of the CO molecule involves transitions between the ground X $^1\Sigma^+$ state and the high Rydberg states, and the perturbations observed in the spectrum need to be interpreted in terms of the interaction between the Rydberg states and some valence states.

The strong interaction between the first Rydberg state $3s\sigma$ $B^1\Sigma^+$ and the $D'^1\Sigma^+$ valence state has been previously studied in details[4]. A two-state diabatic model was built and vibrational term values, predissociation widths, as well as the relative intensities of the B-X transitions were calculated by close coupling method. The model reproduces satisfactorily the experimental results[5] concerning the B-X transitions.

In the present work, we investigate the predissociation of the second Rydberg state $3p\sigma$ C $^1\Sigma^+$ and some $ns\sigma$ $^1\Sigma^+$ belonging to the same Rydberg series as the B $^1\Sigma^+$ state, by their interactions with the D' $^1\Sigma^+$ valence state, The previous B-D' interaction model was used as a starting point for modelling these interactions.

SUMMARY OF AVAILABLE EXPERIMENTAL DATA

The predissociation features observed in the C state[6] show a weaker interaction with the D' state. Three levels (v = 0, 1, 2) are observed in emission. The v = 3 and 4 levels are observed in absorption only. The C-X (3-0) lines have a width of about 1-2 cm^{-1}, while the C-X (4-0) band appears diffuse. Compared with the ground state ion CO$^+$, the vibrational terms are downward shifted by smaller amounts than in the B sate, and presents a discontinuity between the v = 3 and 4 levels. The higher Rydberg states belonging to the $ns\sigma$ series should also be perturbed by the D' state. However, rotational structure was observed for several of them[7], e.g., $4s\sigma$ (v = 1) and $5s\sigma$ (v=1), showing a weak broadening by predissociation. These experimental evidences were already used for modelling the D' state diabatic potential in the previous work[4]. Indeed, the repulsive branch of the D' potential was chosen so that it crosses the C sate potential between the outer turning points of v = 3 and 4, and its crossing points with the $ns\sigma$ potentials are localized to the right of the outer turning points of the observed levels. Furthermore, for the $ns\sigma$ levels, different predissociation widths were observed in different isotopes. For instance, the $4s\sigma$-X and $5s\sigma$-X (0-0) bands were found to be diffuse in $^{12}C^{16}O$ and rotationally resolved in $^{13}C^{16}O$, while the contrary was observed for the $4s\sigma$-X (1-0) band. These differences put constraints on the model.

THEORETICAL METHODS AND MODEL ASSUMPTIONS

Two theoretical approaches have been adopted for calculating the predissociation widths of the higher Rydberg state levels : a nonperturbative close coupling calculation and a perturbation calculation.

The close coupling method applied to the two state interaction in the B-D' model has been described in detail in numbers of papers and was summarized in Reference 4.

It can be generalized to three interacting states, with B, D' and another Rydberg state, where the B state and the second Rydberg state are two closed channels and the predissociating D' state is an open channel. The photodissociation cross section is

$$\sigma(\lambda) = \sigma(E) = (8\pi^2 / 3\lambda)|\tau(E)|^2 \qquad (1)$$

where $\tau(E)$ is the single open channel component of the bound-free matrix element

$$\tau(E) = \langle 0|M_{op}|\Psi^-_{open}(E,R)\rangle \qquad (2)$$

and M_{op} is the dipole operator, $|0\rangle$, the ground rovibronic wavefunction. In the calculation, we used a three component dipole moment vector $[D_1, D_2, D_3]$ where D_i represents the transition dipole between the ground state and the respective D', B and C states integrated over rotational and electronic coordinates.

In the perturbation theory, when a Rydberg state is weakly coupled to a continuum state, the probability of predissociation is given by the Fermi Golden Rule:

$$\varpi(E) = \frac{2\pi}{\hbar}|\langle\phi_{i,v}|V_{i-\varepsilon}|\varepsilon(E)\rangle|^2 \qquad (3)$$

where $|\phi_{i,v}\rangle$ is the rovibronic wavefunction describing a bound level v of the Rydberg state i considered, $|\varepsilon(E)\rangle$ is a two-component vector representing the predissociating continuum state of energy E, which, in the present case, results from the close coupled B $^1\Sigma^+$ and D' $^1\Sigma^+$ states. The coupling operator $V_{i-\varepsilon}$ is also a two-component quantity where only the coupling strength between states i and D' is nonzero. The predissociation width of a level v at energy E_v is then given by

$$\Gamma = \hbar\varpi(E_v) \qquad (4)$$

In this preliminary work, the B and D' diabatic potentials are taken from the previous model[4]. For the C $^1\Sigma^+$ diabatic potential, we used a Rydberg-Klein-Rees (RKR) potential generated from the molecular constants ω_e, $\omega_e x_e$, r_e fitted[6] with the experimental data from v = 0 and 1, assuming these low levels remain unperturbed. As for the nsσ diabatic potentials, we assumed they are parallel to the potential of the ground state ion CO$^+$. The coupling strength between the C state and the D' state was ajusted so as to best reproduce the level shifts in the C state. The coupling strength between a $ns\sigma$ state and the D' state was scaled from the B-D' coupling strength found previously, following the $(n^*)^{-3/2}$ behaviour for the Rydberg series (n^* : effective principal quantum number). The transition dipoles for D'-X and B-X are taken from the previous B-D' model, the C-X transition dipole is derived from the B-X transition dipole by the ratio of the respective oscillator strengths[1].

PRELIMINARY RESULTS

All the calculations show that the results are very sensitive to the model, i.e., the choice of diabatic potentials, the coupling strength, the transition dipoles etc. The amount of experimental data available puts very restrictive constraints upon the model.

The three state close coupling calculation reproduces photodissociation resonances due to the C state vibrational levels at correct position, except for v = 4, and with reasonable widths, superimposed to other resonance features due to the B state, which are located in the B-D' predissociation continuum.

The table shows comparison of preliminary results on the level shifts δG_v (relative to the ground state ion) and predissociation widths Γ calculated with the two methods quoted above, for the most perturbed vibrational levels of the C state v = 2, 3, 4. The coupling operator used here is defined (in cm^{-1}) as :

$V_{C-D'} = 300$ for $R < R_x$
$V_{C-D'} = 300 \exp[-\ln 2(R-R_x)^2/0.20^2]$ for $R > R_x$

where R_x is the internuclear distance of the C-D' potentials crossing point (in Å).

Preliminary results on level shifts δG_v (cm^{-1}) and predissociation widths Γ (cm^{-1}) for the C $^1\Sigma^+$ state

v	δG_v exp.	δG_v Calc. (Close-coupling)	Γ exp.	Γ Calc. (Close-coupling)	Γ Calc. (Perturbation)
2	- 96.3	- 100.1	—— (emission)	0.055	0.052
3	- 152.05	- 149.3	1-2	0.36	0.35
4	- 97.89	- 221.8	> 5	12.5	12.1

The figure represents the overlap integral which appears in eq.(3), assuming a constant coupling $V_{i-\varepsilon}$:

$$I^2 = |\langle \phi_{i,v} | \varepsilon(E) \rangle|^2 \qquad (5)$$

calculated for a $n s\sigma\ ^1\Sigma^+$ v = 0 state and the B-D' continuum state versus energy, calculated for two isotopes $^{12}C^{16}O$ and $^{13}C^{16}O$. It illustrates the possible differences of predissociation widths for different isotopes. The predissociation width for a given level v is given by :

$$\Gamma = 2\pi |V_{i-\varepsilon}|^2 I^2(E_v) \qquad (6)$$

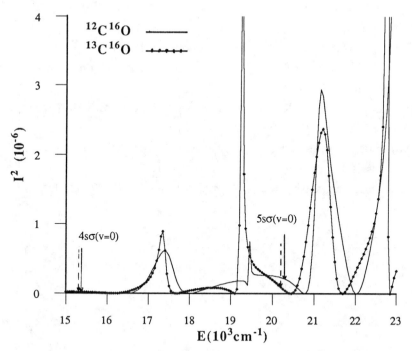

Overlap integral defined by eq. (5), where $|\phi_{i,v}\rangle$ is the rovibronic wavefunction describing a $n s\sigma$ (v=0) state, $|\varepsilon(E)\rangle$, the B-D' coupled continuum state of energy E. Full arrows: level position for $^{12}C^{16}O$, dashed arrows: level position for $^{13}C^{16}O$.

In the present calculation, we found a larger width in $^{12}C^{16}O$ (Γ = 1.32 cm^{-1}) than in $^{13}C^{16}O$ (Γ = 0.26 cm^{-1}) for 5sσ v = 0, in qualitative agreement with the experimental results[6]. However, the differences observed for 4sσ v=0 are not reproduced (calculated $\Gamma \approx$ 0.27 cm^{-1} for both isotopes).

The present calculations account for some main features of the predissociating interaction observed in higher Rydberg states of the CO molecule. However, not all interactions are taken into account. The amount of experimental data should allow further improvement of the modelling and bring more detailed understanding of the predissociation process.

REFERENCES

[1] C. Letzelter, M. Eidelsberg, F. Rostas, J. Breton, and B. Thiebelmont, Chem. Phys. **114**, 273 (1987).

[2] Y.P. Viala, C. Letzelter, M. Eidelsberg, and F. Rostas, Astron. Astrophys. **193**, 265 (1988).

[3] E.F. van Dishoeck and J.H. Black, Astrophys. J. **334**, 771(1988).

[4] W.-Ü L. Tchang-Brillet, P.S. Julienne, J.-M. Robbe, C. Letzelter and F. Rostas, J. Chem. Phys. **96**, 6735 (1992).

[5] M. Eidelsberg, J.Y. Roncin, A. Le Floch, F. Launay, C. Letzelter and J. Rostas, J. Mol. Spectrosc. **121**, 309 (1987).

[6] M. Eidelsberg, A. Le Floch, F. Launay, J. Rostas, J.Y. Roncin and M. Vervloet, to be published.

[7] M. Eidelsberg and F. Rostas, Astron. Astrophys. **235**, 472 (1990).

THE $A\,{}^2\Pi_i - X\,{}^2\Sigma^+$ AND $B\,{}^2\Sigma^+ - A\,{}^2\Pi_i$ ELECTRONIC TRANSITIONS OF CS^+ BY HIGH RESOLUTION FOURIER TRANSFORM SPECTROSCOPY.

D. Cossart, M. Horani and M. Vervloet.
Laboratoire de Photophysique Moléculaire du CNRS, Bât 213,
Université de Paris-Sud, 91405 Orsay, France.

ABSTRACT

The emission spectrum of the carbon monosulfide cation has been recorded under high resolution by Fourier transform spectrometry. Rotational analyses of some bands of the A - X and B - A electronic transitions have been carried out. From these analyses, prelimary molecular constants have been obtained for the vibrational levels v = 0, 1 and 2 of the $A\,{}^2\Pi$ state and v = 0 and 1 of the $B\,{}^2\Sigma^+$ state.

INTRODUCTION

The sulphur compounds are considered to be of importance in the identification of shock chemistries in star-forming regions [1,2]. Several sulphur-containing molecules, including CS, have been already observed in interstellar clouds [3] but the ion CS^+ has not been detected so far. This unsuccessful search [4,5], based on the data [6] obtained from the electronic transition $A\,{}^2\Pi - X\,{}^2\Sigma^+$, may be due to a lack of spectroscopic information on its electronic transitions. Moreover, its astrophysical or laboratory detection has never been reported in the millimeter wavelengths region; this may be caused by its low predicted dipole moment [7]. In order to obtain more accurate expected microwave transitions, it was decided recently to analyse the near infrared transiton A - X recorded by Fourier transform spectrometry. This analysis [8] has yieded precise molecular parameters for the six lowest vibrational levels of the ground electronic state $X\,{}^2\Sigma^+$.

The present work is devoted to the study of the two electronic states, $A\,{}^2\Pi$ and $B\,{}^2\Sigma^+$, respectively located at about 11806 cm^{-1} and 36210 cm^{-1}. This study is made by analysing the rotational structure of bands involving different vibrational levels of those two electronic states.

EXPERIMENT

Earlier experiments were achieved by mixing metastable excited helium, created in a high speed flowing system, with a low pressure of carbon disulfide vapor. The light emitted from the collisional zone was focused onto the entrance iris of a Fourier transform spectrometer (Bomem DA3.002) of the Herzberg Institute of Astrophysics (Ottawa, Canada). A total of nine bands of the A - X electronic transition were recorded in the range 14000 - 5800 cm^{-1} at an apodized resolution of 0.04 cm^{-1} or 0.05 cm^{-1}. Figure 1 shows a part of this spectrum. With this source, we have observed that, for each band, the intensity of the $^2\Pi_{3/2} - {}^2\Sigma^+$ sub-band is about 10 times larger than the intensity of the $^2\Pi_{1/2} - {}^2\Sigma^+$ sub-band[8]. We have also noticed that no emission of other molecular species was overlapping this spectral range.

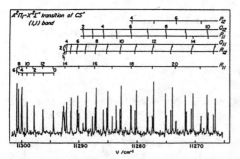

Fig. 1: A part of the sub-band (1,1) $A\,^2\Pi_{3/2} - X\,^2\Sigma^+$.

The intensity of the weaker sub-bands mentioned above was too low to perform a rotational analysis of them. In order to get these sub-bands with higher intensity, we have used another emission source producing spectra of higher rotational temperature. This source is similar to that one described in the previous study of this electronic transition recorded on photographic plates[6]. Basically, a low pressure of carbon disulfide was flowing through a d.c.

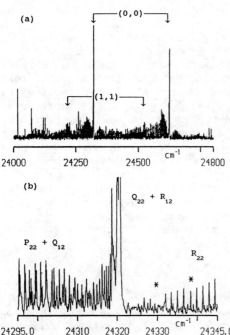

Fig. 2: The emission $B\,^2\Sigma^+ - A\,^2\Pi$ of CS^+ at around 24500 cm^{-1}.
(a) Vertical arrows point to band heads of the (0,0) and (1,1) bands.
(b) Asterisks show irregularities in the R_{22} branch of the (0,0).

discharge produced between a hot tungsten filament and an anode. The spectral range 14000 - 9000 cm^{-1} was recorded by means of a Fourier transform spectrometer (Bruker IFS 120 HR) at an apodized resolution of 0.07 cm^{-1}. Under these conditions, each band shows sub-bands of about the same intensity; however, some of them are overlapped by the emission of the electronic transition $d\,^3\Delta - a\,^3\Pi$ of the neutral species CS.

The $B\,^2\Sigma^+ - A\,^2\Pi$ transition, located at around 24500 cm^{-1}, was recorded by using another emission source which seemed to give a stronger intensity than the two others mentioned before. Carbon disulfide vapor was flowing through a microwave discharge (2450 MHz) at low pressure. The products of this discharge were allowed to mix with excited helium generated by a d.c. discharge. The light emitted from the mixing zone was focused onto the entrance iris of the Fourier transform spectrometer (Bruker IFS 120

HR). The spectrum was recorded from 26000 cm^{-1} to 21000 cm^{-1} at an apodized resolution of 0.1 cm^{-1}. Figure 2 shows the region of interest where the spectrum of the B - A transition occurs. The B $^2\Sigma^+$ - X $^2\Sigma^+$ transition, expected at around 30000 cm^{-1}, was tried unsuccessfully to be recorded.

RESULTS AND DISCUSSION

The determination of the molecular constants of the v = 0, 1 and 2 levels of the A $^2\Pi_i$ state has been made by analysing the (0,3), (1,0), (1,1) and (2,0) bands. The rotational analysis of the (0,3) and (2,0) bands was straightforward and gave constants reported in Table 1. The rotational analysis of the (1,0) and the (1,1) bands was not as easy as for the other bands because of irregularities in some branches. Unequivocal assignments of the branches were obtained from agreement with the ground state combination differences obtained in the other analysed bands [8]. The irregularities are located around J'=19.5 in the P_{22}, R_{22} and $^RQ_{21}$ branches, indicating a perturbation of the f component of the $^2\Pi_{1/2}$ (v = 1) sub-state.

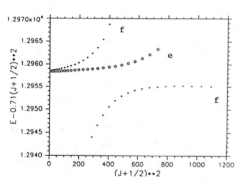

Fig. 3: Reduced term values of the A $^2\Pi_{1/2}$ (v = 1) sub-level perturbed by the high level v = 10 of the X $^2\Sigma^+$ state.

In order to deperturb the A $^2\Pi$ v=1 level, we have worked with the upper state term values. These term values were derived by adding the observed line frequencies to the appropriate X $^2\Sigma^+$(v = 0 and 1) term values calculated from the precise rotational constants [8]. In figure 3, the observed term values for the A $^2\Pi$ v=1 level of CS$^+$ have been plotted against $(J+1/2)^2$. This figure shows that the perturbing state lies below the A $^2\Pi_{1/2}$ v = 1 sub-level. As predicted in the earlier analysis work [6], the perturber is the v = 10 vibrational level of the X $^2\Sigma^+$ ground state.

The deperturbation was performed using an Hamiltonian matrix for a $^2\Pi$ - $^2\Sigma$ interaction [9]. The off-diagonal spin-orbit and rotational-orbit parametres (ξ and η) have been derived by an iterative procedure which also gives a set of partially deperturbed molecular constants for both interacting levels. This set of constants was found to best fit the observed rotational levels of the A $^2\Pi_i$ v = 1 state and are reported in Table 1.

Table I: Rotational constants (in cm^{-1}) for the A $^2\Pi_i$ (v') and X $^2\Sigma^+$ (v") levels of CS$^+$.

A$^2\Pi_i$	v' = 0	v' = 1	v' = 2
T$_0$	11804.866(1)	12805.174(21)	13791.571(6)
B	0.715463(9)	0.70903(6)	0.70284(3)
10^{6}D	1.446(9)	[1.5]	1.506(30)
A	-302.318(3)	-302.163(42)	-301.001(12)
10^{4}A$_D$	0.32(9)	0.36(40)	-5.03(30)
10^{2}p	0.73(2)	1.08(51)	1.64(9)

X$^2\Sigma^+$		v" = 10	
T$_0$		12909.41(55)	
B		0.7981(90)	
10^{6}D$_e$		[1.2]	
10$^2\gamma$		[1.7]	
10$^3\eta$		-3.34(19)	
ξ		8.588(75)	
σ	0.015	0.12	0.024

Note: Uncertainties (3σ) are in units of the last significant figure.
Bracket figures were held constant in the least squares.

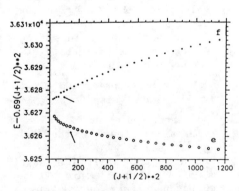

Fig. 4: Reduced term values of the B $^2\Sigma^+$ (v = 0) state. The arrows indicate local perturbations.

The B $^2\Sigma^+$ - A $^2\Pi$ transition is shown in figure 2. There are two very strong features which are not resolved. These two features were widely and contradictorily discussed [6] and eventually assigned to the (0,0) band of the B - A transition of CS$^+$. We have been able to carry out unambiguously the rotational analysis of the resolved branches of this Fourier transform spectrum by using combinaison differences of term values in A $^2\Pi$ v = 0, which have been precisely determined from the analysis of the A - X system reported above. This rotational analysis confirms the earlier assignment. The two strong features, located at around 24320 cm^{-1} and 24625 cm^{-1}, are respectively given by the Q_{22} and $R_{11} + Q_{21}$ branches. The molecular constants for the B $^2\Sigma^+$ v = 0 and 1 levels were determined by using a non linear least square fit on the level term values. Only the term values unaffected by local perturbations have been taken into account in this calculation. For the v = 0 level, a local perturbation is detected in the two components e and f as shown in figure 4. This perturbation was not analysed in detail as for the A state but is certainly due to a high vibrational level of the A $^2\Pi$ state as observed in the analogous case of the (0,0) band of the B $^2\Sigma^+$ - X $^2\Sigma^+$ transition of the isovalent molecule CN [10]. For the v = 1 level, strong perturbations have also been detected in both the components e and f.

Table II: Preliminary rotational constants in (cm^{-1}) for the B $^2\Sigma^+$ (v') levels of CS$^+$.

B $^2\Sigma^+$	T_0	B	10^6D	γ
v' = 0	36271.09(6)	0.69795(21)	1.47(15)	-0.0153(12)
v' = 1	37178.9(15)	0.6715(40)	[1.47]	-0.11(12)

Note: Uncertainties (3σ) are in units of the last significant figure.
Bracket figures were held constant in the least squares.

CONCLUSION

In this work, we have determined for the first time the molecular constants of v = 0 and 1 and improved those of v = 2 of the first electronic excited state A $^2\Pi_i$ of CS$^+$. It is an extension of the earlier work [6] where v = 2, 3, 4 and 5 of the A state have been analysed. The perturbing state in $^2\Pi_{1/2}$ of v = 1 has been clearly identified as v = 10 of X $^2\Sigma^+$, which fully agrees with an earlier prediction[6]. We succeeded in the rotational analysis of the two bands (0,0) and (1,1) of the B $^2\Sigma^+$ - A $^2\Pi$. Precise term-values for the two lowest vibrational levels of the B state can be obtained by adding line frequencies to the energies of the levels of the ground state calculated from precise molecular parameters[8]. At the moment, the perturbations observed in v = 0 and 1 of the B state are not understood and the identification of the perturbing states is not yet established. The study of these perturbations are presently in progress.

REFERENCES

1. D. A. Neufeld, A. Dalgarno, ApJ., 340, (1989).
2. B. E. Turner, ApJ., 396, L107, (1992).
3. D. Smith, N. G. Adams, K. Giles, E. Herbst, A&A, 200, 191, (1988) and references cited therein.
4. R. Ferlet, E. Roueff, M. Horani, J. Rostas, A&A, 125, L5, (1983).
5. R. Ferlet, E. Roueff, J. Czarny, P. Felenbok, A&A, 168, 259, (1986).
6. D. Gauyacq, M. Horani, Can. J. Phys., 56, 587, (1978).
7. J. H. Blöcker, E.-A. Reinsch, P. Rosmus, H.-J. Werner P. J. Knowles, Chem. Phys., 147, 99, (1990).
8. M. Horani, M. Vervloet, A&A, 256, 683, (1992).
9. H. Lefebvre-Brion, R. Field on Perturbations in the Spectra of Diatomic Molecules (Academic Press, Inc., 1986), p. 124.
10. N. H. Kiess, H. P. Broida, J. Mol. Spectrosc., 7, 194, (1961).

A LASER VAPORIZATION SOURCE FOR SPECTROSCOPIC CHARACTERIZATION OF REFRACTORY COMPOUNDS

J. Chevaleyre, C. Bordas, A. M. Valente, V. Boutou, J. Maurelli, B. Erba
and J. d'Incan

Laboratoire de Spectrométrie Ionique et Moléculaire (URA n°171), CNRS et
Université Lyon 1
Bat 205, 43, boulevard du 11 Novembre 1918
69622 Villeurbanne Cedex, France

ABSTRACT

In classical ovens, vapour pressure of small molecules including refractory compounds is generally too weak to undertake their spectroscopic characterization in the gas phase. Laser sources as they allow any material to be vaporized whatever its melting point, bring a satisfactory answer to this problem.

We designed a source of this type with the aim to undertake the spectroscopic characterization of small molecules including carbon or nitrogen together with a metal of large abundance (Al, Mg, Ca, Fe). Indeed, such molecules which could be present at the early stage of grain formation are still to be characterized.

INTRODUCTION

The interplay between astronomers and molecular physicists has been very stimulative to reach the best knowledge of phenomena associated to molecular species in the astophysical media. This dynamics led to very successful results for molecules containing the more abundant elements such as H, C, O, N, S and new developments are permanently expected. The situation is rather different for metal-bearing molecules where observations as well in astrophysics as in the laboratory are less developped. They are mainly limited to molecules where metal atoms are associated with O, H, S and, except some noticeable exemples like MgCN discussed in this conference [1], very less is known about molecules including metal and N or C atoms. In views of the large abundance of C in carbon-rich stars and more generally of N and C in astrophysical media, this lack of knowledge leads to rather speculative assertions on the carriers of metals of large abundance like Na, Mg, Al or Fe. In his wide review on the observations and chemistry of refractory elements in the interstellar medium, Turner [2] came to the conclusion that a large laboratory effort is awaited to characterize carbides but especially nitrides of Al, Na or Mg that could appear preferentially than the corresponding oxides in these media. Moreover, after the observations of SiC in IRC +10 216 [3], mono or polycarbides of Al, Mg or Fe could be observed too.

Those nitrides or carbides are difficult to characterize in the laboratory because of their refractory character; the vapour pressure that can be reached with classical ovens is generally too weak for spectroscopic investigations in the gas phase. Moreover, theoretical calculations, as they must include a large number of active electrons are difficult; this is particularly true for transition elements like Fe where an open d-shell is involved. For this reason, reliable theoretical calculations have long been unavailable. Recently, Bauschlicher et al [4] using MRCI and CASSCF methods calculated electronic levels of the AlC molecule. These results have just been partly confirmed with the observation of a few bands of the AlC ($B^4\Sigma^-$ - $X^4\Sigma^-$) system in the 22,000 cm^{-1} region and we can expect in the next future new calculations related to C or N metal-bearing molecules to be available.

We present in this paper a laser vaporization source, especially designed to produce small molecules of refractory materials. As it can vaporize any material, whatever its melting point, this source could be particularly suited for gas phase spectroscopic investigatons of refractory compounds.

Our preliminary results, presented in this paper, concerns the capability of this source to produce AlC and Al_nC molecules.

EXPERIMENT

A variety of laser vaporization sources have been designed in the last decade; generally to produce clusters of various metals. Their efficiency to produce large amounts of refractory compounds in the gas phase has been demonstrated for any metal, up to the most refractory like molybdenum and tungsten [5]. For this reason, these sources could be well suited to undertake spectroscopic investigations on refractory molecules which could be present in appreciable quantities in molecular beams downstream the nozzle.

Briefly, a target rod is located in the throat of a pulsed nozzle. It is shot by the second harmonic of a pulsed YAG laser leading to metal vaporization and formation of a high temperature plasma. Before laser pulse, helium was injected into this source through a pulsed valve in such a way that the laser shot coincide with the maximum pressure of the helium carrier gas; the injection pressure can be varied in the 1-10 bar range.

The subsequent expansion of the gas mixture allows recombination and thermalization of the plasma with production of small molecules and clusters.

After the nozzle, these molecules and clusters expand into the vacuum (10^{-3} torr), forming a cool molecular beam. A skimmer placed 1-2 cm downstream allows the supersonic molecular beam to enter the low pressure region (10^{-6} torr) where various excitation processes will be undertaken with different lasers.

At that point, it is worthwhile to notice that in our earlier experiments and in order to favour the production of small molecules, we reduced considerably the nozzle volume: it consisted simply of a channel of 2 mm in length and 1 mm in diameter. By this way, we expected to reduce the number of thermalizing collisions with helium to avoid the production of large clusters. Indeed, in that case, the molecular beam was very poor and unstable and the velocity of the molecules emerging from this nozzle was about twice larger than the sound velocity. Moreover, it appeared that the production of these molecules was not influenced by the providing of the carrier gas. Indeed, they were directly ejected from the plasma without any expansion in the helium carrier gas. Similar behaviour was also observed by Cai et al [6].

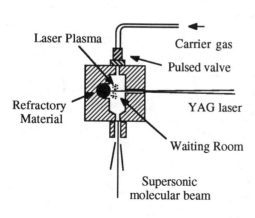

Fig. 1. The vaporization source.

Subsequently, the vaporization source was modified as shown in Fig.1, essentially by addition of a 0.5 cm³ waiting room where the plasma has time enough to expand into the carrier gas before the final expansion into vacuum. With this new geometry, considerable improvements in stability and yields of molecular species were obtained. In the low pressure region, downstream the skimmer, the supersonic molecular beam allows various spectroscopic investigations to be undertaken, all of them via laser excitation and resulting in the yield of ions. These ions are produced into the extraction region of a time of flight mass spectrometer (TOF), that permits mass characterization.

The simplest excitation is the direct ionization by one UV photon. Increasing the energy of this photon, new peaks of given masses appear as the photon energy reaches the ionization potential. By this way, ionization potentials can be easily determined. Moreover, photoionization efficiency as well as its evolution near the threshold can also be studied. In another scheme, depicted in Fig 2, one photon excites the molecule to an electronically excited state while a second photon brings it beyond the ionization limit. By scanning the exciting photon over a given electronic state an ion signal is produced only when the exciting energy matches a v, J level of this state, allowing its spectroscopic investigation. A general block diagram of the experimental device is presented in Fig. 3.

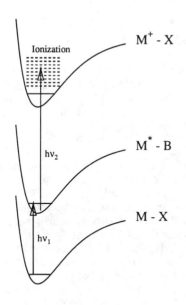

Fig. 2. Two photon ionization.

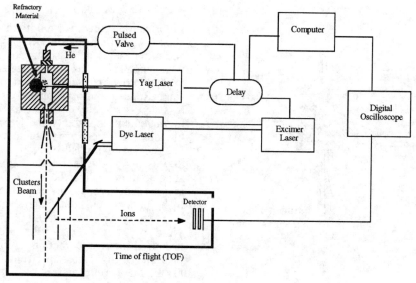

Fig. 3. Schematic of the experimental device.

PRELIMINARY RESULTS

We started the experimental program by the mass characterization of molecular species existing in the molecular beam. This was achieved by direct ionization with one UV photon provided by a frequency doubled excimer pumped dye laser. A typical spectrum is shown in Fig 4 with a photon energy: 5.8 ev. Unsurprisingly this spectrum exhibits a large series of Al_n clusters, up to n=40, as they are usually produced in this kind of source [7]. Nevertheless, larger clusters could also be present in the beam but, in our apparatus, as the TOF is perpendicular to the molecular beam, the extraction voltage is too low to allow them to enter the TOF. Taking into account that the higher intensities observed for n=7 and n=14 correspond to magic numbers in the shell jellium background model, the regular envelop of the mass spectrum for n>7 suggests a unique one photon ionization. From the ionization potential measurements by Schriver et al [8], IP (Al_n)<5.8 ev only for n>14; while the classical metallic sphere model would lead to IP<5.8 ev for n>8. This deviation was attributed to a not yet complete s-p hybridization in this n-range. It would be of interest to compare our results with Schriver's; but, at the present time, no conclusion can be drawn on this point as long as new mass spectra, using various near-by frequencies of the ionization laser, are not yet recorded. In particular, we must carefully control the laser fluence in order to get sure that no multiphoton ionization processes pollute the expected one photon ionization process. Indeed, a multiphoton ionization whose efficiency would decline from larger clusters to vanish towards n=7 could be partially responsible for the observation of clusters in this mass range. This remark could also explain why Al_{4-6} do not appear in the spectrum as two photons are

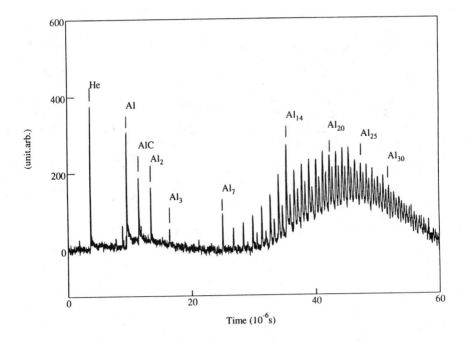

Fig. 4. Mass spectrum obtained by laser vaporization of an aluminium rod combined with direct one photon ionization : hν = 5.8 eV.

necessary to ionize n<7 clusters. As a consequence, Al_{2-3} peaks are only present owing to the very large abundances of these clusters as compared to the larger ones.

Nevertheless, regarding the purpose of this experiment, the most exciting feature of this spectrum is the appearance of mass peaks for the Al_nC molecules. At that point we are not yet sure whether carbon is an impurity of the current aluminium we used or a degradation product from the carbonated oil pump. The large intensity of the AlC mass peak gives us the opportunity to start with spectroscopic investigation of this molecule using the one photon resonant-two photon ionization technique and scanning the resonant photon over AlC electronically excited states. Moreover, we intend to undertake a systematic study of the ionization potential and photoionization efficiency of Al_nC molecules. We expect valuable information regarding the C-bonding with metal particles to be deduced from these experiment.

REFERENCES

1. L. M. Ziurys, this conference
2. B. E. Turner, Astrophys. J. 376, 573 (1991)

3. J. Cernicharo, C. A. Gottlieb, M. Guelin, P. Thaddeus and J. M. Vrtilek, Astrophys. J., 341, L25 (1989)
4. C. W. Bauschlicher, S. R. Langhoff and L. G. M. Petersson, J. Chem. Phys. 89, 5747 (1988)
5. J. B. Hopkins, P. R. R. Langridge-Smith, M. P. Morse and R. E. Smalley J. Chem. Phys., 78, 1627 (1983)
6. M. F. Cai, T. P. Dzugan and V. E. Bondybey, Chem. Phys. Letters, 155, 430 (1989)
7. M. Pellarin, B. Baguenard, M. Broyer, J. Lermé and J. L. Vialle, J. Chem. Phys., 98, 944 (1993)
8. W. E. Schriver, J. L. Persson, E. C. Honea and R. L. Whetten Phys. Rev. Letters, 64, 2539, (1990)

DISCUSSION

LELEYTER — Do you have other data on the $A\ell_n$ C molecules especially a curve similar to the curve IP($A\ell_n$) against n ?

CHEVALEYRE — The mass spectrum showing $A\ell_n$ C molecules is the first and preliminary result obtained with this experimental device. A possible next step could be to record different mass spectra as a function of the increasing frequency of the ionizing laser. This procedure allows to determine very easily the ionization potential of the different masses existing in the molecular beam as they appear in the mass spectrum when the photon energy reaches the ionization potential value.

LELEYTER — Are there papers about MgC on FeC ?

CHEVALEYRE — The best known metal carbide is, at the present time, $A\ell C$. Very reliable calculations have been published by the Langhoff and Bauschlicher group and the first observed bands of the $B^4\Sigma^- - X^4\Sigma^-$ system have just been published by Brazier. MgC and FeC are not less exciting molecules than $A\ell C$, in view of their relatively large abundances. A few group are working calculations and their results are expected in a very near future. They are also good candidates for spectroscopic characterization in supersonic beam generated by laser vaporization sources.

RADIATIVE RELAXATION OF SPIN-ORBIT STATES IN Xe^+ AND Kr^+ AND OF VIBRATIONAL AND SPIN-ORBIT STATES IN HBr^+ AND DBr^+

M. Heninger, S. Jullien, J. Lemaire, G. Mauclaire, R. Marx and S. Fenistein
Laboratoire de Physico-Chimie des Rayonnements (URA 75)
Bâtiment 350, Université Paris-Sud, 91405 Orsay cedex

ABSTRACT

To illustrate the capabilities and limitations of the "monitor ion technique" using a triple cell ICR spectrometer to measure radiative lifetimes in the ms to s range, two sets of results are presented. For the spin-orbit transitions $^2P_{1/2} \rightarrow {}^2P_{3/2}$ in Xe^+ and Kr^+ the experimental lifetimes: 48.7 ms and 320 ms respectively are in very good agreement with previous theoretical calculations. For HBr^+ and DBr^+ vibrational and spin-orbit state relaxations have been studied. The vibrational lifetimes, 10 ms and 56 ms respectively, show an important isotopic effect like $H(D)Cl^+$. Calculations performed in collaboration with G. Chambaud and P. Rosmus are in good agreement with experimental results. The $^2\Pi_{1/2} \rightarrow {}^2\Pi_{3/2}$ spin-orbit relaxation: (890 ± 170) ms for both isotope is much slower. The experimental uncertainty is large since for such long lifetime there is an important, but very difficult to quantify, contribution of collisional deactivation. There is to date no calculation available for spin-orbit relaxation in molecular ions.

INTRODUCTION

Experimental determination of radiative lifetimes in the millisecond to second range is extremely difficult, especially for charged species, and most of our information originates from theoretical calculations.

A few years ago we developed a non spectroscopic method, based on the so called "monitor ion technique" and using a triple cell Ion Cyclotron Resonance spectrometer, to study the radiative relaxation of ions.

Although there are severe limitations to this technique, it turns out that the few other spectroscopic or non spectroscopic methods proposed to date are also restricted to a very limited category of ions.

Recent improvements of our experimental set up allowed the extension of the monitor ion technique to several new systems.

We present here two sets of results which illustrate the capabilities and the limitations of this technique.

1. EXPERIMENTAL

The triple cell ICR spectrometer used in these experiments has already been described [1,2]. The reactant ions are produced by a pulsed electron beam in a low pressure ion source (P~10^{-6} torr). They are drifted into the relaxation cell where they are stored for variable times. They are then drifted into the third cell to react with the monitor gas at constant pressure (typically 10^{-6} - 2×10^{-6} torr) and constant reaction time (10 - 20 ms). Finally, the reactant and product ions are drifted back into the relaxation cell to be detected via a Fourier Transform data system.

Differential pumping ensures a pressure ratio on the order of 250 between the relaxation-detection cell and the two adjacent ones.

To determine the radiative decay rate of an ion, its storage time in the relaxation cell is varied. Since all the other parameters (ion production, reaction and detection) are kept constant during an experiment, the variation of the monitor ion intensity as a function of storage time reflects only the processes occurring in the relaxation cell, i.e. radiative relaxations and collisions with the background gas. To account for the ion losses by diffusion to the walls, the monitor ion intensity is normalized to the total ion signal.

For long lived ions the experimental lifetimes must be corrected for collisional quenching and reactions occurring in the relaxation cell. As an example, for a radiative lifetime of 500 ms, a background pressure of 2×10^{-8} torr, and a collision rate constant of 10^{-9} cm^3 mol^{-1} s^{-1}, the measured lifetime would be only 367ms. Unfortunately, in most cases, the quenching and reaction rate constants of excited ions are not known; moreover, the pressure and composition of the background gas, mainly H_2O desorbing from the walls and monitor gas leaking from the reaction cell, are very difficult to assess.

2. SPIN FORBIDDEN TRANSITIONS $^2P_{1/2} \rightarrow {}^2P_{3/2}$ IN Xe$^+$ AND Kr$^+$

In these simple two level systems, the monitor ion technique is easy to use: For Xe$^+$, charge transfer with N_2O^+ is energetically allowed for the $^2P_{1/2}$ state only. Fig. 1 shows the corresponding energy diagram and Fig. 2 a log plot of the decay curve of N_2O^+ corrected for a small constant contribution of stable ions to the monitor ion signal. The slope of the best fit straight line gives:

$1/\tau_{measured} = 1/\tau_{rad} + \Sigma k_M [M]$,

where $\Sigma k_M [M]$ is the collisional decay rate in the relaxation cell. An upper limit of $\Sigma k_M[M] = 1.2$ s^{-1} is determined with $p_{H2O} = 10^{-8}$ torr, $p_{N2O} = 2\times10^{-8}$ torr and the collision rate constant calculated with the parametric formula of Su and Chesnavich [3]. This gives $\tau_{rad} = (48.7 \pm 5)$ ms

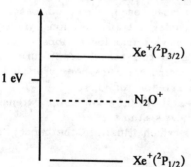

Fig. 1. Energy diagram for Xe$^+$ and the monitor ion

in good agreement with the calculated value τ = 47.2 ms [4].

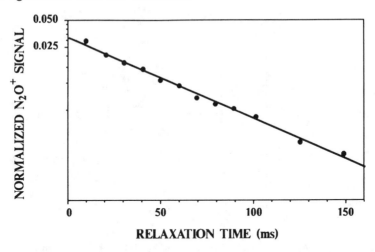

Fig. 2. log plot of normalized N_2O^+ signal as a function of relaxation time

For Kr+ charge transfer with CH_4 and CO has been used as monitor.
With CH_4 only $Kr^+(^2P_{1/2})$ produces CH_3^+ and the radiative lifetime is extracted from the exponential decay of $[CH_3^+]$ as a function of the relaxation time.

With CO, both spin states react to give CO^+, but the rate constant is much smaller with $Kr^+(^2P_{1/2})$ than with $Kr^+(^2P_{3/2})$. Therefore, $[CO^+]$ increases with the relaxation time t_2 as $A - B \times \exp(-t/\tau)$. No detail will be given here on the way to correct the measured lifetimes for collisional decay due to the background gas.

The radiative lifetime determined with the two different monitor reactions are 300 ms and 340 ms respectively, in good agreement with the theoretical value: 357 ms [5,6].

3. VIBRATIONAL AND SPIN-ORBIT STATES RELAXATION IN HBr+ and DBr+

Fig. 3 shows an energy diagram of the first levels of HBr+ together with the thermodynamic threshold for charge transfer with O_2, and proton (or deuteron) transfer with CO_2.

The energy of the ionizing electron beam has been set at 14 eV, well below the appearance potential of Br+ and of the first excited state $HBr^+(A^2\Sigma^+)$. If the populations were Franck-Condon, there would be 96% of v=0 and 4% of v=1 with an equal amount of the two spin states. In the absence of any information on ionization by low energy electrons, populations close to F.C. are assumed.

As shown on fig. 4, the O_2^+ signal as a function of the relaxation time

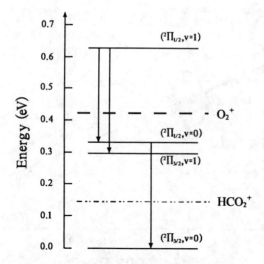

Fig. 3. Energy diagram of the first level of HBr$^+$ and the monitor ions.

may be fitted to a sum of two exponentials:
$A_1 \times \exp(-t/\tau_1) + A_2 \times \exp(-t/\tau_2)$, with $\tau_1 = (10 \pm 2)$ ms and $\tau_2 = (490 \pm 100)$ ms.
The HCO$_2^+$ signal also exhibits a double exponential decay:
$A_3 \times \exp(-t/\tau_3) + A_4 \times \exp(-t/\tau_4)$, with $\tau_4 = (600 \pm 120)$ ms and $\tau_3 = (5 \pm 1)$ s

According to the energy diagram, O$_2$ can react only with HBr$^+(^2\Pi_{1/2}$, v=1). But although the reaction with HBr$^+(^2\Pi_{1/2}$, v=0) is endothermic by 0.092 eV, a small fraction of these ions can also react.

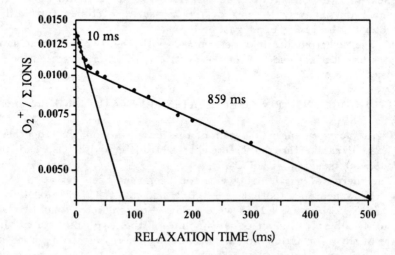

Fig. 4. Log plot of normalized O$_2^+$ signal as a function of relaxation time.

As the calculated vibrational lifetime for v=1 is 1.5 ms for HF$^+$, and 4.6 ms for HCl$^+$ [7], τ_1 = 10 ms is a likely lifetime of HBr$^+$($^2\Pi_{1/2}$,v=1). Then τ_2 would correspond to the spin transition ($^2\Pi_{1/2}$, v=0) $\rightarrow$ ($^2\Pi_{3/2}$, v=0).

Proton transfer with CO_2 is energetically allowed for HBr$^+$($^2\Pi_{1/2}$, v=0) and also for the ($^2\Pi_{1/2,3/2}$, v=1) states, but these are at least 10 times less populated, and the corresponding decay (10 ms) cannot be seen. Therefore, τ_3 must be the lifetime of the ($^2\Pi_{1/2}$, v=0) state, equal, within the experimental uncertainties, to τ_2 determined with O_2 as monitor.

Finally, although endothermic by 0.143 eV, reaction of a small fraction of the HBr$^+$($^2\Pi_{3/2}$, v=0) state is possible and the very slow decay τ_4 is probably due to reaction of these ions with the background gas in the relaxation cell.

The same experiments with DBr$^+$ give: τ'_1 =(51 ± 10) ms, $\tau'_2 \approx \tau'_3$ =(650 ± 130) ms, τ'_4 =(6 ± 1.2) s.

Correction for quenching and reaction with the monitor gas in the relaxation cell has been evaluated using partial pressures p_M = p_3/250 (where p_3 is the pressure measured in the reaction cell) and an upper limit of the rate constant (the collisional rate constant) k_c = 6.17×10^{-10} cm^3 mol^{-1} s^{-1} with O_2 and 7.07×10^{-10} cm^3 mol^{-1} s^{-1} with CO_2.

For H_2O, an upper limit is obtained with a partial pressure of 7×10^{-9} torr determined in previous experiments, and k_c = 2.33×10^{-9} cm^3 mol^{-1} s^{-1}. The lower limit is the reaction rate of HBr$^+$ with H_2O determined by measuring [H_2O^+] and [H_3O^+] as a function of the relaxation time. The upper and lower limits for the overall correction are 0.52 s^{-1} and 0.14 s^{-1} for H(D)Br$^+$.

The experimental corrected radiative lifetimes are presented in table I together with theoretical calculations performed in collaboration with G. Chambaud and P. Rosmus. As expected, an important isotope effect is observed for the v=1 radiative lifetimes.

There is, to our knowledge, no theoretical calculation for spin transitions in molecular ions.

Table I Radiative lifetimes

Ions	Experiment (ms)	Theory (ms)
HBr$^+$($^2\Pi_{1/2}$, v=1)	10 ± 2	9.3
HBr$^+$($^2\Pi_{1/2}$, v=0)	859 ± 136	?
DBr$^+$($^2\Pi_{1/2}$, v=1)	56 ± 10	53.5
DBr$^+$($^2\Pi_{1/2}$, v=0)	924 ± 157	?

REFERENCES

1. S. Fenistein, M. Heninger, R.Marx, G. Mauclaire and Y.M. Yang Chem. Phys. Letters, **172**, 89 (1990).
2. S. Jullien, J. Lemaire, S. Fenistein, M. Heninger, G. Mauclaire and R. Marx, accepted in Chem. Phys. Letters.
3. T. Su and W.J. Chesnavich, J. Chem. Phys., **76**, 5183 (1982).
4. P. Armentrout, D.W. Berman and J.L. Beauchamp, Chem. Phys. Letters, **53**, 255 (1978).
5. R.H. Garstang, J. of Research of the NBS-A Physics and Chemistry, **Vol. 68A**, n°1, 61 (1964).
6. B. Edlen, Phys. Rev., **65**, 248 (1944).
7. H.-J. Werner, P. Rosmus, W. Schätzl and W. Meyer J. Chem. Phys., **80**, 831 (1984).

PHOTODISSOCIATION AND ROTATIONAL EXCITATION OF CO AND ITS ISOTOPOMERS IN INTERSTELLAR CLOUDS

S. Warin, J.J. Benayoun
Observatoire de Grenoble, BP 53X, F-38041 Grenoble Cedex 9, France

Y.P. Viala
DEMIRM, Observatoire de Meudon et Radioastronomie, ENS, F-92195 Meudon Cedex, France.

ABSTRACT

We have developed a model of an interstellar cloud that treats the chemistry among simple carbon and oxygen bearing molecules, including ^{13}C and ^{18}O isotopic compounds together with the rotational excitation of H_2, CO, ^{13}CO and C^{18}O. The rotational population of the CO isotopomers is controlled by chemical processes, including selective photodissociation of rotational levels, collisionnal processes and photon trapping of the millimeter and submillimeter lines. We concentrate this study on the abundance and rotational population of the isotopomers of CO in a "standard" transluscent cloud model, compare our results with those obtained in lte calculations and briefly address the problem of UV penetration.

I. DESCRIPTION OF THE MODEL

We have developed an interstellar cloud model dealing with chemistry and rotational excitation of molecules. For the moment, the model treats the population of the rotational levels of only H_2 (HD), CO and its isotopomers ^{13}CO and C^{18}O. Assuming steady state equilibrium, chemical abundances and rotational population are computed as a function of position within the cloud.

The level population is controlled by chemical reactions (including photodissociation by UV photons which, for H_2 and CO, occurs through absorption in rotational lines), collisional processes, UV pumping through excited electronic levels (for H_2 alone), radiative processes including emission and absorption of millimeter and submillimeter lines connecting the rotational levels (for CO and its isotopes alone). The transfer problem for these lines is resolved using an escape probability formalism.

The statistical equilibrium equations are solved simultaneously with the chemical balance equations, each rotational level being considered as a separate chemical species.

Because H_2 and CO are photodissociated through rotational line absorption towards electronic states, self-shielding is important. Furthermore, because the photodissociation of both molecules occurs in the same wavelength range (912-1150 Å), shielding of CO and its isotopomers by H_2 is also important. A particular attention has hence been paid to the calculation of the photodissociation rates which are computed separately for each individual rotational line using molecular data from Eidelsberg et al[1] (1991). The radiation field within each line is obtained by solving a transfer equation which takes into account absorption by both grains and gas. This allows to treat explicitly line overlap (including the Lyman series of H) and continuum absorption by gas (i.e. photoionisation of C).

The chemical model involves the simplest molecules formed from H, C and O as well as isotopic species containing ^{13}C and ^{18}O. The reactions involving isotopic species include isotopic fractionation reactions and reactions obtained by duplicating those between the main isotopomer. The library of chemical reactions contains 3113 reactions among 159 species.

At last, the model can compute the temperature of the gas together with its chemical composition. This is done by solving a thermal balance equation together with the chemical balance equations and a transfer equation for UV photons. The transfer problem is solved by assuming a plane parallel geometry.

In the results presented here we do not solve thermal problem. We assumed a constant temperature, of 50K in most models and constant density of 10^3 cm^{-3}. The visual optical depth to the mid plane of the cloud is $\tau_v = 1$. The UV external radiation field is that of Mathis et al[2] (1983). The cosmic ray ionisation rate is $\zeta_H = 6.8 * 10^{-18} s^{-1}$. The elemental abundances are those adopted by Le Bourlot et al[3] (1992) and are as follows : H:He:C:O:Metals = 1:9.75(-2):3.62(-5):8.53(-5):1.1(-7). The isotopic ratios are [^{12}C/^{13}C]=90 and [^{16}O/^{18}O]=500.

II. ROTATIONAL POPULATION OF CO. COMPARISON WITH LTE CALCULATIONS.

Fig.1 shows the photodissociation rates and the fractional abundances of the first rotational levels of CO, as function of position within the cloud.

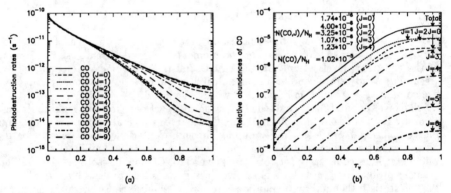

Figure 1: - *Photodestruction rates and abundances of CO and its rotational levels as a function of visual optical depth (τ_v) from the edge of the cloud.*

Photodissociation is strongly dependent on the rotational level because of the importance of the self-shielding. At low optical depth, the photodissociation rates are identical for all rotationals levels. As the optical depth increases, the photodissociation rates of low-lying levels decrease more rapidely than those of upper levels ($J \geq 5$) because the former are more abundant than the latter, at the temperature under consideration. The main consequence of self-shielding is that it leads to a selective photodissociation among rotational levels : the more populated the level, the lowest its photodissociation rates. The net effect of self-shielding is to overpopulate low-lying levels and to underpopulate high levels with respect to an LTE calculation.

This is clearly shown in fig.2 where the abundance of the six first levels of CO are plotted for the exact calculation and by assuming LTE (the total abundance of CO being the same in the two cases).

The consequences on the profiles of emergent rotational lines J=1-0, 2-1, 3-2 are shown on fig.3 where the Rayleigh-Jeans brightness temperature is plotted as a function of velocity assuming a microturbulent broadening of the line. In the

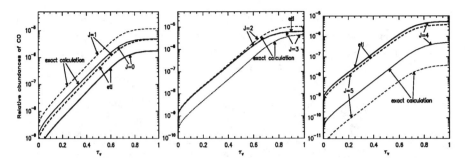

Figure 2: - *Relative abundances of the rotational levels of CO as a function of visual optical depth, in the case of the exact calculation and of the LTE hypothesis. Normal line : exact calculation, bold line : LTE.*

exact calculation the brightness temperature decreases as excitation increases. It is the opposite in the LTE calculation, at least up to the 4-3 transition. The most spectacular difference between the two sets of calculation is obtained for the 2-1 and 3-2 lines whose excitation temperature are inversed.

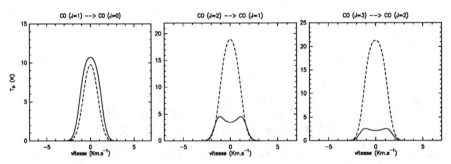

Figure 3: - *Line profiles emergent for the standard model assuming microturbulent broadening of mean velocity $v_{turb} = 1$ kms^{-1}, ^{12}CO $(J = 1 - 0)$, $(J = 2 - 1)$ and $(J = 3 - 2)$. Rayleigh-Jeans brightness temperature is plotted as a function of the velocity. The contribution from the 2.7 K background radiation has been removed. Full line : exact calculation, dashed line : LTE calculation.*

III. INFLUENCE OF THE TREATMENT OF THE UV RADIATION TRANSFER

For this study, we choose a plane parallel geometry with : i) finite thickness and isotrope radiation flux (4a). ii) semi-infinite thickness and isotropic radiation flux (4a with $\tau_{tot} = \infty$). iii) finite thickness and collimated incident radiation flux. The radiation attenuation varies as exp($-\tau$), i.e. takes into account absorption from only the shorter distances of the current point to the two edges of the cloud (4b).

The abundance distribution of C, C^+ and CO for the three geometries are plotted on the fig.5.

Plane parallel semi-infinite geometry neglects the radiation coming from the opposite side of the cloud compared to the finite thickness situation : the consequence

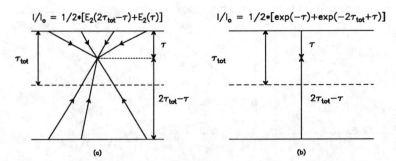

Figure 4: - *Schematic representation of the different treatment of the UV radiative transfer. (a) : plane parallel geometry with finite thickness and isotropic radiation flux. (b) : plane parallel geometry with finite thickness and a collimated incident flux.*

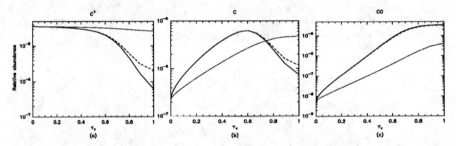

Figure 5: - *Relative abundances of C^+ (a), C (b) and CO total (c) as a function of visual optical depth (τ_v) from the edge of the cloud, in the three different geometries used. Full line : plane parallel geometry, semi-infinite thickness; long-dashed line : plane parallel geometry, finite thickness; short-dashed line : plane parallel geometry, finite thickness with collimated incident UV flux.*

is that it underestimates in each point the radiation field, the difference reaches its maximum of a factor 2 at the center. The plane parallel semi-infinite geometry underestimates relative abundance of C and C^+ by a factor 1.6 and 3.4 respectively, while there is practically no change for CO.

The difference is much more sensitive between an isotrope and a collimated radiation flux. The collimated flux overestimates the capacity of penetration of the radiation from the nearest two edges of the cloud to the current point. The effect on the abundances of C, C^+ and CO oversteps the order of magnitude. We can see that C^+ is strongly increased and remains dominant across the cloud. Conversely CO is strongly decreased due to enhanced the photodestruction rates. The effect on C is more complex. It is less abundant at the surface because it is more strongly ionized and it is more abundant at the center because it is more easily produced by the photodissociation of CO.

For isotropic flux the transition C, C^+, CO occurs at $\tau_v = 0.7$. And the column density ratio $N(X)/N(H+2H_2) = 2.2*10^{-5}$ for C^+, $3.2*10^{-6}$ for C and $1.0*10^{-5}$ for CO. A collimated flux leads to an increase of the column density of C^+ by a factor 1.5 and a decrease in the column density of C and CO by factor 1.3 and 13 respectively.

IV. PHOTODISSOCIATION AND ABUNDANCES OF THE ISOTOPOMER ^{13}CO AND $C^{18}O$

The photodissociation rates and the rotational population of ^{13}CO and $C^{18}O$ are plotted on fig.6.

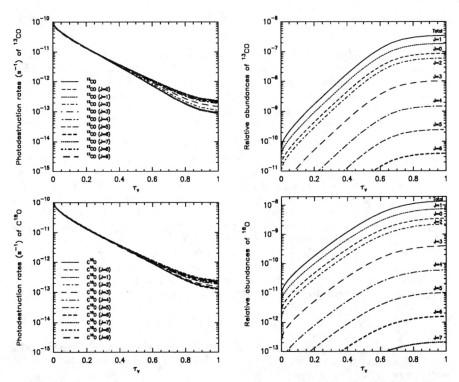

Figure 6: - *Photodestruction rates and relative abundances of the isotopomer ^{13}CO and $C^{18}O$ and their ratational levels as a function of optical depth (τ_v) from the edge of the cloud.*

The self-shielding for the rare isotopes is much less efficient than for CO, leading to much larger photodissociation rates, of the same order for all rotational levels and with a smoother variation with optical depth. This seems to indicate that the rare isotopes are not efficiently shielded by the main one.

Because the photodissociation is more efficient for the rare isotopomers, we expect a $^{12}CO/^{13}CO$ and $^{12}CO/C^{18}O$ ratio larger than the cosmic isotopic ratio $[^{12}C/^{13}C]=90$ and $[^{16}O/^{18}O]=500$. In the case of ^{13}CO, this effect is counterbalanced by the isotopic fractionation reaction :

$$^{13}CO^+ + ^{12}CO \rightleftharpoons ^{13}CO + ^{12}CO^+ \ (\Delta E = -35K)$$

which favours the transformation of ^{12}CO in ^{13}CO at low temperature. Indeed, as can be seen on fig.7, the ratio $^{12}CO/C^{18}O$ is increased with respect to regard the cosmic ratio, by a factor of 5 towards the center of the cloud.

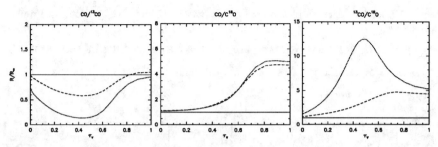

Figure 7: - *Abundance ratios of various isotopic forms of CO as function of visual optical depth (τ_v) from the edge of the cloud. R/R_∞ is the computed ratio (R) divided by the ratio that would occur in the absence of any isotopic fractionation effects (R_∞)- that is the ratio of the isotopic forms of the nuclei in the gas. The solid line represents the isotopic ratio (equal to 1). Full line : isotopic ratio equal to 1, short-dashed line : model with $T = 10$, long-dashed line : standard model.*

This is of course the consequence of selective photodissociation. The enhancement of ^{12}CO over C^{18}O is more important than obtained in the simplified model of Chu and Watson[4] (1983). The isotopic fractionation reaction is efficient to convert a fraction of ^{12}CO in ^{13}CO, resulting in an isotopic ratio smaller than the cosmic ratio by nearly a factor 2 in the external parts of the clouds, around $\tau_v = 0.4$. The selective photodissociation does not play any role since it would have produced an opposite effect, an enhancement of ^{12}CO over ^{13}CO.

It is worth noting that these results strongly depend on the temperature, since the efficiency of isotopic fractionation for ^{13}CO increases as the temperature decreases. As shown on fig.7, reducing the temperature down to $T = 10K$, leads to a decrease of the ^{12}CO/^{13}CO ratio to nearly an order of magnitude at $\tau_v = 0.4$.

V. CONCLUSION

These preliminary calculations emphasize the importance of self-shielding for CO photodissociation wich result in a very selective photodissociation between the isotopes and among rotational levels.

The isotopic fractionation of ^{12}CO in ^{13}CO overwhelms this effect of selective photodissociation especially at low temperature.

The geometry of the cloud and the way the penetration of UV radiation is accounted for play an important role in the abundances of C$^+$, C and CO.

The last effect of accounting for selective rotational photodissociation is a drastic change in rotational population and emitted line profiles of CO and its isotopes compared to LTE calculation.

REFERENCES

1. M. Eidelsberg, J.J. Benayoun, Y. Viala and F. Rostas, A&AS **90**, 231 (1991).
2. J.S. Mathis, P.G. Mezger and N. Panagia, A&A **128**, 212 (1983).
3. J. Le Bourlot, G. Pineau des Forêts, E. Roueff and D.R. Flower, A&A **267**, 233 (1993).
4. Y.H Chu and W.D. Watson, ApJ **267**, 151 (1983).

CONSTANTS FOR ASTROPHYSICAL GAS-PHASE MOLECULES: PHOTODISSOCIATION RATES, MICROWAVE AND INFRARED SPECTRA

J. Crovisier
Observatoire de Paris-Meudon, F-92195 Meudon, France
MESIOB::CROVISIE or 17671::CROVISIE *(SPAN)*
CROVISIE@mesioa.obspm.circe.fr

A compilation of references to molecular data relevant to infrared and radio spectra was made for species (molecules, radicals and ions) of astrophysical interest. This work was originally undertaken in the frame of the working group "Laboratory Spectroscopy for ISO"[1]. Its scope, however, should interest all radio and infrared observers, both planetary and interstellar.

The compilation presently includes 211 species, representing all molecules already detected in the interstellar medium, in comets and in planetary atmospheres, as well as several molecules believed to be of important astrophysical interest. It is intended to be regularly updated. Needs for further laboratory data are pointed out.

The compilation is not distributed in printed form (which would be more than 250 pages), but as computer files. On an experimental base, these files may be obtained through *anonymous ftp* to 130.84.200.6, in the \pub\molecule directory. The files are in the RTF format; they are intended to be downloaded to a Macintosh or a PC and treated with the WORD word processor. Further instructions are given in a "readme" file.

REFERENCE

1. T. Encrenaz, J. Crovisier, L. d'Hendecourt, P. Lamy, J.A. Tully, Laboratory spectroscopy for ISO. Report of the ISO working group. In *Infrared Astronomy with ISO*, edts Th. Encrenaz and M.F. Kessler, Nova Science Publishers, Inc, pp. 141-170 (1992).

PHOTODYNAMICS OF ACETYLENE

J.H. Fillion, N. Shafizadeh and D. Gauyacq
Laboratoire de Photophysique Moléculaire, Bât.213
Université de Paris-Sud 91405 ORSAY CEDEX (France)

ABSTRACT

The aim of this work is to study the role of the excited Rydberg states in the photofragmentation process of acetylene in order to understand the formation of the C_2H and C_2 radicals in the circumstellar medium. The np Rydberg series show a good example of fast predissociation through a strong valence-Rydberg interaction in the ^1A_g geommetry. These np series are probed by two-color, (3+1') photon excitation via the resonant $G^1\Pi_u(4s\sigma)$ Rydberg state. A preliminary experiment shown in this paper demonstrates that selective photoionization of this G state is very efficient by using an infrared photon. By scanning the infrared laser below the IP limit, one should be able to probe even more efficiently the high lying np states and study their relaxation mecanism.

INTRODUCTION

The acetylene molecule is an important constituant of some stellar atmospheres, as has been shown by the first infrared detection in IRC+10 216 [1]. In carbonated stars envelopes, C_2H_2 and its fragmentation product C_2H are the main carbon reservoirs from which carbon containing interstellar dust is formed [2]. The photodestruction of C_2H_2 by the interstellar UV radiation has been established as the dominant mecanism leading to the radical C_2H [3-6]. On the other hand the fragmentation mecanisms of this molecule are still very poorly understood, albeit the large number of laboratory experimental work carried out. Among these previous studies, dissociation cross section measurements between 153 nm and 193 nm have stressed out the role of the resonant molecular structure and of the discrete states in this process, especially at low temperature [7]. In addition, one-color Resonant Multiphoton Ionization (REMPI) studies have shown that, above 9.4 eV, the Rydberg states of acetylene undergo very fast predissociation [8,9].

Although they cannot be excited by a direct one photon excitation from the gerade ground state, the np series present a typical example of a fast dissociation channel above an energy threshold. Ab initio calculations have recently suggested the important role of Rydberg-valence interaction involving these Rydberg np states and trans-bent valence states [10]. Experimental studies by using one-color REMPI [9] have shown an efficient decay channel above 9.4 eV, precluding the observation of any ionization signal from these resonant states. Nevertheless, this experimental search by REMPI has been limited to the energy region below 90 000 cm^{-1}, i.e. to Rydberg states from 3p to 7p. We therefore have undertaken the study of the higher members of this np series between 7p and the ionization limit, i.e. around the IP energy, i.e. 11.406 eV, in order to study the energy dependant dissociation mecanism in this series.

© 1994 American Institute of Physics

EXPERIMENTAL

The two-color REMPI experiment was performed in a "magnetic bottle" time-of-flight photoelectron spectrometre [11]. In the present experiment two separate Nd:YAG and Excimer pumped dye laser systems were used to generate the pump and probe beams in a countrepropagating geommetry. The output of the first dye laser (QUI dye solution) was used to pump individual rotational lines of the three-photon $G^1\Pi_u \leftarrow X^1\Sigma_g^+$ transition (0_0^0 and 2_0^1 bands). The pump laser output was of about 1 mJ and focused in the ionization region through a 150 mm focal length lens. The second dye laser (probe beam) was operated with a red dye (DCM dye solution) and Raman shifted by passing through a high pressure hydrogen cell. The resulting infra-red beam (between 840 nm and 930 nm) was focused together with the attenuated red beam into the ionization volume through a 75 mm focal length lens. The two lasers were synchronized by a home made delay generator with an approximate arrival time overlap within a time jitter of 10-20 ns. The pulse energy of the infra-red beam was typically 2-3 mJ.

Acetylene was introduced in the ionization chamber without any purification, at a pressure of 10^{-3} mbar.

The electron signal resulting from the the one-color (3+1) and the two-color (3+1') ionization via the G state was detected on micro-channel plates, amplified and recorded both by a digital oscilloscope and a CAMAC acquisition system.

RESULTS AND DISCUSSION

Figure 1 displays the one- and two-color excitation mecanism. The lowest ionization limits corresponding to the ground state $^2\Pi_u$ of the ion $C_2H_2^+$ at the energy 11.41 eV above the neutral molecule ground state, and to the vibrationnaly excited ionic ground state in the ν_2 (C-C stretching) mode are shown by hatched areas. The pump laser frequency (blue photon of energy $h\nu_1$) is resonant either with the 0_0^0 band or with the 2_0^1 band of the three-photon transition $G^1\Pi_u \leftarrow X^1\Sigma_g^+$. In all cases the G state (v=0 or ν_2=1) is further ionized by a fourth blue photon as indicated in Fig.1a. When the second laser is operating, the G state can also be ionized by either a red photon (energy $h\nu_2$) or an infra-red (IR) photon (energy $h\nu_3$), as shown in Fig.1a. The intensity of the probe laser has to be relatively high in order to compete with the very efficient one-color (3+1) process with the blue photons only. This explains the unusual tight focus used for the probe laser beam (see section above). On the other hand, since no time delay was set between the probe and pump lasers, the intensity of the probe laser has to be kept low enough in order to avoid unwanted depopulation of the ground state of acetylene by non resonant multiphoton ionization with 6 red photons (or 9 IR photons).

The energy diagram of Fig.1b illustrates the excitation mechanism when the IR photon energy is scanned below the ionization limit (the excitation processs involving a red photon is very similar to the case of Fig.1a and is not shown here for the sake of clarity). Then a double resonance signal occurs in the ionization channel every time the IR photon is energy resonant with a one-photon transition between the intermediate G state and one excited np (v=0 or $v_2=1$) Rydberg state converging to the ground state of the ion ($v^+=0$ or $v_2^+=1$, respectively). The np,v=0 states are ionized by a second IR photon while the np,$v_2=1$ states are rather ionized through rapid vibrational autoinization into the ground state ion continuum. Both vibrational series can also undergo predissociation (a loss mechanism with respect to the ionization detection) into the dissociation continuum of $C_2H + H$.

The two-color REMPI process illustrated in Fig.1.b allows for a detailed study of the fragmentation dynamics of the np Rydberg series of acetylene, by recording

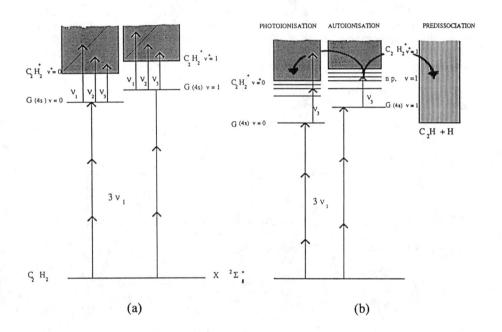

Fig.1 (a) Energy diagram for the REMPI of acetylene via the G state (v=0 and $v_2=1$) by using a one- or a two-color technique (in the later case, the second photon is either in the red or in the infra-red) ; (b) Energy diagram for the Double Resonance REMPI of acetylene probing the high np Rydberg states from the G state (v=0 or $v_2=1$). The three relaxation channels (ionization, autouionization and dissociation) are indicated.

intensities and linewidths of the double resonance ionization signal via these Rydberg states.

The recent ab initio calculation of the valence C' state (trans-bent equilibrium geommetry) has demonstrated the existence of at least one type of efficient valence-Rydberg interaction in the ^1A_g symmetry which could lead to predissociation above the $C_2H + H$ dissociation limit of the valence state. This mechanism is certainly responsible for the fast decay channel opened above the 4p states, leading to a sudden disappearance of this np series in the ionization channel, as observed in one-color REMPI experiments [9]. However, since any Rydberg-valence interaction approximately decreases as $1/n^{*3}$ (where n* is the effective principal quantum number), one expects a slower predissociation process with increasing energy and, therefore, the reappearence of the high np Rydberg series in the ionization channel. This means that the high lying np states should be observed, at least just below the IP, by double resonance two-color REMPI by using the excitation mechanism of Fig.1b.

Up to now, none of the cross sections involved in the competing mechanisms schematized in Fig.1b are known.

In order to check the possibility of observing the high np series by using the experimental procedure described above, we have performed a preliminary experiment corresponding to the situation of Fig.1a. In the case of this "three-color" REMPI of acetylene, each of the three photons (blue, red and infra-red) has enough energy to contribute to the direct ionization of the G state. Under some laser intensity conditions, it has been possible to observe ionization in the three channels as shown in Fig.2, both for the v=0 and $v_2=1$ levels of the G state.

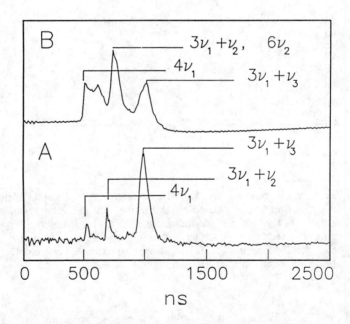

Fig.2 Photoelectron spectrum of acetylene corresponding to the excitation scheme of Fig.1a. The laser intensities of the three beams are 0.7 mJ (blue, v_1), 4 mJ (red, v_2) and 2 mJ (IR, v_3) (A) via G,v=0 ; (B) via G, $v_2=1$.

This means that the one-photon discrete-to-continuum transition is not saturated by the blue laser beam in the (3+1) ionization one-color process. Fig.2 also shows that, under certain laser intensity conditions, the ionization probability from the G state by using an infra-red photon is significant by comparison to the other competing mechanisms. Therefore, one expects a similar or even a larger order of magnitude for the discrete-to-discrete transition probability from the G state to the Rydberg states converging to the ionization continua, as shown in Fig.1b. The corresponding double resonance experiment is in progress and will be described in a further paper.

CONCLUSION

This work has proposed a very selective experimental technique in order to probe the relaxation dynamics of highly excited Rydberg states of Acetylene, which have not been observed yet. The double resonance experiments are in progress and should yield a quantitative estimate of the predissociation rate of the np states as a function of excitation energy. Parallel to this work, ab initio calculations are performed in order to determine the strength of the valence-Rydberg coupling as a function of the principal quantum number (limited to low n values) and of the Rydberg orbital symmetry (pσ or pπ) [12]. These calculations can then be extrapolated to higher energy regions, where predissociation dominates. An alternative to this experiment is to detect the fragmentation channels instead of the ionization channel, by using an ion mass analyser in addition to the previous set-up. Such a photofragment analysis experiments needs a third laser in order to perform the multiphoton ionization of C_2H, H or C_2. Although such a three-color experiment is not easy, it would yield new informations in the photodynamics of these Rydberg states even when they cannot be observed by ionization detection techniques.

The last step of this work is to extend the present study to ungerade Rydberg states which seem play an important role in the photofragmentation of acetylene in circumstellar atmosphere.

AKNOWLEDGEMENTS

We are grateful to Prof. Ph. Brechignac for his help and for stimulating discussions. We also thank M. Bonneau and and D. Furio for their technical assistance. This work has been supported by the CNRS through a GDR "PCMIG" contract and by a EEC Science collaboration program (n°SC1*-ct91-0711).

REFERENCES

1. S.T. Ridgway, D.N.B. Hall, S.G. Kleinmann, D.A. Weinberger and R.S. Wojslaw, Nature **264**, 345 (1976).
2. A. Omont "Chimie dans les enveloppes circumstellaires", Colloque du GDR PCMI (Paris, 1991).
3. S. Lafont, R. Lucas and A. Omont, Astron.Astrophys. **106**, 201 (1982).
4. P.J. Huggins and A.E. Glassgold, Astrophys.J. **252**, 201 (1982) ; P.J. Huggins, A.E. Glassgold and M. Morris, Astrophys.J **279**, 284 (1984).
5. Truong-Bach, Nguyen-Q-Rieu, A. Omont, H. Olofsson and L.E.B. Johansson,

Astron.Astrophys. **176**, 285 (1987).
6. J.H. Bieging and Nguyen-Q-Rieu, Astrophys.J. **329**, L107 (1988).
7. C.Y.R. Wu, T.S. Chien, G.S. Liu and D.L. Judge, J.Chem.Phys. **91**, 272 (1989).
8. T.M. Orlando, S.L.Anderson, J.R. Applingand M.G. White, J.Chem.Phys. **87**, 852 (1987).
9. M.N.R. Ashfold, B. Tutcher, B. Yahg, Z.K. Jin and S.L. Anderson, J.Chem.Phys. **87**, 5105 (1987).
10. J. Lievin, J.Mol.Spectrosc. ,**156**, 123 (1992).
11. P. Kruit and F.H. Read, J.Phys. e Sci.Instr.,**16**,313 (1983).
12. J. Lievin, private communication.

DISCUSSION

LEACH — If it is possible to detect C_2H by Laser Induced Fluorescence excitation this would give important information concerning the internal energy constant of the nascent fragment species.

GAUYACQ — This is a very good suggestion. Unfortunately, our experimental set up doesn't allow for a selective detection of the fragments coupled to a Laser Induced Fluorescence (LIF) detection. Another experimental set up has to be used for such a detection.

NENNER — If photodissociation proceeds via a "isomeric" state (for ex. :

$$C_2H_2 \rightarrow \begin{array}{c} CH_2\text{-}C \\ \end{array} \rightarrow C_2 + H_2),$$

which experiment do you intend to perform to show the existence of such a mechanism ?

GAUYACQ — We will set up an experiment including both fragment detection and ionization detection. The fragments will be ionized by a further laser and detected by an ion mass analyzer. This detection will give us branching ratios for the different fragments (C_2H, C_2, H, H_2, ...) formation. The ionization channel will be observed via a TOF photoelectron spectrometer. The PES spectrum tells up which cation internal vibrations are formed following photoionization of the high lying Rydberg states, bringing some information on the possible isomeric geometry of the valence state responsible for the Rydberg state predissociation. Theoretical models for this kind of interaction are of course highly needed !

ROSTAS — When you observe the REMPI spectrum of the np Rydberg states, which is the photon which produces the ionization step ?

GAUYACQ — The np Rydberg states are ionized by an infrared photon as shown in the photoelectron spectrum. Actually, this photoelectron spectrum indicates a strong non Frank-Condon behaviour in the ionization process of these Rydberg states. This behaviour is not yet well understood. It most probably corresponds to a very strong Rydberg-valence mixing in the neutral molecule.

COLLISIONS AND REACTIVITY

RATE CONSTANT FORMULAE FOR FAST REACTIONS

D. C. Clary

Department of Chemistry, University of Cambridge, Cambridge CB2 1EW, UK

ABSTRACT

This paper presents formulae that we have derived recently for the rate constants of some chemical reactions important in interstellar chemistry. First of all, the application of a simple classical capture theory to the rate constants for the fast reactions of ground-state carbon atoms with organic molecules is described. It is found that the rate constants scale with the size of the molecule as $N^{1/3}$, where N is the number of carbon atoms in the molecule. This is found to be in remarkable agreement with experimental results for the reactions of alkenes and alkynes, even up to N=16. Rate constant formulae at very low temperatures for the fast reactions of molecules in $^1\Sigma$ and $^2\Pi$ electronic states with various ionic and molecular partners, including those in $^2\Pi$ states, are also summarised. A new formula is also derived for the rate constant of a reaction that has a scattering resonance for a very low kinetic energy. This rate constant is found to be temperature independent.

INTRODUCTION

The modelling of dense interstellar clouds requires rate constants for many chemical reactions[1]. These can sometimes be obtained from experimental measurements made in the laboratory[2], but such experimental data is not always available, especially at the very low temperatures of interstellar clouds. Therefore, theory has a valuable role in providing estimates for the rate constants in the cases that experiments are difficult to perform. Such theoretical estimates have already been shown to be particularly useful in the important case of the reactions of ions with closed-shell molecules[3]. In this paper we present reaction rate constant formulae for various types of reactions involving neutral species that could be important in interstellar clouds.

REACTIONS OF C(^{3}P) ATOMS WITH ORGANIC MOLECULES

Recent measurements by Husain and co-workers demonstrate that the reactions of ground-state carbon atoms with 1-alkenes and 1-alkynes have surprisingly large rate constants (>10^{-10} cm^3 molecule^{-1} s^{-1})[4,5]. It is, therefore, of interest to examine if a classical capture theory[6] combined with a simple treatment of the long-range intermolecular forces might be appropriate for these reactions as this type of theory has been demonstrated to work so well for ion-molecule reactions. A simple approximation is to assume that the main term in the potential energy surface is the isotropic dispersion factor

$$V(R) = -\frac{C_6}{R^6} \tag{1}$$

where R is the distance between the C atom and the centre of mass of the molecule and the C_6 coefficient depends on the reactants. In classical capture theory, the centrifugal potential b^2E/R^2 is added to $V(R)$, where b is the impact parameter for the collision and E is the collision energy. The assumption that reaction occurs with unit reaction probability for all impact parameters provided the reaction is classically allowed gives a maximum impact parameter b_{max} from which the reaction cross section

$$\sigma = \pi(b_{max})^2 = \pi \frac{3}{2^{2/3}} \frac{C_6^{1/3}}{E^{1/3}} \tag{2}$$

is obtained. Maxwell-Boltzmann averaging this cross section over collision energy gives the rate constant

$$k = 8.56 \frac{C_6^{1/3}}{\mu^{1/2}} (k_B T)^{1/6} \tag{3}$$

where μ is the collisional reduced mass and k_B is the Boltzmann constant. If the parameters in this expression are all in atomic units, the factor 0.613 x 10^{-8} converts it into units of cm^3 molecule^{-1} s^{-1}.

A simple approximation for the C_6 coefficient is the formula[7]

$$C_6 = \frac{3}{2}\left(\frac{E_a E_b}{E_a + E_b}\right)\alpha_a \alpha_b \quad (4),$$

where E_a and E_b are ionization potentials and α_a and α_b are polarisabilities of monomers a and b, respectively. The ionization potentials of $C(^3P)$ and the alkenes and alkynes considered here are all close to 11eV and it is the polarisabilities of the organic molecules that have more variation with the size of the organic molecule. Experimental results[8] show that the isotropic polarisabilities of hydrocarbons depend approximately on the number N of carbon atoms in the molecule. This suggests that the polarisabilities of both the 1-alkenes, C_NH_{2N}, and the 1-alkynes, C_NH_{2N-2}, will be approximately proportional to N. It follows from equations (3) and (4) that the rate constants for the reactions of ground state C atoms with these organic molecules will vary as

$$k \propto N^{1/3} \quad (5).$$

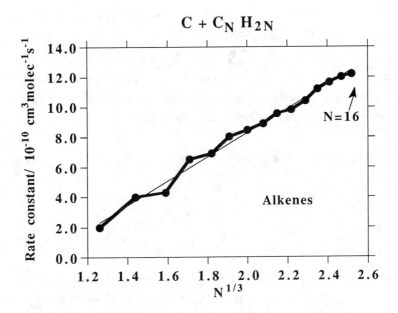

Figure 1. Plot of the experimental[5] rate constant for the reaction of C atoms with 1-alkenes C_NH_{2N} against $N^{1/3}$.

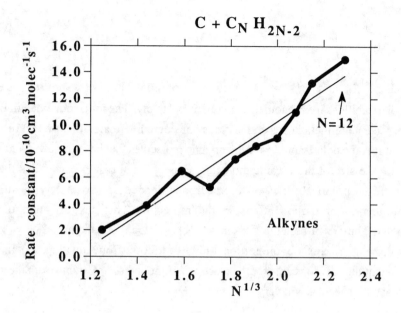

Figure 2. Plot of the experimental[5] rate constant for the reaction of C atoms with 1-alkynes C_NH_{2N-2} against $N^{1/3}$.

Figures 1 and 2 are plots of the experimental rate constants[5] at room temperature for the 1-alkenes, C_NH_{2N}, and the 1-alkynes, C_NH_{2N-2}, against $N^{1/3}$. The fits are very good in both cases. This is strong evidence that these reactions are controlled largely by long-range intermolecular forces even for reactions with as many as 16 carbon atoms. The plots also give the following simple formulae for the rate constants

$$k = 8.0 N^{1/3} - 7.8 \tag{6},$$

for the 1-alkenes, and

$$k = 11.9 N^{1/3} - 13.6 \tag{7},$$

for the 1-alkynes, with units of 10^{-10} cm^3 molecule^{-1} s^{-1}. The results strongly suggest that these formula can be used for the reactions of ground state C atoms with alkenes and alkynes with N even larger than 16. As the theoretical rate constant suggests only a very slight temperature dependence of $T^{1/6}$ (see equation (3)) it is likely that these formulae will also be applicable for the low temperatures of interstellar interest.

It is important to examine if the absolute values of the rate constants can be computed accurately from first principles. Use of the ionization potentials of 11.76 eV and 10.5eV together with the polarisabilities of 1.76 x 10^{-24} cm^3 and 4.25 x 10^{-24} cm^3 for carbon and ethene, respectively[9] enables the coefficient C_6=104 a.u. to be estimated for the C + C_2H_4 reaction via equation (4). As the rate constant depends on $(C_6)^{1/3}$, a very accurate value for C_6 is not essential. Equation (3) then gives a rate constant of 8x 10^{-10} cm^3 molecule^{-1} s^{-1} for C + C_2H_4. This is much larger than the experimental[4] value of 2.0 x10^{-10} cm^3 molecule^{-1} s^{-1}. However, this simple calculation of the absolute value of k for C + C_2H_4 would suggest that the constant multiplying $N^{1/3}$ in equation (6) should be 6.3, which is close to the experimentally determined value of 8.0.

The main weakness of this theory for calculating an absolute value for k is the neglect of shorter range chemical forces, and other long-range terms such as quadrupole-quadrupole interactions and the anisotropy of C_6. Furthermore, non-adiabatic transitions and spin-orbit effects are neglected. The good fits to the $N^{1/3}$ dependence of the rate constant obtained in equations (6) and (7) suggest, however, that the contributions to the rate constant of the dispersion term and other terms in the potential energy surface are additive. Also, the contributions to the rate constant from terms in the potential surface not treated explicitly in our treatment do not scale with the size of the organic molecule. Clearly, *ab initio* calculations of the potential energy surfaces for these reactions would be very useful, but we feel that the simple classical theory described here is a useful first step in understanding the rates of these reactions.

REACTIONS OF DIPOLAR MOLECULES IN $^2\Pi$ STATES

The simple theory described above is for reactions of atoms with closed-shell organic molecules which are assumed to be non polar. For reactions involving polar molecules, the theory becomes more complicated but, in the limit of very low temperatures, analytical formulae for rate constants can still be derived[10-13]. This has been done for reactions of molecules in both $^2\Pi$ and $^1\Sigma$ electronic states with various ionic and neutral molecular partners. Here we briefly describe the method for deriving the rate constants of $^2\Pi$ molecules[11,12].

It is assumed that there are no barriers in the potential energy surface and the reaction cross section is controlled by the long-range region of the reactant channel. The calculated rate constants of the exothermic reactions are expected to be upper bounds to the true values unless an ion-pair forms in the long-range region of the potential[6,14]. For low collision energies, the centrifugal barriers to the reaction occur at very large intermolecular distances so that the long-range expansion of the potential energy surface is applicable. Therefore, the classical capture theory for fast reactions should become more accurate as the temperature of the reaction is lowered. Fortunately, it is the region of low temperatures that are most relevant for interstellar chemistry.

The interaction potential V is a function of the distance R between the centres of mass of the two reacting partners and their respective orientations. The simple dispersion term in the potential is considered in the first section of this paper and, in this section, ion-dipole, dipole-dipole and dipole-quadrupole interactions are treated. For ion-molecule reactions it is assumed that the ion is a structureless particle, while neutral linear molecules are treated as rigid rotors. Molecular vibrations are normally not important in these fast reactions and are neglected. Several closely related rotationally adiabatic theories to that emplyed by us have been used by other groups[10,15-19], especially for ion-molecule reactions.

A basis set Φ_i, where i labels a **pair** of quantum states appropriate for each of the reactant molecules, is used to form matrix elements over V for a fixed value of R. Diagonalisation of this matrix gives a set of adiabatic energy curves $V_{eff}^i(R)$, each of which can be identified for large R with a pair of quantum states i of the isolated molecules. For each of the $V_{eff}^i(R)$, the maximum impact parameter $b_{max}^i(E)$ that gives reaction for a given collision energy E is then calculated and the capture cross section is obtained as $\sigma_i(E) = \pi [b_{max}^i(E)]^2$. This is state-selected in the initial pair of states labelled by i. Maxwell-Boltzmann averaging then gives the rate coefficient k(T) for temperature T.

A molecule in a $^2\Pi$ electronic state has a non-zero projection Ω of its molecular angular momentum along the molecular axis, while molecules in $^1\Sigma$ states have $\Omega=0$. Thus, $^2\Pi$ and $^1\Sigma$ molecules are described by different basis functions and will not have the same effective potentials $V_{eff}^i(R)$. The cross sections and rate constants will be different and are labelled by the total angular momentum quantum numbers of each molecule, j and Ω. The value of Ω appropriate for the electronic ground state depends on the sign of the spin-orbit coupling constant in the molecule and different j and Ω states can become populated as the temperature is increased. For interstellar chemistry, OH and CH are two of the most important radicals and the ground state of OH($^2\Pi$) has (j=3/2, Ω=3/2) while that for CH($^2\Pi$) has (j=1/2, Ω=1/2).

For very low E, the centrifugal barrier in the effective potential

$$\frac{Eb^2}{R^2} + V^i_{eff}(R) \tag{8}$$

occurs for very large R and perturbation theory can be used to determine analytical formulae for $V^i_{eff}(R)$. This enables analytical formulae for the cross sections and rate constants to be derived in the limiting cases of very low energy or temperature, respectively[10-13]. First-order perturbation theory is applicable when calculating the long-range part of $V^i_{eff}(R)$ for an interaction involving a $^2\Pi$ molecule, while second-order perturbation theory is normally needed for reactions involving only $^1\Sigma$ molecules. This results in interesting differences in the temperature dependencies of the rate constants for these two types of reaction at very low temperatures.

At higher temperatures, a larger number of rotational states can be populated in the reactant molecules. The infinite-order-sudden (IOS) approximation can then be applied[14]. This theory, which assumes an infinite manifold of rotational states, involves computing capture cross sections for fixed orientation angles of the reacting partners, and then averaging over all such angles. The analytical formulae obtained in this way are appropriate for high collision energy (and high temperatures). It is important to emphasise that these particular formula are only very approximate as the capture theory should eventually break down at high collision energies when short range intermolecular forces become important.

A clear example of the differences in the formulae obtained for reactions of $^2\Pi$ and $^1\Sigma$ molecules at low temperatures is the ion-dipole interaction. Here, the leading term in the electrostatic potential is

$$V(R,\theta) = -\frac{\mu_D q}{R^2}\cos\theta - \frac{\alpha q^2}{2R^4} \tag{9}$$

where θ is the angle between the molecular axis and the line joining the ion to the centre of mass of the molecule, μ_D is the dipole moment of the molecule, q is the charge of the ion and α is the isotropic polarisability of the molecule. Application of the perturbation theory described above to the reaction between a structureless ion and a $^2\Pi$ molecule in ground state (j,Ω) gives the rate constant[11] in the limit of very low temperature

$$k(T \to 0) = \frac{1}{2(2j+1)} \sum_{P_j, \pm \Omega} \left[\sqrt{\frac{8\pi}{\mu k_B T}} \frac{\mu_D q \Omega P_j}{j(j+1)} + 2\pi q \sqrt{\frac{\alpha}{\mu}} \right] \quad (10),$$

where P_j is the projection of j on **R** and the sum only contains positive products of ΩP_j.

For the ground state of OH with j=3/2 and Ω=3/2, this formula reduces to

$$k(T \to 0) = \frac{1}{5} \mu_D q \sqrt{\frac{8\pi}{\mu k_B T}} + \pi q \sqrt{\frac{\alpha}{\mu}} \quad (11)$$

and the appropriate formula for the ground state of CH with j=1/2 and Ω=1/2 is[13]

$$k(T \to 0) = \frac{1}{6} \mu_D q \sqrt{\frac{8\pi}{\mu k_B T}} + \pi q \sqrt{\frac{\alpha}{\mu}} \quad (12)$$

For the reaction of an ion with a molecule in a $^1\Sigma$ state the low-temperature formula, first derived by Troe[10], is

$$k(T \to 0) = 2\pi q \sqrt{\left[\frac{\alpha}{\mu}(1 + \frac{\mu_D^2}{3\alpha B})\right]} \quad (13),$$

where B is the rotor constant of the molecule. The obvious difference is that molecules in $^2\Pi$ states give a rate constant that has a temperature dependence of $T^{-1/2}$ in the limit of very low temperature while that for $^1\Sigma$ molecules tends to a constant value.

The same principles can be applied to the derivation of rate constants in the limit of low temperatures for other interactions such as $^2\Pi$–$^2\Pi$ and interesting temperature dependencies are again obtained[12,13]. A summary of these temperature dependences of the rate constants for various molecular partners in the low and high temperature limits are given in the Table below.

TABLE I. Summary of the temperature dependencies of the rate constants for various molecular partners in the low and high temperature limits.

Reaction	T dependence of rate constant	
	Low T	High T
Ion-$^1\Sigma$ dipole	T^0	$T^{-1/2}$
Ion-$^2\Pi$ dipole	$T^{-1/2}$	$T^{-1/2}$
Ion-$^2\Pi$ quadrupole(j≤ 0.5)	T^0	$T^{-1/6}$
Ion-$^2\Pi$ quadrupole(j> 0.5)	$T^{-1/6}$	$T^{-1/6}$
$^1\Sigma$ dipole-$^1\Sigma$ dipole	$T^{1/6}$	$T^{-1/6}$
$^1\Sigma$ dipole-$^1\Sigma$ quadrupole	$T^{1/4}$	T^0
$^2\Pi$ dipole-$^1\Sigma$ dipole(j=0)	$T^{1/6}$	$T^{-1/6}$
$^2\Pi$ dipole-$^1\Sigma$ dipole(j≠0)	T^0	$T^{-1/6}$
$^2\Pi$ dipole-$^2\Pi$ dipole	$T^{-1/6}$	$T^{-1/6}$
$^2\Pi$ dipole-$^1\Sigma$ quadrupole(j=0)	$T^{1/4}$	T^0
$^2\Pi$ dipole-$^1\Sigma$ quadrupole(j≠0)	T^0	T^0

The coefficients of the rate constants depend on the properties of the molecular partners such as dipole and quadrupole moments and, in some cases, rotor constants. More accurate rotationally adiabatic capture calculations[11,13] that do not use perturbation theory have demonstrated that these rate constant formulae are only accurate for very low temperatures, typically less than 3K. However, for temperatures of interest to interstellar chemistry up to 20K, they still provide quite useful estimates.

Experimental measurements have shown that the rotationally adiabatic capture theory works well for ion-molecule reactions for temperatures down to 20K[20], and it would be very useful to have experimental results for even lower temperatures to test the theoretical predictions. Preliminary measurements of this kind are now being reported[21]. In the case of neutral reactions, there have been some very promising recent experimental results down to quite low temperatures[22]. The best test of the theoretical predictions will be for reactions between highly polar molecules with large rotor constants. It would also be of interest if the experimental measurements could be done for the reactions of ions with $^2\Pi$ molecules such as OH at low temperatures so that our theoretical prediction of a $T^{-1/2}$ dependence in the rate constant can be examined.

RATE CONSTANTS FOR REACTIONS THAT GO VIA A SCATTERING RESONANCE

There are several reactions of neutral molecules of importance to interstellar chemistry that have unexpected temperature dependencies in the experimentally measured rate constant. For example, the

$$OH + CO \rightarrow CO_2 + H$$

reaction has quite a small rate constant[23] at room temperature (10^{-13} cm^3 s^{-1} molecule^{-1}) that remains constant for temperatures as low as 80K. With a small rate constant such as this, the Arrhenius formula for the rate constant

$$k(T) = A \exp(-\frac{\Delta E}{k_B T}) \qquad (14)$$

might be expected to be applicable, but this is clearly not the case. Many reactions, including OH+CO, are known from spectroscopic measurements to have long-lived complexes as intermediates[24] and this raises the question as to whether excited states of these complexes above the dissociation limit, which are called resonances in the language of scattering theory, can influence significantly the reaction cross sections and rate constants. This would seem to be the case for the OH+CO reaction as quantum reactive scattering calculations[25,26], using an *ab initio* potential energy surface[27], show that the reaction probabilities are very small except at resonance energies corresponding to excited vibrational states of the HOCO complex formed during the reaction. This has been found to be the case for quantum scattering models that explicitly treat both two[25] and three[26] degrees of freedom.

The *ab initio* potential energy surface for the OH+CO→CO$_2$+H reaction has small barriers in both the entrance and exit channels and shows a narrow bottleneck that leads to reaction[27]. It is known that the exit channel barrier is a little too large to give realistic rate constants[26,28], but the potential still serves as a useful guide in examining the dynamics of a reaction that goes via scattering resonances. Figure 3 presents the calculation of the reaction probabilities for energies close to threshold for a two-degree of freedom model that explicitly accounts for the OH vibration in the entrance channel and a local CO vibration in the exit channel.

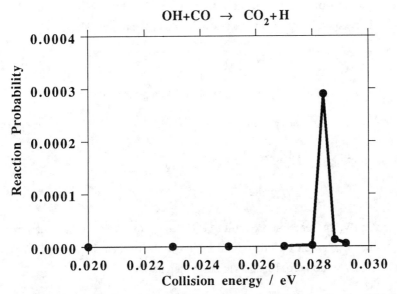

Figure 3. Plot of quantum mechanical reaction probabilities for the OH+CO reaction for low collision energies.

A particular feature of this figure is that there is a scattering resonance at an energy very close to threshold. It is interesting to examine the consequences of this for the rate constant at low temperatures and an analytical formula is now derived.

The reaction cross sections for a chemical reaction for a given initial state is given by

$$\sigma(E) = \frac{\pi}{2\mu E} \sum_J (2J+1) P^J(E) \tag{15}$$

where J is the total angular momentum and P^J is the probability for reaction at collision energy E. For a reaction that only goes via a resonance, this probability will be very small for all values of J except for the one that tunes into the resonance. The effective centrifugal term in the potential

$$\frac{J(J+1)}{2\mu R^2} \tag{16}$$

can be written approximately as $BJ(J+1)$ as the excited states of the collision complex will ensure[26] very restricted values for the average of the wavefunction over $1/R^2$. If a single resonance occurs at an energy E_r with a reaction probability P for an energy E and J=0, then the reaction probability will only be significant for J values which satisfy the relationship

$$E - BJ(J+1) = E_r \tag{17}.$$

If E_r is very close to threshold then we have, to a good approximation,

$$J(J+1) = E/B \tag{18}.$$

Approximating $J(J+1)$ as J^2 for large J gives

$$J = (E/B)^{1/2} \tag{19}.$$

Accordingly, the reaction probability $P^J(E)$ is set to zero for all J except for the value of J given by this equation at which it takes the value P. This enables the reaction cross section to be written as

$$\sigma(E) = \frac{\pi P}{B^{1/2} \mu E^{1/2}} \tag{20}.$$

The reaction rate constant is then obtained by Boltzmann averaging to give

$$k = \frac{\pi P \, 2^{1/2}}{B^{1/2} \mu^{3/2}} \tag{21}.$$

The most interesting feature of this formula is that the rate constant is temperature independent. This suggests that rate constants obtained from measurements on this type of reaction that show no temperature dependence might also be applicable at the very low temperatures of interstellar clouds. This is true even though the rate constant can be quite small as *only one partial wave* contributes to the cross section for each value of the energy E.

Of course this analysis involves several approximations, including neglecting the possibility of more than one resonance and the effect of internal states. A realistic treatment of reactions such as these probably requires accurate quantum scattering calculations on an accurate *ab initio* potential energy surface. However, calculations such as these are very time consuming and difficult to carry out, especially for the many reactions of interest to interstellar chemistry. Therefore, our simple analysis is useful in giving some justification to the extrapolation of experimentally measured rate constants to lower temperatures for certain kinds of reaction.

CONCLUSION

The modelling of interstellar clouds requires a large database of rate constants for a wide variety of chemical reactions at low temperatures. The rate constant formulae derived or discussed here enable useful estimates to be easily made with no computation for a wide variety of reactions. In several cases, these rate constants should serve as upper bounds to the exact values[6] and, therefore, their use in interstellar models should enable predictions to be made on the maximum effect a certain chemical reaction can have. Theoretical data for ion-molecule reactions has already been used widely in such models and this paper also demonstrates that some neutral reactions might also be important in interstellar clouds.

ACKNOWLEDGEMENTS

The work on the reactions of carbon atoms was done in collaboration with Dr David Husain of this laboratory. The formulae for the rates of dipolar molecules were derived in collaboration with Mr Andrew Wickham and Dr Thierry Stoecklin. This work was supported by the Science and Engineering Research Council.

REFERENCES

1. T.J. Millar and D. A. Williams, eds., Rate Coefficients in Astrochemistry, (Kluwer, 1988).
2. See articles by B. R. Rowe, p135, and I. W. M. Smith,, p103, in Ref 1.
3. D. C. Clary, D. Smith and N. G. Adams Chem. Phys. Lett., 119, 320 (1985); D. C. Clary, C. E. Dateo and D. Smith, Chem. Phys. Lett., 167, 1 (1990).
4. N. Haider and D. Husain, J.Chem.Soc. Faraday Trans. 89, 7 (1993); D. Husain, J. Chem. Soc. Faraday Trans., 89, 2164 (1993)
5. D. C. Clary, N. Haider, D. Husain and M. Kabir, Ap. J., in press.
6. D. C. Clary, Ann. Rev. Phys. Chem., 41, 61 (1990) ; D. C. Clary, Ref 1, P1.
7. J. O. Hirschfelder , C. F. Curtiss and R. B. Bird, Molecular Theory of Gases and Liquids, (John Wiley, New York, 1954).
8. K. G. Denbigh , Trans. Faraday Soc., 36, 936 (1940).
9. D. R. Lide , ed., Handbook of Chemistry and Physics(CRC, Boca Raton, 1992).
10. J. Tröe, J. Chem. Phys., 87, 2773 (1987).
11. A. G. Wickham, T. Stoecklin, and D. C. Clary, J. Chem. Phys., 96, 1053 (1992).
12. A. G. Wickham and D. C. Clary, J. Chem. Phys., 98, 420 (1993).
13. D. C. Clary, T. S. Stoecklin and A. G. Wickham, J. Chem. Soc. Faraday Trans., 89, 2185 (1993); T. Stoecklin, C. E. Dateo and D. C. Clary, J. Chem. Soc. Faraday Trans., 87, 1667(1991).
14. D. C. Clary, Molec. Phys., 53, 3 (1984); D. C. Clary, Molec. Phys., 54, 605 (1985); D. C. Clary and J. P. Henshaw, Faraday Discuss. Chem. Soc., 84, 333 (1987).
15. K. Sakimoto and K. Takayanagi, J. Phys. Soc. Jpn, 48, 2076 (1980).
16. D. R. Bates and W. L. Morgan, J. Chem. Phys., 87, 2611 (1987).
17. J. Turulski and J. Niedzielski, Chem. Phys., 137, 191(1989).
18. M. L. Dubernet and R. McCarroll, 1989, Z. Phys. D., 13, 255 (1989); M. L. Dubernet, M. Gargaud and R. McCarroll , Astron. and Astrophys., 259, 373 (1992).
19. N. Markovic, G. Nyman and S. Nordholm, Chem. Phys. Lett., 159, 435 (1989).

20. B. R. Rowe, Ref 1, p135; C. Rebrion, J. B. Marquette, B. R. Rowe and D. C. Clary, Chem. Phys. Lett., 143, 130 (1988).
21. T. L. Mazely and M. A. Smith, Chem. Phys. Lett., 144, 563 (1988).
22. B. R. Rowe, A Canosa and I. R. Sims J. Chem. Soc. Faraday Trans., 89, 2193 (1993).
23. M. J. Frost, P. Sharkey and I. W. M. Smith, Faraday Discuss. Chem. Soc., 91, 305(1991).
24. N. F. Scherer, L. R. Khundkar, R. B. Bernstein and A. H. Zewail, J. Chem. Phys., 87, 1451 (1987).
25. G. C. Schatz and J. Dyck, Chem. Phys. Lett., 188, 11 (1992).
26. D. C. Clary and G. C. Schatz, J. Chem. Phys., in press (1993).
27. G. C. Schatz, M. S. Fitzcharles and L. B. Harding, Faraday Discuss. Chem. Soc., 84, 359 (1987).
28. K. Kudla, G. C. Schatz and A. F. Wagner, J. Chem. Phys., 95, 1635 (1991).

DISCUSSION

ZIURYS — Do you expect reactions of open-shell species with ground electronic states other than $^2\Pi$ states to have the same $T^{-1/2}$ rate dependence ?

CLARY — Molecules in Σ states will not have the $k(T) \to T^{-1/2}$ dependence as $T \to 0$, but molecules in Δ states will.

OMONT — Could you comment upon reaction rates of polar neutral molecules with neutral long chains ?

CLARY — If the long chain is polar then the rates might have the temperature dependences discussed in my paper for reactions between polar molecules. If the chain is not polar, the rates might be similar to those for the $C(^3P)$ atoms with alkenes and alkynes.

NENNER — You have predicted, for C + hydrocarbons, a $N^{1/3}$ dependence of the rate. What is your prediction for $N > 16$. Is there any asymptote ?

CLARY — The experiments cannot be done for $N > 16$ so we do not know what the asymptote is.

GUELIN — What would be the upper limit at low temperatures for the rate of a reaction between two (moderately - e.g. few D) polar neutral radicals ? few 10^{-10} cm^3 s^{-1} ?

CLARY — This has been discussed by T. Stoecklin and D.C. Clary, J. Phys Chem., 96 (1992) 7346. In this case, the rate constants could be as high as 10^{-} cm^3 s^{-1}.

Mc CARROL — Your predicted temperature dependence for neutral-neutral reactive rate constants are really quite different from those measured by E Rowe and co-workers? There is no way a long-range multiple interactions can explain the observed $1/T$ dependences. Is there some additional mechanism occurring which your model does not account for ? What do you think ?

CLARY — We do not consider non-adiabatic collisions and electronic degeneracy effects (e.g. of the OH molecule), these could play a role.

FIELD — Is it possible to calculate the rate coefficients of singlet excited C [^{1}D (say)] since these could be important, perhaps, in astrophysics or in planetary atmospheres ?

CLARY — I would expect these rates to be similar to those for C(^{3}P) when these reactions are very fast.

REACTIONS BETWEEN NEUTRAL SPECIES WITHOUT A SIGNIFICANT BARRIER: THE C + NO AND C + N$_2$O REACTIONS

M. Costes, C. Naulin, N. Ghanem and G. Dorthe
Laboratoire de Photophysique et Photochimie Moléculaire, URA 348 du CNRS,
Université Bordeaux I, 33405 Talence Cedex, France

ABSTRACT

The relative reactive cross-sections of the neutral reactions C + NO → CN + O and C + N$_2$O → CN + NO have been determined in crossed molecular beams. Both cross-sections are found to decrease when the relative translational energy of reactants is increased in the ranges experimentally scanned, i.e. 0.038 - 0.163 eV for C + NO and 0.035 - 0.130 eV for C + N$_2$O. Low energy barriers, if any, may be inferred from the data.

INTRODUCTION

The chemical reactivity in interstellar medium is usually described through mechanisms where ionic species play the major role. However, it clearly appears now that a better fit to experimental observations can be obtained by including contributions of neutral species reactions which were previously discarded or ignored. In particular, it could be the case for the C + NO and C + N$_2$O reactions, as far as the CN production is concerned.

Observations of CN in diffuse clouds have been reported by Dickman et al.[1] and Federman et al..[2] They showed that CN is primarily produced by neutral-neutral reactions N + CH → CN + H and N + C$_2$ → CN + C. This was confirmed by Viala[3], who showed that including contributions of all possible ion-molecule reactions in his model could not account to the actual CN yield.

Other neutral-neutral reactions can yield CN. In particular, the title reactions might be important, due to the relative abundance of neutral atomic carbon in diffuse clouds.

EXPERIMENTAL

The experimental set-up has been detailed elsewhere.[4] Main characteristics only are recalled here. Oxidant-molecule beam is obtained from a first pulsed nozzle, whereas ground-state carbon-atom beam is produced by laser-ablation from a graphite rod at the exit of a second pulsed nozzle. Both reactant molecular beams, truly supersonic, cross at right angles in a vacuum chamber.

Under such conditions, the internal energy of reactants is negligible. The collision energy thus reduces to the relative translational energy of reactants, ε_{tr}, which is selected by adjusting reactant velocities, v_C and v_{NxO} (x = 1 or 2):

$$\varepsilon_{tr} = \tfrac{1}{2} \mu \left(v^2_C + v^2_{NxO} - 2 v_C v_{NxO} \cos\alpha \right) \quad (1)$$

Here, μ stands for the reactant reduced mass, and α for the beam crossing angle. The beam velocity depends on mean molecular weight, mean heat capacity and initial temperature of the expanding gas. Adjusting the mean molecular weight by changing either the carrier gas (He, Ne, Ar, Kr) or the dilution (generally referred to as seeding technique), allows a wide range of relative translational energy values to be scanned.

CN produced in its ground state $X^2\Sigma^+$ by reactive collisions of C with NO or N_2O is probed at the beam crossing point by laser-induced fluorescence (LIF) on its $B^2\Sigma^+$-$X^2\Sigma^+$ vibronic transitions, on $\Delta v = 0$ and $\Delta v = 1$ sequences, respectively near 388 and 422 nm.

RESULTS AND DISCUSSION

Excitation spectra of the $CN(X^2\Sigma^+)$ product obtained for the C + NO (at $\varepsilon_{tr} = 0.08$ eV) and C + N_2O (0.26 eV) reactions are given in Figures 1 and 2.

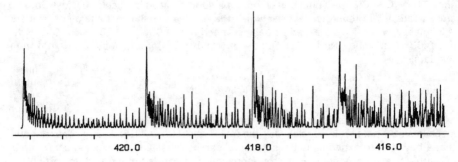

Fig. 1. Excitation spectrum of CN produced by the reaction C + NO → CN + O at 0.08 eV. $B^2\Sigma^+$-$X^2\Sigma^+$ transitions of the $\Delta v = -1$ sequence.

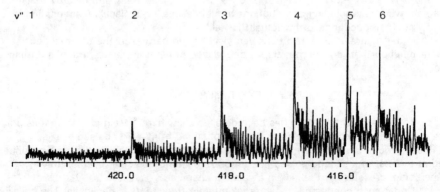

Fig. 2. Excitation spectrum of CN produced by the reaction C + N_2O → CN + NO at 0.26 eV. $B^2\Sigma^+$-$X^2\Sigma^+$ transitions of the $\Delta v = -1$ sequence.

The strong rovibrational excitation results in prominent bandheads for the P branches. However, the two spectra are markedly different. The vibrational excitation extends up to v" = 4 with a monotonically decreasing distribution for the C + NO reaction[5]; it extends up to v" = 6 or 7 with a bell-shaped distribution for the C + N_2O reaction[6].

For a given relative velocity of reactants, u_r, the reaction rate is given by

$$d[CN]/dt = \sigma \, u_r \, [C] \, [N_xO], \qquad (2)$$

where [M] stands for the density of the relevant species. The reactive cross-section, σ, depends on the translational energy. In particular, the determination of this dependence, $\sigma = f(\varepsilon_{tr})$, referred to as the excitation function, allows the energy threshold for reactive collisions to be derived.

Under the present experimental conditions, the CN flux, $d[CN]/dt$, was derived from the LIF probing of CN.

The C atoms and N_xO species were not probed. Nonetheless, for each N_xO-beam condition, reference CN data were recorded at high collision energy, i.e. entraining C atoms with helium. Assuming that the variation of σ is negligible at high energy in the range scanned while varying the N_xO beam allows for N_xO-beam density calibration (with C in He, ε_{tr} ranges from 0.20 to 0.29 eV for extreme NO-beam conditions). Once this calibration had been achieved, it was then possible to lower collision energies using heavier carrier gases for C beam. Although there is no calibration procedure for C atom densities, [C] remains constant for constant beam conditions, i.e. the same carrier gas and the same delay between the pulsed valve and the ablation laser triggerings. For a set of experiments with variable N_xO-beam and constant C-beam conditions, it was thus possible to obtain $\sigma \, [C] = (d[CN]/dt)/(u_r \, [N_xO])$ as a function of collision energy over the range scanned when varying N_xO beam conditions. This procedure was repeated for different C-beam conditions to broaden the energy range scanned, yielding a set of relative excitation functions, $\sigma \, [C] = f(\varepsilon_{tr})$, for each reaction.

However, all these curves reveal a similar ε_{tr} dependence, which can be described by the following function, derived from the simple model of Levine and Bernstein,[8]

$$\sigma = \sigma_0 \, \{\varepsilon_{tr} / |\Delta\varepsilon_0|\}^{-2/s} \qquad (3)$$

where s characterises the reactant long-range interaction potential. Fitting the experimental excitation functions with a constant s value allows the σ_0 to be determined, and hence the normalised data, σ/σ_0, to be plotted as a function of ε_{tr} at the same scale. This yields the excitation functions displayed in Fig. 3 (for C + NO) and 4 (C + N_2O). This procedure has been described for the C + NO reaction.[7]

The C + NO data were fitted using s = 6, corresponding to a dispersion (London) interaction potential. Note that such a s = 6 value results in a 10 % variation of σ in the 0.20 - 0.29 eV range spanned in the NO-density calibration procedure, which justifies a posteriori the assumption made (see above). No threshold energy can be pointed out from the data, since σ is always decreasing when ε_{tr} is increased from the lower (0.038 eV) to the higher (0.16 eV) value scanned. However, introducing a barrier, ε_{thr}, in the model results in the following excitation function:

$$\sigma_R = \sigma_0 \, \{\varepsilon_{tr} / |\Delta\varepsilon_0|\}^{-2/s} \, \{1 - \varepsilon_{thr}/\varepsilon_{tr}\} \, ; \text{ for } \varepsilon_{tr} \geq \varepsilon_{thr}$$
$$\sigma_R = 0 \, ; \text{ for } \varepsilon_{tr} < \varepsilon_{thr} \qquad (4)$$

This function appears as solid lines in Fig. 3 for several ε_{thr} values, the maxima of which lying at:

$$\varepsilon_{tr} = (1 + s/2) \, \varepsilon_{thr} \qquad (5)$$

i.e. $\varepsilon_{tr} = 4 \, \varepsilon_{thr}$ for s = 6.[7] If the maximum would occur at the lower limit of the data ($\varepsilon_{tr} = 0.038$ eV), the maximum compatible threshold would be $\varepsilon_{thr} < 0.010$ eV.

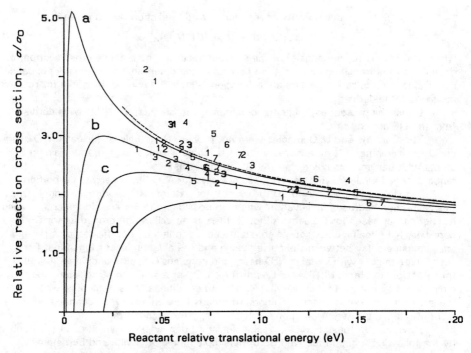

Fig. 3. Normalised excitation function of the reaction C + NO → CN + O. Data point labelling corresponds to different C-atom beam conditions; dashed line curve, eqn. (3) with s = 6; solid line curves, eqn. (4) with s = 6 and ε_{thr} = 0.001 (a), 0.005 (b), 0.010 (c) and 0.020 eV (d).

The C + N_2O data were obtained with a signal-to-noise ratio for CN detection considerably lower than in the C + NO case. Indeed, the detection efficiency is lowered because (i) the CN flux is spread over a much wider set of internal energy states, and (ii) the product separation velocities (in the laboratory frame) can reach much higher values due to the higher reaction exoergicity. It is worth recalling that the LIF signal is proportional to the product density, which do not actually correspond to the flux and furthermore depends on the velocity in the laboratory frame: clearly, products recoiling at low velocities are detected more efficiently than those recoiling at high velocities.

The data, which exhibit a greater scatter especially at low collision energies, (Fig. 4) could only be fitted with s = 2. Such a s value corresponds to an ion - permanent dipole interaction, which seems meaningless in this case. Using a similar procedure as for C + NO (eqn (5) with s = 2) would allow a maximum threshold value, ε_{thr} < 0.02 eV, to be inferred.

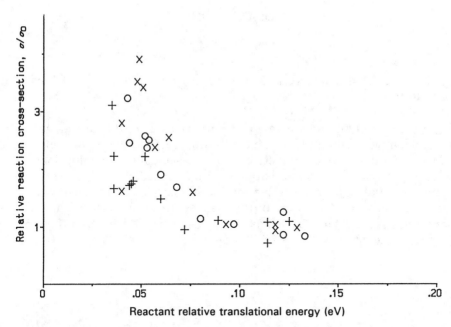

Fig. 4. Normalised excitation function of the reaction C + N_2O → CN + NO. Data point symbols correspond to different C-atom beam conditions.

REFERENCES

1. R. L. Dickman, W. B. Sommerville, D. C. Whittet, D. Mc Nally and J. C. Blades, Astrophys. J. Suppl. Ser. 53, 55 (1983).
2. S. R. Federman, A. C. Danks, D. L. Lambert, Astrophys. J. 287, 219 (1984).
3. Y. P. Viala, Astron. Astrophys. Suppl. Ser. 64, 391 (1986).
4. M. Costes, C. Naulin, G. Dorthe, G. Daleau, J. Joussot-Dubien, C. Lalaude, M. Winckert, A. Destor, C. Vaucamps And G. Nouchi, J. Phys. E. Sci. Instrum., 22, 1017 (1989).
5. C. Naulin, M. Costes and G. Dorthe, Chem. Phys. 153, 519 (1991).
6. M. Costes, C. Naulin, G. Dorthe and Z. Moudden, Laser. Chem. 10, 367 (1990).
7. M. Costes, C. Naulin, N. Ghanem and G. Dorthe, J. Chem. Soc. Faraday Trans. 89, 1501 (1993).
8. R.D. Levine and R.B. Bernstein, Molecular Reaction Dynamics and Chemical Reactivity (Oxford University Press, N. Y., 1987) pp. 58-61.

DISCUSSION

<u>LEACH</u> — At the fluences you use there must be also creation of ions. What happens to them ?

<u>COSTES</u>— Most of the ions created are neutralised by the carrier gas in the extension channel of the pulsed valve and afterwards in the supersonic expansion. Neutralization is not complete but if necessary, the remaining ions can be easily removed from the beam by applying an electric field between two deflections plates situated immediately after the extension channel. However, this procedure is only used for time-of-flight measurements performed with fast-ionisation gauges. So far, no difference has been noticed in the LIF signals of reaction products, with the electric field on or off.

PRELIMINARY EXPERIMENTAL AND THEORETICAL RESULTS ON THE DYNAMICS OF THE REACTION N + CH → CN + H

G. Dorthe*, P. Caubet*, N. Daugey*,
M. T. Rayez**, J. C. Rayez** and P. Halvick**.
Photophysique et Photochimie Moléculaire, URA 348 (CNRS)*
And Physico-Chimie Theorique, URA 503 (CNRS)**,
Université Bordeaux I 33405 Talence, France.

P. Millié
Physique Générale, CEN Saclay, 91191 Gif-Sur-Yvette, France

B. Levy
Physico-Chimie des Rayonnements, URA 75 (CNRS),
Université Paris-Sud, 91405 Orsay, France.

ABSTRACT

Owing to its exoergicity, the reaction N + CH → CN + H might produce CN in 3 electronic states: $X^2\Sigma^+$, $A^2\Pi_i$ and $B^2\Sigma^+$. However, the correlation diagram rules out the production of $B^2\Sigma^+$ state for any symmetry of the reaction complex, either linear or planar. It has been checked experimentally. A vibrational population inversion in $A^2\Pi_i$ state has been observed. The excitation limit at v' = 16 agrees with the total energy available to the products. Theoretical calculations found no potential energy barrier as expected for an exoergic atom + radical reaction.

INTRODUCTION

Observation of CN in diffuse clouds have been reported by Dickman et al.[1] and Federman et al.[2] The latter discussed the CN chemistry and concluded that CN was primarily produced by neutral-neutral reactions of N with both CH and C_2. More recently, Viala[3] confirmed this result since all ion-molecule reactions which could finally lead to CN were included in his model for diffuse clouds and were found to have negligible contribution. Moreover, he found that the N + CH and N + C_2 reactions were also the source of CN in dense dark clouds.

Although a single experimental determination of the room temperature rate constant of the N + CH reaction has been performed from reactant decay[4], no experimental observation of the expected products has been reported. Owing to its strong exoergicity, the reaction could produce CN in 3 electronic states, $X^2\Sigma^+$, $A^2\Pi_i$ and $B^2\Sigma^+$. It was thus very appealing to study the dynamics of such a reaction, both experimentally and theoretically.

EXPERIMENT

A fast-flow reactor at room temperature was used. $CH(X^2\Pi_r)$ was produced through successive abstractions of Br atoms in $CHBr_3$ with H atoms, which result in the overall process $3H + CHBr_3 \rightarrow CH + 3HBr$. H and N were obtained from the microwave dissociation of H_2 and N_2 diluted in He. $CHBr_3$, H_2 and N_2 pressures were typically 1 mTorr while He carrier gas pressure ranged from 0.5 to 4 Torr.

The energy available to the products is the sum of the exoergicity, taken between zero-point energy levels of reactants and products (4.30 eV), the mean thermal energy of reactants (0.06 eV) and the activation energy for the involved pathway. For a negligible value of the latter, the energy available sets an excitation limit at $v'=5$ in $CN(B^2\Sigma^+)$ and $v'=16$ in $CN(A^2\Pi_i)$. Such excitation limits could be easily checked from chemiluminescence spectra.

Origin of $CN(B^2\Sigma^+)$ chemiluminescence.

Only emission from vibrational level $v'=0$ was detected, with an unusual band contour since the typical monotonous envelope of a thermalized rotational structure was not observed. This feature has been cleared up at a spectral resolution allowing to distinguish rotational lines. At a low carrier gas pressure (Fig. 1a), a set of intense double lines appears superimposed over weaker single lines. The double lines arise from the interaction of zero-order, near-resonant, rotational levels of $A^2\Pi_i, v'=10$ and $B^2\Sigma^+, v'=0$. These interactions result in series of actual two rotational levels of mixed $A^2\Pi_i$ and $B^2\Sigma^+$ character instead of the fictive zero-order single levels with pure $B^2\Sigma^+$ character. The rotational level retaining a major $B^2\Sigma^+$ character provides a rotational line appearing as a perturbed line in the sequence of true $B^2\Sigma^+$ rotational lines. The rotational level with a major $A^2\Pi_i$ character nevertheless provides, thanks to its small $B^2\Sigma^+$ character, a rotational line appearing as an extra-line in the sequence of regular rotational lines of the vibronic transition. Despite much lower transition probabilities, these extra-lines appear more intense than perturbed lines and even more intense than pure $B^2\Sigma^+$ rotational lines. This shows that the rovibronic levels corresponding to the extra-lines are much more populated than the levels corresponding to other lines. An increase in the carrier-gas pressure (Fig. 1b) results in an enhancement of perturbed and regular rotational lines demonstrating the filling of the corresponding levels by collisional transfer from extra-lines levels. This shows that the reaction $N + CH$ does not produce CN in its $B^2\Sigma^+$ state and that the observed emission from $B^2\Sigma^+, v'=0$ stems from the filling of that state by collisional transfer from $A^2\Pi_i, v'=10$.

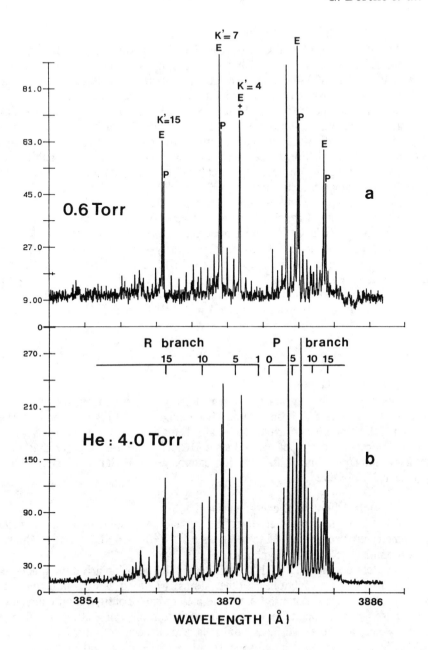

FIGURE 1. $CN(B^2\Sigma^+)$ chemiluminescence.

$CN(A^2\Pi_i)$ vibrational distribution.

An intense chemiluminescence from $CN(A^2\Pi_i)$ was observed throughout the visible. It extends in the infrared but our detector was only sensitive for wavelengths smaller than 800 nm so that vibrational distribution could be determined for $v' \geq 4$. Overlapping bands prevented to determine the distribution directly from the chemiluminescence spectrum. Synthetic spectra were used with a routine adjusting the rotational temperature, the ratio between 1/2 and 3/2 components of $A^2\Pi_i$ state and the vibrational distribution until identity between experimental and calculated spectrum was obtained. For the calculation of rovibronic transition probabilities, we introduced the Einstein coefficients calculated by Knowles et al.[5] The spectral detectivity was determined between 400 and 800 from the reference spectral distribution of the chemiluminescence $O + NO + M \to NO_2 + h\upsilon$ [6].

The chemiluminescence spectrum between 500 and 600 nm is given in Fig. 2 since it shows clearly the emission from the highest vibrational levels. In particular, the unoverlapped (15-8) band is clearly seen as the weak (16-9) band superimposed on the tail of the overlapping (11-5) and (6-1) bands. The vibrational distribution is given in Fig. 3. The highest populated level is $v'=16$, in agreement with a negligible activation energy. The hollow in the distribution at $v'=10$ confirms the collisional transfer between $A^2\Pi_i, v'=10$ and $B^2\Sigma^+, v'=0$.

THEORY

Correlation diagram

The correlation diagram is given in Fig. 4. Energy values for exoergicities of different pathways are those obtained from experimental dissociation energies of CH[7] and CN.[8] Energies of triplet HCN are theoretical values given by Schwenzer et al. with respect to singlet ground-state HCN.[9] It appears clearly that for any symmetry, planar or linear, the reaction cannot produce CN in its $B^2\Sigma^+$ state as experimentally observed.

$^3A'$ and $^3A''$ potential-energy surfaces

As seen in the correlation diagram, $^3A'$ and $^3A''$ potential-energy surfaces of C_s symmetry connect ground-state reactants to $CN(^2\Sigma^+)$ and $CN(A^2\Pi_i)$. They must play a prominent role for the dynamics.

Pople's double zeta 6-31** basis set (including p polarisation functions on hydrogen atoms and d functions on other atoms) was used. As the number of open electronic shells varies along the pathways a large multiconfiguration calculation is required to get a comparable evaluation of electronic correlation all along the surface. Potential energy surfaces points have then been computed at the valence CAS/MCSCF level in which only the 1s orbitals of carbon and nitrogen are kept double occupied. All the geometries of stationary points (saddle-points and minima) have been fully optimized at the valence CAS/MCSCF level. The pathways have been obtained keeping fixed either CN or CH distance, the other degrees of freedom being optimized at the same level of theory

Figures 5a and 5b show the relative energies corresponding to the

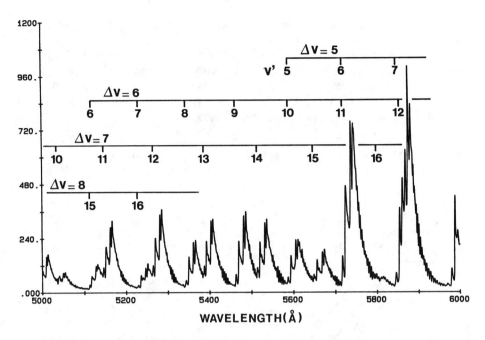

FIGURE 2. $CN(A^2\Pi_i)$ chemiluminescence.

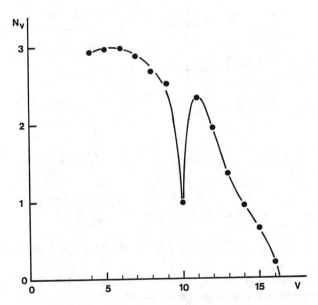

FIGURE 3. $CN(A^2\Pi_i)$ vibrational distribution

characteristic features of the A' and A" surfaces. The results can be summarized as follows:
- The reactions issued from reactants in their ground state and leading to $CN(X^2\Sigma^+)$ and excited $CN(A^2\Pi_i)$ are both highly exoergic and their exoergicities are in a reasonable agreement with experiment: respectively, -4.13 eV (exp. -4.30 eV) and -2.79 eV (exp. -3.15 eV). Therefore, these reactions can lead to vibrational excitation of CN in both electronic states.

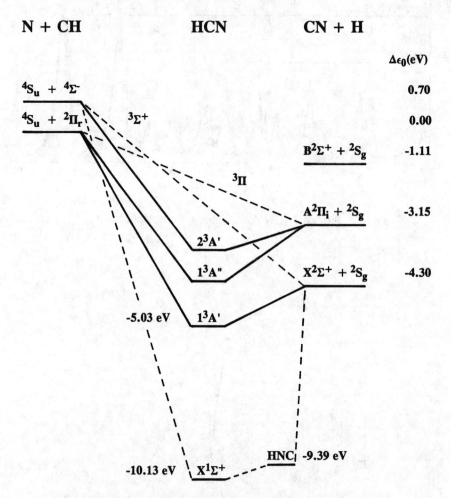

FIGURE 4. Correlation diagram of the N + CH → CN + H reaction.

- Both surfaces, $^3A'$ and $^3A''$ have no barrier in the entrance valley (the HCN angle being optimized all along the reaction path)
- Each surface has two wells, one for the HCN isomer (HCN($^3A'$): r_{CN}=1.315Å, r_{CH}=1.116Å, HCN=2.3°; HCN($^3A''$): r_{CN}=1.332Å,

$r_{CH}=1.134$Å, HCN$=118.9°$) and the other for HNC (HNC(^{3}A'): $r_{CN}=1.284$Å, $r_{NH}=1.051$Å, HNC$=110.2°$; HNC(^{3}A"): $r_{CN}=1.416$Å, $r_{NH}=1.050$Å, HNC$=110.4°$). The energetics are shown in Fig 5a and 5b. The triplet isomers HCN and HNC are separated by a barrier, of energy far below the energy of the reactants.

- The potential wells ^{3}A' and ^{3}A" associated with HNC and HCN are respectively related to the products, via a col, the height of which is mentionned on Fig.5a and 5b.

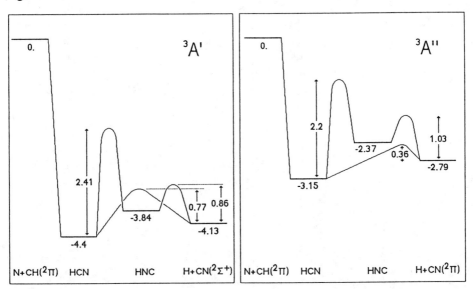

Fig. 5a. Relative energies for stationary points on ^{3}A' potential energy surface.

Fig. 5b. Relative energies for stationary points on ^{3}A" potential energy surface.

REFERENCES

1. R. L. Dickman, W. B. Somerville, D. C. B. Whittet, D. Mc Nally and J. C. Blades, Astrophys. J. Suppl. Ser. 53, 55 (1983).
2. S. R. Federman, A. C. Danks, D. L. Lambert, Astrophys. J. 287, 219 (1984).
3. Y. P. Viala, Astron. Astrophys. Suppl. Ser. 64, 391 (1986).
4. I. Messing, S. V. Filseth, C. M. Sadowski and T .Carrington, J. Chem. Phys. 74, 3874 (1981).
5. P. J. Knowles, H. J. Werner, P. J. Hay and C. Cartwright, J. Chem. Phys. 89, 7334 (1988).
6. G. R. Bradburn and H. V. Lilenfeld, J. Phys. Chem. 92, 5266 (1988).
7. H. Okabe, Photochemistry of Small Molecules (John wiley & Sons, N. Y., 1978), p. 193.
8. M. Costes, C. Naulin and G. Dorthe, Astron. Astrophys. 232, 270 (1990).
9. G. M. Schwenzer, S. V. O'Neil, H. F. Shaeffer III, C. P. Baskin and C. F. Bender, J. Chem. Phys. 60, 2787 (1974).

ASTROPHYSICALLY IMPORTANT REACTIONS INVOLVING EXCITED HYDROGEN

J.M.C. Rawlings
Department of Mathematics, UMIST, Manchester, UK

J.E. Drew
Department of Physics, Astrophysics, University of Oxford, Oxford, UK

M.J. Barlow
Department of Physics and Astronomy, University College London, London, UK

ABSTRACT

The associative ionization reaction $H(n=2) + H \rightarrow H_2^+ + e^-$ is found to be a greater contributor to the H_2 formation rate than the direct radiative association reaction $H(n=2) + H \rightarrow H_2 + h\nu$ in most regions of astrophysical interest. The endothermicity ($\simeq 1.1 eV$) of the reaction and the high departures from LTE that are required for the H I (n=2) level to be sufficiently populated restrict its significance to regions of high excitation. The reaction $H(n=2) + H^+ \rightarrow H_2^+ + h\nu$ may be significant in highly excited ionized regions, such as planetary nebulae and shocks. Chemical models of circumstellar regions have been reassessed in the light of this information. A critical examination reveals that excitation effects are, in general, very important in many astrophysical situations. Only exceptionally, will reactions involving the higher excited states (n>2) be as significant as those involving $H(n=2)$.

INTRODUCTION

The chemistry of circumstellar and nebular regions, where temperatures and densities are high and the radiation field strong, differs greatly from that which is applicable to the general interstellar medium. In particular, the formation of H_2 would be expected to be severely inhibited. Yet there are examples of such environments where it has been detected (e.g. in certain planetary nebulae).

The well-established pathways to H_2 formation in the gas phase are as follows. First, there is the direct three-body reaction:

$$H + H + H \longrightarrow H_2 + H \qquad (1)$$

For this reaction to be significant the density must typically be greater than about 10^{11} cm^{-3}. At lower densities and moderate ionization levels a much more efficient route is via the H^- ion:

$$H + e^- \longrightarrow H^- + h\nu \qquad (2)$$

$$H^- + H \longrightarrow H_2 + e^- \qquad (3)$$

However, the H^- ion can be lost by mutual neutralisation and is susceptible to photodetachment by the radiation field even at near-infrared wavelengths.

Finally, as a 'last resort' there is the much less efficient H^+ route:

$$H^+ + H \longrightarrow H_2^+ + h\nu \qquad (4)$$

$$H_2^+ + H \longrightarrow H_2 + H^+ \qquad (5)$$

Again, the intermediate, H_2^+ is susceptible to photodissociation and dissociative recombination.

Previously, it has not been common practise to take into proper account the internal excitation of the reacting chemical species. If the relevant downward transitions are highly optically thick then trapping may occur (as is often the case for H I Lyα) and the higher states may be well populated. A critical examination reveals that excitation effects are, in general, very important in many astrophysical situations.

RADIATIVE ASSOCIATION

Latter and Black[1] proposed that H_2 formation may occur as a result of direct radiative association;

$$H + H(n = 2) \longrightarrow H_2 + h\nu \qquad (6)$$

The rate coefficient for this reaction was calculated to be $1.2/1.1 \times 10^{-14}$ cm^3 s^{-1} at 1000/10000 K for the 2s channel and $3.4/4.2 \times 10^{-14}$ cm^3 s^{-1} at 1000/10000 K for the 2p channel. Latter and Black investigated the recombination era in the early Universe ($z > 1000$), diffuse HII regions and cool protostellar winds (as modelled by Glassgold et. al.[2,3]). Only in the first of these did the new process make a significant difference to the production of H_2.

ASSOCIATIVE IONIZATION

Associative ionization is the most elementary collisional process by which molecular bonds can be formed. The cross-section for the associative ionization reaction:-

$$H + H(2s) \longrightarrow H_2^+ + e^- \qquad (7)$$

has been measured by Urbain et al.[4]. The measured threshold for this reaction is about 1.1 eV. The 2p channel cannot, as yet, be measured directly owing to the short radiative lifetime of the state. We have assumed that the sum of the 2s and the 2p processes results in a smooth, featureless cross-section.

The calculated rate coefficient for the reaction (from Rawlings, Drew and Barlow[5]) are given in Table 1 for the 2s channel using the measured cross-sections, for the 2p channel using our inferred cross-sections and for the n=2 weighted mean cross-section (assuming statisitical weight populations).

The temperature dependences of the rate coefficients are smooth and fits to them are given by:

$$k_{2s} = 1.16\times10^{-10} T^{-0.07} e^{-16941/T} \text{ cm}^3 \text{ s}^{-1} \qquad (3800 \leq T \leq 21800 \text{ K})$$

$$k_{2p} = 1.19\times10^{-8} T^{-0.5} e^{-19065/T} \text{ cm}^3 \text{ s}^{-1} \qquad (3400 \leq T \leq 22000 \text{ K})$$

$$k_{2s,p} = 2.41\times10^{-9} T^{-0.35} e^{-17829/T} \text{ cm}^3 \text{ s}^{-1} \qquad (3000 \leq T \leq 22000 \text{ K})$$

These are accurate to within 10% over the specified temperature ranges.

Table 1: Values of the rate coefficient (cm^3 s^{-1}) for the associative ionization reaction for several temperature values

Temperature (K)	H(2s)	H(2p)	H(2s,p)
3000	3.1×10^{-13}	4.6×10^{-13}	4.2×10^{-13}
5000	2.1×10^{-12}	3.7×10^{-12}	3.3×10^{-12}
10000	1.11×10^{-11}	1.78×10^{-11}	1.61×10^{-11}
15000	1.69×10^{-11}	2.73×10^{-11}	2.51×10^{-11}

The reaction is followed by charge exchange (re. 5) to form H_2. Although this route again involves the unstable intermediate H_2^+, the measured cross-sections are substantially larger than those for the radiative association reaction. Hence, except at lower temperatures (<6000 K) this reaction is of wider significance.

Associative ionization is unlikely to compete with the negative ion route in circumstances in which H^- easily survives. However, there are situations in which associative ionization is favoured over the positive ion radiative association route. Over the probable temperature range of interest, $5000 \lesssim T \lesssim 10000$ K, n_2/n_+ must be in excess of 1–3 $\times 10^{-5}$ for associative ionization to dominate.

At the densities typical of the situations where H_2 formation is an issue, large departures from LTE are therefore required before associative ionization dominates. Such environments are typically less than half-ionized, cooler than 10000 K and with a high Lyman-α optical depth.

EXCITED HYDROGEN CHEMISTRY

If a true assessment is to be made of the importance of these reactions and related pathways, we must consider the effects that the excitation of molecular as well as atomic species has on the chemistry.

To describe a pure hydrogen gas experiencing a significant degree of excitation in a non-equilibrium environment, the effect of radiation transport in at least a two-level atom must be treated. The generation and transfer of H I Lyα and two-photon emission must be included, both in the treatment of the hydrogen atom and in the photochemistry of the more complex molecular species. Listed below are some of the reactions whose high energy behaviour has been analysed in some detail as part of these investigations.

- $H + H^+ \longrightarrow H_2^+ + h\nu$
- $H^- + e^- \longrightarrow H + e^- + e^-$
- $H_2(v'' \geq 0) + h\nu \longrightarrow H + H$
- $H_2^* + h\nu \longrightarrow H_2^+ + e^-$
- $H_2^{+*} + h\nu \longrightarrow H^+ + H$
- $H_2^{+*} + e^- \longrightarrow H + H$
- Collisional dissociations of CO, H_2, N_2, C_2, O_2 and other molecules which have a small or zero dipole moment

FURTHER REACTIONS INVOLVING EXCITED ATOMIC HYDROGEN

The corresponding associative ionization reaction for H(n=3)

$$H + H(n=3) \longrightarrow H_2^+ + e^- \qquad (8)$$

is exothermic by $\simeq 1.14$eV. However, the n=3 level is $\simeq 1.89$ eV above n=2 and will have a correspondingly lower population in nearly all regions of interest. Calculations of the cross-section by X.Urbain (private communication) imply a rate-coefficient of;

$$k \simeq (2.1 \times 10^{-7}/T)e^{-7137/T} \text{ cm}^{-3} \text{ s}^{-1} \quad \text{for } 4400 \leq T \leq 22000 \text{ K}$$

Over the temperature range 4500–8000 K this is never more than about four times as large as the corresponding rate for the n=2 reaction and to exceed the corresponding n=2 rate, marked population inversion is required. The reaction:

$$H(n=3) + H \longrightarrow H_2 + h\nu \qquad (9)$$

may also need to be considered in certain circumstances, but as yet no data are available.
Urbain, Cornet & Jureta[6] have measured the cross-section for the reaction:

$$H(2s) + H(2s) \longrightarrow H_2^+ + e^- \qquad (10)$$

while Poulaert et al.[7] have measured, and Urbain et al.[8] have calculated, the cross-section for the reaction:

$$H^+ + H^- \longrightarrow H_2^+ + e^- \qquad (11)$$

The values are found to be some 10 to 100 times as large as the calculated values for the H(n=2) associative ionization reaction at the collision energies of interest. However, it is not expected that these reactions will be important since they will be in direct competition with other reactions discussed above.

Of greater potential interest is the excited state analogue of the radiative association reaction (4):

$$H^+ + H(n=2) \longrightarrow H_2^+ + h\nu \qquad (12)$$

Preliminary estimates indicate that k$\sim 4 \times 10^{-13}$cm^3 s^{-1} in the temperature range of interest, in which case the reaction may be significant in regions of high ionization (such as recombination zones, old PN etc.).

APPLICATIONS TO ASTROPHYSICAL MODELS

Excited hydrogen chemistry has so far been considered only in immediate circumstellar environments, focussing on the winds of young stellar objects[5]:

1. winds from massive young stellar objects

2. T Tauri winds

3. cool neutral outflows

We are currently working on a variety of other environments[9] including;

1. old planetary nebulae

2. supernovae (and remnants)

3. shocks

4. the recombination epoch of the early Universe

The basic details of the T Tauri and cool neutral outflow models can be found in Rawlings, Williams & Canto[10] and Glassgold, Mamon & Huggins[3]. Particular attention is paid to the radiative transfer and the relevant continuum opacities, and their effects on the photorates. Because of the high H(n=2) population in the regions of interest, the time-dependence of the Balmer continuum opacity is especially important.

1. Winds from massive young stellar objects.

   The winds associated with massive YSOs are likely to be warm out to large radii, and are certainly dense and fairly well-ionized. The low outflow velocities ($\sim$100 km s^{-1}) and high mass loss rates can yield significant Balmer as well as extreme Lyman continuum optical depth in the outflow. In these circumstances both the H(n=2) population and the H:H$^+$ ratio will be enhanced. In addition, many molecular species (e.g. H$_2$ and, to an extent, H$_2^+$) are protected against photodissociation.

   The model assumes spherical symmetry and uses a velocity law following Castor & Lamers[11]. We use the wind temperature profile based on the OB star wind models of Drew[12] (fit presented in Bunn & Drew[13]). The emergent radiative energy distribution of the YSO was represented by the calculated spectrum from a Kurucz[14] model atmosphere.

   The Balmer continuum opacity is extremely sensitive to the details of the velocity law as well as to mass loss rate. Results indicate that for high mass loss rates ($\sim 10^{-5}$ M$_\odot$ yr^{-1}) and slow wind accelerations the associative ionization is the dominant H$_2$ formation channel out to about 80R$_*$ despite the rapid photodissociation rate for H$_2^+$. In these circumstances a high H(n=2) population can be maintained over a large radius range. Reaction (6) accounts for less than 18% of the H$_2$ formation rate throughout the wind. The H$_2$ abundance peaks at 1.4×10^{-7} at a radius of about 4.3R$_*$. This is a factor of about six times larger than would be the case if the associative ionization reaction (7) were omitted from the chemistry.

2. T Tauri winds.

   T-Tauri winds are both hot (10^4 K) and excited (with an excitation/ionization temperature peaking at $\geq$ 15000 K). The model of Rawlings et al.[10] has been adapted to accommodate the new chemistry. The physical basis of this model is described in Hartmann, Edwards & Avrett[15]; Terminal velocity and maximum temperatures are reached at about 3-5 R$_*$. The mass loss rate is small ($\simeq 5 \times 10^{-9}$ M$_\odot$ yr^{-1}) and the cooling rate is large. The Balmer continuum optical depth is therefore low at all times.

Associative ionization is found to have a marginal significance and only within $0.1R_*$ of the starting radius ($5R_*$). The reason for this is straightforward – because of the high excitation temperature and the (relatively) low densities, H_2^+ is more rapidly destroyed by the IR radiation field than by charge exchange to form H_2. Direct excited state radiative association (6) is more important and results in a significant enhancement of H_2 (by a factor of about 7 within a few R_*).

3. Cool neutral outflows.

In the dense, cool, neutral winds associated with low mass YSOs, such as SVS 13/HH 7-11, there is no significant Lyman continuum radiation. Thus, in the inner wind, where Lyα is extremely optically-thick, detailed balance ($b_1 = b_2$) holds.

In order to test the importance of the associative ionization reaction and the other reactions discussed above we have investigated a model for the outflow from SVS 13 that is much like that (Case 2) considered by Glassgold, Mamon & Huggins[3].

At the lower temperatures (5000 K) H(n=2) reactions tend to be unimportant, since detailed balance holds; but at 7500 K the H(n=2) fractional abundance is over three orders of magnitude higher at radii where Lyα remains in detailed balance ($b_1 = b_2 \simeq 90$) and a large Balmer continuum optical depth accumulates. In this setting the associative ionization H_2 formation route is dominant out to about $35R_*$, where the H^+ route takes over (re. 4,5). The H_2 abundance is higher by a factor of about 3 (within $20R_*$) as compared to the cooler model. This higher H_2 abundances may be sustained in the outer parts of the wind if a realistic cooling law is employed - essentially "freezing out" the chemistry.

The outflow from SVS 13 is apparently too cool for excited state routes to H_2 to be significant. However, the strong temperature dependence of the associative ionization implies that it is important in slightly warmer environments.

REFERENCES

1. Latter, W.B. and Black, J.H., Astrophys. J. **372**, 161 (1991)
2. Glassgold, A.E., Mamon, G.A. and Huggins, P.J., Astrophys. J. **336**, L29 (1989)
3. Glassgold, A.E., Mamon, G.A. and Huggins, P.J., Astrophys. J. **373**, 254 (1991)
4. Urbain, X., Cornet, A., Brouillard, F. and Giusti-Suzor, A., Phys. Rev. Lett. **66**, 1685 (1991)
5. Rawlings, J.M.C., Drew, J.E. and Barlow, M.J., Mon. Not. of the Roy. Ast. Soc. (in press) (1993)
6. Urbain, X., Cornet, A. and Jureta, J., J. Phys. B. **25**, L189 (1992)
7. Poulaert, G., Brouillard, F., Claeys, W., McGowan, J.W. and Van Wassenhove, G., J. Phys. B. **11**, L671 (1978)
8. Urbain, X., Giusti-Suzor, A., Fussen, D. and Kubach, C., J. Phys. B. **19**, L273 (1986)
9. Barlow, M.J., Rawlings, J.M.C. and Drew, J.E., Mon. Not. of the Roy. Ast. Soc. (to be submitted)
10. Rawlings, J.M.C., Williams, D.A. and Canto, J., Mon. Not. of the Roy. Ast. Soc. **230**, 695 (1988)

11. Castor, J.I. and Lamers, H.G.J.L.M., Astrophys. J. Supp. **39**, 481 (1979)
12. Drew, J.E., Astrophys. J. Supp. **71**, 267 (1989)
13. Bunn, J.C. and Drew, J.E., Mon. Not. of the Roy. Ast. Soc. **255**, 449 (1992)
14. Kurucz, R.L., *Stellar Atmospheres: Beyond Classical Models*
    (NATO ASI series C Vol. 341, Kluwer Academic Publishers, 1991) p.441
15. Hartmann, L., Edwards, S. and Avrett, E., Astron. J. **261**, 279 (1982)

EXPERIMENTAL STUDIES OF GAS-PHASE REACTIONS AT EXTREMELY LOW TEMPERATURES.

B.R. Rowe, I.R. Sims, and P. Bocherel.
Département de Physique Atomique et Moléculaire, URA 1203 du CNRS,
Université de Rennes I, Campus de Beaulieu, 35042 RENNES Cedex, FRANCE.

I.W.M. Smith.
University of Birmingham, School of Chemistry, Edgbaston,
Birmingham B15 2TT, UNITED KINGDOM

ABSTRACT

The use of uniform supersonic flows as chemical reactors (the technique known as CRESU) has been extremely successful for the study of ion-molecule reactions at ultra-low temperatures. Recently a new apparatus, designed to allow kinetic experiments on neutral-neutral reactions, has been commissioned at Rennes University. This paper describes some of the features of these new CRESU facilities, together with the most recent results obtained on reactions of CN, OH and CH radicals with various molecules.

INTRODUCTION

In the early seventies the first gas phase models[1-4] highlighted the key role of ion-molecule reactions in the synthesis of molecules in interstellar clouds. Since that time it has become apparent that a number of more complex processes cannot be ignored in the pathways leading to molecule formation.[5] These include grain surface chemistry followed by desorption, as well as the effects of heating by shock waves and turbulence. Astrochemists have certainly not suffered from a lack of imagination in this field, although many of the data concerning the elementary processes that are invoked owe more to guesswork than to laboratory studies at the relevant temperatures. The merit of the so-called gas phase models is that the data that they require are clearly identified and can be obtained in the laboratory or by theoretical calculations. From year to year these models have been refined, and now include hundreds or thousands of reactions with more than a hundred species.[6] Very few neutral-neutral reactions have generally been included in these models, partly due to the common prejudice that many of them will be less efficient in the ultra-low temperature environment of the clouds. However, some authors[7] have pointed out the critical importance of a few neutral-neutral reactions for the predicted abundances of several molecules.

To place our understanding of interstellar chemistry on a firm basis, it is clear that the various processes included in the models have to be studied at ultra-low temperatures. Such measurements are very challenging from an experimental point of view but the last decade has seen remarkable breakthroughs in the field of ion-molecule reactions for which there now exist a bulk of data at temperatures well below that of liquid nitrogen.[8] These results have been obtained using a variety of techniques some of which rely heavily on the ease with which ions can be manipulated by electrical and magnetic fields. Of this kind are the cryogenically cooled static drift tube of Bohringer and Arnold[9] as well as ion traps.[10,11] These two techniques are mainly restricted to molecular hydrogen as neutral reactant since most other species would readily condense on the ultra-cold walls of the apparatus. The merged-beam apparatus of Gerlich[11] also uses fields to merge an ionic and a neutral beam in conditions of very low centre-of-mass-energy. Such an experiment provides cross sections which can be integrated over a Boltzmann energy distribution in order to obtain a rate coefficient. Another kind of experiment uses supersonic expansion to cool down a buffer gas at ultra-low temperature. The free-jet flow reactor of M. Smith[12] uses an expansion of the kind widely used in molecular spectroscopy. The strong density gradient exhibited by the flow is a serious disadvantage of the technique when it has to be applied to rate coefficient measurements. Also in such expansions the velocity and internal state distributions are no longer thermal and in many cases the rotational distribution is unknown, a problem which also occurs in the merged-beam experiment. In this connection it is important to note that some theoretical studies[13] indicate that the rate constants for fast reactions between neutral species, as well as for ion-molecule reactions, at ultra-low temperatures depend strongly on the rotational states of the reagents. The CRESU (Cinétique de Réaction en Ecoulement Supersonique Uniforme, or Reaction Kinetics in Uniform Supersonic Flow) experiment first devised by Rowe and co-workers[14] has the great advantage of achieving fully thermal conditions. It uses the uniform flow generated by a specially contoured Laval nozzle as a flow reactor. The uniformity of the flow makes rate coefficient measurements especially simple since the density changes of the various species are only linked to chemical reactions. Using this technique a large number of ion-molecule reactions has been studied. It has been shown[15] that normally the rate coefficient is very close to that calculated assuming capture due to the long-range part of the potential. The versatility of the CRESU technique in using very condensable species has been extremely useful in obtaining a wealth of data for reactions of ions with polar molecules. From these results it can be concluded that the rate coefficients at ultra-low temperature for most ion-molecule reactions can be predicted by theory with a good degree of confidence.

In contrast to ion-molecule reactions there exist very few data on neutral-neutral reactions below 200 K.[16] Recently a new CRESU apparatus has been built in Rennes with one chamber devoted to neutral chemistry. It has been designed in order to allow the use of the powerful pulsed laser photolysis (PLP), laser-induced fluorescence (LIF) technique,[17] widely used previously by two of the present authors at Birmingham University. Numerous reactions of CN, OH and CH radicals with various molecules have already been studied, demonstrating the strength of the technique. The aim of the present paper is to describe the CRESU facilities at Rennes and to review the results obtained. The technique employed in the neutral-

neutral experiments has recently been described in full,[18] and only the main features and special points will be outlined below. Several papers[18-21] giving measured rate coefficients for different reactions have been published or will be submitted. Therefore the detailed discussion of the results presented can be found elsewhere and the principal purpose of the present review is only to show that the rates of neutral-neutral reactions without activation energy often increase as the temperature is lowered.

EXPERIMENTAL

The various kinds of supersonic expansion that can be used for studies in chemical physics are shown in figure 1. Although it is clear from this figure, it has to be emphasised that the nature of the CRESU flow is very different from that in a molecular beam: in the first case, the flow is in the continuous regime and, even though at the lowest temperature (high mach numbers) atoms or molecules in the flow travel at essentially the same velocity, there is always a large number of collisions, ensuring local thermodynamic equilibrium; in the second case the flow is free of molecular collisions, equilibrium is not established, and the notion of temperature is meaningless. As a consequence the experiments that can be conducted with each system are very different in principle. Shown in figure 1 is one of the way in which the uniform CRESU flow has been used for ion-molecule studies: a high energy electron beam ionises species in the flow and reactant and product ions are detected further downstream using a mass spectrometer. The rate coefficients are deduced using standard flow reactor analysis.

The CRESU facilities at Rennes have been designed to allow studies of either ion-molecule or neutral-neutral reactions. In figure 2 a general sketch shows the two chambers dedicated respectively to each kind of experiment. They can be pumped by several large roots pumps yielding pumping speeds of ca. 19,500 m^3/h for argon and nitrogen and ca. 22,500 m^3/h for helium. Two additional roots pumps can separately evacuate the ion-molecule chamber upstream of the mass spectrometer, which has been designed with a supersonic diffuser. In principle, this arrangement permits one to work with a volume flow rate larger than the total pumping speed, due to the pressure rise at the diffuser entrance. This ion-molecule apparatus is not yet operational.

The neutral-neutral apparatus is also shown in figure 2. On a moveable reservoir various nozzles can be easily flange-mounted, allowing numerous flow conditions (temperature and pressure) to be obtained. It is also possible to work at room temperature by increasing the pressure in the chamber to such a value that a standing shock wave develops in the nozzle, resulting in a subsonic flow of low mach number. The reservoir and the nozzle can be cooled with liquid nitrogen in order to obtain an extremely low temperature at a moderate mach number. Two pulsed, Nd:YAG-pumped dye laser beams are combined on a dichroic mirror and co-propagate along the axis of the flow. The first, photolysis laser pulse produces a given density of radicals by photodissociation of a suitable precursor. A second

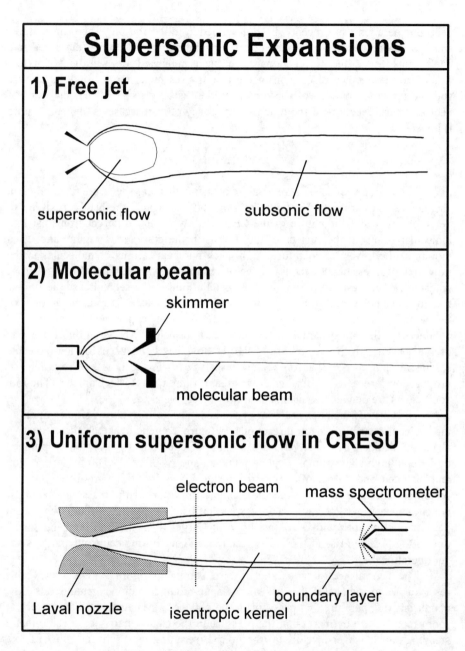

Fig.1: various kinds of supersonic expansions.

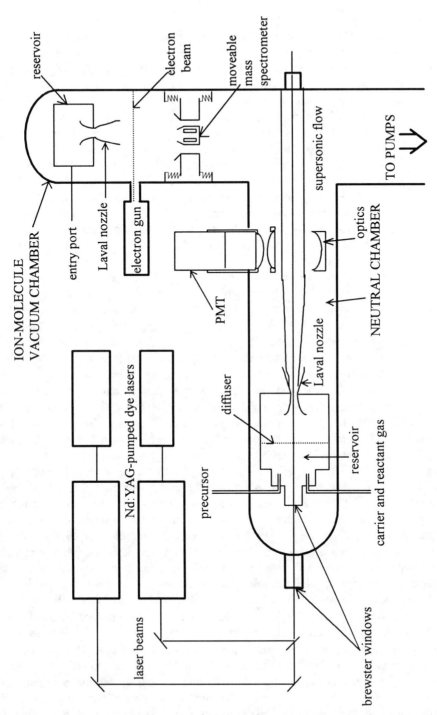

Fig 2: CRESU apparatus at RENNES University.

"probe" laser pulse, arriving after a given delay, enables the detection of the radicals by laser-induced fluorescence. The buffer, reactant and precursor gases are introduced together in the reservoir. The fluorescence light is gathered at a given distance (4–45cm) downstream of the nozzle exit with highly efficient optics and directed onto a photomultiplier tube. With the PLP-LIF technique the supersonic flow is used only to provide constant ultra-low temperature and constant density for a sufficient time delay between the two laser pulses. Figure 3 shows a typical variation of the LIF signal versus time delay in absence of reactant. The initial nearly constant plateau corresponding to the uniform part of the flow is followed by a peak resulting from photolysis within the throat of the Laval nozzle. By varying the nozzle-photomultiplier distance, this maximum can be used for time of flight measurement of the flow velocity. Excellent agreement was found with the value deduced from impact pressure measurements; however the flow velocity is not very sensitive to the mach number and is not used as a parameter in the rate coefficient calculations. A full LIF spectrum was most valuable as a check of the rotational equilibrium and as a measurement of the temperature.

Three radicals have been studied to date in the CRESU neutral-neutral apparatus: CN, CH and OH. CN was produced rotationally and vibrationally cold by the near-threshold photolysis of NCNO at 582 nm, and detected by off-resonance LIF in the $CN(B^2\Sigma^+ - X^2\Sigma^+)$ system, exciting in the (0,0) band at ca. 388 nm and detecting fluorescence from the (1,0) band at ca. 420 nm. OH was produced by 266 nm photolysis of H_2O_2 and detected by off-resonance LIF in the $OH(A^2\Sigma - {}^2\Pi)$ system, exciting at ca. 282 nm in the (0,1) band and detecting at ca. 310 nm in the (1,1)/(0,0) bands. CH was generated by 266 nm sequential multiphoton photolysis of $CHBr_3$ and detected by off-resonance LIF in the $CH(A^2\Delta - X^2\Pi)$ system, exciting at ca. 430 nm in the (0,0) band and detecting at ca. 490 nm in the (1,0) band. For each transition rotationally resolved spectra have been obtained. Figure 5 shows typical spectra obtained for the CN radical, which is an excellent "thermometer" with a rotational line spacing of 5.4 K.[22] The details of the spectroscopic analysis can be found elsewhere:[18] it was shown that the spectra correspond to a Boltzmann distribution of the rotational level populations corresponding to a temperature in very good agreement with the value deduced from impact pressure measurements performed using a Pitot tube, confirming the excellent uniformity of the flow, as can be seen in figure 4. It is quite interesting to note that there is now a substantial body of results, using either LIF[18] or electron-beam-induced fluorescence,[14] which all confirm the validity of the isentropic hypothesis, used for the analysis of the impact pressure measurements, as well as for the calculation of the core of the flow.

In contrast to the $CN(X^2\Sigma^+)$ radical which can be produced from NCNO without excess of internal energy, the photolysis of H_2O_2 at 266 nm leads to $OH(X^2\Pi)$ fragments with considerable excess rotational energy. This is shown in figure 6 which compares spectra taken at 23 K in He buffer for different time delays between the two laser pulses. It can be seen that the initial rotational distribution is very "hot" but that relaxation is rapid, leading to a very simplified spectrum at this temperature, corresponding essentially to one single rotational level. Recording on

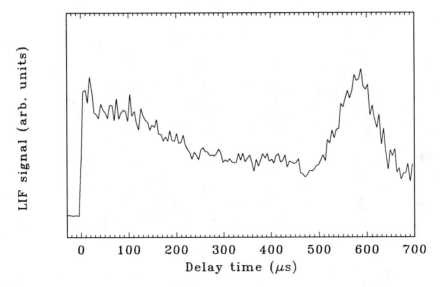

Fig.3: variation of the CN LIF signal versus time delay in absence of reactant.

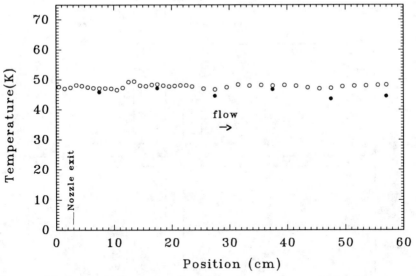

Fig.4: comparison of temperature measurements calculated by an aerodynamic method (open circles) and by relative line intensities of the CN spectra (closed circles).

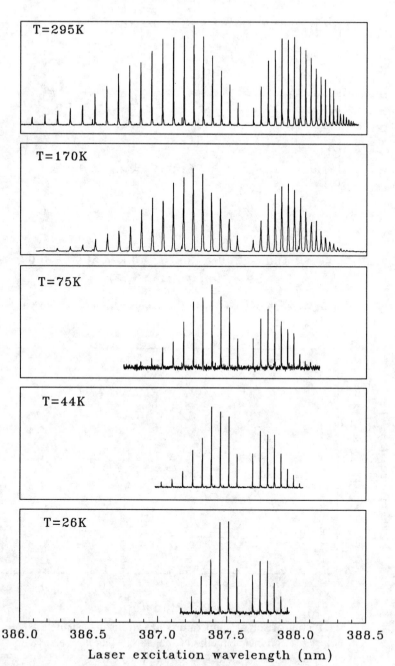

Fig.5: LIF spectra of CN from different nozzles with temperature calculated by an aerodynamic method.

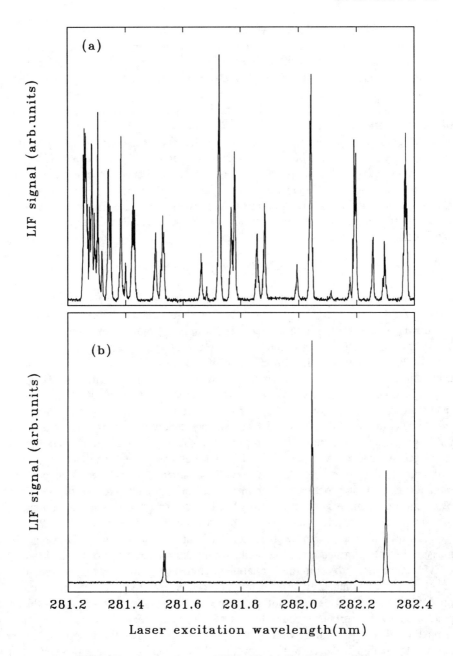

Fig.6: OH LIF spectra taken at 23K in He buffer for two different time delays:
(a) delay = 0.1µs; (b) delay = 40µs.

one of the "cold" lines of figure 6 versus the time delay yields the exact time delay after which reaction kinetic measurements can be performed. It has to be emphasised that in contrast with ion-molecule reactions, where the total ion density is measured by a mass spectrometer, the relative radical density in the present PLP-LIF measurement is inferred from a single rotational line. It is therefore essential that rotational thermalisation prevails in the flow in order to avoid the convolution of relaxation and reaction kinetics.

When a neutral reactant is added to the flow the LIF signal decreases exponentially, as shown in figure 7(a) for the reaction of CN with C_2H_2. This exponential decay yields a pseudo-first-order rate coefficient. The values of this coefficient obtained at different flow rates of the added reactant are plotted against reactant concentration as shown in figure 7(b). The slope of the linear plot obtained yields the second-order rate coefficient of the reaction. Details of the error analysis can be found elsewhere.[18]

RESULTS AND DISCUSSION

Using the Rennes CRESU apparatus the bimolecular reactions of the CN radical with O_2, NH_3, C_2H_2, C_2H_4 and C_2H_6 have been studied down to 13 K for O_2 and down to 25 K for the other species. The results are summarised in figures 8 and 9 together with other measurements performed at Birmingham University. The bimolecular reactions of OH with HBr and of CH with O_2, NH_3, NO and D_2 have also been studied. Figure 10 shows the results of the measurements for CH which should be considered as preliminary: the scatter in some of these data at the lowest temperature, although not very large, has not yet been satisfactorily explained. It should be noted that the values which appear slightly low were obtained in the argon buffer gas, which is the most likely to give problems with dimerisation. Data for the termolecular reaction of OH with NO and butenes have also been obtained. Although studies of termolecular association may improve our understanding of the radiative association processes which may occur in interstellar clouds, this last kind of reaction is much less significant than fast bimolecular reactions in the context of interstellar chemistry and results will not be presented in this paper.

Figures 8 and 9 show the excellent agreement obtained between the measurements performed at Rennes and Birmingham on reactions of the CN radical with O_2 and hydrocarbons in the overlapping temperature range. The other reactions have not been studied previously below room temperature. The determination of product branching ratios was not performed in the present experiments: this is possible only in certain favourable cases using LIF techniques. At very low temperatures the reactions would not proceed at a significant rate for any appreciably endothermic channel, restricting the number of possible products. For example in the $CH + D_2$ reaction the only products expected on energetic grounds are CD + HD. Previous work at room temperature can give some insight into the nature of the products, although when several exothermic channels exist the branching ratios could change with temperature.

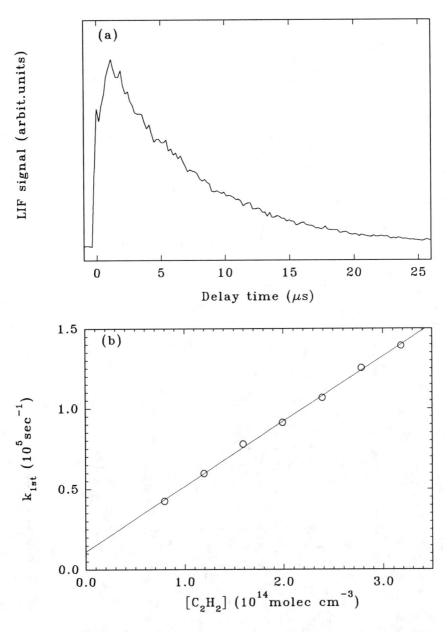

Fig.7: (a) First-order decay of CN LIF in the presence of C_2H_2 at 25K.
(b) First-order decay constants plotted against the concentration of C_2H_2 at 25K.

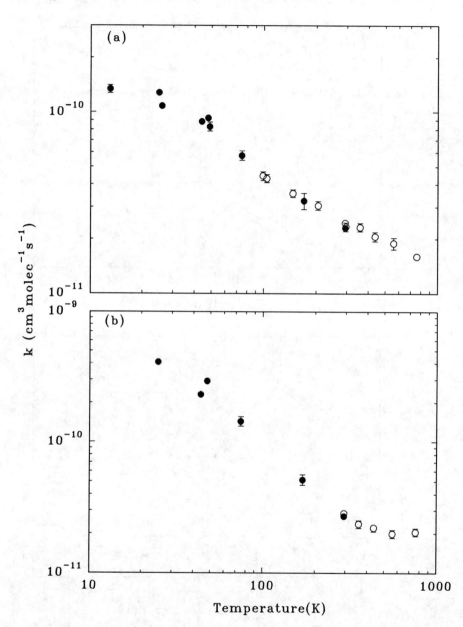

Fig.8: rate constant as a function of temperature for: (a) $CN+O_2$; (b) $CN+NH_3$. The open circles represent Birmingham measurements and closed circles represent CRESU measurements.

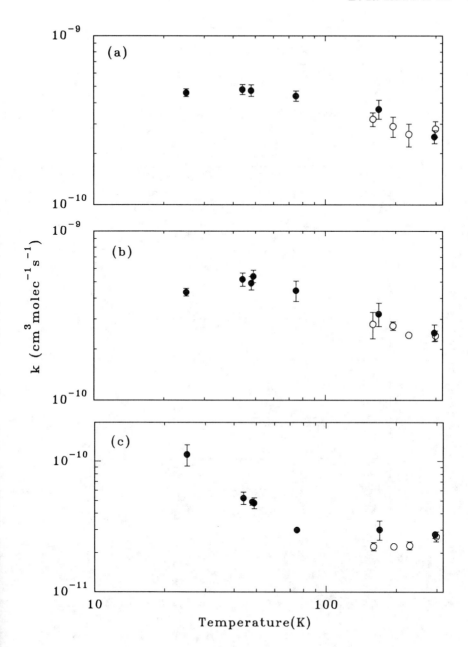

Fig.9: rate constant as a function of temperature for:(a) CN+C_2H_2;(b) CN+C_2H_4; (c) CN+C_2H_6. The open circles represent Birmingham measurements and closed circles represent CRESU measurements.

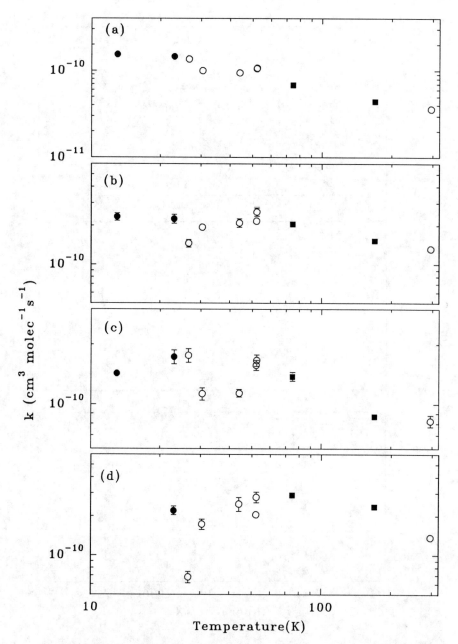

Fig.10: rate constant as a function of temperature for:(a) $CH+O_2$;(b) $CH+NO$; (c) $CH+D_2$;(d) $CH+NH_3$. The open circles represent measurements in Ar, the closed circles measurements in He and the squares measurements in N_2.

The possibility that the measured rate-constants result from the onset of a termolecular associative channel to form stabilised van der Waals complexes at the lowest temperature may be ruled out. Measurements were performed at various total densities and with different buffer gases in order to check the bimolecular nature of the reactions. Furthermore, stabilised van der Waals complexes are more likely to form either between the radical species and the argon buffer gas which is present at a much greater concentration, resulting in a rapid decrease in radical concentration even in the absence of reactant. No evidence of this was ever observed.

It is clear from the results presented that widely varying behaviour of thermal rate constants with temperature is observed at low and ultra-low temperatures for the reactions studied, and no general rule can be proposed. The interesting case of the $CN + C_2H_6$ reaction illustrates that rate coefficients at the lowest temperatures cannot be inferred from behaviour at higher temperatures. Such a temperature dependence is not easy to explain and could involve the formation of a *transient* van der Waals complex.[20] Clearly there is a need for advances in theory in order to understand reaction behaviour at ultra-low temperatures and, eventually, to be able to make quantitative predictions for the temperature dependence. Those reactions whose rate coefficients increase or remain approximately constant as the temperature is decreased to extremely low values, must be occurring over potential energy surfaces without significant barriers. Such reactions have recently attracted much theoretical attention.[23-25] Attempts have been made to calculate the rate coefficients of the $CN + O_2$, $OH + HBr$, and $CH + D_2$ reactions at ultra-low temperatures. Calculations by Clary and co-workers[23] use various forms of adiabatic capture theory, assuming that the reaction rates are equal to the capture rates calculated using the long-range part of the potential between the molecular reactants. In the case of ion-molecule reactions, this kind of calculation has given rate coefficients in good agreement with theory.[15] For the neutral-neutral reactions investigated in the present study satisfactory agreement is found only for the $CH + D_2$ reaction for which theoretical data exist below 40 K. For $OH + HBr$ the experimental values remain well below the calculated ones[23] which do not reproduce the strong negative temperature dependence found in the experiments.[21] In this case it seems that capture calculations provide only an upper limit for the rate constant. However, for the $CN + O_2$ reaction, the experimental value of the rate coefficient is larger than the theoretical one at temperatures below 50 K. Also the capture rate exhibits a positive temperature dependence instead of the experimental $T^{-0.63}$ dependence. Recent calculations by Klippenstein and Kim[25] using variational transition state theory reproduce quite well the observed behaviour at temperatures over 50 K. These results suggest that the behaviour of neutral-neutral reactions in the 13–300 K temperature range cannot be fully reproduced using only the long-range intermolecular potential and that shorter range, chemical forces are important.

ASTROCHEMICAL CONCLUSION

For twenty years gas-phase models of interstellar chemistry have relied heavily on ionic chemistry to explain the formation of molecular species observed in the interstellar medium. Although these models have included a few neutral-neutral reactions, the results presented above show that the role of neutral chemistry in interstellar molecule synthesis has hitherto been underestimated. Several of the stable neutral reactants in this study are known to be abundant in interstellar clouds and recent observations[26] have shown that this is the case for molecular oxygen. Coupled with the findings reported here, this implies that reactions of such species with CN and CH cannot be neglected, underlining the importance of these measurements. Indeed, following on from the recent studies on neutral-neutral reactions at ultra-low temperatures Herbst *et al.*[27] have recently shown that the inclusion of several fast neutral-neutral reactions in their model has a dramatic effect on the predicted abundances of a number of molecules. However, owing to a lack of experimental data, these new predictions have to rely on a large number of assumptions regarding unmeasured rate coefficients and product branching ratios at ultra-low temperatures. This clearly shows that a wide range of experimental and theoretical quantum chemical studies are urgently needed in this field.

[1] E. Herbst and W. Klemperer, Astrophys. J. **185**, 505, (1973).

[2] P.M. Solomon and W. Klemperer, Astrophys. J. **178**, 389, (1972).

[3] J.H. Black and A. Dalgarno, Astrophys J. **184**, L101, (1973).

[4] A. Dalgarno and J.H. Black, Rep. Prog. Phys. **39**, 573, (1976).

[5] D.A. Williams, 1988, in Rate Coefficients in Astrochemistry, eds T.J. Millar and D.A. Williams, Kluwer Academic Publishers, see also the present proceedings.

[6] E. Herbst and C. M. Leung, Astrophys J. **310**, 378, (1986).

[7] T.J. Millar, C.M. Leung and E. Herbst, Astron. Astrophys., **231**, 466, 1990.

[8] V.G. Anicich, Astrophys. J. Suppl. Ser. **84**, 215, (1993).

[9] H. Bohringer and F. Arnold, J. Chem. Phys. **77**, 5534, (1982).

[10] S.E. Barlow, G.H. Dunn and M. Schauer, Phys. Rev. Lett. **52**, 902, (1984).

[11] D. Gerlich, J. Chem. Soc. Far. Trans., **89**, 2199, (1993).

[12] M. Hawley, T.L. Mazely, L.K. Randeniya, R. S. Smith, X.K. Zeng and M.A. Smith., Int. J. Mass Spectr. Ion. Proc. **97**, 55, (1990).

[13] D. C. Clary, Mol.Phys., **53**, 3, (1984).

[14] B.R. Rowe, G. Dupeyrat, J.B. Marquette and P. Gaucherel, J. Chem. Phys. **80**, 4915, (1984).

[15] C. Rebrion, J.B. Marquette and B.R. Rowe, Chemistry and Spectroscopy of Interstellar Molecules, Edited by D.K. Bohme, E. Herbst, N. Kaiju and S. Saito, University of Tokyo Press (1993).

[16] R. Atkinson, D.L. Baulch, R.A. Cox, R.F. Hampson, J.A. Kerr and J. Troe, J. Phys. Chem. Ref. Data, **18**, 881, (1989).

[17] I.R. Sims and I.W.M. Smith, J. Chem. Soc. Faraday Trans. 2, **84**, 527, (1988).

[18] I.R. Sims, J.L. Queffelec, A. Defrance, C. Rebrion-Rowe, D. Travers, P. Bocherel, B.R. Rowe and I.W.M. Smith, J. Chem. Phys., Submitted.

[19] I.R. Sims, J.L. Queffelec, A. Defrance, C. Rebrion-Rowe, D; Travers, B.R. Rowe and I.W.M. Smith, J. Chem. Phys. **97**, 8798, (1992).

[20] I.R. Sims, J.L. Queffelec, D. Travers, B.R. Rowe, L.B. Herbert, J. Karthauser and I.W.M. Smith, Chem. Phys. Lett, **211**, 461, (1993).

[21] Several papers will be submitted concerning OH and CH reactions

[22] C.V.V. Prasad, P.F. Bernath, C. Frum and R. Engleman Jr, J. Mol. Spectrosc. **151**, 459, (1992).

[23] D.C. Clary, J. Chem. Soc. Far. Trans., **89**, 2185, (1993).

[24] S. Zabarnick, J.W. Fleming, M.C. Liu, J. Chem. Phys. **132**, 407, (1989).

[25] S.J. Klippenstein and Y.W. Kim, J. Chem. Phys. in press.

[26] L.Pagani, Astron.Astrophys., in press.

[27] E. Herbst, H. Lee, D.A. Howe and T.J. Millar, Mon. Not. R. Astron. Soc., submitted.

AMMONIA IN THE INTERSTELLAR MEDIUM

C.M.Walmsley

Max Planck Institut für Radioastronomie, Bonn

ABSTRACT

I discuss the astrophysical importance of ammonia in the interstellar medium with special reference to it's use as a "thermometer". A very brief overview is given of the observations of interstellar ammonia in cold dark clouds and in hot star forming regions. The corrections needed when attempting to derive the gas kinetic temperature from observed level populations are discussed as well as their sensitivity to collisional rates. Observations of maser action in interstellar ammonia as well as observations of circumstellar ammonia are also briefly reviewed.

INTRODUCTION

Ammonia is one of the most commonly observed interstellar molecules. It was initially detected by Cheung et al. [4] in 1968 and since then has been found to be present in a variety of astronomical situations including circumstellar shells [2] and other galaxies [16] as well as in galactic molecular clouds. A useful review of both observations and interpretation is that of Ho and Townes [11].

Observations of the low lying inversion transitions are important for our understanding of the properties of star forming regions. The fact that there are many such transitions which are concentrated in a narrow frequency band around 23 GHz has made possible detailed studies of the population distribution among the different rotational levels in interstellar molecular clouds. This population distribution has been interpreted with the help of models of the statistical equilibrium in order to infer density, temperature, and ammonia abundance. In this contribution, I briefly describe such models and the extent to which they are dependent upon current knowledge of collisional rates.

AMMONIA OBSERVATIONS IN STARS AND CLOUDS

Below, I give a brief summary of the characteristics of the different regions in interstellar and circumstellar space which have been found to contain ammonia. I neglect in so doing the distribution of ammonia within our solar system.

In the Galaxy, ammonia has been most extensively observed in dense molecular clouds and I give a brief summary of the observational characteristics of such clouds in the following section . Most observations of ammonia in molecular clouds are consistent with "normal" or subthermal population distributions of the inversion levels. That is to say, the temperature corresponding to the ratio of the populations in the upper and lower inversion levels (i.e defined in analogous manner to equation 4 below) is less than or of the same order as the gas kinetic temperature. On the other hand, in very occasional "abnormal" situations, there is evidence for population inversion or maser action in ammonia inversion transitions. One believes this to be the exception rather than the rule. Nevertheless, as abnormalities always arouse interest, I give a brief summary of work on ammonia masers . Finally, circumstellar ammonia is a subject of considerable current interest , albeit with few positive results, and I give a brief resumé of circumstellar NH_3 observations .

AMMONIA IN MOLECULAR CLOUDS

Ammonia is prevalent in molecular clouds with an abundance relative to hydrogen which is typically of order 10^{-8} to 10^{-7}. There have been a number of surveys (e.g Refs 1,9) which have mainly been aimed at demonstrating a correlation between "high density" molecular concentrations seen in NH_3 and the presence of extremely young (less than 10^6 years old) newly formed stars. The premise behind this is that , when observing ammonia emission, one will select higher density gas than in lower dipole moment species such as CO. It is believed that the observed molecular emission lines from interstellar clouds are collisionally excited. Then one expects, when observing a transition with Einstein A-value A_{ul}, that much of the emitting gas will have a density of order $A_{ul}/\langle\sigma_{ul}v\rangle$ where the quantity in the denominator is the collisional rate for deexcitation of the upper state. Thus, for ammonia inversion transitions with $A_{ul} \sim 10^{-7}$ s^{-1} and $\langle\sigma_{ul}v\rangle$ of order 10^{-11} in cm^3 s^{-1}, one tends to see gas with hydrogen number density 10^4 cm^{-3}. As the ammonia observations themselves show moreover, this gas is usually cold with temperatures in the range 10-20 K. Very often, only the (1,1) and (2,2) transitions can be detected (see level diagram in fig.1) . Thus the ammonia observations are thought to pick out cold dense clumps (10^4 cm^{-3} which is astrophysically speaking a high density even if low for laboratory physicists) embedded in lower density material (density of a few hundred cm^{-3}) which is seen on larger size scales in the transitions of CO and it's isotopically substituted forms. Such dense clumps are often associated with young stars which are still in their contraction phase on the way towards becoming a solar type star. As a consequence, such ammonia "cores" are thought likely to be the sites of the next generation of young stars.

The vast majority of observational work on interstellar ammonia has concerned itself with the inversion transitions and in particular the low lying metastable lines (i.e the levels with J=K for which radiative transitions to lower rotational states are forbidden). Much less time and ink has been devoted to the far infrared and submm lines although the first transitions of this type were detected roughly a decade ago [13,26]. This is in the first place due to the fact that the inversion lines are readily observable from ground-based telescopes whereas observation of the rotational transitions requires the telescope to be above the greater part of atmosphere. Indeed , the detections referred to above were made with the Kuiper Airborne Observatory or KAO. Although much important work has been carried out with the KAO, ground-based observations are much more convenient and are to be preferred where possible. Secondly however, most of the regions observed at radio wavelengths have densities which are too small

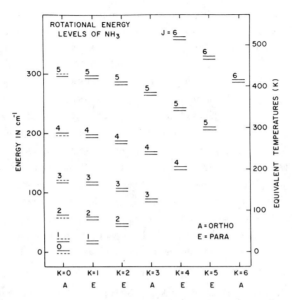

Figure 1: Ammonia level diagram showing level excitation as function of angular momentum quantum number J and the component along the molecular symmetry axis K

to excite the rotational transitions (and often also temperatures which are too small). This can be deduced from the densities given above and the fact that Einstein A-values for rotational transitions are roughly five orders of magnitude higher than the corresponding values for the inversion lines. Thus, in analogy to the argument outlined above, one can conclude that the "critical density" of the gas seen in these lines is of the order of 10^9 cm^{-3}. In other words, rotational lines should only be seen in hot high density compact regions.

In fact, the two detections of rotational lines mentioned above were both made towards the luminous young cluster of infrared sources embedded in the Orion molecular cloud and known (after their discoverers) as the Kleinmann-Low (or KL) nebula. These infrared sources are thought to be due to cocoons of hot dust particles surrounding young newly formed stars. The dust grains prevent us from seeing the stars directly but allow us to infer their presence. The dust passes on its heat to surrounding molecular gas and hence one finds that in this locality, the kinetic temperature of the gas is much higher than the value of 10-20 K quoted above for general molecular cloud material. In fact, temperatures as high as 200 K have been derived from the ammonia inversion line observations in this direction [10]. The hydrogen density is also thought to be very high (very high is defined for the current purpose as above 10^6 cm^{-3}) and thus it is not surprising that the first measurements of ammonia rotational transitions were made in this direction.

One reason why we believe the density is above 10^6 cm^{-3} in regions similiar to Orion Kleinmann-Low is that not only the metastable J=K lines are detected but also the non metastable inversion transitions with J > K. Non-metastable states are simply the upper levels of rotational transitions and hence one might conclude that the rotational lines should be excited wherever the non-metastable lines are detected. In fact, it seems at first surprising that we detect non-metastable NH$_3$ from regions of much lower density than the 10^9 cm^{-3} which was "naively" estimated above as the critical density for exciting a rotational transition. The reason for this is twofold. In the first place, the strong far infrared radiation field in regions such as Orion-KL is capable of exciting the rotational transitions of ammonia. Secondly, a photon emitted in an NH$_3$ rotational transition can be reabsorbed and reemitted many times by the surrounding dense molecular gas before escaping into "real" interstellar space. One can show that this reduces the "critical" density for excitation by an amount proportional to the number of scatterings (reabsorptions). The net result of all of this is that non-metastable ammonia is a useful marker for regions such as the Orion Kleinmann-Low where massive stars are currently forming.

Such hot dense regions appear to mark the birthplace of hot hot massive young stars. One hopes therefore, by studying the properties of "hot cores " to infer the physical conditions which pertained prior to the "switch-on" of the stars. Several studies (e.g Ref.3) have therefore been carried out aimed at determining typical densities, temperatures, and abundance distributions. One interesting result is that the molecular abundance distribution in the hot regions differs from that found in "typical" clouds in that saturated species such as NH$_3$ appear to be much more abundant than radicals and molecular ions. This appears to be the consequence of reactions on dust grain surfaces and subsequent evaporation of the "ices" thus formed. A review of the characteristics of these "hot core" regions is given by Walmsley and Schilke [29] and an overview of molecular abundance measurements is given by Walmsley [30].

One of the by-products of the studies of hot dense regions is that transitions from vibrationally excited states of a variety of molecules have been observed. In fact, in the case of Orion-KL, emission from the $v_2 = 1$ state of vibrationally excited ammonia has been detected [18,23]. The evidence suggests that the "vibrational" temperature computed from the ratio of populations in the excited and ground vibrational states is similiar to the "rotational" temperature obtained from the relative population in different rotational levels. Thus, one is in these regions approaching an LTE situation where all level populations are Boltzmannian. However, departures from LTE do occur as the observations of ammonia masers makes clear.

AMMONIA MASERS

Ammonia masers are the exception and not the rule. In fact, the reason why ammonia is useful as a temperature probe (see below) is that the population distribution between levels is usually "well behaved". Nevertheless, a few examples of "maser emission" have been found and thus population inversion (over-population of the upper inversion level relative to the thermal value) can occur.

The most prevalent ammonia maser appears to be the $NH_3(9,6)$ maser which was found by Madden et al. [15] towards several star forming regions similiar to Orion-KL but not towards Orion-KL itself. The most prominent of the (9,6) masers was detected towards a region known as W51. A subsequent VLBI (Very long baseline Interferometry) experiment towards this source [21] showed that the observed (9,6) emission originates in a region less than 0.1 milli-arcsec. in angular dimensions. This corresponds to less than the earth-sun distance which is tiny in astronomical terms. The line intensity expressed in terms of an equivalent black body is above 10^{13} K. Clearly some non-thermal maser like process must be responsible but it is presently unclear what. Also unclear is to what extent the emission which we observe has been beamed in our direction.

Other somewhat less spectacular masers have been found in the (9,8) line [31] and in the (5,5) line [3]. All of these appear to be rare which makes their analysis difficult. Also rare but of interest in the present context is the maser in the (3,3) line of $^{15}NH_3$ towards the region NGC7538 [12].

In this latter case, an attractive feature is that a mechanism has been proposed to invert the (3,3) levels [6,8,28] based upon the collisional selection rules for transitions out of the ground 0_0 state (see fig.2). The precise nature of these selection rules is somewhat controversial (see contribution by Flower) but the effect is that collisions out of the ammonia ground state with para-H_2 or He preferentially populate the upper (3,3) state relative to the lower one. Collisions with ortho-H_2 on the other hand do not show such a selection and hence the maser efficiency depends upon the abundance ratio of ortho to para molecular hydrogen. Flower et al. [6] use this characteristic to put limits upon the ortho-to-para H_2 ratio which is difficult to determine by direct means. The ortho-to-para H_2 ratio can in turn be used to estimate cloud "age" and hence one can imagine using the $^{15}NH_3(3,3)$ measurements to "date" the NGC7538 cloud. However, the feasibility of this depends upon the validity of the proposed inversion mechanism.

The (3,3) maser in this "collisional" model becomes quenched at high density and certainly above 10^6 cm^{-3}. One would thus expect it to be associated with regions where the non-metastable NH_3 lines are not excited. Towards NGC7538 however, many non-metastable transitions have been detected. Moreover, further analysis [22] shows that the (3,3) line is not the only transition of $^{15}NH_3$ which appears to be inverted towards NGC7538. In fact, the region where the masers form appears to be another "hot dense core" of the same general type as Orion-KL and the regions where the (9,6) ammonia masers have been found. At such high densities, the collisionally pumped (3,3) maser should be quenched. Also other pump mechanisms involving the excited vibrational state become plausible. These facts throw doubt upon the

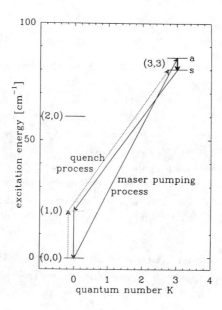

Figure 2: Level diagram illustrating proposed inversion mechanism for ammonia (3,3) line

collisional selection mechanism for explaining the observed $^{15}NH_3$ (3,3) population inversion in NGC7538. On the other hand, the competing pump mechanisms also are unconvincing. Thus, one can conclude that controversy about the pump mechanism for this and other ammonia masers will persist for some time to come.

AMMONIA IN CIRCUMSTELLAR ENVELOPES

Ammonia has been found in a very limited number of circumstellar envelopes [2,17,19,20]. In all cases, the objects in which ammonia has been detected are losing mass rapidly and appear to be in the last stages of their evolution prior to becoming planetary nebulae. The ammonia observations are of interest both because they allow estimates of the gas temperature in the outflowing gas and because the ammonia is a useful tracer of the kinematics of the wind. The abundances estimated for NH_3 vary between 10^{-5} and 10^{-8} relative to hydrogen. Ammonia has been found in both carbon and oxygen-rich circumstellar envelopes.

TEMPERATURE DETERMINATIONS USING AMMONIA

The main astrophysical application of ammonia observations has been to probe the temperature in dense condensations in star forming regions. Here, one is making use of radiative selection rules which forbid $\Delta K=1$ transitions. It follows that the relative population of levels in different K ladders is governed by collisions and hence is sensitive to the temperature (see fig.1). In fact, for hydrogen densities below 10^6

cm^{-3}, one finds that only the lowest J=K metastable levels of these ladders are populated and hence the effective rate for transfer of population from a ladder K to a ladder K$'$ is

$$R(K, K') = \sum \langle \sigma(K, K, J', K') v \rangle \qquad (1)$$

where the sum is over J'. Here, I am ignoring the distinction between the upper and lower inversion levels and just consider the sum of the two. The population in level (K,K) n(K) is then determined by a series of equations of the type:

$$n(K) \sum R(K, K') = \sum n(K') R(K', K) \qquad (2)$$

The statistical equilibrium between K-ladders is thus determined purely by collisions and in particular by K-changing collisions. One can get some more insight by looking at the particular case of the balance between the (1,1) and (2,2) levels which for many applications is all that matters. In this case at low temperatures (below 40 K), one can reasonably restrict ones attention to the (1,1), (2,2), and (2,1) levels. If one labels these levels 1,2,3 in order of energy, one can write down the condition for population balance for (2,2) as:

$$n_2 (C_{21} + C_{23}) = n_1 C_{12} \qquad (3)$$

where C_{12} is the collisional rate $\langle \sigma v \rangle$ from (1,1) to (2,2), C_{23} from (2,2) to (2,1) etc. By applying the condition of detailed balance and rearranging, one can express the ratio of (2,2) to (1,1) populations in terms of the ratio of collisional rates C_{23}/C_{21}. The quantity which one can determine observationally is the population ratio n_2/n_1 or equivalently the column density ratio $N(2,2)/N(1,1)$. One usually expresses the population ratio in terms of the rotation temperature T_{12} between (2,2) and (1,1). This is defined in terms of the Boltzmann equation.

$$n_2/n_1 = (5/3) \exp -(41/T_{12}) \qquad (4)$$

In the above, 5/3 is the ratio of statistical weights and 41 K is the energy difference between (2,2) and (1,1). Applying detailed balance, one can then show that

$$T_{12} = \frac{T}{1 + (T/41) ln(1 + C_{23}/C_{21})} \qquad (5)$$

where T is the kinetic temperature. Thus deviation from a Boltzmann distribution comes about because collisional depopulation of (2,2) to (2,1) is comparable with that to (1,1). Moreover, at low T, the rotation temperature and kinetic temperatures coincide but at higher T, T_{12} considerably underestimates the kinetic temperature. Finally, one should note that the relationship between rotational and kinetic temperature is in this approximation purely a function of the collisional rates and is independent of density.

These results can be generalised in various ways taking into account the effects of more highly excited levels. Figure 3 shows the results of computations by Danby et al.(Ref.5) for T_{12} as a function of T. The results are in reasonable agreement with those which could be deduced from equation 5 and thus the dominant effect determining deviations from a Boltzmann distribution is the depopulation of (2,2) into the non-metastable level (2,1). Expressed in other terms, the under-population of (2,1) causes the (2,2) to (1,1) population ratio to be smaller than the thermal equilibrium value.

As the temperature becomes higher, a larger number of levels need to be taken into consideration and a larger number of transitions become observable. Thus in hotter regions, $NH_3(4,4)$ and $NH_3(5,5)$ (respectively 200K and 300K above ground) become observable and one can measure the rotation temperatures T_{14} between (1,1) and (4,4), T_{45} between (4,4) and (5,5) etc. As with (1,1) and (2,2), one finds that the rotation temperatures underestimate the kinetic temperature and that the extent of this underestimate is determined by ratios analogous to the ratio C_{23}/C_{21} of importance for T_{12}. The currently available collisional rates consider levels up to (5,5) and for more highly excited levels, extrapolations of dubious reliability have been used.

Fortunately, there exist also other astrophysical temperature indicators which one can use to compare with the ammonia results. One such is CO which, because of it's large relative abundance, is thought to be "optically thick" in most galactic molecular clouds. Thus the mean free path for reabsorption of photons emitted in the rotational transitions is much smaller than the cloud size and radiation at frequencies within the line is within an essentially black enclosure. CO has moreover a relatively small permanent dipole moment of the order of 0.1 debye and thus one expects the population distribution of at least the lower rotational levels to be Boltzmannian. Then on grounds of classical thermodynamics, one expects the radiation intensity at the line frequency within the cloud to approximate the black body value. This may be a reasonable approximation also for the line radiation emergent from the cloud and thus a measurement of the line intensity is simultaneously a measure of temperature.

Hence CO measurements can be used to infer temperatures and compare with values obtained using NH_3. For a variety of reasons, one should not expect such a comparison to yield excellent agreement. Most obviously, for reasons outlined earlier, one expects the observed CO line radiation to form in more extended low density regions than the ammonia lines. One can confirm this with detailed maps. Nevertheless, the comparison (figure 4) does show qualitative agreement between ammonia and CO temperatures. This is despite the fact that the corrections between the ammonia rotational and kinetic temperatures are occasionally quite large. The conclusion seems to be that, to the present level of accuracy, ammonia temperatures corrected in the manner described above are reasonable (although see references 14,24,27 for a more precise discussion).

CONCLUSIONS

I have attempted to give a "lightning tour" of the various ways in which interstellar ammonia observations have been used to gain information about the physical

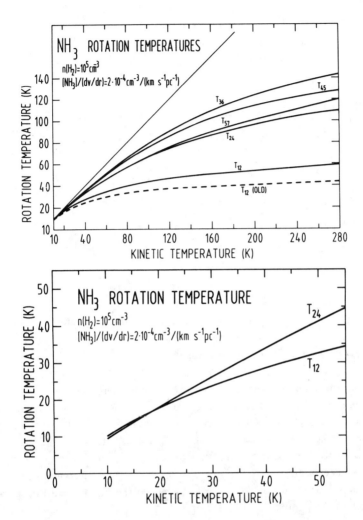

Figure 3: Results of computations by Danby et al. [5] showing ammonia rotational temperature between different metastable levels as a function of kinetic temperature. T_{12} denotes the rotation temperature (see text) between (1,1) and (2,2) ; T_{24} denotes the rotation temperature between (2,2) and (4,4) etc. Calculations were carried out for a molecular hydrogen density of 10^5 cm^{-3} and an ammonia abundance to velocity gradient ratio of $2\,10^{-4}$cm^{-3}/(km s^{-1} pc^{-1}) but the results are insensitive to this choice of parameters. The dashed curve marked T_{12} (old) in the upper panel shows results computed using older collisional rates due to Green [7]

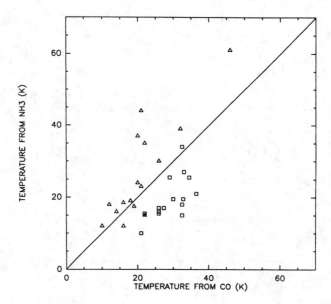

Figure 4: Comparison of temperatures derived from ammonia with temperatures derived from CO (from ref.5). Triangles refer to observations taken from Takano [25] and boxes to observations from Walmsley and Ungerechts [28].

parameters in star forming regions. One may pose the question ; to what extent do the detailed collisinal rate calculations and measurements carried out by physicists and chemists affect the astrophysical conclusions. Equation 5 and figure 3 give one answer to that question. The tentative model for the (3,3) maser outlined in the text (even if it turns out not to correspond to reality) gives another. I think one may conclude that, given the uncertainties placed in our path by the unruly behaviour of molecular clouds, it is important for astrophysicists to reduce the uncertainties in relevant collisional rates to a minimum.

REFERENCES

1 Benson P.J., Myers P.C. 1989 *Astrophys.J. Suppl.* , **71**,89.
2. Betz A.L., McClaren R.A. , Spears D.L. 1979 *Astrophys. J.* **229**,L97.
3. Cesaroni R., Walmsley C.M., Churchwell E. 1992 *Astron. Astrophys.* ,**256**,618.
4. Cheung A.C., Rank D.M., Townes C.H., Thornton D.D., Welch W.J. 1968 *Phys. Rev. Lett.* **21**, 1701.
5. Danby G., Flower D.R., Valiron P., Schilke P., Walmsley C.M. 1988 *Monthly Notices Roy. Astron. Soc.* ,**235**,229.
6. Flower D.R., Offer A., Schilke P. 1990 *Monthly Notices Roy. Astron. Soc.* **244**, 4p.
7. Green S. 1983 NASA Technical Memorandum 83869.

8. Guilloteau S., Wilson T.L., Martin R.N., Batrla W., Pauls T.A. 1983 *Astron. Astrophys.* ,**124**,322.
9. Harju J., Walmsley C.M., Wouterloot J. 1993 *Astron. Astrophys. Suppl.* **98**,51.
10. Hermsen W., Wilson T.L., Walmsley C.M., Henkel C. 1988 *Astron. Astrophys.* ,**201**,285.
11. Ho P.T.P., Townes C.H. 1983 *Ann. Rev. Astron. Astrophys.* ,**21**,239.
12. Johnston K.J., Stolovy S.R., Wilson T.L., Henkel C., Mauersberger R. 1989 *Astrophys. J.* ,**343**,L41.
13. Keene J., Blake G.A., Phillips T.G. 1983 *Astrophys. J.* , **271**, L27.
14. Kuiper T.B.H. 1987 *Astron. Astrophys.* ,**173**,209.
15. Madden S.,Irvine W.M., Matthews H.E., Brown R.D., Godfrey P.D. 1986 *Astrophys. J.* ,**300**,L79.
16. Martin R.N., Ho P.T.P. 1986 *Astrophys. J.* ,**308**,L7.
17. Martin-Pintado J., Bachiller R. 1992 *Astrophys. J.* ,**391**,L93.
18. Mauersberger R., Henkel C., Wilson T.L. 1988 *Astron. Astrophys.* **205**,235.
19. Morris M., Guilloteau S., Lucas R. ,Omont A. 1987 *Astrophys. J.* ,**321**,88.
20. Nguyen-Q-Rieu, Graham D.A., Bujarrabal V. 1984 *Astron. Astrophys.* , **138**,L5.
21. Pratap P., Menten K.M., Reid M.J., Moran J.M. , Walmsley C.M. *Astrophys. J.* ,**373**,L13.
22. Schilke P., Walmsley C.M., Mauersberger R. 1991 *Astron. Astrophys.* ,**247**,516.
23. Schilke P., Güsten R., Schulz A., Serabyn E., Walmsley C.M. 1992 *Astron. Astrophys.* ,**261**,L5.
24. Stutzki J.,Winnewisser G. 1985 *Astron. Astrophys.* ,**148**,254.
25. Takano T. 1986 *Astrophys. J.* ,**303**, 349.
26. Townes C.H., Genzel R., Watson D.M., Storey J.W.V. 1983 *Astrophys. J.* ,**269**,L11.
27. Ungerechts H., Walmsley C.M., Winnewisser G. 1986 *Astron. Astrophys.* ,**157**,207.
28. Walmsley C.M., Ungerechts H. 1983 *Astron. Astrophys.* ,**122**, 164.
29. Walmsley C.M., Schilke P. (1992) in *Dust and Chemistry in Astrophysics*, (edited Millar T.J., Williams D.A. , publ. IOP).
30. Walmsley C.M. (1993) in Proceedings of the Faraday Symposium 28, *Journal of Chem. Society*, **89**,2119.
31. Wilson T.L., Johnston K.J., Henkel C. 1990 *Astron. Astrophys.* ,**229**,L1.

DISCUSSION

FIELD — In your very interesting presentation of the importance of NH_3 as a thermometer in the ISM, you stressed the role of collisions in establishing the relative populations of NH_3 energy levels. I should like to emphasise how important it is to consider the importance of line and continuum radiation fields in establishing these relative populations. With relevance to this meeting, the form of the dust emission as a function of wavelength will influence the results obtained. In addition, the treatment of radiative transport will also be important and the use of the large velocity gradient approximation will introduce systematic errors into the results. At least for simpler systems, approximate Λ iteration should be used if possible.

WALMSLEY — I agree that in the hot regions close to strong infrared sources, radiative pumping will become important. In relatively low temperature gas far from regions of star formation, this seems improbable. I don't think as a rule that one incurs large errors by using the large velocity gradient approximation if the focus is merely on the relative populations in different metastable levels.

THADDEUS — A collisional maser pump in a rare isotopic species ($^{15}NH_3$) which is absent in the normal species is strange. If optical depth is responsible for quenching the pump in the normal species, one wonders why it isn't observed on the edge of the cloud where the optical depth is lower. If there is a nearby infrared source, I would suspect that it may be the key.

WALMSLEY — I agree that in the case of the NGC 7538 $^{15}NH_3$ maser, infrared pumping may be the explanation. However, it is the only object of it's kind that we know of and so one should keep an open mind. The maser may be amplifying the continuum radiation from the background HII region and hence be only found along the line of sight to this source.

GIARD — Does one have any idea of what would be the effect on an $NH_3(3,3)$ maser of taking into account the radiative pumping by warm dust. Would it favor or not the inversion ?

WALMSLEY — In the purely collisional model which I discussed, any process which increases the population of the (1,0) level will act to (quench the maser. Hence, if there is strong dust emission at the wavelength of the (1,0)-(0,0) transition (500 microns roughly), this would not favor the inversion. However, this is a wavelength where under most circumstances, dust emission is not very intense. If one on the other hand considers the possibility that the maser region is close to hot dust radiating at 10 microns, then the situation is different because the ftrst vibrational state can be excited. Schilke et al (1991) have considered this latter possibility.

DESPOIS — Do you think it is possible to have a maser effect in the cometary $NH_3(3,3)$ line ? Could collisions with water molecules lead to such an effect?

WALMSLEY — In the cometary situation, it is possible that pumping by solar infrared radiation into excited vibrational states is important for the relative populations of rotational levels. Also, H_2O is the most abundant collision partner in a cometary atmosphere and I would presume that collisional selection rules are different for NH_3 with H_2O than for NH_3 with H_2.

THE ROTATIONAL EXCITATION OF NH_3 BY ORTHO- AND PARA-H_2: A STATUS REPORT

D. R. Flower
Physics Department, The University, Durham DH1 3LE

Alison Offer
School of Chemistry, The University, Bristol BS8 1TS

ABSTRACT

The results of recent theoretical and experimental studies of NH_3-H_2 collisions are summarized. Particular attention is paid to the possibility that the collisional propensity rules differ for excitation by ortho- and para-H_2.

The existence of different propensity rules for rotationally inelastic collisions between ortho- and para-H_2 and other molecules would have interesting consequences for studies of the interstellar medium. In particular, it might be poossible to deduce the relative abundance of ortho- and para-H_2 in dark clouds, where it is not directly observable. The value of this ratio can shed light on the chemical history and even the age of molecular clouds [1,2].

The work of Green [3] on NH_3- He collisions laid the foundations for subsequent studies of the rotational excitation of NH_3. The cross-sections, σ, calculated by Green show propensity rules favouring the excitation, from the ground state, of one component of an inversion doublet relative to the other component. For example,

$$\sigma(0\ 0\ +\ \rightarrow 3\ 3\ +) \ll \sigma(0\ 0\ +\ \rightarrow 3\ 3\ -)$$

$$\sigma(0\ 0\ +\ \rightarrow 4\ 3\ +) \ll \sigma(0\ 0\ +\ \rightarrow 4\ 3\ -)$$

in ortho-NH_3, and

$$\sigma(1\ 1\ +\ \rightarrow 2\ 2\ +) \ll \sigma(1\ 1\ +\ \rightarrow 2\ 2\ -)$$

$$\sigma(1\ 1\ -\to 2\ 2\ -) \ll \sigma(1\ 1\ -\to 2\ 2\ +)$$

in para-NH$_3$. We use the notation $\sigma(j\ k\ \varepsilon \to j'\ k'\ \varepsilon')$ for the transition between the states j k ε and j' k' ε', where j is the rotational quantum number and k its projection on the symmetry axis of the molecule; ε is an index which is related to the usual 'symmetric' (s) and 'asymmetric' (a) labels of the components of an inversion doublet through

$$\varepsilon = (-1)^j \text{ for the a component,}$$

and

$$\varepsilon = -(-1)^j \text{ for the s component.}$$

In molecular clouds, excitation of ammonia occurs principally in collisions with H$_2$ molecules, rather than He atoms, owing to the higher abundance of the former. It is, therefore, relevant to ask:

(i) do the above propensity rules apply to collisions with para-H$_2$? Both H$_2$ and He possess two electrons, and para-H$_2$ (j = 0) is spherically symmetric, as is He in its ground state. Consequently, para-H$_2$ is often supposed to exhibit behaviour similar to that of He in collisions with molecules.

(ii) Do the same or different propensity rules apply to collisions with ortho-H$_2$, in which the ground state has non-zero angular momentum?

Two sets of calculations have been performed in recent years in order to answer the above questions. The first was by Offer & Flower [4,5] and the second by Rist, Alexander & Valiron [6]. The latter extended and refined the computations of the NH$_3$-H$_2$ interaction potential and incorporated this improved potential in their collision calculations. Both groups employed the quantum mechanical coupled channels method to determine rotational excitation cross-sections, using independent computer codes.

In Tables I and II, we compare the theoretical results for excitation by para-H$_2$ and ortho-H$_2$, repectively, for transitions out of the 0 0 + ground state of ortho-NH$_3$. Offer & Flower [5] reported results obtained using two potentials ('1' and '2' in the Tables) which differed in the method used to calculate the contribution of the dispersion energy to the total interaction. Potential 2 should be more accurate than potential 1.

Table I Cross sections (in units of 10^{-16} cm^2) for transitions out of the 0 0 + ground state of ortho-NH$_3$, induced by collisions with para-H$_2$ at an energy of 125 cm^{-1}. Potentials '1' and '2' differ in the method used to calculate the dispersion energy (second order and many body perturbation theory, respectively) but have the same SCF part.

j' k' ε'	Rist et al [6]	Offer and Flower [5]	
		Potential 1	Potential 2
1 0 +	8.20	25.5	13.0
2 0 +	4.15	2.30	4.91
3 0 +	0.29	0.36	0.51
3 3 +	0.01	0.05	0.06
3 3 −	3.19	4.71	5.04

Referring first to Table I, which pertains to excitation by para-H$_2$, we see that all three calculations predict a strong propensity for excitation of the 3 3 − component of the inversion doublet, rather than 3 3 +. At higher energies, where the 4 3 ± levels can be excited, the calculations predict that σ(0 0 + → 4 3 −) >> σ(0 0 + → 4 3 +). These properties are equivalent to those discovered by Green [3] in his study of NH$_3$- He scattering. We see further from Table I that the differences between the results of Rist et al. [6] and Offer & Flower [5] are of similar magnitude to those between potentials 1 and 2. Overall, the results of Rist et al. agree better with those that we obtained using potential 2, which we believe to be more accurate than potential 1. We conclude that, for collisions with para-H$_2$, the discrepancies between our work and that of Rist et al. probably relate to differences in the potentials employed.

Table II   Cross sections (in units of 10^{-16} cm^2) for transitions out of the 0 0 + ground state of ortho-NH$_3$, induced by collisions with ortho-H$_2$ at an energy of 125 cm^{-1}. Potentials '1' and '2' differ in the representation of the dispersion energy but have the same SCF part.

j' k' ε'	Rist et al [6]	Offer and Flower [5] Potential 1	Potential 2
1 0 +	31.2	36.8	40.5
2 0 +	9.17	10.9	9.93
3 0 +	0.28	0.44	0.48
3 3 +	1.30	3.05	3.15
3 3 −	2.10	1.77	1.50

We now turn our attention to Table II. On the basis of a comparison of the results obtained using potentials 1 and 2, Offer & Flower [5] concluded that cross-sections for excitation by ortho-H$_2$ are less sensitive to the detailed form of the interaction potential than for para-H$_2$. This conclusion is supported by the further comparison with the calculations of Rist et al. The only substantial discrepancy occurs for σ(0 0 + → 3 3 +), which Rist et al. predict to be *smaller* than σ(0 0 + → 3 3 −) at this energy, whereas we predict the reverse.

Rist et al. attribute the persistence of the propensity rule favouring the 0 0 + → 3 3 − transition to a 'quasisymmetry' of the interaction potential. The symmetry to which they refer applies strictly only to the long-range multipolar interaction between molecules and not to induction or dispersion forces (cf. Leavitt [7]), nor at short range. Nevertheless, Rist et al. incorporate this quasisymmetry in the potential matrix elements which appear in the collision calculations. A consequence of treating the quasisymmetry as if it were exact is to eliminate the direct coupling between the 0 0 + and 3 3 + states. The omission of this coupling may at least partly account for the smaller value of σ(0 0 + → 3 3 +) which they obtain.

Table III Cross sections (in units of 10^{-16} cm^2) for transitions out of the 0 0 + ground state of ortho-NH$_3$, induced by collisions with ortho-H$_2$ at an energy of 600 cm^{-1} (Rist et al. [6]) and 605 cm^{-1} (Offer & Flower [4]).

j' k' ε'	Rist et al [6]	Offer and Flower [4] Potential 1
3 3 +	1.35	2.39
3 3 −	4.27	3.07
4 3 +	0.45	1.12
4 3 −	2.07	1.65

Experimental measurements of state-to-state cross-sections for rotational transitions of ortho- and para-NH$_3$, induced by ortho- and para-H$_2$, have been reported by Schleipen, ter Meulen & Offer [8]; the collision energy was 605 cm^{-1}. Rist et al. published cross-sections for four transitions from the ground state of ortho-NH$_3$ at an energy of 600 cm^{-1}. Their results for excitation by ortho-H$_2$ are compared with those obtained by Offer & Flower [4], using potential 1, in Table III. We see from the Table that, once again, the computations of Rist et al. predict a stronger propensity in favour of transitions in which ε' ≠ ε. Their omission of the direct coupling between states for which ε' = ε, mentioned above, could partly account for the strength of this propensity rule.

The experimental evidence (Schleipen et al. [8]) does not lend support to the calculations of Rist et al. The measured value of the ratio σ(0 0 + → 3 3 −)/σ(0 0 + → 3 3 +) is 0.71 [i.e. σ(0 0 + → 3 3 +) is found to be *larger* than σ(0 0 + → 3 3 −)] and is closer to that computed by Offer & Flower (1.3) than by Rist et al. (3.2). For the ratio σ(0 0 + → 4 3 −)/σ(0 0 + → 4 3 +), the corresponding numbers are 2.5 (Schleipen et al.), 1.5 (Offer & Flower), and 4.6 (Rist et al.).

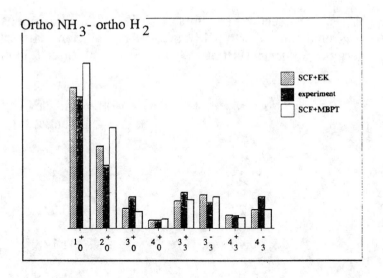

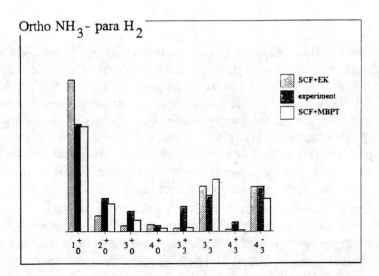

Fig. 1. Relative cross-sections for transitions out of the ground state of ortho-NH_3, induced by ortho- and para-H_2. The theoretical results are denoted 'SCF+EK' and 'SCF+MBPT', corresponding to potentials '1' and '2', respectively, in the text. (From Schleipen et al.[8].)

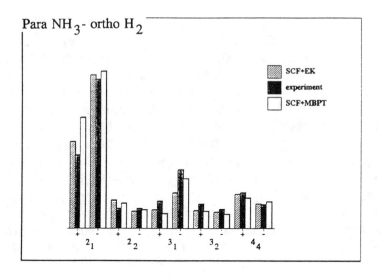

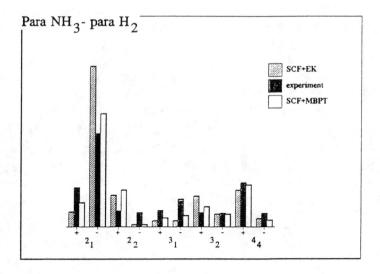

Fig. 2. As Fig. 1, but for para-NH₃.

Schleipen et al.[8] compare their measured, relative cross-sections with those computed (by Alison Offer) using potentials 1 and 2. This comparison is shown for ortho-NH_3 in Fig. 1 and for para-NH_3 in Fig. 2. These Figs. show that the agreement between theory and experiment for the excitation of both ortho- and para-NH_3 by ortho-H_2 is satisfactory. The results obtained using the two forms of the potential differ from each other by about the same amount as the theory differs from the experiment. However, for collisions involving para-H_2, the agreement is less good. In particular, the strong propensity in favour of certain ε-changing transitions, discussed above and evident in *all* the theoretical results, is apparently not confirmed by the experiments. In collisions with ortho-NH_3, the measurements show that the tendency to preferential excitation of the 3 3 - and 4 3 - components is stronger for para-H_2 than for ortho-H_2, but not as strong as predicted by theory. In collisions with para-NH_3, no clear propensity towards either the 2 2 + or 2 2 - states is seen.

Finally, we briefly discuss NH_3- He scattering. The experiments of Schleipen & ter Meulen[9] show the collisional propensity rules to be *more* pronounced than for NH_3- para-H_2 collisions (see Fig. 3), in the sense of the theoretical results for the latter system. Apart from the early calculations by Green[3], which were based on a potential which no longer represents the state-of-the-art, the only other quantal calculations for NH_3- He scattering which have been published are those performed by Schinke (Meyer et al.[10]). However, Schinke used the coupled states approximation to the coupled channels equations, which yields identically zero cross-sections for transitions such as 0 0 + → 3 3 + , 0 0 + → 4 3 +, and 1 1 - → 2 2 - . As yet unpublished work by Rist (1991, Thesis, University of Grenoble) suggests that the discrepancies between theory and experiment, discussed above for NH_3- para-H_2 scattering, are as severe for NH_3- He collisions.

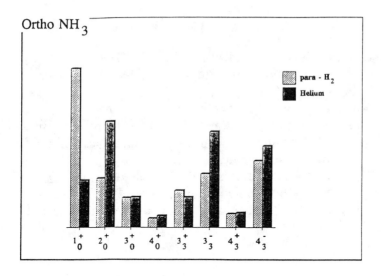

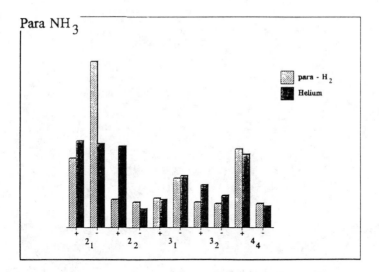

Fig. 3. A comparison of measured relative cross-sections for NH_3-He and NH_3-para-H_2 scattering. (From Schleipen et al.[8].)

REFERENCES

1. D. R. Flower, G. D. Watt, MNRAS, 209, 25 (1984).
2. D. R. Flower, A. Offer, P. Schilke, MNRAS, 244, 4P (1990).
3. S. Green, J. Chem. Phys., 64, 3463 (1976).
4. A. Offer, D. R. Flower, J. Phys. B, 22, L439 (1989).
5. A. Offer, D. R. Flower, J. Chem. Soc. Faraday Trans., 86, 1659 (1990).
6. C. Rist, M. H. Alexander, P. Valiron, J. Chem. Phys., 98, 4662 (1993).
7. R. P. Leavitt, J. Chem. Phys., 72, 3472 (1980).
8. J. Schleipen, J. J. ter Meulen, A. R. Offer, Chem. Phys., 171, 347 (1993).
9. J. Schleipen, J. J. ter Meulen, Chem. Phys., 156, 479 (1991).
10. H. Meyer, U. Buck, R. Schinke, G. H. F. Diercksen, J. Chem. Phys., 84, 4976 (1986).

DISCUSSION

FIELD — Calculations show clearly that the collisional transition $(0,0,+) \to (3,3,+)$ for NH_3 colliding with para-H_2 is of very low cross section compared with $(0,0,+) \to (3,3,-)$. Is there a sum rule which would tend to dictate that some other collisionally induced transition would tend to compensate for this and tend to ensure that the total cross section for scattering from $(0,0,+)$ with p-H_2 is about equal to the total scattering cross section (i) from $(0,0,-)$, (ii) for o-H_2 in collision with $(0,0,+)$?

FLOWER —
  (i) The stale $(j,k,\varepsilon) = (0,0,-)$ does not exist. Doublets occur only for non-zero values of k.
  (ii) The close coupling calculations show that the collisional propensity rules differ for para- and ortho-H_2. These calculations should take account of both direct and indirect transitions between the initial and final states.

SCAPPINI — For the collisional excitation of the upper sub level of the (3,3) doublet of NH_3, you propose $\Delta J = 3$ collisionally induced transitions from the ground (0,0) level. It seems to me that the $\Delta J = 3$ quantum jump violates both dipole and quadrupole selection ("propensity") rules. Is it possible that instead, the collision process goes in steps of $\Delta J = 1$ from the (0,0) to the upper (3,3) sub-level ?

FLOWER — The step-by-step proces to which you refer will proceed only within the k=0 ladder. Transitions from $(j,k) = (0,0)$ to (3,3) require that $\Delta k = 3$.

RECENT PROGRESS IN EXPERIMENTAL STUDIES OF ION–MOLECULE REACTIONS RELEVANT TO INTERSTELLAR CHEMISTRY

D. Gerlich
Fakultät für Physik der Universität, D–79104 Freiburg, Germany

ABSTRACT

This review is concerned with recent developments in the study of ion–molecule reactions at low temperatures using a cryo cooled 22–pole ion trap. Experimental progress was achieved by efficiently thermalizing the ions by injecting a very intense, short duration (ms) pulse of He into the trap and by using laser methods to characterize the ion velocity distribution. New results, measured at a nominal temperature of 10 K, are presented for the radiative association reactions of C^+ and CH_3^+ with H_2. The controversy concerning the low temperature formation of protonated acetylene in the $C_2H_2^+ + H_2$ reaction is briefly summarized and new experimental results are presented which corroborate the earlier conclusion that this reaction is endothermic. For all the above reactions isotope fractionation with HD and D_2 was studied, and the influence of $H_2(j=1)$ was examined by utilizing $p-H_2$ instead of $n-H_2$.

INTRODUCTION

The use of chemical composition as a diagnostic probe for the interstellar medium or other astrophysical environments requires a detailed understanding of the formation, excitation, and destruction of molecules. As a consequence of observational advances and the development of more realistic models, very detailed information on the physical and chemical processes occurring on interstellar grains and in the gas phase are needed. This has lead to a variety of theoretical activities and to the development of a variety of experimental setups for measuring cross sections or rate coefficients for processes such as photoionization, inelastic collisions, chemical reactions between radicals, ions and neutrals, and electron–ion recombination.

The specific importance of ion–molecule reactions in the chemistry of diffuse and dense interstellar clouds and an overview of the different routes to the synthesis of molecules was discussed in several recent review papers (see for example Ref. 1 and references therein). In general, the physical conditions of interstellar and circumstellar objects vary widely in temperature, number density, radiation fields, and are often far from thermal equilibrium. Therefore cross sections for reactions of state–selected ions at collision energies ranging from meV to several eV are required for describing nonequilibrium environments. A short overview of guided ion beam instruments, which are suitable for this purpose, was given recently.[2] Typical examples of astrophysical importance where kinetically excited ions may play a role include the $N^+ + H_2$ hydrogen abstraction reaction[3] and the $C^+ + H_2$ reaction.[1]

An experimental challenge is to study ion–molecule reactions especially at very low temperatures. Results are not only important for modeling the chemistry of cold interstellar clouds but they are also of fundamental interest for several reasons. At a total energy of a few meV only, the outcome of a collision can be changed significantly by excitation of the lowest rotational or fine structure state of the reactants or by slight differences in zero point energies if isotopes are exchanged. Significant energy dependencies are due to the fact that in very slow collisions only a small number of total orbital angular momenta are involved, although the electrostatic long range interaction can lead to capture at very large impact parameters. Another consequence of low energy collisions is that for systems with even a few atoms, intermediates with very long lifetimes can be formed; this can lead to improbable processes such as tunneling or radiative association.

In recent years remarkable progress was achieved in the development of experimental methods which allow the study of ionic reaction processes with ions at temperatures way below 80 K. A detailed comparative discussion of the various techniques can be found in a recent article by M. Smith.[4] In the majority of these experiments the reactant species are contained by a dense buffer gas which is cooled cryogenically or by supersonic expansion. This has the advantage of many thermalizing collisions but often leads to higher order kinetics. In the following an experiment is described which makes use of a 22—pole radio frequency ion trap and which has the unique features of wide accessible number density range, high sensitivity, versatility and reliability. From recent experimental results which include laser studies of cold stored ions and the observation of the dynamics of the growth of H_n^+ clusters on a time scale of seconds, three reactions were selected for this paper which are important in the formation of hydrocarbon molecules in interstellar clouds.

EXPERIMENTAL

Today various instruments are in use which utilize inhomogeneous radio frequency (rf) fields for mass and energy selection of ions, for guiding simultaneously charged reactants and ionic products or for trapping charged particles. Very fundamental keywords which are related to the principle of particle confinement with fast oscillating forces and which are needed for characterizing the features of suitable experimental devices, are the *adiabatic approximation*, the *effective potential*, and the *adiabaticity parameter*. Applications of rf fields include not only the well—known Paul trap, quadrupole mass spectrometers and the octopole ion beam guides, but also early developments in accelerator and plasma physics, the interaction of microwave and laser fields with electrons and the storage of droplets and other macroscopic charged particles in electrical fields alternating at a few Hz. Many theoretical and experimental aspects of these techniques and some specific applications have been discussed thoroughly in a recent review.[5]

For trapping ions a variety of rf—based electrode arrangements have been developed.[5] The most prominent example is the quadrupole or Paul—trap which was used for many fundamental experiments, especially in spectroscopy. Since the effective potential of an rf—quadrupole is harmonic, this type of trap is the ideal device for sideband laser cooling and for the precise localization of a single ion. For the study of collision processes between stored ions and neutrals, electrode structures with wide field—free regions are not only superior but mandatory if one intends to operate at low ion temperatures. Traps consisting of stacks of ring electrodes have been used successfully in recent years for collision experiments with slow stored ions.[6] Another more recent approach is the construction of a linear 22—pole trap which is shown in Fig. 1. Here the effective potential has a very flat plateau in the inner region and extremely steep walls towards the poles.

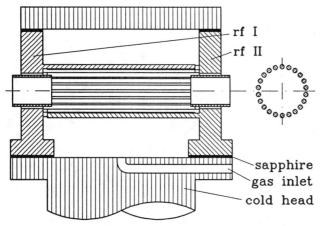

Fig. 1. Variable temperature 22–pole ion trap (rod diameter 1 mm, rod length 36 mm, inscribed radius r_0=5 mm). The rf electrodes I and II are mounted onto the cold head using sapphire for electrical insulation and good heat conductance. In the axial direction the ions are confined by electrostatic voltages applied to the two cylindrical tubes. These electrodes are operated in a pulsed mode for injecting selected ions and for extracting the content for analysis.

Some technical details of the 22–pole ion trap and a selection of recent results have been published elsewhere.[7] As indicated in Fig. 1, the electrodes and the surrounding walls are thermally connected to the cold head of a closed cycle refrigeration system and the nominal temperature can be varied between T_n=10 K and 300 K. The translational and internal degrees of freedom of the ions are coupled to the cold environment by inelastic collisions with a suitable buffer gas or via radiation. The actual temperature was determined applying different methods. Below 50 K temperature–dependent rate coefficients were used for reactions such as $N^+ + H_2 \rightarrow NH^+ + H$ or the formation of He^+_2–dimer ions and H_n^+ clusters and at present the coldest ion cloud (e.g. T_n=10 K) is characterized by 15±5 K.

In a recent experiment[8] the translational temperature of N_2^+ ions stored in Ar gas was derived from the Doppler width of selected spectral lines. In this study a single–mode diode laser (783–787 nm) excited the N_2^+ ions into the A,v=2 state. Radiative decay of these ions leads in most cases to vibrationally excited $N_2^+(X,v>0)$ ions. They have sufficient internal energy to undergo efficient charge transfer with Ar; so the excitation spectrum of N_2^+ is recorded with a very good signal–to–noise ratio by counting the Ar^+ ions as a function of the laser wave length. Between 300 K and 50 K the temperatures derived from the Doppler profile were in good overall accordance with the nominal temperature of the walls of the trap.

The ion trap depicted in Fig. 1 is the central part of an apparatus which was described elsewhere.[5] Ions are prepared in a separate ion source, mass selected and then injected via the pulsed entrance electrode. Ions can be stored for many minutes or even hours. In the absence of target gas the mean decay time is only determined by reactions with the background gas, the pressure of which is estimated to be below 10^{-11} mbar at T_n=10 K. For the study of reaction process, target gas is added at a number density varying from below 10^9 cm^{-3} to above 10^{15} cm^{-3}. By choosing a suitable combination of number density and interaction time, the rates of rather fast as well as very slow bimolecular reactions can be measured. The accessible number density range allows us both to determine reliable ternary association rate coefficients and to operate under conditions where the rate for three–body collisions becomes smaller than 1 day^{-1} and where radiative association prevails.

Fig. 2. Time dependence of the He—gas pulse injected into the ion trap for thermalizing the ions using a piezoelectric pulsed valve. The FWHM of the pulse is 5 ms. As can be seen from the fit, the decay of the number density after the pulse is described by two time constants, 3.5 ms and 23 ms. The first one is determined by the small volume (~ 15 cm^3) of the box surrounding the trap electrodes and the conductance of the entrance and exit tube.

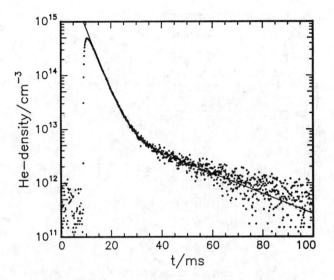

At low target densities the thermalization of internally excited ions is a problem. In some cases it is possible to wait long enough for the trapped ions to be cooled by radiation. However, for many applications such as for the study of slightly endothermic reactions, collisional cooling is superior since it avoids formation of background products by the reactions of initially excited ions. For this purpose an additional pulsed gas inlet was included into the trap apparatus. Using a piezo electric valve a very intense pulse of He buffer gas (peak number density up to 10^{15} cm^{-3}) can be injected. The resulting temporal change of the number density has an half width of 5 ms and during the pulse, ions make in the order of 10^4 collisions. As can be seen from Fig. 2 the number density drops (with a time constant of 3.5 ms) by 2 orders of magnitude in 20 ms. The subsequent slower decay has a larger time constant of 23 ms. After 100 ms the rates of ternary processes involving the He buffer gas atoms are negligible. This He pulse not only leads to efficient thermalization of the ions but also to more efficient trapping of those which otherwise would escape during the filling period.

The rates of attenuation of the primary ions in reactions with a target gas and/or for forming specific products are determined from the temporal change of the ion composition which is measured by periodically filling the trap, extracting its content after various times, and analyzing its mass composition with a quadrupole mass spectrometer. Depending on the reaction process, repetition periods between ms and a min are chosen. Two typical examples are shown in Fig. 3. Both panels show an exponential decline of the primary ions (CH_3^+ and $C_2H_2^+$ reacting with H_2) recorded in a time interval between 2 s and 60 s. In both cases the dominant reaction is radiative association; due to the low H_2 number density three—body collisions contribute less than 10 % in the case of $CH_3^+ + H_2$ and are negligible in the $C_2H_2^+ + H_2$ association reaction. In addition to association, both primary and product ions also exchange an H—atom for a D—atom in collisions with the natural traces of HD mixed with the hydrogen. Due to the large abundance of H_2 in comparison the HD the slightly endothermic back—reactions, i.e. D—H exchange in collisions of the partially deuterated ions with the H_2, are also important.

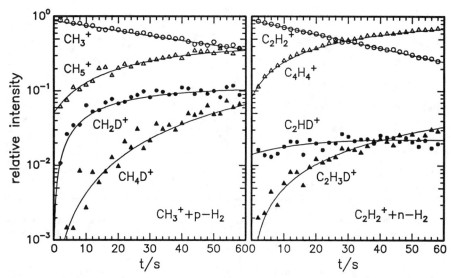

Fig. 3. Time evolution of ions stored in hydrogen target between 2 s and 60 s in the 22–pole ion trap at $T_n=10$ K (left panel: CH_3^+ ions, $[p-H_2]=8.7\times10^{10}$ cm^{-3}, right panel: $C_2H_2^+$, $[n-H_2]= 1.6\times10^{10}$ cm^{-3}). The dominant reaction in both cases is radiative association. The products plotted with the solid symbols are formed in collisions with HD impurities. The lines are the solutions of coupled rate equations which include in addition to association reactions the H–D exchange in collisions with HD and the D–H exchange with H_2. In the case of CH_3^+, the hydrogen used had the natural abundance, $[HD]/[H_2]=3\times10^{-4}$, for the reaction with $C_2H_2^+$ purified hydrogen, $[HD]/[H_2]=4\times10^{-5}$, was used.

RESULTS

Collisions of C^+ ions with hydrogen atoms or molecules are possible steps leading to the formation of small hydrocarbons in interstellar chemistry. However, since the reaction $C^++H_2 \rightarrow CH^++H$ is endothermic by almost 0.4 eV and since radiative association of C^+ with H atoms plays a negligible role, the only process occurring at low temperatures and densities is the radiative association reaction

$$C^+ + H_2 \longrightarrow CH_2^+ + h\nu. \quad (1)$$

It is important to note that the CH_2^+ complex can be formed without a barrier and that there is a chance that it may be effectively stabilized by a rapid electronic transition between the low lying excited $^2B_1$ state and the $^2A_1$ ground state. The various theoretical estimates bracket the rate coefficient for radiative association for this process, k_r, between 10^{-15} cm^3/s and 10^{-16} cm^3/s and predict only a weak temperature dependence below 100 K (for a discussion see Ref. 6). The first experimental study was performed in a 10 K Penning ion trap apparatus[9] but no CH_2^+ product was detected. Based on an estimate of the sensitivity of the method used, an upper limit of $k_r < 1.5 \times 10^{-15}$ cm^3/s was derived. Further experimental studies have been performed in this laboratory using two different ring electrode ion traps[6] and an more precise estimate for k_r was reported, $k_r(80$ K$) < 7\times10^{-16}$ cm^3/s.

	$k_r/\text{cm}^3\text{s}^{-1}$	$k_3/\text{cm}^6\text{s}^{-1}$	τ_{dis}/ns	τ_r/ms
$C^+ + n\text{-}H_2$	6.8±0.9 (-16)	1.1±0.1 (-28)	0.2	0.47
$C^+ + p\text{-}H_2$	1.7±0.2 (-15)	1.8±0.2 (-28)	0.4	0.37
$C^+ + D_2$	1.5±0.2 (-15)	4.5±0.5 (-28)	1.6	1.3
$CH_3^+ + n\text{-}H_2$	5.0±2.0 (-14)	6.0±1.2 (-26)	120.	3.9
$CH_3^+ + p\text{-}H_2$	1.1±0.1 (-13)	1.9±0.2 (-25)	390.	5.7
$C_2H_2^+ + n\text{-}H_2$	1.3±0.1 (-12)	3.6±0.4 (-25)	770.	0.88
$C_2H_2^+ + p\text{-}H_2$	4.7±1.2 (-12)	1.5±0.3 (-24)	3200.	1.
$C_2D_2^+ + D_2$	7.9±0.9 (-12)	1.9±0.2 (-24)	7700.	1.1

Tab. I. Summary of ternary and radiative association rate coefficients for the indicated collision systems. The lifetimes were evaluated on the basis of a simple model (see Ref. 6) using a stabilization efficiency $f = 0.2$.

New experiments have been performed at $T_n=10$ K in the 22–pole trap. The He pulse was used for capturing and thermalizing the translational energy of the C^+ ions; however, there is not yet any experimental evidence whether or not the excited spin orbit state (excitation energy 8 meV) relaxes. Storage times have been extended up to 60 s. Since the CH_2^+ and CH_3^+ products undergo fast secondary reactions the dominant product is CH_5^+. Also other subsequent reactions (e.g. the formation of CH_7^+ and H–D exchange) have been accounted for in the determination of the rate coefficients.

In order to derive reliable values for the ternary and the radiative rate coefficients, experiments have been performed at H_2 number densities ranging from 7×10^{11} cm^{-3} to 2×10^{14} cm^{-3}. In this number density range the measured apparent second–order rate coefficient can be approximated by

$$k^* = k_3 [H_2] + k_r . \tag{2}$$

Here k_3 is the termolecular, $[H_2]$ the number density, and k_r the bimolecular radiative rate coefficients. The dependence of measured effective rate coefficients on $[H_2]$ is shown in Fig. 4 for $C^++p\text{-}H_2$ association at $T_n=10$ K. The results can be nicely fitted with the linear function given by Eq. (2). In the log–log representation, contributions from the additive term k_r become apparent by the curvature of the fit, which is obvious at $[H_2]$ below 10^{13} cm^3/s. Similar results for collisions of C^+ with n–H_2 have been presented recently[2]. For clearness these data are omitted in Fig. 4 but represented by the fit (dashed curve). The resulting values for k_r and k_3 are given in Tab. I. It is interesting to note that radiative association with rotationally excited $H_2(j=1)$ is about 5 times slower. This can be seen from the dash–dotted curve which was derived from the two other fits.

In view of the rather small rate coefficients for radiative association it is it may be necessary to consider also the above mentioned endothermic hydrogen abstraction reaction in models of interstellar clouds. As discussed recently[1], it cannot be excluded that those C^+ ions, which are formed by dissociative charge transfer in collisions with He^+, have sufficient kinetic energy to produce CH^+ ions with a rate comparable to that of radiative association.

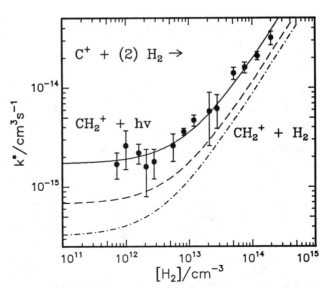

Fig. 4. Ternary and radiative association of C^+ with p–H_2 at a nominal temperature of 10 K, measured over a wide range of [H_2]. The error bars indicate exclusively the statistical errors. The solid line represents a fit of Eq. 2 to the measured effective rate coefficients with the parameters
$k_3 = 1.8 \times 10^{-28}$ cm^6/s,
$k_r = 1.7 \times 10^{-15}$ cm^3/s.
The dashed line shows the corresponding fit to data determined with n–H_2 (see Ref. 2), the dash–dotted line is for pure $H_2(j=1)$.

Based on several assumptions (Langevin rate coefficients, third–body stabilization efficiency $f = 0.2$, for details see Ref. 6) one obtains the collision complex and radiative lifetimes given in Tab. I. Note, that the model, used for evaluating k_3 and k_r, characterizes the complexes with only one single mean value for both the radiative and the complex lifetime. Therefore the following conclusions should not be overestimated. The derived values for τ_r are much larger than those used in the theoretical estimates and their magnitude raises the question whether exclusively electronic transitions contribute to radiative stabilization, or whether vibrational transitions are also involved. In order to shed some more light onto this problem experiments with D_2 reactant gas have also been performed. The results are included in Tab. I. As expected, the higher density of states of the $CD_2^+ \cdot D_2$ complex leads to an increase of the complex lifetime. In contrast to a pure electronic transitions, the evaluation with the simple model results also in a larger value for the radiative lifetime of the CD_2^+ complex. This indicates that nuclear motion is involved; however, it is not clear, whether this is due to vibrational coupling between the involved potential energy surfaces or due to infrared transitions. More insight into this "simple" triatomic system may be obtained by performing spectroscopic studies on CH_2^+ in the vicinity of the dissociation limit.

After formation of CH_2^+ via radiative association, a fast hydrogen abstraction with H_2 leads to CH_3^+. The next step in the chain of formation of small hydrocarbons, formation of CH_4^+, is endothermic and cannot occur under interstellar conditions. Therefore the only reaction of CH_3^+ at low temperatures is again a radiative association process,

$$CH_3^+ + H_2 \rightarrow CH_5^+ + h\nu. \qquad (3)$$

This process has attracted the attention of many theorists and experimentalists. However, despite considerable effort on both parts, there are still some conflicting interpretations, partly caused by the fact that it was not straightforward to use theoretical models to compare 80 K high number density SIFT results with 13 K low number density ion trap measurements.[10-14] In addition, there was an apparent discrepancy between the radiative rate coefficient measured in the pioneering Penning ion trap experiment by Barlow et al.[14] at a temperature of 13 K, $k_r=1.1\times10^{-13}$ cm^3/s, and in the 80 K ring electrode ion trap, $k_r=6\times10^{-15}$ cm^3/s.[12] For a temperature change from 80 K to 13 K, theories predict an increase of the rate coefficient by a factor varying between 2.5 and 4 whereas the ratio between the two experimental values is 17.[6]

Additional experiments were performed at $T_n=10$ K by determining effective cross sections at [H$_2$] between 10^{11} cm^{-3} and 10^{13} cm^{-3} and by using the fitting procedure described above. The results for ternary and radiative association are given in Tab. 1. The deviation of only a factor 2 from the Penning trap experiment must be regarded as good, especially if one considers the various difficulties in determining radiative association rate coefficients. A thorough discussion of the possible sources of error in both experiments was given in Ref. 6. Using the simple one–complex model the mean radiative and complex lifetimes have also been determined. The result that τ_r has several ms clearly indicates that the CH$_5^+$ complex is stabilized by the spontaneous emission of infrared photons as already concluded in Ref. 12. This result it is in serious conflict with the radiative lifetime given by Bates, $\tau_r=2.9\cdot10^{-5}$ s.[11] The role of ortho and para hydrogen, will be discussed in a separate publication.

Another process which plays an important role in the chain of formation of small hydrocarbons at low temperatures are the collisions between acetylene ions and hydrogen. In contrast to the other two systems discussed above, the energetics in this case are not so clear and the competition between the hydrogen abstraction reaction

$$C_2H_2^+ + H_2 \longrightarrow C_2H_3^+ + H \qquad (4)$$

and radiative (and, at high [H$_2$], ternary) association

$$C_2H_2^+ + H_2 \longrightarrow C_2H_4^+ + h\nu \qquad (5)$$

has attracted a lot of interest in the last two years and various experimental and theoretical studies have focused on the question whether the formation of the C$_2$H$_3^+$ products is endothermic and/or hindered by a barrier.[15-17] The possibility that reaction (4) prevails at low temperatures is of importance for astrophysical applications since C$_2$H$_3^+$, which has a strong microwave emission spectrum due to its dipole moment, is a good candidate for being detected and thus as an indicator of the carbon chemistry.

Measurements performed with low temperature swarm[15] and ion trap techniques[6] have revealed that the rate coefficient for C$_2$H$_3^+$ formation via reaction (4) is less than 5×10^{-13} cm^3/s at 80 K and that it increases with increasing temperature. The strong positive temperature dependence was attributed to an endothermicity of about 4 kJ mol^{-1}.[15] The question whether this assumed endothermicity is in reality a barrier was raised by measurements performed in a low temperature free–jet flow reactor[16] in which a significant C$_2$H$_3^+$ signal was found. From these data a rate coefficient of 10^{-12} cm^3/s was deduced at $T_n=10$ K with an increasing tendency towards 10^{-11} cm^3/s if the temperature was lowered to 1 K (see dots in Fig. 5). This negative temperature dependence was qualitatively explained by Hawley and Smith by a mechanism which involves low temperature tunneling. Based on a corrected *ab–initio* energy reaction pathway, this hypothesis was supported by a phase space calculation[17] the results of which are included in Fig. 5 as a solid line.

In a series of low temperature ion trap studies it was tried to reproduce the low temperature behavior seen in the free–jet experiment, but a significant production of $C_2H_3^+$ at temperatures below 80 K has never been detected.[6] The first reactivity measurements were confused by $C_2H_3^+$ background ions which were formed during the thermalization of the primary ions and which undergo association with H_2 to form $C_2H_5^+$; however, this process is very slow as determined in a separate study with directly injected $C_2H_3^+$. In recent experiments this problem was significantly reduced by utilizing the intense He–pulse for thermalization described in Fig. 2, by operating at very low H_2 number densities and by extending the storage time up to 1 min.

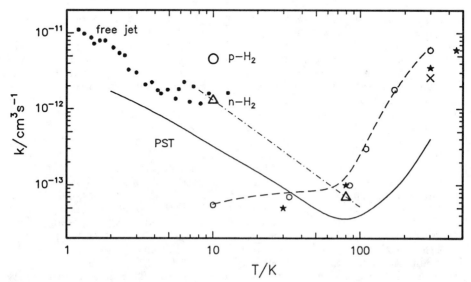

Fig. 5. Temperature dependence of bimolecular rate coefficients for formation of ions with mass 27 in collisions of $C_2H_2^+$ with $n-H_2$. The dots at low temperatures are data from the free jet experiment[16], the cross at 300 K is a recent SIFT result[18], the stars are ring electrode ion trap results[6], and the open circles (connected with the dashed line) are current results measured in the 22–pole ion trap (assuming that all the mass 27 u products are $C_2H_3^+$). The phase space prediction[17] is given as solid line. For comparison, the rate coefficients for radiative association of $C_2H_2^+$ with $n-H_2$ are shown as open triangles; the dash–dotted line emphasizes the strong temperature dependence. A surprisingly large value for k_r is measured at 10 K if one uses $p-H_2$ (large open circle).

A selection of previous ring electrode trap[6] measurements and new 22–pole ion trap data (connected with a spline fit) are included in Fig. 5. The slight discrepancies of the 300 K values with that from a recent SIFT measurement[18] (cross) will be discussed elsewhere. The steep decrease of the rate coefficients seems to level off at temperatures below 80 K; however, this change in slope is only due to the fact that in the evaluation of the data is was assumed that all ions with mass 27 u are $C_2H_3^+$ products (i.e. disregarding C_2HD^+, see below). Therefore, the rate coefficients plotted below 80 K can be regarded as safe upper limits for hydrogen abstraction reaction (4), i.e., this process must be slower than 5×10^{-14} cm^3/s.

A closer analysis of other possible reactions, e.g. also the influence of ions containing ^{13}C has revealed the importance of H–D exchange reaction in

collisions with HD present in the hydrogen, especially the exothermic reaction
$$C_2H_2^+ + HD \rightarrow C_2HD^+ + H_2 . \tag{6}$$
which also leads to the formation of ions with mass 27 u as recently discussed.[19] For a quantitative evaluation, the rate coefficient for this process, $k(10\ K) = (7\pm3)\times10^{-10}$ cm^3/s, was determined directly in the ion trap using HD with a purity of >97%. The $C_2H_4^+$ association products become also partially deuterated by
$$C_2H_4^+ + HD \rightarrow C_2H_3D^+ + H_2 . \tag{7}$$
In experiments with HD—purified hydrogen ($[HD]/[H_2]<4\times10^{-5}$) the formation of C_2HD^+ and $C_2H_3D^+$ was carefully studied as a function of storage time. Some typical data are shown in the right panel of Fig. 3 together with the solutions of a rate equation analysis. In this analysis, radiative association, H–D–, and D–H–exchange was included, but no $C_2H_3^+$ formation. It is important to note that both deuterated products, C_2HD^+ and $C_2H_3D^+$, are nicely fitted by using just one set of parameters for both reactions (6) and (7), i.e., the number density $[HD] = 6.2\times10^5$ cm^{-3} and $k = 7.5\times10^{-10}$ cm^3/s. Similar observations have been made for other hydrocarbon ions (see for example left panel in Fig. 3). Since in the analysis $C_2H_3^+$ formation was completely disregarded one can conclude that the rate coefficient for reaction (4) at 10 K is less than 10^{-14} cm^3/s. The largest uncertainty in deriving a final value is the actual value of [HD] since it has not yet been determined directly *in situ*. To obtain a precise temperature dependence of the rate coefficient for $C_2H_3^+$ formation below 80 K, experiments with hydrogen, containing almost no HD, would be necessary.

These conclusions are in obvious contradiction to the free—jet flow reactor rate coefficients which are also included in Fig. 5. For a better judgment of these controversial results which are also partly discussed in Ref. 19 and 20 a few additional points are briefly mentioned:
(i) There is no doubt that the ions stored in the 22—pole trap are really thermalized. Various measurements lead to the conservative estimate that, at $T_n=10$ K, the actual temperature is 15 ± 5 K.
(ii) The temperature dependence of the radiative association rate coefficient clearly shows that the complex lifetime increases strongly with decreasing temperature; therefore, a tunneling process, if occurring, should become observable, especially for p–H_2 where the mean complex lifetime reaches 3.2 ms (see Tab. I).
(iii) Tests with directly injected $C_2H_3^+$ prove that there is no loss of these ions, e.g. by subsequent reactions.
(iv) The rate coefficient for the isotopic variant of reaction (4), i.e. $C_2D_2^+ + D_2 \rightarrow C_2D_3^+ + D$, is also less than 10^{-14} cm^3/s.
(v) The 10 K ion trap rate coefficient for ternary association with n–H_2 (see Tab. I) is significantly larger than that determined in the free—jet experiment.[16] The may partly be explained with saturation effects which are discussed in detail by Herbst and Yamashita.[21] Note that at 3 K and at the high densities prevailing in the free—jet (up to 10^{15} cm^{-3}) the effective two—body rate coefficient for $C_2H_4^+$ formation reaches values above 10^{-10} cm^3/s.[21]

In conclusion it seems to be necessary to analyze in more detail the experimental conditions of the multiple collision environment of the free—jet to obtain a proper understanding of the origin of the $C_2H_3^+$ ions. What are the implications for astrophysical models? In these applications it is advisable to use the ion—trap measurements and to assume that the bimolecular reaction (4) is completely negligible at temperatures below 30 K and that the radiative association process (5) plays the dominant role in low temperature collisions of acetylene ions with hydrogen. Recently published[19] low temperature results for k_r and k_3 are given in Tab. I.

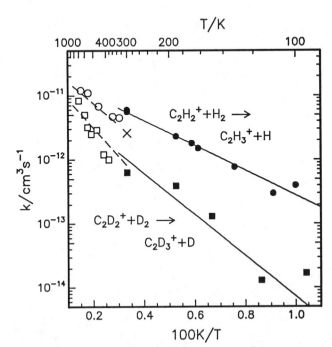

Fig 6. Arrhenius plot of bimolecular rate coefficients for the formation of $C_2H_3^+$ in collisions of $C_2H_2^+$ with n–H_2 (circular symbols, squares are for deuterated reactants). The open symbols show the drift field dependent SIFDT results,[18] the solid ones thermal ion trap results. Note that the energy distributions in the SIFDT experiment are non–thermal. The indicated fits lead to the following Arrhenius energies:
H_2 SIFDT: 5.9 kJ mol^{-1}
  Trap: 4.6 kJ mol^{-1}
D_2 SIFDT: 11.6 kJ mol^{-1}
  Trap: 6.0 kJ mol^{-1}.
The cross at 300 K is a recent SIFT result.[18]

There remains finally the question whether or not the hydrogen abstraction reaction $C_2H_2^+ + H_2 \rightarrow C_2H_3^+ + H$ is really endothermic or whether the low temperature behavior is due to a barrier which is large enough to hinder tunneling. The only clear experimental answer which is presently available is the measurement and evaluation of both the forward and reverse rate coefficients of reaction (4) in a SIFT experiment which indicates that formation of $C_2H_3^+$ is endothermic by 7.5 kJ mol^{-1}.[18] A very recent high–level *ab initio* quantum–chemical calculation[22] has corroborated this result, indicating a value of 8.4 kJ mol^{-1}. This is also in reasonable agreement with the Arrhenius energies derived from the slopes of both the SIFDT and ion trap data plots shown in Fig. 6. The obvious deviations in the slopes of these plots are most probably caused by the differences in populations amongst the different energy states of the reactants in the two experiments. More details, including a discussion of the reaction of the deuterated species (see Fig. 6) and also results from state specific measurements, will be published later.

CONCLUSIONS

Remarkable experimental progress was achieved in the last years in the field of ion–molecule reactions. This contribution has focused predominantly on recent advances in low temperature experiments; however, there are also various other experimental developments such as crossed and merged beams[2] which can provide information needed to understand reactions of ions at higher energies and under non thermal conditions prevailing in many astronomical environments. Further experimental progress relevant to interstellar chemistry is to be expected by combining existing techniques with laser methods for preparation of the reactants, for analysis of the products, for spectroscopy and for particle detection. For example, the modification of the low temperature equilibrium of stored ions by exposing them to a continuous infrared laser field can provide both spectroscopic information and specific reaction rate coefficients. The use of rf traps for storing charged microparticles or experiments with atomic targets such as N, C or O are also within the reach of the existing methods. These few examples reveal that one can expect more progress in experimental studies of ion–molecule reactions relevant to interstellar chemistry.

ACKNOWLEDGMENTS

The major contribution to this work is due to A. Sorgenfrei. Many thanks are due to Prof. M. Smith and especially to Prof. D. Smith for stimulating discussions and helpful advice. The author is indebted to Prof. Ch. Schlier for his continuous interest in this work. Support of the Deutsche Forschungsgemeinschaft (SFB 276) is gratefully acknowledged.

REFERENCES

1   D. Smith, *Chem. Rev.*, **92**, 1471 (1992).
2   D. Gerlich in: ICPEAC 1993, *Book of Invited Papers*, eds. T. Anderson et al. (1993).
3   E. T. Galloway and E. Herbst, *Astron. Astrophys.* **211**, 413 (1989).
4   M. A. Smith in: *Current Topics in Ion Chemistry and Physics*, Vol. 2, C. Y. Ng, T. Baer, and I. Powis, eds., Wiley, New York (1993).
5   D. Gerlich, *Adv. in Chem. Phys.* **LXXXII**, 1 (1992).
6   D. Gerlich and S. Horning, *Chem. Rev.* **92**, 1509, (1992).
7   D. Gerlich, *J. Chem. Soc., Faraday Trans.* **89**, 2199 (1993).
8   W. Paul and D. Gerlich in: ICPEAC 1993, *Book of contributed papers*, eds. T. Anderson et al. 807, (1993).
9   J. Luine, G. Dunn, *Ap. J.* **299**, L67 (1985).
10   N. G. Adams, D. Smith, *Chem. Phys. Lett.* **79**, 563 (1981).
11   D. R. Bates, *Ap. J.* **312**, 363 (1987).
12   D. Gerlich and G. Kaefer, *Ap. J.* **347**, 849 (1989).
13   D. R. Bates, *Ap. J.* **375**, 833 (1991).
14   S. E. Barlow, G. H. Dunn, M. Schauer, *Phys. Rev. Lett.* **52**, 902 (1984).
15   D. Smith, N. G. Adams, E. E. Ferguson, *Int. J. Mass Spectrom. Ion Proc.* **61**, 15 (1984).
16   M. Hawley and M. A. Smith, *J. Chem. Phys.*, **96**, 1121 (1992).
17   K. Yamashita, E. Herbst, *J. Chem. Phys.* **96**, 5801 (1992).
18   D. Smith J. Glosik, V. Skalsky, P. Spanel and W. Lindinger, *Int. J. of Mass Spectrom. Ion Proc.* in print (1993).
19   D. Gerlich, *J. Chem. Soc., Faraday Trans.* **89**, 2210 (1993).
20   M. Smith, *J. Chem. Soc., Faraday Trans.* **89**, 2209 (1993).
21   E. Herbst, K. Yamashita, *J. Chem. Soc. Faraday Trans.* **89**, 2175 (1993).
22   S. A. Maluendes, A. D. Mc. Lean, and E. Herbst, *J. Chem. Phys.* submitted (1993).

QUESTION

SCAPPINI — You have mentioned the following reaction:
$$CH_3^+ + CO \rightarrow H_3CO^+ + C$$
I would like to know if the thermicity of this reaction is known or, if not, whether an experiment to determine it is feasible.

GERLICH — I do not know the details about the thermicity of this reaction; however, there is no doubt that atom abstraction with CO is endothermic due to the large C—O bond strength. Since, in addition, the proton affinity of CO is very small only ternary association or, at low densities, the radiative association
$$CH_3^+ + CO \rightarrow CH_3CO^+ + h\nu$$
has been observed experimentally at room temperature and below. An experimental value for the thermicity of the O—atom abstraction reaction can be determined by measuring the kinetic energy dependence of the H_3CO^+ formation, using for example a Guided Ion Beam apparatus.

ROWE —
1. Do you know the reason why there was a problem on $C_2H_2^+ + H_2$ in Mark Smith's experiment?
2. In your merged beam experiment do you plan to work with condensable species such as H_2O or NH_3? If yes, how well defined are the rotational distributions of these species?

GERLICH —
1. I have discussed with Mark Smith many different possibilities, including fragmentation of $C_2H_4^+$ association products during the extraction, the role of HD, excitation of $C_2H_2^+$ etc. We didn't find a clear evidence for a problem in that experiment and concluded that more experimental tests are needed. I personally have not yet analyzed in detail the evaluation procedures of the measurements, but they seem to be rather complicated due to the high density environment (up to 10^{15} cm^{-3}) and the non-equilibrium conditions. In contrast, the evaluations of our experiments are straightforward since they have been performed at very low densities and under quite different conditions.
2. In our merged beam experiment, we have performed successfully first experiments with H_2O, seeded in He. As you know, the rotational temperature can be varied significantly by using different expansion conditions and we plan to determine cross sections for collisions with polar mole-

cules as a function of the rotational state. For this goal we intend to measure the rotational populations using in situ laser methods.

LEACH — In radiative association, what is known about the levels actually emitting and the population of these levels as a function of temperature in the experiment? If the emitting levels have a variety of emission rates, any temperature dependence of their populations would affect the competition of the radiative association reaction with respect to other channels.

GERLICH — In many ion—molecule collisions such as $H^+ + H_2$ or $CH_3^+ + H_2$ the collision takes place on a single potential energy surface, which often has a well several eV deep. Therefore the intermediate complex can be regarded as a highly excited molecule and the radiative stabilization occurs by an infrared transition from a continuum state into a rovibrational quasicontinuum. It is unlikely that different continuum states decay with very different radiation rates. For a few systems we have performed some correlated tests by using an infrared fragmentation method and found an exponential decay with one single time constant. In other situations, where several electronic potential energy surfaces can become involved (e.g. $C^+ + H_2$) the situation is less clear and significant differences in radiation lifetimes can be expected. However, the dominant temperature dependence of the radiative rate coefficients is due to the temperature dependence of the lifetime of the complex. Here it is very important to remember that these lifetimes are mean values and that they strongly depend on the total energy, the orbital angular momentum and also on other (approximate) constants of the motion. The only experimental way I see at the moment to obtain some quantitative answers to your questions is to use the method of laser stimulated radiative association.

SHALABIEA — Do you have a good estimate for the rate coefficient for the reaction
$$H_3^+ + HD \rightarrow H_2D^+ + H_2 ?$$

GERLICH — As discussed in detail during the 28. Faraday Symposium, the experimental determination of this rate coefficient at low temperatures is complicated by problems such as thermalization of H_3^+, formation of clusters, and by various secondary reaction with H_2 and D_2 admixtures, especially if we use in addition to the HD target gas p—H_2 for thermalizing the H_3^+ ions. The most reliable information we can provide at the moment in relation to your question is the measured H_3^+/H_2D^+ equilibrium fractional abundance. If we store H_3^+ ions in natural hydrogen ($[HD]/[H_2] = 3 \times 10^{-4}$) at 15 ± 5 K we obtain for n—H_2 $[H_2D^+] / [H_3^+] = 0.002 \pm 0.001$ and for p—H_2 $[H_2D+] / [H_3^+] = 0.15 \pm 0.1$.

Note that the actual HD/H_2 ratio in the trap is not precisely known. The small value measured in the case of n—H_2 is most probably also influenced by the fact that the stored H_2D^+ ions may be rather "hot" due to energy pooling in subsequent inelastic collisions with $H_2(j=1)$.

The best estimate for the D—H exchange rate coefficient we can propose presently is $(3 \pm 2) \times 10^{-10}$ cm^3/s This value has been determined from a measurement with pure HD. The largest uncertainty (not included in the error bars) is the internal energy of the H_3^+.

ION–TRAP EXPERIMENTS ON $C_3H^+ + H_2$: RADIATIVE ASSOCIATION vs. HYDROGEN ABSTRACTION

A. Sorgenfrei, D. Gerlich

Fakultät für Physik der Universität, D–79104 Freiburg, Germany

ABSTRACT

Various reaction products, formed in collisions of C_3H^+ with $n-H_2$, $p-H_2$ and D_2, have been studied over a wide range of temperatures and densities using a 22–pole ion–trap. The rate coefficients for forming $C_3H_2^+$ have been found to increase with falling temperature and, between 100 and 10 K, they remain almost constant at a value close to 10^{-10} cm^{3}s^{-1}. This proves that hydrogen abstraction is not endothermic, as concluded from earlier swarm measurements and from *ab initio* calculations. However, all reaction steps involved are so slow that, at 10 K, a large fraction of the collision complexes are stabilized via emission of a photon. This competition among radiative association, abstraction reaction, and decay back towards reactants is tentatively explained with a model in which potential barriers exist in both the entrance and exit channels.

INTRODUCTION

In the last years, there has been considerable interest in the association reaction

$$C_3H^+ + H_2 \rightarrow C_3H_3^+ + h\nu \qquad (1)$$

and the competing hydrogen abstraction reaction

$$C_3H^+ + H_2 \rightarrow C_3H_2^+ + H . \qquad (2)$$

One of the reasons for this interest is that both sets of ionic products most probably play an important role as possible precursors for forming the cyclic neutral molecules, c–C_3H_2 and c–C_3H, both of which have been observed in interstellar clouds.[1-3]

In a selected ion flow tube (SIFT) experiment, saturated ternary association with He; i.e.,

$$C_3H^+ + H_2 + He \rightarrow C_3H_3^+ + He , \qquad (3)$$

has been identified as the only detectable channel at 80 K whereas products from reaction (2) have been found only at higher temperatures.[4-6] This led to the assumption that formation of $C_3H_2^+$ is slightly endothermic and to the conclusion that only reaction (1) is of astrophysical significance.[6] In an attempt to measure directly the rate coefficient for the radiative process in an ion–trap apparatus it has been discovered that at low densities reaction (2) is the dominant channel;[7] however, this channel is suppressed by reaction (3) even at densities much lower than those prevailing in flow tubes.

The overall situation is illustrated in Fig. 1 which shows a selection of effective binary rate coefficients as a function of He buffer gas density. The SIFT results[5] (open squares) are independent of the helium gas density at both 80 K and 300 K, and only at 550 K is a slight increase evident. The 80 K ion–trap results from Ref. 7 are reproduced in the lower panel of Fig. 1. The competition between the two products $C_3H_2^+$ and $C_3H_3^+$ is indicated by the fact that the sum of both rate coefficients (solid dots) remains constant. The significant density dependence has been explained quantitatively with a rather elaborate model, the results of which are in very good agreement with both the

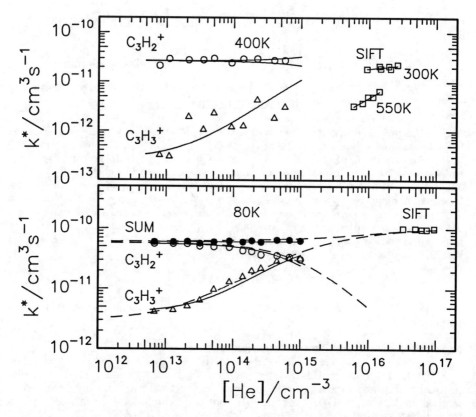

Fig. 1: Effective rate coefficients for formation of $C_3H_2^+$ and $C_3H_3^+$ products in $C_3H^+ + H_2$ collisions as a function of the He buffer gas density and at various temperatures. The SIFT measurements (open squares) have been performed at such high densities, that almost all collisions lead to $C_3H_3^+$ formation via saturated ternary association.[5] The ion–trap experiments clearly show that at low densities the $C_3H_2^+$ channel becomes dominant and that its rate coefficient increases with falling temperature. The theoretical model which has been used to calculate the depicted lines, has been explained in detail in Ref. 7. Some remarks concerning the resulting model parameters can be found in the text.

SIFT and the ion–trap results (dashed lines, for details see Ref. 7). In Fig. 1 the 80 K and the 400 K ion–trap results are fitted with a simpler model (solid lines), which is based on the assumption that only one class of complexes is formed with a rate coefficient k_c. Subsequent steps are described by the rate of unimolecular redissociation to reactants, $1/\tau_{dis}$, the rate of dissociation into reaction products, $1/\tau_{rxn}$, the radiative emission rate, $1/\tau_r$, and the rate of stabilizing collisions with He, $1/\tau_s = k_s[\text{He}]$. The fitted parameters are:

400 K: $k_c = 5.7 \times 10^{-11}$ cm^{3}s^{-1}, $k_s = 5 \times 10^{-10}$ cm^{3}s^{-1}
            $\tau_{dis} = 9 \times 10^{-7}$ s, $\tau_{rxn} = 1 \times 10^{-6}$ s, $\tau_r = 1 \times 10^{-4}$ s

80 K: $k_c = 6.2 \times 10^{-11}$ cm^{3}s^{-1}, $k_s = 5.6 \times 10^{-10}$ cm^{3}s^{-1}
            $\tau_{dis} = 8 \times 10^{-5}$ s, $\tau_{rxn} = 2.5 \times 10^{-6}$ s, $\tau_r = 3.5 \times 10^{-5}$ s

Note that these numbers are based on rather crude model assumptions and are only mean values. Nonetheless, there is no doubt that extremely long time constants play a role since they are directly related to the experimentally observed competition between reaction and association and also to the early saturation of threebody stabilization. The new experimental results, presented below, will indicate that at temperatures below 80 K all reaction steps are even slower, and that the mean lifetime of the $C_3H_3^+$ collision complexes becomes so long that a significant fraction is stabilized via emission of a photon. In addition, the influence of the J=1 rotational energy of the hydrogen target and first results from experiments with deuterated reactants will give some further guidance for understanding the dynamics of the $C_3H^+ + H_2$ reaction.

Besides the mentioned experimental efforts, a variety of calculations have been undertaken on aspects of the $C_3H_3^+$ system (see Refs. 8, 9, and references therein). The very recent *ab initio* minimum energy pathway calculated by Maluendes *et al.*[8] shows that the C_3H^+ and H_2 reactants can form directly the strongly bound propargyl ion, which can subsequently undergo fast isomerization to the lower energy cyclic isomer $C_3H_3^+$. These authors also performed phase space dynamics calculations to determine rate coefficients for radiative and ternary association and to get an estimate for the H_2CCCH^+ : c–$C_3H_3^+$ product branching ratio. Unfortunately, the calculations fail to reproduce a large fraction of the experimental observations. The most evident problem is that in the *ab initio* calculations[8] the c–$C_3H_2^+ + H$ product channel lies 7.3 kJ mol^{-1} above the energy of the reactants (HCCCH$^+ + H$: 27 kJ mol^{-1}), whereas ion–trap results clearly show that reaction (2) is the dominant process at 80 K.[7] The argument, used in some discussions, that the ions are not fully thermalized in the ion–trap experiment, will be ruled out by the present study. Other experimental results which are in conflict with the theory are that the measured association rate coefficients saturate at effective rate coefficients well below the Langevin limit. There are also some discrepancies concerning complex and radiative lifetimes, although the calculated and measured rate coefficient for radiative association are within a factor of a few of one another.[8] In summary, the *ab initio* minimum energy pathway is inconsistent with previous and also with new experimental results and, therefore, we are forced to invent a modified schematic energy profile for interpreting the measurements.

EXPERIMENTAL

The experiments, presented in this contribution have been performed on an improved variable temperature 22–pole ion–trap (T_n=10 K − 300 K). The principle of several radio frequency (rf) storage devices and the apparatus have been discussed in a comprehensive review.[10] More details concerning recent improvements and experimental tests can be found in another contribution in this book of proceedings.[11]

The C_3H^+ ions were generated by electron bombardment of methyl acetylene in a separate ion source, mass selected, and then injected into the trap. A modification which is most important for the present studies is the use of a piezo–electric valve, which allows us to inject an intense pulse of buffer gas for efficient thermalization of the C_3H^+ ion within a few ms.[11] Rate coefficients for forming $C_3H_2^+$ and $C_3H_3^+$ in collision with H_2 target gas are derived from the temporal change of the ion composition, two typical examples of which are depicted in Fig. 2. The number of ions of each individual mass, in total typically only a few hundred per filling, is determined by periodically injecting the primary ions into the trap, extracting its content after various times, and counting them after mass analysis. To ensure that the ions are really thermalized, we have not only used the intense He–pulse but we have also operated with storage times between 100 ms and 60 s. To reduce the reaction rates, hydrogen densities as low as 10^9 cm^{-3} have been utilized. For a detailed evaluation of the observed time dependence, the data are fitted in a suitable time range (see Fig. 2). The fits are based on solutions of a rate equation approach which accounts not only for reactions 1–3 but also for minor contributions from secondary reactions such as $C_3H_2^+ + H_2 \rightarrow C_3H_3^+ + H$, which is exothermic but very slow, or for H–D exchange with HD impurities. Additional experiments have been performed by directly injecting C_3^+ or $C_3H_2^+$, by taking p–H_2 ([$H_2(j=0)$]>99%) instead of n–H_2, and by using deuterium instead of hydrogen.

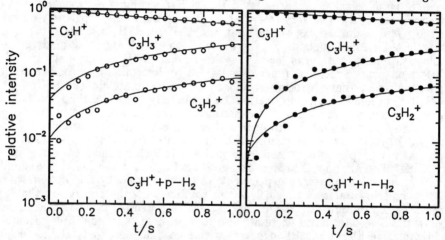

Fig. 2: Time dependence of the indicated ion intensities at T_n=10 K and [H_2]=1.5×10^{9}cm^{-3}. The lines are solutions of a rate equation approach using k_r= 2.5×10^{-10} cm^{3}s^{-1} and k_{rxn}=4×10^{-11} cm^{3}s^{-1} for p–H_2 and k_r= 1.9×10^{-10} cm^{3}s^{-1} and k_{rxn}=5×10^{-11} cm^{3}s^{-1} for n–H_2. The slight deviations at short times arise because the initially hot C_3H^+ ions react differently.

RESULTS

As can be seen from the upper panel of Fig. 1, the rate coefficient for forming $C_3H_2^+$ at 400 K is only k_{rxn}=2.8×10^{-11} cm^{3}s^{-1} but it increases if one goes to 80 K (see lower panel). If one further decreases the temperature, a further increase is observed but, below 50 K, k_{rxn} remains more or less constant at about 10^{-10} cm^{3}s^{-1}, as can be seen from the right panel of Fig. 3. The results for p–H_2 are somewhat smaller and there seems to be also some indication that, below 30 K, the rate coefficients have a tendency to fall. These two observations could be taken as a hint that the hydrogen abstraction channel is endothermic, but, in contrast to earlier conclusions from swarm measurements, the endothermicity must be less than a few tenths of a kJ mol^{-1}. There is, however, another reason for this decline; in this case formation of $C_3H_2^+$ competes with association reactions. But in contrast to the competition by ternary collisions, as depicted in Fig. 1, only radiative association can play a role in the present

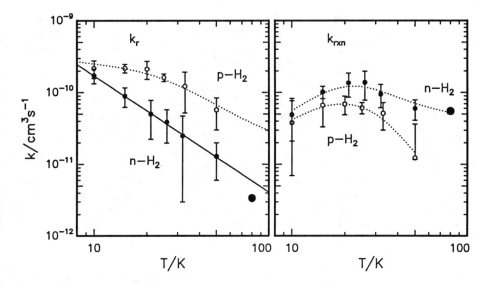

Fig. 3: Temperature dependence of the rate coefficient for the radiative association reaction $C_3H^+ + H_2 \rightarrow C_3H_3^+ + h\nu$ (left panel) and the hydrogen abstraction reaction $C_3H^+ + H_2 \rightarrow C_3H_2^+ + H$ (right panel). Collisions with p–H_2 (open circles) lead apparently to longer complex lifetimes and therefore to larger values of k_r than with n–H_2 (filled circles); k_{rxn} shows the opposite trend. The dotted lines are splines for guiding the eyes, the line fitting the n–H_2 results decreases proportional to $T^{-1.6}$. The larger symbols are the 80 K results from Ref. 7.

ion–trap experiment, since there is no He–buffer at all and the hydrogen target density is below 10^{11} cm^{3}s^{-1}. The temperature dependence of the rate coefficient for stabilizing $C_3H_3^+$ by photoemission is shown in the left panel of Fig. 3. For n–H_2 target, k_r increases very steeply and reaches a value close to 2×10^{-10} cm^{3}s^{-1} at 10 K. The fact that the use of p–H_2 target leads to an increase of k_r is in qualitative accordance with statistical calculations.[8] The j–dependent decrease of k_{rxn} and also the differences in the temperature dependencies will be discussed below.

Comparison of the two panels of Fig. 3 corroborates the result reported in Ref. 7 that at 80 K hydrogen abstraction is the dominant channel, but it also reveals that the trend is reversed and that at low temperatures radiative association prevails. From this behavior one can conjecture that the slight decline of k_{rxn} below 30 K is caused by competition, i.e., a large fraction of the collision complexes emit a photon before they get a chance to decay to $C_3H_2^+$. This idea is supported by the fact that the sum $k_r + k_{rxn}$ is a function which increases monotonously with falling temperature for n–H_2. For p–H_2 and below 30 K, this sum reaches a saturation value of about 20% of the Langevin limit. A careful verification of the density independence between 10^9 cm^{-3} and 10^{11} cm^{-3} has proven that both products are formed in a bimolecular process. Saturated ternary association with a second H_2 partner would require a ternary rate coefficient larger than 10^{-19} cm^{6}s^{-1} which is completely unreasonable.

Interesting supplementary information has been obtained by experiments with deuterated modifications. A selection of rate coefficients determined at 10 K is listed in Tab. I. Reaction of C_3H^+ with D_2 has been observed to be significantly slower; in particular, the rate coefficient for radiative association is 50 times smaller than in collisions with H_2! In this case one has in addition to the three channels listed in Tab. I the exothermic proton–deuteron exchange leading to C_3D^+ which has been found to be slower than 10^{-12} cm^{3}s^{-1}.

Rate coefficients for the fully deuterated reaction $C_3D^++D_2$ have been derived from measurements where primary C_3^+ ions were injected into the trap containing D_2 at a density of 2.4×10^{10} cm^{-3}. In this case C_3D^+ ions are formed in a first step via an exothermic deuterium abstraction process (1.3×10^{-9} cm^{3}s^{-1}). Although the mean residence time of these most probably excited products is many ms before they undergo further reactions to $C_3D_2^+$ or $C_3D_3^+$, it is not clear whether they lose all their internal energy by radiation or inelastic collisions with D_2. Therefore the rate coefficients given in the table for the fully deuterated system may slightly deviate from the thermalized 10 K values. Further studies (comparison with $C_3^++H_2$ or injection of C_3D^+ into the He–pulse) are in progress.

Tab. I. Rate coefficients for hydrogen abstraction and radiative association for different isotopic combinations of $C_3H^++H_2$. The column at right provides information on the rate coefficient as a fraction of the Langevin limit.

			$k/10^{-11}$cm^{3}s^{-3}	$\Sigma k/k_L$
$C_3H^+ + H_2$	$\rightarrow$	$C_3H_2^+ + H$	5.	
	$\rightarrow$	$C_3H_3^+ + h\nu$	20.	17 %
$C_3H^+ + D_2$	$\rightarrow$	$C_3D_2^+ + H$	2.7	
	$\rightarrow$	$C_3HD^+ + D$	1.	
	$\rightarrow$	$C_3HD_2^+ + h\nu$	0.4	4 %
$C_3D^+ + D_2$	$\rightarrow$	$C_3D_2^+ + D$	17.	
	$\rightarrow$	$C_3D_3^+ + h\nu$	13.	28 %

DISCUSSION

There are a variety of experimental observations on the $C_3H^++H_2$ system which have to be explained by a theoretical model. These observations include:
1. The measured ion–trap rate coefficients k_r+k_{rxn} and also the saturated SIFT results are significantly smaller than the Langevin value ($k_L=1.5\times10^{-9}$ cm^{3}s^{-1});
2. The ratio $\Sigma k/k_L$ increases from below 1% at 550 K to 17% at 10 K.
3. There is a class of collision complexes which live extremely long. At 80 K, competition with ternary stabilization by He reveals that their lifetime is above 10^{-5} s and, at 10 K, longer than the radiative lifetime.
4. The decay of the collision complex back towards reactants or into $C_3H_2^+$ products is so slow that it competes with ternary association at the experimen-

tal conditions of the 80 K SIFT experiment and with radiative association at 10 K.
5. There are some significant isotope effects (see Tab. I).

As already mentioned in the Introduction, these observations are in serious conflict with recent *ab initio* calculations. For all three isotope combinations the ion–trap experiments provide evidence that the hydrogen abstraction reaction occurs at a significant rate whereas Wong and Radom[9] predict an endothermicity of 4 kJ mol^{-1} and Maluendes et al.[8] 7.3 kJ mol^{-1}. This discrepancy is not so serious since, due to the uncertainty of the *ab initio* calculations, thermoneutrality or even a small exothermicity are not ruled out by these results. More problematic is that there are presently no other experiments which can provide independently support for the exothermicity of reaction (2) since the conclusions from Prodnuk et al.[12] have been shown to be based on an incorrect assumption.[9] Therefore we would like to add a few arguments here which provide additional support for the reliability of our results. One of the most obvious proofs of fully relaxed ions is the steep increase of the radiative association channel with decreasing temperature. Similar observations have been made for the growth of hydrogen clusters. Many of the possible artifacts of

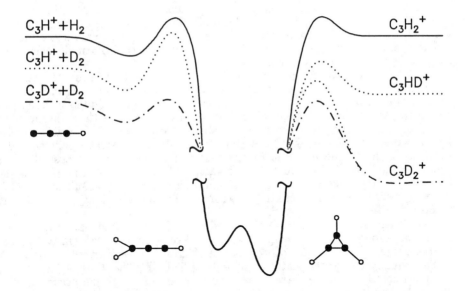

Fig. 4: The depicted schematic energy profile for the $C_3H^+ + H_2 \rightarrow C_3H_2^+ + H$ reaction we have assumed in order to interpret experimental results is shown. Note especially the barriers in the entrance and the exit channel, which are needed to slow down the transition rates. The relative positions for the deuterated analogs are shifted by their zero–point energies. Note; however, the change in scale between the long range part and the strongly bound intermediate. This plot is not in accordance with recent high level *ab initio* calculations[8] which predict that the reactants can form directly H_2CCCH^+ (−390 kJ mol^{-1}) and that a transition state (−184 kJ mol^{-1}) allows fast isomerization to the cyclic $C_3H_3^+$ (−501 kJ mol^{-1}).

our machine can be excluded by comparing the $C_3H^+ + H_2$ results with those for $C_2H_2^+ + H_2$ discussed in this book.[11] It is also important to note that we determine the exothermic reaction $C_3H_2^+ + H_2 \rightarrow C_3H_3^+ + H$ to be very slow at 10 K ($k = 5 \times 10^{-13}$ cm^{3}s^{-1}) in accordance with the calculated barrier.[9]

Besides the controversy concerning the enthalpy change, a serious discrepancy between experiment and theory is that the calculations predict minimum energy pathways which lead directly to the various isomers of the collision complex and which directly connect the cyclic $C_3H_3^+$ isomer with the c–$C_3H_2^+ + H$ product channel. In contrast to that, any explanation of the results presented in Fig. 1 and 3 requires potential barriers or dynamical bottlenecks, both in the entrance and the exit channels. From this point of view, the following interpretation of our results, based on a multi–barrier model, is not only speculative but also in conflict with the available theoretical information.

Fig. 4 shows symbolically the experimentalist's view of the minimum energy pathway. In accordance with the sequential model invoked by Smith and Adams[5] we assume that the long range polarization potential leads to formation of a weakly–bound entrance complex which is connected via a barrier to the strongly–bound intermediate. Maluendes et al.[8] have found such a pathway but disregarded it due to the dominance of the direct one. Competition between the slow transition towards the strongly bound intermediate via the bottleneck and the decay back to reactants leads to a rate coefficient for product formation which is only a fraction of the Langevin limit. The dependence of the lifetime of the weakly bound complex on its total energy can not only explain the measured temperature dependence but also the influence of the rotational excitation in the case of n–H_2 shown in Fig. 3.

The fraction of complexes that has undergone modification to the very stable complex has then three possibilities, (i) back dissociation, (ii) radiative or ternary stabilization, or (iii) decay towards the $C_3H_2^+ + H$ product channel. For simplicity, we ignore here other alternative pathways such as those discussed in Ref. 7 in connection with the two "parallel" channel model. The low barrier between the H_2CCCH^+ and the c–$C_3H_3^+$ intermediates illustrates schematically that, according to the *ab initio* result[8] that fast isomerization is possible. The energy of the $C_3H_2^+ + H$ product has been assumed to be at the same level as the entrance channel, but the exit channel must be hindered by a suitable mechanism in order to reproduce the experimentally derived long complex and reaction lifetimes.

Using estimated zero point energies for the indicated deuterated reactants and products, their energy positions have been plotted relative to the $C_3H_3^+$ curves. The heights of the barriers have been estimated in order to explain qualitatively the data presented in Tab. I. The $C_3H^+ + D_2$ reaction (dotted lines) has a large entrance channel barrier since reaction occurs only at 4% of the Langevin limit, but the exit height has been assumed to be smaller to become consistent with the fact, that fewer than 10% of the strongly bound complexes undergo radiative association, while 24% decay to C_3HD^+ and 66% to $C_3D_2^+$. The fully deuterated reactants have the largest reaction probability, i.e. the smallest entrance channel barrier, and the measured branching ratio between radiative association and abstraction reaction requires an exit channel barrier slightly below the entrance barrier. It is clear that this discussion of the isotope effects is rather speculative; however, it demonstrates that the measured changes in branching ratios will provide a very critical test for any *ab initio* potential surface and for statistical phase space calculations.

A final remark must be made concerning the radiative lifetime, τ_r, of the highly excited $C_3H_3^+$ collision complexes. In Ref. 7 a value of 2.9×10^{-5} s was derived from a fit based on a "two—complex" model. This short lifetime would require an electronic transition, but a suitable excited electronic state has not been found by Maluendes et al.[8]. Therefore these authors used a rate of only 500 s^{-1}. In Ref. 7 it was suspected that the large measured radiative rate coefficient could be in error due to other interfering reactions. This can now be excluded because, going from 80 K to 10 K, we have measured a radiative rate coefficient as large as 2×10^{-10} cm^{3}s^{-1}! This is only possible if the complex lifetime, τ_{dis}, has the same magnitude as the radiative lifetime, τ_r. Unfortunately, under these conditions, we cannot determine a reliable experimental value for τ_{dis} and τ_r by adding He buffer gas, as in the case of other reaction systems.[7]

CONCLUSIONS

This contribution gives an overview of recent low temperature ion—trap results for the $C_3H^+ + H_2$ system. The discussion clearly reveals that more theoretical work is needed to explain all the complicated dynamics leading to the various competing exit channels. For astrophysical implications and especially concerning the question how the c—C_3H_2 molecules are formed under interstellar conditions, one has to note that at 80 K hydrogen abstraction prevails and therefore c—$C_3H_2^+$ formation is the dominant product whereas at 10 K the prevailing reaction mechanism is radiative association, which leads both to H_2CCCH^+ and c—$C_3H_3^+$. The ratio which is finally obtained between these two stable isomers depends in a complex way on isomerization and relaxation, but one can expected roughly equal abundances.[8] Further experimental studies are in progress, and it is planned especially to determine experimentally the ratio of the two $C_3H_3^+$ isomers by utilizing their different reactivities.[5]

REFERENCES

1   E. Herbst, N. G. Adams, D. Smith, *Ap. J.* **285**, 618 (1984).
2   S. A. Maluendes, A. D. Mc. Lean, and E. Herbst, *Ap. J.* **417**, 181 (1993).
3   H. E. Matthews, W. M. Irvine, *Ap. J,* **298**, L61 (1985).
4   D. Smith, N. G. Adams, E. E. Ferguson, *Int. J. Mass Spectrom. Ion Proc.* **61**, 15 (1984).
5   D. Smith, N. G. Adams, *Int. J. Mass Spectrom. Ion Phys.* **76**, 307 (1987).
6   N. G. Adams, D. Smith, *Ap. J.* **317**, L25 (1987).
7   D. Gerlich and S. Horning, *Chem. Rev.*, **92**, 1509 (1992).
8   S. A. Maluendes, A. D. Mc. Lean, K. Yamashita, and E. Herbst, *J. Chem. Phys.* **99**, 2812, (1993).
9   M. W. Wong, L. Radom, *J. Am. Chem. Soc.* in press.
10 Gerlich, D. *Adv. in Chem. Phys.* **LXXXII**, 1 (1992).
11 Gerlich, D. *Recent progress in experimental studies on ion—molecule reactions relevant to interstellar chemistry*, Invited lecture, this book of proceedings.
12 S. D. Prodnuk, S. Gronert, V. M. Bierbaum, and C. H. DePuy *Org. Mass Spectrom.* **27**, 416, (1992).

ABOUT THE FORMATION OF INTERSTELLAR SiN

O. Parisel
M. Hanus
Y. Ellinger
Equipe d'Astrochimie Quantique
Ecole Normale Supérieure, 24 rue Lhomond, F-75231 Paris CEDEX 05, France
DEMIRM, Observatoire de Paris, 5 place Jules Janssen, F-92195 Meudon, France

ABSTRACT

The recent detection of the SiN molecule in the interstellar medium has given a renewed interest in the silicon chemistry. Following the carbon chemistry involved in the formation of CN, this molecule has been assumed to come from a dissociative recombination of a complex ion formed by the action of Si^+ onto NH_3. However, our preliminary calculations confirm that such an analogy, which is *a priori* attractive, is, in fact, abusive.

INTRODUCTION

Interest for the interstellar silicon chemistry in gas phase has been recently renewed by the detection[1] of SiN in outer envelope of IRC +10216 : silicon compounds were usually expected to reside in grains[2]. The reactive radical $X^2\Sigma^+$ SiN has been the first silicon-nitrogen molecule observed in space and thus is the first observed example of a link between silicon and nitrogen chemistry, although such a link had already be considered in modeling chemistry in interstellar clouds[3,4].
To our knowledge, no theoretical studies have been done on the possible ways to form this diatomic molecule in interstellar conditions. Although some hypothetical reaction mechanisms have been proposed[1,4], we will only be concerned here with the one involving a dissociative recombination of a complex ion formed by the action of Si^+ onto NH_3.

STRENGTH AND WEAKNESSES OF THE CARBON ANALOGY

Since the Si and C atoms are valence-isoelectronic, it is an appealing prospect to predict some properties of silicon chemistry from carbon chemistry.
HCN and HNC, both observed in interstellar medium, are supposed to be formed according to :

$$C^+ + NH_3 \longrightarrow HCNH^+ + H$$

followed by the dissociative recombinations :

$$HCNH^+ + e^- \longrightarrow (HCN + H) + (HNC + H)$$

Both steps have been shown[5,6] to proceed without an activation barrier through the linear structure $HCNH^+$. The Si/C analogy would then lead to the following reactions for silicon :

$$Si^+ + NH_3 \longrightarrow HSiNH^+ + H$$

followed by the dissociative recombinations :

$$HSiNH^+ + e^- \longrightarrow (HSiN + H) + (HSiN + H)$$

In fact, the analogy seems abusive : if HCN and HNC have been detected in space, their silicon analogs SiNH and NSiH have not. Only SiN has been observed, which suggests that these compounds are not issued from a similar complex. This argument might be questionable since calculations[7,8] and experiments[9] have shown that the dipole moment for HNSi is about 0.2 D, which means that the detection of this molecule by microwave spectroscopy is unlikely due to the actual sensitivity of detectors. However, silicon compounds of astrophysical interest have been already shown to have some surprising properties relative to their carbon analogs : the exotic structure[10] of HSi_2H (acetylene analog) has been one of the first example of such differences, another one is the ground state of the SiH_2 which is a singlet state[11] while that of methylene CH_2 is known to be a triplet. Concerning the problem we are interested in, the definite proof of the unrelevancy of the analogy will be exposed in the following sections.

COMPUTATIONAL DETAILS

Some calculations on the $(Si, N, H, H)^+$ have already been performed[8,12] but the level of theory used was not sufficient to answer all the questions the present problem brings : previous studies for example do not include electronic correlation in the determination of the geometries. We will report here our preliminary results concerning this system : the calculations presented here are all obtained using the MP2 perturbation theory to account for both core and valence electronic correlation, using a triple-zeta basis set extended with polarization and diffuse functions. This leads to the standard 6-311++G** basis set included in the GAUSSIAN92[13] code which was used to optimize geometries using an UHF wavefunction as zero-order function for the perturbation.

We have studied this way all possible structural isomers, triplets and singlets, arising from the action of $Si^+(^2P)$ on NH_3 ($^1A_1$) and from the action of $N(^4S$ and excited $^2P)$ on $SiH_3^+(^1A_1')$. Details of these studies, including geometrical and spectroscopic parameters for the species studied, will be published elsewhere.

BREAKDOWN OF THE ANALOGY

These calculations show unambiguously that the $^1A_1$-C_{2v} structure of $SiNH_2^+$ is the most stable species among all the structures that can be formed, lying 49.1 kcal/mol (51.9 kcal/mol if zero-point energies are not taken into account) below the $^1A'$-C_s quasi-linear $HSiNH^+$. The situation is then completely reversed compared to the carbon chemistry in which the $^1A_1$-C_{2v} CNH_2^+ isomer lies about 46 kcal/mol above the linear $HCNH^+$ isomer[14].

Such results definitely rule out a $HSiNH^+$ structure as a starting point for a dissociative recombination to HSiN and HNSi : further studies must then be focused on the $SiNH_2^+$ formation and reactions. It should however be noticed that the calculated dipole moment for this compound is found to be 0.11 D : this will complicate its detection by microwave spectroscopy in the interstellar medium.

ON THE WAY TO $SiNH_2^+$

We have begun the investigation of the potential energy surface of the $Si^+ + NH_3$ reaction using the method described above in order to determine whether this reaction proceeds with an activation barrier. We will be concerned in this preliminary report with only one of the possible path to the products $SiNH_2^+(^1A_1$-$C_{2v})$ and $H(^2S)$. This path involves the formation of the C_{3v} complex $SiNH_3^+$ followed by the abstraction of an hydrogen atom through a transition state $(SiNH_2...H)^+$.
This reaction is found to be exothermic by 25.8 kcal/mol (17.6 kcal/mol if zero-point corrections are not included), the relaxed energy being transferred to the ejected hydrogen in the form of kinetic energy or to the molecular product $SiNH_2^+(^1A_1$-$C_{2v})$ in the form of vibrational energy.
At this level of theory, the reaction proceeds without activation barrier by -2.6 kcal/mol or with a barrier of 3.0 kcal/mol if zero-energy corrections are not considered. Previous studies[8] reported no

barrier by -5.6 kcal/mol. Further refinements are thus needed to determine unambiguously the height of an eventual activation barrier on that pathway : as well as the zero point energies, the influence of the correlation energies may have a crucial importance along the whole surface and higher-level correlation techniques than MP2 treatments must be used to ensure an homogeneous description of all structures involved in the potential energy surface. Furthermore, a second path has to be investigated : it has indeed been found that the C_{3v} complex $SiNH_3^+$ could isomerize easily to the planar quasi-isoenergetic C_s-$HSiNH_2^+$ compound, confirming previous reports[8]. This possible second pathway is presently under active investigation.

CONCLUSIONS AND PROSPECTS

We have reported our preliminary results concerning possible ways to interstellar SiN molecule refining and extending the results obtained by previous authors to whom the theoretical methods used in this communication were not available. It is confirmed that linear $HSiNH^+$ can not be considered as a starting point to the formation of HSiN and HNSi : $SiNH_2^+$ is unambiguously a much more attractive candidate. Dissociative recombination on $SiNH_2^+$ could then lead to HNSi (and not HSiN) or SiN directly. However, the ways through $SiNH_2^+$ do not definitely rule out the formation of the HSiN isomer. Further investigations are in progress to determine :

1- the isomerization barrier from HNSi to HSiN
2- the barrier for the isomerization of neutral SiNH2 to neutral HSiNH

We would like to emphasize that a careful description of all the reactions involved in this chemistry can only be reached using highly correlated wavefunctions since energy differences involved are small : even if the MP2 level we have used gives relevant results, they have to be refined using more sophisticated methods such as first- and second-order configuration interactions in order to get definitive theoretical answers to the formation of SiN in interstellar conditions.

REFERENCES

1. B. E. Turner, Astrophys. J., 388, L35 (1992)
2. B. E. Turner, BAAS, 23, 933 (1991)
3. W. D. Langer and A. E. Glassgold, Astrophys. J., 352, 123 (1990)
4. E. Herbst, T. J. Millar, S. Wlodek and D. K. Bohme, Astron. Astrophys., 222, 205 (1989)
5. T. L. Allen, J. D. Goddard and H. F. Schaefer III, J. Chem. Phys., 73(7), 3255
6. D. Talbi and Y. Ellinger, to be published
7. R. T. Luke, J. A. Pople, M. B. Krogh-Jesperon, Y. Apeloig, M. Karni, J. Chandrasekhar and P. von Rague-Schleyer, J. Am. Chem. Soc., 108, 270 (1986)
8. J. R. Flores, F. Gomez-Crespo and J. Largo-Cabrerizo, Chem. Phys. Lett., 147, 84 (1988)
9. M. Bogey, C. Demuynck, J. L. Destombes and A. Walters, Astron. Astrophys., 244, L47 (1991)
10. M. Cordonnier, M. Bogey, C. Demuynck and J.- L. Destombes, J. Chem. Phys., 97(11), 7984 (1992)
11. C. Winter and P. Millié, Chem. Phys., 174, 177 (1993) and references cited
12. J. R. Flores and J. Largo-Cabrerizo, J. Mol. Struct. (Theochem), 183, 7 (1989)
13. Gaussian92, Revision B, M. J. Frisch, G. W. Trucks, M. Head-Gordon, P. M. W. Gill, M. W. Wong, J. B. Foresman, B. J. Johnson, H. B. Schlegel, M. A. Robb, E. S. Replogle, R. Gomperts, J. L. Andres, K. Raghavachari, J. S. Binkley, C. Gonzalez, R. L. Martin, D. J. Fox, D. J. DeFrees, J. Baker, J. J. P. Stewart and J. A. Pople, Gaussian, Inc, Pittsburgh PA, 1992
14. D. J. DeFrees and W. J. Hehre, J. Phys. Chem., 82, 391 (1978) and references cited

REACTION OF C+ IONS WITH MOLECULES AT LOW TEMPERATURES

M. Ramillon and R. McCarroll
Laboratoire de Dynamique Moléculaire et Atomique,
CNRS et Université P. et M. Curie,
4, place Jussieu T12-B75, 75252 Paris Cedex 05, France

M. Gargaud
Observatoire de l'Université de Bordeaux I, F 33270 Floirac, France

ABSTRACT

Adiabatic rotational state methods are developed for the study of reactions of open shell ions with linear molecules at low temperatures. Multiple potential energy surfaces, some of which are non-reactive, participate in the collision process and when the fine structure splitting is comparable with the energy separation of the rotational states and with the thermal collision energy, non adiabatic effects in the non reactive region can become important. Results for C+ ions reacting with HCl and HCN are presented.

INTRODUCTION

Adiabatic capture models[1,2] of ion/molecule reactions compare favourably with experiment in the low temperature range ($\leq$ 300K) for a wide variety of systems[3-6]. In these models, it is generally assumed that a detailed knowledge of the electronic potential energy surfaces is not required since the probability of a reactive collision is largely controlled by long-range interactions. For most systems, it suffices to take account of the leading terms of a multipole expansion.

However, there is now considerable experimental evidence[7,8] indicating that these simple adiabatic capture models do not always work for ions with open-shell p electrons. For example, the measured rate constants for reactions of C+ and N+ ions with simple polar molecules such as HCl or HCN are considerably lower than the predicted theoretical values of the adiabatic capture model. This would suggest that it is incorrect to assimilate such ions as structureless point charges. Ab-initio calculations[9] on the C^+/HCl system provide an interesting illustration of the problem. It is found that, of the three potential energy surfaces correlated to the $C^+(^2P)$/HCl entry channel, the upper $^2A'$ state is repulsive for all orientations of the HCl molecule whereas the other $^2A'$ and the $^2A''$ are both attractive (C^+/HCN has similar SCF surfaces). If the effect of spin-orbit interaction is neglected, a corrective factor of 2/3 should then be applied to take account of the non-reactive character of the repulsive upper $^2A'$ potential energy surface[7].

But at low temperatures where kT is of the order of, or smaller than $^2P_{1/2}$ - $^2P_{3/2}$ fine-structure splitting in C^+ (63.42 cm^{-1}), the neglect of spin-orbit interaction is unjustified. At low astrophysical temperatures, it is inappropriate to apply an overall corrective factor of 2/3.

DESCRIPTION OF THE METHOD

The adiabatic capture methods like ACCSA[1] or SACM[2] compute the eigenvalues of an R-dependent hamiltonian (R is the ion/molecule distance). When the ions and the molecules are of closed-shell type, it is sufficient to perform the calculation in a basis of free rotor states. These rotor states can be used to label the adiabatic curves obtained.

The C^+ ion is an open-shell atom with one unpaired 2p electron. In this case the corresponding electronic states must be taken into account. Accordingly, for the C^+/molecule system one basis state will be defined using the four quantum numbers $\{f, m_f, j, m_j\}$ where f designates the fine structure states of C^+ (f = 1/2, 3/2) and j the rotation states of the molecule.

A rotating (BF) reference frame is used in preference to a space fixed (SF) frame for the description of the collision dynamics. The polar axis of the BF frame follows the direction of the collision vector **R** which joins the center of mass of the ion to the center of mass of the molecule. In the BF frame the orientation of the position vector of the C^+ (2p) electron is given by the angles (θ_1, φ_1) while the orientation of the linear molecule is given by (θ_2, φ_2). The use of the BF frame is of prime importance. This choice of coordinates enables us to exploit the Ω-conserving symetry (Ω is the projection of the total angular momentum along the collision axis).

A major part of the method involves diagonalizing the hamiltonian H_R

$$H_R = Al(\theta_1, \varphi_1) \cdot s + Bj^2(\theta_2, \varphi_2) + V(R, \theta_1, \varphi_1, \theta_2, \varphi_2) \qquad (1)$$

where A is the fine structure constant of C^+ and B the rotation constant.

The ion-molecule interaction potential V is expanded as a sum of the charge-dipole, charge-quadrupole, charge-induced dipole and quadrupole-dipole interactions. This series expansion up to $(1/R)^4$ is expected to reproduce the SCF surfaces down to about 7 a.u. . The quadrupole-dipole interaction, which will be denoted by $W^{(\Theta_1-P_2)}$, is the only term which can couple the motion of the 2p electron to the rotation of the molecule.

In order to determine the electronic properties induced by the quadrupole-dipole interaction, we have formed the 6x6 matrix representing the operator $Al.s + W^{(\Theta_1-P_2)}$ in the $\{f, m_f\}$ basis of electronic states. This matrix which we shall call the quadrupole-dipole matrix shows two kinds of terms, in addition to the diagonal terms for the fine structure splitting. There are $\cos\theta_2$ terms which are in general responsable for a global deviation of the rotational adiabatic curves while $\sin\theta_2$ terms cause avoided crossings in place of simple crossings.

In the limit of a fine structure separation taken to be equal to zero the quadrupole-dipole matrix can be reduced to a 3x3 matrix. This 3x3 matrix can be easily diagonalized analytically. We have checked that the three subsequent eigenvalues show qualitative electronic characters identical to those of the SCF surfaces given by Dateo and Clary[9].

At short distances ($\approx$ 7 a.u.) it is possible to associate an electronic surface with each of the adiabatic curves obtained from the diagonalisation of H_R. This way each rotational adiabatic curve $\varepsilon_i(R), i \equiv \{f, m_f, j, m_j\}$, will be attributed a reactive factor r_i equal to 1 or 0 depending on whether it correlates to a reactive or non-reactive surface.

The addition of the centrifugal term $J(J+1)/2\mu R^2$ to each adiabatic energy $\varepsilon_i(R)$ yields the set of channels $\{\varepsilon_i^J(R)\}$. The assumption of a classical capture for each channel is correct with regard to the large number of partial waves that generally contribute to the cross section. It can be shown that the formula for the cross section (J partial wave expansion), obtained by the ACCSA theory, can be thermally averaged in a analytical way over the collision energy. This leads to the state-selected rate constant formula of the SACM theory[10].

$$k_i(T) = \left(\frac{2\pi}{\mu^3 KT}\right)^{1/2} \sum_{J=0}^{\infty} (2J+1) \exp\left(-E_i^{J*}/KT\right) \qquad (2)$$

where E_i^{J*} is the maximum value of the radial potential $\varepsilon_i^J(R) - \varepsilon_i^J(\infty)$. The state-selected rate constants are thermally averaged over the distribution of initial population.

$$k(T) = \frac{\sum_i r_i \, k_i(T) \, e^{-\varepsilon_i(\infty)/KT}}{\sum_i e^{-\varepsilon_i(\infty)/KT}} \qquad (3)$$

PSEUDO-ADIABATIC CURVES

The main advantage of the ACCSA or SACM theories resides in the fact that the full quatum calculations are avoided and there is no need to solve coupled differential equations. In the non-reactive region, the adiabatic energy curves govern the dynamics of the ion/molecule system.

In the case presented in figure 1 all the visible curve crossings are in reality avoided crossings from a rigorous mathematical point of view. The quadrupole-dipole interaction is responsible for these avoided crossings, but beyond a distance of 20 a.u this interaction is too weak to make the avoidings discernible. It is therefore evident that the system will no longer follow the purely adiabatic curves. On the other hand, a Landau-Zener study of avoided crossings occuring at shorter distances shows that for distances shorter than 10 a.u. transition probabilities attain intermediate values. In that case the exchange of energy between the rotation of the molecule and the fine structure of C$^+$ can be of considerable importance.

It is of some interest to point out some computational aspects. The numerical diagonalization of H_R is performed on a mesh of R values ranging from R_{max} down to R_{min}. It yields a series of columns of eigenvalues. Each diagonalization at a given distance R introduces a new line of eigenvalues arranged in ascending order. In this way it can be considered that the nth column represents a pile of values of the nth adiabatic rotor state.

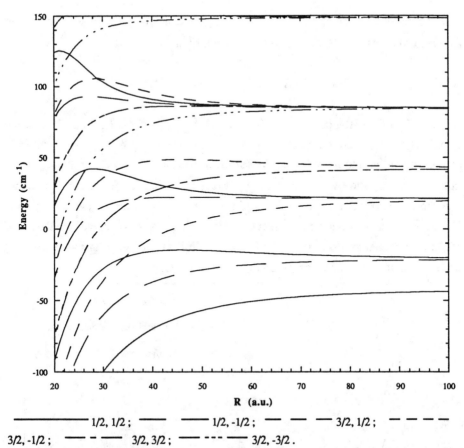

Figure 1 : C$^+$/HCl rotationally adiabatic curves at large distances, from diagonalization of (1). In this case $\Omega=1/2$. A pseudo-adiabatic curve follows a definite legend whereas as a purely adiababatic does not. The legends refer to the asymptotic electronic states $\{f,m_f\}$, the rotor states are not indicated.

We have developed a computer software which overcomes the problem of non-significant avoided crossings. This program follows the eigenvalues data from R_{max} to R_{min} and performs the relevant data permutations, using a continuity criterion on the first derivative. After this treatment each column of data represents a *pseudo-adiabatic* curve. Each pseudo-adiabatic curve is therefore supposed to be a possible path for the dynamics of the system. It should be pointed out that a suitable mesh size must be chosen in order to obtain reliable pseudo-adiabatic curves.

We shall examine two different cases for the quadrupole-dipole interaction, the *adiabatic limit* for which the complete quadrupole-dipole matrix is used and the *diabatic limit* for which the $\sin\theta_2$ terms (responsable for most of the significant avoided crossings) of the quadrupole-dipole matrix are removed.

RATE CONSTANTS FOR C⁺/HCN AND C⁺/HCl

When the distance between the ion and the molecule is less than 10 a.u. the electric field of the ion is very strong and it can be considered that the polar molecule axis is constrained to lie close to the collinear orientation $\theta_2 = 0$ (for small j). For this orientation, the two attractive SCF surfaces are almost degenerate. It is likely that the two surfaces are strongly mixed and both can be reactive (even if only the lowest $^2A'$ leads to reaction products). At even shorter distances, the orientation of the molecule changes to near perpendicular (optimized geometry for the reaction[11]) and the two attractive surfaces are no longer degenerate.

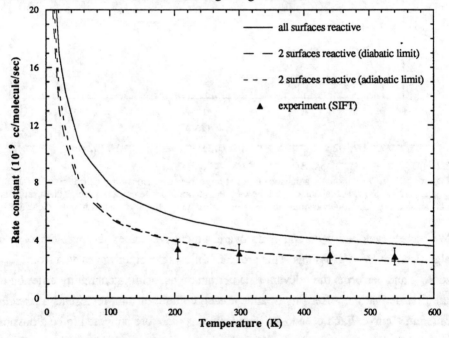

Figure 2 : Thermally averaged rate constant for C⁺/HCN system.

On figure 2 are presented the thermally averaged rate constants for the C⁺/HCN system. The full curve represents the result of the simple adiabatic capture theory (open-shell nature of C⁺ neglected). They clearly overestimate the experimental data[3]. Our results are the broken curves. The adiabatic and diabatic limits, when averaged over the distribution of initial states, give sensibly the same results. They are in good agreement with experiment between 205 and 540 K. In this range of temperature the dependence of the rate constant is smooth. Experimental data at lower temperatures, where the theory predicts steeper slopes, would therefore be of great interest.

In figure 3 are presented the thermally averaged rate constants for the C^+/HCl system. The legends used here are similar to those for the figure 2. The simple adiabatic capture results are of course too large, but even the rate constants considering two surfaces reactive still overestimate the experimental data[4,5]. However, we may note that the temperature dependence of the rate constant is very well reproduced by the two surfaces reactive model and if we multiply our results by a global factor of 0.7 a perfect agreement with experiment is obtained. Calculations assuming only one reactive surface yield absolute rate constants, which on average are in reasonable agreement with experiment, but the temperature variation is much less satisfactory than for the two surfaces model. We conclude that the two-surface model is basically correct, but a justification of the corrective factor must await a more detailed study of the capture probability in the reactive region.

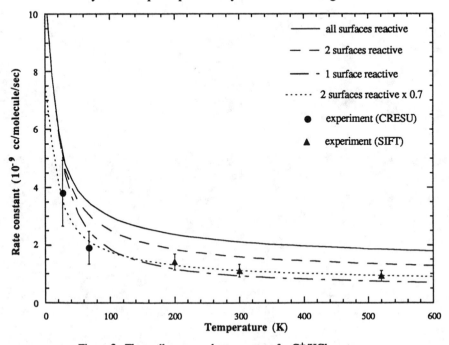

Figure 3 : Thermally averaged rate constant for C^+/HCl system.

REFERENCES

1. D.C. Clary, Mol. Phys. 54, 605 (1985)
2. J. Troe, J. Chem. Phys. 87, 2773 (1987)
3. D.C Clary, D. Smith and N.G. Adams, Chem. Phys. Lett. 119, 320 (1985)
4. C. Rebrion, J.B. Marquette, B.R. Rowe and D.C. Clary, Chem. Phys. Lett. 143, 130 (1988)
5. D.Smith and N.G. Adams, Astrophys.J.Lett. 298, 827 (1985)
6. M.L. Dubernet and R. McCarroll, Z. Phys. D 15, 333 (1990)
7. D.C. Clary, C.E. Dateo and D. Smith, Chem. Phys. 167, 1 (1990)
8. D.C. Clary, Ann. Rev. Phys. Chem. 1990, 41, 61 (1990)
9. C.E. Dateo and D.C. Clary, J. Chem.. Phys. 90, 7216 (1989)
10. M. Ramillon and R. McCarroll, J. Chem. Phys. (1993) to appear
11. C. Barrientos, A. Largo, P. Redondo, F. Pauzat and Y. Ellinger, J. Phys. Chem. 97, 173 (1993)

DISCUSSION

CLARY — Do you think that it will be important, in a very accurate treatment, to account for the dependence of the spin-orbit coupling constant of C^+ on the geometry of the HCℓ molecule ?

RAMILLON — I am not sure of what you mean exactly by the geometry of the HCℓ molecule. If you mean the dependence of the energy separation between the fine structure levels of C^+ on the orientation of the HCℓ molecule, it is automatically taken into account at intermediate and large distances with the inclusion of the quadrupole-dipole interaction. On the other hand, inside the chemical region the interaction energies are of course much stronger and it is likely that the changes in the geometry of the HCℓ molecule will affect the spin-orbit interaction of C^+. In this work we are not concerned with the detailed treatment of the dynamics in the reactive region.

ROUEFF — Have you considered the influence of the closed rotational channels of HCℓ ?

RAMILLON — The rotationally adiabatic states correlate at infinite separations with the free rotor states. In this way the states for which free rotor energy is very large compared to kT do not contribute to the thermally averaged rate constant. But these closed rotational channels are nevertheless indirectly involved. They are included in the basis set of free rotor states used in the representation of the adiabatic rotor states.

McCARROL — Just a comment or a question to the astronomers. In astrophysical situations, there will not necessarily be a thermal distribution neither of the C^+ fine structure states nor of the molecular rotational states. Under these conditions, extrapolation of experimental results to low temperatures may be dangerous. The theoretical model can investigate the detailed state selected reactions.

DALGARNO — In many astrophysical environments, significant departures occur from thermal equilibrium in the populations of fine-structure levels and in the populations of rotational levels.

GAS PHASE REACTION OF ATOMIC HYDROGEN, ATOMIC NITROGEN RADICAL WITH TRANSITION METAL CARBONYL CATIONS

M. Sablier, L. Capron, H. Mestdagh, C. Rolando
UA 1679 du CNRS, Processus d'Activation Moléculaire
Ecole Normale Supérieure, Département de Chimie
24 rue Lhomond, 75231 Paris Cedex 05, France

N. Billy, G. Gouédard, J. Vigué
UA 18 du CNRS, Laboratoire de Spectroscopie Hertzienne
Ecole Normale Supérieure, Département de Physique
24 rue Lhomond, 75231 Paris Cedex 05, France

ABSTRACT

The reactivity of transition metal carbonyl cations $M(CO)_n^+$ (n = 0-N, where N is the number of CO ligands for the stable neutral precursor) with molecular and atomic hydrogen and nitrogen has been studied, since no information was available on the reactivity of metal-containing ions with free atoms. Reaction of $M(CO)_n^+$ with H or N atoms led respectively to $HM(CO)_{n-1}^+$ and $NM(CO)_{n-1}^+$. The efficiency of the reaction is strongly dependant of the number of CO ligand and is shifted toward higher value of n going downwards periodic table. Except for MH^+ all these products are new species.

INTRODUCTION

Whereas a large amount of ion-molecule reactions, as well as various atom-molecule reactions, have been studied in the gas phase, there is a scarcity of experimental data concerning the reactivity of ions with free atoms such as hydrogen or nitrogen, or with organic radical species[1].

The knowledge of such processes has an astrophysical interest: free atoms, the most abundant of which being H˙, are important components of interstellar medium, and electropositive elements such as transition metals are likely efficiently photoionized to positive ions; therefore metal ions might react with free atoms or with radicals under interstellar conditions and maybe play the role of catalysts[2].

We report here our results concerning the reactivity of bare and transition metal carbonyl cations with molecular, and atomic hydrogen or nitrogen[3].

EXPERIMENTAL

A multiquadrupole MS/MS/MS triple analyzer[4] (figure 1) was used for all the studies. The spectrometer includes a source S, three quadrupole analyzers, referred to sequentially as Q_1, Q_2, Q_3, two collision cells (C_1 between Q_1, Q_2 and C_2 between Q_2, Q_3) and an off-axis detector D. Ions are generated by a dual EI/CI source or by a FAB source using a MScan company neutral gun. Samples are introduced via an ultravacuum leak or via the FAB target probe using a glycerol matrix[5]. The source compartment is differentially pumped by a 700 $l.s^{-1}$ diffusion pump.

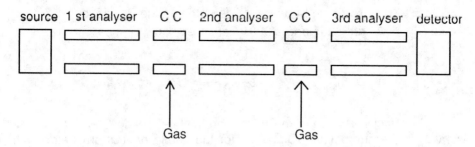

Figure 1: Experimental apparatus for the study of ion-atom reactions

Quadrupole analyzers Q_1, Q_3 (1.56 cm diameter rods, 35 cm long) and Q_2 (1.56 cm diameter rods, 25 cm long) can be tuned to unit mass resolution throughout the entire mass range 4-1500 amu. A variable DC offset can be applied independently on each quadrupole in the range 0, 50 V for Q_1, Q_2 and 0, 250 V for Q_3. The two collision cells are RF only quadrupoles (1.56 cm diameter rods, 12 cm long) encased in a closed stainless steel tube with 0.4 cm only diameter entrance and exit holes. The cells can be pressurized up to 0.1 Torr of any standard gas and the absolute pressure is measured by a MKS company Baratron gauge. The collision and analysis sections are pumped by two diffusion pumps (135 and 700 $l.s^{-1}$). The upper limit pressure in the collision cell is determined by the lack of differential pumping between the cells and the analyzers. The tension of the lenses delimiting the cells can be modulated allowing trapping of ions inside for a variable time up to several hundred milliseconds. This ion confinement technique allows access to thermodynamic and kinetic data[6]. Alternatively an RF field can be added allowing ICR-like experiments by mass selective ion rejection in the quadrupole reaction chamber[7]. The detector is a modified 21 stage CuBe dynode electron multiplier mounted off axis.

In all the reactivity experiments an average kinetic energy of ca 2eV was used; the corresponding a center-of-mass collision energies, depending on reactant masses, vary between 0.01 and 0.07 eV (0.2-1.6 kcal/mol) for atomic and molecular hydrogen, between 0.1 and 0.7 eV (3-15 kcal/mol) for atomic and molecular nitrogen. For collisionnally activated dissociation an average kinetic energy of 25 eV was used.

The discharge was effected in a 7 mm diameter Pyrex tube, with a frequency of 2.45 GHz and a power of 60 W at a pressure of ca 1 Torr. About 10 cm separated the discharge zone from the end of the tube, from which a 0.4 mm diameter hole, followed by a 30 cm length, 4 mm diameter teflon tube, led to the collison cell.

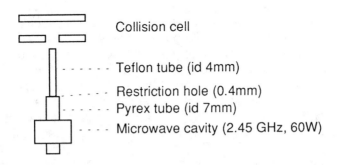

Figure 2: Detailed view of the microwave discharge fitting to the collision cell.

The measured pressure in the collison cell is in the range 0.2-2 mTorr and is not affected by switching the discharge on or off. The flow through the hole, estimated either from hole diameter or from pump delivery, is in the range 10-100 ml s^{-1}, corresponding to a 0.05-0.5 s average residence time of the gas in the tube downstream from the discharge.

This allows quenching of metastable electronic states of H_2, all of which have radiative lifetimes shorter than a few milliseconds[8], and collisional quenching of vibrationally excited H_2, (except the v=1 level, for which only partial quenching can be expected)[9]. In the case of nitrogen the long-lived excited state A $^3S_u^+$ is probably formed but as it is very efficiently quenched by nitrogen atoms, this state is actually not detected downstream from a microwave discharge[10]. Important recombination is expected to occur in the Pyrex tube. Therefore hydrogen atoms production is expected to be relatively inefficient compared to experiments using a larger flow rate[11]. Fractional dissociations measured under conditions similar to the present experiment (although with a slightly longer residence time) are actually in the 3-4% range both for hydrogen[12] and nitrogen[13].

RESULTS AND DISCUSSION

We report here our results concerning the reactivity of transition carbonyl cations with molecular and atomic hydrogen or nitrogen. The reactions involving atomic species can be summarized in the following equations :

$$M(CO)_n^+ + H^\cdot \rightarrow HM(CO)_{n-1}^+ + CO \quad (1)$$

$$M(CO)_n^+ + N^\cdot \rightarrow NM(CO)_{n-1}^+ + CO \quad (2)$$

Non excited bare transition metal cations do not react with molecular hydrogen, nor with molecular nitrogen[14]. For example the cross section reported for the reaction of hydrogen with Fe^+ ions produced by EI on $Fe(CO)_5$ is ca $2 \cdot 10^{-18}$ cm^2 at low collision energy,[15,16] which is too low to be detected under our reaction conditions. In the presence of atomic hydrogen or nitrogen (discharge on) no reaction was observed either; it is not surprising since collisional stabilisation of short-lived diatomic intermediates such as FeH^+ or FeN^+ is very unlikely at the pressure used, while radiative association is far too slow[2].

With the discharge on, in addition to minor peaks due to contaminants like water or oxygen observed also discharge off, a new product appeared, corresponding to replacement of a CO ligand by a H atom (equation 1) or a N atom (equation 2). The observed reaction extents amount to several percents, which corresponds to quite efficient reactions since the fractional dissociation of molecules to atoms is probably small, as discussed above, and the atom beam is introduced perpendicularly to the axis of the collision cell, so that ion-atom reactions are likely to occur only in a small fraction of the total cell length (about 1 cm instead of 12 cm), contrary to ion-molecule reactions.

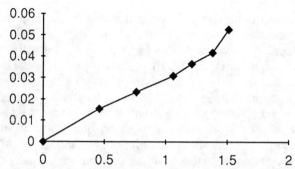

Figure 3: Dependance of FeH^+ branching ratio with hydrogen pressure (mTorr)

As shown in figure 3 the yields of FeH^+ ion increases linearly with hydrogen pressure which should be roughly proportional to hydrogen atom concentration. This behaviour demonstrates a reactivity of $Fe(CO)^+$ with hydrogen atoms mainly with the bulk constituant of the ion beam and not with a minor excited state.

Both with atomic hydrogen and atomic nitrogen the reaction efficiency decreased with the number n of CO ligands, as shown in Figure 4. This trend is more pronounced in the case of nitrogen atoms, which give no detectable product for n=3 to 5.

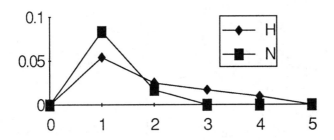

Figure 4: Extent of formation of $HFe(CO)_{n-1}^+$ and $NFe(CO)_{n-1}^+$ from $Fe(CO)_n^+$

Although we have yet no experimental evidence of their structure, it seems likely that in all these compounds the hydrogen or nitrogen atom is bound directly to iron, by analogy with FeH^+ and FeN^+ obtained for n = 1. Similarly tetracarbonyl iron anion was reported to yield also $Fe(CO)_3H^-$:

$$Fe(CO)_4^- + H^\cdot \rightarrow Fe(CO)_3H^- + CO$$

as a primary reaction product by reaction with hydrogen atom[17,18]:

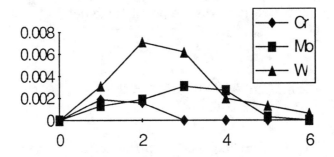

Figure 5: Extent of formation of $HM(CO)_{n-1}^+$ from $M(CO)_n^+$ for group VI family

This study has been extended to heavier transition metal carbonyl ions. As shown in figures 4 and 5 the nature of the most reactive species is shifted towards higher number of CO ligand when going downards the periodic table.from chromiumcarbonyl to tungstencarbonyl cations or from ironcarbonyl to rutheniumcarbonyl cations.

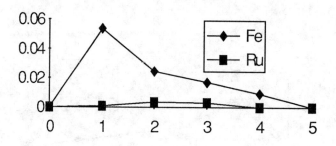

Figure 6: Extent of formation of $HM(CO)_{n-1}^+$ from $M(CO)_n^+$ for Fe, Ru

CONCLUSION

This study is being extended either to other free organic radicals like cyanogen radical $CN^{\cdot}$, methyl radical $CH_3^{\cdot}$, and to complex mixture resulting of dissociation of polyhalogenated hydrocarbons like freons in a a microwave plasma discharge. Preliminary experiments[19] on the latter systems shows that ion-atom reaction also allows the possibility to get useful information on complex mixtures.

ACKNOWLEDGEMENT

We thank the GDR "Physicochimie des Molécules Interstellaires" and the GDR "Dynamique des Réactions Moléculaires" for financial support of this work.

REFERENCES

1. C. Rolando, M. Sablier, Gas Phase Ion-Atom Reactions, Mass Spectrometry Review in press, and references cited therein.

2. K.K. Irikura, W.A. Goddard III, J.L. Beauchamp Int. J. Mass Spectrom. Ion Processes 99, 213 (1990).

3. H. Mestdagh, C. Rolando, M. Sablier, N. Billy, G. Gouedard, J. Vigué J. Amer. Chem. Soc. 114, 771 (1992).

4. C. Beaugrand, G. Devant, D. Jaouen, H. Mestdagh, N. Morin, C. Rolando Adv. Mass Spectrom. 11A, 256 (1989).

5. H. Mestdagh, N. Morin, C. Rolando Tetrahedron Letters 27, 33-36 (1986).

6. C. Beaugrand, H. Mestdagh, D. Jaouen, C. Rolando Anal. Chem., 61, 1447 (1989).

7. J. T. Watson, C. Beaugrand, H. Mestdagh, D. Jaouen, C. Rolando Intern. J. Mass Spectrom. Ion Processes, 93, 225 (1989).

8. K. P. Huber, G. Herzberg Constants of Diatomic Molecules (Van Nostrand Reinhold Company, New York, 1979) pp 240-253 and references cited therein.

9. M. Cacciatore, M. Capitelli, G.D. Billing Chem. Phys. Letters 157, 305 (1989).

10. H. Vidaud, R. P. Wayne, M. Yaron, A. J. Von Engel Chem. Soc. Faraday Trans. II, 72, 1185 (1976).

11. D. Spence, O. J. Steingraber Rev. Sci. Instrum. 59, 2464 (1988).

12. C. Tsai, D. L. Fadden J. Phys. Chem. 93, 2471 (1989).

13. C. Tsai, S. Belanger J. T. Kim, J. R. Lord, D. L. McFadden J. Phys. Chem. 93, 1916 (1989).

14. K. Eller, H. Schwartz Chem. Rev. 91, 1121 (1991) and refrences cited therein.

15. L. F. Halle, F. S. Klein, J. L. Beauchamp J. Am. Chem. Soc. 106, 2543 (1984).

16. J. L. Elkind, P. B. Armentrout J. Phys. Chem. 90, 5736 (1986).

17. R. N. McDonald, A. K. Chowdhury, P. L. Schell Organometallics 3, 644 (1984)

18. R. N. McDonald, A. K. Chowdhury Organometallics 5, 1187 (1986)

19. H. Mestdagh, C. Rolando, M. Sablier Proceedings of the 40th ASMS Conference on Mass Spectroscopy and Allied Topics, 77 (1992).

REACTION OF ANTHRACENE WITH HE⁺ AND AR⁺ AT ROOM TEMPERATURE

H. Abouelaziz, J.C. Gomet, D. Pasquerault, L. Nedelec, A. Canosa, C. Rebrion and B.R. Rowe.
Département de Physique Atomique Moléculaire, URA 1203 du CNRS,
Université de Rennes I, Campus de Rennes Beaulieu, 35042 RENNES Cedex,
FRANCE.

P. Lukac
Plasma Physics Department MFF, Commenius University, Mylnska Dolina F2,
842 15 Bratislava, SLOVAQUIA.

ABSTRACT

This paper deals with the measurement of the rate coefficient of anthracene ($C_{14}H_{10}$) with atomic ions such as He⁺ and Ar⁺. The experiment was carried out at room temperature in a flowing afterglow Langmuir probe apparatus equipped with a movable mass spectrometer. We obtained a rate coefficient of 6.4×10^{-9} cm³s⁻¹ and 1.8×10^{-9} cm³s⁻¹ for He⁺ and Ar⁺ respectively.

INTRODUCTION

Polycyclic aromatic hydrocarbon (PAH) molecules and ions are thought to be present in interstellar and circumstellar environments. It is more particularly believed that they are responsible for the diffuse infrared emission bands[1] and there are also some arguments that suggest they could be carriers of the visible diffuse interstellar bands[2,3]. Astrochemical models are now attempting to reproduce the chemistry of PAH compounds[4]. Therefore, ion-PAH rate coefficients are clearly needed. The neutralization of PAH ions by electron recombination also appears to be one of the significant processes that must be included in the models. There is presently a lack of data concerning the dissociative recombination (hereafter DR) of PAH ions as well as ion-molecule reactions involving PAH.

In our laboratory, a flowing afterglow experiment, including a movable Langmuir probe and a movable mass spectrometer (hereafter FALPMS), allows us to measure DR rate coefficients and ion-molecule reactions at room temperature. We recently measured[5] the DR rate coefficient for several cyclic ions including benzene ($C_6H_6^+$) and naphthalene ($C_{10}H_8^+$). We presently concentrate our efforts on the study of anthracene DR as well as ion-molecule reactions such as He⁺ + anthracene or C⁺ + anthracene. This paper presents our results concerning the reaction He⁺ + $C_{14}H_{10}$ as well as the additional reaction Ar⁺ + $C_{14}H_{10}$.

© 1994 American Institute of Physics

EXPERIMENTAL TECHNIQUE

A sketch of the FALPMS apparatus can be seen in Figure 1. As it was extensively described in previous publications[5,6] only the main features and improvements will be discussed here. As with the FALP apparatus previously used at Rennes, a helium buffer gas (P=0.5 Torr) flows through a glass tube and is ionized by a microwave discharge. The glass tube can be connected to various stainless steel tubes of different lengths. When pure helium is used, the main species in the plasma are He^+ and He_2^+ ions together with metastable helium He^M and electrons. Anthracene was introduced into the flow by means of a needle entry port G_3. As anthracene $C_{14}H_{10}$ is in solid state at room temperature, it was set into a cylindrical oven which can be heated up to 600 K by means of a heating wire connected to a temperature regulation device. A helium buffer was also introduced into the oven to transport the vaporized anthracene towards the 8 needle entry port where the helium-anthracene mixture merged with the plasma. A flow controller was used to measure the helium flow rate and the pressure in the oven was measured with a needle manometer. A precise determination of the anthracene flow rate was then possible. To avoid condensation of anthracene, the stainless steel line connecting the oven to the needle entry port and the under-vacuum-torus that supports the 8 stainless steel needles were also heated. Two temperature regulation devices allowed us to maintain these to a higher temperature than into the oven. The 8 needles themselves were electrically heated by a current of about 1.7A. It was observed indeed that in some plasma conditions, the main helium buffer gas could cool the needles so that they were rapidly coated by solid anthracene and stuffed up eventually. The potential difference between the exit of the needles and the torus was very small so that the plasma was not disturbed: Langmuir probe measurements were carried out with and without electrical heating and lead to the same electron density. By using such a technique, it became possible to introduce large flow rates of vapor anthracene (several tenth of scc) into the plasma. This flow could be changed either by heating (or cooling) the oven or by changing the oven pressure by closing (or opening) a little bit the valve that insulates the oven from the experiment.

The measurement zone is located in a stainless steel chamber downstream of the needle entry port. This design allows one to move a quadrupole mass spectrometer along the flow axis as well as an off-axis Langmuir probe which can be set on-axis by rod rotation. It is possible therefore, to measure the densities of both ions and electrons as a function of distance z along the flow.

This arrangement allows one to study ion-molecule reaction between the parent ions and the neutral reactant gas by measuring the ion density decrease along the flow axis.

Provided that no He^+ source species were present in the flow and neglecting diffusion, the decrease of He^+ density is given by:

$$\frac{d[He^+]}{dz} = -\frac{k}{v}[He^+][C_{14}H_{10}] = -k_{fst}[He^+] \tag{1}$$

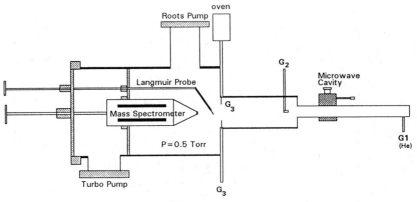

Figure 1. Sketch of the FALPMS

where k is the rate coefficient of $He^+ + C_{14}H_{10}$ and v is the flow velocity. Using time-of-flight techniques, v was obtained ($v=1.7\times10^4$ cm/s) and it was also checked that the flow in the recombination zone had the same characteristics as that in the mixing tube. The plasma velocity did not depend on z and was the same in both regions. The rate coefficient k was determined by plotting k_{fst} as a function of anthracene density. The slope of the line that was obtained being equal to the ratio k/v.

RESULTS

As He^M may be a source of electrons in the measurement chamber by Penning ionization of anthracene:

$$He^M + C_{14}H_{10} \rightarrow C_{14}H_{10}^+ + He + e$$

it was firstly removed by Penning ionization of argon by adding argon in the flow through the entry port G_2 that is far downstream of the needle entry port. He_2^+ was also removed by:

$$He_2^+ + Ar \rightarrow Ar^+ + 2He$$

so that the only present ions were Ar^+ and He^+ at the anthracene entry port location. A typical mass spectra obtained, with a low resolution, at this position, in the absence of anthracene, can be seen in figure 2. When $C_{14}H_{10}$ is introduced, the two primary ions are partly destroyed and several products are formed among which $C_{14}H_{10}^+$ was usually found to be largely dominant. Figure 3 illustrates a situation where other minor products could be observed in a non negligeable amount.

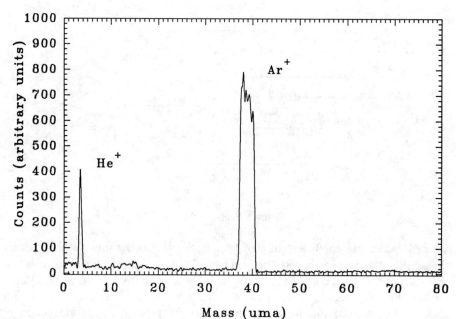

Figure 2. Typical mass spectra close to the needle entry port when no anthracene is introduced in the plasma.

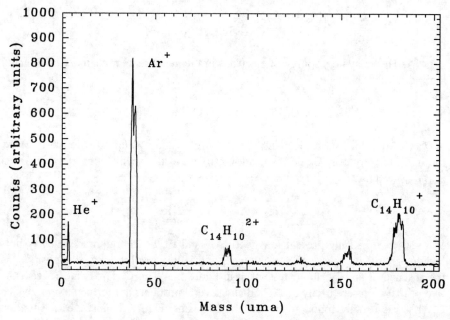

Figure 3. Mass spectra at z=65mm with a large anthracene flow rate.

Note the presence of $C_{14}H_{10}^{++}$ coming from the $He^+ + C_{14}H_{10}$ reaction. Further work is needed to properly identify all the products.

Different anthracene flow rates were used to determine the rate coefficient k as well as the one which corresponds to the $Ar^+ + C_{14}H_{10}$ reaction. A typical ion density decrease is shown in figure 4 for a given anthracene flow rate. As it was previously specified, the rate coefficients were then obtained by plotting the variation of the slope (that is k_{fst}) of these decreases as a function of anthracene density. Figure 5 shows the results that were obtained for He^+ and Ar^+. The rate coefficients were respectively found to be equal to 6.4×10^{-9} cm^3s^{-1} and 1.8×10^{-9} cm^3s^{-1}. It is interesting to notice that the ratio of these two rate coefficients is close to the square root of the ratio of the atomic masses. This suggests that reactions should occur at rate coefficients close to the Langevin rate given by:

$$k_L = 2\pi q \sqrt{\frac{\alpha}{\mu}} \quad (2)$$

where μ is the reduced mass of anthracene and the reacting ion, q the ion charge and α the polarizability of anthracene. An estimation of α gave a large value: $\alpha = 25$ Å^3 that is more than twice the polarizability of benzene.

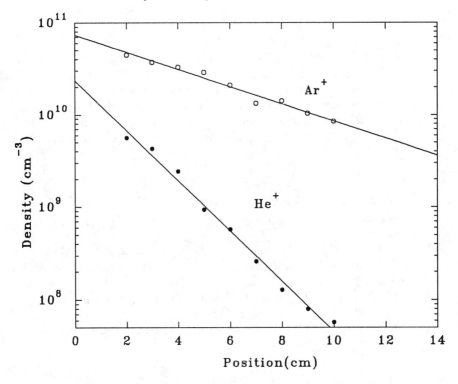

Figure 4. He^+ and Ar^+ density decrease as a function of distance z for a given anthracene flow rate, the needle entry port is located at z = -2mm.

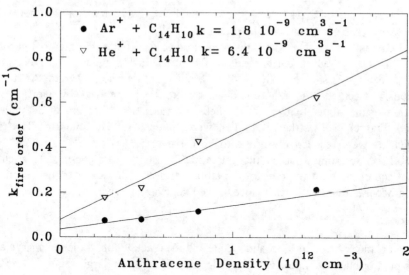

Figure 5. Plot of $k[C_{14}H_{10}]/v$ as a function of anthracene density

PERSPECTIVES

These results should be considered as preliminary, more work is planned on the topic. The nature of the products should be interesting to be determined. Other atomic ions are also supposed to be studied. We more specially think of C^+, N^+ and O^+. The present study gave also some qualitative indications concerning the recombination of anthracene ions with electrons. A value much greater than 10^{-6} $cm^3 s^{-1}$ is probably to be expected. Further work is in progress on this topic.

REFERENCES

[1] A. Leger and J. L. Puget, Astron. Astrophys. 137, L5 (1984).
[2] G.P. Van der Zwet and L.J. Allamandola, Astron. Astrophys. 146, 76 (1985)
[3] A. Léger and L.B. D'Hendecourt, Astron. Astrophys. 146, 81 (1985).
[4] G. Pineau des Forets, D.R. Flower and A. Dalgarno, Mon. Not. Roy. Astr. Soc. 235, 21 (1988).
[5] H. Abouelaziz, J.C. Gomet, D. Pasquerault, B.R. Rowe and J.B.A. Mitchell, J. Chem. Phys. 99, 237 (1993).
[6] B.R. Rowe, J.C. Gomet, A. Canosa, C. Rebrion and J.B.A. Mitchell, J. Chem. Phys. 96, 1105, (1992).

IS STRIPPING OF POLYCYCLIC AROMATIC HYDROCARBONS A ROUTE TO MOLECULAR HYDROGEN ?

P. Cassam-Chenaï
F. Pauzat
Y. Ellinger
Equipe d'Astrochimie Quantique
Ecole Normale Supérieure, 24 rue Lhomond, F-75231 Paris CEDEX 05, France
DEMIRM, Observatoire de Paris, 5 place Jules Janssen, F-92195 Meudon, France

ABSTRACT

Though molecular hydrogen is the most abundant species in space, no global explanation for its formation in the interstellar medium has been proposed. The actual understanding of the process relies upon the recombination of hydrogen atoms in presence of a third body, generally an interstellar grain. In the present note we report the results of quantum mechanical calculations which suggest that molecular hydrogen can be formed by reaction between hydrogen atoms and positively charged PAHs, resulting in a simultaneous dehydrogenation of the aromatic ions. Since regeneration of the original cation by addition of atomic hydrogen to the dehydrogenated positive ion is thermodynamically allowed, the process can be cycled. The present study shows another possible aspect of the PAHs puzzling role. It suggests that the positively charged species may be an important partner in the catalytic formation of molecular hydrogen in the interstellar medium.

INTRODUCTION

Molecular hydrogen H_2, by far the most abundant molecule in the universe, is also the starting point of reaction chains in interstellar chemistry. In addition, it can act as a shield against ionizing radiations for other molecules, and participates actively in the cooling processes through collisions with less abundant molecules. It is usually assumed that H_2 molecules form by recombination of H atoms on grains[1]. In this process, hydrogen atoms are captured, adsorbed on the grain, and, by hopping from one site to another eventually react and form H_2, which is vaporized in the surrounding space. However, to the best of our knowledge, models have failed so far to obtain the required rate coefficient[2], even when triggering by cosmic rays is considered[3]. A scenario where an incident H atom is chemisorbed on a carbon grain, then migrates and reacts with peripheral chemically bonded hydrogen has also been proposed[4]. Actually, if such a process exists for graphite-like grains, it could also work for smaller polycyclic aromatic hydrocarbons (PAHs). Spectroscopic arguments have led to the hypothesis[5] that a large part of the interstellar carbonaceous matter could be in the form of compact PAHs, those being partially dehydrogenated, partially ionized. Consequently, the reaction

$$PAH^{+\bullet} + H^{\bullet} \rightarrow DPAH^{+} + H_2$$

where DPAH stands for dehydrogenated PAH, is a logical chemical process to be investigated. Such a reaction that involves positively charged species had never been considered, although it appears to be a unique route leading at the same time to molecular hydrogen and DPAH cations whose existence is supported by the close resemblance between their theoretical IR spectra and the intensity pattern of the UIR bands[6]. In parallel with experimental studies[7], which are carried out in the laboratories of the Ecole Normale Supérieure, we have performed semi-empirical and *ab initio* calculations to investigate the thermodynamical balance of the reaction. The results are presented in this paper and the astrophysical consequences discussed.

© 1994 American Institute of Physics

THEORETICAL APPROACH

a small size example

The smallest PAH system to test our hypothesis is the naphthalene radical cation $C_{10}H_8^+$. Calculations have been performed using the GAUSSIAN 88 computer program[8], at the restricted open-shell Hartree-Fock (ROHF) level, with the built-in 3-21G basis set. D_{2h} symmetry has been used for $C_{10}H_8^+$ and C_s symmetry for $C_{10}H_7^+$. Results (see Tab.I) show that the most favorable product would be the dehydro-2-naphtalene cation $^3A"$ state. Nevertheless, the reaction is found endothermic ($\Delta H = 47.5$ kJ.mol^{-1} after zero-point correction).

Table I : HF/3-21G energies (atomic units)

Cation	State	Energy
Naphthalene	^2A_u	-380.94924
Dehydro-1-naphthalene	$^3A"$	-380.29973
	$^1A'$	-380.27077
Dehydro-2-naphthalene	$^3A"$	-380.29539
	$^1A'$	-380.26860

Thus, the formation of H_2 by stripping the naphthalene cation appears impossible in the interstellar medium. However, the preceding result, obtained for a small molecule, should not be generalized because interstellar PAHs most probably contain more than 10 carbons and reactivity is known to vary with the size of the species involved. In order to obtain a better view of the problem, larger systems had to be considered (see Figure 1). Since *ab initio* calculations are not tractable for those systems, we switched to a semi-empirical method.

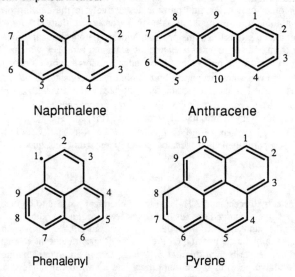

Figure 1 : Nomenclature of PAHs referred to in the text and standard numbering

medium size examples

The AM1 method[10], which is parametrized using heats of formation, was found particularly relevant to our problem. To assess the quality of the method, the heats of formation for the singlet states of the dehydro-1-naphthalene and dehydro-2-naphtalene cations have been calculated at the restricted Hartree-Fock (RHF) level. The results: 1243.8 kJ.mol^{-1} and 1238.3 kJ.mol^{-1} respectively give a difference of 5.5 kJ.mol^{-1}, in perfect agreement with the *ab initio* value of 5.7 kJ.mol^{-1}.

Since the PAHs are thought to be compact[5] in the interstellar medium, we have concentrated our effort on the pyrene cation. Results of the calculations in C_S symmetry are given in Tab.II. Attention is drawn to the fact that the doublet and triplet states calculations have been performed at the ROHF level, with the "half-electron" approximation[10], and some caution must be taken for the comparison with the singlet states done at the RHF level. Nevertheless, the reaction is found *exothermic* in all cases.

Results for the anthracene cation, with no symmetry condition imposed, show that the singlet is too high. This state has an aromatic π system of 4n+2 electrons and two missing electrons; in non compact systems, aromaticity seems too weak to stabilise sufficiently the DPAH+ singlet. The triplet, which is lower, has not been studied. In the case of the phenalenyl radical, the same reaction can occur in a neutral form:

$$PAH^{\bullet} + H^{\bullet} \rightarrow DPAH + H_2$$

Resuts in C_S symmetry show that the reaction is exothermic with a triplet state but not with a singlet. Here again, as suggested by mesomer formulas, aromaticity is weak and having 4n+2 electrons in the π system is not sufficient to allow exothermicity with a singlet state.

Table II : AM1 Heats of formation (kJ.mol^{-1}).

PAH	$\Delta H_f(PAH^+ + H)$	$\Delta H_f(DPAH^+ + H_2)$	DPAH
Pyrene	1326.7	1319.2 (singlet)	Dehydro-1-pyrene
		1257.3 (triplet)	
		1316.3 (singlet)	Dehydro-2-pyrene
		1262.7 (triplet)	
		1309.2 (singlet)	Dehydro-4-pyrene
		1258.1 (triplet)	
Anthracene	1232.6	1292.0 (singlet)	Dehydro-1-anthracene
		1296.6 (singlet)	Dehydro-2-anthracene
		1296.6 (singlet)	Dehydro-9-anthracene
Phenalenyl	514.2	564.0 (singlet)	Dehydro-1-phenalenyl
		509.6 (triplet)	
		628.4 (singlet)	Dehydro-2-phenalenyl
		508.8 (triplet)	

To summarize, our theoretical calculations show that the reaction of stripping PAHs+ as small as the pyrene cation is a thermodynamically allowed process. Of course, this is not sufficient for this chemical reaction to play a part in interstellar chemistry. Complementary quantum chemical studies, now in progress, are needed to determine whether or not there is an activation barrier on the reaction path.

ASTROPHYSICAL CONSEQUENCES

Taken as an isolated mechanism, the following reaction of stripping

$$PAH^{+\bullet} + H^{\bullet} \rightarrow DPAH^{+} + H_2 \qquad (k_2)$$

would at most transform all PAH^+ in positively charged graphite clusters and its contribution to the density of H_2 in interstellar clouds would be negligible. However, the situation looks different if we consider that the PAH^+ can be regenerated by the following exothermic reaction ($\Delta H \cong 400$ kJ.mol^{-1} in the average for the species considered):

$$DPAH^{+\bullet} + H^{\bullet} \rightarrow PAH^{+} \qquad (k_2)$$

In this way, the overall process becomes a catalytic cycle which transforms hydrogen atoms into molecular hydrogen.

Taking for the rate coefficients k_1 and k_2 their Langevin values :
$$k_1 \cong k_2 \cong 2.10^{-9} \text{ cm}^3.\text{s}^{-1},$$
we find, in the steady state approximation
$$N(PAH^+) = N(DPAH^+)$$
where $N(X)$ denotes the relative abundance of species X with respect to that of H nuclei. Following Omont[11], d'Hendecourt and Leger[12], Lepp and Dalgarno[13], Lequeux and Roueff[14], we can assume
$$10^{-7} \leq N(PAH) \leq 10^{-6}$$
Such values can be obtained by taking, for example: $N(C) = 4.10^{-4}$, ten per cent of carbon atoms locked in the PAHs, and an average of 50 carbon atoms per PAH; taking x for the unknown percentage of ionized PAHs, we find
$$x\ 10^{-9} \leq N(PAH^+) + N(DPAH^+) = x\ N(PAH) \leq x\ 10^{-8}$$
and thus,
$$x/2\ 10^{-9} \leq N(PAH^+) \leq x/2\ 10^{-8}$$
Denoting $n(X)$ the abundance of species X in a typical diffuse cloud, and n the abundance of H nuclei, we get:
$$n(H_2) = k_1\ n(H)\ n(PAH^+)$$
$$= k_1\ n(H)\ n\ N(PAH^+)$$
which gives
$$x\ 10^{-18} \leq n(H_2) / n\ n(H) \leq x\ 10^{-17}$$
Comparing this result with the rate of formation of H_2 (3.10^{-17} cm^3.s^{-1}) according to Duley and Williams[1] we get 3 to 30 percent of ionized PAHs. Considering that the present process is not the only one efficient in the formation of H_2, the values obtained here can be considered as an upper limit of ionized PAHs which could be active in such a process.

At the level of this study, there are still a number of questions that need a deeper insight, in particular we cannot give an answer as to the distribution of the energy released in the process (IR emission of the aromatic substrate, rotationnal excitation of the H_2 fragment..). It is clear that more theoretical and experimental work is necessary for a complete understanding of what appears to be a promising route to molecular hydrogen.

ACKNOWLEDGEMENTS

We are particularly indebted to professor V. Barone for fruitful discussions and to P.J.E. Encrenaz for stimulating exchanges on the astrophysical consequences of the chemistry presented here.

REFERENCES

1. W.W. Duley and D.A. Williams, Interstellar Chemistry, (Acad. Press Inc., London, 1984).
2. S. Aronowitz and S. Chang, Astrophys. J. **293**, p.243, (1985).
3. V. Pironello and D. Averna, Astron. Astrophys. **196**, p.201, (1988).
4. S. Klose, Astron. Nachr. **310**, p.409, (1989).
5. A. Léger, L.B. d'Hendecourt, L. Verstraete and P. Ehrenfreund, in *Chemistry in Space* (Edited by J.M. Greenberg and V. Pirronello, Kluwer Academic Publishers, Dordrecht, 1991).
6. F. Pauzat, D. Talbi and Y.Ellinger, 'UIR bands : computational experiments on model PAHs', communication n° 78 in this book
7. G. Gouedard, N. Billy, J. Vigué, M. Sablier, H. Mestdagh and C. Rolando, in Colloque G.D.R. 1991: Physico-chimie des molécules interstellaires, (Institut d'Astrophysique de Paris, 1991).
8. *Gaussian 88* , M.J. Frisch & al., Gaussian, Inc., Pittsburg PA, (1988).
9. M.J.S. Dewar, E.G. Zoebisch, E.F. Healy, and J.J.P. Stewart, J. Am. Chem. Soc. **107**, 3902, (1985).
10. M.J.S. Dewar and N. Trinajstic, J. Chem. Soc. **A**, p.1220, (1971).
11. A. Omont, Astron. Astrophys. **164**, 159, (1986).
12. L.B. d'Hendecourt and A. Léger, Astron. Astrophys. **180**, L9, (1987).
13. S. Lepp and A. Dalgarno, Astrphys. J. **324**, p.553, (1988).
14. J. Lequeux and E. Roueff, Physics Reports **200**, p241, (1991).

DISSOCIATIVE RECOMBINATION OF H_3^+

A. E. Orel
University of California, Davis, Livermore, CA 94550

K. C. Kulander and B. H. Lengsfield III
Lawrence Livermore National Laboratory, Livermore, CA 94550

ABSTRACT

Recent experiments by Larsson et al.[1] have confirmed the prediction[2] of a high energy (~ 9.5 eV) peak in the cross section for dissociative recombination of H_3^+. This peak is caused by four doubly excited resonance states of H_3. Electron scattering calculations using the complex Kohn method provide resonance positions and widths as functions of the internuclear geometry. This information was used as input to a wave packet calculation for the dissociation dynamics on a fit to the resonant state potential energy surfaces. The resulting cross sections agree well with this experiment.

INTRODUCTION

A great deal of theoretical work exists on the dissociative recombination (DR) of diatomics.[3] In contrast there is little work on polyatomic systems. We have initiated a study on the resonant dissociative recombination of H_3^+[4,5]:

$$e^- + H_3^+ \rightarrow H_3^* \rightarrow H_2(v, J) + H$$
$$\rightarrow H + H + H$$

To our knowledge this is only *ab initio* treatment of DR that includes more than one nuclear degree of freedom. There has been considerable interest in the dissociative recombination of H_3^+ due to its importance in low-energy plasmas, particularly in the modeling of Jovian atmospheres and the interstellar media. In recent measurements of the DR cross section a high energy resonance peak was reported near 9.5 eV.[1] In earlier theoretical studies[2] a resonance in this energy range was found and predicted to contribute to the DR process. In resonant or 'direct' DR, the electron is captured into a resonance state of the neutral molecule. After capture, the molecule begins to dissociate. During this time, the molecule can autoionize, leaving the ion in a vibrationally or rotationally excited state. Once the resonance curve crosses the ionic curve and becomes bound with respect to the emission of an electron, autoionization can no longer occur, and the system evolves into asymptotic final states. Thus to describe the direct DR process one must know the energies and widths of the resonant states and describe the dynamics of the dissociative process. We will first describe how the potential surfaces and widths were determined, then how the dynamics was performed, and finally summarize our results.

CALCULATION OF RESONANCE POSITIONS AND WIDTHS

The ground state equilibrium geometry of H_3^+, an equilateral triangle[4], (D_{3h} symmetry) with bond lengths 1.65 a_0 is shown in Figure 1.

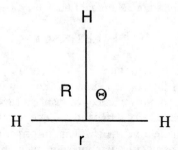

Figure 1: Coordinate system used in calculation. The equilibrium geometry is defined by $r_{eq} = 1.65 a_0$, $R_{eq} = 1.4289 a_0$, $\Theta = 90°$.

The electronic configuration is $1a_1'^2$. The resonance states correspond to the capture of an electron into the low-lying doubly degenerate e' orbital, with the simultaneous promotion of one of the a_1' electrons to the same orbital:

$$H_3^+ (1a_1'^2) + e^- \rightarrow H_3 (1a_1' 1e'^2)$$

Because of the degeneracy of the doubly occupied orbital, four distinct molecular configurations are possible.

If we allow the molecule to distort keeping two of the bond distances equal, the D_{3h} symmetry is broken. In the resulting C_{2v} geometry there are four distinct states, two $^2A_1$ states:

$$c_1 1a_1 2a_1^2 + c_2 1a_1 1b_2^2$$

and two $^2B_2$ states:

$$c_3 1a_1 (2a_1 1b_2)^1 + c_4 1a_1 (2a_1 1b_2)^3$$

where the superscript in the last two cases represents the spin coupling of the two electrons in the parentheses. The coefficients, c_i, vary with nuclear geometry. We have found that all four states contribute to the observed peak in the DR cross section.

We carried out both restricted configuration-interaction (CI) calculations and electron-H_3^+ scattering calculations using the complex Kohn variational method[5] to map out the four resonance potential energy surfaces from the Franck-Condon region to the points where they cross into the bound state manifolds. These latter calculations are more complete because they correctly include the coupling of the resonance states to the background and allow the determination of the resonance widths as well as positions.

The basis set used was the same as in our calculations[6] for dissociative excitation of H_3^+. At the equilibrium geometry of the ion (shown in Figure 1) six natural orbitals were selected, using the procedure outlined in Reference 6. At each new geometry, calculations on H_3 were carried out with these orbitals in the extended basis, with the restrictions that the $1a_1$ orbital remain singly occupied and not more than one electron is excited to the virtual space. This eliminates contributions from the background $1a_1^2 ka_1$ and $1a_1^2 kb_2$ states which represent a free electron in the field of the ion core. The results for these resonance states near the equilibrium geometry of H_3^+ are shown in Figure 2. The two $^2B_2$ states differ only by the spin-coupling of the two excited electrons, either singlet or triplet, and the curves are roughly parallel. By comparison there is a strong interaction between the two $^2A_1$ states with an avoided crossing exactly at the D_{3h} geometry. At these same points, the lowest $^2B_2$ and $^2A_1$ states become the degenerate components of the $^2E'$ state, leading to a *triple* intersection of electronic states along the symmetric stretch seam of the potential energy surface.

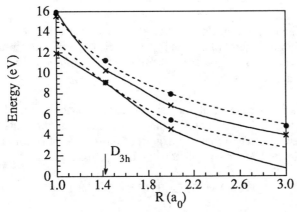

Figure 2: Cut through the H$_3$ potential energy surfaces. Θ is fixed at 90°, r is fixed at 1.65 a$_0$. The solid curves are for A_1 symmetry, the dotted curves are B_2 symmetry, derived from bound-state calculations. The symbols (X A_1, • B_2) are derived from electron scattering calculations.

In order to validate these bound state resonance calculations and obtain the resonance (autoionization) widths, electron scattering calculations from H_3^+ were carried out at a small number of geometries using the complex Kohn variation method.[5] The details of this method and its application to the dissociative excitation of H_3^+ have been described elsewhere.[6] The scattering calculation explicitly includes the effects of the continuum background. Carrying out scattering calculations at a series of energies, we can fit the eigenphase sums to a Breit-Wigner form to obtain the resonance widths and positions. For both symmetries we found it necessary to fit the two resonances simultaneously. A typical example can be seen in Figure 3, where the eigenphase sum for elastic scattering of an electron from H_3^+ in A_1 symmetry for the molecule at its equilibrium geometry. The results for the potential energy surfaces of the resonance are shown in Figure 2. The two methods give the same shape for the surfaces with a shift of no more than ~ 0.5eV due to the coupling to the background.

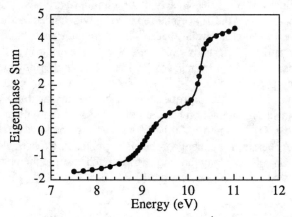

Figure 3: Eigenphase sums for electron - H_3^+ elastic scattering at the equilibrium geometry of the ion in A_1 symmetry. Solid circles are the calculated points, and the line is the fit to the sum of two Breit-Wigner resonance lineshapes.

In addition we carried out scattering calculations for Jacobi angle of 80°. The surfaces was found to be relatively flat near the C_{2v} geometries in this antisymmetric stretch degree of freedom. Therefore, the neglect of this degree of freedom should cause little change in the width or magnitude of the DR cross section.

DYNAMICS OF THE DISSOCIATION

Our calculation was carried out in C_{2v} geometries, thereby restricting the problem to two degrees of freedom. The dynamics was calculated using a wave packet method which involves the direct integration of the time dependent Schrödinger equation.

$$i\hbar \frac{\partial \Psi}{\partial t} = H\Psi \qquad (1)$$

In Jacobi coordinates, shown in Figure 1, the initial wave function is defined on an r, R, grid, where r is the bond distance, and R is the distance from the remaining H to the center of mass of the H_2 bond. θ the angle between r and R was held fixed at 90°. In these coordinates, the Hamiltonian for the nuclear motion is given by,

$$H(r,R) = -\frac{1}{2\mu_1}\frac{\partial^2}{\partial r^2} - \frac{1}{2\mu_2}\frac{\partial^2}{\partial R^2} + V(r,R) \qquad (2)$$

where μ_1 is the reduced mass of H_2, μ_2 is the reduced mass of $H + H_2$. The potential energy is:

$$V(r,R) = V_0(r,R) + \frac{i\Gamma(r,R)}{2} \qquad (3)$$

where V_0 is the real part of the resonance energy, and $\Gamma(r,R)$ is the complex portion (the resonance width). This is necessary since the resonant state can autoionize during dissociation.[7] The kinetic energies are evaluated by finite difference, and the time propagation was carried out using the Chebsyshev polynomial method.

Our treatment of dissociative recombination is a generalization of the time dependent treatment of photodissociation given by Kulander and Heller[8]. At t=0, the wave packet is defined to be

$$\Psi_0(r,R) = \sqrt{\frac{\Gamma(r,R)}{2\Pi}} \Phi_0(r,R) \qquad (4)$$

$$= \sqrt{\frac{\Gamma(r,R)}{2\Pi}} \phi_i(Q_s)\phi_j(Q_b) \qquad (5)$$

where $\phi_0(r,R)$ is the initial vibrational wave function on the resonant state surface, taken to be the simple product wave function of harmonic oscillator wavefunctions, $\phi_i(Q_s)$ and $\phi_j(Q_b)$ in the symmetric stretch and bending normal modes of H_3^+ for each initial vibrational state of interest.

From the time propagation of the wave packet on the dissociative surfaces, we can determine the total capture probability and the resonant dissociative recombination cross section. The total dissociative recombination cross section is calculated by projecting the final wave packet onto final states.[9] For example, when dissociation produces to an atom and a diatomic one can define:

$$S_2(E) = \sum S_i(E) \qquad (7)$$

where the $S_i(E)$ are the final state probability distributions given by:

$$S_i(E) \propto \left| \iint drdR \chi_i(r)\phi_T(R)\Psi_t(r,R) \right|^2 \qquad (8)$$

where $\chi_i(r)$ is a vibrational state eigenfunction of the diatomic, and $\phi_T(R)$ is the translation function describing the motion of the atom from the molecular center-of-mass, t is chosen large enough that the dissociation is complete, i.e.-the fragments are no longer interacting. This expression, when summed over all bound vibrational final states, yields the branching ratio into the two-body channel, $S_2(E)$. Once the sum in Eq.7 is extended to include the continuum H_2 states the result is the total dissociation cross section.

In Figure 4 we show the individual contributions of the four resonance states and the total DR cross section as a function of impact energy. These results are in quantitative agreement, both for the position of the peak and its shape and magnitude, with the ~ 9.5 eV peak reported by high energy cross section measured by Larsson et al.[1]

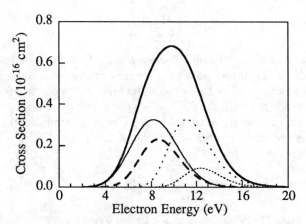

Figure 4: Total (heavy solid line) and partial resonance DR cross sections as a function of incident electron energy: 1^2A_1 (solid line), 1^2B_2 (long dashed line), 2^2A_1 (medium dashed line) and 2^2B_1 (short dashed line).

ACKNOWLEDGEMENTS

We wish to acknowledge very helpful discussion with Sheldon Datz and Jim Peterson regarding the experimental results. This work has been carried out under the auspices of the U. S. Department of Energy at the Lawrence Livermore National Laboratory under contract number W-7405-ENG-48. A. E. Orel acknowledges support of the National Science Foundation, grant No. PHY-90-14845. Computer time was supplied by the San Diego Supercomputer Center.

REFERENCES

1. M. Larssen, H. Danared, J. R. Mowat, P. Sigray, G. Sundström, L. Broström, A. Filevich, A. Källberg, S. Mannervik, K. G. Rensfelt and S. Datz, Phys. Rev. Lett.
2. K. C. Kulander and M. F. Guest, J Phys. B **12**, L501 (1979).
3. see Dissociative Recombination: Theory, Experiment and Applications, edited by J. B. A. Mitchell and S. L. Guberman (World Scientific, Singapore, 1989) p. 97, and references therein.
4. G. Herzberg, *Electronic Spectra of Polyatomic Molecules*, (Van Nostrand Reinhold Company, New York), pg. 289 (1966).

5. T. N. Rescigno, *The Physics of Electronic and Atomic Collisions, ICPEAC XVII, Invited Papers* edited by W. R. MacGillivray, I. E. McCarthy and M. C. Standage (IOTP: London) pg 283 (1992)
6. A. E. Orel, Phys. Rev. A, **46** 1333 (1992).
7. C. W. McCurdy and J. L. Turner, J. Chem. Phys. **78**, 6773 (1983).
8. E. J. Heller, J. Chem. Phys. **68**, 3891 (1978).
9. K. C. Kulander and E. J. Heller, J. Chem. Phys. **69**. 2439 (1978).

PHOTODISSOCIATION OF SMALL POLYATOMIC MOLECULES

Reinhard Schinke
Max-Planck-Institut für Strömungsforschung
D-37073 Göttingen, Germany

ABSTRACT

We will discuss the photodissociation of small polyatomic molecules in the gas phase. The main theoretical tools, namely the time-independent and the time-dependent quantum mechanical approaches as well as classical trajectory calculations, are briefly reviewed. A few examples will be discussed and especially the relationship with the multi-dimensional potential energy surface will be emphasized.

INTRODUCTION

Photodissociation of polyatomic molecules through the absorption of single UV photons is of central interest for many topics in physical chemistry or related fields.[1] It is particularly important for the chemistry in the atmosphere where many chain reactions are ignited through the release of radicals like OH and NO in the fragmentation of diatomic or polyatomic molecules.[2,3] The sun light provides the necessary radiation in the visible or the UV. Likewise, photodissociation is a prominent process in the interstellar space;[4] the production and destruction of molecules in interstellar clouds often involve a photochemical process. Besides its practical importance, photodissociation of small molecules is also ideally suited to study the intra- and intermolecular dynamics.[5] With dynamics we mean the coupling or communication among the different bonds of the polyatomic molecule, the energy transfer from one mode to the others, the breaking of bonds and the formation of new ones, or the quenching of electronic states by non-adiabatic coupling. The advantage of "half collisions" in investigating the change of a chemical system, in contrast to "full collisions", is the principal ability of preparing the system in an unique quantum mechanical state. This allows one to follow the evolution of the system in a most detailed way, not blurred by the averaging over many angular momentum states as in full collisions.

During the last two decades or so sophisticated experimental methods, which were made available by advances in the molecular beam technique in combination with tunable, narrow-band lasers or ultra short laser pulses, have contributed unprecedented details of photodissociation processes in tri- and polyatomic molecules.[6-9] Complete specification of the initial state and complete resolution of the fragment states is today possible,[10,11] providing the pieces to construct a clearest picture of the evolution of a molecular system from its preparation all the way to the products.

Influenced by the overwhelming success of modern experimental techniques, theorists have greatly advanced their computational methods in order to treat photodissociation processes on the basis of *ab initio* calculations, avoiding unrealistic models and/or unjustified approximations. They have demonstrated this for quite a number of small molecules of which only some prominent examples are mentioned here: H_2O,[12,13] H_2S,[14] ClNO,[15-17] and HCO.[18] The power of modern computers gives access to study the dissociation dynamics of triatomic molecules by essentially exact quantum mechanical methods and calculated potential energy surfaces. In the field of photodissociation the close interplay between experiment and theory has greatly contributed to our understanding of photochemical and photophysical processes, in particular, and molecular reaction dynamics, in general.

Figure 1 illustrates the photodissociation of a triatomic molecule ABC into products A and BC. Prior to the absorption process the molecule is assumed to be in a particular initial state $|\Psi_{gi}\rangle$ in the ground (g) electronic state where the index i represents a complete set of vibrational-rotational quantum numbers. Absorption of a photon promotes the molecule to the excited (e) electronic state in which the nuclei start to move subject to the classical or quantum mechanical equations of motion. The corresponding potential energies will be denoted by V_g and V_e, respec-

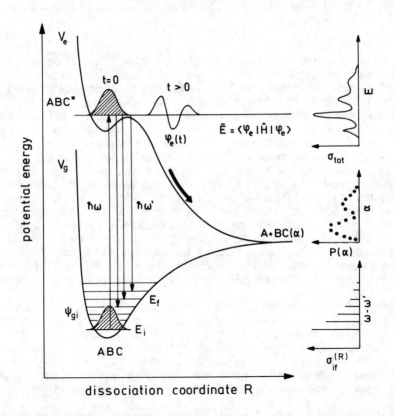

Fig. 1: Schematic representation of the UV-photodissociation of a triatomic molecule ABC into products A and BC(α). The horizontal line in the excited state indicates the expectation value $\bar{E}$ of the excited-state Hamiltonian H_e calculated with the wave packet $\Phi_e(t)$; it is constant in time. The right-hand side shows the total absorption cross section $\sigma_{tot}(E)$, a final product state distribution P(α), and the emission (Raman) cross section $\sigma_{if}^{(R)}$, respectively. The heavy arrow illustrates the dissociation path. For further details see the text.

tively. If the upper-state potential is purely repulsive along the coordinate R the molecule dissociates immediately into the fragments A+BC (or any other variation: AB+C, AC+B, or A+B+C). If, on the other hand, V_e has a barrier which blocks immediate or direct dissociation the bond rupture will be delayed.

The standard quantities measured in a conventional experiment with long pulse duration and narrow frequency resolution are the *total photodissociation cross section* $\sigma_{tot}(\omega)$ as a function of the laser frequency ω, and the *partial photodissociation cross sections* $\sigma(\omega;\alpha)$. The total absorption spectrum measures the probability for absorbing a photon with specific frequency irrespective of the fate of the excited molecule, i.e., independent of the particular quantum states the fragments are produced in. The absorption spectrum depends primarily on the shape of the PES in the inner region extending from the Franck-Condon (FC) point (equivalent to the equilibrium geometry in the ground electronic state) to the barrier, if there exist any. Possible structures in the spectrum reflect the structure of the excited complex while their widths yield information about the lifetimes of the compound states and hence the coupling with the exit channel.[5,19,20]

The partial photodissociation cross sections $\sigma(\omega;\alpha)$, which represent the probability of absorbing a photon of frequency ω and producing the fragments in the particular quantum state α, contain information about the *entire* dissociation process starting from the excitation point and proceeding all the way to the product channels. It is therefore sensitive to a larger part of the PES than the absorption spectrum. In general, one measures final state distributions $P(\omega;\alpha) = \sigma(\omega;\alpha)/\sigma_{tot}(\omega)$ for several frequencies ω and tries to relate the shape of the distributions to the forces acting during the break up. Provided the fragment is a diatomic molecule the index α usually comprises the vibrational (n) and rotational (j) quantum numbers as well as the particular electronic state, for example $^2\Pi_{1/2}(A')$ in NO, in which the molecule is produced. Finally, we note that the sum of all partial cross sections,

$$\sigma_{tot}(\omega) = \sum_{\alpha} \sigma(\omega;\alpha), \qquad (1)$$

yields the total photodissociation cross section.

During dissociation there exists a small probability that the molecule emits radiation (with frequency ω') and thereby recurs to a particular vibrational-rotational state $|\Psi_{gf}\rangle$ in the ground electronic state.[21] Because the lifetime of the dissociating molecule is usually in the subpicosecond region and hence much shorter than the lifetime for spontaneous emission, which is of the order of nanoseconds, the intensity of the emitted light is extremely small. Nevertheless, it can be measured with sensitive detectors and yields through the Raman shift $(\omega-\omega')$ information about the energy levels of the excited vibrational states (f) and therefore indirectly about the ground-state potential energy surface V_g. The intensities $\sigma_{if}^{(R)}(\omega)$ reflect the motion of the dissociating complex in the upper electronic state and therefore they can be used to extract information about V_e. Because the emission (or Raman) spectrum involves the overlap of the wave function in the upper state with excited vibrational states in the ground electronic state, it is sensitive to a wider region of the upper PES than the absorption spectrum σ_{tot}.

Measurement of the absorption spectrum $\sigma_{tot}(\omega)$, the Raman spectrum $\sigma_{if}^{(R)}(\omega)$, and the partial cross sections $\sigma(\omega;\alpha)$ usually provides sufficient information to fully reveal the dissociation dynamics, especially if they are complemented by theoretical studies, either performed in the framework of classical mechanics or by solving the quantum mechanical equations of motion. In recent years it has been impressively demonstrated for a variety of small molecules that with the help of laser pulses in the sub-picosecond regime ($\sim$ 100 fs) the temporal evolution of the molecular system can be *directly* probed in the time domain.[22,23] Pump-probe experiments with femtosecond resolution opened a new window in the study of molecular spectroscopy and dynamics. They are particularly profitable in cases where, because of spectral congestion, high-resolution spectroscopy fails to unravel the time-dependence.

TIME-INDEPENDENT AND TIME-DEPENDENT THEORY

Photodissociation processes can be studied in the time-independent or in the time-dependent framework of quantum mechanics (see, for example, chapters 2-4 of Ref. 5). Both approaches are related by a Fourier transformation between the energy and the time domain. While yielding the same photodissociation cross sections they provide, however, different views of the fragmentation process.

In the time-independent picture one solves the time-independent Schrödinger equation for a particular energy $E = E_i + \hbar\omega$ in the excited state,

$$H_e \Psi_{e,E}^\alpha = E \Psi_{e,E}^\alpha , \qquad (2)$$

where α indicates the particular product channel. The different wave functions $\Psi_{e,E}^\alpha$ are distinguished by the boundary conditions at large intermolecular distances R. According to first-order perturbation theory for the light-matter interaction the partial dissociation cross sections are given by[5,24]

$$\sigma(\omega;\alpha) \propto \omega |<\Psi_{e,E}^\alpha | \mu_{eg} | \Psi_g >|^2, \qquad (3)$$

where μ_{eg} is the coordinate-dependent transition dipole moment function between the ground and the excited electronic states. The total photodissociation cross section is then given by summation over all final product states α. By varying the frequency ω (i.e., the energy E) one calculates the full spectrum point by point.

In the time-dependent picture, on the other hand, the time-dependent Schrödinger equation

$$i\hbar \frac{\partial}{\partial t} \Phi_e(t) = H_e \Phi_e(t), \qquad (4)$$

is solved, where $\Phi_e(t)$ is a wave packet evolving in the excited electronic state. In formal terms, a wave packet is a coherent superposition of all stationary eigenstates $\Psi_{e,E}^\alpha$ in the excited state and therefore does not correspond to a particular energy.

Since the wave packet is not an eigenstate of H_e it moves in time on the excited-state PES. In order to describe the absorption process and the subsequent fragmentation one starts the wave packet at time $t = 0$ with the initial condition

$$\Phi_e(0) = \mu_{eg}\Psi_{gi}, \qquad (5)$$

i.e., we assume that the initial state in the ground electronic state, Ψ_{gi}, multiplied by the transition dipole function μ_{eg}, is *instantaneously* promoted by the photon to the upper state.[19,20] There it immediately starts to move under the action of H according to Eq. (4). In this idealized picture of the absorption process the molecule is excited by an infinitely narrow pulse.

The advantage of the time-dependent approach is that the motion of the wave packet, such as trapping in the inner region of the potential, vibrational motion along the various modes, or recurrences to its origin, can be followed directly in real time. At least for short times the center of the wave packet closely follows a classical trajectory which helps to make the molecular dynamics more transparent. The solution of the time-dependent Schrödinger equation is an initial value problem and as such it is much easier to visualize than the time-independent Schrödinger equation which has to be solved with particular boundary values.

Several efficient methods have been developed for the propagation of multi-dimensional wave packets and for more details we refer the reader to the appropriate literature.[25-28] Basically, the wave packet is discretized on a multi-dimensional grid and the kinetic energy is evaluated by means of the Fourier transformation. The angular degree of freedom, which requires some special attention because it is not a linear coordinate, is most conveniently treated in the discrete-variable-representation.[29] Instead of discussing a particular method we illustrate the motion of a wave packet using a typical example, the photodissociation of FNO by excitation into the S_1 state (Fig. 2). The motion in the S_1 state starts at $t = 0$ with a two-dimensional Gaussian function in R and r located at the FC point. Initially the wave packet follows the gradient of the PES which in the FC region points towards the shallow barrier. In the proximity of the transition state it begins to separate into two parts, one entering the exit channel and immediately leading to dissociation without ever recurring to the place of birth, and a second one which is trapped in the shallow potential well. The latter performs one oscillation in the well and after about 30 fs it recurs for the first time to the FC region where a second cycle begins: some part of $\Phi_e(t)$ escapes from the inner region and the remaining part performs a second oscillation. This continues until the entire wave packet has left the inner zone and travels freely in the product channel.

The motion of the wave packet is reflected by the autocorrelation function

$$S(t) = <\Phi_e(0)|\Phi_e(t)>, \qquad (6)$$

which is simply the overlap of the evolving wave packet with the initial wave packet at $t = 0$. The autocorrelation function reveals how fast the wave packet leaves the FC region, how often it recurs to its place of birth, or how rapidly it leads to dissociation. It reveals many details about the motion of the evolving wave packet. When S(t) has decayed to zero, the absorption spectrum is finally calculated as the Fourier transform of the autocorrelation function,[19,20,31,32]

$$\sigma_{tot}(\omega) \propto \omega \int_{-\infty}^{+\infty} dt\, S(t)\, e^{iEt/\hbar}, \qquad (7)$$

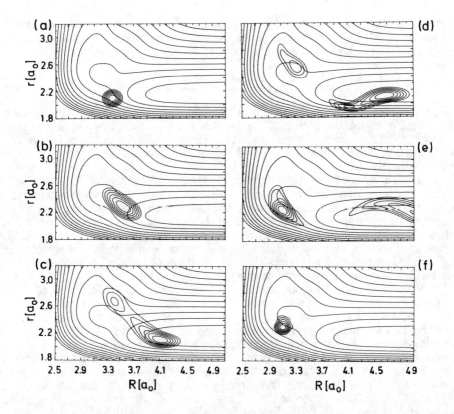

Fig. 2: Evolution of a wave packet $\Phi_e(t)$ in the dissociation of FNO in the S_1 state. The bending angle is fixed at the equilibrium value in the ground electronic state, $\gamma = 128.4^0$. The times (in femtoseconds) are (a) 0, (b) 7.26, (c) 14.52, (d) 21.78, (e) 29.04, and (f) 36.30. Reprinted (with permission of American Institute of Physics) from Ref. 30.

where $E = E_i + \hbar\omega$. Equation 7 is central to the time-dependent picture of spectroscopy; it constitutes the relation between the dynamics in the time domain on the one hand and the energy-resolved spectrum on the other. It should be noted that the wave packet calculation also yields all partial photodissociation cross sections by projecting $\Phi_e(t)$ in the limit of large times onto the stationary eigenstates of the fragments. Thus, all cross sections for the entire range of energies can be extracted from a single calculation which illustrates the power of the wave packet method.

Finally, we emphasize that in many cases classical trajectory methods are also well suited to explore photodissociation processes, especially when the fragmentation dynamics is direct and fast. In order to mimic the motion in the ground electronic state as quantum mechanically as possible, it is recommended to weight the initial coordinates and momenta by the Wigner distribution function. Classical trajectories are easy to implement and can be extended to several degrees of freedom without great difficulty (see, for example, chapter 5 of Ref. 5). Very often few trajectories (10-20 or so) are sufficient to make the general dissociation dynamics transparent and therefore it is always recommended to start a theoretical investigation at the classical level. Moreover, classical mechanics might yield simple pictures of interpretation, especially of the final state distributions, where quantum mechanical calculations appear to be complicated and not very informative.[34-36]

DISCUSSION

The dynamics of photodissociation is determined by the shape of the multidimensional potential energy surface (PES) $V(Q_1,Q_2,...)$ in the excited state, where Q_1 etc. are the coordinates or internal bonds necessary to describe the system. Because the complex breaks apart normal coordinates are by no means useful to describe the system. Let us consider a triatomic molecule ABC dissociating into A and BC. The appropriate coordinates are then the Jacobi coordinates R, the distance from A to the center-of-mass of BC, r, the vibrational coordinate of the diatomic fragment BC, and γ, the orientation angle between the two vectors **R** and **r**. The excited-state PES depends on all three coordinates and in general the dependence of V on all of them must be known. Only in very special cases it is allowed to ignore e.g. the dependence on the bending angle γ because it is weak along the entire dissociation path.[13]

The total absorption spectrum is mainly determined by the forces $\partial V/\partial Q_i$ in the region of the Franck-Condon point where the molecule accesses the excited-state PES. The lifetime of the excited complex is essentially determined by the slope of V in the direction of the exit channel, i.e., the force $\partial V/\partial R$. The systems RNO with R = Cl, F, OH, or CH_3O, for example, are well suited to illustrate the relationship of the PES on the one hand and the lifetime on the other.[37-39] If the upper-state PES is steeply repulsive in R the A-BC bond breaks immediately leading to products A and BC. The resulting absorption spectrum is broad and completely free of so-called vibrational structures (or resonances in the language of scattering theory). With decreasing slope $\partial V/\partial R$ the bond ruptures on a longer time scale and the absorption spectrum becomes narrower, possibly showing very diffuse vibrational structures due to the temporary excitation of an internal bond "perpendicular" to the dissociation path.[19,20] The photodissociations of ClNO via excitations in the S_1 and the T_1 states are illustrative prototypes.[16,17] If a small barrier hinders immediate dissociation the lifetime becomes substantially longer depending on the barrier height and the coupling between the dissociation coordinate on one hand and the other degrees of freedom on the other. The dissociations of CH_3ONO[40] and FNO[30] both via the S_1 state nicely illustrate the occurrence of relatively narrow resonance structures in absorption cross sections. Figure 3 illustrates the gradual transition from fast and direct dissociation to resonance scattering.

The distribution of quantum states of the products are essentially determined by the forces "perpendicular" to the dissociation mode, namely r and γ for the triatomic molecule. If the coupling with R is weak relatively few vibrational or rotational levels of the diatom will be populated after the fragmentation and the distributions can be well described by so-called FC mapping (see, for example, chapters 9 and 10 of Ref. 5). When, however, the coupling between R on one hand and r respectively γ on the other hand is large dynamical effects are more important. In such cases a semiclassical approach, in which the initial state is treated quantum mechanically and the dynamics in the exit channel is described classically, gives a simple but yet realistic explanation of the final state distributions (rotational[41] and vibrational[42] reflection principle (see chapter 6 of Ref. 5).

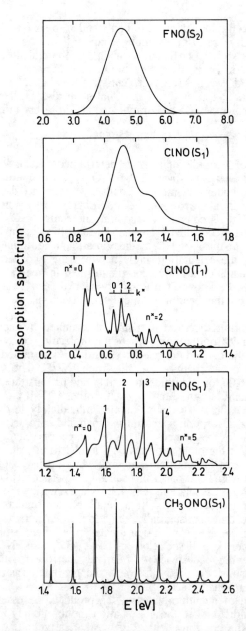

Fig. 3: Calculated absorption spectra as functions of the energy in the excited state for five RNO systems as indicated. Note that except for $FNO(S_2)$ the scale of the energy axis is the same in all cases. n^* and k^* denote the NO stretching and the RNO bending quantum numbers in the excited state and the star indicates resonance states in distinction to true bound states. Reprinted (with permission of American Chemical Society) from Ref. 39.

A realistic description of photodissociation processes requires knowledge of the excited-state PES as a function of all internal coordinates. One- or two-dimensional cuts might be helpful to illustrate the main mechanism but, in general, variation of all coordinates is needed when cross sections and product state distributions are to be determined on an accurate level. Let us consider the dissociation of a symmetric triatomic molecule ABA into A and BA. H_2O^{13} and $CH_2^{43,44}$ are typical examples. In order to rigorously describe the fragmentation it is mandatory to know the PES as a funtion of both AB bond distances; variation of only the bond which finally breakes but fixing the other one is usually unrealistic because in the very first moments of the motion in the excited state the molecule performs preferentially symmetric stretch motion. Thus, keeping one of the bonds fixed might yield an unrealistic picture. In both cases mentioned above, however, activity in the bending angle is very weak and consequently, freezing the bending angle in a sudden-type appxorimation (see, for example, chapter 3 of Ref. 5) is appropriate. If the dissociation of a molecule with more than three atoms is considered, the corresponding PES is a function of more than three coordinates. On the other hand, the calculation of an accurate PES in four or even more dimensions is extremely time-consuming and probabily impossible except for special cases. The main question, which requires a great deal of experience and chemical insight (which, nevertheless, can be often misleading) is which coordinates must be definitely taken into account and which coordinates can be ignored.

Most photodissociation examples which have been treated by *ab initio* theory up to now involve fragmentation on a single PES without coupling to other electronic states. That is certainly the exception rather than the rule. In most cases, especially when higher excited states are involved and the density of (electronic) states is larger, the dissociation proceeds on two or even several PESs and the non-Born-Oppenheimer coupling between them must be incorporated. Needless to say, that such calculations are still very demanding and time-consuming. What requires the most computer time is not the propagation of two or even three coupled wave packets but the determination of the potential energy surfaces and the coupling elements between the various electronic states. To our knowledge a rigorous theoretical treatment of non-adiabatic transitions during the breakup of a polyatomic molecule has been performed only for H_2S in the first absorption band.[14] Despite the many model calculations the study of non-adiabatic transitions in polyatomic molecules is still in its infancy and more realistic examples are required for a deeper understanding.

REFERENCES

1. M. Klessinger and J. Michl, Lichtabsorption and Photochemie organischer Moleküle (VCH Verlagsgesellschaft, Weinheim, 1989).
2. H. Okabe, Photochemistry of Small Molecules (Wiley, New York, 1978).
3. R. P. Wayne, Principles and Applications of Photochemistry (Oxford University Press, Oxford, 1988).
4. K. Kirby and E. F. van Dishoek, Adv. At. Mol. Phys. 25, 437 (1988).
5. R. Schinke, Photodissociation Dynamics (Cambridge University Press, Cambridge, 1993).

6. M. N. R. Ashfold and J. E. Baggott (Eds.), Molecular Photodissociation Dynamics (Royal Society of Chemistry, London, 1987).
7. J. P. Simons, J. Phys. Chem. 91, 5378 (1987).
8. R. N. Dixon, Acc. Chem. Res. 24, 16 (1991).
9. M. N. R. Ashfold, I. R. Lambert, D. H. Mordaunt, G. P. Morley and C. M. Western, J. Phys. Chem. 96, 2938 (1992).
10. D. Häusler, P. Andresen, and R. Schinke, J. Chem. Phys. 87, 3949 (1987).
11. F. F. Crim, Ann. Rev. Phys. Chem. 44, 397 (1993).
12. P. Andresen and R. Schinke. In: Molecular Photodissociation Dynamics, edited by M. N. R. Ashfold and J. E. Baggoott (The Royal Society of Chemistry, London, 1987).
13. V. Engel, V. Staemmler, R. L. Vander Wal, F. F. Crim, R. J. Sension, B. Hudson, P. Andresen, S. Hennig, K. Weide, and R. Schinke, J. Phys. Chem. 96, 3201 (1992).
14. B. Heumann, K. Weide, R. Düren, and R. Schinke, J. Chem. Phys. 98, 5508 (1993).
15. R. Schinke, M. Nonella, H. U. Suter, and J. R. Huber, J. Chem. Phys. 93, 1098 (1990).
16. A. Untch, K. Weide, and R. Schinke, J. Chem. Phys. 95, 6496 (1991).
17. D. Sölter, H.-J. Werner, M. von Dirke, A. Untch, A. Vegiri, and R. Schinke, J. Chem. Phys. 97, 3357 (1992).
18. E. M. Goldfield, S. K. Gray, and L. B. Harding, J. Chem. Phys. 99, 5812 (1993).
19. E. J. Heller, Acc. Chem. Res. 14, 368 (1981).
20. E. J. Heller. In: Potential Energy Surface and Dynamics Calculations, edited by D. G. Truhlar (Plenum Press, New York, 1981).
21. D. G. Imre, J. L. Kinsey, A. Sinha, and J. Krenos, J. Phys. Chem. 88, 3956 (1984).
22. A. H. Zewail, Science 242, 1645 (1988).
23. A. H. Zewail, Faraday Discuss. Chem. Soc. 91, 207 (1991).
24. R. Loudon. The Quantum Theory of Light (Oxford University Press, Oxford, 1983).
25. R. B. Gerber, R. Kosloff, and M. Berman, Computer Physics Reports 5, 59 (1986).
26. R. Kosloff, J. Phys. Chem. 92, 2087 (1988).
27. K. C. Kulander (Ed.). Computer Phys. Comm. 63, 1-577 (1991).
28. J. Broeckhove and L. Lathouwers (Eds.). Time-Dependent Quantum Molecular Dynamics (Plenum Press, New York, 1992).
29. J. C. Light. In Time-Dependent Molecular Dynamics, edited by J. Broeckhove and L. Lathouwers (Plenum Press, New York, 1992).
30. H. U. Suter, J. R. Huber. M. von Dirke, A. Untch, and R. Schinke, J. Chem. Phys. 96, 6727 (1992).
31. R. G. Gordon. In: Advances in Magnetic Resonances, Vol. 3, edited by J. S. Waughn (Academic Press, New York, 1968).
32. L. S. Cederbaum and W. Domcke, Adv. Chem. Phys. 36, 205 (1977).
33. S. Goursaud, M. Sizun, and F. Fiquet-Fayard, J. Chem. Phys. 65, 5453 (1976).
34. R. Schinke, Ann. Rev. Phys. Chem. 39, 39 (1988).
35. R. Schinke, Comments At. Mol. Phys. 23, 15 (1989).
36. R. Schinke, A. Untch, H. U. Suter, and J. R. Huber, J. Chem. Phys. 94, 7929 (1991).

37. C. X. W. Qian, A. Ogai, J. Brandon, Y. Y. Bai, and H. Reisler, J. Phys. Chem. 95, 6763 (1991).
38. C. X. W. Qian and H. Reisler. In: Advances in Molecular Vibrations and Collision Dynamics, Vol. 1B, edited by J. M. Bowman (JAI Press Inc., Greenwich, 1991), p. 231.
39. J. R. Huber and R. Schinke, J. Phys. Chem. 97, 3463 (1993).
40. A. Untch, R. Schinke, R. Cotting, and J. R. Huberpage, in print (1993).
41. R. Schinke, J. Chem. Phys. 85, 5049 (1986).
42. A. Untch, S. Hennig, and R. Schinke, Chem. Phys. 126, 181 (1988).
43. R.A. Beärda, M.C. van Hembert, and E.F. van Dishoeck, J. Chem. Phys. 97, 8240 (1992)
44. R.A. Beärda, G.-J. Kroes, M.C. van Hemert, B. Heumann, R. Schinke, and E.F. van Dishoeck, J. Chem. Phys., in press (1994).

CLUSTERS AND LARGE HYDROCARBONS

ISOMERIZATION OF PURE CARBON CLUSTER IONS: FROM RINGS TO FULLERENES

J. M. Hunter, J. L. Fye, E. J. Roskamp, and M. F. Jarrold
Department of Chemistry, Northwestern University,
2145 Sheridan Road, Evanston, IL 60208

ABSTRACT

Laser vaporization of graphite generates medium sized carbon cluster ions (30-100 atoms) which are either fullerenes or a mixture of polyyne ring isomers. Studies of the isomerization of the polyyne ring isomers have been performed using injected ion drift tube techniques. These experiments reveal the existence of giant carbon rings (monocyclic rings containing up to 90 atoms) and show that for some cluster sizes it is possible to convert the polycylic polyyne rings into spheroidal fullerenes. The mechanism of this remarkable structural transformation is discussed.

INTRODUCTION

Fullerenes are spheroidal shell molecules of pure carbon containing 12 pentagons and (n-20)/2 hexagons [1]. Probably one of the most significant scientific developments so far this decade has been the discovery of a method to generate macroscopic quantities of these unusual species [2]. Macroscopic quantities have only been isolated for a few particularly stable fullerenes such as C_{60} and C_{70}. However, fullerenes with as few as 30 atoms have been observed in gas phase experiments [1]. Studies of the geometries of medium-sized gas-phase carbon cluster ions, generated by pulsed laser vaporization of graphite, show that fullerenes are not the only geometry that can exist for carbon clusters containing 30-100 atoms. In addition to the fullerenes there is a mixture isomers which are believed to be roughly planar polyyne rings [3-9]. These non-fullerene isomers, which are direct products of cluster growth from smaller carbon fragments, are relatively unstable compared to the fullerenes. In this article we describe recent experimental studies of the structural interconversions that the non-fullerene isomers undergo when heated [5-8]. The experiments reveal the existence of giant carbon rings (monocyclic rings containing up to 90 atoms) and show that it is relatively easy to convert the polycyclic polyyne rings into spheroidal fullerenes. The implications of these results for understanding how large carbon clusters grow, and how fullerenes are formed, will be discussed.

EXPERIMENTAL METHODS

The experiments were performed using injected ion drift tube techniques. A schematic diagram of the experimental apparatus is shown in Fig. 1. The carbon cluster ions were generated by pulsed laser vaporization of a graphite rod in a continuous flow of helium buffer gas. Laser vaporization of graphite generates a broad

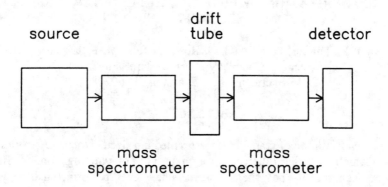

Fig.1 Schematic diagram of the experimental apparatus

distribution of cluster sizes so a mass spectrometer is used to select a particular cluster size for detailed study. Cluster ions which exit the source were focussed into a quadrupole mass spectrometer where the size selection occurs. The size selected clusters are then focussed into a low energy ion beam and injected (at various energies) into a 7.6 cm long drift tube which contains helium buffer gas at around 5 torr. As the clusters enter the drift tube they are rapidly heated by collisions with the buffer gas [10,11]. After their injection energy is thermalized, further collisions with the buffer gas cool the clusters down again. Thus the clusters experience a rapid transient heating and cooling cycle as they enter the drift tube. At high injection energies the clusters may become sufficiently hot that they fragment [10]. At lower injection energies they may still become hot enough that they anneal (isomerize to a lower energy structural isomer) [11]. After injection into the drift tube the clusters travel across the drift tube under the influence of a weak electric field (around 13 V/cm). The mobility of the cluster ions (how rapidly they travel across the drift tube) depends on their geometry [3]. Compact isomers, such as fullerenes, travel across the drift tube more rapidly than less compact isomers such as the roughly planar polyyne ring isomers. After traveling across the drift tube a small fraction of the cluster ions exit through a small aperture, they are then focussed into a second quadrupole mass spectrometer which can be set to transmit either the injected cluster or one of its fragments (thus it is possible to probe the geometry of the injected cluster and its fragments). At the end of the quadrupole the ions are detected by an off-axis collision dynode and dual microchannel plates. The mobilities of the cluster ions are measured by injecting a short pulse (around 50 µs) into the drift tube and recording the arrival time distribution at the detector.

GIANT CARBON RINGS

Fig. 2 shows arrival time distributions recorded for C_{60}^+ cluster ions with a number of different injection energies. These arrival time distributions show the

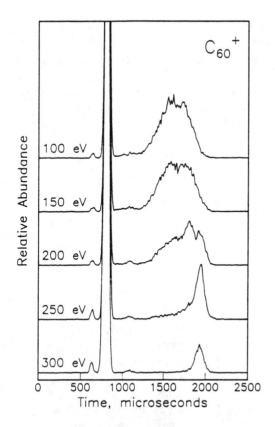

Fig. 2 Arrival time distributions recorded for C_{60}^+ as a function of injection energy

amount of time it takes for the clusters to travel across the drift tube. There is no significant change in the arrival time distributions for injection energies less than 100 eV, so the arrival time distribution at the top of Fig. 2 represents the distribution of isomers coming from the source. The sharp peak at around 800 μs is due to the relatively compact fullerene. The broad distribution at longer times is due to a mixture of polycyclic ring isomers. These isomers consist of polyyne or cumulene rings (the precise nature of the bonding is not yet known [12]) fused together to give a variety of bicyclic, tricyclic, and quadracyclic rings. For smaller clusters it is possible to resolve some of these isomers and compare their mobilities with those calculated for the various possible ring isomers [4]. However, for the larger clusters such as C_{60}^+ the large variety of different isomers leads to the broad distribution observed experimentally.

As the injection energy is increased, and the clusters are heated as they enter the drift tube, the broad distribution centered at ~1500 μs gradually disappears as the polycylic ring isomers are converted into the fullerene and an isomer which grows in at ~1900 μs. The isomer at ~1900 μs has a mobility which suggests that it is a large monocyclic ring. We have measured the mobility of this isomer for clusters containing as few as 10 atoms (the smaller clusters including C_{10}^+ are known to be monocyclic rings [4,13,14]) and the observed variation in the mobility as a function of cluster size is in excellent agreement with that expected for a monocyclic ring [5]. Note that for the larger clusters it is not possible to provide a definitive structural assignment from the mobility alone. Thus the calculated mobility of a 59 atom ring with a one atom "tail" stuck on the outside is almost identical to that of a sixty atom ring. However, the 59+1 geometry is expected to be considerably less stable than the sixty atom ring, so the monocyclic ring is the most plausible geometry that is consistent with the experimental observations.

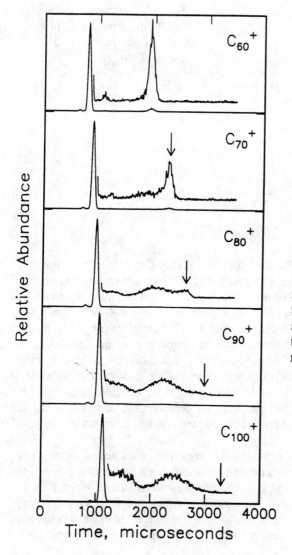

Fig. 3 Arrival time distribution recorded for carbon clusters containing 60, 70, 80, 90, and 100 atoms with an injection energy of 300 eV. The intensities to the left of the fullerene peaks have been multiplied by a factor of 25 (and offset for clarity). The arrows show the expected arrival times of the monocyclic rings.

If a ring of sixty carbon atoms is a reasonably stable entity, what is the largest carbon ring that can be observed in these experiments? As the results shown in Fig. 3 demonstrate, we have now observed this isomer for clusters containing as many as 80-90 atoms [15]. The smallest carbon clusters (n<10) are known to be linear chains [16]. However, the monocyclic ring becomes the most stable isomer for relatively small carbon clusters [4,13,14]. Thus cluster growth by coalescence of small rings is probably an important process. As shown in Fig. 4, this could occur by a [2+2] cycloaddition of two rings to give a bicyclic ring: two polyyne rings linked by a four membered ring. Our experiments show that for the smaller clusters the bicyclic rings convert into a monocyclic ring when heated [5,7]. As shown in Fig. 4, this isomerization can easily be accomplished by a retro [2+2] reaction (perpendicular to the [2+2] which generated the bicyclic ring). The strain energy of a monocyclic ring is less than for a bicyclic ring and so this reaction is driven by the strain energy relief that occurs on going from the bicyclic ring to the monocyclic ring. A cluster growth sequence of ring coalescence through a [2+2] cycloaddition followed by a retro [2+2] to open the resulting bicyclic ring to a large monocyclic ring provides an efficient means of cluster growth. This is probably the dominant cluster growth mechanism in hot environments, such as carbon arcs, where the temperature is high enough to drive the ring opening step.

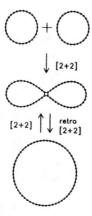

Fig. 4 Diagram showing the growth of giant monocyclic rings by coalescence of two smaller rings followed by a retro [2+2] to open up the bicyclic ring. For large clusters, where the strain relief is small, opening of the bicyclic ring becomes less favorable.

As noted above the opening of the bicyclic ring to form a large monocyclic ring is driven by the lower strain energy of the monocyclic ring. On the other hand, new chemical bonds are formed when the large monocyclic ring is converted into a bicyclic ring (the reverse of the retro [2+2] in Fig. 4). For a sufficiently large cluster, formation of the new chemical bonds compensates for the increased strain energy and the bicyclic ring becomes lower in energy than the monocyclic ring. From simple considerations of the strain energy and the energy associated with formation of the new chemical bonds we estimate that the bicyclic ring should become lower in energy than the monocyclic ring for clusters with around 35-40 atoms [15]. This is obviously considerably smaller than the largest carbon ring observed in the experiments described above. The reason for this difference is that the clusters in our experiments are heated to high temperatures and then rapidly quenched. At high temperatures an equilibrium is rapidly established. The monocyclic ring persists to larger cluster sizes that expected from simple energetic considerations because it's more floppy and its lower frequency vibrations result in a larger entropy. The rapid quenching that occurs in our experiments freezes in the high temperature isomer populations. Thus the monocyclic ring persists to larger cluster sizes than expected from energetic considerations alone. Simulations incorporating these ideas predict that with our experimental conditions the giant monocyclic ring should persist to cluster with around 80-90 atoms in reasonable agreement with our experimental observations [15].

FROM RINGS TO FULLERENES

Fig. 5 shows the relative abundance of the various C_{60}^+ isomers (determined from the arrival time distributions) plotted against injection energy. Not all the clusters survive injection into the drift tube. Two main groups of products are

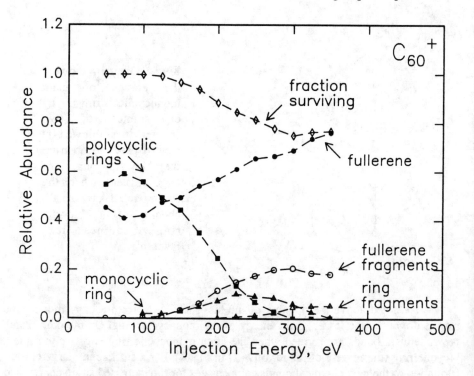

Fig. 5 Plot of the relative abundances of the various isomers and fragments observed in the annealing of C_{60}^+.

observed. These are labelled fullerene fragments and ring fragments in Fig. 5. The fullerene fragments result from the loss of C_2 species to give fragments such as C_{58}^+ and C_{56}^+. This is the characteristic fragmentation pattern of the fullerenes [17], and the arrival time distribution of the products show that they are almost entirely fullerenes. However, these products emerge at injection energies that are too small to fragment pre-formed C_{60}^+ fullerene ions from the source. As can be seen from Fig. 5, the appearance of the fullerene fragments appears to be correlated with the disappearance of the polycyclic ring isomers. Thus we believe that the fullerene fragments result from conversion of the polycylic ring isomers into fullerenes followed by dissociation. Conversion of the polycyclic polyyne rings into fullerenes is a highly exothermic process and this will provide most of the energy needed to fragment the strongly bound fullerene.

At higher injection energies a second group of products emerges (labelled ring products in Fig. 5). These products appear to result from breaking the cluster into two roughly equal parts to give fragment ions such as C_{26}^+, C_{30}^+, C_{34}^+, C_{38}^+, and C_{42}^+. The arrival time distributions for these products show that they are predominantly monocyclic rings. The appearance of these products appears to be correlated with the disappearance of the monocyclic ring suggesting that these smaller rings may be fragments of the large monocyclic ring. Fragmentation into two smaller rings is probably the lowest energy dissociation pathway of the large monocyclic rings since this process conserves the number of chemical bonds. Thus for an infinitely large ring this fragmentation process would be thermoneutral. For a finite ring the additional strain energy associated with making two smaller, more tightly wrapped rings, makes this process endothermic.

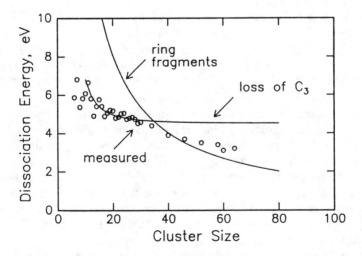

Fig. 6 Dissociation energies for the monocyclic rings. The points are the experimental data and the lines are the results of calculations described in the text

Fig. 6 shows dissociation energies estimated from our measurements for the monocyclic rings [15]. The results shown in the figure for the smaller clusters are in reasonable quantitative agreement with the previous measurements of Sowa, Hintz, and Anderson [18]. For the small clusters significant size dependent oscillations in the dissociation energies are observed with a periodicity of four. These oscillations can be attributed to aromatic stabilization. As can be seen from Fig. 6 the oscillations rapidly diminish with increasing cluster size, and essentially vanish for clusters with around 25-30 atoms. The solid line labelled ring fragments in the figure shows the calculated increase in the strain energy on going from one large ring to two smaller equally sized rings. As expected, this strain energy increases rapidly with decreasing cluster size. Note that the increase in the strain energy closely tracks the increase in the dissociation energies of the monocyclic rings with decreasing size. For clusters with 30-40 atoms a fundamental change in the fragments resulting from the monocyclic rings is observed. Over this size range the products change from fragmentation into two smaller rings to loss of a C_3 species. The implication of this observation is that loss of C_3 becomes the lowest energy dissociation pathway for the smaller clusters. Unlike fragmentation into two rings, loss of C_3 does not conserve the number of bonds, so this process is intrinsically endothermic. However, as shown in Fig. 6 the additional strain energy associated with loss of a C_3 species is relatively small, and so loss of C_3 becomes the lowest energy dissociation channel for the smaller carbon rings. From the results shown in Fig. 6 we estimate that loss of C_3 from an infinitely large ring is endothermic by around 4.5 eV. This can be compared with the energy required to remove a C_3 species from graphite which is 8.42 eV. Thus the large monocyclic rings have a cohesive energy which is approximately 1.3 eV/atom less than graphite. In comparison C_{60} fullerene is approximately 0.4 eV/atom less stable than graphite [19].

In addition to the fullerene fragments which appear to result from conversion of the polycyclic polyyne ring isomers into fullerenes followed by dissociation, it is apparent from the results shown in Fig. 5 that the relative abundance of the intact C_{60}^+ fullerene also increases as the injection energy is raised. These results suggest that some of the polycyclic polyyne ring isomers are converted into an intact fullerene. Conversion of these roughly planar ring isomers into spheroidal fullerenes is clearly a remarkable structural transformation. We will discuss a plausible mechanism for this process below. Close inspection of the results shown in Fig. 5 reveals that the abundance of the ring fragments observed at high injection energies does not equal the abundance of the monocyclic ring observed at intermediate injection energies. Furthermore, the relative abundance of the intact fullerene appears to increase as the monocyclic ring disappears. These results, and results for other cluster sizes, suggest that at high levels of excitation some of the monocyclic ring may be converted into a fullerene. Note that the activation energy for this process is larger than for the polycyclic rings because it requires higher injection energies.

The results described above are for C_{60}^+. C_{60} is special. In addition to its high symmetry, it is the smallest fullerene that has isolated pentagons (C_{70} is the next largest). C_{60} is one of the few clusters for which macroscopic quantities can be made by the arc synthesis method. An obvious question is how does the annealing behavior of the other carbon clusters compare to that of C_{60}^+? We find that the behavior of the even numbered clusters close to C_{60}^+ (such as C_{58}^+ and C_{64}^+) is very similar to that of C_{60}^+ [7]. However, with decreasing cluster size the behavior changes completely. For the smaller clusters the isomer distribution is dominated by the monocyclic ring at intermediate injection energies, and at higher energies the monocyclic ring fragments (as described above). Fig. 7 shows the relative abundance of the fullerene (intact and fragments) and the monocyclic ring plotted against cluster size. As can be seen from the figure there is a threshold for fullerene formation which occurs at around C_{34}^+. With increasing cluster size the relative abundance of the fullerene generated by annealing of the polycyclic rings increases slowly reaching around 80% at C_{60}^+.

The observations described above have important implications for understanding how fullerenes are synthesized from small carbon fragments. As described above, coalescence of two monocyclic rings to give a bicyclic ring followed by opening of the bicyclic ring to yield a monocyclic ring is probably an important cluster growth cycle. This growth cycle can continue unimpeded until the cluster is larger than around 40 atoms. At this size opening up to form a giant monocyclic ring becomes less favorable (there is less strain relief), and a second process - conversion to a spheroidal fullerene - begins to compete. As will become clear below, an important factor in determining when conversion to a spheroidal fullerene becomes possible is probably that the rings need to be sufficiently floppy that they can begin to wrap-up and form a partial fullerene, and so begin recovering some of the fullerene stabilization energy.

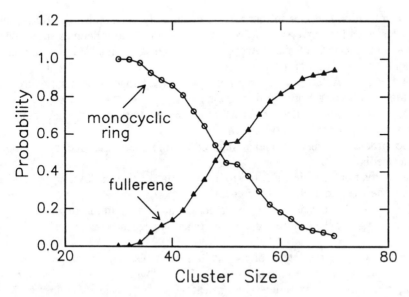

Fig. 7 The relative abundance of the monocyclic ring and the fullerene (intact and fragments) resulting from the annealing of the polycyclic ring isomers plotted against cluster size

An important issue that we have not yet addressed is how much energy is required to promote the conversion of the polycyclic rings to the fullerene. An estimate of the activation energy associated with this process can be obtained from simulations of the measured injection energy thresholds for these processes [6,8]. This is done using a modified impulsive collision model to estimate the fraction of the clusters injection energy that is converted into internal energy (as the clusters enter the drift tube), and a statistical model to describe the subsequent isomerization or dissociation of the excited cluster. From these simulations (which are described in detail elsewhere [6,8]) we estimate that the activation energy associated with conversion of the polycyclic ring isomers to the spheroidal fullerene is ~2.4 eV for C_{60}^+. This is less than a typical C-C bond energy. Considering the magnitude of the structural transformation, this activation energy is remarkably small! Studies of other cluster sizes show that the activation energies are relatively insensitive to cluster size (they decrease slightly with increasing size). This insensitivity to cluster size is significant because it suggests that the critical steps along the pathway from rings to fullerenes occur early - before the cluster can sense the stability of the final fullerene product.

THE MECHANISM

How do the planar polycyclic polyyne rings convert into spheroidal fullerenes? Our experimental results provide a number of important clues about the detailed mechanism of these remarkable structural transformations:

1. The activation energies for conversion of the polycyclic rings into fullerenes are less than for conversion of the monocyclic rings into fullerenes suggesting that some feature of the polycyclic rings acts as a nucleation point and promotes formation of a fullerene.
2. Formation of fullerenes from the polycylic rings competes with formation of a monocyclic ring from the polycyclic rings.
3. The efficiency of fullerene formation increases with cluster size (smaller clusters forming mainly monocyclic rings).
4. The process is very efficient, suggesting that there may be a common intermediate.
5. The activation energies are low indicating that there is a low energy route from the rings to a partial fullerene.
6. The process is insensitive to the precise number of atoms in the cluster, suggesting that the critical steps in the reaction sequence occur early.

Fig. 8 shows the first few steps of a reaction mechanism which appears to be consistent with these experimental observations. We start with a bicyclic polyyne ring with the four-membered ring acting as a nucleation point. The four-membered ring and the attached polyyne chains are configured to undergo a Bergman enediyne cyclization [20] to give the first hexagon. Further Bergman cyclizations are possible, but these generate a chain of hexagons (such as (3) in Fig. 8) and do not lead to a

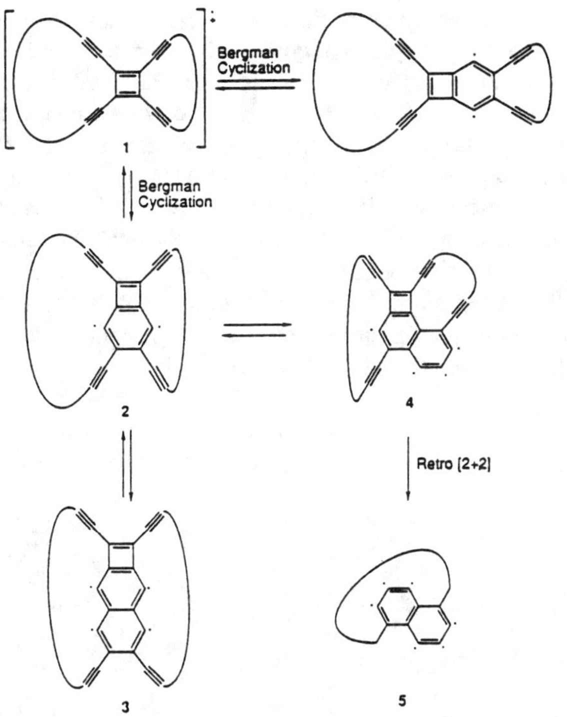

Fig. 8 Diagram illustrating the first few critical steps in the mechanism proposed for conversion of the bicyclic polyyne ring isomers into fullerenes.

fullerene. In order to start forming a fullerene it is necessary to induce a twist in the cyclization process. This is readily accomplished by a radical induced ring closure to give (**4**) in Fig. 8. A retro [2+2] releases the polyyne chain yielding a common intermediate (**5**). This retro [2+2] is probably the critical step in the reaction sequence. Note that if the retro [2+2] occurs before the Bergman cyclization it would lead to a large monocyclic ring. At the energies required to induce the retro [2+2] the first few steps can be assumed to be in equilibrium. For a small cluster with short polyyne chains the relative abundance of (**1**) will be larger than (**4**) (because of ring strain) and when the critical retro [2+2] occurs this will lead mainly to monocyclic rings. On the other hand for large clusters, the ring strain will be less, the relative abundance of (**4**) will be larger than (**1**), and the dominant product will be the fullerene precursor (**5**). Note that (**5**) is a common intermediate, all bicyclic rings, regardless of the size of the individual rings, will be channeled through this intermediate. This is also true for larger ring systems such as the tricyclic rings where the first step will probably be to open up to a bicyclic ring.

From the common intermediate (5) further radical induced ring closures can occur as the two ends of the polyyne chain start to spiral around the growing fullerene. This is illustrated in Fig. 9 for C_{60}. In this way it is possible to assemble a perfect fullerene without breaking a single carbon-carbon bond. In order to assemble a perfect fullerene from the spiralling polyyne chains it is necessary to place all the hexagons and pentagons in their correct positions. The probability of this occurring on the first try is probably relatively low. However, if a mistake is made and the fullerene cannot close-up then the badly formed partial fullerene will simply unravel and try again. When all the hexagons and pentagons are in their correct positions the fullerene shell can close up. Five new chemical bonds are formed when the last hexagon is locked into the fullerene shell. So once the spheroidal shell of the fullerene is complete, it is unlikely that it will unravel again.

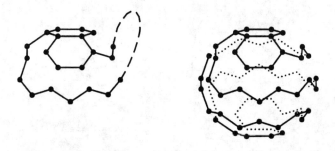

Fig. 9 Diagram illustrating how the spiralling polyyne chains can zip up to form a perfect C_{60}. The two six-membered rings formed in the first two cyclization steps (described in Fig. 8) are at the top of the fullerene. The bonds that zip up the spiralling polyyne chains to form the fullerene are not shown for clarity.

The mechanism described above for the synthesis of fullerenes is very different from previous proposals where it is generally assumed that fullerene growth occurs through precursors which are fullerene fragments [21-24]. The icospiral growth models of Smalley, Kroto and their coworkers [21] and the ring stacking model of Wakabayashi and Achiba [24] are examples. Further evidence that fullerenes can be synthesized by coalescence of monocyclic rings followed by a dramatic structural rearrangement can be found in the important work of McElvaney, Diederich and coworkers [25] who examined the products of laser desorption of large substituted ring compounds such as $C_{18}(CO)_3$, $C_{24}(CO)_4$, and $C_{30}(CO)_5$ using Fourier transform ion cyclotron resonance mass spectrometry and showed that fullerenes are formed.

SUMMARY

In this article we have provided a brief summary of recent experimental studies of the structural dynamics of large carbon clusters. In addition to the work described here for the pure carbon clusters, understanding the behavior of contaminated or doped carbon clusters is also of considerable importance. For example, how does the presence of contaminates such as hydrogen, oxygen, or nitrogen influence the geometries of the clusters and the structural interconversions that the clusters can undergo. If the clusters are doped with metals such as lanthanum, then how does the metal atom get inside to form metallofullerenes. Experiments addressing these issues are in progress.

ACKNOWLEDGEMENTS

We grateful acknowledge support of this work by the National Science Foundation (Grant No. CHE-9306900). JLF acknowledges partial support as a U.S. Department of Education GANN/Dow Chemical Graduate Fellow.

REFERENCES

1. H. Kroto, J. R. Heath, S. C. O'Brien, R. F. Curl, and R. E. Smalley, Nature, **318**, 162 (1985); R. F. Curl and R. E. Smalley, Science, **242**, 1017 (1988).
2. W. Kratschmer, L. D. Lamb, K. Fostiropoulos, and D. R. Huffman, Nature, **346**, 354 (1990).
3. G. von Helden, M.-T. Hsu, P. R. Kemper, and M. T. Bowers, J. Chem. Phys., **95**, 3835 (1991).
4. G. von Helden, N. G. Gotts, and M. T. Bowers, J. Phys. Chem. **97**, 8182 (1993).
5. J. M. Hunter, J. L. Fye, and M. F. Jarrold, J. Phys. Chem., **97**, 3460 (1993).
6. J. M. Hunter, J. L. Fye, and M. F. Jarrold, Science, **260**, 784 (1993).
7. J. M. Hunter, J. L. Fye, and M. F. Jarrold, J. Chem. Phys., **99**, 1785 (1993).
8. J. M. Hunter, J. L. Fye, E. J. Roskamp, and M. F. Jarrold, J. Amer. Chem. Soc., (submitted).
9. G. von Helden, N. G. Gotts, and M. T. Bowers, Nature, **363**, 60 (1993).
10. M. F. Jarrold and E. C. Honea, J. Phys. Chem., **95**, 9181 (1991).
11. M. F. Jarrold and E. C. Honea, J. Amer. Chem. Soc., **114**, 459 (1992).
12. See, for example, M. Feyereisen, M. Gutowski, J. Simons, and J. Almlof, J. Chem. Phys., **96**, 2926 (1992).
13. R. Hoffmann, Tetrahedron, **22**, 521 (1966).
14. S. W. McElvaney, B. I. Dunlap, and A. O'Keefe, J. Chem. Phys., **86**, 715 (1987).
15. K. Shelimov and M. F. Jarrold, (to be published).
16. N. Moazzen-Ahmadi, A. R. W. McKellar, and T. Amano, J. Chem. Phys., **91**, 2140 (1989); J. R. Heath and R. J. Saykally, J. Chem. Phys., **94**, 1724 (1991); A. Van Orden, H. J. Hwang, E. W. Kuo, and R. J. Saykally, J. Chem. Phys.,

98, 6678 (1993).
17. L. A. Bloomfield, M. E. Geusic, R. R. Freeman, and W. L. Brown, Chem. Phys. Lett., **121**, 33 (1985); S. C. O'Brien, J. R. Heath, R. F. Curl, and R. E. Smalley, J. Chem. Phys., **88**, 220 (1988).
18. M. B. Sowa, P. A. Hintz, and S. L. Anderson, J. Chem. Phys., **95**, 4719 (1991).
19. H.-D. Beckhaus, C. Ruchardt, M. Kao, F. Diederich, and C. S. Foote, Angew. Chem Int. Ed. Engl., **31**, 63 (1992).
20. R. G. Bergman, Acc. Chem. Res., **6**, 25 (1973).
21. Q. L. Zhang, S. C. O'Brien, J. R. Heath, Y. Liu, R. F. Curl, H. W. Kroto, and R. E. Smalley, J. Phys. Chem., **90**, 525 (1986); H. Kroto, Science, **242**, 1139 (1988).
22. A. Goeres and E. Sedlmayr, Chem. Phys. Lett., **184**, 310 (1991); M. Broyer, A. Goeres, M. Pellarin, E. Sedlmayr, J. L. Vialle, and L. Woste, Chem. Phys. Lett., **198**, 128 (1992).
23. T.-M. Chang, A. Naim, S. N. Ahmed, G. Goodloe, and P. B. Shevlin, J. Amer. Chem. Soc., **114**, 7603 (1992).
24. T. Wakabayashi and Y. Achiba, Chem. Phys. Lett., **190**, 465 (1992).
25. Y. Rubin, M. Kahr, C. B. Knobler, F. Diederich, C. L. Wilkins, J. Amer. Chem. Soc., **113**, 495 (1991); S. W. McElvaney, M. M. Ross, N. Goroff, and F. Diederich, Science, **259**, 1594 (1993).

DISCUSSION

THADDEUS — I will discuss some of the astrophysical implications of these remarkable experiments tomorrow morning. They are obviously of a astrophysical interest because the 2 - 3 ev required for the spontaneous isomerization of large mono or polycyclic rings to fullerenes are readily supplied by the interstellar radiation field. The long discussed heating of large molecules are thought to account for the unidentified infrared bands. It is noteworthy too that in the progression from one to two to three dimensional carbon via spontaneous isomerization, the two dimensional graphitic structure is positively avoided, because small graphitic plates are not the most stable carbon clusters owing to all the unsatisfied bonds at the plate edges.

JARROLD — This is a comment. No response.

LEACH — It would be of interest to do experiments with buffer gases other than Helium and see whether the simulated mobilities reproduce the conversion processes consistent both with experiment and with the He results.

JARROLD — We have performed some experiments with buffer gases other than helium. A basic assumption in our interpretation of the mobility measurements is that the mobility of the ion is determined by the shape of the cluster and thus by hard sphere interactions between the cluster ion and the buffer gas. We were naturally concerned about how ion-induced dipole interactions influence the mobility. We have performed measurements of the mobilities of small silicon cluster ions in He, Ne, and Ar to examine this issue [1]. The mobilities measured in all three buffer gases show the same trends, but with argon (which has a relatively large polarizability) there was evidence that the ion induced dipole interaction starts to become important. The evidence for this is a reduction in the difference between the mobilities of different isomers. On the other hand these measurements suggest that the polarizabilities of helium and neon are sufficiently small that (at room temperature) the mobilities measured in these buffer gases are dominated by hard sphere interactions. The nature of the buffer gas also influences the amount of energy deposited into the cluster as it is injected into the drift tube. The heavier the buffer gas, the larger the fraction of the clusters injection energy that is converted into internal energy. For silicon we have found that the fraction of the clusters injection energy that is converted into internal energy appears to follow the predictions of a simple impulsive model [2].

(1) M. F. Jarrold and J. E. Bower, J. Chem. Phys., 96, 9180 (1992).
(2) M. F. Jarrold and E. C. Honea, J. Phys. Chem., 95, 9181 (1991).

SCHUTTE — How would the presence of hydrogen influence your results ?

JARROLD — This is clearly an important issue. Nobody has yet examined how contaminants such as hydrogen, oxygen or nitrogen influence the geometries of the clusters and their structural dynamics.

ELLINGER — In your experiments you were able to determine an upper limit for the number of atoms in your carbon rings (80-90 atoms). Do you have any estimation of the lower limit for the small rings.

JARROLD — We have not attempted to examine this issue. Chemical reactivity studies by McElvaney, Dunlap and O'Keefe [1], and the mobility measurements of von Helden et al [2] suggest that the change from linear chains to monocyclic rings occurs at around C_{10}^+. For the anions the linear chains appear to persist to larger cluster sizes.

(1) S. W. McElvaney, B. I. Dunlap, and A. O'Keefe, J. Chem. Phys., 86, 715 (1987).
(2) G. von Helden, M. T. Hsu, P. R. Kemper, and M. T. Bowers, J. Chem. Phys., 95, 3835 (1991).

BAUMGARTEL —

1. Can large monocyclic carbon systems be postulated unambiguously from m/e ratio and a "retention time".
2. Can mechanistic considerations developed for reactions in condensed phase (solutions) at $T \leq 400$ K be used to explain high temperature reactions (~ 2000 K) in a gaseous medium of low density ?

JARROLD—
1. For the monocyclic rings we have performed measurements down to clusters as small as C_{10}^+ and compared these results with the calculated mobility of a rigid monocyclic ring. The agreement between the measured and calculated mobilities is exceedingly good. However, the mobility does not permit an unambiguous assignment of the geometry. For large clusters such as C_{60}^+, for example, a 59 atom ring with an atom stuck on the ouside has roughly the same mobility as a monocyclic ring. The monocyclic rings can, however, be distinguished from bicyclic rings.
2. I do not think that there is a problem using the ideas of mechanistic organic chemistry to understand gas phase processes. Gas phase studies have played an important role in physical organic chemistry for many years. While the temperatures involved are rather high, I believe that the most reasonable approach to understanding these process is to use the ideas of mechanistic organic chemistry as a starting point. If these ideas cannot explain all our observations then it will be necessary to invoke different processes.

SCHREEL— Your method of separation of the fullerenes and the clusters is based upon differences in cross-section. Have you calculated those cross-sections and do they agree with the observed travelling times through the drift-tube ?

JARROLD— We have performed calculations for the fullerenes and the giant carbon rings. The results are consistent with our experimental measurements.

LEACH—
1. The fullerene formation mechanism described is for cations. What about the mechanism for neutral fullerenes ?

2. The suggested intermediates in the fullerene ion synthesis are polyradicals. Can you give some information about such highly unstable reactive species ?

JARROLD—
1. The results we have obtained are for cluster ions. At this point we are assuming that the charge does not strongly influence the results. Unfortunately, it is not possible to perform analogous measurements for neutral clusters. So the best approach to this issue will probably be theoretical studies where calculations are performed for both the charged and neutral clusters.
2. We do not expect that the polyradicals will exist as shown schematically in the reaction mechanism. We expect a significant bonding interaction between the neighboring biradicals as is observed for example in o-benzyne [1].

J. A. Blush, H. Clauberg, D. K. Kohn, D. W. Minsek, X. Zhang, and P. Chen, Acc. Chem. Res., **25**, 385 (1992).

EXCITED STATE LIFETIME BROADENING EFFECTS ON ROTATIONAL BAND CONTOURS OF C_{60} CALCULATED FOR COMPARISON WITH THE DIFFUSE INTERSTELLAR BANDS

S.A.Edwards and S.Leach
DAMAP et URA 812 du CNRS, Observatoire de Paris-Meudon, 92195 Meudon, France.
and
Laboratoire de Photophysique Moléculaire du CNRS, Bât. 213, Université Paris-Sud, 91405 Orsay, France.

ABSTRACT

Recent calculations of some rotational band contours expected for electronic spectra of the C_{60} molecule and related species have been extended to include the effects of rotational line broadening induced by excited state decay mechanisms. Some typical calculated contours are presented here for the purpose of comparison with the observed profiles of the unassigned diffuse interstellar bands. Some of the mechanisms leading to such reduced excited state lifetimes are discussed, and it is suggested that under certain combinations of conditions, which could be present in the interstellar medium, polyhedral carbon species can give rise to absorption band profiles similar to those of observed interstellar bands.

INTRODUCTION

In a previous publication[1] we have presented calculations of rotational band contours predicted for electronic spectra of the icosahedral molecule C_{60}. These calculations were made using a value for the rotational constant B" of ground state C_{60} calculated from the observed geometry of the molecule in the gas phase[2] and assuming a spherical structure. A range of possible excited state rotational constants B' was estimated by introducing reasonable increments and decrements of the carbon-carbon bond lengths measured for the ground state. The calculations were carried out at a variety of rotational temperatures corresponding to those expected in various regions of the interstellar medium, with the aim of making comparisons between the simulations and the observed diffuse interstellar bands (DIBs). It was suggested that the band contours calculated for C_{60}, in that work, would also be relevant to those expected for similar species such as neutral or ionized polyhedral carbon species (PCs) and their partially or completely hydrogenated counterparts (fulleranes).

The comparisons between the calculated contours and some of the observed DIBs led us to conclude that for rotational temperatures above about 100 K the species in question could produce profiles similar to some of the observations. It was noted, however, that the ultraviolet and visible spectrum of C_{60} itself, as observed in the laboratory[3,4], does not reveal peaks corresponding to any of the main DIBs.

In an attempt to continue this study we have enlarged on our previous calculations by introducing the effects of rotational line broadening such as would be observed if there were mechanisms leading to reduced excited state lifetimes taking place.

These additional calculations are part of an ongoing study to test many of the various groups of molecules and ions proposed as candidates for the carriers of the DIBs and we mention here previous such studies for some polycyclic aromatic hydrocarbons[5] (PAHs).

CALCULATIONS

All the calculations were carried out on the VAX 4500 computer at Meudon using a program written specifically for this work. As in our previous study[1] we use a value of $B'' = 0.0027679$ cm^{-1} for the ground state rotational constant of C_{60} and we adopt the range of excited state rotational constants B' proposed in that work, giving as values for the change of rotational constant on excitation $\Delta B = B' - B'' = (-5.69, -3.82, -1.92, 1.62, 3.88$ and $5.86) \times 10^{-5}$ cm^{-1}. In the present calculations we neglect the effects of Coriolis coupling on the spectra, as the introduction of such effects did not alter the conclusions obtained from our previous calculations. Similarly the present calculations have only been carried out for a general transition for which the electronic orbital angular momenta $\Lambda' = \Lambda'' = 0$. Finally we also adhered to the same rotational temperatures as used previously, that is 3, 10, 50, 100 and 500 K.

The effects of reductions in the excited state lifetimes are included by increasing the full width at half maximum (FWHM) of the (gaussian) convolution of the contours. The FWHM (in cm^{-1}) are calculated from the excited state lifetimes τ (in s) from the formula:

$$\text{FWHM} \approx 5\times10^{-12}\times\tau \qquad (1)$$

We introduce the lifetimes 10^{-11}, 10^{-12} and 10^{-13} s and corresponding FWHM of 0.5, 5.0 and 50.0 cm^{-1}. It should be noted that, in the absence of efficient radiative or non radiative decay mechanisms, excited state lifetimes of the order of 10^{-9} s are to be expected, corresponding to FWHM of 0.005 cm^{-1}. Our previous calculations[1] used linewidths of 0.15 cm^{-1} for the calculated spectral convolution, corresponding to that of the best available observational resolution[6].

RESULTS AND DISCUSSION

Some typical results from our calculations are presented in figures 1(a) to 1(i). These contours result from calculations with a moderate reduction in rotational constant on excitation, that is $\Delta B = -3.82 \times 10^{-5}$ cm^{-1}, and the rotational temperatures 3, 10 and 100 K using the enlarged linewidths given above. As in our previous work the contours, plotted as relative absorbance against frequency (in cm^{-1}), are normalized to the maximum absorbance and the band origins are set to zero.

In our previous publication[1] we suggested that the forms of some of the observed DIBs were either apparently symmetrical or unsymmetrical with a steep edge to either the low or high frequency side. The earlier calculations performed at 0.15 cm^{-1} resolution were able to reproduce the forms of unsymmetrical bands only when relatively high temperatures (above 100 K) were included and in no cases were symmetrical contours produced. Here we see that when very short excited state lifetimes (10^{-13} s) are introduced into the calculation (figures 1(g) to 1(i)) broad, symmetrical bands are obtained, even at low rotational temperatures. We also note that for lifetimes of the order of 10^{-12} s, narrower but still featureless contours are seen, at low rotational temperatures (figure 1(d)), which are apparently symmetrical. At higher temperatures

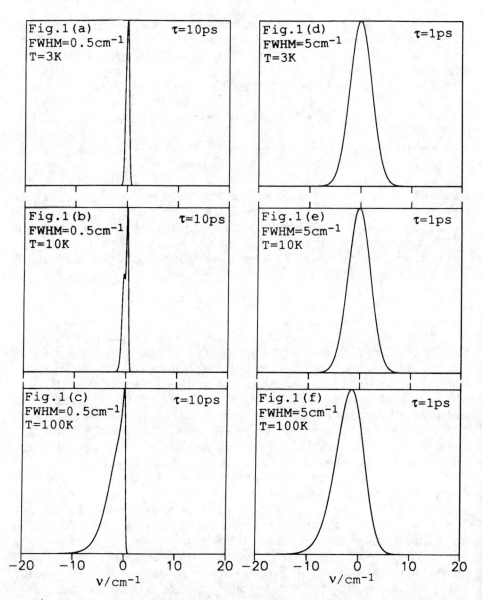

Figure 1. Calculated rotational band contours of C_{60}

(figure 1(f)) the calculated contours begin to show some asymmetry. The longer lifetime used for figures 1(a) to 1(c) gives contours similar to those obtained previously in that they are relatively sharp at low temperatures and have some tendency to exhibit multipeaked structure

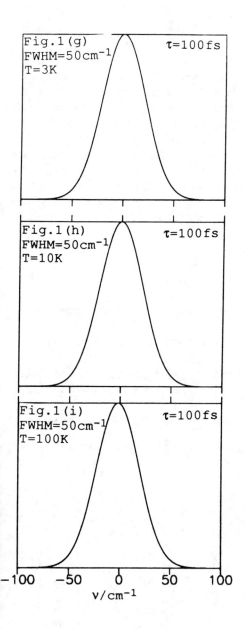

Figure 1. continued

(figure 1(b)).

From the calculated rotational band contours presented here we suggest that the C_{60} molecule and similar species, neutral or ionized, can exhibit absorption bands profiles similar to many of the DIBs if mechanisms leading to rotational line broadening are present. The mechanisms leading to such broadening have been discussed in relation to PAHs[5] and here we reiterate that mechanisms due to collisions, the Doppler effect and pre-dissociation can probably be ruled out for these types of species in the interstellar medium. Two possible processes that could give rise to line broadening are radiationless intramolecular transfer caused by coupling of the excited electronic state to lower lying states and intramolecular vibrational redistribution in the excited state. It is, however, uncertain that these effects, if indeed, present in the excited states of the species under consideration, would lead to the extremely short lifetimes neccessary to reproduce the broad, symmetrical peaks shown in figures 1(g) to 1(i).

CONCLUSION

The new calculations discussed here, of which some of the results are shown in figure 1, demonstrate that polyhedral carbon species, neutral or ionized, could, under certain temperature conditions give rise to absorption bands whose profiles are similar to those of many of the observed diffuse interstellar bands. A requirement for this is that the excited electronic states of the species are subject to efficient decay processes, leading to rotational line broadening as a consequence of Heisenberg's uncertainty principle. For many of the DIBs, however, the lifetimes of the excited states of PCs would have to be exceptionally short if these species are indeed the carriers of the diffuse interstellar absorption bands.

REFERENCES

1. S.A.Edwards and S.Leach, Astron. Astrophys., 272, 533 (1993)
2. K.Hedberg, L.Hedberg, D.S.Bethune et al., Science, 254, 410 (1991)
3. S.Leach, M.Vervloet, A.Després et al., Chem. Phys., 160, 451 (1992)
4. S.Leach, J. Chem. Soc. Faraday Trans., 89, 2305 (1993)
5. C.Cossart-Magos and S.Leach, Astron. Astrophys., 233, 559 (1990)
6. B.E.Westerlund and J.Krelowski, Astron. Astrophys., 203, 134 (1988)

ELECTRONIC STRUCTURES AND STABILITIES OF M_pC_n MICROCLUSTERS. I. Si_pC_n ($n+p \leq 6$)

J. Leclercq, G. Pascoli, M. Leleyter and M. Comeau

GSINPA, Faculté des Sciences d'AMIENS, 33 rue Saint-Leu,
80039 AMIENS CEDEX (France)
and CURI, 5 rue du Moulin Neuf, 80000 AMIENS

ABSTRACT

$Si_pC_n^q$ clusters (n+p<6, q=±1) produced by various experimental techniques (SIMS, SSMS, LAMMA, etc) often show alternations in their emission intensities $I(Si_pC_n^q)$ according to the parity of n (resp.p), the number of carbon (resp. silicon) atoms. For instance, if p≤3 and if they are plotted versus n, they show maxima for n+p odd (q=+1).

It is well known (correspondence rule) that such a parity effect in the emissions arises from alternations in the ions stabilities. These data are analysed within a model built from the Hückel approximation with hybridization sp^ν (ν = 1-3). We compute the electronic energies for various shapes, then compare the geometries at given (n+p) in the 3 kinds of hybridizations, and finally derive the relative stabilities of the clusters. The evolution of the relative stabilities of the most stable clusters is studied as a function of n or p. With the selected geometries, we find that the relative stabilities actually show the same behaviours as the experimental data. Some of these results are confirmed by *ab initio* computations at the HF-SCF level in the DZ+P 6-31G* basis.

INTRODUCTION AND EXPERIMENTAL RESULTS

Silicon carbides are very important compounds since they are encountered as industrial materials known for applications in high-temperature ceramics or semi-conductors. Silicon carbon species are also very useful to understand the chemical processes in the atmospheres of carbon-rich stars or in interstellar and circumstellar environments.

The aim of this paper is to show that a quasi systematical study of the Si_pC_n clusters can be carried out by means of a model built within the framework of Hückel approximation which takes into account hybridization, a very fundamental phenomenon for IV-B elements such as C and Si. We begin with recalling the experimental data and then we shall interpret and discuss them

596 Stabilities of M_pC_n Microclusters. I

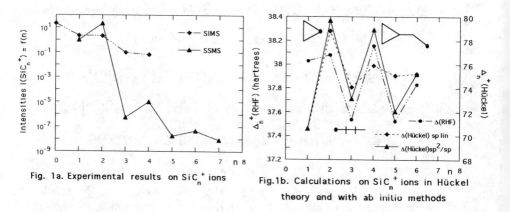

Fig. 1a. Experimental results on SiC_n^+ ions

Fig. 1b. Calculations on SiC_n^+ ions in Hückel theory and with ab initio methods

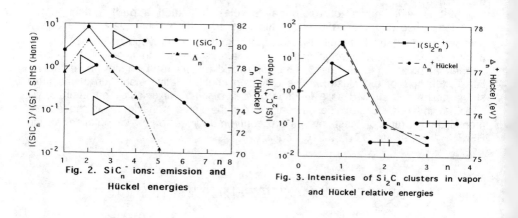

Fig. 2. SiC_n^- ions: emission and Hückel energies

Fig. 3. Intensities of Si_2C_n clusters in vapor and Hückel relative energies

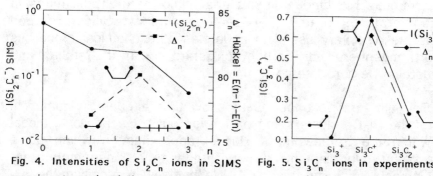

Fig. 4. Intensities of $Si_2C_n^-$ ions in SIMS experiments and relative Hückel energies

Fig. 5. $Si_3C_n^+$ ions in experiments and Hückel calculations

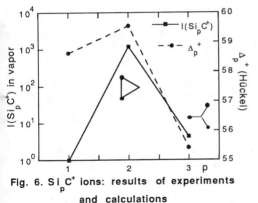

Fig. 6. Si_pC^+ ions: results of experiments and calculations

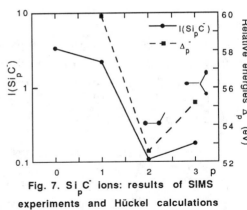

Fig. 7. Si_pC^- ions: results of SIMS experiments and Hückel calculations

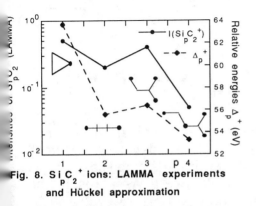

Fig. 8. $Si_pC_2^+$ ions: LAMMA experiments and Hückel approximation

from our model. We also compare our results to those of the litterature.

In the laboratory, a number of silicon-carbon clusters Si_pC_n are produced by various experimental techniques (SIMS, SSMS, LAMMA, vaporization, etc..). The intensities of emission $I(Si_pC_n^q)$ (q= ±1) (in solid lines in figures 2 to 8) often show alternations according to the parity of n, the number of carbon atoms, or p, the number of silicon atoms.

- We first present the experimental data at constant p versus n, the number of carbon atoms, for positive or negative ions (figures 1 to 5). For p=1, in Fig.1a, the positive ions are found in 2 different kinds of experiments, SIMS [1] and spark source mass spectrometry SSMS [2] (curve in solid line) (maxima for even n); in Fig.1b, the negative ions have been got in SIMS experiences [3] and there is only a clear maximum for n=2, the other ones being very less pronounced. For p=2, the positive ions (Fig.3) are observed in vapor in thermodynamical equilibrium at 2316 K with solid SiC [4] and their maxima occur for odd n; the negative ions (Fig.4) observed in SIMS experiments [3] are maxima for even n. For p=3 (Fig.5) the positive ions detected in other SIMS experiments [5] have maximum intensities for odd n.

In most of the cases, that is if p is not too large (p=1,2), the intensities of the positive ions are the largest ones when n+p is odd, exactly as in the very known case of C_m^+ ions [6] (Fig.1a and 2) Conversely, for p≥3 (Fig.5), the ions are more frequently detected when n+p is even, as in the case of pure Si_p^+ ions [7,8].

• Data can also be shared at n constant versus p, the number of silicon atoms. For n=1, the positive ions in vapor at 2316 K [4] (Fig.6) show a sharp maximum at p=2 while the negative ones observed in SIMS [5] (Fig.7) are maximum for odd p. Finally, for n=2, the positive ions found in LAMMA experiments [9] are more frequently detected for odd p, that is , for n=1 or 2, they have always maximum intensities for odd n+p; it is the opposite result for the negative ions.

INTERPRETATION AND DISCUSSION : HÜCKEL MODEL WITH sp^v HYBRIDIZATION

Let us now present our model. At the present time, it is well known that such a parity effect in the detection of small clusters is due to an even-odd effect in the stabilities of the clusters themselves (the "correspondence rule"). As a result, the above experimental data mean that generally, the $Si_pC_n^q$ aggregates are likely to be more stable when the corresponding measured intensities are the highest ones. We have thus to determine the stabilities of $Si_pC_n^q$.

We shall use a model built in the Hückel approximation, but since the 2 elements belong to IV-B group, we have to take hybridization into account ; it is derived from the Friedel-Lannoo model [10] for the bulk, which we already made suitable elsewhere [6] for microclusters of pure IV-B elements. Here, we only modified our preceding calculations by introducing 2 different elements in the 2 Hamiltonians (the sp Hamiltonian and the π one which can be separetely treated in the Hückel theory since it is a "one-electron" approximation). The sp and π energy levels are then calculated by diagonalizing the corresponding operators in the three kinds of hybridization sp^v (v =1,2 or 3) and the total electronic energies E of the various ions are derived by filling in the levels with the total number of valence electrons (let us recall that the origin of the energies is taken at the vacuum level). The corresponding program was written in the C language and the parameters are those used in our preceding papers [6,11]. However an important restriction has to be brought : in our model, indeed, neither the repulsive energies between the nuclei nor the dielectronic correlations are taken into account. Consequently, we cannot get absolute conclusions, but we can only do comparative studies in every series of aggregates.

In a first stage, we had respected the angles between the hybrid bonds (linear shapes of cumulene-like type for sp, angle of 120° for sp^2, that is 2D shapes, and 3D shapes for sp^3). However, since the geometries of small silicon-carbon clusters with 3 or 4 atoms, are known to be closed rings in some cases [12,13], in the second stage, we also considered such shapes in sp or sp^2 hybridizations which are then "forced".

In order to determine the most relatively stable geometry at given m=n+p, we compare the electronic energies of the shapes having the same number of bonds, keeping the one with the highest energy. When the number of bonds is not the same, we retain the geometry with the highest energy only if this value is at least 3 eV per extra bond higher than those of the other shapes with the same number of atoms (the 3 eV value is necessary as we already showed it in an early paper [6], to take into account the extra repulsive nuclear energy due to every extra bond between the atoms). Owing to these selection rules, we are able to find the most stable shapes for the various m=n+p. The comparisons will be then made by considering the quantities $\Delta_m=|E(m)-E(m-1)|$ which measure the increases in electronic energies when going from a (m-1)-atom cluster to a m-atom one, that is when a further atom (C or Si) and hence an extra bond are added.

As far as sp^3 hybridization is concerned, our preceding studies on Si_n microclusters [11] showed that the highest occupied molecular orbital ("HOMO"), the filling of which governs the relative stability of the cluster, is always lying in a flat band of pure p degenerate levels at the energy E_p of the atomic p level of silicon, due to the existence of dangling bonds in the cluster; as a result, since the energy of HOMO keeps the same value whatever the total number of atoms is, the sp^3 structure is quite unable to explain the existence of the alternations in the stabilities of the aggregates. Moreover, the sp³ geometries are never the most stable ones, that is these microclusters never have three-dimensional shapes in our model, and consequently they never have more than three nearest neighbours. Therefore, in the following, we shall only present the studies in sp or sp² hybridizations.

RESULTS OF THE CALCULATIONS

The main results are summed up by the different figures (Fig.2-8) where we have plotted the variations of the quantities Δ_m (m holds here for n or p according to the graphs). The geometries of the corresponding clusters are showed too (the silicon atoms are represented by black points in the sketches of the molecules).

Let us examine the first series of results, the case of SiC_n^+ ions plotted as a function of n (Fig.1b). The graph presents 3 curves. In dashed line with full rhombuses are shown the Δ_n^+ got in Hückel theory by assuming strictly linear structures whatever n (sp) and the impurity atom (Si) bound to one end of the carbon chain (we studied the influence of the position of the silicon atom along the chain and it appears that the preferred one is at the end of the chain, due to the fact that C-C bonds are energetically more favourable that Si-C bonds); the resulting curve seems quite similar to that of SSMS experiment in Fig.1a with a very strong maximum at n=2 and the other ones being lessened as in experiments. Thus, we can see from this curve that $\Delta_{2k}^+ > \Delta_{2k\pm1}^+$, which means that the energy gain is larger when going from an "odd" ion to an "even" one than in the opposite case and that consequently the "even" ions are more stable than the "odd" ones.

However, if we search for the most stable structures, we find that all these aggregates seem to prefer sp² hybridization with triangular form at n=2, four-membered ring with a transannular C-C bond (cross ring) at n=3 (exactly as predicted by *ab initio* calculations for neutral SiC_3 [4,12], C_4 in a rhombic structure with transannular C-C bond, bound to the silicon atom for n=5, all in C_{2v} symmetry. But if we compute the various Δ_n for these structures, the variation is first monotonically increasing from n=0 to n=3, then decreasing at n=4, which is quite different from the experimental data, since there are no maxima for even n. This suggests that it is very likely that the aggregates take other geometries which are not necessarily the most stable ones in an absolute way, and in order to obtain a behaviour of the Δ_n^+ more consistent with the data, we are lead to choose these shapes which are a little less stable, but seem to be favoured over the most stable ones, under the conditions of the experiences. These geometries are sketched in Fig.1b and the corresponding Δ_n^+ are plotted in solid line. As a result, SiC_3^+ would prefer the linear shape with Si bound at one end of the C_3^+ chain, SiC_4^+ the sp^2 hybridization as SiC_2^+, with a Si-C submolecule bound to a cyclic C_3, the next clusters being in sp hybridization (but their study is not yet complete at the present time).

The 3rd curve on the graph shows the Δ_n^+ derived from the total energies obtained in *ab initio* calculations carried out on SiC_n^+ (n=1-6) at the Hartree-Fock-SCF level (program MONSTERGAUSS) in the standard DZ+P (double-zeta + polarization) 6-31G* basis (restricted open-shell Hartree-Fock, RHF, in order to avoid spin contamination). These computations were done in order to get an idea of the shapes of the clusters. Geometry optimizations, however, always lead to linear geometries at this level of theory. It is known indeed that the cyclic forms can be obtained only at a higher level of the theory, that is with configuration interaction calculations [14], since the cyclic isomers become more stable only when correlation effects are included. In any case, for SiC_2^+, the linear and the cyclic isomers lie very close in energy [14]. The HF computations also lead to a saw-toothed curve quite similar to the experimental data. Moreover, this study shows that the adiabatic ionization potentials of SiC_n (n=1-3) at the first stage of our computations also present alternating values: 11.4 eV (n=1), 8.9 eV (n=2) and 11.0 eV (n=3), which enables to understand that it is easier to get an "even" ion as SiC_2^+ than an odd one as SiC^+ or SiC_3^+ in the ionization of the neutral species in the experiences [15]. All these results will be detailed in a forthcoming paper [16].

The other series of $Si_pC_n^q$ clusters were treated from the same point of view. For instance, negative ions SiC_n^- have the same more stable shapes as the corresponding positive ones, but a discussion similar to the above one, lead us to keep the geometries which are drawn in Fig.2, which gives a theoretical curve quite similar to the experimental one. For $Si_2C_n^+$, the preferred

$Si_2C_2^+$ structure in our Hückel model is a rhombus with a transannular C-C bond as in the study of Trucks et al.[17] and the one for $Si_2C_3^+$ is the rhombic form of the above SiC_3 with the extra Si atom bound to the apex of the rhombus opposite to the first Si atom, due to the fact that the Si-C bond is energetically more favourable than the Si-Si one. Nevertheless, the shapes which are finally retained here are the linear carbon chains with one Si at each end. A same kind of reasoning can be hold for the other figures.

CONCLUSION

We can see that with a good choice of the various geometries of the $Si_pC_n^q$ clusters, we are able to get theoretical curves which are quite similar to the experimental ones. We have determined geometries of these clusters which are not necessarily the most stable ones in an absolute way, but are likely to be the shapes encountered in the different experiences. It is interesting to remark that with a model as simple as this one, built in Hückel approximation, one can arrive to the same kinds of most stable geometries as with very sophisticated *ab initio* computations. However, we will have to improve our study for taking into account in one hand the nuclear repulsion energies in a more satisfactory way, and in other hand the influence of the correlations to explain in particular from where comes the discrepancy between the "absolute" most stable geometries and those which account for the experimental results for the best.

Acknowledgments : All the calculations were carried out on the computers of the CURI (Centre Universitaire Régional Informatique d'Amiens).

REFERENCES

1. A. Benninghoven, W. Sichtermann and S.Storp, Thin Solid Films, 28, 59-64 (1975).
2. F.N. Hodgson, M. Desjardins and W. Baun, Mass Spectrom. Conference, San Francisco, paper 78, 444-9 (1965).
3. R.E. Honig, Adv. Mass Spectrom., 2, 25-37 (1962).
4. J.D. Presilla-Marquez and W.R.M. Graham, J. Chem.Phys., 96, 6509-14 and references therein.
5. C.E. Richter and M. Trapp, Int. J. Mass Spectr. Ion Phys., 40, 87-100 (1981).
6. M. Leleyter and P. Joyes, J. Physique (France), 36, 343-55 (1975).
7. R.E. Honig, J. Chem. Phys., 22, 1610-11 (1954).
8. In some graphs, the maxima can be attenuated, but the general evolution of the curves allows us to consider that they are actual maxima, even if they appear to only be slope breakings.

9. D. Consalvo, A. Mele, D. Stranges, A. Giardini-Guidoni and R. Teghil, Int. J. Mass Spectrom. Ion Proc., 91, 319-25 (1989).
10. J. Friedel and M. Lannoo, J. Physique, 34, 115-121 and 483-93 (1973).
11. M. Leleyter, J. Microsc. Spectr. Electron., 14, 61-72 (1989).
12. C.M.L. Rittby, J. Chem. Phys., 96, 6768-72 (1992).
13. R.S. Grev and H.F. Schaeffer III, J. Chem. Phys., 80, 3552-5 (1984); ibid., 82, 4126-30 (1985); D.L. Michalopoulos, M.E. Geusic, P.R.R. Langridge-Smith and R.E. Smalley, ibid, 80, 3556-60 (1984); K. Raghavachari, ibid., 83, 3520-5 (1985); C.W. Bauschlicher,Jr. and S.R. Langhoff, ibid., 87, 2919-24 (1987); A.D. McLean, B. Liu and G.S. Chandler, ibid., 97, 8459-64 (1992).
14. J.R. Flores and A. Largo, Chemical Phys., 140, 19-26 (1990).
15. We also found that the Si-C distance slowly increases from n=1 to 3 in the neutral species (1.624, 1.671 and 1.700 Å) while it shows alternations for the ionized ones, with small values for n=1 or 3 (about 1.53 or 1.55 Å) and larger values (about 1.78 or 1.76 Å) for n=2 or 4.
16. G. Pascoli, M. Leleyter, J. Leclercq and M. Comeau, to be published.
17. G.W. Trucks and R.J. Bartlett, J. Molec. Struct. (Theochem), 135, 423-8 (1986).

ELECTRONIC STRUCTURES AND STABILITIES OF M_pC_n MICROCLUSTERS. II. B_pC_n (n < 6, p=1,3)

M. Comeau, M. Leleyter, J. Leclercq and G. Pascoli

GSINPA, Faculté des Sciences d'AMIENS, 33 rue Saint-Leu,
F-80039 AMIENS CEDEX
and CURI, 5 rue du Moulin Neuf, 80000 AMIENS

ABSTRACT

$B_pC_n^+$ ions were observed in laser vaporization (Becker-Dietze) and their emission intensities show very strong alternations against the number n of carbon atoms: they are maximum for even n if p=1, and for odd n if p=3. In addition, they show a four-fold periodicity with strong enhancements of $I(BC_{4k+2}^+)$ (k=0-4) and deep minima of $I(B_3C_{4k+2}^+)$ (k=0-3).

Such an oscillating behaviour of the emission intensities corresponds to alternations in the stabilities of the observed ions (correspondence rule). Now, these alternations arise from the electronic structures of the clusters. So we studied the B_pC_n clusters (p=1,3) within our Hückel model with hybridization sp^ν (ν =1-3) by determining the relative stabilities of the clusters. The calculations show that the four-fold periodicity arises from closed-ring shapes and a good agreement is found with the experimental data. *Ab initio* computations have also been carried out on BC_n^q (q=0,1, n<6) at the HF-SCF level in the DZ+P 6-31G* basis in order to get further informations on the distribution and the transfers of electronic charges along the BC_n chain.

INTRODUCTION AND EXPERIMENTAL RESULTS

Boron carbide is a material, the structure of which is not yet very well known, and it is obvious that its electronic properties are the main factor governing all its physical properties. From this point of view, it could be interesting to study the first steps of the formation of the solid by considering the microclusters B_pC_n and their likely structures. In this paper, we first recall the experimental results on these small aggregates. In the second section, we present our model built in the Hückel approximation with hybridization and adjust the suitable parameters for boron by comparison to bulk data. Then, in the following section we interpret and discuss the data, and finally, in the last section, we report on results of *ab initio* calculations carried out on the BC_n^q clusters (q=0,1, n<6) which give further informations on the distributions and transfers of electronic charges and the degree of hybridization inside the aggregates.

© 1994 American Institute of Physics

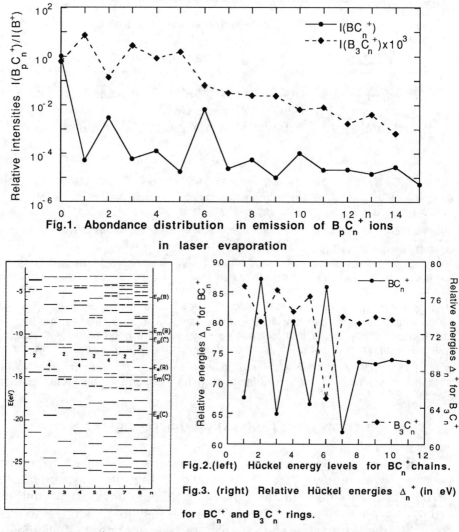

Fig.1. Abondance distribution in emission of $B_pC_n^+$ ions in laser evaporation

Fig.2.(left) Hückel energy levels for BC_n^+ chains.

Fig.3. (right) Relative Hückel energies Δ_n^+ (in eV) for BC_n^+ and $B_3C_n^+$ rings.

Such $B_pC_n^+$ microclusters are encountered for instance in laser ionization mass spectrometry experiments (Becker-Dietze) [1] and they have a very peculiar behaviour (Fig. 1); on the one hand, they show a very markedly parity effect, already well known for a number of polyatomic ions [2], with very strong alternations in the emission intensities as a function of n, the number of carbon atoms. For p=1, the intensities $I(BC_n^+)$ are maximum for even n (curve in solid line) while for p=3, the maxima of $I(B_3C_n^+)$ occur for odd n (curve in broken line), but on the other hand, one can observe in addition a four-fold periodicity with strong enhancements of $I(BC_{4k+2}^+)$ (k=0-4) and deep minima of $I(B_3C_{4k+2}^+)$ (k=0-3). A very simple reasoning enables us to expect that the four-fold periodicity is due to closed-ring shaped clusters, so we are going to use a model built in the Hückel approximation in order to determine the most stable clusters and check this hypothesis, and finally interpret the experimental data.

HÜCKEL MODEL WITH sp^v HYBRIDIZATION

As in the preceding paper (paper I) [3], we know that such a saw-toothed behaviour in the emission of polyatomic ions derives from a parity effect in the stabilities of the clusters themselves ("correspondence rule"), i.e. in their electronic properties.

The model is built in the framework of the Hückel approximation in which the hybridization phenomena are taken into account. They are indeed very important not only for carbon, but also for boron which has a well-known tendency to form hybrids [4]. We use the same kinds of calculations as in I, but we have to fit the parameters for boron. The method has been detailed in our earlier paper [5]. Three parameters of the Hückel hamiltonians have to be adjusted: β_σ and β_π, the resonance integrals respectively for the sp and the π hamiltonians, and $\Delta_\sigma = (E_S-E_p)/(v+1)$ (v order of the sp^v hybridzation), the promotion integral between hybrid orbitals on the same site. β_σ and Δ_σ are derived from bulk data and atomic energy levels E_S and E_p. β_σ =-4.84 eV is then deduced from the gap (1.39 eV) between the valence and conduction bands in the band structure of β-rhombohedral boron [6] with the further condition that $\Delta_\sigma/\beta_\sigma$ <1 (fulfilled for a covalent material [7]). $E_S(B)$=-14 eV and $E_p(B)$=-5.71 eV are taken from the Slater tables [8]. The value of β_π =-1.9eV is adjusted from the distribution of the sp and π energy levels in 2 simple cases, the linear B_n or BC_n chains (thus in sp hybridization). Fig. 2 shows the BC_n molecules : the sp levels (solid lines) are shared out in 2 bands, the bonding (bottom of the figure) and the antibonding (top) bands, and the π levels (in broken lines) are distributed around $E_p(B)$ and $E_p(C)$. β_π is chosen in order that a gap subsists between the π band and the sp bonding band since boron is not a conducting material; moreover the highest π levels remain below the highest sp antibonding ones [9]. The Hückel parameters for carbon are those already used elsewhere [3,5]

RESULTS OF THE CALCULATIONS AND DISCUSSION

In addition to the above hypotheses, we have kept the same method and selection rules as in I [3] in order to compare the various geometries of the $B_pC_n^+$ ions. The relative stabilities are well represented by the quantities $\Delta_n^+ = |E(n)-E(n-1)|$ which measure the increases in electronic energies when going from a (n-1)-atom cluster to a n-atom one, that is when a further C atom and hence an extra bond are added. Our results are plotted Figure 3 as a function of n: BC_n^+ ions are in solid line and $B_3C_n^+$ ions in broken line.

• **BC_n^+ ions** : Numerous isomers were studied and the preferred geometries seem to be the triangular form at n=2 (in sp hybridization), a four-membered ring (rhombus) with a transannular C-C bond at n=3 (sp^2) exactly as for the SiC_3^+ ion in I, a hexahedron with B at one top at n=4 (sp^3) (quite similar to the structure of the boron skeleton in $(B_5H_5)^{2-}$ quoted in (4)) and a B-C submolecule bound to the rhombus C_4 at n=5. Let us notice that BC_2 was observed in a $^2\Sigma^+$ ground state, hence a linear shape consistent with the structure ·B=C=C: [10]; nevertheless, other results would let to think that BC_2 is rather in the C_{2v} symmetry .

However, these most stable shapes do not lead to values of Δ_n^+ giving the good behaviour on the graphs. The most suitable geometries in the present state of our study (all the various isomers for n>6 have not yet been studied), are simple closed-ring forms, either in sp hybridization (n=1,2,4,6) , or in sp^2 (n=3,5, 7-11). Under these conditions, the theoretical curve for n<9 shows 2 main maxima at n=2 or 6 and 2 weaker maxima at n=4 and 8 , which fairly well agrees with the experimental curve in Fig.1.

• **$B_3C_n^+$ ions**: the only isomers we studied till now are the linear and cyclic shapes with various positions of the 3 boron atoms. It appears that for instance a CBCBCB chain seems to be preferred to a BBBCCC one because the B-C bond is energetically more favourable than the B-B bond. In the present state of things, we keep the relative stability values of the cyclic $B_3C_n^+$ ions (with the 3 adjacent boron atoms) in sp hybridization if n<6 and in sp^2 hybridization for n≥6 (curve in broken line Fig.3) to get a behaviour similar enough to the experimental data. We found then large minima at n=2 or 6 and maxima at n=1,3,5 as in the corresponding curve Fig.1. The Δ_n^+ have to be improved for n>6 in order to obtain a better agreement. Our work on this question is still in progress.

RESULTS OF HARTREE-FOCK SCF CALCULATIONS

In another point of view, we undertook *ab initio* calculations in order to get further informations on the BC_n. We used the MONSTERGAUSS program at the HF-SCF level (without configuration interaction) in the standard DZ+P (double-zeta + polarization) 6-31G* basis (RHF to avoid spin contamination which is very severe in these cases). Geometry optimizations always lead at this level to linear shapes except for BC_2 where B seems to be in a bridge position above the C_2 fragment. Let us remark that in our results, BC^+ appears to be stable with respect to the dissociation to $B+C^+$, but not to B^++C. The main results are summed up in Tables I and II

Table I: Total energies, relative stabilities, adiabatic ionization potentials and atomic populations of BC_n^+ ions

BC_n^+	Total energy	$\|\Delta_n^+\|$	IP (eV)	Atomic popul.		
n	hartrees	hartrees		s	p	d
0	-24.234042	-	7.76	2	1	0
1	-61.903361	37.669319	11.22	1.71	0.43	0.02
2	-99.849697	37.946336	9.20	1.11	1.08	0.03
3	-137.678103	37.828406	7.89	1.10	1.28	0.03
4	-175.535179	37.857076	8.48	1.14	1.18	0.03
5	-213.373688	37.838509	7.54	1.14	1.23	0.03

Å	n	1	2	3	4	5
BC_n^+	BC_1	1.6948	1.4282	1.4393	1.4155	1.4105
	C_1C_2		1.2014	1.2393	1.2207	1.2230
	C_2C_3			1.3320	1.3403	1.3355
	C_3C_4				1.2944	1.2269
	C_4C_5					1.3572
BC_n	BC_1	1.4328	1.3769*	1.5585	1.3570	1.5550
	C_1C_2		1.2686 *	1.2230	1.2586	1.2055
	C_2C_3		(* non	1.3655	1.2931	1.3629
	C_3C_4		linear)		1.2605	1.2191
	C_4C_5					1.3578

Table II: Distances in Å between the atoms in the BC_n^+(top of the table) and BC_n (bottom) chains. BC_2, though non linear, is reported too.

We find again (Table I) that the relative stabilities are larger for even n (italic) than for odd n. Moreover the atomic populations show a sign of a significant sp hybridation as it was expected, which supports our above Hückel model. The energy level distribution quite agrees with our preceding CNDO calculations [2]; in particular, we found again the bonding σ_B(B-C) MO localized on the B-C bond and almost especially arising from the pσ atomic orbital of boron. As far as the distances are concerned, we observe that the BC distance is nearly unchanged for the ions (except n=1) contrary to the neutral species which show alternations for n=3 to 5. It would seem that the ions and BC_4 are in a "cumulene type" shape while BC_3 and BC_5 would be under polyacetylenic forms. The question remains to be more completely studied. Moreover, as we wrote in I, the cyclic isomers which are the most likely shapes, are only found when correlation effects are included.

A last remark can be made about the electronic charge distribution (Fig. 4 and 5). Fig. 4 reports on the charge variation along the chain for n=4 and 5. We find here the so-called "charge oscillations" known as "Friedel oscillations", due to the existence of 2 dangling bonds at each end of the chain [11]. They are the answer of the system to the local perturbation to the periodicity of the linear "crystal" represented here by the ends of the chain. The behaviour of these oscillations is however different according to the n parity : for BC_4 (broken line), they are similar to "stationary waves" in phase with the "lattice", while it is not the case for n=5 (solid line). Fig. 5 shows the evolutions of the charge on the terminal boron atom (solid line) and on the terminal carbon (dashed line) and the carbon first neighbour of boron atom (broken line). Boron appears as always being an electron donor as it could be expected from its electronegativity, and the same result also holds for the terminal carbon atom, a foreseeable result since the sign of the terminal atom charge is a direct consequence of the filling of the sp band [12]. On the other hand, it is then obvious that the carbon 1st neighbour of the boron has an excess of electron. Let us notice that, by analyzing the nature (σ ou π) of the charge in the clusters, it appears that the ionization of the BC_n chain is mainly due to the removing of a σ electron except for n=4.

As a conclusion, we showed that the observed four-fold periodicity in the emissions of B_pC_n clusters can be very well understand from a simple Hückel model. On the other hand, the ab

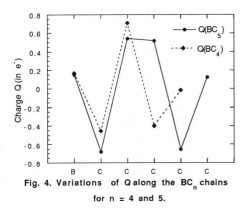

Fig. 4. Variations of Q along the BC_n chains for n = 4 and 5.

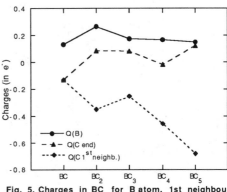

Fig. 5. Charges in BC_n for B atom, 1st neighbour and terminal C atoms.

initio calculations enable to study the influence of an "impurity" such as boron atom bound to one end of a C_n chain, a kind of calculation which can be interesting in order to determine for instance the doping effects in a matrix by an impurity.

Acknowledgments : All the calculations were carried out on the computers of the CURI (Centre Universitaire Régional Informatique d'Amiens).

REFERENCES

1. S. Becker and H.J. Dietze, Int. J. Mass Spectrom. Ion Process., 82, 287-98 (1988).
2. M. Leleyter, J. Physique II France, 1, 1179-96 (1991).
3. J. Leclercq, G. Pascoli, M. Leleyter and M. Comeau, preceding paper.
4. P. Joyes, Les agrégats inorganiques élémentaires (Les Éditions de Physique, Les Ulis, France 1990) p 234 .
5. M. Leleyter and P. Joyes J. Physique (France), 36, 343-55 (1975).
6. R. Franz and H. Werheit, 10th Intern. Symp.on Boron, Borides and Related Compounds, AIP Conf. Proc. (USA) n° 231, 29-36 (1991).
7. J. Friedel and M. Lannoo, J. Physique, 34, 115-21 and 483-93 (1973)
8. J.C. Slater, Phys. Rev., 98, 1039-45 (1955).
9. Figure 2 also shows the number of electrons (2 or 4 according to the parity of n) in the highest occupied molecular orbital (HOMO) of the BC_n^+ ions. One can see that for even n, the HOMO which is generally a π level, is complete with 4 electrons (which corresponds to a more stable cluster) while, for odd n, there are only 2 electrons and the HOMO is either half-filled, or higher in energy, which corresponds to a less stable ion.
10. W.C. Easley and W. Weltner Jr, J. Chem. Phys., 52, 1489-93 (1970). They state that the odd electron in BC_2 is in a highly mixed s-p hybrid, so that the observed BC_2 molecule cannot be the symmetric CBC, so that the question is still to be discussed.
11. M. Leleyter, J. Physique, 48, 1975-80 (1987).
12. M. Leleyter, J. Physique, 42, 1115-28 (1981). When the sp band is indeed less than half-filled , one finds a lack of electron at the "surface" of the "linear crystal", i.e. a positive charge.

PHOTOPHYSICAL STUDIES OF JET-COOLED PAHs : EMISSION SPECTRA, LIFETIMES AND van der WAALS CLUSTERS ASTROPHYSICAL IMPLICATIONS

Philippe BRECHIGNAC and Patrice HERMINE
Laboratoire de Photophysique Moléculaire, C.N.R.S.
Université de Paris-Sud, Bâtiment 213, F91405 ORSAY CEDEX

ABSTRACT

The supersonic free jet experimental apparatus which has been used to study the photophysical properties of cold isolated PAHs in conditions which simulate the astrophysical environments is described. Selected results obtained with naphtalene, anthracene, phenanthrene and pyrene are presented. The efficiency and spectral composition of the light emitted by these molecules when submitted to solar radiation are discussed. LIF excitation spectra of aminopyrene-$(H_2)_n$ clusters are also shown. They exhibit spectrally resolved features which are assigned to structural isomers differing by the occupation number and the nature of the bindidng sites.

1. Introduction.

The intramolecular energy conversion mechanism in isolated polycyclic aromatic hydrocarbons (PAHs) leading to transformation of UV photons emitted by the stars into mid infrared photons (so-called UIR emission features) is now widely considered as an efficient and ubiquitous process in interstellar and/or circumstellar medium.

It has been recently suggested that some of the carriers of the diffuse interstellar absorption bands (so-called DIBs) which are observed in the visible and near infrared part of the spectrum, and which remain after more than 50 years the longest standing mystery of modern astrophysics, could be PAHs or PAHs cations [1].

Finally correspondences in terms of carriers have very recently been established between narrow visible emission features observed from the Red Rectangle reflection nebula and well-identified absorption DIBs [2,3].

These three remarks support our idea that PAHs should be considered not only for their properties as absorbers, but also as emitters, in the visible-near-UV part of the spectrum, in addition to the mid infrared region. Indeed, most of these polycyclic molecules have non-zero fluorescence and/or phosphorescence quantum yields, so that at least one part of the energy of the photons absorbed thanks to the allowed optical transitions in the UV is re-emitted in the near-UV or in the visible. The rest of the energy is of course mostly radiated in the infrared after non radiative decay has taken place.

The possibility to observe isolated cold PAHs in the laboratory, by seeding PAHs vapors in a carrier gas which expands supersonically through a heated nozzle, has been shown to be an efficient spectroscopic tool. The very low temperature and the collision-free regime achieved in a free jet provide a unique way to valuably simulate most of the common astrophysical situations, when cold large molecules are involved. Thus it is under this kind of conditions that we have investigated the photophysical properties of a number of small (up to 4 benzenic rings) PAHs under UV laser excitation. The experimental procedure and the results are described in Section 2.

The goal is to obtain the relevant data needed to predict (by synthetic simulation) the spectral composition and photon efficiency of the light re-emitted by a PAHs population under stellar and/or solar irradiation in a given astrophysical situation. Such a prediction has been done for the specific astrophysical conditions corresponding to the coma of comet Halley on the day of encounter with the VEGA2 probe, since astronomical spectra to compare with are available. This is presented in Section 3.

An interesting aspect of our experimental set-up is its capability to produce weakly-bound van der Waals clusters formed by binding an adjustable number of small molecules to a large PAH. This is the very small size analogue of the ices physisorbed at the surface of a carbonaceous grain. Some results of our investigations on such clusters are presented in Section 4.

2. Experimental.

2.1. *Set-up and procedure.*

The experimental set-up consists of four parts :

i) the supersonic free jet apparatus is made as follows. The PAH / rare gas mixture is expanded through a heated pulsed nozzle (0.5 mm diameter) into a vacuum chamber evacuated by a pumping system made of a Roots pump (1200 m^3/h) backed by a rotary pump. The adiabatic expansion of the mixture produces a fast and very efficient cooling of the molecules. At about 10 mm downstream from the nozzle the kinetic temperature, determined by the carrier gas collisions, has dropped below 10 K, and the PAH molecules travelling with the gas flow velocity have been rotationally cooled to a temperature close to the kinetic temperature. At this distance where the laser beam crosses the jet, the molecular density has become small enough for the average time between collisions being larger than the radiative lifetimes of the electronically excited states.

ii) the laser systems include two tunable dye lasers. The first one, being pumped by the second harmonic (λ=532 nm) of a Nd^{3+}:YAG laser, covers the red portion of the visible spectrum. After frequency-doubling by a KDP crystal it provides tunable UV radiation whose wavelength is between 280 nm and 400 nm. The second one, being pumped by an XeCl excimer laser, covers the blue part of the visible spectrum. After frequency-doubling by a BBO crystal it provides tunable UV radiation whose wavelength extends down to 220 nm. Also available is a beam of 266 nm wavelength obtained by frequency-doubling the 532 nm radiation of the YAG laser.

iii) the spectral analysis is made by a stepping motor scanned grating slit monochromator of 320 cm focal length, after collection of the light emitted from the laser-free jet interaction region by means of an efficient (f:1 aperture) optical assembly, and imaging onto the entrance slit.

iv) the data collection system processes the signal of a fast broadband photomultiplier placed behind the exit slit of the monochromator. Time-gated detection and analog to digital converter are used before averaging by a microcomputer and storage of data files.

As it will appear from the content of Section 3 below, the properties of interest are the following : absorption spectra from cold species, absolute and relative intensities (oscillator strengths) of the electronic transitions, lifetimes and quantum yields, non radiative rates, **emission spectra from photo-excited cold isolated species**, *as a function of excitation wavelength*. Then three kinds of measurements have been performed : laser-induced fluorescence excitation spectra, lifetimes of single vibronically excited levels, dispersed emission spectra following monochromatic fluorescence excitation. In the first kind the laser wavelength is scanned while the monochromator wavelength is kept fixed with a wide bandpass. In the second kind, both wavelengths are kept fixed and a digital oscilloscope provides recordings of the temporal shape of the fluorescence decay. In the third kind the laser wavelength is kept fixed while the monochromator is scanned with a narrow bandpass (typically 5 Å).

2.2. *Selected results.*

The species studied include four small PAHs : naphtalene, anthracene, phenanthrene, pyrene, which have been selected for their capability to emit in the 340 to 390 nm spectral range; and one monosubstituted species, aminopyrene.

Figure 1 shows a typical energy levels scheme for large molecules and the main radiative and/or non radiative processes which can be efficient in the relevant species. Excitation to the first electronically excited state S_1 can give rise to intersystem crossing transitions to lower triplet states and to fluorescence decay to the ground state S_0. When sufficient excess vibrational energy resulting from the excitation is available, intramolecular vibrational redistribution becomes active that will affect the shape of the emission spectrum. Excitation to the second electronically excited state S_2 opens a new non radiative channel which will often be very fast : internal conversion from S_2 to S_1. When excess vibrational energy in redistributed S_1 is large enough internal conversion from S_1 to S_0 also takes place. Then the only way for the molecules to cool down energetically is to radiate in the infrared range, which process is at the origin of the PAHs identification in reflection nebulae, but not our scope in this work.

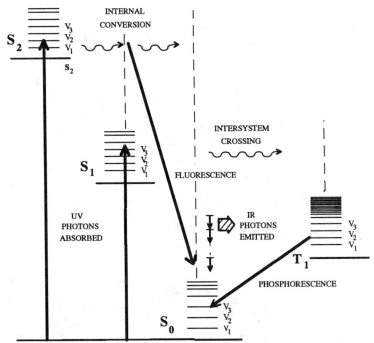

Fig. 1. Typical energy levels scheme of a large molecule showing the main radiative and non radiative processes.

Possible lifetime broadening due to fast non radiative rates can appear in the fluorescence excitation spectra. Typical examples are shown in figure 2. The bands of the S_1-S_0 transition of phenanthrene are narrow, their width (about 3 cm^{-1}) being explained by the unresolved rotational contour at low temperature, which indicates that no fast non radiative process is present. On the contrary the bands of the S_2-S_0 transition of phenanthrene are much wider. Their width, from which the S_2-S_1 internal conversion rate can be derived, is vibrational energy dependent, going from about 12 cm^{-1} at the origin to 6.2 cm^{-1} at 677 cm^{-1}. The third spectrum shown in Fig.2, showing the region of the origin of the S_2-S_0 transition of pyrene, corresponds to the "intermediate case" where the observed bandshape reflects the details of the couplings between the initially excited optically-allowed "doorway" state (S_2 origin) and the manifold of "dark" (final) states (vibrational states of S_1 at about the same energy)[4].

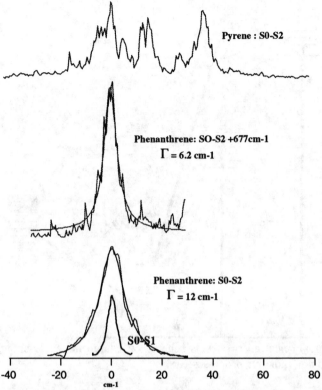

Fig. 2. LIF excitation contours of : **a-** S_1-S_0 and S_2-S_0 origins of phenanthrene; **b-** S_2-S_0 + 677 cm^{-1} vibronic band of phenanthrene; **c-** S_2-S_0 origin of pyrene.

In all cases the fluorescence emitting levels are highly excited vibrational levels of S_1. Then the fluorescence emission spectra exhibit red shifts with respect to the origin of the S_1-S_0 transition. When the excitation energy is such that IVR in S_1 is basically complete, the emission spectrum is esssentially stable against excitation wavelength. The exact position and shape is then characteristic of the specific PAH which has been excited. Higher lying electronic states like S_3 or S_4 can also be involved, but the kind of intramolecular redistribution processes will be similar. Nonetheless the increase of the rate of internal conversion to S_0 will gradually decrease the fluorescence quantum yield. The characteristic spectra of the four PAHs cited above are displayed in Figure 3.

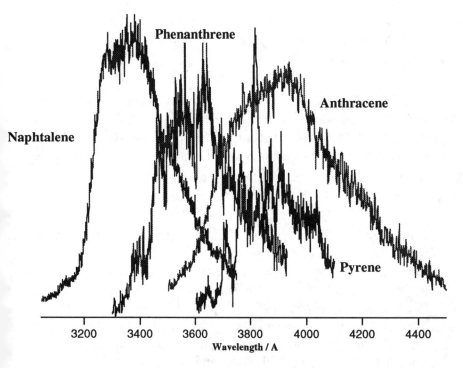

Fig. 3. Stable characteristic dispersed emission spectra of naphtalene, phenanthrene, anthracene and pyrene under UV excitation at sufficiently large photon energy

3. Astrophysical simulation and discussion.

We are now interested in the prediction in terms of emission rates and spectral composition (number of photons emitted per unit time - second- and per wavelength interval -Å-) of the light which would be re-emitted by a low density gas of cold PAHs in given concentrations submitted to a given radiation flux coming from the surrounding stars. The explicit prediction will be done for the case of the solar radiation field, which is relevant for cometary conditions.

Making such a prediction quantitative requires the knowledge of two important quantities : the pumping efficiency and the fluorescence quantum yield. Both quantities must be known for each of the PAHs present in the distribution and as a function of the excitation wavelength. For a given molecule the pumping efficiencies can be calculated from the absorption cross sections $\sigma_j(\lambda)$ of the various electronic transitions S_j-S_0

and the spectral composition and absolute magnitude of the solar flux $F(\lambda)$. Then if the fluorescence quantum yields $\Phi_j(\lambda)$ are known, the total emision rate for this molecule is :

$$K = \sum_j \int \Phi_j(\lambda) \, \sigma_j(\lambda) \, F(\lambda) \, d\lambda$$

The photons are distributed in wavelength according to the stable spectrum which has been measured in the free jet (see Fig.3).

The absorption cross sections can be derived from the interconnected data on oscillator strengths f_j, molar absorptivities $\varepsilon_j(\lambda)$, radiative lifetimes τ_R. Effective lifetimes which are measured also connect the fluorescence quantum yields and the non radiative rates k_{nr} by :

$$\Phi = 1/(1+k_{nr}\tau_R)$$

The solar flux data have been taken from Simon [5] and Labs et al. [6] for a sun to molecule distance of 1 astronomical unit (AU).

The resulting emission rates are listed in Table I below. The naphtalene and pyrene molecules, which both absorb mainly thanks to their S_2-S_0 transitions, appear to be respectively the weaker and the stronger emitters among the four molecules under study. On the contrary the anthracene molecule absorbs mainly thanks to its S_1-S_0 transition, while phenanthrene absorbs on both S_2-S_0 and S_3-S_0 transitions.

Table I- *Values of the emission rates for one PAH molecule placed at 1 AU from the Sun (in photons per second and per molecule)*

PAH	Naphtalene	Anthracene	Phenanthrene	Pyrene
Transition	S_2-S_0	S_1-S_0	S_2-S_0 and S_3-S_0	S_2-S_0
Emission rate (ph/s)	2.10^{-3}	65.10^{-3}	9.10^{-3}	250.10^{-3}

In the specific case of phenanthrene, we have shown that the energy conversion mechanism for UV photons presented above does occur in astrophysical objects, which are believed to have been formed from interstellar material in the primitive solar nebula, that are comets [7]. Indeed the stable dispersed fluorescence emission spectrum of jet-cooled phenanthrene when laser-pumped to its second excited singlet state S_2 and above, made of three main broad features, properly accounts for the spectral features which have been observed from the Halley's inner coma by the tricanal spectrometer (TKS) on board the VEGA2 probe during the encounter with comet Halley on March 9, 1986 [8]. The production rate of phenanthrene molecules outgassed from the comet nucleus has been estimated to be about 10^{27} molecules/s, which is as high as 10^{-3} the water vapor production rate.

As a matter of fact this suggestion must be considered as the first proposed identification of a specific PAH in space. It is worth pointing out that such a proposition has been made possible thanks to the specificity of the UV signatures, by contrast with the IR signatures which are extremely similar for a whole family of molecules. It is of interest for the interstellar medium chemistry to consider the consequences of such an identification in this kind of primitive astrophysical objects. On the other hand future astronomical observations may tell us whether stellar excitation of the visible-UV emission spectra of PAHs and/or their cations is also active in different kinds of environments.

4. Van der Waals clusters involving aromatics and small molecules.

Recent models of interstellar medium including distribution of grains of various sizes have recognized that the total area available to incoming particles offered by PAHs is comparable to that offered by the classical grains surfaces. Consequently the proper account of the physical and/or chemical processes taking place on these microsurfaces might reveal to be of primary importance for the realistic modelling of interstellar chemistry. Moreover the van der Waals microclusters, made by binding one or several small abundant interstellar molecules onto an aromatic molecule, which can easily be formed in a supersonic jet, provide a valuable model to investigate the energetics and the dynamics of the physisorption step at carbonaceous surfaces. They are accessible to experiment by UV spectroscopy and to theoretical modelling by Molecular Dynamics (MD) simulations, as is shown below.

Hydrogen is by far the most abundant molecule in interstellar clouds. A previous study of the aniline-H_2 cluster by laser induced fluorescence (LIF) spectroscopy [9] has given the binding energy of the hydrogen molecule at a sp^2 carbon surface and shown that the rotation of the adsorbed molecule remains nearly free. Here the aminopyrene molecule, shown in Figure 4, offers a wider microsurface and the amino substituent provides enhancement of S_1-S_0 oscillator strength and site-specific van der Waals spectral shifts.

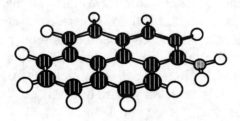

Fig. 4. Skeleton of the aminopyrene molecule.

The LIF excitation spectra of this electronic transition, obtained in the free jet for different seeding ratios of H_2 in He expanded from a backing pressure of 5 bars, are drawn in Figure 5. Besides the monomer 0^0_0 band which saturates the detection (and is chosen as reference for energies), all the peaks to the blue of the main one at -15 cm^{-1} can be assigned to the aminopyrene-(H_2) cluster with a single molecule on the microsurface (origin and the two Van der Waals bending vibrations). The three peaks to the red, which grow with the H_2 concentration, could be due to aminopyrene-(H_2)$_2$ clusters. But the presence of three closely-spaced peaks needs some interpretation. Indeed a search of the minimum energy structures for the aminopyrene-(H_2)$_2$ cluster by the "quenching" method in MD trajectories has revealed the presence of three different isomers [10]. The most stable isomer has one H_2 molecule bound on either side of the planar surface. The other two isomers have a (H_2)$_2$ dimer bound either along the long axis or the short axis of the aminopyrene molecule. Since, as we have shown in the case of aniline-Ar_n clusters [11], the van der Waals shift depends on the distance of the adsorbate to the NH_2 substituent, it explains why the three isomers are specifically shifted in frequency. Thus it is possible to spectrally discriminate clusters of the same size in which adsorbed hydrogen molecules occupy different sites on the microsurface. This finding opens a way for a direct real time measurement of the migration of the adsorbed molecules on their substrate.

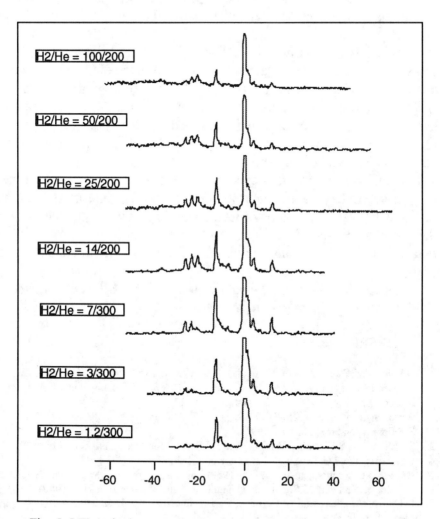

Fig. 5. LIF excitation spectra of aminopyrene-$(H_2)_n$ clusters in the region of the S_2-S_0 origin, for increasing concentrations of H_2 in the Helium carrier gas.

5. Conclusion.

We have shown that the possibility to simulate in a supersonic free jet the low temperature and collision-free conditions which prevail in most "cold" astrophysical environments, like interstellar clouds or comets, allow to study the photophysical properties of this important and ubiquitous component of interstellar matter that are polycyclic aromatic hydrocarbons.

We have in particular studied the emission characteristics of a series of four molecules selected for their characteristic spectral range in the near UV, with emphasis on the yield and spectral composition of the UV radiation re-emitted after illumination by star light. Confrontation of the results with cometary data has led to the first proposition of identification of a specific PAH, namely phenanthrene, in space (see the corresponding paper in this book).

Finally LIF excitation spectra of the S_2-S_0 origin of aminopyrene-$(H_2)_n$ clusters have been reported and theoretically interpreted, showing that selective laser excitation of isomers differing by the site occupation of the adsorbates is possible. Such results are expected to be useful for the detailed modelling of the formation of ices at the interstellar grains surfaces.

Acknowledgement: The autors wish to thank Pascal Parneix for the MD calculations on aminopyrene-$(H_2)_2$.

REFERENCES

1. G.P.van der Zwet and L.J.Allandola, Astron.Astrophys. 146, 76 (1985); A.Léger and L.d'Hendecourt, Astron.Astrophys. 146, 81 (1985); F.Salama and L.Allamandola, J.Chem.Phys. 94, 6964 (1991); Nature 358, 42 (1992); A.Léger and L.d'Hendecourt, Astron.Astrophys. 259, 257 (1992).
2. S.F.Fossey, Nature 353, 393 (1991).
3. S.M.Scarrott, S.Watkin, J.R.Miles, and P.J.Sarre, Mon.Not.R.Astr.Soc. 11p (1992).
4. P.Avouris, W.M.Gelbartand M.A.El-Sayed, Chem.Rev. 77, 793 (1977)
5. P.C.Simon, Aeronomica Acta A211 (1980), Institut d'Aéronomie Spatiale de Belgique.
6. D.Labs, H.Neckel, P.C.Simon and G.Thuillier, Solar Physics 107, 203 (1980).
7. G.Moreels, J.Clairemidi, P.Hermine, Ph.Bréchignac and P.Rousselot, Astron.Astrophys. (december 1993).
8. G.Moreels et al., Astron.Astrophys. 187, 551 (1987).
9. B.Coutant and Ph.Bréchignac, J.Chem.Phys. 91, 1978 (1989).
10. P.Parneix, unpublished.
11. P.Hermine, P.Parneix, B.Coutant, F.Amar and Ph.Bréchignac, Z.Phys. D22, 529 (1992); P.Parneix, F.Amar and Ph.Bréchignac, Z.Phys. D26, 217 (1993)

DISCUSSION

LEGER — What is the quantum yield of the fluorescence you find in phenanthrene? Could a similar process explain the observed ERE in ISM?

BRECHIGNAC — The fluorescence quantum yield of phenanthrene is of the order of 10%. Starting from 15% when excited at the S 1 origin it drops slightly when one reaches the S_2 state, and stays nearly constant.

Concerning the possibility that such a process could contribute to the extended red emission (ERE), I wish to make the following comment. The typical object in which the ERE has been observed is the Red Rectangle. It is also the one in which some of the diffuse interstellar bands (DIBs) have been seen in emission. The gas phase character of the carriers of the DIB s is rather well established. The so-defined carriers necessarily offer a whole set of vibronic transitions for the stellar photons to be absorbed. When the low-lying transitions are excited there is a good chance of nearly resonant emission to take place, producing sharp lines if the gas kinetic temperature is low. On the contrary, when higher-lying transitions get excited (which happens because of the presence of larger energy photons in the stellar radiation field), intramolecular dynamics becomes active at redistributing the population in vibrationally-excited S_1 or T_1, the result of which is a broad, red-shifted emission spectrum (which may contain fluorescence and/or phosphorescence).

Such a picture is basically consistent with the observations, and would lead to conclude that emission in the DIBs bands would not be possible without ERE. This suggestion strongly constrains the observational search of new sources showing DIBs in emission. It is, however only qualitative: the wavelengths and relative intensities of DIBs versus ERE obviously depend on the exact electronic structure of the carrier(s).

Finally one can say that if such carriers are neutral PAHs the ERE would involve, in order to fit the spectral range either phosphorescence processes (as we observe in pyrene) or molecules much larger than those we have studied. If the carriers are PAH cations the ERE would rather involve fluorescence emission.

CROVISIER — We have tried to derive the PAH abundances in half a dozen comets from their 3.28 μm emission (Bockelée-Morvan, Brooke and Crovisier; to be submitted to Icarus). The retrieved abundances are of the order of a few 10^{-5}, which is much smaller than what you found for phenanthene in comet Halley.

BRECHIGNAC — Apparently the discrepancy between your retrieved abundances and ours is between one and two orders of magnitude. I would only make two points both in the direction that one should be very careful in the comparison of such numbers. Indeed what we are both measuring is the concentration of aromatic molecules in the gas phase outflows from the comet nuclei.

My first remark is that the spatial structure of these outflows as compared to the spatial resolution of the observing instrument might be very important to take into account. The TKS data we have used correspond to probing the coma very close to the nucleus, down to 500 km, along a line of sight in the direction of jets. It is clearly different from what can be seen from ground.

My second point concerns the distance from the Sun. Indeed the molecules of interest are solids with small vapor pressure in normal conditions. The only reason why we see them in the gas phase is their evaporation from the comet nuclei. The evaporation rate, and consequently the gas phase concentration, is crucially dependent on the nucleus surface temperature. Beyond the possible spatial variations of this temperature due to the local thermal and radiative properties of a given area, we expect large temporal changes in the evaporation rates provoked by small changes of the surface temperature, which are a consequence of the variation of the comet-to-Sun distance along the orbit. For the TKS data taken on March 9, 1986, Halley was very close to the perihelion.

Ideally the only meaningful comparison between IR and UV data would be by observing the same comet along the same line of sight, with the same field of view and at the same date.

As a final remark, one should consider the uncertainties in both cases coming from the model of the excitation mechanism which is used to retrieve abundances. I believe that in the IR emission case the model itself must be molecule dependent. But I think your remark is interesting .

LEE — I would like to add two comments. First, PAHs generally have strong fluorescence emission, as indicated by such as anthracene's and pyrene's high fluorescence quantum yields; however, such pronounced blue emission is no observed in space. Dr. Leach reported in 1987 that most of electronically excited PAH cations would undergo rapidly internal conversion to vibrationally excited levels of electronic, doublet ground states, therefore emit IR photons. Experimental examination of PAH ion fluorescence is certainly needed. Second, you mentioned about ice formation at PAHs surface, if it is the case, the changes of PAH emission spectra, both in fluorescence and in phosphorescence, would be expected due to energy transfer through/to the icy medium.

BRECHIGNAC — I perfectly agree with you that the near UV emission from PAHs is not observed in interstellar space, at least up to now. But I argue that it has been observed from Halley comet, which is also in space. Certainly the electronically excited PAHs, whether neutral or ionized, will radiate part of their energy in the IR range, as I mentioned. But I also wish to point out that the possibility for the same excited PAHs to also radiate part of their energy in the near UV or visible has probably been hastily forgotten and neglected.

Concerning the change of the emission spectra of "icy" PAHs as compared to free PAHs, I also agree with you since it is specifically this spectral change that we use to study them. But the kind of "redistributed" emission which I am concerned with will not be greatly affected by the coverage. On the contrary it will be crucial in the case of nearly resonant emission in narrow bands, as it may be the case for the carriers of the emitting DIBs.

MOREELS — The quantitative evaluation of phenanthrene in Halley's comet deduced from its UV fluorescence is higher than the evaluation of aromatic compounds from the 3.28 µm X-CH stretch band. This discrepancy may be explained by the fact that the geometrical field of view of the UV spectrometer (Vega 2 TKS) and of the IR instruments (Vega 1,2 IKS) were not the same. There are also calibration uncertainties. However, it should emphasized that the idendification of at least one aromatic compound (in fact probably several) in Halley's coma is shown by three independent observations: in the UV and the IR from the Vega space probe and from ground-based telescopes.

BRECHIGNAC — I have nothing to add to your comment.

LEACH — Are there reasons why phenanthrene but not other PAHs should be seen in cometary comas?

BRECHIGNAC — This is a very important point. I have two pieces of information to give in place of an answer.

I have already mentioned the first one in pointing out the fact that we do not see the actual composition of the comet nucleus, but only the gas phase species in the coma. We have estimated roughly from the vapor pressure data that at a temperature of 300 K (reasonable as a surface temperature) the relative concentrations of the four PAHs would be: very large for naphtalene, very small for pyrene, and ten times larger for phenanthrene than for anthracene.

The second point refers to measurements of PAHs abundances in meteorites. Phenanthrene is usually found among the most abundant species, together with pyrene. But anthracene is nearly two orders of magnitude weaker.

Combining the two points one may understand why anthracene and pyrene are not seen. Naphtalene should be abundant but its fluorescence yield is weak.

A subsidiary reason might be the fortuitous occurence of a particular emission spectrum in a "clean" region, i.e. free of strong radical emission. That was the case for phenanthrene, but not for naphtalene which is superimposed on the NH 336nm feature.

IR SPECTROSCOPY OF LABORATORY-SIMULATED INTERSTELLAR PAHS
ROLE OF TEMPERATURE ON THE BAND POSITIONS

C. Joblin
GPS, Université Paris 7, 2, place Jussieu, 75231 PARIS Cedex 05
NASA Ames, Moffett Field, CA 94035

L. d'Hendecourt, A. Léger
IAS, bât 121, Université Paris Sud, 91405 ORSAY

D. Défourneau
GPS, Université Paris 7

ABSTRACT

The IR absorption spectra of polycyclic aromatic hydrocarbons have been measured in gas phase for temperatures up to 920 K. The band positions clearly depend on the vibrational content of the molecules, which is interpreted as an anharmonic effect.
From the laboratory data, we have deduced that the position of the interstellar 3.3 μm band is consistent with the carriers being free PAHs at high temperatures.

INTRODUCTION

In the PAH model as proposed by Léger and Puget [1], polycyclic aromatic hydrocarbons (PAHs) are responsible for the "Unidentified" IR (UIR) bands observed in emission in many astronomical regions. The emission mechanism requires the molecules to be isolated; after the absorption of a UV photon, there is a fast conversion into the electronic ground state with a high vibrational energy, and finally the cooling takes place by slow IR emission (Léger et al. [2]). Although the IR spectra of PAH molecules can in first approximation account for the UIR bands, the spectral fit has to be improved. However, most laboratory spectra have been obtained for condensed molecules at low temperatures, whereas the PAH model states that the interstellar molecules are isolated and at high temperatures (T~1000 K). Here, we discuss the IR absorption spectra measured for high-temperature gas-phase PAHs. In that experiment, molecules are evaporated and thermally excited in an oven. In the following, we will focus on the dependence of the band positions with temperature, especially for the 3.3 μm band.

Léger et al. [2,3] and Schutte et al. [4] have shown that most of the IR emission, for an isolated photoexcited molecule, is correctly described by a simple thermal model with a vibrational temperature (T_{vib}). The spectral information derived from our experiment, in which molecules are thermally excited, should then be relevant to interpret the astronomical spectra.

RESULTS

Figure 1 displays the 3.3 μm band measured at several temperatures for different PAHs. The right plots give the variation of the band positions with temperature. Our results are compared to those obtained with similar techniques (Kurtz [5], Flinckinger et al.[6]), and to those available for photoexcited PAHs (Shan et al. [7], Brenner and Barker [8]). The band positions measured in neon matrices at 4 K are also given (Joblin et al. [9]).

Figure 1 shows that the bands shift towards longer wavelengths and broaden as the temperature increases. Whereas the band broadening results from a combination of vibrational and rotational effects (Fig. 2), the band shift is mostly due to a vibrational (anharmonic) effect.
This is suggested by the fact that the linear extrapolation of our data (Fig. 1) can account for the positions measured for photoexcited molecules at high vibrational (~2000 K) and low rotational temperatures (~300 K).

The linear dependence of the band frequencies with temperature can be explained by including the first anharmonic terms into the vibrational energy.

$$\frac{E_v^{anh}}{hc} = \sum_i \omega_i (v_i + \frac{1}{2}) + \sum_i x_{ii} (v_i + \frac{1}{2})^2 + \sum_{i;k\ k\neq i} x_{ik} (v_i + \frac{1}{2})(v_k + \frac{1}{2}) + ... \quad (1)$$

ω_i : harmonic frequency of the v_i mode
x_{ii} : anharmonicity constant of the v_i mode
x_{ik} : coupling constant between v_i and v_k modes
v_i, v_k : number of quanta in the v_i and v_k modes

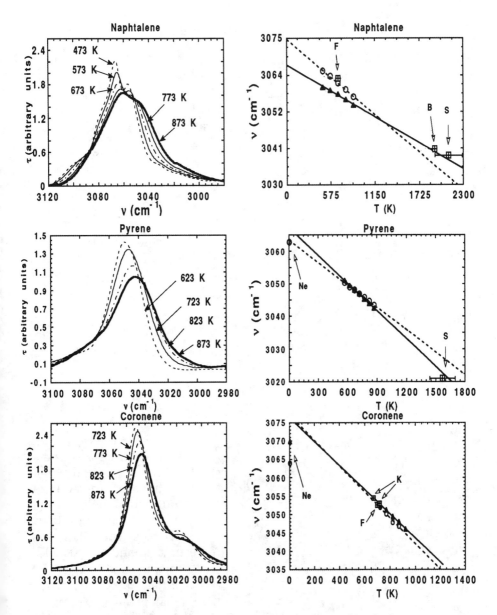

Fig. 1. Evolution of the 3.3 μm band of naphtalene ($C_{10}H_8$), pyrene ($C_{16}H_{10}$) and coronene ($C_{24}H_{12}$) with the excitation temperature of the molecules.
Right diagrams: band positions at the peak (triangles), and at half maximum (circles) versus temperature. Ne: data in neon matrices (Joblin et al. [9]). The published data in gas phase are also reported: K (Kurtz [5]), F (Flinckinger et al.[6]), S (Shan et al. [7]), and B (Brenner and Barker [8]).

At the temperature T, the mean population of the ν mode is given by:

$$\langle n \rangle = \frac{1}{\exp(hc\nu/k_BT) - 1} \quad (2)$$

For a transition involving one quantum of the ν_i mode, the mean band position at the temperature T can be estimated by:

$$hc\nu_i(T) = E_\nu^{anh}(\nu_i = \langle n_i \rangle + 1, \nu_k = \langle n_k \rangle) - E_\nu^{anh}(\nu_i = \langle n_i \rangle, \nu_k = \langle n_k \rangle)$$

$$= hc\,(\nu_i(0) + 2x_{ii}\langle n_i \rangle + \sum_{k;k \neq i} x_{ik}\langle n_k \rangle) \quad (3)$$

By developing (3) in the high-temperature range ($\frac{hc\nu}{k_BT} < 1$), and using (2), a linear dependence of the band positions with temperature is found.

$$\nu_i(T) = \nu_i(0) + \frac{k_BT}{hc}\left[\frac{2x_{ii}}{\nu_i} + \sum_{k;k \neq i} \frac{x_{ik}}{\nu_k}\right] \quad (4)$$

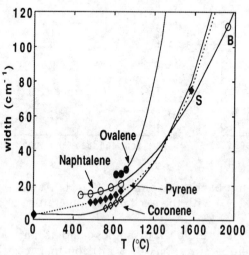

Fig. 2. Expected width of the 3.3 μm feature for several PAH molecules at different vibrational temperatures. In our experiment, the contribution of the rotation to the band width has been estimated and subtracted (see Joblin et al. [10]). The measurements from S (Shan et al. [7]), and B (Brenner and Barker [8]) are also reported. The data points have been extrapolated by using polynomial fits.

ASTROPHYSICAL IMPLICATIONS

Why does the 3.3 µm interstellar band usually occur at 3.289 µm?

Tokunaga et al [11] have shown that the position of the interstellar 3.3 µm band is rather constant at 3.289 µm (3040 cm^{-1}). Moreover the PAH model states that the carriers of this feature are isolated PAHs which absorb UV photons of mean energy 10 eV (Rouan et al [12]). Using the calculations from Léger et al. [2], we have estimated the mean emission temperature over the cooling for molecules of different sizes : naphtalene (T = 1915 K), pyrene (T = 1440 K), coronene (T = 1140 K), and ovalene (T = 970 K). The mean positions and widths of the 3.3 µm bands associated to these temperatures can be estimated using Figures 1 and 2. In the case of naphtalene, which is the smallest and then the hottest molecule, the band width has been measured in the Brenner and Barker's experiment [8], and is large (~110 cm^{-1}) compared to the interstellar case (typically 40 cm^{-1}). In Figure 3, we have then only considered the case of larger molecules (pyrene $C_{16}H_{10}$, coronene $C_{24}H_{12}$, and ovalene $C_{32}H_{14}$) which are cooler and therefore associated to narrower 3.3 µm bands. We did not make any assumptions on the relative abundances of the different species, and we have simply displayed the different bands assuming a lorentzian shape and an arbitrary intensity (normalized to unity).
The sum of the three bands gives a feature which is in good agreement with both the position and the intensity of the 3.3 µm band (type 1 from Tokunaga et al [11]).

This result is probably not a mere chance, and it suggests that some other PAH mixtures will also give a good fit of the 3.3 µm feature. We deduce that the observed 3.3 µm band is then consistent with the carriers being free PAHs at high temperatures. Although, the position of this feature does not allow to identify any specific PAH molecule, it depends on the radiation field, and then could be a way to probe this field.

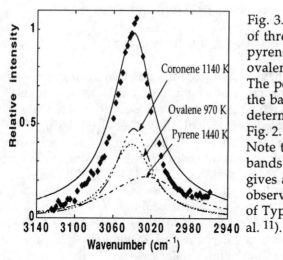

Fig. 3. Plausible contribution of three different PAHs: pyrene, coronene, and ovalene to the 3.3 μm band. The positions and widths of the bands have been determined using Fig. 1 and Fig. 2.
Note that the sum of these 3 bands (continuous line) gives a reasonable fit to the observations (3.3 μm band of Type 1 from Tokunaga et al. [11]).

REFERENCES

1. Léger, A., & Puget, J. L., Ast. & Ap., 137, L5 (1984).
2. Léger, A., d'Hendecourt, L., & Défourneau D., Ast. & Ap., 216, 148 (1989).
3. Léger, A., Boissel, P., Désert, F.X., & d'Hendecourt, L., Ast. & Ap., 213, 351 (1989).
4. Schutte W.A., Tielens A.G.G.M, & Allamandola L.J., Ap. J. in press (1993).
5. Kurtz, J., Ast. & Ap., 255, L1 (1992).
6. Flinckinger, G.C. & Wdowiak, T.J.& Gomez, P.L., Ap. J. 380, L43 (1991).
7. Shan, J., Suto, M., Lee, L.C., Ap. J. 383, 459 (1991).
8. Brenner, J.D., Barker, J.R., Ap. J., 388, L39 (1992).
9. Joblin, C., d'Hendecourt, L., Léger, A., & Défourneau D., Ast. & Ap. in press (1993).
10. Joblin, C., Boissel P., d'Hendecourt, L., Léger, A., & Défourneau D., in preparation (1993).
11. Tokunaga, A., Sellgren, K., Smith, R., Nagata, T., Sakata, A., & Nakada, Y., Ap. J. 380, 452 (1991).
12. Rouan, D., Léger, A., Omont, A., & Giard, M., Ast. & Ap., 253, 498 (1992).

UIR BANDS:
COMPUTATIONAL EXPERIMENTS ON MODEL PAHS

F. Pauzat
D. Talbi
Y. Ellinger
Equipe d'Astrochimie Quantique
ENS, 24 rue Lhomond, F-75231 Paris Cedex 05, France
DEMIRM, Observatoire de Paris, F-92195 Meudon, France

ABSTRACT

A number of ab-initio calculations have been performed these last years in our group around the hypothesis of PAHs being at the origin of UIR bands in space. We present here a brief account of our progress and conclusions and the questions still opened at this point of our work.

INTRODUCTION: WHY COMPUTE ON PAHS?

The starting point of the story lies somewhere in 1984 when comparison of experimental absorption spectra of neutral PAH-type molecules with well-known Unindentified IR emission bands (at 3.3, 6.2, 7.7, 8.6, 11.3μ), showed nice striking coincidences to researchers'eyes [1,2]. Links could obviously be done with CH stretching at 3.3μ, CC stretching at 6.2μ, CC cycle in-plane deformation at 7.7μ, CH in-plane bending at 8.6μ, out-of-plane CH bending at 11.3μ. **Could Polycyclic Aromatic Hydrocarbons be the emitters of UIR bands?**
      Though it was not unique [3,4], the hypothesis was attractive. However, when analyzing more precisely the comparison, discrepancies and difficulties showed up. The main ones are presented below:
- If a reasonably good agreement could be considered for positions of the IR bands between experimental and observed spectra, it was not at all the case for the band intensities. Tentative explanations through dehydrogenation of emitters, temperature effects or differences between emission (observed spectra) and absorption (experimental spectra) were laborious and unsatisfactory.
- The physical and energetical conditions in the regions where UIR bands are most observed, imply that numbers of molecules are probably ionized. On the other hand, there is no argument allowing to state that IR spectra of two parent molecules, neutral and ionized, are analogous. Thus, the initial comparison might be partly erroneous.
      Consequently a lot of questions were to be answered before any final conclusion, positive or negative, about the PAHs hypothesis could be given. In particular, how will dehydrogenation, ionization, attachment and substitution affect the IR spectra of PAHs? Will any of these modifications improve or make worse the agreement? Laboratory experiments with any kind of molecule which is not the fundamental neutral fully hydrogenated one are difficult to realize for obvious reasons as the uncertainty about the nature and quantity of the species really formed and the influence of the matrix environment. Then, **another source of information can be ab-initio quantum mechanical calculations.**

PART1:
ARE COMPUTATIONS ON PAHS RELIABLE
AND COMPARABLE TO EXPERIMENTS AND OBSERVATIONS ?

The molecules to consider are non saturated molecules and for half of them at least, when ionized or dehydrogenated, radicals; such molecules are "difficult" molecules for ab initio calculations, which might imply rather sophisticated treatments to get reliable results. Moreover, in this particular problem, the size of the real molecules may become too large for this type of calculations, and we may have to deal with the problem of the choice of representative models. The last eventually tricky point is about the types of observables to be calculated (frequencies and intensities); they are not the most usual ones, i.e. energies, dipole moments, etc... They depend on derivatives of energies, which, introducing one more degree of uncertainty, implies a higher degree of precision at the level of the determination of energies and wavefunctions. All these reasons show that some tests of reliability were necessary to decide the proper level of sophistication to be used for such calculations.

a) Level of computation

Taking into account our group's experience and litterature available on the subject, we have chosen to determine the infrared emission spectra at the HF/3-21G level of theory, using the Gaussian 88 and 90 computer codes [5].

For each molecule considered, the frequencies and normal modes were determined; they were correlated to specific types of nuclear displacements and the relative weights of the different local motions such as stretching, bending, or ring torsion were identified for each normal mode.

In order to improve the numbers calculated, we used a correcting procedure in which the frequencies are adjusted with scaling factors depending on the type of vibration. The scaling factors used were determined on neutral fully hydrogenated molecules for which experimental data exist. This was done by minimizing, for each type of vibration (i.e. CH stretching, CC stretching, CH in-plane bending, CC in-plane bending, CH out-of-plane bending and CC out-plane bending), the root mean square deviation between the calculated frequencies and the corresponding experimental ones; then, these factors were transferred to the cation.

b) Test for "internal" coherence

The smallest representative of PAH type is benzene, but considering its high symmetry and the fact that the cation presents a Jahn-Teller effect, it is not a good choice as a test. So, we turned our attention to pyridine (see figure1) which is of the same size, aromatic too, but where the presence of nitrogen in place of a CH group removes the preceding problems due to symmetry degeneracy.

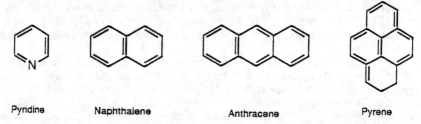

Pyridine   Naphthalene   Anthracene   Pyrene

Figure 1 - Structure of some model PAHs

We computed IR spectra of neutral and ionized pyridine with increasing levels of sophistication for method (from SCF which does not take into account electronic correlation to MP2 which calculates part of the correlation with a second order pertubative development) and basis set (from a simple split valence basis set 3.21G to a double zeta basis including polarization functions on hydrogen as well as heavy atoms, 6-31G**) [6].

The results are illustrated in figure 2 where simulated IR spectra are reported at various levels of wave functions for comparison. Obviously, they are stable and meet our criteria of internal coherence established according the level of agreement needed, i.e. less than 10% for the frequencies and a factor smaller than 2.5 for intensities, allowing us to procede with a rather standard level of calculation.

Figure 2 - Computed transmittance function of frequency (cm^{-1}) for neutral pyridine. Computation levels are respectively RHF/3.21G, RHF/6.31G**, RHF/6.31+G*, MP2/6.31G** for graphs I, II, III and IV.

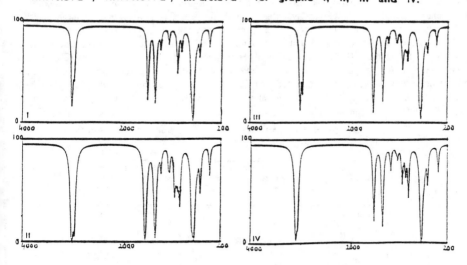

Table 1- Computed and Experimental IR spectra for naphthalene cation

Symmetry		Computed[7]			Experimental[8]	
		Frequency (cm^{-1})	Intensity (Km/mole) absolute(relative)		Frequency (cm^{-1})	Intensity relative
b1u	r(CH)	3088	1			
b1u	r(CH)	3072	2			
b1u	R(CC)+β(CH)	1497	147	(0.13)	1519	0.08
b1u	β(CH)	1429	19	(0.02)	1401	0.04
b1u	R(CC)+β(CH)+α(CCC)	1292	5			
b1u	R(CC)+β(CH)+α(CCC)	1090	22	(0.03)	1023	0.06
b1u	α(CCC)	782	0			
b1u	α(CCC)	350	0			
b2u	r(CH)	3100	1			
b2u	r(CH)	3071	10			
b2u	R(CC)+β(CH)	1498	180	(0.16)	1525	0.16
b2u	R(CC)+β(CH)	1324	3			
b2u	R(CC)+β(CH)	1173	31			
b2u	R(CC)+β(CH)	1093	1123	(1.0)	1218	1.0
b2u	R(CC)+α(CCC)	955	662	(0.59)	1016	0.2
b2u	α(CCC)	559	324			

Computed frequencies are scaled to 0.91 for r(CH) (CH stretching vibrations), 0.903 for R(CC) (CC stretching vibrations) and β(CH) (CH in-plane bending vibrations), 0.88 for α(CCC) (CCC cycle vibrations), 0.84 for ε(CH) (CH out-of-plane bending vibrations) and 0.86 for τ(CCC) (CC out-of-plane bending vibrations).
Only dominant components are reported in attributions for each vibration when mixed with others.
We give only the active vibrations of b2u and b1u symmetries which are the ones we can compare with experiments.

c) Test for "external" coherence

Our second series of test using the level quoted above, implies comparison with corresponding experimental results. Some results were available, but only for neutral fully hydrogenated molecules. An interactive collaboration with Vala's experimental group gave numbers to be compared for cations.

An extensive study on naphthalene in both neutral and ionized states [7] has been carried out. It shows that a HF treatment (RHF and ROHF respectively) in a split valence basis set (3-21G) is a reasonable compromise between accuracy and computational effort since it gives coherent enough orders of magnitude for the intensities to account for the effects of ionization on the IR spectra. As can be seen in table 1, the spectra calculated at this level of theory were in very good agreement with the experimental one of Szczepanski et al.[8].

Table 2- Computed and Experimental IR spectra for neutral anthracene

Symmetry		Computed[9]		Experimental[9]	
		Frequency (cm^{-1})	Intensity (Km/mole)	Frequency (cm^{-1})	Intensity (Km/mole)
b1u	r(CH)	3051	56	3062,3055	35.2
b1u	r(CH)	3033	7	3032	7.3
b1u	r(CH)	3027	4	3022,3017	8.2
b1u	R(CC)	1678	16	1627	16.3
b1u	R(CC)	1487	1	1450	6.4
b1u	R(CC)	1347	1	1346	2.0
b1u	β(CH)	1288	8	1272	6.9
b1u	β(CH)	1185	6	1149,1151	5.3
b1u	α(CCC)	911	4	908	3.2
b1u	α(CCC)	642	2	652	1.3
b1u	α(CCC)	230	2		
b2u	r(CH)	3065	62	3068,3067	59.5
b2u	r(CH)	3036	0		
b2u	R(CC)	1576	7	1542,1540	5.3
b2u	R(CC)	1495	6	1460	5.3
b2u	β(CH)	1398	1	1400	2.7
b2u	β(CH)	1301	7	1318	17.0
b2u	β(CH)	1187	1	1167,1169	4.8
b2u	R(CC)	1082	2		
b2u	R(CC)	974	2	1001	9.6
b2u	R(CC)	783	0		
b2u	α(CCC)	607	9	603	20.0
b3u	ε(CH)	985	13	955,958	10.1
b3u	ε(CH)	904	87	878.5	95.8
b3u	ε(CH)	721	127	729,726	139.9
b3u	τ(CCC)	480	38	470,468	40.5
b3u	τ(CCC)	385	0		
b3u	τ(CCC)	90	2		

Computed frequencies are scaled to 0.906 for r(CH) (CH stretching) vibrations, 0.922 for R(CC) (CC stretching vibrations), 0.912 for β(CH) (CH in-plane bending vibrations), 0.893 for α(CCC) (CCC cycle vibrations), 0.854 for ε(CH) (CH out-of-plane bending vibrations) and 0.896 for τ(CCC) (CC out-of-plane bending vibrations).
Only dominant component is reported in attributions for each vibration when mixed with others.

Another comparative study about anthracene [9] is illustrated in table 2. The excellent agreement can be noticed for both band positions and intensities. The shifts between scaled quantum mechanical values and experimental band positions are less than 50 cm^{-1} in the worst cases, less than 20 cm^{-1} in most cases. It is well within the error bar usually accepted for this kind of calculations, and confirm the validity of the scaling procedure by type of vibration. The results concerning the intensities are more than satisfying, the differences being negligeable compared to the usual error bar known for computed intensities at this level of calculation. A good agreement is observed in band positions for anthracene cation between quantum mechanical values and experimental ones. Direct comparison of the computed intensities individually was difficult in this case where some of the vibrations are delocalized (when they are of same type and same symmetry). For a more realistic comparison, we summed up the band intensities of same vibration type i.e. the C-H in-plane bending ones, between 1350 cm^{-1} and 1050 cm^{-1} and calculated the intensity of the others relative to this band interval. These ratios can be compared to the corresponding experimental one. The agreement is satisfactory, the largest discrepancy being observed for the C-H out-plane bending vibration for which the experimental relative intensity is 3 times higher than the computed one. This is acceptable taking into account the usual error bar for this kind of calculations and the fact that intensity from this band might have been transferred to that of same type at 748cm^{-1} because of a delocalization effect.

It can be concluded that **such ab initio calculations are reliable enough to be much helpful to give information about PAHs spectra.**

PART2:
WHAT IS THE EFFECT OF IONIZATION ON THE IR SPECTRA OF PAHS ?

Three molecules have been considered as models for PAHs: naphthalene as the smallest possible one, anthracene and pyrene, respectively as the first representatives of the linear and compact series (figure 1). The spectra of the corresponding neutral and ionized species have been compared carefully through bands positions and absorption intensities.

a) Effect on band positions

The frequencies are only slightly changed when ionizing, whatever the molecule considered, the effects are similar for all types of molecules. The differences between the neutral and the corresponding onized systems are most of the time less than 50 cm^{-1}, which is within our error bar; however, these differences are systematic and can be rationalized from the calculated geometry modifications for most of them. A precise analysis of our results can be found in publication by DeFrees et al.[10]. Briefly, we found that:
- The CH stretching vibrations are slightly displaced towards higher frequencies by 20 cm^{-1}(0.02μ) and still correlate to the observed band centered at 3.28μ.
- The aromatic CC stretching vibrations are concentrated and systematically displaced towards higher wavelengths, roughly from 6.2 to 6.4μ, due to bond relaxation. However, this band which is somewhat too high, compared to the 6.2μ observed, which it is supposed to be correlated with, will be shifted back when increasing the size of the PAHs and thus the increasing rigidity of the structure.
- The broad band related to the 7.7μ observed in space does not change significantly; then ionization cannot help for discriminating ionized from neutral small PAHs which present this feature at slightly smaller wavelengths.
- The other broad band at 8.6μ does not shift significantly.
- The components of the last main interstellar band, between 1000 and 700 cm^{-1} (10-14μ), are shifted towards lower wavelengths for most of them, by 10 cm^{-1} at most.

640 UIR Bands

At this point, the conclusion may be that, **concerning positions, if things are not really worsened by ionization, they are not improved too.** Anyway, agreement was already rather good.

b) Effect on intensities

This effect is much more interesting while unexpected by its scale. IR intensities are very sensitive to ionization and their variations are strongly related to the type of vibration implied. Moreover, these variations are rather similar for all models considered, even if specific differences appear in a closer analysis, which argues for an averaging of the emission over the different types of PAHs possible.

In order to provide direct comparison with the observed spectra in space, we summed the intensities per type of vibration; each frequency interval corresponds to a different type of movement of the nuclei, identified from the normal coordinate analysis. Histograms were realized, using bands widths deduced from those of the corresponding interstellar bands (see figure 3). The ratios between the major features of the spectra are also given in table 3.

Table 3 - Integrated intensities normalized to the "6.2+7.7μ" band

Molecule	I(3.3μ)	I(6.2μ)	I(11.3μ)
C10H8 naphthalene	2.45	0.64	
C14H10 anthracene	3.31	0.60	2.23
C16H10 pyrene	2.56	0.53	
average for neutrals	2.77	0.59	
C10H8+ naphthalene	0.04	0.92	
C14H10+ anthracene	0.06	0.38	0.17
C16H10+ pyrene	0.066	0.72	
average for cations	0.04	0.67	
observ.*	0.05	0.35	0.19

*These ratios have been deduced from the ratio to the 7.7μm band reported by Cohen et al. [13].

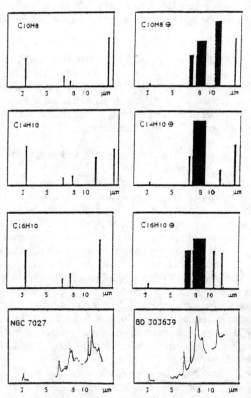

Figure 3 - Comparison of computed IR spectra for neutrals and cations with IR emission spectra of BD 303639 and NGC 7027.

The experimental spectra were provided courtesy of Schutte, Tielens and Allamandola.

The main effects are the following:
- The CH stretching vibrations (3.3μ) decrease by one order of magnitude, while the CC stretching vibrations (6.2-7.7μ) increase by one order of magnitude; thus, comparing our CH/CC intensities ratio with the observed one, it is obvious that it is consistent with the cations of our model PAHs, contrary to the neutral ones.
- The CH in-plane at 8.6μ is strongly enhanced too by ionization while the CH out-of-plane is unaffected.
- Among the three models we studied, only anthracene has an hydrogen "solo"; we found that its intensity at 11.3m, relatively to the CC stretching one, is consistent only for the cation.

At that point, the obvious conclusion is that PAHs, if they exist, are predominantly ionized, which is consistent with estimates of ionization fractions in the regions where these spectra are observed, and which relieves the necessity of the highly questionable intensive stripping of the hydrogens.

PART3:
WHAT IS THE EFFECT OF DEHYDROGENATION ON THE IR SPECTRA OF PAHS ?

Dehydrogenation does not affect band positions, except for two points which are very positive:
The band corresponding to the CH out-of plane vibrations splits in several peaks according to the relative positions of the hydrogens, solo, duo or trio; which is what was expected and is in accordance with the observations. At least to get this effect, some dehydrogenation of the PAHs is required.
A vibration for a triple bound CC in naphthalene appears at 5.2μ, position where a weak emission has been recently observed and associated with the family of the IR emission features by Allamendola and coworkers [11] (for more precise discussion of this point see Pauzat et al. [12]).

a) Dehydrogenation of neutral PAHs

This study has become necessary because of the rule of thumb used hypothesis of proportionality between the number of hydrogens surrounding the PAHs and the intensities of CH vibrations. This assumption was first made in order to account for the too high intensity of the CC stretching band in regards to the CH one. Our computational experiment was done on the naphthalene taken as a model, which was progressively dehydrogenated to 50% (which means 4 hydrogens left out of the initial 8). Different possibilities of dehydrogenation have been taken into account. The influence over all types of CH vibrations is given in table 4. Except for the out-of-plane bending, it is evident that no proportionality of any kind can be applied. Moreover, if a decrease in the intensity is observed, as expected, with the diminution of the number of hydrogens for the CH stretching, it is not even the case for the CH in-plane bending whose intensities seem to increase. Vibrations linked to CC displacements vary too, but in a chaotic way which cannot be rationalized.

This, definitively, rules out the possibility of using proportionality between the number of hydrogens attached on the PAHs and intensities of the CH vibrations to get rid of one of the main discrepancy with observed spectra.

b) Dehydrogenation of ionized PAHs

The same study was done on the ionized corresponding molecules and results are even more unexpected. It seems that the effect of decreasing the intensity by ionizing is temporized by the dehydrogenation and especially that, in an apparent contradictory way, the CH stretching intensity is slowly increasing again when the number of hydrogens decreases (see table 4). The combined effect of ionization and some dehydrogenation (for our limited model, naphthalene, that means didehydrogenation) leads to rather satisfying relative intensities for all bands (see table 5 and figure 4). However, we have to keep in mind that, even if the result is satisfying, it has been obtained on a model and on a rather small one.

In conclusion the behaviour of our model compounds is a very encouraging support for the existence of PAHs in space: PAHs might be ionized and partially doubly dehydrogenated on adjacent carbons, leading to a few triplet bonds spread at the periphery of the carbon skeleton.

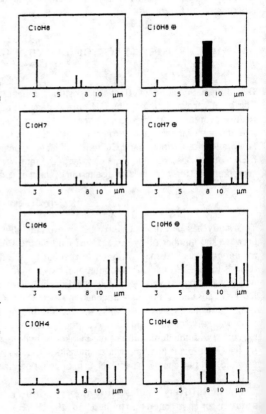

Figure 4 - Computed IR spectra for neutral and cation dehydrogenated naphthalene.

Table 4 - Evolution of IR intensities of the CH vibrations with dehydrogenation for naphthalene neutral and cation

molecule	C10H8	C10H7	C10H6	C10H4
%H	100	87.5	75	50
CH stretching				
neutral	100	67	51	17
cation	100	140	175	380
CH in-plane				
neutral	100	79	195	650
cation	100	75	64	22
CH out-of-plane				
neutral	100	90	77	51
cation	100	90	67	30

Intensities are referred to 100% for the molecule with all 8 hydrogens.
A statistical average of the different possible conformations has been calculated for each molecule.

Table 5 - Integrated intensities normalized to the "6.2+7.7μ" (except when indicated)

Molécule	I(3.3μ)	I(5.3μ)	I(6.2μ)	I(11.3 μ)	I(11.3μ)(a)	I(11.3μ)(b)
$C_{10}H_8$	2.5		0.64	4.0	3.4	7.6
$C_{10}H_7$	1.5		0.60	3.5	3.0	6.5
$C_{10}H_6$	0.97	0.03	0.52	2.3	1.9	3.5
$C_{10}H_8^+$	0.04		0.92	0.46	0.10	0.13
$C_{10}H_7^+$	0.06		0.51	0.45	0.12	0.14
$C_{10}H_6^+$	0.077	0.207	0.497	0.330	0.100	0.120
observ.	0.04-0.06	0.03-0.05	0.31-0.39	0.15-0.23		0.125*

The "11.3μ" intensities are relative to all the CH out-of-plane bending vibrations i.e. solo-duo-trio.
Observations are from Cohen et al.[13], excepted for * which is from Zavagno et al.[14].
(a) ratio to the (6.2+7.7+8.6μ) band intensities. (b) ratio to the (7.7+8.6μ) band intensities.
A statistical average over the different possible isomers has been done for each molecule.

UP DATE CONCLUSIONS AND QUESTIONS

a) What are the parts of the PAHs puzzle?

If we consider PAHs as being responsible of the UIR bands, we have first to keep in mind two points when comparing our results with the observed spectra:
- some corrections should be introduced to deal with the emission-absorption differences
- our results are obtained for models of PAHs which are rather small and no statistical averaging has been done because the number of species considered is obviously too small.

Up to that point the conclusions are:

i) Intensities should certainly be those of cations while positions should be those of neutral molecules which at first sight seems unreconciliable.

ii) Some dehydrogenation is needed to get some precise bands as the 5.2μ and the trilogy 11.2, 11.9, 12.7μ. Furthermore, we have to remember that all species have to be in their electronic ground state, no excited state being able to survive long enough in space.

All these features taken together seem to be parts of a puzzle and somewhat contradictory. The main feature pointed out by our studies is the change in the intensities when ionizing the PAHs. One may question the nature of such a change and the conditions required to get it. The ions we studied are all π radicals and the neutrals are σ states (radicals or not) with a true aromatic electronic structure. **The characteristic difference between the two species is a hole in the π system.**

Then, the logical following question is: what about **neutrals with a hole** in the π system or **cations with no hole** in the π system? Two computational experiments have been carried out on naphthalene to get the beginning of answer to that question:

- From our preceding calculations on dehydrogenated naphthalene we had the spectra of the cation ground state dehydrogenated naphthalene $C_{10}H_7^+$, which has a hole in its π system and thus is a π state, $^3A"$; if we fill this hole with an electron from the σ system, we get an excited state of σ type, $^1A'$. We computed the spectrum of this molecule in this state and found that the intensities ratios were very close to those of the neutrals in their σ ground states, $^2A'$, in particular the crucial CH/CC vibrations ratio.

- In the same way, we calculated the spectrum of the neutral π excited state, $^2A''$, obtained from the neutral ground state σ of $C_{10}H_7$, $^2A'$, by filling the σ hole with an electron from the π system and we found that the resemblance was obvious with the corresponding cations in their π ground states.

These preliminary calculations seem to indicate that, **among the large family of PAHs, neutrals as well as cations could be good candidates for UIR emitters as long as they are of π symmetry.**

b) Is it a way to put parts of the puzzle together ?

Taking one by one the conditions required to rationalize the PAHS hypothesis, we arrive to the following scheme:

- Dehydrogenation is mandatory but limited and probably depending on the type of species. This condition raises no difficulty of its own.

- Considering ionized PAHs, intensity ratios between the different observed bands are very satisfactory; but some positions are to be shifted back to interstellar values, eventually through size effect, attachment or stacking. We have not yet reliable numbers about these last two effects, which have certainly to be considered for small PAHs.

- Considering neutral PAHs, the main problem is that intensities must be balanced to the interstellar values some way or anorther. If we consider the same possibilities as for the ions, then the size effect is the most appealing possibility. Indeed, neutral PAHs in π elctronic states would do it. But, we have to remember that an absolute requirement is that molecules have to be in their electronic ground state, which is not the case for our small neutral π systems. However, the problem is only apparent. If small neutral PAHs are obviously of σ symmetry, when increasing the size, we will observe that the energy gap with the first π excited state will progressively decrease, leading to a crossover of the two states. Ultimately, the dehydrogenated neutral PAH will present a π ground state. The question is to determine at what size the crossing occurs. Calculations on the subject are actually in progress.

Thus, contrary to simple consideration of ionization potentials, solutions with small PAHs as cations and large PAHs as neutrals, could be considered. Different regions with different conditions could give rise to such distinct populations.

REFERENCES

1. A.Leger & J.L.Puget, A&A, **137**, L5,1984.
2. K.Sellgreen, Ap.J., **277**, 623, 1984.
3. R.Papoular, J.Conard, M.Guiliano, J.Kister & G.Mille, A&A, **217**,204,1989.
4. A.Borghesi, E.Bussoletti & L.Colangeli, Ap.J., **314**, 422, 1987.
5. M.J.Frisch et al., 1988, Gaussian 88, Gaussian, Inc., Pittsburg.
   M.J.Frisch et al., 1990 , Gaussian 90, Gaussian, Inc., Pittsburg.
6. Y.Ellinger, D.Talbi & F.Pauzat, to be published
7. F.Pauzat, D.Talbi, M.D.Miller, D.J.DeFrees, and Y.Ellinger, J.Phys.Chem., **96**, 7882, 1992.
8. J.Szczepanski, D.Roser, W.Personnette, M.Eyring, R.Pellow & M.Vala,
   J. Phys.Chem., **96**, 7876, 1992.
9. J.Szczepanski, M.Vala, D.Talbi, O.Parisel and Y.Ellinger, J.Chem.Phys., **98**, 4494, 1993.
10. D.J.DeFrees, M.D.Miller, F.Pauzat, D.Talbi & Y.Ellinger, Ap.J., **408**, 530,1993.
11. L.J.Allamendola, A.G.Tielens & J.R.Baker, Ap.J.Suppl., **71**, 733, 1989.
12. F.Pauzat, D.Talbi and Y.Ellinger to be published
13. M.Cohen,A.G.Tielens,A.G.Bregman,F.C.Witteborn,D.M.Rank,L.J.Allamendola, D.H.Wooden & M.DeMuizon, Ap.J., **341**, 246,1989.
14. A.Zavagno, P.Cox & J.P.Baluteau, A&A, **259**, 241, 1992.

DISCUSSION

BOTSCHWINA— Did you analyse the changes in the IR intensities of the CH stretching vibrations occurring upon ionization in terms of contributions from individual internal or symmetry coordinates ? Do the significant reductions result mainly from reduced first derivatives of the electric dipole moment with respect to the CH stretching coordinates or do contributions from the carbon ring play a major role ?

PAUZAT— Up to now, the analyses we attempted showed no clear cut evidence of a dominant role, the effects being delocalized.

LEACH— You have calculated the Einstein coefficients for the PAHs and its cationic and dehydrogenated variants but compare with observed UIR emission spectra, whose intensity depends on a convolution of these coefficients and level population factors. Have you attempted to apply a photophysical model so as to estimate populations and then make more realistic comparisons with observations ?

PAUZAT— The IR emission is linked to the instantaneous temperature of the emitter - which is not known - but may be a-posteriori simulated by applying a Planck correction on the IR absorption. The spectra are only affected for low temperatures, the 3.3µ band decreasing below 600°K.

LEE— One of the results deduced from your computational calculations is that the strength of the 3.3µm band of neutral PAH species is decreased by a factor of 10 when they are in the ionic state, from which you concluded that the relative strengths of the CH stretching bands to the CC vibration of PAH cations match better those of observational data. However, both the Florida group and d'Hendecourt and Léger in their IR mesurements of cationic PAHs embedded in cryogenic matrices have found that the 3.3µm band almost totally vanishes; this is true for several different cationic PAH species. If this is an intrinsic property of PAH cations and not a matrix effect, it is more likely that the 3.3µm VIR band is mainly contributed by neutral PAHs. Would you, d'Hendecourt or Léger like to say some words about this problem ?

LEGER— Our experimental measurements on PAH cations are qualitatively compatible with the calculation of F. Pauzat and coworkers. We find that the 3.3µm absorption is strongly reduced to value ≤ 20% of that for the neutral. Because of the residual presence of neutrals in our experiments, we cannot be more precise.

D'HENDECOURT— More a comment: PAH cations now observed in matrices (Pyrene, Coronene, Ovalene) by ourselves as well as the Vala's group do indeed show strong differences in their IR spectra, compared with the spectra of the neutrals. We agree that the reduction of the 3.3.µm band is compatible with an almost total disappearance of this band (a factor of 10 for example is compatible with our experiments). A strong band appears at 635 µm and another one (very strong) at 7.25 µm, bands wich are not compatible with those observed in the ISM. We really confirm your calculations on ions with one important difference: in the laboratory, the position of the ion bands (at 6.35 and 7.25 µm) is obtained with probably a better accuracy than in your computations and pose a problem to interpret the observed IS spectra with PAH ions.

PAUZAT— Our calculations predicted a strong decrease of the 3.3µ which has been confirmed by the recent experiments, a reduction factor of an order of magnitude being found in both theoretical and experimental determinations

As to the comparison with the interstellar spectra, we have also found that the bands calculated for the small PAHs around 6.4 and 7.4μ do not exactly match the observed positions, though the intensity profile appears satisfactory after some dehydrogenation. Substitution, stacking and size effects may account for this discrepancy; if not, we have to find the ideal carrier with the frequencies of the neutrals and the intensities of ions. Our preliminary calculations on the π states of open-shell dehydrogenated radicals are encouraging in this sense. However, for these states to be electronic ground states, PAHs have to be larger than a critical size that remains to be determined.

SARRE — Most of the ions we know about in space are protonated species. Have any calculations been done on protonated PAH's

Could Protonation be related to the satellite band problem ?

PAUZAT— Only protonated naphthalene has been calculaled. This ion shows interesting features, in particular at 3.4μm we found a band corresponding to the CH_2 stretching coming from the addition of the proton. This band whose position corresponds to the most important satellite of the 3.3μ feature is even stronger than those corresponding to the aromatic ring protons.

LEGER — If the ionization cross section measurements made at LURE are correct, considering the balance ionisation/recombination, we find that PAH would be basically neutral in diffuse ISM and ionized in reflection Nebulae. IR spectra are known only for the latter. Are you making a prediction of what they would look like in the diffuse ISM ?

GIARD — About the question of the observation of unidentified infrared emission bands in the diffuse interstellar medium by A. Léger: the 3.3μm and 12μm bands have been observed in the diffuse medium towards the inner Galaxy where the interstellar radiation field is about 10 times the solar neighborhood value (see poster by Ristorcelli et al. in this meeting). 3.3μm and 6.2μm intensities relative to the total infrared flux are in this medium the same as in UV excited nebulae where the intensities of the radiation fields are 10^3 to 10^5 times higher.

PAUZAT — Our calculations suggest two possible types of carriers to accomodate the frequency - intensity dilemma. Small ionized PAHs or large neutral PAHs, both partly dehydrogenated. The spectra in the interstellar medium will then critically depend on the size distribution of the available PAHs.

COMPUTATIONAL CHECKING OF AROMATIC MODELS FOR INTERSTELLAR VUV ABSORPTION

O. Parisel
Y. Ellinger
Equipe d'Astrochimie Quantique
Ecole Normale Supérieure, 24 rue Lhomond, F-75231 Paris CEDEX 05, France
DEMIRM, Observatoire de Paris, 5 place Jules Janssen, F-92195 Meudon, France

ABSTRACT

There has been in the last few years a large interest in the study of the Polycyclic Aromatic Hydrocarbons (PAHs) which have been proposed as possible carriers of the Unidentified Infra-Red bands (UIR), in an ionized or dehydrogenated form. This hypothesis has first been supported by *ab initio* quantum chemistry calculations and the same features are now observed in matrix experiments.

A similar situation prevails for the Visible and Ultra-Violet (VUV) spectra of PAHs and derivatives. However, since most data concerning ionized PAHs have been recorded in polar or polarizable solvents or matrices, substantial shifts in the transition energies have to be expected relative to the gas phase spectra. Furthermore, no experimental data are available on dehydrogenated PAHs. Quantum chemistry appears thus to be an attractive alternative.

INTRODUCTION

The origin of about 200 DIBs observed between 4400 and 8700 Å in many objects remains unknown since their discovery about 70 years ago[1-10]. Many hypothesis have been formulated in the past to identify the carriers of these bands[11-12], but no one is definitevely convincing, and none of them leads to an unambiguous one-to-one attribution.

There has been a large interest in the last few years in the study of the Polycyclic Aromatic Hydrocarbons which have been supposed to be the carriers of the UIR bands [13,14]. However, as pointed out by the NASA group[15], "although the IR emission band spectrum resembles what one might expect from a mixture of PAHs, it does not match in details such as frequency, band profile, or relative intensities predicted from the absorption spectra of any known PAH molecule". Another point which is often assumed is that PAHs, in the region where the UIR and DIBs are observed, might be partly ionized (positively charged)[14] and partly dehydrogenated[16-17]. Recent *ab initio* calculations performed in our group since 1990 on the IR spectra of PAHs and their ionized and/or dehydrogenated forms have supported these assumptions. They have shown that these species, if relatively small, are much more attractive carriers than neutrals[18-20], even if some discrepancies remain. These theoretical results on IR absorption spectra of ionized PAHs are in good agreement with the laboratory experiments developed in Florida by M. Vala and coworkers[21-23] and those of Léger's group in Paris.

Furthermore, it has been shown recently that some PAHs (especially those like phenanthrene) could account for the satellites of the IR 3.3 μm band[20, 24]. It is another argument in favour of PAHs and derivatives as possible carriers for the UIR emission features in space.

It is then obvious that if PAHs are responsible for the IR emission, they must absorb energy leading to absorption bands in the visible or ultra-violet spectrum : the previous results are encouraging enough to stimulate analogous studies on the VUV absorption spectra of PAHs and derivatives and correlate them to the DIBs as proposed by Crawford et al.[25].

PAHs AND DERIVATIVES AS CARRIERS FOR DIBs ?

Only recently, have jet-cooled experiments been performed on neutral PAHs; these experiments, in which interstellar conditions are best reproduced, supported the detection of phenanthrene in the Halley comet[26-27]. On another hand, it is the ionized PAHs which have been proposed as possible carriers for the DIBs. Such cationic species can now be studied as isolated molecules, using the 'ion trap' coupled to a mass- and a VUV-spectrometer recently developed by P. Boissel, which offers a promising tool for the analysis of the ionized PAH's photochemistry in simulated interstellar conditions[28]. However, manipulations are extremely difficult and dehydrogenated species are far from reach. Quantum chemistry calculations remain here the only possible alternative to laboratory experiments.

The situation prevailing for VUV spectra of ionized spectra of PAHs was the following when we began this work in 1991 : except those coming out from the very recent 'ion trap', all data concerning ionized PAHs had been obtained in polar or polarizable solvents or matrices. Although considerable amount of experimental works and efforts have now been done by several groups (Allamandola and Salama, Vala and Szczepanski, Léger, d'Hendecourt and Joblin, Lei and Wdowiak) to avoid the problems induced by environmental perturbations on the recorded spectra, it must be emphasized that none of these matrix-experiments reproduces the interstellar conditions : even if rare gas matrices are used, host-guest interactions between the molecule under study and the matrix remain. These interactions can induce shifts, broadening, enhancement of intensities with respect to an isolated molecule. For example, energy shifts have been reported to be as large as 2000 cm^{-1} for some states of neutral naphtalene[29]. Even using neon and argon matrices shifts up to 800 cm^{-1} can occur[30].

This lack of unquestionable experimental data explains why only few detailed comparisons are available for DIBs. Most arguments for and against PAHs have been based on indirect considerations. The low ionization potential of PAHs, the well known visible absorption of the correcponding ions[31] together with the high stabiliy of the polyaromatic frame to survive UV photon irradiation[32, 33] are all pro arguments. In addition, it has been observed that the intensities of some DIBs are enhanced in regions where the infrared emission has been ascribed to ionized PAHs[34]. On the other side, it has been pointed out that calculated rotational band contours of vibronic transitions of neutral and ionized PAHs are incompatible with the intrinsic profiles for some DIBs[35] while in agreement for others[32]. Finally, it should also be reminded that the molecular nature of the DIB's carriers has long been a matter of discussion because no structures were observed on the spectra until the recent high resolution observations[36] of the strong bands at 5780 and 6284 Å.

Comparison of experimental VUV spectra of various ionized PAHs[31] with interstellar absorption has been done by Crawford et al.[25] who have shown that, providing an energy shift of 1500 cm^{-1} to correct matrix effects, the agreement between the DIBs and the experimental vibronic spectra is "striking enough to justify serious consideration". The values used by Crawford et al. were the only ones available in 1985 and had been obtained in a very polar environment leading to serious energy shifts relative to the gas phase. The aim of the present work is to report calculated transition energies and oscillator strengths for ionized PAHs using quantum chemistry methods whose results correspond to a gas phase experiment. Many of the recent high-level experiments on PAHs ions were developed simultaneously. It will be seen that comparison of the observed DIBs with both theoretical predictions and experimental results is much more striking than it was with the data available to Crawford et al.

THEORY AND COMPUTATIONAL DETAILS

The electronic spectra of PAHs and their ions (especially anions) have been widely studied using various theoretical methods. Although most of them have been semi-empirical approaches, benzene has been studied by several groups using *ab initio* methods[37-42]. Despite these efforts, calculated and experimental electronic transitions have been found to differ by more than several tenths of an electron volt. Only very recently have Roos and coworkers succeeded to describe accurately *some* of the electronic states of the benzene molecule[43]. However, the extension of their high-level *ab initio* treatment to larger molecules such as PAHs remains problematic, even unrealistic, due to the computational efforts required.

To remedy the actual computational limitations of *ab initio* methods for such species, the use of properly parametrized semi-empirical methods is an attractive alternative. Numerous such studies have been reported in the past : the most famous of them is the use of the non-self-consistent Hückel's Molecular Orbital (HMO) theory[44], followed by limited Configuration Interaction (CI)[58]. This level of theory has been widely used and usually gives a first qualitative starting point for the discussion of electronic spectra[45]. An improvement in the description of the molecular electronic structure can be achieved by the self-consistent-field PPP/CI method[31, 46, 47]. However, if one wants to go beyond the σ-π separation, all-valence electron approximations such as CNDO[48] or INDO[49] coupled to variational-perturbative[43, 50, 51] post-SCF treatments seem to be the most attractive compromise between accuracy and computational cost and task, keeping in mind that no theoretical method is able to account for the entire VUV spectra of species as large as PAHs with the precision needed for a direct comparison with astronomical observations.

We report here the frame of the methods used in our theoretical studies of the VUV spectra of neutral and ionized PAHs.

1- Getting the molecular SCF orbitals :

Two types of semi-empirical hamiltonians have been used to obtain the molecular orbitals.
In the PPP approach, which is a π-electron method, we have used a restricted open-shell Hartree-Fock (ROHF) scheme, including overlap[46,52].
To obtain the molecular orbitals within the all-valence electron approach, we have used a modified version of the INDO framework which was developed to account simultaneously for both molecular electronic and conformational properties. The so-called CS-INDO[53] method differs from the common version of INDO by the use of hybrid orbitals so that the non-diagonal terms of the one-electron core hamiltonian can be written :

$$h_{pq} = S_{pq} K_{ij} \beta°_{AB}$$

where i and j are hybrid types (σ, n or π), p and q are hybrids of the type i and j, centered on atoms A and B, and $\beta°_{AB}$ is an atomic parameter depending of the nature of both atoms A and B.

Here, the so-called "screening constants" K_{ij} appears to be the only molecular parameters injected in the method : the use of hybrid orbitals allows thus a parametrization that reflects the bonding properties of the structure studied. However, for such compounds as planar PAHs, only the $K_{\pi\pi}$ constant plays a fundamental role.

The description of ionized PAHs, which all have one unpaired electron, was achieved using a Longuet-Higgins-Pople (LHP) type hamiltonian[54]: the Fock operator is thus written :

$$F = H + \sum_{i=1}^{m} \frac{n_i}{2} (2 J_i - K_i)$$

where H is the one-electron core hamiltonian, J_i and K_i are the Coulomb and exchange operators, and n_i is the occupation number (0, 1 or 2) of the ith molecular orbital. The advantage of this formulation is its simplicity[55, 56] compared to the more rigorous Restricted Open shell Hartree Fock (ROHF) method (for a system with only one unpaired electron in a non-degenerated orbital, see ref. 57).

2- Getting a good description of excited states :

The most efficient way to go beyond the SCF approximation, is to use a CI method[58]. However, due to the size of the species studied, truncation of the CI matrix is necessary. A commonly used approach consists in including in the CI space only (some of) the mono-excitations with respect to the reference SCF determinant (Single Configuration Interaction : SCI[59]). However, such a treatment is generally inadequate to obtain a relevant description of the highest excited states because the weight of higher-order excitations increases[43].

In order to avoid the use of large scale CI (including at least simple, double, triple and dominant quadruple excitations) which would be impracticable for large molecules , we have taken into account a Multi-Reference CI in the CIPSI[50] (Configuration Interaction by Perturbation with multiconfigurational zeroth-order wave functions Selected Iteratively) framework : the CI space is built iteratively by selecting the most important configurations using second order perturbation theory.

Thus, the final CI space contains only the most important excitations among all the simple, double, triple and quadruple excitations. The final CI wavefunction obtained by diagonalization of the so-built CI matrix is developed on a basis of a few hundred symmetry- and spin-adapted determinants. This multiconfiguration wave function is then perturbated by all configurations generated by single and double excitations from all the S^2 eigenfunctions used in the CI expansion. So, excited states are described by functions containing up to 6 excitations with respect to the reference open-shell SCF determinant. The total number of Slater determinants included in the perturbative treatment is of the order of a few hundred millions.

3- Choice of geometry :

The following additional assumptions were made : at the PPP level, standard geometries were chosen (all C-C distances equal to 1.40 Å and all C-C-C angles equal to 120°). In the CS-INDO, approach, however, we have taken the geometries optimized at the ROHF/3-21G *ab initio* level of theory. In all cases, species were assumed to be and to remain planar upon ionization and VUV excitation, so that all calculated electronic transitions are vertical. This last assumption agrees with the analysis of the experimental vibronic absorption spectra[21-23, 60], in which vibrational frequencies vary only slightly relative to the level of electronic excitation.

4- Parametrization :

It can be shown that for planar PAHs, it is sufficient to introduce only one adjustable molecular parameter describing the non-diagonal term of the one-electron core hamiltonian : the so-called β_{pq} of the PPP method and the $K_{\pi\pi}$ screening constant in the CS-INDO formalism. This parameter has been optimized for both σ and $\sigma-\pi$ hamiltonians at the CIPSI level so that the first calculated transition matches the first observed intense band on the corresponding experimental spectrum[31].

At the CIPSI/CS-INDO level, it has been found that the two optimized parameters for naphtalene and its cation were transferable to other neutral and cationic PAHs. The optimized values are: $K_{\pi\pi}$ = 0.57 for neutrals and $K_{\pi\pi}$ = 0.41 for cations ; the other parameters were taken as : $K_{\sigma\sigma}$ = 1.00 and $K_{\sigma\pi}$ = 0.65, as previously used for other aromatic compounds[61, 62].

At the CIPSI/PPP level, it has been found that the β_{pq} parameter varies from one compound to another, the value for the cation being always larger than the one for the corresponding neutral. Transferability was not applicable and the parameter has been optimized for every neutral and ionic systems.

5- Test of the method :

Our calculations are expected to reproduce experimental absorption spectra in the gas phase, and since no experimental data are available for ionized PAHs in that phase, the photoelectronic specrta (PES) are the only experimental results comparable to our calculations. This comparison is given in scheme 1 for the anthracene cation : the agreement is good.

6- Results :

The following neutral and cationic species were studied : naphtalene (na), anthracene (an), tetracene (te), phenanthrene (ph), pyrene (py) and chrysene (ch). Coronene (co), initially considered was omitted in the present work because of the large number of state degeneracies induced by the high symmetry of this molecule and because of the perturbations expected by a Jahn-Teller effect in the cationic form.

The spectra obtained with the method described above are in excellent agreement with the available experimental ones, so long as an optimization of the β_{pq} parameter in the CIPSI/PPP method is made for each compounds. The general features of the experimental spectra recorded in freon[31] are correctly reproduced with energy shifts lying between 300 and 2000 cm^{-1}. Such shifts are in the order of magnitude that can be expected with such calculations: it should be emphasized that : 1) our calculations do not include any coupling between electronic and vibrational levels, 2) no correction has

DIB cm^{-1}	calculated transition energy cm^{-1}	oscillator strength	species	ΔE cm^{-1}
3048	3694	0.006	ph+	646
8474	8711	0.145	ch+	237
11561	10501	0.156	ph+	1060
11600	11453	0.198	te+	147
13160	12647	0.030	py+	513
13210	13308	0.030	te+	98
13224	13810	0.476	an+	586
14568	13969	0.020	py+	599
14572	14437	0.015	ch+	135
14927	14950	0.146	na+	23
15009	15082	0.107	an+	73
15153	15155	0.0001	ph+	2
17090	17075	0.546	ph+	15
18498	18631	0.0004	na+	133
18645	19196	0.055	na+	551
	19575	0.056	ph+	930
20478	19680	0	te+	798
	19841	0.014	ph+	637
20916	20889	0.051	ch+	27
20989	21212	0.235	an+	223
21154	21293	0.190	te	139
22207	21777	0.076	an+	430

Table 1. Tentative assignment using calculated energies. Calculated oscillator strengths and deviation from related DIB are also reported (see text for further details)

Scheme 1. Photoelectron spectrum of vapor phase anthracene compared to the PPP and CS-INDO + CIPSI calculations[22].

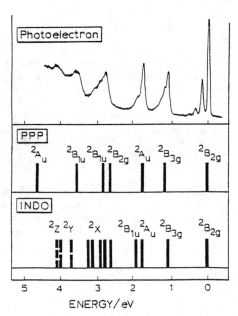

been added to account specifically for matrix environment, 3) no vibrational structure has been evaluated.

At the CIPSI/PPP level, oscillator strengths account well for the relative intensities of transitions. At the CIPSI/CS-INDO level, the values are merely indicative since only the variational part of the correlated wavefunction has been considered in the calculation. A detailed comparison of our calculations with experimental results can be found in references 21-23 and reference 62.

CORRELATION TO DIBS

1- Tentative assignments :

We considered all DIBs reported between 885 and 22577 cm^{-1} (about 80 DIBs : see references 1-9) and compared them to the set of transitions we have obtained at the highest level of theory for neutral and ionized PAHs : naphtalene (neutral and cation : CIPSI/CS-INDO), anthracene and phenanthrene (neutral : CIPSI/PPP, cation : CIPSI/CS-INDO), tetracene, chrysene and pyrene (neutral and cation : CIPSI/PPP). It can be seen, from the values reported in table 1, that 20 DIBs can be correlated to our calculated transition energies showing that ionized PAHs must be considered as possible carriers. It must be emphasized that *all calculated bands* are related to a DIB. A number of bands, however are missing this way : they can be related either to species we haven't considered here (larger PAHs, dehydrogenated aromatics, or completly different carriers) or to vibronic bands. Since no vibronic structure can be obtained at the present level of theory, an estimation of these bands can only be recovered from the corresponding accurate experimental data in matrix as Salama and Allamandola did for the naphtalene cation[60a].

This approach leads to the values and assignments presented in table 2 where the computed data concerning the naphtalene[60b], anthracene[22] and pyrene[23] cations have been replaced by experimental vibronic transitions. In addition, we have included experimental results for coronene(co) and ovalene(ov)[63]. For each of these molecules, the experimental spectrum selected is that whose recording conditions were supposed to be the closest to the interstellar conditions.

In both cases, assignments were made so as to minimize the global root-mean-square (RMS).

2- Comments :

First of all, it must be noticed that any calculated or available experimental band occuring in the 4400-8700 Å can be connected to a DIB. Apart from a single band which can be related to the neutral tetracene, *all the electronic transitions matching DIBs come from ionized PAHs.*

The deviation of each experimental or calculated value from the DIB it is related to is always less than 1100 cm^{-1}. This value is of the order of magnitude of vibrational energies : vibronic couplings and application of Franck-Condon rules could account for such a shift. Furthermore, it should be kept in mind that vibronic couplings may be important (they are responsible, for instance, for the observation of the electronically forbidden first transitions of benzene and coronene). We can thus expect some of our calculated values to present a shift of a few tens of cm^{-1} due to such a coupling. Concerning the values observed in laboratory, such a shift falls well within matrix environment effects. Furthermore, it has been known for a long time that substitution(s) on the aromatic ring also can induce slight shifts on the VUV spectra : we will discuss briefly this point in the conclusion.

The quality of the two assignments can be estimated from the corresponding Root-Mean-Square deviation : we have 103 cm^{-1} for the first one (only theory) and 54 cm^{-1} for the second one (mixing theory and experiments) : such values are very encouraging since RMS deviation decreases as the number of assigned DIBs raises from 20 to 31 when vibronic structures are included.

DIB cm^{-1}	transition energy cm^{-1}	species	ΔE cm^{-1}
3048	3694	ph+	646
8474	8711	ch+	237
11561	10501	ph+	1060
	10565	co+	996
11600	11453	te+	147
13160	12779	py+	381
13210	13308	te+	98
13224	13843	an+	619
14568	14010	py+	558
	14225	an+	343
14572	14437	ch+	135
14599	14598	an+	1
14712	14706	co+	6
14928	14815	na+	113
15153	15155	ph+	2
	15157	na+	4
	15174	an+	21
15558	15322	na+	236
15671	15664	na+	7
15735	15808	na+	73
16184	16234	co+	50
	16238	na+	54
	16265	an+	81
16631	16600	an+	31
	16739	na+	108
17090	17050	na+	40
	17075	ph+	15
17245	17246	na+	1
17348	17458	na+	110
17523	17666	na+	143
	17730	an+	207
18062	18123	an+	61
18197	18186	na+	11
18645	19091	na+	446
	19179	an+	534
	19481	ph+	836
20478	19574	na+	904
	19680	te+	798
	19920	co+	558
	20542	py+	62
20916	20721	an+	195
	20889	ch+	27
20990	21190	ov+	200
21154	21293	te	139
	21588	ov+	434
22207	21786	co+	421
	21829	py+	378
	21911	an+	296
	22099	te+	108
22577	22533	py+	44

Table 2. Tentative assignment using calculated energies. Calculated oscillator strengths and deviation from related DIB are also reported
(see text for further details)

CONCLUSIONS AND PROSPECTS

We first would like to emphasize the fact that the determination of the VUV spectra of species as large as PAHs remains a long and exacting task on both the computational and experimental sides. The methods we used cannot be considered routine work : the use of the CIPSI procedure to get an accurate description of the excited states amounts to including correlation effects progressively and must be carefully analyzed at each step of the iterative process. In that way, it differs fundamentally from the much simpler S/INDO method which has been recently used for a first evaluation of the spectra of dehydrogenated naphtalene in its neutral or ionized forms[64]. The transferability of the parameters within families of PAHs derivatives (same state of hydrogenation; same state of charge;...) is a strength of the present approach, even though calibration on at least one experimental value is desirable for a better precision. When this work was undertaken, no experimental data were available for ionized PAHs in the gaz phase. The calibration has been done on the spectra reported by Shida and Iwata[31] who studied a large number of PAHs and corresponding cations using the same experimental conditions.

Although the reference spectra were not free from environment effects, we think that the present study is one more step towards a better understanding of the puzzling role of PAHs in the interstellar medium. Even if a complete one-to-one assignement of all DIBs to ionized PAHs is hardly possible, the results presented above and the now available experiments give more than a strong presumption. The recent experiments by Léger and d'Hendecourt on the cationic radical of methylene pyrene and the correlation to 4 DIBs[65a,b], confirming a previous intuition of Salama and Allamandola[65c] are encouraging. However, it is also a proof of the extreme complexity of the problem and we think that, within the PAHs model, derivatives have of all sorts have to be considered, such as substituted or heteroaromatic PAHs for instance... possibly neutral, ionized or (not exclusif) dehydrogenated !

There are so many structural possibilities and so many DIBs : PAHs and PAH-like species are one of the most attractive family at the present time, however, other models should also be considered (role of non-free-flyer species, carbonated long chains[66], role of coals[67] ...). The fact that the carriers could belong to several different families or that the intensities of the signatures could vary dramatically according to the physical and chemical conditions of the surrounding medium is supported by several observations, among them are :

* some UIR bands have been detected without the detection of DIBs in the same region[68]
* some DIBs have been observed without UIR signatures in the same region[69]
* DIBs have been observed in N-rich environments[69]

The consequences of all that is obvious to us : theoreticians, observers and experimentalists have to work with each other so that the search for the DIBs carriers does not turn to a speculative angling when nothing is known about a possible type of carriers.

REFERENCES

1. M. L. Heger, Lick Obs. Bull., 10, 146 (1921)
2. P. W. Merrill, Pub. A. S. P., 46, 206 (1934)
3. P. W. Merrill, R.F. Sanford & O. C. Wilson, Astrophys. J., 86, 274 (1937)
4. G. H. Herbig, Astrophys. J., 196, 129 (1975)
5. F. Sanner, R. L. Snell & P. A. van den Bout, Astrophys. J., 226, 460 (1978)
6. R. L. Snell & P. A. van den Bout, Astrophys. J., 244, 844 (1981)
7. G. H. Herbig, Astrophys. J., 331, 999 (1988)
8. A. Gammelgaard, Astron. & Astrophys., 135, 177, (1984)
9. C. Joblin, J. P. Maillard, L. d'Hendecourt & A. léger , Nature, 346,729 (1990)
10. a- G. H. Herbig & T. H. Leka, Astrophys. J., 382, 133 (1991)
    b- P. Jenniskens & F.-X. Desert, accepted in Astron. Astrophys., 1993
11. G. E. Bromage, QJRAS, 28, 294 (1987)

12. G. van der Zweet, 'Possible carriers of the diffuse interstellar bands', in 'Polycyclic Aromatic Hydrocarbons and Astrophysics', A. Léger, L. d'Hendecourt & N. Boccara, editors, Dordrecht, Reidel (1986)
13. A. Léger & J. L. Puget, Astron. Astrophys., 137, L5 (1984)
14. L. J. Allamandola, A. G. G. M. Tielens & R. J. Barker, Astrophys. J., 290, L25 (1985)
15. L. J. Allamandola, A. G. G. M. Tielens & R. J. Barker, in 'Physical Processes in Interstellar Clouds', G. E. Morfill & M. Scholer editors, Dordrecht, Reidel (1987)
16. A. Omont, Astron. Astrophys., 164, 159 (1986)
17. J. L. Puget & A. Léger, ARA&A, 27, 161 (1989)
18. F. Pauzat, D. Talbi, M. Miller, D. J. DeFrees & Y. Ellinger, J. Phys. Chem., 96(20), 7882 (1992)
19. D. J. DeFrees, M. D. Miller, D. Talbi, F. Pauzat & Y. Ellinger, Astrophys. J., 408, 530 (1993)
20. F. Pauzat, 'UIR bands : computational experiments on model PAHs', communication n° 78 in this book. To be published in the Astrophys. J.
21. J. Szczepanski, D. Roser, W. Personnette, M. Eyring, R. Pellow & M. Vala, J. Phys. Chem., 96(20), 7876 (1992) (ionized naphtalene)
22. J. Szczepanski, M. Vala, D. Talbi, O. Parisel & Y. Ellinger, J. Chem. Phys., 98(6), 4494 (1993) (ionized anthracene)
23. J. Szczepanski, M. Vala, D. Talbi, O. Parisel & Y. Ellinger, submitted to the J. Phys. Chem. (ionized pyrene)
24. D. Talbi, F. Pauzat & Y. Ellinger, Astron. Astrophys., 268, 805 (1993)
25. M. K. Crawford, A. G. G. M. Tielens & L. J. Allamandola, Astrophys. J., 293, L45 (1985)
26. P. Hermine & P. Bréchignac, 'Photophysical studies of jet-cooled PAHs', communication n° 40 in this book
27. a- G. Moreels, J. Clairemidi, P. Hermine, Ph. Bréchignac & P. Rousselot, to be pulished in Astron. Astrophys., and communication n° 55 in this book
    b- G. Moreels et al., Astron. & Astrophys., 187, 551 (1987)
28. P. Boissel, G. Lefèvre & Ph. Tiébot, 'Photofragmentation of PAH ions : laboratory experiments on long time scale', communication n° 14 in this book
29. L. Andrews, B. J. Kelsall & T. Blankenship, J. Phys. Chem., 86, 2916 (1982)
30. cited by P. Bréchignac, in 'General discussion', J. Chem. Soc. Far. Trans., 89(13), 2314 (1993)
31. T. Shida & S. Iwata, J. Am. Chem. Soc., 95, 3473 (1973)
32. G. P. van der Zwet & L. J. Allamandola, Astron. Astrophys., 146, 76 (1985)
33. A. Léger & L. d'Hendecourt, Astron. Astrophys., 146, 81 (1985)
34. M. Cohen & B. F. Jones, Astrophys. J., 321, L151 (1987)
35. C. Cossart-Magos & S. Leach, Astron. Astrophys., 233, 559 (1990)
36. P. Jenniskens & F.-X. Desert, Astron. Astrophys., 274, 465 (1993)
37. R.J. Buenker, J.L. Whitten & J.D. Petke, J. Chem. Phys., 49, 2261 (1968)
38. S.D. Peyerimhoff & J.R. Buenker, Theor. Chim. Acta, 19, 1 (1970)
39. P.J. Hay & I. Shavitt, J. Chem. Phys., 60, 2865 (1974)
40. J.M.O. Matos, B. Roos & P.-A. Malmqvist, J. Chem. Phys., 86(3), 1459 (1987)
41. O. Kitao & H. Nakatsuji, J. Chem. Phys., 87(2), 1169 (1987)
42. M.H. Palmer & I.C. Walker, Chem. Phys., 133, 113 (1989)
43. B.O. Roos, K. Andersson & M.P. Fülscher, Chem. Phys. Letters, 192(1), 5 (1992)
44. E. Hückel, Z. Phys., 70, 206 (1931) ; 72, 310 (1931) ; 76, 628 (1932)
45. a- L. Salem, The Molecular Orbital Theory of Conjugated Systems, W.A. Benjamin (New York, 1966)
    b- E. Heilbronner & H. Bock, 'Das HMO-Modell und seine Anwendung', Verlag Chemie (Weinheim, 1968)
46. a- R. Pariser & R.G. Parr, J. Chem. Phys., 21, 456 (1953) ; 21, 767 (1953)
    b- J.A. Pople, Trans. Faraday Soc., 49, 1375 (1953)
47. R. Zahradnik & P. Carsky, J. Phys. Chem., 74(6), 1235 (1970)
48. J.A. Pople & G.A. Segal, J. Chem. Phys., 44, 3298 (1966)
49. J.A. Pople, D.L. Beveridge & P.A. Dobosh, J. Chem. Phys., 47, 2026 (1967)

50. B. Huron, J.P. Malrieu & P. Rancurel, J. Chem. Phys., 58(12), 5745 (1973)
51. K. Andersson, P-.A. Malmqvist & B.O. Roos, J. Chem. Phys., 96(2), 1218 (1992)
52. G.Berthier, J.Baudet & M. Suard, Tetrahedron, 19 suppl. 2, 1 (1963)
53. a- F. Momicchioli, I. Baraldi & M.C. Bruni, J. Chem. Soc. Faraday II, 68, 1556 (1972)
    b- F. Momicchioli, I. Baraldi & M.C. Bruni, Chem. Phys., 82, 229 (1983)
54. H.C. Longuet-Higgins and J.A. Pople, Proc. Phys. Soc. (London), A68, 591 (1955)
55. M.J.S. Dewar & N. Trinajstic, Chem. Commun., 646 (1970)
56. F.O. Ellison and F.M. Matheu, Chem. Phys. Letters, 10, 322 (1970)
57. C.C.J. Roothaan, Rev. Mod. Phys., 32, 179 (1960)
58. E.A. Hylleraas, Z. Phys., 48, 469 (1928)
59. J. B. Foresman, M. Head-Gordon, J.A. Pople & M.J. Frisch, J. Phys. Chem., 96, 135 (1992)
60. a- F. Salama & L. J. Allamandola, Atrophys. J., 395, 301 (1992)
    b- F. Salama & L. J. Allamandola, J. Chem. Soc. Far. Trans., 89(13), 2277 (1993) and references cited
61. I. Baraldi & G. Ponterini, Gazz. Chim. Ital., 118, 109 (1988)
62. A. Després, V. Lejeune, E. Migirdicyan, A. Admasu, M. S. Platz, G. Berthier, J.-P. Flament, O. Parisel, I. Baraldi & F. Momicchioli, J. Phys. Chem., in press
63. P. Ehrenfreund, L. d'Hendecourt, L. Verstraete, A. Léger, W. Schmidt & D. Defourneau, Astron. Astrophys., 259, 257 (1992)
64. P. Diu, F. Salama & G. H. Loew, Chem. Phys., 173, 421 (1993)
65. a- L. d'Hendecourt, 'Interstellar grains', communication n°6 in this book
    b- A. Léger & L. d'Hendecourt, 'Toward the identification of the (single) carrier of the 4430 Å and 7564 Å DIBs', communication n° 32 in this book
    c- F. Salama & L. J. Allamandola, Nature, 358, 42 (1992)
66. a- A. E. Douglas, Nature, 269, 130 (1977)
    b- P. Thaddeus, C. A. Gottlieb, R. Mollaaghababa & M. Vrtilek, J. Chem. Soc. Far. Trans., 89(13), 2125 (1993)
    c- P. Thaddeus, 'Free carbenes in the interstellar gas', communication n° 79 in this book
67. a- R. Papoular, J. Breton, G. Gensterblum, I. Nenner, R. J. Papoular and J. J. Pireaux, accepted for publication in Astron. Astrophys. (1993)
    b- O. Guillois, R. Papoular, C. Reynaud, K. Ellis & I. Nenner, 'Coal model for the UV-visible interstellar exctinction curve', communication n° 24 in this book
    c- K. Ellis, O. Guillois, R. Papoular, C. Reynaud & I. Nenner, 'Recent measurements on coal in the Mid-IR as a model for interstellar dust', communication n° 25 in this book
68 L. B. F. M. Walters, H. J. G. L. M. Lamers, T. P. Snow, E. Mathlener, N. R. Trams, P. A. M. van Hoof, C. Waelkens, C. G. Seab & R. Stanga, Astron. Astrophys., 211, 208 (1989)
69. T. Le Bertre & J. Lequeux, Astron. Astrophys., 274, 909 (1993)

EFFECTS OF MOLECULAR SIZE ON THE DISSOCIATION RATES OF PAH CATIONS

H.W.Jochims[1], E.Rühl[1], H.Baumgärtel[1], S.Tobita[1,2] and S.Leach[2,3]

[1] Institut für Physikalische und Theoretische Chemie der Freien Universität Berlin, Takustr.3, W-1000 Berlin 33, F.R. Germany
[2] Laboratoire de Photophysique Moléculaire du CNRS, Bât.213, Université Paris-Sud, 91405-Orsay, France
[3] Département Atomes et Molécules en Astrophysique, CNRS-URA812, Observatoire de Paris-Meudon, 92195-Meudon, France

ABSTRACT

The photostability of a series of polycyclic aromatic hydrocarbons has been studied experimentally by determining the internal energy E_{int} of their monocations at which the dissociation rate is $10^4 s^{-1}$. The results on the hydrogen atom loss reaction, fitted to an RRK model calculation, were then scaled to determine the internal energy E_{crit} at the astrophysically critical dissociation rate $10^2 s^{-1}$. Data were also obtained on H_2 and C_2H_2 loss channels. The quasi-linear dependence of E_{int} and E_{crit} on three PAH size parameters was demonstrated and modelled. The results indicate that in HI regions, photoexcited regular PAHs containing less than 30 - 40 carbon atoms will dissociate rather than relax by infrared emission, whereas for $N_C \geq 30 - 40$, and for analogous PAH photoions of any size, the principal relaxation channel will be infrared emission. Some other implications are discussed concerning the photophysics and photochemistry of PAHs in the interstellar medium.

INTRODUCTION

Following absorption of an ultraviolet photon and subsequent internal conversion, polycyclic aromatic hydrocarbons and their monocations in the interstellar medium would be in highly excited vibrational states whose infrared radiative rates are considered to be greater than or equal to $10^2 s^{-1}$. The only process competitive with the PAH infrared emission is dissociation. The present study provides, for the first time, experimental data on the PAH size effect on the internal energy E_{crit} at which the rate of dissociation is of the order $k_d = 10^2 s^{-1}$, which can be considered as the astrophysically critical dissociation rate above which the yield of the competitive infrared emission channel would rapidly become extremely small.

EXPERIMENTAL

The experimental technique is that of Photoion Mass Spectrometry (PIMS) using monochromatized synchrotron radiation as photon excitation source in the energy range 7-35 eV. Measurements were made of the appearance potentials for ionic fragmentation of a set of PAHs (Figure 1), including both alternant and non-alternant hydrocarbons [1], ranging from benzene C_6H_6 up to coronene $C_{24}H_{12}$. We remark that coronene is often considered as the model PAH in theoretical studies of interstellar polycyclic aromatic hydrocarbons [2,3].

RESULTS AND DISCUSSION

Three unimolecular dissociation reactions were studied, corresponding to loss of a hydrogen atom, a hydrogen molecule and acetylene, respectively. The results have implications on the photostabilitiy of PAH monocations which are expected to be generally valid also for neutral PAHs.

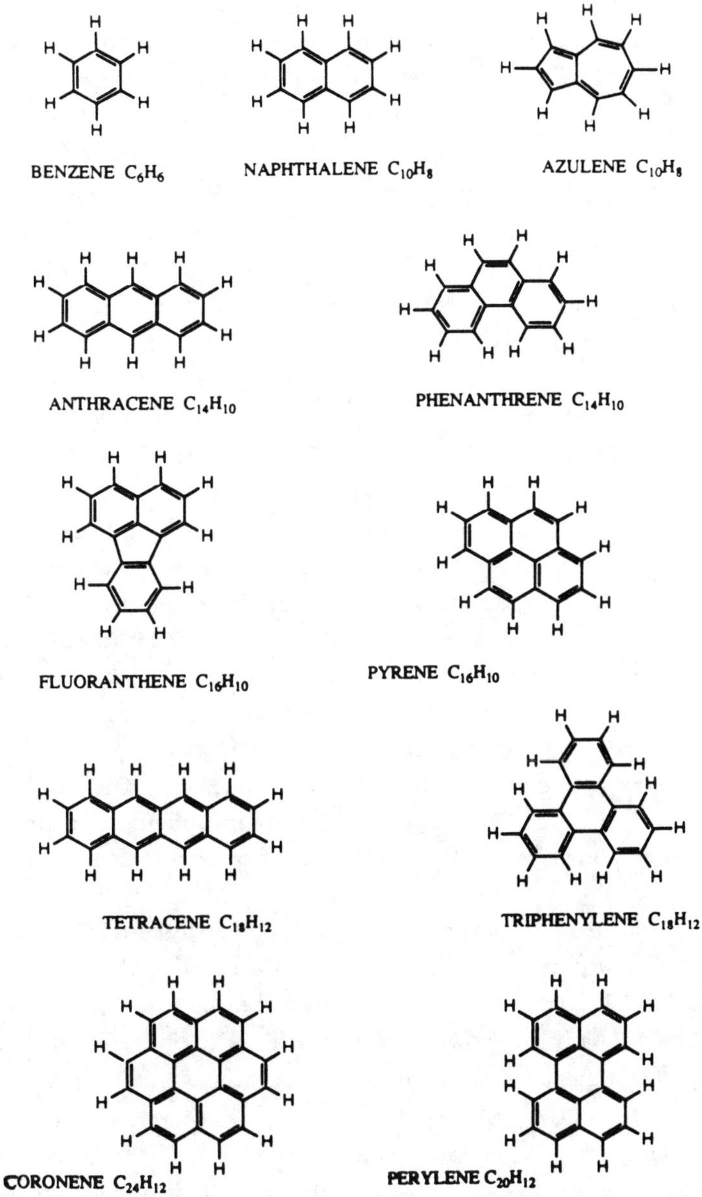

Figure 1. Polycyclic aromatic hydrocarbons studied in the present work on size effects on dissociation rates of their monocations

For the –H process a RRK fit to the dissociation rates of the PAH monocations was established as a function of the number of their vibrational degrees of freedom. A scaling procedure enabled an estimate to be made of the internal energy content, E_{crit}, corresponding to the astrophysically critical dissociation rate $k_d = 10^2 s^{-1}$ of the PAH monocations. The dependence of E_{crit} on N_A, the number of carbon atoms in the PAH, was compared (Figure 2) with the theoretical values obtained previously on the basis of quantal RRK calculations [3,4]. The latter predict E_{crit} values over 1 eV greater than our experimentally based results. The observed linear dependence of E_{int} (and E_{crit}) on N_A and on N_C (number of carbon atoms) was shown to be consistent with the RRK model [5].

The upper limit of PAH size below which neutral PAHs in HI regions would mainly relax by photodissociation is shown to be a species containing 30-40 carbon atoms (Figure 3). Our results also indicate that PAH monocations of any size formed by ionization in HI regions should be stable and relax essentially by emission of infrared photons. Dissociation of PAH monocations thus formed would require absorption of a second photon of appropriate energy. We remark that our results do not lend support for models of PAH fragmentation by direct photodissociation mechanisms in HI regions.

In this study, most of the polycyclic aromatic hydrocarbons were regular pericondensed or catacondensed [6] species (Figure 1). Some indication was obtained that E_{crit} values are low for PAHs which have mutually perturbing H atoms, leading to critical N_C values greater than 30 - 40 for these species. This suggests that photodissociation will tend to eliminate the less regular PAHs at a greater rate than the more regular ones. The resulting distribution of PAHs in the interstellar medium would then tend to be heavily weighted in favour of the more regular species. Future laboratory work will be carried out to provide means of substantiating this conjecture.

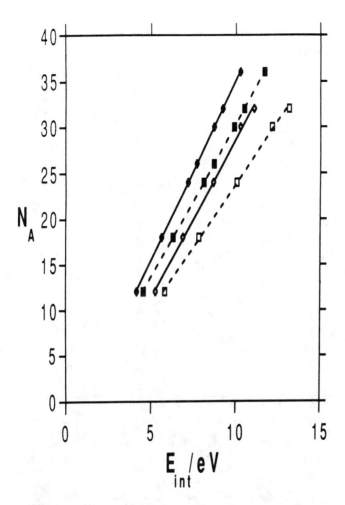

Figure 2. Effects of size of PAH cations on the energy required for hydrogen atom loss. Calculated values of the internal energy of the parent cation, E_{int}, at the appearance potential of this reaction, as a function of N_A, the total number of atoms. RRK calculation values (present study) at $k_d = 10^2 s^{-1}$ (filled diamonds) and $k_d = 10^4 s^{-1}$ (filled rectangles). Quantal RRK calculation values [3] at $k_d = 10^2 s^{-1}$ (open diamonds) and $k_d = 10^4 s^{-1}$ (open rectangles).

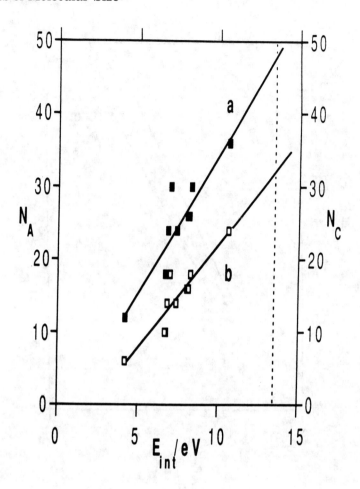

Figure 3. Effects of size of PAH cations on the energy required for hydrogen atom loss. Critical internal energy, E_{crit}, of the parent cation for a dissociation rate $k_d = 10^2 s^{-1}$ as a function of (a) N_A, the total number of atoms and (b) N_C, the number of carbon atoms. The vertical dotted line at $E_{int} = 13.6$ eV corresponds to the upper limit of photon energy in H I regions of the interstellar medium.

From the discussion in previous sections, it is clear that at particular loci in the interstellar medium there can exist both neutral PAHs and ionic PAHs, the size and number distribution of which, as well as the degree of dehydrogenation, will depend on the incident radiation field and its spectrum. This distribution should be reflected in the frequencies and relative intensities of the UIR bands which originate in these species. Satisfactory modelling of the infrared emissions depends on improvement in the quality and quantity of photophysical and photochemical properties of neutral and ionic PAHs as well as further understanding of the incident radiation and its spatial variations.

This work is reported more fully elsewhere [5].

REFERENCES

1. S.Leach, in IAU Symposium 135, *Interstellar Dust*, ed. L.J.Allamandola and A.G.G.M.Tielens, (Kluwer, Dordrecht) 1989, p.221.
2. A.Léger and J.L.Puget, Astron.Astrophys.137, L5,(1984).
3. L.J.Allamandola, A.G.G.M.Tielens and J.R.Barker, Astrophys.J.Suppl. 71,733,(1989).
4. A.G.G.M.Tielens, L.J.Allamandola, J.R.Barker and M.Cohen, in *Polycyclic Aromatic Hydrocarbons and Astrophysics*, ed. A.Léger, L.d'Hendecourt and N.Boccara, (Reidel, Dordrecht), 1987, p.273.
5. H.W.Jochims, E.Rühl, H.Baumgärtel, S.Tobita and S.Leach, Astrophys.J, in press, Jan.1, 1994 issue.
6. S.Leach, in *Polycyclic Aromatic Hydrocarbons and Astrophysics*, ed. A.Léger, L.d'Hendecourt and N.Boccara, (Reidel, Dordrecht), 1987, p.99.

PHOTOFRAGMENTATION OF PAH IONS:
LABORATORY EXPERIMENTS ON LONG TIMESCALES

P. Boissel, G. Lefèvre and Ph. Thiébot
Laboratoire de Photophysique Moléculaire du CNRS,
Bat. 213, Université Paris Sud   91405 Orsay Cedex, FRANCE.

ABSTRACT

Photofragmentation of PAH ions is studied under high isolation conditions, using an ion trap. Dissociation threshold is reached by gradual increase of the ion internal energy following the absorption of several low energy photons. This experiment provides a tool for a first spectroscopic approach of truly isolated ions. Different fragmentation paths are observed, depending on ion geometry and isotopic substitution. The history of the primary and secondary fragments can be followed over long periods. These results provide guide-lines to infer which type of ionic species can be encountered in the interstellar medium.

INTRODUCTION

The possible existence of large aromatics molecules (PAHs) as a third component of the interstellar medium, between small molecules and grains, opens the question of the stability of such "medium sized" species (20 to 200 atoms) under UV irradiation. Experiments concerning the photofragmentation of PAHs are quite scarce. Owing to the short detection times generally used, they probe an energy range well above threshold. In these experiments, the fragments obtained are detected by impact on a target and cannot be subsequently studied. All these reason have urged us to devise a series of experiments using an ion trap to approach some of the conditions encountered in the interstellar medium: long timescales, collisionless environment.

EXPERIMENTAL

A detailed description of the trap will be given in a forthcoming paper[1]. We only recall here the features that are useful for the understanding of the present experiment.

The trapping principle is that of the Penning trap, widely used in Ionic Cyclotronic Resonance (ICR) cells[2]. Ions are confined through the conjugated action of a magnetic field and a static quadrupolar potential. The nature of the ions that are present in the trap at a given instant can be probed by Fourier Transform Mass Spectroscopy (FT-MS) with impulse excitation[3]. The only difference with usual ICR cells is that the four plates used for excitation and detection of the ion motions are replaced with four rods, in order to leave large optical access. This modification appears to have no influence on the ion trapping. The main drawback of our system is due to the small size of the trap rather than its geometry. Indeed, when more than one type of ion is present, the different ions clouds interact and the peak heights in the mass spectrum are no longer an exact image of the population. With two type of ions of nearly the same mass, but quite different populations, the minor peak may be completely washed out in the mass spectrum. For instance, the peak corresponding to the ^{13}C isotope is never detected although it should reach 17% of the main peak in the case of anthracene.

The whole device, placed in an UHV quality vacuum cell, is attached to a closed cycle helium cryostat. The temperature of the different parts ranges from 12 K to 30 K, leading to efficient cryopumping. This insures a very low rate of ion-neutral collisions, confirmed by the long trapping times obtained: more than 10 minutes for anthracene cation (mass 178) and even longer for ions of lower mass.

Ions are produced via laser ablation of a solid PAH pellet, located on the trap axis, behind one end plate. Neutral species, evaporated by the laser shot, are immediately condensed on the outer parts, while ionic ones, guided by the magnetic field, enter the trap through a small hole. Due to space charge limitation, only a small fraction of the ions produced are trapped. This saturation effect leads to reproducible initial conditions. Typically, 10^5 to 10^6 unfragmented PAH cations are obtained, with a dispersion lower than 20% from shot to shot. For a given mass range, the natural losses of the trap are also quite reproducible. The mass spectra, taken at given time delays after the laser shot, can then be normalized to account for these losses.

Once trapped, the ions can be irradiated by a focussed cw xenon arc lamp. A shutter allows irradiation during defined periods and color filters are used to check which spectral range is efficient.

For the photofragmentation experiments described here, a typical run is conducted as follows:
- ejection of remaining ions from the preceding run
- laser shot,
- dead time (10 to 15 seconds)
- initial mass spectrum and opening of the shutter (time = 0 on the figures)
- mass spectra at regular time intervals.

Figure 1 presents the result of such a run, performed on fully deuterated anthracene cation, the filter used being a simple glass plate. Three points can be noticed: i: a very efficient fragmentation occurs, since the parent ion completely disappears after 15 s. ii: the only fragment obtained corresponds to the loss of a neutral acetylene molecule (C_2D_2, mass 28). iii: this fragment is not destroyed under the same irradiation.

The two last points are important. However, before discussing their possible astrophysical implications, it is necessary to understand the first one. So, let us explain how an anthracene ion can be broken by such low energy photons.

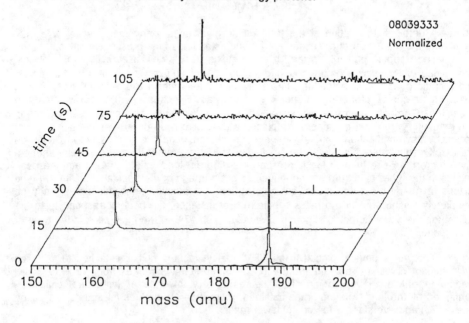

Figure 1: Fragmentation of fully deuterated anthracene cation

MECHANISM OF THE FRAGMENTATION

The glass plate in front of the lamp only transmits wavelengths longer than 330 nm, corresponding to photon energies lower than 3.8 eV. In the spectroscopic survey described below, we show that dissociation can be obtained with photons of still lower energy (1.73 eV). In comparison, in a different experiment[4] on deuterated naphthalene, a very similar ion, the same dissociation channel (C_2D_2 loss) only appears above 8 eV of internal energy. Then, in the present experiment, dissociation must result from accumulation of the energy brought by successive photon absorptions.

Let us describe the mechanism, in the simple case where photons corresponding to the first electronic transition are only used. After absorption of a photon, the ion does not stay in the excited electronic state. A competition takes place between reemission of the energy (fluorescence) and redistribution of this energy between the vibrational degrees of freedom of the electronic ground state (internal conversion). For large ions, the fluorescence quantum yield is not very high, so that, in most cases, we are left with an ion in the electronic ground state, but with an increased vibrational energy content. In the absence of collisions, this energy can be only evacuated radiatively. This radiative cooling, either through vibrational fluorescence or electronic delayed fluorescence (Poincaré fluorescence[5]), is highly unefficient. During the cooling time (typically a few seconds), the ion can absorb another photon before having lost the energy brought by the preceding one. Then, its energy content increases gradually up to the region (6-8 eV) where fragmentation becomes important.

A numerical simulation of the whole mechanism has been performed. The model used and the detailed calculation will be exposed in another paper. Starting with a population of initially cold ions, the competition between heating by absorption and radiative cooling leads to a stable distribution of internal energies after a few seconds (fig 2).

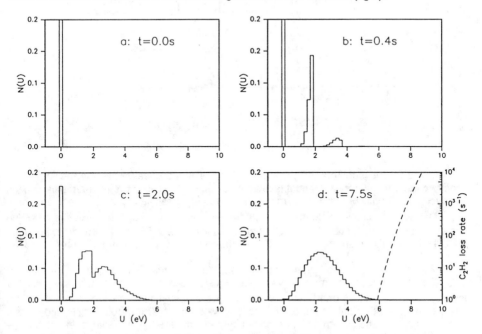

Figure 2: Evolution of the internal energy distribution among a population of isolated anthracene ions, irradiated by 1.7eV photons.

The fragmentation processes are treated using the Inverse Laplace Transform theory (ILT)[6]. The C_2D_2 loss rate, calculated with energy barriers and preexponential factors deduced from reference 4, is reported on the last frame of figure 2. Fragmentation occurs when ions in the high energy tail of the distribution absorb an extra photon. This leads to an exponential decrease of the parent ion population when the stationary distribution is established..

DETAILED RESULTS

Fragmentation path

With the pyrene cation, the fragmentation path corresponding to the loss of C_2H_2 is completely absent. The fragment remaining corresponds to the loss of two hydrogen atoms (figure 3).

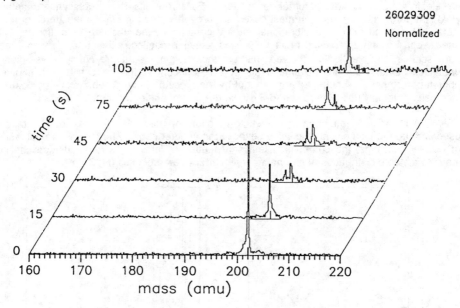

Figure 3: Fragmentation of Pyrene cation

Due to the problems in the discrimination of nearby peaks, the intermediate mass spectra cannot be fully trusted and it is difficult to know if the two hydrogen atoms have been removed sequentially or inside a H_2 molecule. However, a transient peak seems to appear at mass M-1, which could indicate that the final fragment is obtained by a two step process. The same indication comes out of the result of an unpublished "MIKES" study[7]. In this experiment the only dissociation path observed for the pyrene cation is the loss of one H atom.

With fully deuterated anthracene cation, the main reaction observed is the loss of C_2D_2 (fig. 1). A peak at the mass M-4 is sometimes observed, but marginally, at a very low level. In contrast, with normal anthracene (fig. 4), the acetylene loss remains the dominant channel, but the other path does exist. Peaks at M-26 and M-2 are both visible and remain after 50 s, when the parent ion has completely disappeared. This isotope effect can be explained qualitatively by the shift in the zero point energy of C-H oscillators when H is replaced by D. This shift increases the dissociation barrier for H ejection, but leaves unaffected the barrier for the other channel, which does not break a C-H bond. An improvement of the numerical simulation is currently in progress to check if this effect can be reproduced correctly.

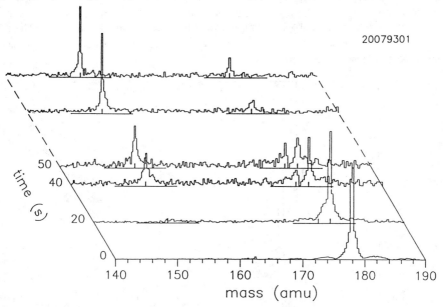

Figure 4: Fragmentation of anthracene cation

Stability of the fragments

One of the main advantages of the ion trap is that, FT-MS being a non-destructive test, the fragments remain in the trap and can be subsequently studied.

We have already seen the case of hydrogen ejection from pyrene and normal anthracene. The absence of the M-1 peak at long times indicates that the primary fragment is rapidly dissociated, a second H atom being removed to form a triple bond. The secondary fragment with mass M-2 seems no longer affected by the irradiation. However, in some long runs on pyrene, a peak at M-4 seems to appear above the noise level, indicating further dehydrogenation.

As already noticed, $C_{12}H_8^+$ and $C_{12}D_8^+$ fragments, resulting from the acetylene loss, are not dissociated under irradiation with wavelength longer than 330 nm (fig. 1 and 4). In longer runs, we have check that this is still true for irradiation times longer than 5 minutes. The difference in behaviour between the parent ion, dissociated within a few seconds, and the fragment may be due to two different reasons. i: absorption lines in the spectral range concerned are less intense or absent. The internal energy distribution is then shifted downwards. ii: the energy barrier for the lowest dissociation channel is too high and cannot be reached. Further studies are needed to check which of this two reasons is dominant.

Spectroscopic survey

A crude spectroscopy of the isolated ions can be made with colored glass filters. In the spectral region between 300 and 800 nm, the xenon lamp spectrum is smooth and looks like that of a blackbody at 6000 K. Using a series of long pass filters (Schott), the evolution of the fragmentation rate with cut-off wavelength has been measured. The difference observed between two successive glass in the series gives an indication on the contribution of the spectral region comprised between the two cut-off. This has been done with the deuterated anthracene cation. The fragmentation efficiency is quantified by determination of the crossing time t_c at which the parent and fragment peak heights are equal. The evolution of $1/t_c$ with the cut off wavelength λ_c of the filters is presented in figure 5 and compared with the absorption spectrum of anthracene cation in argon matrix[8]. Starting from the right, no fragmentation occurs with λ_c=780 nm, beyond the fist electronic transition. Then, for decreasing λ_c, the efficiency increases step by step, each time a new absorption line is concerned by the irradiation.

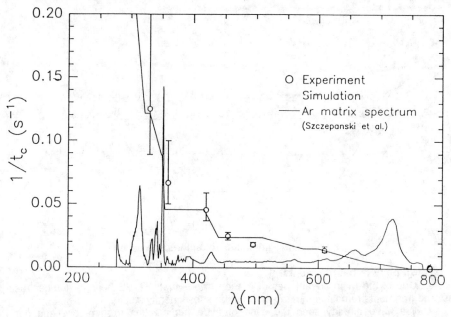

Figure 5: Spectral evolution of the fragmentation efficiency of anthracene d_{10} ion.

A quantitative comparison can be done with the help of our numerical simulation. The spectrum is modelized using 7 transitions with energies corresponding to the center of the lines. The absorption rates in these transitions are calculated with the oscillator strength given in reference 8 and a 6000 K blackbody spectrum for the unfiltered lamp. The relative spectral variations of the calculated values of $1/t_c$ are in good agreement with the measured ones. However, concerning the absolute value, the calculated illumination differs from the experimentally measured one by a factor of 20. Verifications on this point are currently in progress, but this could indicate that oscillator strength in reference 8 are overestimated.

The 720 nm absorption band region has been studied with more details (figure 6). In this limited wavelength range, the fragmentation efficiency only depends on the power absorbed by the ion. Conversely, after calibration with neutral attenuators, the crossing time can be used to measure the absorbed power. This value can be compared to that given by an integration of the absorption spectrum. The spectral intensity of the lamp being constant, the absorbed power is proportional to the integral of the absorptivity, weighted by the transmission of the filter. The agreement between the two determination is good up to 665 nm and at 780 nm but, at 695 and 715 nm, the energy actually absorbed is higher than the one predicted from the spectrum. This indicates a red shift of the absorption line or the appearance of a wing on the long wavelength side. This effect seems too large to be merely due to the deuteration or to the matrix environment. It can rather be explained by the fact that, once the stationary distribution is established, the absorption measured it that of "hot" ions, with internal energies around 2 eV.

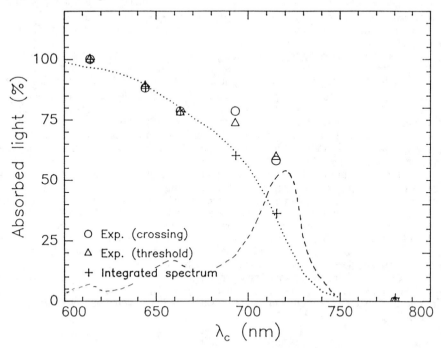

Figure 6: Fragmentation of deuterated anthracene cation: detail of the evolution inside the fist electronic absorption.

Looking back to figure 5, it can be noticed that a similar shift is observed for the 350 nm line. With a cut-off wavelength of 360 nm, this transition seems at least partly excited since the efficiency observed is higher than expected.

The same study has been done on pyrene cation. The fragmentation efficiency remains constant for λ_c between 280 and 400 nm, indicating the absence of absorption lines in this region. The fall-off is entirely located between 400 and 480 nm. This result is compatible with the rare gas matrix spectra [9]. However, here again the band seems to extend farther towards long wavelength.

CONCLUSION - ASTROPHYSICAL IMPLICATIONS

Most of the results of the present experiment are transferable to the study of astrophysical ions, although the irradiation used here is quite different from that encountered in the interstellar medium (ISM).

First, observation, over several minutes, of PAH ions having lost 2 or even 4 hydrogen atoms, gives a confirmation of the exixtence of stable dehydrogenated species, an hypothesis which have been used for a while without clear experimental confirmation.

Beyond this qualitative observation, the important point is that, owing to the long timescale used, this experiment probes the energy region where fragmentation is in competition with radiative cooling. Now, this is just the key region for the study of the survival of PAHs in the ISM. Indeed, the energy scales with the size of the species. For PAHs expected in the ISM, larger than those studied here, this competition will occur in the 9-13 eV range, corresponding to the energy brought by hard UV photons. In other terms, a given interstellar irradiation leads to a turn-over zone for the PAH mass, below which species are destroyed too rapidly. The present results raise several questions concerning the selection among species that might occur around this zone.

Let us first examine the fragmentation channels. In the case of pyrene, a compact PAH, fragmentation only leads to the loss of hydrogen atoms. For anthracene, a non-compact one, the dominant channel is the loss of C_2H_2. If this difference between compact and non compact species is confirmed for other members of the PAH family, and if it extends to larger units, it can induce a natural selection between the two types. Hydrogen atoms being abundant, an equilibrium can be reached between hydrogenated and dehydrogenated forms, while the loss of two carbons atoms is much more irreversible. Such a selection would explain why lines characteristic of the non compact PAHs are absent from the interstellar infrared emission spectra [10].

The great stability of the fragment obtained by ablation of acetylene from anthracene cation opens another question. If this stability not only comes from a decrease of visible absorption, it must be due to an increase of the dissociation barrier. In this case, this ion will be also more resistant under UV irradiation. Here again a selection will occur and, in a given mass range, the fragment can be much more abundant than the parent ion. More generally, it should be beared in mind that ions encountered in the ISM may be not only those directly obtained from stable molecules. The search for the species responsible for the diffuse interstellar bands (DIBs) should be extended to such exotic species.

Concerning the DIBs, this first experiment has proven that it is possible to make absorption spectroscopy of gas phase isolated ions. This method avoids the ambiguity brought by matrix environment and can work with large ions, even those which does not give fluorescence.

NOTE

During the conference, we have noticed that our experiment can also give informations about the absorption continuum of PAHs cations. In the case of pyrene, the fragmentation efficiency does not decrease, within experimental errors, when the irradiation of the spectral zone between 280 and 400 nm is blocked by a colored filter. The decrease is only observed when the main line itself is no longer irradiated. This rules out the existence of a continuum with an integrated absorption intensity equivalent or larger than that of the sharp feature. Observation of this continuum seems then specific to the matrix experiments. One possibility, which has not been considered in reference 9, is the ionisation of preexisting neutral dimers. This would produce dimer cations, with a growth correlated with the monomer one. Due to resonant charge exchange, such dimers are known to exhibit very broad absorption features.

REFERENCES

1 P. Boissel and G. Lefèvre, in preparation.
2 M. B. Comisarow and A.G. Marshall, Chem. Phys. Letters **25**, 282, (1974).
3 R.T. McIver Jr., R.L. Hunter and G. Baykut, Rev. Sci. Instr. **60**, 400, (1988).
4 E. Ruhl, S. D. Price and S. Leach, J. Phys. Chem. **93**, 6312, (1989).
5 A. Léger, P. Boissel and L. d'Hendecourt, Phys. Rev. Letters, **60**, 921, (1988).
6 W. Forst, J. Phys. Chem. **86**, 1771, (1982).
7 S. Tobita, Private communication.
8 J. Szczepanski, M.Vala, D.Talbi, O.Parisel and Y.Ellinger, J. Chem. Phys. **93**, 6312, (93).
9 F. Salama and L.J. Allamandola, Nature **358**, 42, (1992).
10 A.. Léger, L. d'Hendecourt and D. Défourneau, Astron. Astrophys. **216**, 148, (1989).

IONIZED POLYCYCLIC AROMATIC HYDROCARBON MOLECULES AND THE INTERSTELLAR EXTINCTION CURVE

Wei Lee and Thomas J. Wdowiak
Department of Physics
University of Alabama at Birmingham, Birmingham, AL 35294, USA

ABSTRACT

Electronic absorption measurements have been carried out on both the neutral and cationic forms of several individual compact and non-compact PAH species in the wavelength range of 190—820 nm. The pronounced decrease in the strength of the strong near-UV absorption bands of the neutral species upon ionization by gamma radiation is indicated by the laboratory spectra. This provides persuasive experimental evidence for resolution of the conflict between the PAH-UIR hypothesis and the fact that expected features, which are characteristic of neutral aromatics in near-UV, are not observed in the interstellar extinction curve.

INTRODUCTION

The principal infrared interstellar emission features, known as the unidentified infrared (UIR) bands at 3.3, 6.2, 7.7, 8.6, and 11.3 μm (3040, 1615, 1310, 1150, and 885 cm^{-1}), have been observed in the planetary nebulae, bipolar nebulae, reflection nebulae, H II regions, Herbig Ae stars, Wolf-Rayet stars, novae, the plane of the Milky Way, and starburst galaxies.[1] The fact that these five features are usually observed as a "family" and are associated with such a wide variety of celestial objects implies that there exists a relatively specific form of matter which is widely distributed in the interstellar medium (ISM) as a major constituent. Attempts to identify these emission features have led to generally accepted hypotheses that radiative relaxation of vibrationally excited aromatic hydrocarbons (PAHs) is involved. The original suggestion that the bands are associated with aromatic species embedded in small amorphous carbon particles[2] was later challenged by the hypothesis[3] that they originate from vapor-phase, neutral polycyclic aromatic hydrocarbons. Allamandola, Tielens & Barker[4] independently argued that because a large fraction of the PAHs are expected to be ionized, the PAH cations are mostly responsible for the bands. PAH species (ionized and neutral) are attractive as an interstellar component because their aromatic structure provides the level of stability necessary for long lifetime in the interstellar UV radiation and shock environment, and also allows for the conversion of the absorbed UV energy into discrete IR emission bands through the mechanism of IR fluorescence.[5]

The PAH-UIR hypothesis has long been investigated via a number of laboratory studies involved with neutral species because of the inherent difficulties in handling ionized molecules in the laboratory. Theoretical studies have for the most part until recently focused on neutral PAH molecules. It has been thought that the vibration modes of neutral PAHs and those of ionized species should not be much different. While most laboratory studies were done at room temperature, recent study of the temperature-dependent aspects[6] of PAH molecules in the vapor phase and condensed state has shown that the C—H stretch feature of compact PAH species at elevated temperature (~800 K) has a closer match than room-temperature of the peak wavelength and profile to those of the observed 3.3 μm UIR feature. This enhances the belief that PAHs are ubiquitous and abundant throughout the ISM, and are responsible for the UIR by providing a better correlation of their general IR spectral characteristics. However, one of the criticisms of the PAH model raised by Leach[7] and Donn, Allen, and Khanna[8] is that the strong absorption peaks of PAH molecules in the near-UV are not evident in the interstellar UV extinction curve. This apparent conflict with the PAH hypothesis, which should be considered as a serious problem,[9] requires resolution if the PAH hypothesis for the UIR is to remain plausible.

Since a large fraction of the PAHs in the general ISM are expected to be ionized, it is important that an investigation be made of the spectral properties of ionized PAHs. The need to obtain UV/visible spectra of cationic PAHs has impelled this research. In this paper, we report the UV/visible spectral properties of the PAH cations, produced by gamma irradiation of the neutral precursors coronene ($C_{24}H_{12}$), benzo[ghi]perylene ($C_{22}H_{12}$), perylene ($C_{20}H_{12}$), benzo(e)pyrene ($C_{20}H_{12}$), pyrene ($C_{16}H_{10}$), 1,2,5,6 dibenzanthracene ($C_{22}H_{14}$), and chrysene ($C_{18}H_{12}$) dissolved in boron oxide glass ($B_2O_3 \cdot \frac{2}{3}H_2O$) matrices. Details of our spectral measurements of the isolated PAHs in their neutral and cationic forms along with a discussion of astrophysical implications have been published elsewhere.[10,11]

EXPERIMENTAL

All spectra of PAHs studied in both the neutral and cationic states were obtained in a low melting inorganic glass—boron oxide glass at room temperature. The preparation of PAH molecules isolated in the glass matrix is accomplished by dispersing a small quantity (~0.15% by mass) of PAH in boric acid crystals (125 mg) before heating the mixture to 240°C in an oven. On cooling the molten boric acid melt solidifies to a glass in which the PAH is presumably molecularly dispersed. The PAH cations were produced by irradiating our samples, including references without PAH, with gamma radiation from a Gamma-Cell 40 2839-Ci ^{137}Cs source at UAB. A thorough description of the sample preparation and the experimental techniques is reported elsewhere.[10]

RESULTS AND DISCUSSION

Figures 1(a) and (b) show the electronic spectra of the neutral and irradiated 5-ringed perylene and benzo(e)pyrene, respectively. As shown in the figures, these neutral PAHs have few or no visible features but do have very strong absorption peaks in the near-UV. However, the irradiated PAH samples, presumably composed of both neutral and ionized species, exhibit a pronounced weakening of the near-UV features with increase in the dose of gamma irradiation. There is a saturation in the conversion of the neutral species to the cation as the dose is increased. The changes

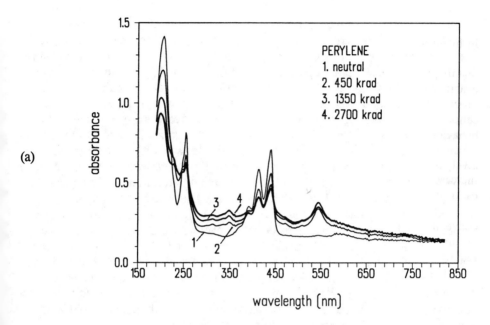

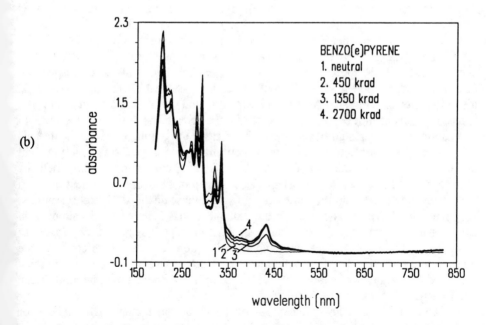

Fig. 1 UV/visible absorption spectra of neutral and cationic PAH species isolated in boron oxide glass vs. boron oxide glass of the same irradiation conditions: (a) perylene and (b) benzo(e)pyrene. PAH:boron oxide glass = 1:1350.

in intensity of the absorption bands with respect to increasing irradiation doses are plotted in Figures 2(a) and (b) for perylene and benzo(e)pyrene, respectively. (Notice that due to the difficult determination of the continuum in the UV, these plots do not exactly present the strength of peaks.) The effect of gamma irradiation on the growth of the cation bands and the correlated depletion of neutral species is obvious. The cation concentration increases sharply at the beginning as represented by the general behavior of the curves. The cation formation rate then reaches a steady state and after it can even decrease. When the population of PAH cations is built up at a certain level, the probability of recombination of a free electron with a singly positive ion increases substantially, resulting in a steady state in which the neutralization of a cation accompanies each creation of another cation. Moreover, the production of PAH dications may take place as the population of monocations increases by the conversion of monocations through a second ionization. The creation of PAH dications can be either direct, with effective photons of $\geq$ 20 eV, or sequential, requiring photons of $\geq$ 7 eV for the first ionization and $\geq$ 13 eV for the second from the readily singly ionized PAHs, as has been shown by a detailed examination of experimental data on the ionization potentials of PAHs.[12] Incomplete removal of the near-UV absorption bands in a highly irradiated sample is attributed to the fact that some of the neutral PAH material is not dissolved in boron oxide glass, remaining as clumps. Ionization would be precluded for these clumps.

The reduction of the strength of the near-UV absorption bands in samples of the matrix-isolated perylene and benzo(e)pyrene when ionized suggests why PAHs are not observed in the ISM at UV wavelengths. Similar results are found for the non-compact 1,2,5,6 dibenzanthracene (5 rings) and chrysene (4 rings) and for the other three compact PAHs studied as we reported previously.[10] Although it is expected that the IR spectra of the cations differ from those of the corresponding neutral species, which might exclude PAH cations as the UIR carriers, recent IR measurements of cationic PAHs by Szczepanski and Vala[13] indicate that the UIR emission bands cannot be explained solely on the basis of neutral PAH species, but that cations must be a significant, and in most cases the dominant, component. This ensures that PAH molecules, expected to be ionized in a large fraction, are evident in the infrared in UIR emission. Theoretical calculations employing *ab initio* methods of the spectra of some cationic PAH species lead to a much better agreement between the calculated CH/CC vibration intensity ratios and those deduced from observations. However, the 3.3 μm CH stretching band of the matrix-isolated PAH cations such as coronene,[14] perylene, pyrene, anthracene, and naphthalene[13] is not found in laboratory studies utilizing low-temperature argon as the matrices. If the disappearance of this band is not a matrix effect of argon ice but an inherent spectroscopic property of cationic PAHs, the 3.3 μm UIR band might be more likely due to neutral PAHs.

Our experiments also indicate, as suggested by Salama and Allamandola for naphthalene[15] and pyrene,[16] that a broad absorption continuum from the near-UV to visible apparently does develop when PAH species are ionized (see Figure 3 for 1,2,5,6 dibenzanthracene and chrysene, and ref. 10 for the other five compact PAH species studied). This continuum absorption, as Salama and Allamandola have

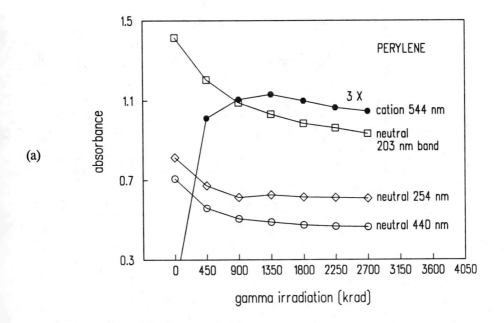

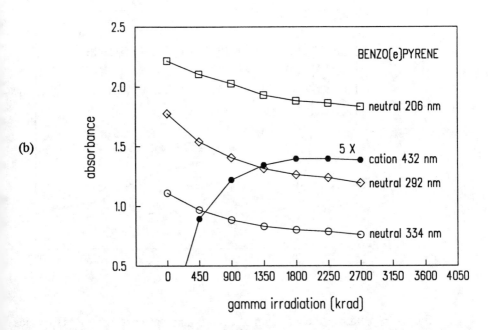

Fig. 2 The decrease of the neutral PAH bands and the growth of the corresponding PAH cation bands as a function of the gamma irradiation doses: (a) perylene and (b) benzo(e)pyrene. These plots are based on the same data selectively used for Fig. 1.

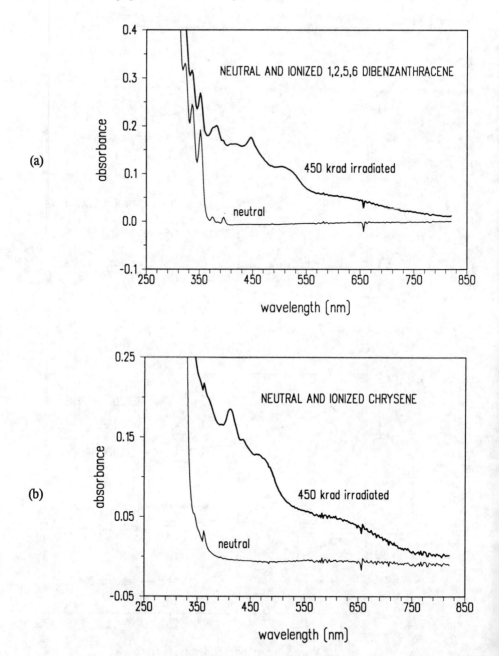

Fig. 3 Formation of a broad continuum of neutral and matrix-isolated PAH species upon irradiation: (a) 1,2,5,6 dibenzanthracene and (b) chrysene. PAH:boron oxide glass = 1:675. Notice that boron oxide glass without PAHs of the same irradiation condition has been used as a reference in measurements.

proposed, may provide the channel by which interstellar UV/visible radiation is efficiently converted to the discrete IR emission. Our work has demonstrated that the continuum phenomenon is general for all PAH species when ionized. The increase in strength with decrease in wavelength is likely not a result of an effect such as electron scattering. As matter as a fact, Treinin in a survey study on inorganic glasses for trapped radicals has pointed out that no spectroscopic indication of trapped electrons could be detected, implying that the electrons are supposed to attach to the solute molecules to produce the mononegative ions.[17] However, there is a chance of this continuum to be due to the matrix effect if Coulomb force perturbs between the constrained neighboring PAH cations and closely spaced suborbitals of higher electronic energy levels overlap more greatly. The progressive increase of the continuum with irradiation doses, compared with the saturation (and the eventual dropping) of the cation bands, suggests the formation of this continuum may be associated with the interaction between the neighboring, positively charged PAH molecules, which would be present even for dications formed in the matrix if there are any. If the broad continuum could be proved as an intrinsic characteristic of cationic PAHs, it will improve our understanding of the how PAH ions convert interstellar UV/visible energy to the IR emission as observed in the UIR.

CONCLUDING REMARKS

The work directed, toward seeking to provide an experimental foundation for resolution of the conflict between PAH-UIR hypothesis and the nature of the interstellar extinction curve, has resulted in several conclusions.

The main result of our work concerns the behavior of UV/visible spectral feature intensities when comparing neutral and positively charged PAHs. Proponents of the PAH hypothesis countering the objection based upon a lack of observed near-UV absorption features have suggested that the mix of PAH molecules likely to exist in interstellar space would result in the "smearing out" of the individual UV features.[18] This "fix" ignores the electronic properties of PAHs affected by ionization. PAH molecules are expected to be ionized in the general ISM and as such they would not exhibit near-UV absorption features between the bluest DIB absorption band at 4428 Å and the UV extinction bump at 2175 Å. Therefore a smearing out is not required.

Ionized PAHs are still to be expected to make their presence known by emissions due to transitions similar to those between vibration states of neutral PAHs, producing the UIR bands. This result, thought to be true[10] and now confirmed by others,[13] along with the lack of the band structure in the near-UV portion of the interstellar extinction curve, on the basis of our experiments, strongly argues for the interstellar PAHs being ionized. This is consistent with the idea that the ejected photoelectrons having excess kinetic energy are heating H I regions.[19]

There does appear to be a broad and intense absorption continuum for the ionized PAHs which extends from the UV to the visible or even the near IR. If this is an intrinsic property of individual PAH cations, this continuum might serve as an efficient channel for pumping the UIR. The structureless feature would also contribute to the visible portion of the interstellar extinction curve and blend but undetectable. Further investigation on the apparent absorption continuum remains as an important goal.

Recognizing the importance of examining other observational conflicts regarding the interstellar PAH hypothesis such as the lack of blue fluorescence as pointed out by Donn et al.,[8] we have started a fluorescence study of PAHs matrix-isolated in boron oxide glass matrices. We have found that when matrix-isolated pyrene is irradiated with a resulting reduction in strength of ~15% of the 275-nm absorption band due to ionization of some of the molecules, the fluorescence bands between 360 and 440 nm weaken by greater than 99%! This preliminary result indicates that the irradiated boron oxide glass matrix serves to quench the fluorescence, suggesting that if PAHs exist in grain mantles they would not emit structured, observable visible radiation. We thank NASA grant NAGW-749.

REFERENCES

1. K. Sellgren, in *Dusty Objects in the Universe*, edited by E. Bussoletti and A. A. Vittone (Kluwer, Dordrecht, 1990), p. 35, and the references therein.
2. W. W. Duley and D. A. Williams, Mon. Not. R. Astron. Soc. **196**, 269 (1981).
3. A. Léger and J. L. Puget, Astron. Astrophys. **137**, L5 (1984).
4. L. J. Allamandola, A. G. G. M. Tielens, and J. R. Barker, Astrophys. J. **290**, L25 (1985).
5. L. J. Allamandola, A. G. G. M. Tielens, and J. R. Barker, Astrophys. J. Suppl. Ser. **71**, 733 (1989).
6. G. C. Flickinger and T. J. Wdowiak, Astrophys. J. **362**, L71 (1990).
7. S. Leach, in *Polycyclic Aromatic Hydrocarbons and Astrophysics*, edited by A. Léger, L. B. d'Hendecourt, and N. Boccara (Reidel, Dordrecht, 1987), p. 99.
8. B. D. Donn, J. E. Allen, and R. K. Khanna, in *Interstellar Dust*, edited by L. J. Allamandola and A. G. G. M. Tielens (Kluwer, Dordrecht, 1989), p. 181.
9. See, for example, A. Evans, *The Dusty Universe* (Ellis Horwood, N. Y., 1993), p. 192.
10. W. Lee and T. J. Wdowiak, Astrophys. J. **420**, L127 (1993).
11. T. J. Wdowiak and W. Lee, J. Chem. Soc. Faraday Trans. **89**, 2308 (1993).
12. A. T. Tokunaga, K. Sellgren, R. G. Smith, T. Nagata, A. Sakata, and Y. Nakata, Astrophys. J. **380**, 452 (1991).
13. J. Szczepanski and M. Vala, Nature **363**, 699 (1993) and the references therein.
14. L. d'Hendecourt and A. Léger, in *Proceedings of the First Symposium on the Infrared Cirrus and Diffuse Interstellar Clouds*, in press.
15. F. Salama and L. J. Allamandola, Astrophys. J. **394**, 301 (1992).
16. F. Salama and L. J. Allamandola, Nature **358**, 42 (1992).
17. A. Treinin, in *Radical Ions*, edited by E. T. Kaiser and L. Kevan (Interscience Publishers, N. Y., 1968), p. 525.
18. C. Joblin, A Léger, and P. Martin, Astrophys. J. **393**, L79 (1992).
19. E. L. O. Bakes and A. G. G. M. Tielens, Astrophys. J., in press.

DISCUSSION

SARRE — Do we know if these PAH cations fluoresce in the gas phase ?

WDOWIAK — It is important to determine if PAH cations in the gas phase emit light when excited with ultraviolet radiation. We have not performed such an experiment to date. However I can tell you the preliminary results of an experiment performed with boron oxide glass matrix isolation. We excited disolved pyrene at its 275 nm band and observed its characteristic structured emission between 360 nm and 420 nm. The sample was then irradiated with gamma rays with the result that the 275 nm absorption band was reduced by 15 percent. We were surprised to find that gamma irradiation had quenched the UV excited emission to a level below one percent of what had been measured previously. Obviously this must be a matrix effect indicating a gas phase experiment is mandatory, even though more difficult to perform.

BUSSOLETTI — Are you able to quantify the irradiation yield? Is this value similar to any actual astrophysical situation ?

WDOWIAK — Our experiment is not intended to be an astrophysical simulation and the technique utilized was carried out because of the ease with which many samples could be investigated simultaneously. Also we did not have access to a powerful ultraviolet source such as a hydrogen lamp, but did have access to one of the many Cs^{137} sources used in biomedical research at our university. The use of a gamma source makes quantification of yield difficult because of the complexity of the process by which gamma rays absorbed in the sample result in ionization. It is not a direct one as would be with an ultraviolet source.

BRECHIGNAC — I wish to comment about the appearance of the UV continuum together with the bands attributed to the cations. If this continuum is indeed due to the cations it is most probably due to higher-lying electronic transitions. But in your experiment, as well as in rare gas matrix experiments, there is an important broadening of the bands due to the matrix. Then it is nothing but obvious whether these electronic transitions would keep a continuum character in the case of gas phase cations, as expected for the carriers of the DIB's.

WDOWIAK — Salama and Allamandola upon observing the apparent continuum in their experiments with naphthalene and pyrene in argon and neon matrices, wondered if it was general for PAH cations. We can now say that it is when they are isolated in matrices. Because of the potential for such a continuum being a channel for pumping at the UIR it is now important to determine if it exists for gas phase cations.

COLANGELI — You have shown the 3.3 μm band produced by Coronene in KBr matrix at high temperature. It fits quite well with astronomical observations. On the other hand, gas phase coronene displays a peak shifted in wavelength. Could you comment about this discrepancy ?

WDOWIAK — Our French friends (d'Hendecourt, Joblin and colleagues) have recently exhibited including at this meeting, data that show that at higher temperatures than in our gas phase experiments the wavelength of the C-H stretch shifts toward that of the UIR band. So now we have two ways of matching the 3.3 μm UIR wavelength. Also the ultraviolet excited emission experiments performed on the smallest PAH's by Barker and colleagues in Michigan also appear to exhibit a shift from our gas phase wavelength towards that of the UIR wavelength.

BOISSEL — In our ion trap experiment, detection of photo-fragmentation allows us to get spectroscopic information about truly isolated ions in the gas phase. In the case of pyrene, preliminary results show that, within the limit of the experimental errors, no absorption is present between 280 and 400 nm, which excludes the existence of a continuum. So, can you exclude that, in condensed phase experiment, the continuum can be due to the presence of electrons in the vicinity of the ion ? That could be the case either in rare gas matrix or in boron oxide environments.

WDOWIAK — That is quite plausible and if so would eliminate the idea of a continuum absorption for PAH cations as an astrophysical channel for pumping of the UIR. Again I'll repeat that our matrix isolation experiments verified the results of the Salama and Allamandola matrix isolation experiments and demonstrated that what they found was general for PAH cations in matrices. Ion trap experiments are very important and I am very happy that you are doing them.

LEACH —
1. It would be of interest to do electron spin resonance measurements on the γ-irradiated samples and correlate with optical observations.
2. It is doubtful whether PAH cations will fluorescence because of the existence of low-lying electronic states which facilitate very efficient internal conversion to the electronic ground state.

A very useful technique for determining whether ions fluorescence is the photoion-fluorescence photon coincidence method (PIFCO) developed by Devoret, Eland and Leach (*Chem. Phys. Letters*, 1976) at Orsay with which it is possible to measure the quantum yield of ion fluorescence of mass selected ions, down to 10^{-5}.

WDOWIAK — This requires no response from me as it is a comment from Leach rather than a question.

FIELD — My question is, what is the ionization balance between PAH and PAH^+ in the interstellar field or in the range of environments encountered in the ISM?

WDOWIAK — I do not know the answer to the question, but I understand that Professor Dalgarno has made some calculations regarding this matter

A LABORATORY INVESTIGATION OF THE DIFFUSE INTERSTELLAR BANDS AND LARGE LINEAR MOLECULES IN DARK CLOUDS

Thomas J. Wdowiak, Wei Lee, and Luther W. Beegle
Department of Physics
University of Alabama at Birmingham, Birmingham, AL 35294

ABSTRACT

A laboratory synthesis via electrical discharge and rare gas cryogenic matrix isolation has produced molecules from a 0.5% methane in argon mixture, many of which exhibit absorption bands at wavelengths close to those of the diffuse interstellar absorption bands (DIBs), including the strongest and widest 4428 Å DIB. The visible spectrum also reveals absorption features due to the HCO molecule which exists in dark clouds. Those laboratory features at DIB wavelengths are stable under ultraviolet radiation emitted from a mercury vapor lamp, while the features attributed to HCO are easily bleached with visible wavelength light. The apparatus consists of a linear discharge tube excited by a high frequency (Tesla) transformer, a sapphire substrate cooled to 10 K by a closed cycle refrigerator, and spectrometers of the diode array and scanning varieties. These experiments demonstrate their potential for investigation of the questions of the general interstellar medium carriers of the DIBs and the kinds of molecules expected to have a role in the chemistry of dark molecular clouds.

INTRODUCTION

The diffuse interstellar bands (DIBs) were first recognized by Paul Merrill in the 1930's when it was noticed that there existed certain spectral absorption features which were common to a large variety and number of stars, thereby indicating their interstellar nature. The elucidation of the carriers of these ubiquitous spectral features has been the goal of many researchers ever since. An early effort on our part resulted in production of absorption bands of species frozen in isolation in an argon matrix at 13 K, that we argued correlated in wavelength with the strongest of the DIBs.[1] Those experiments were inspired by the suggestion of A. E. Douglas that linear carbon chains were responsible for the DIBs.[2] We attempted to produce such free radical species in our original experiment using an electrical discharge of a gas mixture of 0.48% concentration of methane (CH_4) in argon. Such a discharge will produce radical and ion species including CH, C_2^-, and C_3. As the discharge continues additional bands become apparent at wavelengths close to those of the strongest DIBs. Very importantly, a broad band at 4500 Å became the first to be noticed in terms of order of appearance. This wavelength is within matrix shift limits of the bluest DIB at 4428 Å. Matrix shifts occur because the excited state of the molecule interacts differently with the matrix from what the ground state does. This results in a spectral feature being at a wavelength different from what it would be for the absorber in the gas phase. We identified a total of 8 bands at wavelengths close to the strongest DIB wavelengths as then cataloged by Herbig.[3] These bands resisted photobleaching with a quartz envelope mercury vapor lamp.

Kratschmer and colleagues (see ref. 4 and work cited therein) have reported bands at wavelengths near those of DIBs when carbon vapor is isolated in argon and then annealed by elevation of the temperature. We have discussed the possible relationships of our 4500Å band and a similar band they found initially at 4470 Å and which appears to shift upon annealing to 4500 Å.[5]

In this report, we will discuss our return to doing the DIB related gas discharge matrix isolation experiments along with other spectral phenomena we have observed in the past and continue to study now but never reported before. The latter results may have bearing on the question of carbon species in dark molecular clouds.

EXPERIMENTAL TECHNIQUE

Our current experiments are similar to the earlier ones except for the use of an ion pump, different but equivalent spectroscopic instrumentation, and some changes in procedures that will be discussed. A mixture of 0.52% methane in argon (Air Products Custom Grade) is introduced at 8 PSI through a capillary into the central portion of a discharge tube, where it is excited from a metal electrode by a small Tesla coil. The metal surface of a cold finger, cooled by an APD Cryogenics, Inc., Model 202W Displex Expansion Engine refrigerator, serves as the second electrode. A mechanical and ion pump provides an initial vacuum and then the mechanical pump is closed from the system. During deposition, because of the pressure needed to sustain the discharge, the ion pump is turned off. Deposition at 13 K occurs on a sapphire disk appearing as a white, translucent frost.

Spectroscopic measurements were carried out with a tungsten lamp powered at 45 W and a McPherson 1 meter spectrometer, operated by a McPherson step motor controller, with the photomultiplier signal coupled through a Pacific Photometric model 401 photometer to a 10 inch (26 cm) Technicon recorder. A Kodak heat rejection filter was inserted between the lamp and the vacuum shroud giving a smooth continuum spectrum. Bleaching experiments were carried out with a low-pressure mercury vapor lamp and the tungsten lamp. The light source was operated at a lower intensity and greater distance then our earlier experiment. The significance of this will be discussed later.

DIB CANDIDATES

Figure 1 shows the single beam absorption spectrum of reactive species isolated at 13 K in an argon matrix after the sample was exposed for extended time to UV radiation of a quartz-enveloped mercury vapor lamp and visible radiation from a tungsten lamp equipped with a Kodak heat rejection filter. Bands resulting from CH, C_2^-, C_3 and CNN are indicated. The wavelength positions of the four strongest DIBs at 4428 Å, 5778 Å, 6283 Å, and 6613 Å are indicated by the vertical lines intersecting the spectrum. The significance of photobleaching with visible light will be discussed later. UV photobleaching is done to remove most of the C_2^- features at 4730 Å and

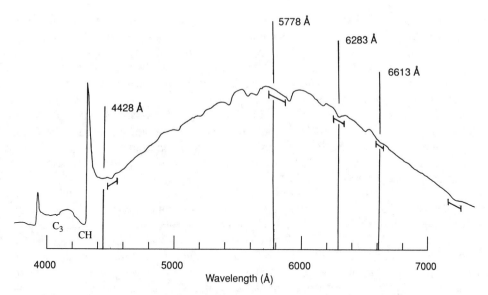

Figure 1. Single beam absorption spectrum of reactive species produced in a gas discharge of CH_4 (0.52%) and Ar (99.48%) and then matrix isolated at 13 K. Afterwards the matrix was photobleached with UV and visible light. Vertical lines indicate wavelengths of 4 strongest DIBs at 4428 Å, 5778 Å, 6283 Å, and 6613 Å and laboratory bands near those wavelengths are also indicated. Other species are identified. See the text and spectrum A in Figure 2 for the wavelengths of photobleachable bands including C_2^- which show some residual absorption in this sample.

5210 Å and demonstrate that the bands near DIB wavelengths are stable against mid-UV (2000 Å- 4000 Å) radiation. This UV bleaching technique was described in our earlier paper.[1]

It can be seen that our recent experiments produce the same features as we reported before[1] and these bands do correlate in approximate wavelength but within matrix shift limits with strong DIBs. The correlation is better for the longer wavelengths than for the blue feature (4428 Å DIB, 4500 Å Lab). In addition, we have been concerned that our bands might be due to atmospheric contaminants including frozen O_2 (the solid state equivalent of the atmospheric or telluric bands). We presented an analysis elsewhere demonstrating that the bands being due to frozen O_2 is unlikely.[5] To reinforce our belief in this regard we have carried out an experiment that duplicates all conditions occurring in a full experiment except omitting electrical discharge. No bands of any sort appeared, including those expected of frozen O_2, indicating that leakage of atmospheric O_2 into our system in sufficient amounts to produce bands does not occur. Also, the bands that occur when the sample is discharged are reactive species, which is demonstrated by their disappearance upon warming the matrix to 45 K. This was done with the ion pump on and it was not overwhelmed by outgassing, which would have been expected if O_2 had been present in sufficient quantities to produce the absorption bands. The increase in pressure from 7×10^{-6} torr to 5×10^{-5} torr upon warming is what would be expected from an increase in vapor pressure of the warmed frozen argon and 0.52 percent methane mixture. We conclude that the experiment produced bands which are not due to the absorption of frozen O_2.

Because of the possibility that bands at wavelengths near those of the DIBs may be due to reactive species produced from trace atmospheric contaminants in the Tesla coil excited discharge, we have done the experiment using argon without methane. This does not result in the formation of a blue band at wavelength of 4500 Å near that of the strong DIB at 4428 Å, indicating that our laboratory band is a product of reactions most likely involving carbon when methane is present along with argon. This reinforces the idea that the 4500 Å laboratory band is a good candidate for being a laboratory analog for the 4428 Å DIB.

The electrical discharge of argon, sans methane, does result in bands near the DIB wavelengths of 5778 Å and 6283 Å. In this case such bands are likely due to reactive species derived from trace atmospheric contaminants such as N_2 and O_2. A candidate for such a contaminant derived constituent is NO_3 which has been observed to have bands near the DIB wavelengths.[6,7] This result indicates that masking by bands of NO_3 derived from atmospheric contamination can be a serious problem for DIB analog generating experiments carried out with electrical discharges and ultraviolet irradiation. This includes experiments with carbon species other than methane, such as acetylene, carbon monoxide, etc. and inert gases different from argon. We are considering possible remedies including doing the experiment in an argon atmosphere as in a glove box. Also being considered as a source of contaminants is outgassing of the ion pump when it is turned off during the discharge because of the 0.5 torr pressure required for the discharge. We intend to install a valve between the ion pump and the vacuum shroud of the closed cycle refrigerator to remove this potential source.

Another solution is to utilize a sealed system having no O-ring seals or valves, and where the optical substrate is cooled with gaseous helium which in turn is cooled by the closed cycle refrigerator. Care would be taken to insure the gases utilized are contaminant free. Using a closed cycle refrigerator has an advantage over use of liquid helium in that a sample can be held indefinitely.

OTHER SPECIES

In addition to the spectral features described in our earlier papers[1,5] including those we have considered as candidates for being DIB laboratory analogs, the gas discharge/matrix isolation experiment produces other spectral features that are of astrophysical interest. As it turns out, their presence was not observed in our initial experiments because of operation of the tungsten light source at full power (80 W) in closer proximity to the sample than we do now. The existence in the matrix of other species became evident when we made measurements using a Cary 14 double beam spectrometer where monochromatic light is passed through the sample rather than polychromatic light, as was done in our single beam measurements. We observed a series of broad bands having a periodic alternation in strength at the wavelengths of 5028 Å (S), 5221 Å (LS), 5426 Å (S), 5647 Å (LS), 5903 Å (S), 6164 Å (LS), 6493 Å (S) and 6825 Å (LS), where S and LS indicate a sequence of "strong" and "less

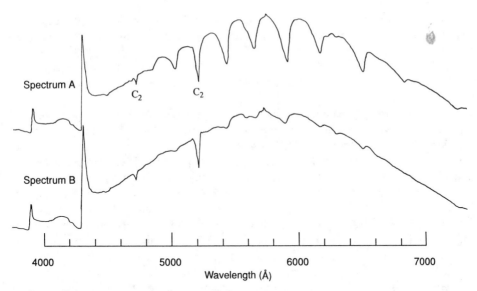

Figure 2. Spectrum A–single beam absorption spectrum of reactive species produced in a gas discharge of CH_4 (0.52%) and Ar (99.48%) and then matrix isolated at 13 K. Note sequence of HCO bands at 5028 Å, 5221 Å, 5426 Å, 5647 Å, 5903 Å, 6164 Å, 6493 Å, and 6825 Å along with C_2^- bands at 4730 Å and 5210 Å. Spectrum B–after photobleaching with visible light there is a pronounced decrease in the strength of the HCO bands while those of C_2^- remain unchanged.

strong". This sequence is most likely due to the HCO molecule[8] formed because of the presence of atmospheric oxygen as a trace contaminant. The reason for our not observing the presence of these bands during our initial experiments[1] became evident when we used a rapid scanning Rofin spectrometer capable of making ten spectra per second in the 220–850 nm range. The HCO bands which were at first very strong would weaken rapidly and disappear within several minutes. It was evident that the species responsible were being photobleached by the light of the tungsten lamp. The phenomenon was not unlike photobleaching of the C_2^- bands at 4730 Å and 5210 Å, except it envolved visible light rather than UV from a mercury vapor lamp as required in the C_2^- case. When the matrix was raised in temperature from 10 K to 35 K or greater the bands returned in original strength and the process of visible light photobleaching followed by thermal annealing could be repeated as many times as desired. Figure 2 shows first the spectrum of a matrix populated with reactive species (A). By operating the tungsten lamp at lower power and further away from the sample we are able to obtain a single beam spectrum with minimal photobleaching and can control photobleaching when it is desired. In our initial experiments, we missed seeing the HCO bands because in operating our lamp at elevated intensity for a period of time prior to a scan to insure a stable condition, we inadvertently bleached the sample. The permanence of the C_2^- bands under visible light irradiation is evident. Spectrum B of Figure 2 shows the effect of extensive visible light photobleaching on the sample initially having spectrum A. Note that the C_2^- bands remain, indicating the energy necessary to photobleach the other species including HCO is less than required to photoeject electrons from C_2^-.

CONCLUSIONS AND FUTURE WORK

In summary, we conclude that gas discharge/matrix isolation experiments can produce reactive species including one that is a good candidate for being a laboratory analog of the carrier of strong DIB at 4428 Å. These same experiments also produce other identified carbon based molecules such as CH, C_2^-, C_3, CNN, and HCO. Species such as HCO are of interest because of their populating dark HI regions and having a role in the chemistry and thermal balance of those regions. Reactive species such as NO_3 formed from trace atmospheric contaminants can be a problem for experiments done for the purpose of producing DIB carrier candidates in the laboratory because they have bands that cover wavelengths regions of interest to the DIB question. Identification of the species discussed remains a high priority. Laser desorption/mass spectroscopy done on the matrices populated with these interesting reactive species should be coupled to optical spectroscopy. Also turnable laser spectroscopy is expected to have utility.

This work was supported by NASA grant NAGW-749.

REFERENCES

1. T. J. Wdowiak, Astrophys. J. **241**, L55 (1980).
2. A. E. Douglas, Nature **269**, 130 (1977).
3. G. H. Herbig, Astrophys. J. **196**, 129 (1975).
4. W. Kratschmer, Chem. Soc. Faraday Trans. **89**, 2285 (1993).
5. T. J. Wdowiak, in *Solid State Astrophysics*, edited by E. Bussoletti and G. Strazzulla (Amsterdam, North Holland, 1991), p. 279.
6. E. J. Jones and O. R. Wulf, J. Chem. Phys. **5**, 873 (1937).
7. W. B. DeMore and N. Davidson, J. Am. Chem. Soc. **81**, 5869 (1959).
8. L. J. van IJzendoorn, *A Spectroscopic Study of Reaction Processes and Molecules in Ices of Astrophysical Interest*, Ph.D. Thesis, Univ. of Leiden (The Netherlands 1985).

THE FORMATION OF THE HYDROCARBON COMPONENT OF CARBONACEOUS CHONDRITES FROM INTERSTELLAR POLYCYCLIC AROMATIC HYDROCARBONS

Wei Lee and Thomas J. Wdowiak
University of Alabama at Birmingham, Birmingham, AL 35294, USA

ABSTRACT

The mid-infrared spectrum of a film-like deposit produced through plasma discharge of a gaseous mixture of hydrogen and the simplest polycyclic aromatic hydrocarbon (PAH), naphthalene ($C_{10}H_8$), is found to be almost identical to that of the benzene-methanol extract of the alkane-predominant hydrocarbon material of an organic-containing meteorite that fell in Murchison, Australia in 1969. This experiment that has evidently resulted in the synthesis of a laboratory analog for the hydrocarbon component of the Murchison carbonaceous chondrite suggests that the meteoritic material originated as PAH species formed in stars during their later stages of evolution. It also implies that the pathway from those formation sites to incorporation into the meteorite parent body probably involved hydrogenation of PAHs into alkanes in plasma of the solar nebula or in H II regions prior to the solar nebula. This model is in agreement with what is known about the meteoritic hydrocarbon including its deuterium to hydrogen (D/H) enrichment, the observed cosmic infrared emission bands known as the UIR and best attributed to PAH molecules, and the intrinsic stability of PAHs that allows their formation in stellar atmospheres and survival under the ultraviolet radiation and shock environment of the interstellar medium.

INTRODUCTION

Organic matter has been found to exist in many apparently primitive chondrites as well as interplanetary dust particles, and seems to be present also in comets, numerous asteroids, and some planetary satellites. Since the organic matter in carbonaceous chondrites (and by inference that in other objects) is believed to have been synthesized abiotically in the forming solar system, investigation of the meteoritic organic matter may reveal the early history of the solar system and aspects related to the origin of terrestrial life. The question of origin of meteoritic organic matter is now posed in the context of the physicochemical processes that are associated with the collapse of an interstellar cloud leading to the formation of a star along with planetary system. The corresponding answer requires considerations of possible formation sites, such as the dense interstellar clouds, the solar nebula, and the surface regions of planetesimals, and of chemically synthetic pathways. Hypotheses suggested for the origin of hydrocarbon compounds in carbonaceous chondrites are restricted by their ability to account for both the molecular and isotopic composition of the compounds and by their general agreement with meteoritic petrology; some laboratory models of proposed processes are evaluated with respect to their ability to form the hydrocarbon species with the meteoritic characteristics of the specific molecular constituents and isotopic abundances. A review of the origin question with emphasis on isotopic analyses has been provided by Mullie and Reisse.[1] The determination of H-isotope ratios for macromolecular carbon and, more recently, for a few soluble hydrocarbon compounds has provided evidence for a genetic relationship between meteoritic organics and interstellar molecules.[2,3] In this paper, we report the synthesis of an exceptional laboratory analog of the hydrocarbon component of the Murchison meteorite, a prototypical CM chondrite, under experimental conditions that simulate those of stellar atmospheres or the solar nebula.[4] A model for the origin of the hydrocarbon component of carbonaceous chondrites resulting from this experiment is discussed as well.

EXPERIMENTAL

The laboratory analog of the hydrocarbon component of the Murchison meteorite is produced when a gaseous mixture of hydrogen and the simplest PAH, naphthalene, is excited by a neon sign transformer in a discharge tube. Fig. 1 schematically shows the experimental configuration. A sapphire tube whose inner

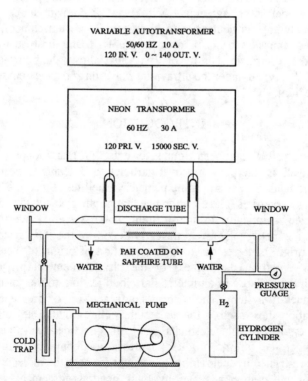

Fig. 1. Experimental apparatus configured for the laboratory synthesis of the hydrocarbon component of the Murchison meteorite.

wall is coated with naphthalene is inserted into the discharge tube to a position halfway between the two electrodes. Hydrogen gas is introduced through one of the small side arms. A mechanical pump holds the pressure at about 0.5 torr as measured with a McLeod gauge. Prior to the introduction of the hydrogen, the composition of the gas mixture of the pumpdown pressure of 0.05 torr is estimated to be 90% of H_2, 8% of N_2, and 2% of O_2. The two latter components are due to backstreaming of atmospheric gases from the pump. A thorough description of the experimental apparatus and and the production of the deposit is reported elsewhere.[4]

The samples are removed from various positions of the discharge tube, washed in benzene (and methanol), or soaked in a benzene-methanol (9:1) solution, and then mounted on a KBr or NaCl disk for infrared measurements. IR spectra are obtained with a Mattson Polaris FT-IR spectrometer, operated at a resolution of 4 cm^{-1} and calibrated using a polystyrene film. For most of the experimental runs, liquid nitrogen is boiled with a ceramic resistor (1Ω, 10 W) for dry nitrogen purge. The nitrogen purge removes atmospheric water vapor and carbon dioxide from the optical compartment of the spectrometer. The input power to the boiler is 100 W providing a purging rate estimated to be 0.4 L/s at room temperature.

RESULTS AND DISCUSSION

Spectrum A in Fig. 2 shows the mid-IR spectrum of the film-like deposit removed from the water-cooled discharge tube surface facing one of the electrodes

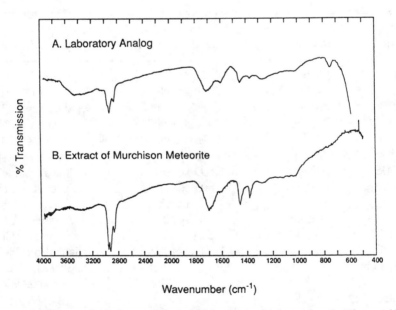

Fig. 2. Comparison of A) the spectrum of laboratory hydrocarbon analog and B) the spectrum of the hydrocarbon component of the Murchison meteorite obtained by Cronin and Pizzarello.[5] Figure taken from ref. 4.

(gas down-stream side), washed in benzene, dried under vacuum, and mounted on a NaCl crystal having a transmission cutoff at 600 cm^{-1}. The band assignments of the infrared features of the spectrum are given in Table I and are based on refs. 6—10. Spectrum B in Fig. 2 is of the hydrocarbon component extracted with benzene-methanol (9:1) mixture as reported by Cronin and Pizzarello,[5] and is considered by those in the field to be the spectrum to match because of the effort expended in obtaining a contamination-free sample. It is obvious that, except for a few minor features in our spectrum which are absent in the Murchison spectrum and can be easily account for by attributing to the naphthalene precursor and its derivatives (see Table I for those attributed to aromatics), detailed comparison of the two spectra indicates that there is a remarkably good match. The overall character of the analog spectrum shows that not only are the same functional groups present as in the meteoritic material, they are present in the same general proportions. The shape of the curve particularly in the "fingerprint region" of 1600—1800 cm^{-1} shows the assemblage of the functional groups into a general molecular configuration is the same in the analog as in the meteoritic material. Our laboratory analog, like the predominant component of the meteoritic organic matter, contains aliphatic alkanes as signatured at 2960 and 2870 cm^{-1} due to —CH$_3$ and at 2925 and 2860 cm^{-1} due to —CH$_2$—. The existence of both —CH$_3$ and —CH$_2$— groups in our residue indicates the rupture and saturation processes of the ring(s) of the naphthalene have taken

Table I. Assignment of absorption features in IR spectrum of the Murchison analog.

Wavenumber (cm^{-1})	Intensity	Classifications	Functional Groups
3453	Medium	O-H stretch	Alcohols
3060, 3030	Weak	C-H stretch	Aromatics
2960	Strong	C-H stretch	Alkanes (-CH$_3$)
2925	Strong	C-H stretch	Alkanes (-CH$_2$-)
2870	Strong	C-H stretch	Alkanes (-CH$_3$)
2860	Strong	C-H stretch	Alkanes (-CH$_2$-)
1706	Strong	C=O stretch	Ketones or aldehydes
1600	Medium	C=O stretch C-C stretch	Ketones or aldehydes Aromatics
1454	Medium	C-H bend	Alkanes (-CH$_2$-)
1377	Weak	C-H bend	Alkanes (-CH$_3$)
1270	Weak	O-H or C-H bend	Alcohol or aldehydes
1163, 1116		C-O or C-C stretch	Alkanes, alcohols, or aldehydes
1030	Weak		
755	Medium	C-H out-of-plane bend	Aromatics

place. There is also a very broad absorption band in both spectra at 1706 cm^{-1} probably due to the C=O stretch in the carboxyl group. We consider the appearance of this functional group in our laboratory analog is a consequence of the atmospheric oxygen in the gas mixture as discussed earlier. The meteorite extract has been found to contain aromatic ketones and several series of alkyl aryl ketones.[3]

Our experimental result demonstrates that a polycyclic aromatic hydrocarbon hydrogenated in ionized hydrogen results in a material with a characteristic IR spectrum which is an excellent match to that of the meteoritic material. It is a highly positive and repeatable result providing insight into a process consistent with the astrophysics involved. The implication best drawn is that our electrical discharge experiment, involving gaseous hydrogen and naphthalene, produced a laboratory analog which is almost identical to the hydrocarbon component of carbonaceous chondrites, suggesting that the complex hydrocarbon matter in the solar system, such as found in carbonaceous chondrites and chondritic interplanetary dust particles,[11] may have originated from a mixture of PAH molecules formed in stellar atmospheres. It has been proposed that PAH molecules exist in the interstellar medium because of the close correlation of their general IR spectral characteristics with the observed celestial IR emission bands known as the UIR at 3.3, 6.2, 7.7, 8.6, and 11.3 μm (3040, 1615, 1310, 1150, and 885 cm^{-1}). These five features, usually observed as a "family" in a wide variety of celestial objects, were first discovered by Gillett, Forest, and Merrill,[12] and have been interpreted as being due to the presence of aromatics.[13-16] The aromatic structure of PAHs also provides the level of stability necessary to survive in the harsh interstellar environment, characterized by high-energy UV photons and energetic shock waves. Before their eventual incorporation into the parent bodies of carbonaceous chondrites, some PAH species, as suggested by this experiment, can become alkanes via hydrogenation in a plasma such as the branched-alkyl substituted mono-, di-, and tricyclic alkanes found to be the predominant aliphatic compounds in the Murchison meteorite.[5] The T Tauri stage of the sun was probably an important factor in the production of the necessary plasma conditions. The idea of an early solar system conversion of aromatic to aliphatic compounds has been recently anticipated on the basis of the deuterium enrichments in the Murchison organic compounds.[3] The fact that the aromatic component of the Murchison contains greater deuterium abundances than the aliphatic does can be explained, according to our experiment, by some of the aromatic interstellar molecules having escaped hydrogenation.[4]

CONCLUSION

The experiment and the relevant astrophysical considerations suggests a model for the pathways that lead to the presence of the complex hydrocarbon material found in carbonaceous chondrites:

(1). PAHs are born in the envelopes (hot stellar atmospheres) of carbon stars,[17-19] and then are ejected (R CrB, planetary nebulae, etc.) into the interstellar medium where they survive interstellar transport;

(2). interstellar PAH species then become deuterated[20-22] in dark clouds;

(3). they are hydrogenated partially into alkanes in a plasma in the solar nebula or in H II regions prior to the solar nebula; some of the PAH molecules may escape the process and retain their interstellar characteristics; and

(4). they are finally incorporated into the parent bodies of carbonaceous chondrites.

We thank John Cronin, Lou Allamandola, and NASA grant NAGW-749.

REFERENCES

1. F. Mullie and J. Reisse, *Topics in Current Chemistry* (Springer-Verlag, Berlin, 1987), Vol. 139, p. 85.
2. E. Zinner, in *Meteorites and the Early Solar System*, edited by John F. Kerridge and Mildred S. Matthews (Univ. Arizona Press, Tucson, 1988), p. 956.
3. R. V. Krishnamurthy, S. Epstein, J. R. Cronin, S. Pizzarello, and G. U. Yuen, Geochim. Cosmochim. Acta **56**, 4045 (1992).
4. W. Lee and T. J. Wdowiak, Astrophys. J. Lett., in press.
5. J.R. Cronin and S. Pizzarello, Geochim. Cosmochim. Acta **54**, 2859 (1990).
6. L. J. Bellamy, *The Infrared Spectra of Complex Molecules* (John Wiley & Sons, N. Y., 1975)
7. H. A. Szymansky, *Infrared Band Handbook* (Plenum, N. Y., 1963)
8. D. J. Pasto and C. R. Johnson, *Organic Structure Determination* (Prentice-Hall, Englewood Cliffs, 1969)
9. N. B. Colthup, J. Opt. Soc. Am. **40**, 397 (1950).
10. G. J. Shugar and J. T. Ballinger, *Chemical Technician's Ready Reference Handbook*, 3rd ed. (McGraw-Hill, N. Y., 1990), p. 749.
11. See, for example, S. J. Clemett, C. R. Maechling, R. N. Zare, P. D. Swan, and R. M. Walker, Lunar Planet. Sci. **24**, 309 (1993).
12. F. C. Gillett, W. J. Forrest, and K. M. Merrill, Astrophys. J. **183**, 87 (1973).
13. W. W. Duley and D. A. Williams, Mon. Not. R. Astron. Soc. **196**, 269 (1981).
14. A. Léger and J. L. Puget, Astron. Astrophys. **137**, L5 (1984).
15. L. J. Allamandola, A. G. G. M. Tielens, and J. R. Barker, Astrophys. J. **290**, L25 (1985).
16. A. Blanco, E. Bussoletti, and L. Colangeli, Astrophys. J. **334**, 875 (1988).
17. R. Keller, in *Polycyclic Aromatic Hydrocarbons and Astrophysics*, edited by A Léger, L. d'Hendecourt, and N. Boccara (Reidel, Dordrecht, 1987), p. 387.
18. M. Frenklach and E. D. Feigelson, Astrophys. J. **341**, 372 (1989).
19. I. Cherchneff, J. R. Baker, and A. G. G. M. Tielens, Astrophys. J. **401**, 269 (1992).
20. J. F. Kerridge, S. Chang, and R. Shipp, Geochim. Cosmochim. Acta **51**, 2527 (1987).
21. L. J. Allamandola, A. G. G. M. Tielens, and J. R. Barker, Astrophys. J. Suppl. Ser. **71**, 733 (1989).
22. P. Ehrenfreund, F. Robert, L d'Hendecourt, and F. Behar, Astron. Astrophys. **252**, 712 (1991).

IRON AROMATICS COORDINATION:
ION TRAP EXPERIMENTS ON Fe^+ $(C_{10}H_8)_n$ COMPLEXATION

P. Boissel
Laboratoire de Photophysique Moléculaire du CNRS,
Bat. 213, Université Paris Sud   91405 Orsay Cedex, FRANCE.

ABSTRACT

This paper reports the first experimental observation of gas phase metal aromatics coordination. Using an ion trap technics, formation and destruction of $Fe(C_{10}H_8)_n^+$ complexes have been studied. These complexes are shown to be stable under collision free conditions. Orders of magnitude for the formation cross section and the dissociation barrier can be drawn from this preliminary experiment.

INTRODUCTION

The possible importance of organometallic chemistry in the interstellar medium (ISM) has been pointed out in a recent paper[1]. In this work, coordination reactions between iron, neutral or ionised, and polycyclic aromatic hydrocarbons (PAHs) were considered. Owing to the predicted abundance of PAHs, such reactions could explain the depletion of atomic iron in the ISM. They could also be important in the accretion of metal particles and for the condensation of PAHs into carbonaceous grains. The kinetics of these processes in molecular clouds has been studied[2]. Concerning the ISM chemistry, complexation of transition metal ions with large aromatics molecules could lead to very efficient catalysis of gas phase reactions. In such complexes, the metal brings its reactivity and PAH provides the thermal bath to stabilize reaction products. Owing to the small size of these entities (compared to grains), a single excitation by an UV photon could be sufficient to release the processed products. These mechanisms might then play a major role, especially for reactions difficult to produce in a simple two body collision. They ought to be taken into account in the study of the ISM chemistry.

However, the relevant quantities concerning the complexation reactions: bond energies, cross sections, activation barriers,... are only derived from calculations[3], or through analogies with condensed phase studies. Up to now, even the existence of iron PAH complexes in the gas phase was hypothetical. This led us to try to observe them in our ion trap apparatus.

EXPERIMENTAL PROCEDURE

The core of our experimental set-up is an ion trap, which has been specifically devised to study medium sized ions of astrophysical interest, in an environment as near as possible as that encountered in the ISM. In this sense, the main interest of an ion trap is to provide high isolation conditions, allowing the study of reactive species. This device has been already used to study the photofragmentation of PAH cations[4]. It will be described in a forthcoming paper[5].

Ions are confined through the conjugated action of a magnetic field and a static quadrupolar potential (Penning trap). The mass spectrum of the ions that are present at a given instant can be obtained by Fourier Transform Mass Spectroscopy (FT-MS) with impulse excitation[6]. The cell is similar to an Ionic Cyclotronic Resonance (ICR) cell, but, in our design, larger openings are provided by the use of rods instead of plates for the transverse electrodes (figure 1). This allows the admission of neutral reactants and the irradiation of the ion cloud through different ports.

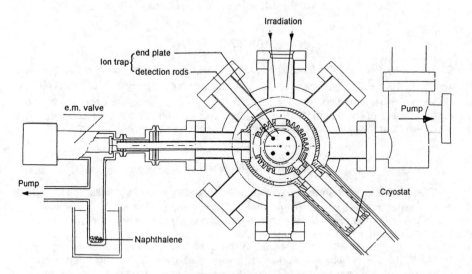

Figure 1: Experimental set-up

An important point, in the present experiment, is the use of cryopumping. The whole device, placed in an UHV quality vacuum cell, is attached to a closed cycle helium cryostat. The temperature of the different parts ranges from 12 K to 30 K. The density of neutral species remaining is very low, allowing long trapping times. Coronene ions, with mass of 300 amu, can be stored for more than ten minutes. For lighter ions, the efficiency is even better. An initial number of 10^6 potassium cations is only reduced by 50% in the same time.

Ions are produced via laser ablation of a solid target, located on the trap axis, behind one end plate. This technics, used to obtain PAH cations[4,5], also works to get iron cations, provided the laser is focussed somewhat tighter. A few millijoules of the fourth harmonics of a Nd/Yag laser (266 nm), focussed with a 300mm lens, are sufficient to saturate the trap with Fe^+ ions. This saturation, due to space charge effect, leads to very reproducible initial conditions from shot to shot.

In a preliminary try to form metal PAH complexes, a composite target was used. The laser, focussed on the metal, evaporated simultaneously the aromatics species. In this way, positively charged adducts of anthracene with iron or copper have been produced. However the amounts obtained were very low and the results hardly reproducible. Then, in the present study, another procedure is used. Complexes are formed through gas phase reactions of metal ions with neutral PAHs. This procedure leads to reproducible results, and it is more relevant for the study of the ISM problem. Up to now, Fe^+ naphthalene complexation has been studied. Iron has been chosen for its astrophysical interest. Naphthalene is the first polycyclic aromatic species, and is easy to handle.

Naphthalene vapor is admitted in the trap through an electromagnetic valve. Owing to the cryogenic pumping, it is immediately condensed on the walls. Short pressure pulses can then be obtained, the pressure falling rapidly after the valve closure. Opening times used range from 0.2s to a few seconds. The stagnation pressure can be fixed by the temperature of a cold arm. Regulation between 0 and 20°C leads to vapor pressures in the range 10^{-2} to 10^{-1} mbar.

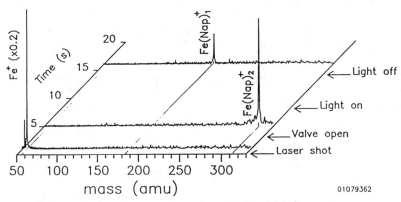

Figure 2: Formation and fragmentation of $Fe(C_{10}H_8)_n^+$ complexes.

The protocol for one run is sketched in figure 2, in the case of a representative experiment. At time 0, Fe^+ ions are produced by the laser shot and a first mass spectrum is taken. A few seconds after, the electromagnetic valve is open for a short time. When it is closed again, a second mass spectrum controls the nature and the amount of products obtained. In the case of figure 2, $Fe(C_{10}H_8)_2^+$, called below complex 2, is only formed and no remaining Fe^+ ions are seen. Now, the main advantage of the ion trap technics is that, once detected, the products are not lost and can be subsequently studied. In the run shown, for instance, $Fe(C_{10}H_8)_2^+$ has been irradiated for a few seconds and, in the last mass spectrum $Fe(C_{10}H_8)_1^+$ (complex 1) appears, as a photodecomposition product. These different steps: formation, storage and decomposition of the complexes are detailed in the following sections.

FORMATION OF THE COMPLEXES, Fe^+ - $C_{10}H_8$ REACTIONS

Reactions of iron cations with neutral naphthalene can lead to different products, depending on the parameters of vapour admission in the trap. A summary of these products can be seen in figure 3.

Concerning the complexation, when the stagnation pressure of naphthalene is low, and when the electromagnetic valve is opened for a short time, monoadducts $Fe(C_{10}H_8)_1^+$ (mass 184 amu) are only formed (figure 3a). The peak around 440 amu, which could correspond to $Fe(C_{10}H_8)_3^+$ has not been reproduced consistently and is probably an artefact. On the other hand, for higher stagnation pressure or longer opening times, $Fe(C_{10}H_8)_2^+$ (312 amu) is the dominant product (figure 3b). For still higher pressures, $Fe(C_{10}H_8)_1^+$ is no longer observed (figure 2). This indicates that $Fe(C_{10}H_8)_2^+$ is formed by addition of a second naphthalene molecule on the monoadduct, rather than by a direct three body reaction.

A peak at mass 128, corresponding to the naphthalene cation is also observed. It is not possible to distinguish wether this charge exchange is a direct process or if the PAH cation results from decomposition of a $Fe(C_{10}H_8)^+$ complex. In any case this charge transfer leads to a neutralisation of the iron cation that can be more efficient than the radiative recombination. This charge transfer reaction has been also observed with pyrene molecule.

The peak at 56 amu monitors the amount of remaining iron cations, those which have not been involved in a reaction or been ejected from the trap. This peak is only observed when small quantities of vapour are admitted. In the conditions where $Fe(C_{10}H_8)_2^+$ is formed, no iron ion remains.

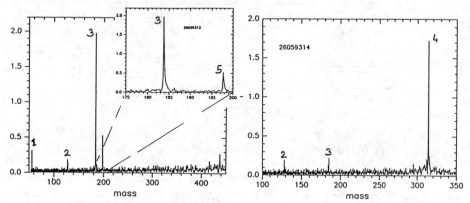

Figure 3: Products obtained in two different experiments. Peaks labels are:
1: Fe^+, 2: $C_{10}H_8^+$, 3: $Fe(C_{10}H_8)^+$, 4: $Fe(C_{10}H_8)_2^+$, 5: $N-Fe-(C_{10}H_8)^+$

In the first experiment, a peak at 198 amu has been obtained, in a reproducible manner, each time complex 1 was formed (figure 3a). This mass might correspond to a ternary complex $NFe(C_{10}H_8)^+$, formed with nitrogen originating from a leak in the naphthalene vessel. More controlled experiments are currently in progress on this point. Observation of this complex would constitute the first step towards the study of catalytic gas phase reactions.

No quantitative measurement of the parameters concerning the complexation has been attempted in this preliminary experiment. Nevertheless, estimates can be drawn from qualitative remarks. The reaction cross section is comparable to the elastic (velocity changing) cross section. Indeed, iron ions trajectories are much affected by the encounters with heavier naphthalene molecules. These ions would then be lost out of the trap after a few collisions if they were not captured to form a complex. Concerning a possible activation barrier for the reaction, the kinetic energy available in the collision cannot be very high. Naphthalene vapor is at room temperature. For iron, the mean velocity of ions in a trap is not easy to appreciate. However, no significant difference in the reaction efficiency has been noticed when Fe^+ cyclotron motion is excited during the vapour admittance. In our usual conditions, this excitation corresponds to an extra kinetic energy of about 0.3 eV. It can then be assumed that an activation barrier, if it exists, cannot exceed 0.5 eV.

LIFETIMES OF THE COMPLEXES IN THE TRAP

Information about the stability is drawn from the time evolution of the mass spectrum. In this way, complex 1 is observed for a few tens of seconds, and complex 2 for more than one minute. These adducts are then true stable species. However, their disappearance is not only due to an escape out of the trap. This natural loss of the trap can be accounted for through a comparison with the decays observed with ions of equivalent mass: anthracene (178 amu) and coronene (300 amu) for complex 1 and 2. The complexes disappear faster, indicating that a kind of decomposition occurs. This self decomposition cannot be due to the energy stored from the initial formation, the observed lifetimes being much longer than the radiative cooling time. Energy from an external source has to be brought in to cause the dissociation. It might come from the collisions with remaining neutral species. Although these collisions are not frequent, energy can be stored between them, owing to the low efficiency of radiative cooling. The gradual internal energy increase leads finally to fragmentation.

Some points are still unexplained in this mechanism. It is not clear, for instance, why the charged fragment resulting from the decomposition is generally not observed. Another question is to determine if the larger stability observed for complex 2 is due to a higher dissociation barrier or merely to a size effect. Experiments are currently in progress to check these points. Anyway, we can conclude that, in the collision free conditions encountered in the ISM, these complexes are stable and can subsist for long times.

PHOTODECOMPOSITION OF $Fe(C_{10}H_8)_2^+$

Photodissociation is a convenient technics to obtain information on a new species. Owing to the long storage times in the ion trap, it can be performed with a simple apparatus only using a cw xenon arc lamp and colored filters[4]. However, in the present case this process is in competition with collisions induced dissociations and the results are not easy to analyse. Then, $Fe(C_{10}H_8)_2^+$, which is more stable, is only studied here.

The experimental procedure has been described in figure 2. To account for the ions escape from the trap and for the collisions induced dissociations, the two mass spectra analysing the content of the trap before and after irradiation are taken at definite instants. Irradiation with different conditions and duration is performed between these two instants, and the result is compared with reference runs during which nothing is done.

The first observation is that, using the same light intensity, the dissociation of the complex is much faster than that of PAH cations of comparable size, anthracene or pyrene[4]. This could be due to a lower dissociation barrier or to a stronger absorption in the visible range. However, the only fragment obtained is $Fe(C_{10}H_8)^+$, resulting from the detachment of a neutral naphthalene. It can then be concluded that intermolecular bonding is weaker than intramolecular ones. A first upper limit on the dissociation barrier for the Fe-PAH bond can then be estimated from the easiest fragmentation path of the naphthalene cation. A value of 3.5 eV has been found for the reaction corresponding to the ablation of a C_2H_2 molecule[7].

The evolution of the fragmentation yield with the exposure duration has been measured, with an irradiation limited to wavelengths longer than 570 nm (figure 4a). An exponential decay of the parent species is observed as soon as the irradiation begins. The intensity dependance of the decay rate seems linear. Qualitatively, this indicates that the fragmentation is obtained for energies comparable to the photon energy, since no accumulation time is needed. More quantitatively, we have tried to reproduce these results with the simulation program developed to study PAHs fragmentation. Although the parameters used in this calculation are not optimized, it leads to a crude evaluation of the dissociation barrier in the range 1 to 2 eV.

Using a set of long wave pass colored filters, the spectral dependence of the dissociation rate has been studied (figure 4b). Assuming that the linear intensity dependence observed is valid on the whole spectral range, this dissociation rate can be used as a measure of the total power absorbed for a given irradiation. It must then decrease when, following an increase in the filter cut off wavelength, an absorption line is left outside the irradiated region. A continuous decrease is indeed observed from 280 to 700 nm, showing that absorption extends over the whole visible range. In the limits of our experimental errors, this decrease is smooth, indicating that this visible absorption is an unstructured continuum. Such a broad, unstructured continuum is generally observed for dimer cations[8]. It is attributed to the resonant charge exchange between the two identical species. Here, this charge exchange seems to occur, despite the presence of iron between the two naphthalene.

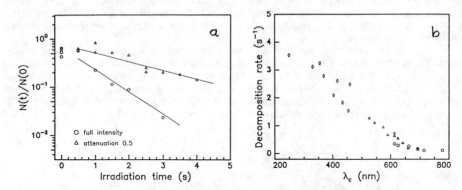

Figure 4: Photofragmentation of $Fe(C_{10}H_8)_2^+$:
a: Time and intensity dependence;   b: Spectral evolution.

CONCLUSION

This experiment demonstrates the existence of stable iron-PAHs adducts. Qualitative results obtained in this preliminary study have been used to get a first estimate of the relevant parameters. Concerning the complex formation, the cross section is comparable to the elastic cross section, and the activation barrier, if it exists cannot be higher than 0.5 eV. Concerning the binding energy, the barrier to dissociation of $Fe(C_{10}H_8)_2^+$ is in the range 1 to 2 eV.

ACKNOWLEDGEMENTS

The experiments described here were performed using the set-up built with the help of G. Lefevre. Obtaining these results and exploiting them over a short lapse of time has been possible owing to of a group of students: F. Amy, B. Grenier, R. Hackert, M. Rouaud and S. Odin whose efficient participation is acknowledged here.

REFERENCES

1 G. Serra, B. Chaudret, Y. Saillard, A. Le Beuze, H. Rabaa, I. Ristorcelli and A. Klotz
   Astron. Astrophys. **260**, 489, (1992)
2 P. Marty, G. Serra, B. Chaudret and I. Ristorcelli, Contribution   in this meeting.
   P. Marty, G. Serra, B. Chaudret and I. Ristorcelli, to be published in Astron. Astrophys.
3 B. Chaudret, A. Le Beuze, H. Rabaa, J.Y. Saillard and G. Serra,   New J. Chem **15**, 791, (1991)
4 P. Boissel, G. Lefèvre and P; Thiébot.   Contribution 14 in this meeting.
5 P. Boissel and G. Lefèvre, in preparation
6 R.T. McIver Jr., R.L. Hunter and G. Baykut, Rev. Sci. Instr. **60**, 400, (1988).
7 E. Ruhl, S. D. Price and S. Leach, J. Phys. Chem. **93**, 6312, (1989).
8 G.P. Smith and L.C. Lee, J. Chem. Phys. **69**, 5393, (1978)

EVALUATION OF THE ROLE OF ORGANOMETALLIC SPECIES IN THE CHEMISTRY OF INTERSTELLAR AND CIRCUMSTELLAR MEDIA

Alain KLOTZ[1,2,3], Isabelle RISTORCELLI[1], Dominique de CARO[2], Guy SERRA[1], Bruno CHAUDRET[2], Jean-Pierre DAUDEY[3], Martin GIARD[1], Jean-Claude BARTHELAT[3], Philippe MARTY[1]

1- Centre d'Etude Spatiale des Rayonnement du CNRS, UPR 8002, associée Université Paul Sabatier, Toulouse, France.

2- Laboratoire de Chimie de Coordination du CNRS, UPR 8241, associée Université Paul Sabatier et Institut National Polytechnique, Toulouse, France.

3- Laboratoire de Physique Quantique, URA CNRS 505, associée Université Paul Sabatier, Toulouse, France.

ABSTRACT

Interactions between metal atoms or ions and unsaturated organic molecules in the interstellar or circumstellar media could play an important role in the chemistry of these media.

Ab-initio calculations were performed to predict bonding energy of the cations [metal-benzene]$^+$ and [Mg-pyrene]$^+$. We showed that the higher bonding energy of [metal-benzene]$^+$ is, the larger is the metal depletion. This result should be extended to bigger PAHs. We conclude that organometallic chemistry can participate to the depletion of metals.

We synthetised grains composed of small cristallized cores of metal (diameter from 20Å to 200Å) embedded in pyrene. the mean diameter of grains depends of the ratio pyrene/metal.

Finaly, we showed that organometallic reactions can be catalytic. Large organic molecules could be synthetised faster in the ISM through such this process than through classical collisions. We propose that the first ring of PAH molecules could be synthetised in this way.

INTRODUCTION

The unidentified infrared emission bands (UIBs) are observed in a wide variety of astrophysical places: planetary nebulae, reflection nebulae, HII regions, external galaxies and the diffuse galactic interstellar medium.

If these bands are due to the vibrational modes of polycyclic aromatic hydrocarbons molecules (PAHs), as proposed by Léger and Puget (1984), then these molecular species are widespread and have to be considered in the study of the physics and chemistry of the interstellar medium.

Transition metals have d shell electrons and can bind to aromatic molecules like PAHs by coordination bonds.

The Red Rectangle shows a strong PAHs infrared emission and a very pronounced depletion of iron. Favorable conditions seem present in this object to invoke molecules composed of iron and PAHs. Another object, IRC+10216, shows in some shells a large abundance of acetylene that can react with metals producing a depletion of such species as observed in other areas.

Our purpose is to know whether it is important to take into account organometallic reactions in interstellar chemistry.

Previous papers on this subject have already been published by Böhme et al.(1989), Chaudret et al.(1991), Serra al.(1992) and Marty et al.(1993). In this paper we present ab-initio calculations on the familly of molecules [metal-arene]$^+$, synthesis of aggregates that contains metals and PAHs and we emphasize the importance of catalytic role of metals.

THEORETICAL CHEMISTRY Study of bonding energy of cationic organometallic molecules [metal-benzene]$^+$

These complexes can be computed by ab-initio methods. The calculation, at self consistent level, of bonding energy and the optimisation of the geometrical structure were performed for aluminium, magnesium and zinc. The bonding energies for other complexes are known from mass spectroscopy or calculations and have been gathered from the litterature.

We have plotted on figure 1, the depletion of the metals observed in interstellar space versus the bonding energy of the [metal-benzene]$^+$ cation.

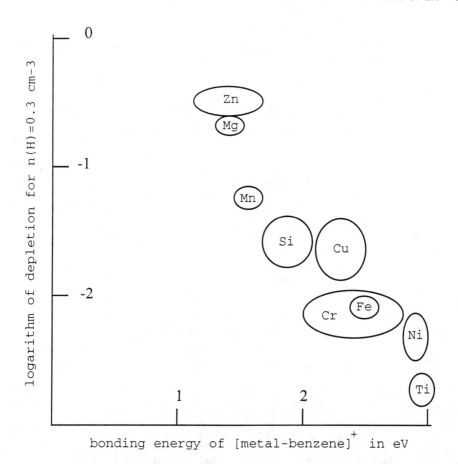

Figure 1: observed interstellar depletion (from Jenkins 1987) of metals versus bonding energy of complexes. The cases of Mg and Zn were calculated. Fe and Ni datas from Hettich et al.(1986). Cu and Cr datas from Hettich et al.(1987). Mn and Ti data from Bauschlischer et al.(1992). Si data from Srinivas et al.(1992).

The correlation between the mean depletion of interstellar free atoms or ions and the bonding energy of [metal-benzene]+ complexes shows that the most stable complexes correspond to the most depleted atoms. This result supports the idea that organometallic interstellar chemistry may occur in interstellar space and could contribute to the depletion of metals.

THEORETICAL CHEMISTRY
Study of cationic fragment [Mg-pyrene]$^+$

There are 2 kinds of cycles in pyrene: external and internals rings. The calculation, at self consistent level, of bonding energy and optimisation of geometrical structure was performed

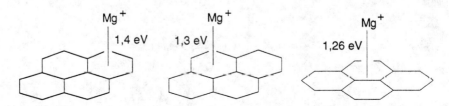

Figure 2: Conformations of magnesium cation approaches to obtain minima of energy.

Three equilibrium structures were found. The more stable, corresponds to the Mg$^+$ approach above the centre of one external ring (bonding energy of 1.4 eV). The second stable conformation corresponds to the Mg$^+$ approach above the centre of one internal ring (bonding energy of 1.3 eV).

The third structure corresponds to the Mg$^+$ approach above the double bond of the centre of pyrene but this is a saddle point (metastable conformation).

These equilibrium structures show poor charge transfer (about 90% of charge remains on magnesium). This implies an electrostatic interaction that explain the small difference of bonding enegy between saddle point and center of rings. The study of charge transfer, out of equilibrium, is more complicated. Preliminary results seem to show that, if an activation barrier exist, it is less than 0.2 eV.

The cationic metal coordination induces a deformation of the PAHs structures. In the cases we studied, the external hydrogen atoms bend out of the carbon plane between 1 to 10°. This fact should modify the γCH vibration of PAHs. Quantitative calculations will be done.

Mg$^+$, which shows a similar electrophillicity as Fe$^+$, has been chosen as a first model of metal PAH interaction. This is particularly valuable for the calculation of the activation barrier to coordination. However, a new pseudo potential for Fe$^+$ will be computed to allow the same calculations for the [Fe-pyrene]⁺ fragment and obtain precise values of the bonding energy.

EXPERIMENTAL CHEMISTRY

Aggregation of metals and PAHs could produce one component of interstellar grains. Laboratory production of model material can reveal the nature and properties of these hypothetical grains

Thermodynamical aggregations were carried out in liquid phase of mixture of transition metal complexe and PAHs. One can obtain homogeneous small grains (about 50Å diameter) probably composed of a crystalised core of metal embedded in PAHs (further analysis are still necessary). In case of small PAH concentrations, big grains are obtained as aggregates of small ones.

It may be possible to obtain thermodynamicaly metastable phases through liquid phase ultrasonic reactions, presently under investigation.

TOWARD A CATALYTIC ASTROCHEMISTRY...

Organometallic reactions can produce complicated organic molecules by catalysis phenomena. The following scheme shows that a catalysis reaction in three steps can be faster than a classical reaction in one step.

However, when the last step is rapid, concentration in organometallic intermediates may be small. This fact implies that organometallic species may not be observable even if they exist! This is generaly the case in well-known catalytic processes

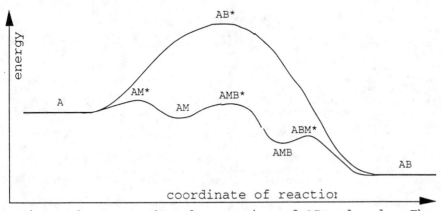

Figure 3: Two paths of creaction of AB molecule. The upper path is a classical collision. It shows a high activation barrier. Lower path is catalytic. Activation barriers are much lower.

Free metal atoms, as well as metal grains if they exist, can participate to catalytic reactions. Free atoms are more abundant than grains, so they are able to increase significantly the rate of creation of organic species.

In the interstellar medium, although metal atoms are not very abundant, compared to H, C, N and O, organometallic reactions may play an important role and may be more efficient than classical organic reactions.

For example, if it is admitted that large PAH structures exist in the ISM (PAHs, coal, graphite, cumulene, etc...) the problem of their synthesis is not solved. Reasonable pathways of multiple ring formation from an initial benzene molecule have been proposed but the formation of the simple benzene molecule is still questionnable (Cherchneff et al. 1992). A cyclotrimerization reaction such as the one described hereafter would thus look attractive. Precedents for such gas phase reactivity has been reported by Schröder et al. (1992). We are presently evaluating the possibilities of existence of such catalytic reactions in the ISM.

$$Fe + C_2H_2 \underset{h\nu}{\rightleftharpoons} Fe(C_2H_2)$$

$$Fe(C_2H_2) + C_2H_2 \underset{h\nu}{\rightleftharpoons} Fe(C_2H_2)_2$$

$$Fe(C_2H_2)_2 + C_2H_2 \longrightarrow Fe(C_2H_2)_3$$

$$Fe(C_2H_2)_3 \overset{h\nu}{\longrightarrow} Fe + C_6H_6$$

BIBLIOGRAPHY

C.Bauschlischer, H.Partridge and S.Langhoff (J. Phys. Chem., 96, 1992, 3273-3278).
D.Böhme, S.Wlodek and H.Wincel (Astrophys. J., 342, 1989, L91)
B.Chaudret, A.Le Beuze, H.Rabaa, J.Y.Saillard and G.Serra, (New J. Chem, 15, 1991, 791-794)
I.Cherchneff, J.Barker and A.Tielens (Astrophys. J., 401, 1992, 269-287)
E.Jenkins ("interstellar Processes", eds. D.J.Hollenbach and H.A.Thronson, 1987, 533-539)
R.Hettich, T.Jackson, E.Stanko and B.Freiser (J.Am.Chem.Soc., 108, 1986, 5086-5093).
R.Hettich and B.Freiser(J.Am.Chem.Soc., 109, 1987, 3537-3542).
A.Léger and J.L.Puget (Astron. & Astrophys., 137, 1984, L5-L8)
P.Marty, G.Serra, B.Chaudret and I.Ristorcelli (Astron. & Astrophys., 1993 in press)
G.Serra, B.Chaudret, J.Y.Saillard, A.Le Beuze, H.Rabaa, I.Ristorcelli and A.Klotz (Astron. & Astrophys., 260, 1992, 489-493).
D.Schröder, D.Sülzle, J.Hrusak, D. Böhme, H.Schwarz (Int. J. of Mass Spectro. and Ion Processes, 110, 1991, 14 5-156)
R.Srinivas, J.Hrusak, D.Sülzle, D.Böhme and H.Schwarz (J.Am.Chem.Soc., 114, n8, 1992, 2802-2806).

ON THE LARGE ORGANIC MOLECULES IN THE INTERSTELLAR GAS

P. Thaddeus
Center for Astrophysics and Division of Applied Sciences,
Harvard University.

ABSTRACT

It is argued that polycyclic aromatic hydrocarbons (PAHs) are not the dominant large molecules in the interstellar gas because (i) the known interstellar molecules are more general in structure and composition; (ii) none of the many infrared and optical interstellar features attributed to large molecules has yet been assigned to a specific PAH; (iii) the PAH hypothesis fails to offer a credible interpretation of the the diffuse interstellar bands, and (iv) PAHs tend to accrete carbon in the diffuse interstellar gas at an unacceptably high rate. An alternate scheme is proposed in which a stable population of long carbon chains and large planar rings is maintained by C^+ insertion, photocleavage, and spontaneous isomerization. Chains with 20-30 atoms are synthesized at about the right abundance to account for the diffuse bands, and it is suggested that these and derivative rings of comparable size or slightly larger are the dominant large molecules in the diffuse gas. It is also suggested that in a slower subordinate cycle spontaneous isomerization further converts some of the larger planar rings to fullerenes, which then add carbon and possibly other atoms to form even larger molecules and amorphous grains.

INTRODUCTION

There is now a good deal of observational evidence for molecules in the interstellar gas which are considerably larger than any of the nearly 100 species which have been identified there or in circumstellar shells (Table 1), the largest of which contain 11-13 atoms; there are strong reasons to think that these, like the smaller known molecules, are largely organic in composition. Although the large molecules may contain a significant fraction of all the carbon in the interstellar gas (perhaps as much as ten percent), none of them has as yet been identified with certainty, and the question of their composition and structure undoubtedly constitutes one of the major unsolved problems in astrophysics. Indeed, there are few problems in natural science which are more interesting and fundamental, because the question bears on the role of organic chemistry and chemical evolution on the largest scale we can address. Here, often following others, I would like to explain why I think the PAH hypothesis[1,2] offers at best a very partial description of what these large molecules are, and to suggest a specific scheme for the synthesis of large molecules in the diffuse interstellar gas, taking as a point of departure the known small species. It is a scheme in which PAHs play at best a marginal role.

Throughout I will confine attention largely to the H I gas, because that is where the diffuse interstellar bands (DIBs), probably the most specific source of information on the large molecules in space, are formed, and because as Omont[3] has stated, "The abundance of PAHs inside molecular clouds is still completely unknown."

OBJECTIONS TO THE PAH HYPOTHESIS

There are at least four reasons for thinking that PAH's are not the dominant component of the large molecules in space:

1. Overspecificity.
2. Failure to achieve definite identifications.
3. Failure to offer a credible explanation for the interstellar diffuse bands.
4. Failure to explain how PAHs are formed, and if they are formed, why they fail to accrete carbon at an unacceptably high rate.

Let us briefly consider each of these objections.

1. <u>Overspecificity</u>. Time and again the temptation to explain the unidentified interstellar features in terms of a single large molecule, or a limited class of molecules, or some highly implausible material, has proven irresistible. A large magnesium porphyrin,[4] cellulose and other polysaccharides,[5] carbyne,[6] and microorganisms[7,8] have all been proposed, without adequate explanation (or any explanation at all) as to why nature has arranged to cast a significant fraction of the chemically active material in the interstellar gas into so special a structure. It may be unfair to compare the PAH hypothesis to these overly specific proposals, because it is more general in its assumptions, worked out in better detail, and more firmly based on the observational data, but it is not entirely free of the charge of overspecificity, as examination of Table 1 shows.

One of the most striking properties of that list is that the great majority (83%) of the molecules included are organic compounds; hydrocarbons although represented are a small minority: only 16%. Most of the organics contain the hetero atoms N, O, S, etc. as well as C, and it is of course the greatly enhanced architectural possibilities permitted by these elements which make organic chemistry enormously richer than hydrocarbon chemistry—as seen on comparing the chemistry of proteins with that of PAHs or any other family of hydrocarbons. If we are guided

NUMBER OF ATOMS

2	3	4	5	6	7	8	9
H$_2$	H$_2$O	NH$_3$	SiH$_4$	CH$_3$OH	CH$_3$CHO	CHOOCH$_3$	CH$_3$CH$_2$OH
OH	H$_2$S	H$_3$O$^+$?	**CH$_4$**	NH$_2$CHO	CH$_3$NH$_2$	CH$_3$CCCN	(CH$_3$)$_2$O
SO	SO$_2$	H$_2$CO	CHOOH	CH$_3$CN	**CH$_3$CCH**		CH$_3$CH$_2$CN
SO$^+$	HN$_2^+$	H$_2$CS	HC≡CCN	CH$_3$NC	CH$_2$CHCN		H(C≡C)$_3$CN
SiO	HNO	HNCO	CH$_2$NH	CH$_3$SH	H(C≡C)$_2$CN		**H(C≡C)$_2$CH$_3$**
SiS	SiH$_2$?	HNCS	NH$_2$CN	**C$_5$H**	**C$_6$H**		
NO	HCN	CCCN	H$_2$CCO	HC$_2$CHO			11
NS	HNC	HCO$_2^+$	**C$_4$H**	**CH$_2$=CH$_2$**			
HCl	HCO	**CCCH**	**C$_3$H$_2$**	**H$_2$CCCC**			H(C≡C)$_4$CN
NaCl	HCO$^+$	**c-CCCH**	CH$_2$CN				
KCl	OCS	CCCO	C$_5$				13
AlCl	**CCH**	CCCS	SiC$_4$				
AlF	HCS$^+$	HCCH	**H$_2$CCC**				H(C≡C)$_5$CN ?
PN	SiCC	HCNH$^+$	HCCNC				
SiN	CCO	HCCN	HNCCC				Total: 96
NH	CCS						
CH	C$_3$						
CH$^+$	MgNC						
CN							
CO							
CS							
C$_2$							
SiC							
CP							

Table 1. Known interstellar and circumstellar molecules. Organic molecules (83%) are in large type, hydrocarbons (16%) are underlined.

therefore by what we *know* to exist in the interstellar gas, we would expect some of the large molecules there to be hydrocarbons, but not all or most, and we would regard any attempt to confine attention to hydrocarbons as a false principle of economy, excluding the most interesting possibilities.

The main features in the unidentified infrared (UIR) bands which have been attributed to PAHs, and on which the PAH hypothesis largely rests, are unfortunately quite non-specific vibrational transitions: aromatic CH and CC stretches and CH wags.[1,2] Such transitions obviously occur in many organic molecules besides PAHs—heterocyclic ring compounds especially—but there has apparently been little attempt to consider these wider possibilities. One would like to know how many of the hetero atoms can be incorporated into the aromatic PAH lattice without spoiling the claimed identifications, and what limits can be set on plausible non-aromatic molecules such as alkanes, polyenes, and polyynes which might be attractive candidates for the interstellar gas. Until such questions are answered and positive identifications are forthcoming, the designation of the large molecules in space as PAHs is quite misleading, because it declares a solution when none exists and deters the search for alternatives. In many or most instances where the term "PAH" is employed in the astronomical literature, a better designation would be simply "large molecule" or "large organic molecule."

2. Failure to achieve identifications

Despite nearly a decade of laboratory investigation and a careful sifting of the spectroscopic literature, not one definite identification of a specific PAH with an IR line or an optical diffuse band has yet been achieved. Since PAHs are stable, readily synthesized compounds which have long been studied by spectroscopists both at IR and optical wavelengths—starting many years before the advent of the astronomical PAH hypothesis—this failure has an ominous ring, hinting that something may be radically wrong with the whole idea, or at least suggesting that the dominant role assigned to PAHs in space is seriously inflated. The explanation that interstellar PAHs are ionized and therefore unfamiliar to the laboratory spectroscopist is small mitigation of this objection, because under ionization equilibrium in the diffuse gas of the order of one-half of PAHs or similar large molecules are neutral.[9] Why has no member of this population of familiar non-ionized PAHs ever been identified? The same kind of objection applies to the explanation that PAHs are unobserved as diffuse bands because they are stripped of their hydrogens. Unless dehydrogenation is extreme, a fraction of the PAHs should retain a full complement of hydrogen, and so be known in the laboratory. Wholesale dehydrogenation might be postulated to produce molecules highly unfamiliar to spectroscopists, but it would introduce a serious problem of stability: bare graphitic plates are not the most stable carbon clusters in the size range 20-60 atoms, so this remedy undercuts the assertion that PAHs are preferred molecules in the interstellar gas because of exceptional stability.

3. Failure to explain the diffuse bands

PAHs were first invoked to explain the UIR bands, but it has been the hope of advocates of the PAH hypothesis that they would provide an explanation for the DIBs as well.[10,11] This hope is founded on the plausible general assumption that the DIBs are not produced by solid state transitions, but are also carried by large molecules, and that these are the same molecules as those which produce the UIR bands—or are closely related to them. This is an assumption which I find sufficiently compelling to adopt throughout the rest of the article.

If it is true, however, the PAH hypothesis is in difficulty, because there are good reasons, already presented by others,[12,13,14] to think that PAHs are *not* the carriers of the DIBs. The reasons are spectroscopic: the optical and ultraviolet spectra of PAHs are markedly different from the pattern displayed by the DIBs. The strongest diffuse band is at 4430 Å, and the rest (now over 100) extend from there in an apparently random pattern well into the infrared; there are none in the well studied octave below 4430 Å to the hump in the interstellar extinction curve at 2200 Å. In striking contrast, PAH's with 20-100 atoms have their strongest electronic transitions below 4430 Å, with some of the most intense in the octave just below. The reason for this distribution to the blue is simply that in a two-dimensional structure of that size the π electrons are confined in both dimensions to too small a length; it is true that there are many organic dyes with 20-100 atoms that have strong transitions throughout the visual, but these are generally not aromatic rings but conjugated chains (e.g., polyenes) where the π electrons can circulate along the entire length of the chain. The advocates of PAHs in fact generally concede that normal PAHs are poor candidates to explain the diffuse bands—hence the emphasis on ionization and dehydrogenation. But I have already indicated the weakness of that line of reasoning.

If a solid state interpretation of the diffuse bands is rejected, there would therefore seem to be two alternatives: (i) the assumption of a single population of large molecules is correct and PAHs are the carriers of neither the DIBs or the UIR bands, or (ii) that assumption is incorrect and a second population of large molecules is required. Neither alternative offers much support to the PAH hypothesis, and both subvert the claim that PAHs are the dominant population of large molecules in the interstellar gas.

4. The accretion catastrophe

The high stability of PAHs has been advanced as one of the main reasons for supposing them an important constituent of the interstellar gas, but it may instead constitute a serious liability. In the diffuse gas a PAH is likely to accrete carbon at a very rapid rate, and to incorporate this carbon tightly into its graphitic lattice. Most of the atomic carbon there is ionized by the interstellar radiation field to C^+, and it is accretion of this very reactive ion followed by non-dissociative electronic recombination that is probably the principal growth mechanism for large organic molecules. Christian, Wan, and Anderson[15] have recently shown in the laboratory by a mass spectrometric study that even the smooth, edgeless surface of C_{60} will react with C^+ at a fast Langevin rate and strongly bond it into its lattice, and that this reaction apparently proceeds without an appreciable activation barrier, and so will remain fast at the low temperature of the diffuse interstellar gas. Presumably by the same or a similar mechanism, linear carbon chains grow very rapidly in a laboratory carbon plasma. There is little reason to suppose that PAHs in space will be immune to similar insertion and growth, and if such growth occurs as generally and as rapidly as these examples and others suggest, the consequences are extremely damaging to the PAH hypothesis.

Consider for example a PAH with about the same number of atoms as C_{60}, and with the same polarizability, $\alpha = 65$ Å^3. The Langevin rate coefficient[16] for reaction with C^+ is then

$$<\sigma v> \equiv r = 2\pi \, (\alpha e^2/\mu)^{1/2} = 5.4 \times 10^{-9} \text{ cm}^3 \text{ s}^{-1} , \quad (1)$$

where $\mu = 12$ amu is the reduced mass, and v is the C^+ velocity, which we take to be 3.7×10^4 cm s^{-1}, corresponding to a normal H I kinetic temperature of 100 K. At the density of a fairly standard H I cloud with $n_H \approx 10$ cm^{-3}, $n_{C^+} \equiv n \approx 3 \times 10^{-3}$ cm^{-3}, the Langevin reaction rate is then $nr \approx 1.6 \times 10^{-11}$ s^{-1}, corresponding to a reaction time $1/nr$ of only about 2000 yr. No mechanism to my knowledge has been proposed which in such short order is able to synthesize PAHs in space or extract them from larger amorphous aggregates via grain evaporation or grain-grain collisions,[17,18] and 2000 yr is clearly orders of magnitude shorter than the time required to transport PAHs into the diffuse interstellar gas from stellar atmospheres (or supernovae). But even if such a production mechanism could be found, the overall depletion of carbon still poses severe difficulties.

Suppose for example that a particular PAH, let us call it P, is produced rapidly enough to be observable, e.g., via one of the stronger interstellar diffuse bands. Even assuming a large oscillator strength for the band (f~1), detectability probably requires that p, the number density of P, be greater than ~ 3×10^{-8} that of H, or greater than $\zeta \sim 1 \times 10^{-4}$ that of C^+. If q is then the production rate of P, equating production with loss via C^+ accretion at some initial time t = 0 yields $q = n_0 p_0 r$, with r the Langevin rate in eq. (1). If it is assumed that at t = 0 only P exists, but that a longer and longer string of descendents each with one more carbon is successively formed at the Langevin rate and each in turn begins to accrete, the rate of change of the C^+ reservoir is readily calculated to be

$$dn/dt = -nr(p_0 + qt) = -nr(\zeta n_0 + \zeta r n_0^2 t), \text{ with the solution}$$

$$n(t) = n_0 \exp(-\zeta r n_0 t) \exp(-\zeta r^2 n_0^2 t^2/2) \tag{2}$$

$$\approx n_0 \exp(-\zeta r^2 n_0^2 t^2/2). \tag{3}$$

Eq. (3) yields C^+ depletion of 50% in an alarmingly short time for $\zeta > 10^{-4}$:

$$t(50\%) = (1/rn_0)\sqrt{(2 \ln 2/\zeta)} < 2000 \text{ yr} \times 117 \approx 230,000 \text{ yr}, \tag{4}$$

which for the diffuse gas is probably unacceptably short by at least one and possibly two orders of magnitude. Essentially the same result can be derived in other ways. When it is recalled that we have considered only accretion by *one* PAH and its descendents, and that many others are presumably required to account for the more than 100 diffuse bands now detected, the conclusion seems inescapable that *unless some efficient mechanism can be found to compete with C^+ accretion and return carbon to the interstellar gas, a detectable population of PAHs will deplete interstellar carbon at a quite unacceptably high rate.* For brevity I will refer to this objection to the PAH hypothesis as the accretion catastrophe.

CHAINS TO RINGS TO FULLERENES TO GRAINS

Is there a way to proceed from the kinds of molecules we know to exist in the interstellar gas to the larger systems with 20-100 atoms which might produce the unidentified infrared bands and the diffuse interstellar bands, while avoiding the limitations of the PAH hypothesis? Here I will suggest a scheme grounded in recent laboratory work on carbon clusters, but with roots that go back a number of years. It is well at the start to emphasize the ubiquitous presence of carbon chains in the table of known interstellar molecules, nearly one-third of which are in that category.

A. E. Douglas[19] suggested in 1977, as the first carbon chains were being discovered by radio astronomers, that these were attractive candidates to explain the DIBs. He pointed out that these conjugated systems were organic dyes, whose strongest electronic transition with increasing chain length would progress from the blue toward the red into the region of the diffuse bands, $\lambda > 4430$ Å; he also noted that fast radiationless transitions could plausibly account for the wide, featureless shapes of the bands. Today as in 1977 the optical spectra of even short carbon chains like those in Fig. 1 are very poorly known, so Douglas's suggestion explains one of the most striking puzzles of the diffuse bands: the total lack of laboratory identification.

If one considers the fate of a long carbon chain in the diffuse interstellar gas, one discovers a crucial difference between the evolution of a chain and that of a large two or three dimensional molecule of high structural stability such as a PAH or C_{60}. The latter as we have seen are subject to rapid accretion, tending to add carbon at an unacceptably high

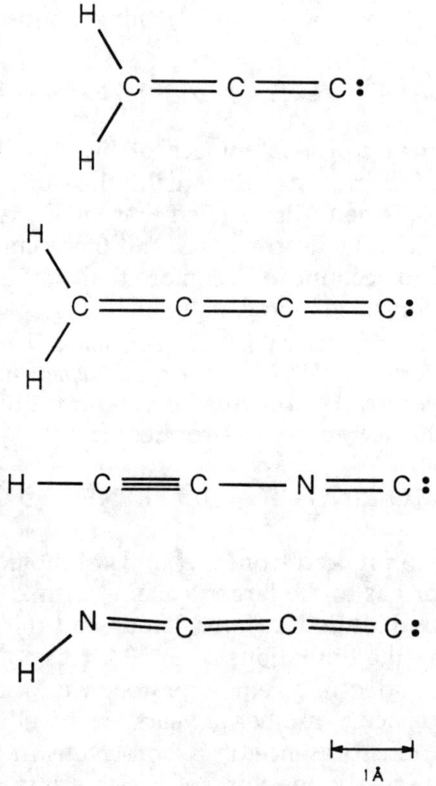

$$H_2C=C=C:$$

$$H_2C=C=C=C:$$

$$H-C\equiv C-N=C:$$

$$HN=C=C=C:$$

|←─1Å─→|

Fig. 1. Four highly polar carbon chains recently detected by radio astronom

rate and becoming larger and larger. A long chain would probably do the same, rapidly growing longer and longer, because there is little reason to think that it is less susceptible to C^+ interaction and insertion, *except that it is much more liable to photocleavage owing to its linear structure: it is only necessary to break a single bond for fission to occur.* It is paradoxically this very fragility that makes long chains interesting candidates for the large molecules in space, because fragmentation solves the problem of returning atomic carbon to the interstellar gas and thereby avoids catastrophic accretion.

Let us consider in detail the very simple steps required to accomplish this recycling of carbon. Assume for simplicity that all chains of length X_n insert C^+ at the Langevin rate—or, if one prefers, at a Langevin rate which increases linearly with the length of the chain when that is greater than the ion-molecule impact parameter of 10-20 Å. The subsequent behavior of the resulting ionized chain X_{n+1}^+ depends critically on length: short chains will dissociatively recombine like other small molecules; long chains, like solids (a graphite electrode is a familiar example) will recombine without fragmentation or sputtering. The transition between these short, fragile or "secondary" chains as I will call them, and the long robust "primary" chains is undoubtedly not sharp, but can be estimated from an RRKM calculation[20] to lie in the vicinity of 12 atoms. By interstellar standards this is not a particularly large number, since it is comparable to the number of atoms in the longest chains actually observed in molecular clouds (HC_9N and $HC_{11}N$ in Table 1). For both short and long chains it will be assumed that subsequent electron recombination is rapid compared to the Langevin ion-molecule reaction rate, because the recombination and Langevin cross sections are comparable and $n_e \sim n_{C^+}$, but the electron velocity is large: $v_e \approx 150\, v_{C^+}$.

As an idealization, let us suppose that the transition will occur precisely between n = 11 and 12 and assume on the basis of the foregoing reasoning that a new link in the chain—C^+ insertion plus rapid recombination—will occur at the Langevin rate, but that growth is opposed by photocleavage in the interstellar radiation field, some of whose uv photons are able to break the strong localized sigma bonds holding the chain together. Just how and at what rate cleavage will occur is not easy to calculate and there is little relevant laboratory data. Let us therefore assume that it occurs randomly at a constant rate irrespective of the chain length or the location of the bond in the chain—fully aware that this is an oversimplification since the triple bonds in acetylenic chains are stronger than the single bonds. The photocleavage rate is the one free parameter in the model; it will eventually be fixed by fitting the observed amount of carbon depletion in the diffuse gas.

When a chain is cut into two fragments, each with 11 atoms or more, each fragment resumes the process of growth by C^+ insertion, and the original chain has essentially reproduced itself and multiplied. If, however, one of the fragments has less than 11 atoms, it continues to react with C^+ at the Langevin rate, and shatters on subsequent recombination, and the fragments in turn shatter on further reaction with C^+ or uv photons until C^+ is returned to the interstellar gas. If both initial fragments are smaller than 11 atoms, the breakup of the original chain is complete. In the present model I have also allowed random photocleavage of small chains at a rate of 3×10^{-11} sec^{-1}, a typical value for small molecules in the interstellar radiation field.[16]

The coupled differential equations which then specify the time derivative of the number density $N_n(t)$ of chains with n atoms are readily written down:

dN_n/dt = gain from collisions – loss from same
       + gain from photocleavage – loss from same

$$= n_{C+} [N_{n-1} r(n-1) - N_n r(n)] + S[2 \sum_{j=n+1}^{\infty} N_j - (n-1)N_n], \quad (5a)$$

when $n \geq n_c = 12$, and

$$= n_{C+} [2 \sum_{j=n+1}^{n_c-2} N_j r(j)/(j-1) - N_n r(n)] + S[2\sum_{j=n_c}^{\infty} N_j] - S'N_n, \quad (5b)$$

Fig. 2. Distribution with time of carbon chains produced by C^+ insertion and destroyed by photocleavage—a numerical solution of the model discussed in the text. At t = 0 it is assumed that all carbon is in the form of C^+ (i.e., n = 1), except for a small amount ($N_{12} = 10^{-4} N_1$) in the form of the smallest primary chain, that with 12 atoms. Within a few $\times 10^4$ years the initial chain population has spread, amplified, and converged on an equilibrium population of long primary chains (n ≥ 12) able to reproduce, and short secondary chains in the process of dissolving to C^+ in the interstellar radiation field but continually replenished by photocleavage of the primary chains. Convergence to the indicated equilibrium distribution is independent of the size and density of the initial seed chains. For the particular value of the photocleavage rate chosen here the equilibrium C depletion is 87%.

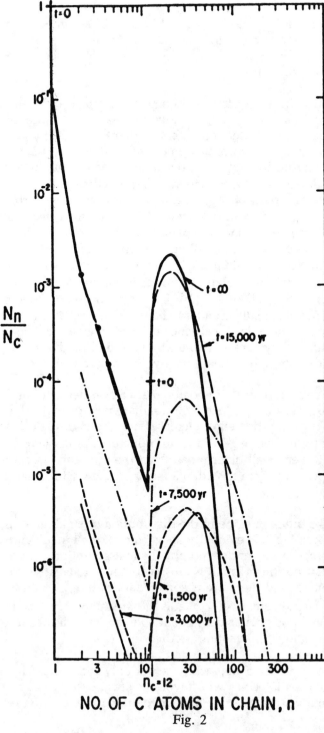

Fig. 2

when $n < n_c$, where S is the (unknown) probability per unit time of bond cleavage in the radiation field, $S' = 3 \times 10^{-11}$ sec^{-1} as assumed for small chains, and provision is made for the Langevin rate r to increase with chain length n. $n_{c_+} = N_1$ is readily determined from conservation of the total number of carbon atoms.

On solving these simple time dependent coupled differential equations numerically, one finds, as shown in Fig. 2, that if the diffuse gas is initially seeded or nucleated with a small quantity of carbon chains of length n = 11 or greater, the population of chains quickly spreads in number and amplifies in total mass, reaching a steady state distribution within ~10^5 years. It is not a coincidence that this time scale is of the same order as that characteristic of the accretion catastrophe. The carbon for this rapid growth is supplied by the reservoir of C$^+$, and when steady state equilibrium is attained, carbon is depleted by an amount which at a given total density depends only on the photocleavage rate S. Carbon depletion of 50%, about that observed in low extinction clouds,[21] fixes S at $\approx 1 \times 10^{-13}$ sec^{-1}, which for a chain with about 20 bonds implies a total photodistruction rate of about 2×10^{-12} sec^{-1}, or about 15 times less than the value adopted for the small secondary chains. One hesitates to state that this is a reasonable value in the absence of experimental or theoretical data, but it qualitatively agrees with the general finding that large polymers and dyes are photostable relative to small molecules.

A positive feature of the model is that carbon depletion is found to increase with decreasing S, which is in qualitative accord with the observation that carbon depletion increases in regions of high density and therefore high uv extinction.[21] The rate of increase of depletion with uv extinction is not very well determined from the available observational data, but it appears to agree to within a factor of a few with what the model predicts.

The sharp break in slope in the distribution of small and large chains at n = 12 in Fig. 2 is evidently an artefact of the model, and would undoubtedly smooth out under more realistic assumptions. Similarly, the slope in the distribution of small chains is somewhat sensitive to the assumptions made, and therefore is somewhat arbitrary. The hump in the distribution of large chains at $n \approx 20$ and the precipitous decline in the distribution when n is greater than about 30 is however a robust feature of the model and its most interesting prediction.

For in the primary chains we have not only synthesized a stable population of large molecules with favorable spectroscopic properties to account for the diffuse bands, we have made them in about the required amount as well. The chains in Fig. 2 near the peak of the distribution at $n \approx 20$ are the ones of major interest and the ones on which we should focus attention. All the reasons Douglas gave for thinking that somewhat shorter chains (5 to 15 atoms) might be the carriers of the diffuse bands apply to these longer ones, probably *a fortiori*, and the optical spectra of these longer chains are even less familiar in the laboratory—in fact totally unknown—explaining again why none of the diffuse bands has yet been identified. The strength of a diffuse band produced by one of these chains is readily calculated; the equivalent width W of an electronic transition in absorption is:

$$W = \pi e^2 \lambda^2 f N_{20} L / mc^2, \qquad (6a)$$

$$= 0.2\text{-}20 \text{ Å}, \qquad (6b)$$

for a range of oscillator strengths $f = 0.01\text{-}1$, $\lambda \approx 5000$ Å, $N_{20} \approx 3 \times 10^{-6}$ cm^{-3} from Fig. 2, assuming the number density of C$^+$ is 3×10^{-3} cm^{-3}, and taking a cloud thickness L along the line of sight of 10 pc. According to Herbig,[22] W for the strongest diffuse bands lies in the range 0.4-4 Å, so there may even be some intensity to spare, accommodating the dilution in intensity that might result from chain variation or embellishment—such as various end groups, insertion of hetero atoms such as N, etc.

A further point to note is that, as a major reservoir of carbon, the primary chains may be observable at other than optical and infrared wavelengths. Radio detection, for example, is not out of the question; polar variants of the chains with $n \approx 20$—e.g., chains with a terminal CN group or highly polar carbene chains like those in Fig. 1.—might be detectable in the cm band in absorption against a strong distant continuum source such as Cas A. Optical depths in the strongest radio lines depend on the unknown dipole moment and degree of rotational excitation, but can be estimated to be possibly as large as 10^{-3}.

To summarize, we have (i) produced a population of large carbon chains in the diffuse gas via explicit processes of formation and destruction, which (ii) exist in stable equilibrium with the reservoir of carbon, (iii) possess favorable spectroscopic properties to account for the DIBs, (iv) have about the abundances required to account for the strength of these bands, and (v) are connected explicitly to the depletion of carbon and the uv extinction. It is further suggested (vi) that these chains may have new observational consequences, e.g., detectable radio lines. The scheme is apparently hermetic, with no connection to classical grains or large two- and three-dimensional molecules, but recent experiments on the origin of fullerenes suggest that this may not be so, and that the large chains proposed here may lead naturally in the interstellar gas to the synthesis of specific two and three dimensional large molecules—although slowly enough to avoid catastrophic accretion.

A major mystery in the fullerene story is the astonishing efficiency with which C_{60} and similar closed carbon cages can be spontaneously produced in a simple carbon arc; the yield of C_{60} under the right conditions can now approach 30%. Two recent experiments[23,24] and previous work by some of the same investigators may provide an explanation. By a technique of gas phase ion chromatography it has been shown that as the number of atoms increases carbon clusters in the gas phase develop from linear chains to planar polyyne ring systems of the kind shown in Fig. 3 to fullerenes. What is new and unexpected is that the transition apparently occurs spontaneously and with high efficiency once a threshold in size is crossed and enough heat is applied to melt or anneal the cluster. These remarkable processes are of astrophysical interest because the heat required to melt or anneal the clusters, of the order of 4 ev for the large monocyclic rings, is comparable to that sporadically supplied in the diffuse gas by the absorption of a uv photon—the very process thought to be responsible for the ubiquitous 3-12 μ emission there which first revealed the presence of large molecules. The size threshold to form rings is also significant: it is apparently as low as 20 atoms which means that all the chains near the peak of the primary distribution in Fig. 2 are probably susceptible to spontaneous isomerization.

The model which I propose for the dominant large molecules in space therefore consists of chains with a distribution similar to that in Fig. 2, complemented by a significant but indeterminate population of planar rings like those in Fig. 3. For the rings the competition between C^+ reaction, photocleavage, and isomerization is clearly fairly complex, as rings are broken, reformed, fragmented, etc., and loops are added or subtracted, and it is not likely that the simple size distribution of chains in Fig. 2 applies to the rings. There is probably a cutoff in the ring distribution at some point above n = 30, but where this occurs, and how sharply, is not yet clear.

The threshold for the spontaneous isomerization of planar rings to fullerenes is according to the laboratory experiments in the vicinity of 40 atoms. Most of the carbon which follows this route is lost to the chain-ring cycle, as illustrated schematically in Fig 3, since the fullerenes and their descendents as highly stable structures are subject to indefinite accretion, until the much slower processes already mentioned—grain evaporation and grain-grain collisions— are able to return the carbon to

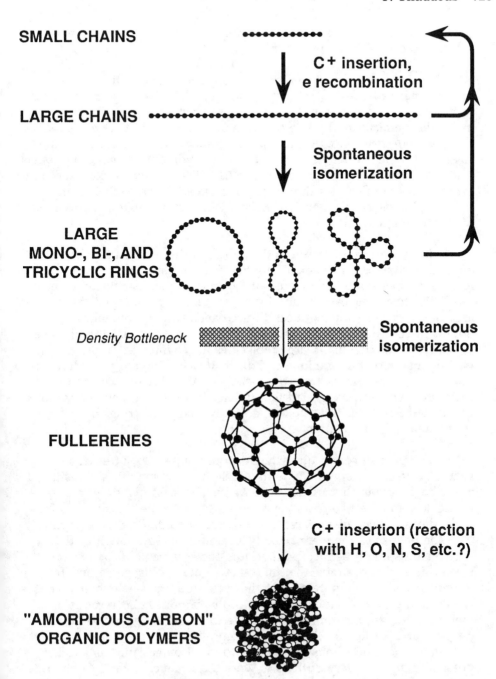

Fig. 3. Schematic ilustration of the growth of carbon clusters via C^+ insertion and isomerization from chains to large monocyclic, bicyclic, and tricyclic rings to fullerenes to amorphous carbon grains.

the interstellar gas. The bottleneck to this loss of carbon is presumably the cutoff in the ring mass spectrum

Notice that we have proceeded from one to two to three dimensional carbon—i.e., from small to long chains to large planar rings to fullerenes to large amorphous carbon clusters—simply with ion-molecule chemistry and photodissociation, *reactions which are universal in the diffuse interstellar gas,* plus an appeal to spontaneous isomerization suggested by recent laboratory experiments—without encountering PAHs. Actually, the present scheme *excludes* PAHs, because isomerization of large planar rings to fullerenes shuns graphitic plates owing to their many dangling bonds; subsequent synthesis of PAHs by hydrogenation is choked off at the source.

To summarize, because of the failure during ten years of concerted effort to assign *any* of the many unidentified interstellar IR and optical features to specific PAHs, because of the tendency of very stable molecules like PAHs to accrete carbon, and for other reasons, it is proposed that PAHs are not the carriers of the optical diffuse bands and do not make a dominant contribution to the unidentified infrared bands. It is proposed instead that the dominant large molecules in the interstellar gas with a well defined structure are long carbon chains, and large planar mono and multicyclic rings formed from these chains. As a side branch of the carbon cycle proposed, it is suggested that fullerenes and from them amorphous grains (but not PAHs) may be formed from such planar rings by spontaneous isomerization.

The present conceptual scheme would appear to suffer from the same charge of overspecificity with which the PAH hypothesis was taxed at the beginning of this article, since we have discussed purely carbon molecules, and the hetero atoms required to make the transition to full-fledged organic chemistry have not been considered. Partly this is simply a question of space. There are actually a number of ways in which the hetero atoms might be incorporated into the scheme, and for several reasons it is very desirable that this be attempted. Nitrogen atoms for example may be incorporated into the cycle from the small chains on; it is worth noting in this context that two of the carbon chains recently identified by radio astronomers in molecular clouds contain interior nitrogen atoms (Fig. 1). Reaction of many different hetero atoms with the fullerenes and amorphous grains that follow isomerization is another possibility, but one very difficult to work out in much detail. At any rate, in elaborating the present theory further, there is no motivation whatever to confine attention to the chemistry of carbon, particularly in view of the evidence for the depletion of many of the hetero atoms in the interstellar gas. Such depletion is a pressing reason why as a working proposition it should be assumed that the large molecules in the interstellar gas draw on the full architecture of organic chemistry, the foundation of the most intricate structures in nature.

Finally, a comment on the crucial question of laboratory confirmation of the unidentified interstellar features. Both the PAH hypothesis and the chain-ring scheme outlined here must be regarded simply as working hypotheses towards *exact* laboratory identification—the last court of appeal, as it has been for all of the molecules in Table 1. Until that is accomplished, all explanations are entirely provisional. I have argued here, often simply echoing the objections of others, that the PAH hypothesis is unlikely to account for the optical diffuse bands; it does provide, however, at least in spectroscopic terms, a fair explanation for the UIR bands—specifically, emission spectra calculated from laboratory absorption data which are "not dissimilar" to the astronomical bands. Can the present chain-ring scheme give as good a description of the UIR bands—or a better one?

That is presently a difficult question to answer, because of the total lack of infrared spectroscopy on long chains and large planar rings. But a partial answer may be possible. C_{60} was first identified spectroscopically[25] on the basis of group theoretical analyses of its infrared active vibrational transitions and theoretical calculations of their frequencies,[26-29] and it might be possible by similar methods to decide whether a reasonable population of large chains and planar rings has a vibrational spectrum which is sufficiently "not dissimilar" to the UIR bands to be viable, or on the contrary is so dissimilar as to be ruled out. Work toward that end is now underway. Also, laboratory investigations of carbon clusters are now being pursued so vigorously, and modern laser and mass spectrometric techniques are becoming so sensitive, that relevant laboratory spectra in the next ten years are not beyond hope.

Acknowledgements

I wish gratefully to acknowledge discussions many years ago with the late A. E. Douglas on the subject of the diffuse bands, and many equally useful conversations over the years with A. Dalgarno and W. Klemperer.

REFERENCES

1. J. L. Puget and A. Léger, Ann. Rev. Astron. Astrophys. **27**, 161 (1989).
2. A. Léger and L. d'Hendecourt, in *Polycyclic Aromatic Hydrocarbons and Astrophysics*, eds. L. D'Hendecourt, A. Léger, and N. Boccara (Dordrecht: Reidel, 1987) p. 223.
3. A. Omont, Astron. and Astrophys. **164**, 159 (1986).

4. F. M. Johnson, D. T. Bailey and P. A. Wegner in *Interstellar Dust and Related Topics*, eds. J. M. Greenberg and H. C. van de Hulst (Reidel, 1973) p. 317.
5. F. Hoyle and N. C. Wickramasingh, Nature **268**, 610 (1977).
6. A. Webster, Mon. Notices Royal Astron. Soc. **192**, 7P (1980).
7. F. Hoyle and C. Wickramasingh, Astrophys. Space Sci. **66**, 77 (1979).
8. S. Yabushita, K. Wada, T. Takai, T. Inagaki, D. Young, and E. T. Arakawa, Astrophys. Space Sci. **124**, 377 (1986).
9. S. Lepp and A. Dalgarno, Astrophys. J. **335**, 769 (1988).
10. A. Léger and L. d'Hendecourt, Astron. and Astrophys **146**, 81 (1985).
11. G. P. van der Zwet and L. J. Allamandola, Astron. and Astrophys **146**, 76 (1985).
12. W. Lee and T. J. Wdowiak, Astrophys. J. **410**, L127 (1993).
13. S. Leach, in *Polycyclic Aromatic Hydrocarbons and Astrophysics*, eds. L. D'Hendecourt, A. Léger, and N. Boccara (Dordrecht: Reidel, 1987) p. 99.
14. B. D. Donn, J. E. Allen, and R. K. Khanna, in *Interstellar Dust*, eds. L. J. Allamandola and A. G. G. M. Tielens (Dordrecht: Kluwer, 1989) p. 181.
15. J. F. Christian, Z. Wan, and S. L. Anderson, J. Phys. Chem. **96**, 3574 (1992).
16. W. D. Watson, Rev. Mod. Phys. **48**, 513 (1976).
17. B. T. Draine, in *The Evolution of the Interstellar Medium*, Astron. Soc. Pacific Conf. Series, **12**, 193 (1990).
18. C. F. McKee, in *Interstellar Dust*, eds. L. J. Allamandola and A. G. G. M. Tielens (Dordrecht: Kluwer, 1989) p. 431.
19. A. E. Douglas, Nature **269**, 130 (1977).
20. W. Forst, *Theory of Unimolecular Reactions* (New York: Academic Press, 1973).
21. E. B. Jenkins, in *Interstellar Processes*, eds. D. J. Hollenback and H. A. Thronson, Jr. (Dordrecht: Reidel, 1987) p. 533.
22. G. H. Herbig, Astrophys. J. **196**, 129 (1975).
23. G. von Helden, N. G. Gotts, and M. T. Bowers, Nature **363**, 60 (1993).
24. J. Hunter, J. Fye, and M. F. Jarrold, Science **260**, 784 (1993).
25. W. Kratschmer, K. Fostiropoulos, and D. R. Huffman, in *Dusty Objects in the Universe*, eds. E. Bussoletti and A. A. Vittone (Dordrecht: Kluwer, 1990) p. 89.
26. Z. C. Wu, D. A. Jelski, and T. F. George, Chem. Phys. Lett. **137**, 291 (1987).
27. S. J. Cyvin, E. Brendsdal, B. N. Cyvin, and J. Brunvoll, Chem. Phys. Lett. **143**, 377 (1988).
28. R. E. Stanton and M. D. Newton, J. Phys. Chem. **92**, 2141 (1988).
29. D. E. Weeks and W. G. Harter, J. Chem. Phys. **90**, 4744 (1989).

DISCUSSION

NENNER — You have considered accretion of C^+ with free carbon chains as a very efficient process. What would be your reasoning if those chains are chemically attached to carbonaceous grains of a polymer type for example ?

THADDEUS — I haven't considered that.

SCAPPINI — Having been associated with the Ottawa group for quite some time I like this revival of Douglas' (1977) idea. I have one concern : if the carbon chains are the carriers of the DIB's, they should be more ubiquitous than they are.

THADDEUS — Nearly-one-third of the known interstellar molecules are carbon chains, and at least the simpler ones are already known to be quite widespread in the dense gas.

WDOWIAK — There is a very strong argument for not excluding PAH's as an interstellar component. It is an argument that does not depend upon photons or computations. It is an argument you can literally hold in your hand and as such it is a heavy argument. I am speaking of the deuteriated PAH's found in meteorites. These are just what you would expect if PAH's are formed in carbon star atmosphere, pass into cold dark cloud when they become deuteriated, and become incorporated into the parent bodies of the meteorite at the time of the formation of the solar system.

THADDEUS — It is very hard from such considerations to show that PAH's are the dominant large molecules in the interstellar gas.

BAUMGARTEL —
1. What is the role of reaction entropy in reactions going from linear system to monocyclic and/or polyclyclic systems ?
2. Why are preferably carbon chains considered consisting of divalent carbon?
   Why not sp^3, sp^2, sp - Valence states of carbon ?
   Why not partially hydrogenated systems ?

730   Large Organic Molecules in the Interstellar Gas

<u>THADDEUS</u> — I am afraid you have to ask Jarrold question 1. With respect to 2 most of the known carbon chains are acetylenic with alternating triple and single bonds, and I believe those are the kind of chains which Jarrold thinks he is observing in the laboratory. Two of the small carbon chain carbenes we have recently detected in the laboratory and in space, H_2CCC and H_2CCCC, are apparently cumulenic in structure, with double bonds.

<u>SCHUTTE</u> — If some types of PAH's would be close to 100% ionized in the diffuse ISM, and C^+ addition would at certain time produce such a PAH, the Coulomb repulsion would prevent the occurrence of further C^+ addition. Thus C^+ addition could perhaps result in a stable population of close to 100% ionized PAH's.

<u>LEACH</u> — Species that are highlighted in models of ISM processes are often "filtered" through observational possibilities. Emphasis on hetero-atom chemistry being important reflects the fact that the presence of heteroatoms in chains creates sizeable dipole moments and so make them observable by radio-astronomy. It would be of value to consider what species would be difficult to observe by present techniques and which, by indirect evidence or reasonable conjecture, could be present in significant amounts in the ISM.

<u>ZIURYS</u> — It is interesting that for many years, radio astronomers have searched unsuccessfully for small aromatic molecules in interstellar gas. This may be in part a partition function effect, but if PAH's are so abundant in the interstellar medium, then one might expect to observe aromatic species in the radio/mm-wave region as well. One readily detects long carbon chains in the dense I.S.M., so having chains account for the diffuse bands/unidentified IR bands nicely fits in with what is already observed.

<u>LEGER</u> — The question is to decide which are the most abundant large molecules in the ISM, aromatic ones or linear ones. There is an almost non-biased way to decide this point which is to consider their IR emission and compare them to the observations. Both classes of molecules are expected to absorb in the UV-visible range and therefore should emit in the IR (transient heating mechanism). Most people agree that the observed UIR band are due to <u>aromatic</u> species. Are you expecting that the linear or cyclic <u>chains</u> that you propose are going to give a similar -or better- fit to the observations ? Do you think that

the presently considered similarity (not fit) between compact PAH spectra and the UIR bands is just chance ?

THADDEUS — The infrared spectra of long carbon chains and large mono and polycyclic rings are quite unknown so it is impossible to say whether they can give as good a fit as the PAH's to the UIR bands- or a better fit, since the PAH fit is fairly good, but not perfect. With respect to the role of chance, I can only say that stranger things have happened in astrophysics!

BUSSOLETTI — I liked very much the evolution path you have presented. I would like to know if you consider it as the only one, or the most likely or what. Actually, according to Frenklach shock-tube studies other evolution path may lead, more or less, to the same end-products.

THADDEUS — Uniqueness is the last thing I would suggest. I do think the present scheme has several features worth exploring, and that it addresses issues neglected by the PAH hypothesis.

BRECHIGNAC — I wish to mention than considering only the number of bonds to break to guess fragmentation pathways is not sufficient for these large molecules. Indeed there is experimental evidence of photo products involving the breaking of several bonds. Secondly I have'nt well seen in your accretion model what is really specific to linear chains. Where is the chain-specificity in your input ingredients ?

THADDEUS — I think that photocleavage is likely to be a specific and crucial property of chains not generally shared by two dimensional molecules like PAHs.

GRAINS, ICES AND CARBONACEOUS MATERIALS

PHYSICAL AND CHEMICAL PROCESSES IN ICY GRAIN MANTLES

Bernard SCHMITT
Laboratoire de Glaciologie et Géophysique de l'Environnement, C.N.R.S.
54 rue Molière, B.P. 96, 38402 Saint Martin d'Hères Cedex, France.

ABSTRACT

Many questions are still largely open concerning the composition and evolution of the icy mantles of interstellar and circumstellar grains. They include the physico-chemical origin of the condensed species, the efficiency of the different chemical processes which may alter the composition of the grains, as well as the various exchange processes between the gas and solid phases and their time constants. We will review the different physical and chemical processes by which icy mantles are formed, evolve and are destroyed. Grain formation involves adsorption and condensation of gas species as well as molecule formation by surface reactions. The chemical evolution of grains includes destruction and synthesis of new species by UV photolysis and by cosmic ray irradiation. The physical evolution and destruction processes consist of thermal or induced desorptions and evaporations, differentiations by selective evaporations, and sputtering. In addition, some specific destruction processes, such as chemical explosions of radical-rich ice mantles, have been proposed.

EVIDENCES OF GAS-GRAIN INTERACTION AND GRAIN CHEMISTRY

a) Accretion and desorption of gas species

Two of the major proofs of exchanges occurring between molecules in the gas phase and molecules condensed or adsorbed on grains come from the simultaneous observations of both solid and gas phase molecules in cold molecular clouds. Indeed, icy grain mantles have been now observed in the line-of-sight towards more than 100 infrared sources [222]. They include protostars and embedded objects in molecular clouds [54, 61, 122, 126, 200, 212], background field stars [201, 223, 225, 226], as well as oxygen-rich circumstellar envelopes of bipolar nebulae [54, 63, 105, 155, 175, 199, 204]. The composition of grain mantles, as derived by infrared spectroscopy, is generally dominated by H_2O ice [43, 54, 182, 200, 201, 209, 223/6, 232], but several other molecules such as CO [54, 61, 121, 122, 126, 139, 214, 224, 226, 235], CO_2 [103], CH_3OH [3, 81, 212], H_2CO [195], H_2S and OCS [62], organic (R-CH_3, R-CH_2-R) [38], and possibly CH_4 [127] and some ions, OCN^- [77], NH_4^+ [80] have been identified. Carbon monoxide, the second widely observed molecule in grains, has been observed yet in the direction of more than 30 infrared sources [227].

All these observations of grain mantles implies either the direct accretion onto the dust particles of molecules formed in the gas phase via ion-molecules reactions or their formation at the surface from adsorbed atoms or radicals or, for some of species (molecules, radicals, ions, ...), their formation inside the grain induced by an external energetic source (UV, cosmic rays).

In fact, it has been accepted well before the observation of the first grain mantle that under the conditions found in cold molecular clouds, gas phase species, other than H_2, colliding with dust grains must stick to them with a high efficiency [48, 131, 187, 218, 229]. If no process of returning the adsorbed species to the gas phase occurs, then all the species will be depleted from the gas phase in a time of about $3 \ 10^9/n_H$ years, where n_H is the number density of hydrogen nuclei (cm^{-3}) [113, 146]. For $n_H \geq 10^4$ cm^{-3} the depletion time is significantly less than the expected lifetime of clouds ($\sim 10^7$ yr). In dense clouds total freeze-out is inevitable, except for young clouds. Since gas phase molecules are widely observed in these regions, some desorption processes must occurs with a sufficient efficiency to explain the observations. Several mechanism, either intermittent or continuous have been proposed. They include thermal desorption, evaporations by pulsed heating due to grain-grain collisions [99], cosmic ray [47, 49, 133, 219, 220, 239], or X-ray interactions [133], sputtering by gas molecules or cosmic rays [16, 116, 216], photodesorption [219], and evaporations induced by surface reactions [51, 159] or radical recombination inside grain mantles [71, 98, 99, 194]. They are rewieved below.

b) molecule formation at grain surfaces or in grain mantles

There is increasing observational evidence that grain chemistry occurs in the interstellar medium [147]. Historically, the evidence that H_2 cannot be efficiently formed by gas phase reactions leaded astrophysicists to seriously consider surface reactions [4/6, 8, 30/34, 41, 50, 65, 75, 92/94, 100, 106/7, 110, 124, 130, 134, 159, 160, 210, 218/220, 228]. The formation of molecular hydrogen by the exothermic reaction of H atoms adsorbed on grain surfaces was the first crucial role of grains in the interstellar chemistry that was widely accepted. The efficiency of the formation mechanism required to explain observations of diffuse clouds must be high [120]. Almost all H atoms sticking onto a grain surface must leave as part of a molecule [50]. Until quite recently, the formation of H_2 was the only contribution of grains included in the theoretical models of dense cloud chemistry.

Recent observations indicate that grain surfaces may play a role in the formation, and subsequent desorption, of some more complex gas phase molecules. In particular, grain chemistry may possibly explain the presence of formaldehyde (H_2CO) and its deuterated isotopes at the edge of dense clouds [57, 215], and the observation of NH in diffuse clouds [145]. In dense regions (hot core sources) and in the outer regions of interstellar clouds some saturated molecules, such as NH_3, NH_2D, H_2O, HDO, H_2CO, HDCO, D_2CO and CH_3OH, are found to have particularly high abundances difficult to be explained with gas phase chemistry only [21, 57, 104, 141/3, 154, 166, 208, 215, 217]. Surface hydrogenation and deuteration of atoms (O, N, C, ...) during the collapse of the cold cloud followed by thermal evaporation, associated with the formation of a nearby star, have been proposed as an additional formation mechanism to explain some of the observed abundances [31, 156]. The release of species formed at the surface of grains during a colder stage, or in a colder part of the cloud and preserved in the grain mantles may also explain the large abundance of saturated molecules, as well as deuterated molecules (HDO, NH_2D, HDCO, D_2CO) [31/33, 40, 42, 57, 104, 142/3, 215, 217].

Finally, the presence in grain mantles of some molecules, such as CO_2 and H_2CO, complex organic molecules and ions (OCN^-, NH_4^+) may require additional energetic formation processes inside the grain [72, 75, 80, 81, 101, 192]. Ultraviolet photolysis [2, 7, 70, 72, 82, 83, 100/1, 192, 233] and cosmic ray induced chemistry [20, 148, 162, 206] are two possible ways of molecular synthesis in the interstellar medium.

INTERSTELLAR GRAIN MANTLES AND THE CHEMISTRY OF THE INTERSTELLAR MEDIUM

The first question about grain mantles is: What are their composition. Infrared spectroscopy is the ideal tool for the identification of the species making up grains and for the derivation of their relative abundances in the solid phase, as well as their column densities. Laboratory IR data of various ices are necessary for the analysis of interstellar infrared spectra.

The second question is: what is the physico-chemical origin of all the species identified in grain mantle. Several different ways have been proposed for the formation and chemical evolution of the composition of grain mantles. The mains are direct condensation of gas phase species, surface reactions of adsorbed atoms, and molecular synthesis inside the grains induced either by UV photolysis or by cosmic ray irradiation. As these different processes have been shown to occurs under laboratory conditions, the question translate into the following: What are the absolute yields and the relative contributions of these processes under various interstellar conditions.

The third question is: What is the contribution of grains to the chemistry of the interstellar medium ? A first observational information may be derived from the knowledge of the relative gas/grain abundance of various molecules in different interstellar environments. The chemical models of interstellar gas may indirectly point out some possible contribution of grain chemistry by its inability to explain some observed abundances. The synthesis and destruction yields of specific molecules by solid state and surface processes must be compared with gas phase models. In addition, gas - grains exchanges may complicate the chemical scheme by selectively removing some species synthesized in the gas phase and releasing others possibly formed at the surface or inside grain mantles.

INFRARED SPECTROSCOPY AS A KEY TO DERIVE THE NATURE AND COMPOSITION OF INTERSTELLAR GRAINS

Mid- and far-infrared spectra of grain mantles (2 - 200 µm) cover the range of the fundamental modes of internal vibrations of the molecules and the lattice vibrations of molecular solids. Thus, the infrared spectral signature of condensed molecules is a powerful means for their identification and quantification, especially for simple molecules (H_2O, CO, CH_3OH, CH_4, CO_2, ...). In addition, the state of the molecules (pure solid, mixed, adsorbed, hydrated, ...), the nature of the solid (amorphous, crystalline phase, ...), its actual temperature and thermal history, as well as possible UV or cosmic ray processing may be derived or constrained from detailed laboratory experiments and analysis of the observations.

Mid- and far-IR laboratory spectra on potential candidates under various astrophysical conditions are necessary to determine the absorption frequency, full width at half maximum (FWHM), absorption coefficient and integrated strength of the absorption features of pure ices and molecular mixtures [9, 52, 79, 84, 85, 102, 109, 150,

[157, 176, 179, 188/90]. For more detailed analysis of the interstellar spectra, Mie's calculations are needed [111]. The knowledge of the optical constants, the real and imaginary parts of the refractive index (n + ik), of solids are then necessary to run theoretical calculations involving small scattering particles. Such data begin to be available for a variety of ices and mixtures [132, 158, 238, 243, 245].

THE VARIOUS TYPES OF GAS-GRAIN INTERACTIONS

As we have seen before, there are numerous physical or chemical interactions between gas and grains that may play a role in the evolution of the interstellar medium. Detailed descriptions and discussions of most of these processes have been published in a few review and key papers on the subject [6, 106, 133, 213, 218/20]. Several additional processes have been proposed or the classical ones have been reconsidered during the last few years. In this paper we qualitatively describe the different grain processes and their estimated efficiencies as proposed by different authors. A brief discussion of the parameters involved follows and most of the relevant references on the subject are cited.

1) What parameters control the interaction between solid and gas phases?

Most of the gas-grain interactions depend on several parameters of the solid phase (the grain), of the gas phase and of the interface between the two (the surface of the grain). On the grain side, its structure (amorphous, crystalline, ...), its composition (pure, mixed, specific compound, ...) and its temperature are the main parameters. On the gas side, the parameters are the total and partial densities (composition) and the kinetic temperature. The interface between the solid and the gas phases determines the interaction potential between a given gas phase species and the surface of a grain of known composition and structure, and therefore defines the type of interaction (physical or chemical). The structure of the surface (amorphous, crystalline, porous, microporous, ...) and its extent (specific area, pore volume-distribution) can significantly affect the interaction potential and its distribution over the surface sites. For example, most of the H_2O-rich interstellar ice mantles show spectral features typical of amorphous material. Amorphous ice condensed at low temperature is found to have a highly porous structure with a fairly high total surface area [138, 183, 185]. This may play an important role on the processes occuring on the surface of interstellar grain mantles.

2) Physical adsorption-condensation - Thermal desorption-evaporation

Physical and chemical adsorption are two basic processes that control the exchanges between the gas and solid phases. In physical adsorption the interaction is due to van der Waals or multipolar forces. The adsorption energy is of the order of the condensation energy. The adsorption process can take place over several layers, is reversible and strongly depends on temperature [73]. In chemisorption, the interaction leads to electron transfer between the solid and the adsorbed molecule. The chemisorption depends on the gas-solid affinity, can occurs at low pressure and high temperature, is generally irreversible, and is restricted to a monolayer or less. The adsorption energy is of the order of the reaction energy [53].

In the case of icy grain mantles at low temperature, chemisorption of molecules and atoms is probably unimportant. Most species will be physically adsorbed. Their surface binding energies, E_b, can be estimated theoretically [106, 213, 218] or determined experimentally [96, 130, 151, 177/8, 180/1, 183, 185]. The nature and the structure of the solid may modify the condensation energy (for a molecule "adsorbed" on its own solid). For example, crystalline H_2O ice seems to have an average surface binding energy about 5% higher than amorphous H_2O ice [177]. In addition, the adsorption energy (for a molecule "adsorbed" on a different solid) and its distribution over the sites depend on the structure of the surface (amorphous or crystalline, porosity). For example, the distribution of the adsorption energy of N_2 on the surface of crystalline H_2O ice range from 800 K to about 1350 K for the more energetic sites [183, 185]. On microporous amorphous H_2O ice the more energetic sites have adsorption energy for N_2 probably exceeding 2000 K. Theoretical calculations of the adsorption energies of H atoms on the surface of a large cluster of amorphous H_2O ice leads to a distribution of binding energy ranging from 300 to 700 K and centered around 500 K [36].

The residence time, τ_{ev}, of a species on a surface determining the evaporation and desorption rates, can be estimated from the binding energy:

$$\tau_{ev} = \nu_0 \cdot \exp(-E_b / kT_{grain})$$

where ν_0 is a characteristic vibrational frequency of the adsorbed species, and E_b is the surface binding energy [60]. ν_0 is frequently estimated from the frequency of lattice vibrations [177/8] and generally ranges between $10^{12} - 10^{13}$ s^{-1}.

A number of experimental and theoretical values of the binding energy are listed in Table I. Large differences or discrepancies still exist between the values found by different authors. Some of these differences come from the different measurement techniques, from different definitions (mean value of the monolayer, ...) and also from different assumptions used to derive the values of E_b.

Another important parameter for the condensation or adsorption processes is the sticking (or condensation) coefficient, α, of a gas species on a surface. Its value corresponds to the probability of adsorption or condensation of a molecule or an atom after hitting a surface. The sticking coefficient depends on the kinetic energy of the gas species, on its ability to transfer this energy to the solid and on its surface binding energy. Therefore it is found to depends on the gas and grain temperatures. Theoretical calculations on the sticking probability of molecules, atoms and ions have been made for several astrophysical gas phase species and grain combinations [37, 106, 118, 134, 219]. At low temperature the sticking coefficient is generally expected to be close to unity for heavy species and for species having large binding energies (H_2O, large molecules, chemisorption, ...). However, it but may have low values even at 10 K - 20 K for molecules with low binding energy such as H_2, CO, N_2 or CH_4. For example Buch & Zhang [36] calculated the sticking coefficient of H and D atoms on a large cluster of amorphous H_2O ice and found values of 0.25 for H and 0.44 for D for a gas temperature of 100 K. This result shows a large isotopic effect that may be important for some grain processes (deuteration by surface reaction, ...). The values increase to 0.83 (H) and 0.91 (D) for a gas temperature of 10 K. Only a few experiments have been performed on the sticking coefficients of molecules or atoms on ices. From recent laboratory experiments we found that the condensation coefficient of N_2 on N_2 ice is about 0.1

near 30 K (Quirico & Schmitt, unpublished). For the sticking of H_2 molecules on H_2O surfaces at 3 K, Govers et al. [66] found a value of 0.07 ± 0.05, but this value increased to about 0.80 when the surface was covered by a layer of H_2 molecules. For the adsorption of H atoms at 350 K on an ice mixture of H_2O, CO_2, O_2 and N_2 at 3 K a sticking coefficient of 0.11 ± 0.05 is found [137]. Similar low values for the adsorption of hydrogen on a variety of ices are estimated by Hunt et al. [112]. The effect of surface temperature on the sticking of H and H_2 has been studied by Brackman & Fite [22]. They found a steep decrease to less than 0.1 below about 18 K. For H_2O on H_2O ice, Haynes et al. [96] derived a condensation coefficient decreasing from unity below about 60 K to about 0.65 near 200 K. Despite these few experiments, the sticking coefficient is still a largely unknown parameter in many important astrophysical cases. In addition, the charge of the grain (expected to be positive in diffuse regions and negative in dark clouds) may have a large effect on the sticking coefficient, especially for ions [205].

TABLE I: Experimental and theoretical values of the binding energies

Solid	Condensation E_b/k (ref.)	Gas / solid	Adsorption E_b/k (ref.)
H_2	230 K (180)	H_2 / H_2O:CH_3OH	550 K (180)
	650 K (130)		
		H_2 / H_2O	450 K (213)
			860 K (130)
		H_2 / CO	350 K (130)
		H / H_2O	350 K (213)
			300-700 K (36)
			750 K (130)
H_2O amorphous	4815 K (177)		
H_2O crystal.	5070 K (177)		
crystal.	6000 K (96)		
CO	960 K (177)	CO / H_2O	1740 K (178)
			1150 K (151)
CO_2	2690 K (178)	CO_2 / H_2O	2860 K (178)
		N_2 / H_2O	1700 K (213)
			1100 K (183, 185)
		Ar / H_2O	950 K (183, 185)
		CH_4 / H_2O	2600 K (213)
NH_3	2985 K (181)		
CH_3OH	4170 K (181)		

3) Gas trapping and release by grain mantles

When mixtures of different molecules stick on a grain, the trapping of a volatile species in a less volatile matrix (such as CO in H_2O ice) may occurs. The trapping efficiency depends on the gas phase composition, on the binding energy of the species, and on the temperature of the grain. The trapping of an adsorbed volatile species occurs when its residence time is larger than the time of monolayer formation of the other species. Experiments have been performed for the trapping of several important molecules (CO, CH_4, N_2, CO_2, H_2) by H_2O at temperatures between 10 K and 100 K [11, 13, 14, 64, 108, 125, 129, 177, 180, 184, 187/90]. For example, the trapping of CO is found to be efficient up to 40 K and to steeply decrease above this temperature under laboratory conditions.

When the temperature of H_2O-rich ice mixtures condensed at low temperature (10 - 20 K) is increased above the sublimation temperature of the trapped species, desorption occurs but at a much slower rate than for simple surface desorption. This is due to the high and narrow porosity of the amorphous ice matrix that substantially increase the adsorption energy of the adsorbed molecules. This desorption stops before all the molecules are desorbed. A small concentration of volatile species (~2% - 7%) remains firmly trapped in the grain mantle. Volume diffusion of these molecules through the grain mantle probably controls the desorption process; the desorption rate is probably very slow (unknown yet). If higher temperatures are reached (> 90 - 100 K) a fast and complete desorption occurs when the H_2O ice matrix crystallizes [125, 187, 188, 191].

These processes of gas trapping and release may apply for slow temperature changes that corresponds for example to temperature cycling of interstellar grain mantles from the inner parts to the outer parts of molecular clouds. They may also apply to fast evaporations induced by pulsed heating.

4) Evaporations by pulsed heating

Evaporations induced by an impulsive heating of the grain are in fact simple thermal evaporations that occur at temperatures higher than the steady state temperature of the grain. Both whole grain heating and local hot spot desorption have been considered [133]. It has been found that for whole grain heating, evaporation is the dominant cooling process above a critical temperature. The number of evaporated molecules depend on the heat stored in the grain above that temperature and on the latent heat of sublimation of the molecules. For grains containing CO, the critical temperature is about 25 K [133]. Below this temperature radiative cooling dominate. Several different sources of impulsive heating have been considered. They are grain-grain collisions, interactions with cosmic rays and with X-rays.

a) Grain-grain collisions

Grain-grain collisions can heat the whole grain by conversion of the of kinetic energy. The increase in temperature depends on the relative velocities of the grains and on its heat capacity. A collision velocity of about 50 m/s leads to a temperature increase of a grain (ice + silicate) from 10 K to about 30 K. At these relative velocities grain-grain collisions are probably not efficient to evaporate grain mantles. However, they may provide a triggering mechanism for chemical explosion of UV irradiated ice mantles [99] (see 6-a).

b) X-rays heating

Heating by X-rays mainly occurs by energy loss of the induced electrons (Auger effect). The temperature increase of the grain mainly depends on X-ray energy and on the grain size and composition. It is found that only moderate but frequent heating may occur for a narrow range of grain sizes (0.02 - 0.04 µm) [133]. Within that range X-rays may desorb molecules with small binding energies such as CO or trigger chemical explosions (see below, 6-a). For smaller and larger grain sizes the heating process is rapidly inefficient.

c) Interaction with cosmic rays

Heating by cosmic rays occurs by energy loss of the electrons excited and ejected on the track of the cosmic ray [56]. The energy deposited is proportional to the square of the atomic number (Z^2), therefore the highest Z cosmic rays are the most efficients. Iron is found to be the dominant source of heating of the grains. Whole grain heating occurs on grains with sizes less than 0.25 µm and is quite efficient for thermal desorption of molecules with low binding energy. More refractory ices (H_2O, NH_3, CH_3OH, ...) are not released to the gas phase by cosmic ray heating. Spot heating, which results from the initial depositing of the energy lost by the cosmic ray into a tube of radius of about 50 angstroms, dominates on larger grain sizes [47, 49, 91, 119, 133, 219].

5) Evaporations by sputtering

The evaporation processes by sputtering corresponds to all erosion processes involving direct transfer of energy between the particle or the ray and a molecule of the grain. They include gas-grain collision, photodesorption and cosmic ray sputtering.

a) Gas-grain collisions

The collision of a gas phase molecule with a grain can transfer the kinetic energy of the impinging molecule to another molecule adsorbed on the surface of the grain. The main gas phase molecule being H_2, high kinetic temperature or high relative velocities (cloud shock, ...) are necessary for the desorption of molecules such as CO. At average interstellar cloud temperature (~100 K) sputtering by gas phase molecules is completely negligible [1].

b) Photodesorption

Electronic excitation of molecules adsorbed on grains leads to the desorption of atoms (photodissociation), ions (photo-ionization) or neutral species (photo-ejection). The process of ejection occurs via excitation and conversion of the electronic excitation to nuclear motion [144]. Since the first model of photo-ejection of adsorbed molecules on grains by Watson and Salpeter [219] suggesting that this ejection process may be efficient, several experiments have been done on ices. However, in most cases the ejection of ions have only been studied for high photon energies (> 20 eV) [46, 58, 97, 135, 167, 173/4, 237]. Photodesorption yields of the order of 10^{-8} - 10^{-6} ion/photon are generally found [174]. Only very few experiments have measured the photodesorption of neutrals from ice surfaces, but no reliable yields have been obtained yet [68, 153]. Experiments in that direction are under way in several laboratories.

c) Cosmic rays

The interaction of cosmic rays with molecular ices leads to its erosion by ejection of atoms, ions and molecules which make up the solid or which are formed by chemical processes induced by the loss of energy of the particles [15, 25/29, 45, 116, 165, 168/9, 216, 239]. Large ionic clusters are also ejected [16, 128, 216]. For example, bombardment of H_2O ice produces clusters with formula $H(H_2O)_n^+$, $O(H_2O)_n^-$ and $OH(H_2O)_n^-$, n varying from 1 to more than 20, in addition to H^+, H_2^+, O^-, and OH^- ions, H_2O molecules and $(H_2O)_n$ clusters. New molecules, H_2 and O_2, are also observed. Erosion yields of neutrals range from a few molecules per incident ion with KeV/amu energies to 10^3-10^6 molecules/ion for incident particles energies of the order of MeV/amu. Erosion yields of ions are about 10^{-3}-10^{-4} that of the neutrals. The yields depend on the total energy loss of the incident ion in the material, i.e., on the thickness of material, on the kinetic energy of the ion (stopping power dE/dx [234]), and on the mass of the ion. Cosmic ray irradiation also induces molecular synthesis (see below).

The stability of ice grains against cosmic ray erosion has been evaluated inside T Tauri nebulae [161, 207].

6) Evaporations induced by chemical reactions

There are mainly two chemical processes that may lead to evaporations: the release of chemical energy stored in grain mantles in the form of stored radicals and the ejection or evaporation induced by reaction energy.

a) Release of chemical energy stored in grains

The sporadic release of chemical energy stored in grain in the form of frozen radicals could lead to the evaporation of part of the grain mantle [69, 70, 98, 99, 133, 194, 213]. This process has been called "explosive desorption". The radicals are produced within the grain mantle by UV irradiation. A triggering event, such as grain-grain collision or cosmic ray heating, is needed to heat the grain to a temperature where diffusion of the reactive species should lead to a chain reaction and the evaporation of the ice mantle. From laboratory experiments the triggering temperature is found to be about 27 K and the radical concentration in the sample to be between 1 to 2.6 % [99, 194]. Léger et al. [133] discussed the efficiency of X-ray and cosmic ray heating for triggering the chemical explosions. D'Hendecourt et al. [99] first deduced that an explosive event could evaporate nearly all the mantle of an interstellar grain. Schutte & Greenberg [194] further studied the process and found that only the most volatile molecules (CO, possibly O_2 and N_2, and partly O_3) have been evaporated during the chemical explosion, and that more refractory molecules are desorbed only very inefficiently.

b) Evaporation induced by reaction energy

Exothermic reactions between adsorbed species may lead directly to the ejection of the product molecules, and possibly of neighboring species by conversion of part of the energy of reaction to kinetic energy [4, 51, 159, 219]. The most well known case is the formation and ejection of H_2 at the surface of grains [106, 136]. There are potential astronomical evidences that NH, formed by hydrogenation of N, is predominantly released upon formation [230] in order to account for the detection of NH in diffuse clouds [145]. On the other hand, OH should have a large probability of retention in order to explain the formation of ice mantles in dark clouds [117, 230]. Duley & Williams [51] recently proposed that H_2 formation on amorphous H_2O ice creates a warm spot on the surface capable of desorbing any nearby adsorbed CO.

SURFACE REACTIONS

The first and most important surface reaction is the formation of molecular hydrogen by reaction of two adsorbed hydrogen atoms. Theoretical calculations show that the mobility of H atoms on different types of surface (silicates, carbonaceous and icy grains) must be very high; adsorbed H-atoms scan the surface of grains many times during their residence time. The binding energy of H-atoms on grains and the interstellar grain temperatures seems to be consistent with a high probability for a H-atom to stick on the surface, diffuse and encounter another physically or chemically adsorbed H-atom and then react to form H_2 before desorption from the surface [10, 36, 50, 65, 106/7, 130, 134, 134b, 202/3].

Considering the high efficiency of H_2 formation at grain surface it is likely that other reactions between atoms and molecules must occur. As a result of the high abundance and mobility of H atoms, hydrogenation of heavier atoms (O, C, N, S, ...) to form H_2O, CH_4, NH_3, H_2S, ..., are the next probable reactions, unless the surface has specific chemical properties that prevent the reaction of heavy atoms with H [117, 219/20]. All exothermic surface reactions between adsorbed species are considered possible if they can approach each other by surface diffusion. The energy barrier for surface diffusion is poorly known and is generally estimated to be a fixed fraction (1/3, 1/2) of the binding energy [210, 213, 218]. Atoms such as C, N, O and S can diffuse on the surface until they find a co-reactant. Molecules, other than H_2, and radicals are generally considered to be trapped in their adsorption site and therefore can only react with migrating co-reactants or with species coming directly from the gas phase. The efficiency of tunneling of H-atoms through substantial activation barriers implies that species such as CO, O_2, H_2CO, H_2S and O_3 may be hydrogenated by surface reaction [210]. The heat of reaction is supposed to be mainly transferred to the grain but a part may be converted into translational energy of the reaction product and may contribute to its ejection from the surface [213, 219] as in the case of H_2 [50, 106]. This initial translational energy or further conversion of vibrational energy into kinetic energy may also allow the newly formed molecule or radical to diffuse over the surface [30, 219]. If the abundance of H atoms on the surface is low, due either to the complete H -> H_2 conversion or to the reduced sticking of H atoms on grains [30], reactions involving radicals and heavy atoms become important. They lead to the formation of larger molecules such as CH_3OH and H_2CO. A large number of surface association reactions without activation energy have been considered by different authors [4/6, 34, 92, 93]. They generally involve reaction between free radicals, but a few radical-radical exchange reactions with a neutral molecule have been considered when the activation energy is negligible [6]. The reaction of CO with O, leading to CO_2, proposed by Tielens & Hagen [210], has been excluded by d'Hendecourt et al. [100] because they found that this reaction possesses some activation energy [75]. In recent models some exothermic reactions with significant activation barriers have been added. They involve molecules such as H_2, N_2, or some stable hydrocarbons and cyanopolyynes [92, 93]. Reactions involving H_2 are very important due to its high surface concentration at low temperature and its high surface mobility. No heterogeneous catalytic reactions (decreased activation energy by physical or chemical adsorption) have been considered yet, but a catalytic cycle that converts H into H_2 via HN_2 and N_2H_2 has been proposed (Hasegawa & Herbst 1993). Direct reactions between gas phase ions and surface molecules seem likely, but have not been yet considered because of uncertainties about ion-surface interactions [218].

MOLECULAR SYNTHESIS BY U.V. PHOTOLYSIS

Due to the presence of an appreciable interstellar ultraviolet flux, the formation of radicals and molecules inside icy grain mantles by photochemical processes has been considered to be an efficient process [55, 67, 68]. Laboratory experiments have largely confirmed these views [72].

Photolysis with vacuum ultraviolet radiation ($h\nu > 6$ eV) of pure or mixed ices at low temperature (10 - 20 K) leads to the formation of radicals and new molecules and the destruction of the initial molecules. The primary step of photolysis occurs by breaking molecular bonds by ultraviolet photons and by subsequent formation of two radicals (H abstraction, for example) or one radical and a molecule (H_2 abstraction, for example). The radical production efficiency per UV photon is found to be about 50% [194]. Each radical either recombines with a surrounding radical, if available, to form a new molecule, or is trapped in the ices if the energy barrier for reaction with the surrounding molecules is too high and if the temperature of the ice is low (less than about 20 K). Above that temperature the radical will diffuse until it encounters another radical and reacts [114]. The maximum concentration of radical trapped in ice ranges between about 1% and 2.6% [99, 194]. These radicals are then available for further reactions. If a transient heating occurs, rapid diffusion and recombination of these radicals will lead to chemical explosion and partial evaporation of the ice mantle [99, 194]. Strong molecules such as CO may be excited by an UV photon and react with another neutral molecule. Further steps in the cycle of radical formation and reaction lead to more and more complex molecules [2, 74, 79, 83, 98, 99, 140, 192, 196, 244]. Average production rates are of the order of 0.01 to 0.1 molecule per absorbed photon [115]. Finally a non-volatile organic residue begins to be produced for UV doses larger than about 10^{20}-10^{21} photon cm^{-2} [7, 24, 72, 113, 140, 193]. Polymeric molecules with atomic masses larger than 400 are found [24, 72]. An additional particular reaction, the proton transfer reaction, has been found to occur in irradiated ice mixtures. This type of reaction leads to the formation of positive and negative ions [77, 78, 80].

A large number of photolysis experiments have been performed with various initial molecules and compositions and various UV photon doses [2, 74, 76/79,

[83, 98, 99, 192, 196, 233, 244]. CO_2, H_2CO, CH_3OH and O_3 are among the main molecules formed by photolysis of most ice mixtures containing carbon and oxygen. The photoproduction rate coefficients of the three first molecules have been estimated from these experiments. They range between $1.5 \cdot 10^{-3}$ molecule per UV photon for H_2CO and CH_3OH to $4.5 \cdot 10^{-3}$ molecule per photon for CO_2 [23, 72, 198].

MOLECULAR SYNTHESIS BY COSMIC RAY IRRADIATION

In addition to erosion, the energy loss of impinging ions in ice mantles leads to chemical effects along the track of the particle in the solid [27, 45, 162]. Chemical reactions occur between the highly concentrated reactive species (excited neutral species and molecular ions) produced along the track either directly by the incident ion or by secondary electrons. Both exothermic and endothermic reactions, even with high activation energy, can occur. "Suprathermal reactions" have also been proposed [170]. Simple molecules (H_2, O_2, CH_4, H_2CO, ...) as well as simple and complex organic molecules are easily produced. The new molecules are formed inside the ice and a part of the more mobile molecules (like H_2 and O_2) can escape the solid by diffusion along the hot cylinder around the ion track. The formation and release of molecular hydrogen has received special attention [12, 164, 236, 242]. They found that in most parts of cold dense clouds the formation rate of H_2 by cosmic rays was more important than that due to the formation on the surface of amorphous ice mantles [202/3].

Experiments on cosmic ray irradiations of pure and mixed ices have been performed by several authors for a variety of ice compositions and temperatures, sample thicknesses, incident ions, and ion energies and doses [12, 16/20, 27, 45, 87/90, 148/9, 152, 162/4, 170/2, 240/1]. High total yield of molecule formation are generally found (about 10-100 molecules per ion). The irradiation of carbon-bearing ices (CH_4, CO and mixtures containing these molecules) leads to a nonvolatile carbon-rich solid at high irradiation doses by preferential release of oxygen and hydrogen. It evolves to a pure carbon layer at still higher doses [39, 44, 59, 115, 148/9, 240/1].

ACKNOWLEDGMENT

This work has been supported by several grants of the GdR "Physico-Chimie des Molécules et Grains Interstellaires" of the French Centre National de la Recherche Scientifique.

REFERENCES

1. Aannestad, P., 1973. *Astrophys. J. Supl. Ser.,* **25,** 505.
2. Allamandola, L.J., Sandford S.A., & Valero G.J., 1988. *Icarus,* **76,** 225.
3. Allamandola, L.J., Sandford S.A., Tielens A.G.G.M., & Herbst T., 1992. *Astrophys. J.,* **399,** 134.
4. Allen, M., & Robinson G.W., 1975. *Astrophys. J.,* **195,** 81.
5. Allen, M., & Robinson G.W., 1976. *Astrophys. J.,* **207,** 745.
6. Allen, M., & Robinson G.W., 1977. *Astrophys. J.,* **212,** 396.
7. Argawal, V.K., Schutte W., Greenberg J.M., Ferris J.P., Briggs R., Connor S., van de Bult C.P.E.M., & Bass F., 1985. In *Origins of Life,* **16,** 21.

8. Aronowitz, S., & Chang S., 1985. *Astrophys. J.,* **293**, 243.
9. Baratta, G.A., Leto G., Spinella F., Strazzulla G., & Foti G., 1991. *Astron. & Astrophys.,* **252**, 424.
10. Barlow, M.J., & Silk J. 1986. *Astrophys. J.,* **207**, 131.
11. Bar-Nun, A., Herman A.G., Laufer D., & Rappaport M.L., 1985a. *Icarus,* **63**, 317.
12. Bar-Nun, A., Herman A.G., & Rappaport M.L., 1985b. *Surf. Sci.,* **150**, 143.
13. Bar-Nun, A., Dror J., Kochavi E., & Laufer D., 1987. *Phys. Rev. B,* **35**, 2427.
14. Bar-Nun, A., Kleinfeld I., & Kochavi E. 1988. *Phys. Rev. B,* **38**, 7749.
15. Bénit, J., 1987. *Ph.D. Thesis,* Université Paris-Sud, Orsay, France.
16. Bénit, J., Bibring J.-P., Della Negra S., Le Beyec Y., & Rocard F., 1986. *Rad. Effects,* **99**, 105.
17. Bénit, J., Bibring J.-P., Della Negra S., Le Beyec Y., Mendenhall M., Rocard F., & Standing K., 1987. *Nucl. instrum. Methods B,* **19/20**, 838.
18. Bénit, J., Bibring J.-P., & Rocard F., 1988. *Nucl. instrum. Methods B,* **32**, 349.
19. Bénit, J., & Brown W.L., 1990. *Nucl. instrum. Methods B,* **46**, 448.
20. Bibring J.-P., & Rocard F., 1984. *Adv. Space Res.,* **4**, 103.
21. Blake, G.A., Sutton E.C., Masson C.R., & Phillips T.G., 1987. *Astrophys. J.,* **315**, 621.
22. Brackmann, R.T., & Fite W.L., 1961. *J. Chem. Phys.,* **34**, 1572.
23. Breukers, R., d'Hendecourt, L.B., & Greenberg J.M., 1992. In: *Chemistry and Spectroscopy of Interstellar Molecules,* Bohme, D.K., Herbst, E., Kaifu, N., & Saito, S., eds. (Univ. of Tokyo Press), p. 273.
24. Briggs, R., Ertem G., Ferris G.P., Greenberg J.M., McCain P.J., Mendoza-Gomez C.X., & schutte W., 1992. In: *Origins of Life and Evolution of the Biosphere* (in press).
25. Brown, W.L. Lanzerotti R.E., Poate J.M. & Augustyniak W.M., 1978. *Phys. Rev. Lett.,* **40**, 1027.
26. Brown, W.L., Augustyniak W.M., Brody E., Cooper B.H., Lanzerotti L.J., Ramirez A., Evatt R., & Johnson R.E., 1980. *Nucl. Instrum. Methods B,* **170**, 321.
27. Brown, W.L., Augustyniak W.M., Simmons E.H., Marcantonio K.J., Lanzerotti L.J., Johnson R.E., Reimann C.T., Boring J.W., Foti G., & Pirronello V., 1982. *Nucl. Instrum. Methods B,* **198**, 8.
28. Brown, W.L., Augustyniak W.M., Marcantonio K.J., Simmons E.H., Boring J.W., Johnson R.E., & Reimann C.T. 1984. *Nucl. Instrum. Meth. B,* **1**, 307.
29. Brown, W.L., & Johnson R.E., 1986. *Nucl. Instrum. Methods B,* **13**, 295.
30. Brown, P.D., 1990. *Mon. Not. R. Astron. Soc.,* **243**, 65.
31. Brown, P.D., Charnley S.B., & Millar T.J., 1988. *Mon. Not. R. Astron. Soc.,* **231**, 409.
32. Brown, P.D., & Millar T.J., 1989a. *Mon. Not. R. Astron. Soc.,* **237**, 661.
33. Brown, P.D., & Millar T.J., 1989b. *Mon. Not. R. Astron. Soc.,* **240**, 25p.
34. Brown, P.D., & Charnley S.B., 1990. *Mon. Not. R. Astron. Soc.,* **244**, 432.
35. Brown, P.D., & Charnley S.B., 1991. *Mon. Not. R. Astron. Soc.,* **249**, 69.
36. Buch, V., & Zhang Q., 1991. *Astrophys. J.,* **379**, 647.
37. Burke, J.R., & Hollenbach D.J. 1984. *Astrophys. J.,* **265**, 223.

38. Butchart, I., McFadzean A.D., Whittet D.C.B., Geballe T.R., & Greenberg J.M., 1986. *Astron. & Astrophys.,* **154**, L5.
39. Calcagno, L., Foti G., & Strazzulla G., 1983. Lett. *Nuovo Cimento,* **37**, 303.
40. Charnley, S.B., Dyson J.E., Hartquist T.W., & Williams D.A., 1988. *Mon. Not. R. Astron. Soc.,* **231**, 269.
41. Charnley, S.B., Tielens A.G.G.M., & Millar T.J., 1992. *Astrophys. J.,* **399**, L71.
42. Charnley, S.B., 1994. In: *Physical Chemistry of Molecules and Grains in Space,* AIP Conf. Proceeding Series, This volume.
43. Chen, W.P., & Graham J.A., 1993. *Astrophys. J.,* **409**, 319.
44. Cheng, A.F. & Lanzerotti L.J., 1978. *J. Geophys. Res.,* **83**, 2597.
45. Ciavola G., Foti G., Torrisi L., Pirronello V., & Strazzulla G., 1982. *Rad. Effects,* **67**, 167.
46. Clampett, R., & Gowland L., 1969. *Nature,* **223**, 815.
47. Duley, W.W., 1973. *Natue Phys. Sci.,* **244**, 57.
48. Duley, W.W., 1974. *Astrophys. Space Sci.,* **26**, 199.
49. Duley, W.W., 1976. *Astrophys. Space Sci.,* **46**, 261.
50. Duley, W.W., & Williams D.A., 1986. *Mon. Not. R. Astron. Soc.,* **223**, 177.
51. Duley, W.W., & Williams D.A., 1993. *Mon. Not. R. Astron. Soc.,* **260**, 37.
52. Ehrenfreund, P., Breukers R., d'Hendecourt L., & Greenberg J.M., 1992. *Astron. & Astrophys.,* **260**, 431.
53. Einstein, T.L., Hertz J.A., & Schieffer J.R., 1980. In *Theory of Chemisorption,* ed. J.R. Smith (Springer Verlag, Berlin), p. 183.
54. Eiroa, C., & Hodapp K.-W., 1989. *Astron. & Astrophys.,* **210**, 345.
55. Ewing G.E., Thompson W.E., & Pimentel G.C., 1960. *J. Chem. Phys.,* **32**, 927.
56. Fano, U., 1963. *Ann. Rev. Nucl. Sci.,* **13**, 1.
57. Federman, S.R., & Allen M. 1991. *Astrophys. J.,* **375**, 157.
58. Feulner, P., Auer S., Müller T., Pushmann A., & Menzel D., 1988. In: *DIET III, Springer Series in Surface sciences,* **13**, R.H. Stulen, & M.L. Knotek, Eds., (Springer-Verlag, Heidelberg).
59. Foti, G., Calcagno L., Sheng K., & Strazzulla G., 1984. *Nature,* **310**, 126.
60. Frenkel, J., 1924. *Z. Physik,* **26**, 117.
61. Geballe, T.R., 1986. *Astron. & Astrophys.,* **162**, 248.
62. Geballe, T.R., Baas F., Greenberg J.M., & Schutte W., 1985. *Astron. & Astrophys.,* **146**, L6.
63. Geballe, T.R., Kim Y.H., Knacke R.F., & Noll K.S. 1988. *Astrophys. J.,* **326**, L65.
64. Ghormley, J.A., 1967. *J. Chem. Phys.,* **46**, 1321.
65. Gould, R.J., & Salpeter E.E., 1963. *Astrophys. J.,* **138**, 393.
66. Govers, T.R., Mattera L., & Scoles G., 1980. *J. Chem. Phys.,* **72**, 5446.
67. Greenberg, J.M. 1963. *Ann. Rev. Astron. Astrophys.,* **15**, 267.
68. Greenberg, J.M. 1973. In: *Interstellar Dust and Related Topics,* J.M. Greenberg & H.C. van de Hulst eds., (Reidel, Dordrecht), p. 413.
69. Greenberg, J.M. 1979. In: Star and star Systems, ed. B.E. Westerlund, (Reidel, Dordrecht), p. 173.
70. Greenberg, J.M., Yencha A.J., Corbett J.W., & Frisch H.L., 1972. *Mem. Soc. Roy. Sci. Liège,* **3**, 425.

71. Greenberg, J.M. & Yencha A.J. 1973. In: *Interstellar Dust and Related Topics*, J.M. Greenberg & H.C. van de Hulst eds., (Reidel, Dordrecht), p. 369.
72. Greenberg, J.M., Mendoza-Gomez C.X., de Groot M.S., & Breukers R., 1993. In: *Dust and Chemistry in Astronomy*, Millar, T.J., & Williams D.A., eds., (IOP Publ. Ltd., Bristol), p. 271.
73. Gregg, S.J., & Sing K.S.W., 1982. *Adsorption, Surface Area and Porosity*, (Academic press, New York).
74. Grim, R.J.A., 1988. *Ph.D. Thesis*, Leiden University, the Netherlands.
75. Grim, R.J.A., & d'Hendecourt L.B., 1986. *Astron. & Astrophys.*, **167**, 161.
76. Grim, R.J.A., & Greenberg J.M., 1987a. *Astron. & Astrophys.*, **181**, 155.
77. Grim, R.J.A., & Greenberg J.M., 1987b. *Astrophys. J.*, **321**, L91.
78. Grim, R.J.A., & Greenberg J.M., 1988. In: *Experiments on Cosmic Dust Analogues*. Eds. E. Bussoletti, C. Fusco & G. Longo (Kluver, Dordrecht), *Astrophys. Space Sci. Lib.*, **149**, 299..
79. Grim, R.J.A., Greenberg J.M., de Groot M.S., Baas F., Schutte W.A., & Schmitt B., 1989a. *Astron. & Astrophys. Supl. Ser.*, **78**, 161.
80. Grim, R.J.A., Greenberg J.M., Schutte W.A., & Schmitt B., 1989b. *Astrophys. J.*, **341**, L87.
81. Grim, R.J.A., Baas F., Geballe T.R., Greenberg J.M., & Schutte W.A., 1991. *Astron. & Astrophys.*, **243**, 473.
82. Hagen, W., Allamandolla L.J., & Greenberg J.M., 1979. *Astrophys. J. Supl. Ser.*, **65**, 215.
83. Hagen, W., 1982. *Chemistry and Infrared Spectroscopy of Interstellar Grains*. Ph.D. Thesis, Leiden University, the Netherlands.
84. Hagen W., Tielens A.G.G.M., & Greenberg J.M. 1981. *Chem. Phys.*, **56**, 367.
85. Hagen, W., Tielens A.G.G.M., & Greenberg J.M., 1983. *Astron. & Astrophys. Supl. Ser.*, **51**, 389.
86. Hanson, D.M., Stockbauer R., & Madey T.E., 1982. *J. Chem. Phys.*, **76**, 5639.
87. Haring R.A., Haring A., Klein F.S., Kummel A.C., & de Vries A.E., 1983. *Nucl. Instrum. Methods*, **211**, 529.
88. Haring R.A., Kolfschoten A.W., & de Vries A.E., 1984a. *Nucl. Instrum. Methods B*, **2**, 544.
89. Haring R.A., Pedrys R., Oostra D.J., Haring A., & de Vries A.E., 1984b. *Nucl. Instrum. Methods B*, **5**, 476.
90. Haring R.A., Pedrys R., Oostra D.J., Haring A., & de Vries A.E., 1984c. *Nucl. Instrum. Methods B*, **5**, 483.
91. Hartquist, T.W., & Williams D.A. 1990. *Mon. Not. R. Astron. Soc.*, **247**, 343.
92. Hasegawa, T.I., Herbst E., & Leung C.M., 1992. *Astrophys. J. Supl. Ser.*, **82**, 167.
93. Hasegawa, T.I., & Herbst E., 1993a. *Mon. Not. R. Astron. Soc.*, **261**, 83.
94. Hasegawa, T.I., & Herbst E., 1993b. *Mon. Not. R. Astron. Soc.*, **263**, 589.
95. Hasselbrink, E., 1992. *Comments At. Mol. Phys.*, **27**, 265.
96. Haynes, D.R., Tro N.J., & George S.M., 1992. *J. Phys. Chem.*, **96**, 8502.
97. Hellner, L., Dujardin G., Ramage M.J., Philippe L., Comtet G., & Rose M., 1994. In: *Physical Chemistry of Molecules and Grains in Space*, AIP Conf. Proceeding Series, This volume.

98. d'Hendecourt, L.B., 1984. *Ph.D. thesis*, Leiden University, The Netherlands.
99. d'Hendecourt, L.B., Allamandola, L.J., Baas F., & Greenberg J.M., 1982. *Astron. & Astrophys.*, **109**, L12.
100. d'Hendecourt, L.B., Allamandola, L.J., & Greenberg J.M., 1985. *Astron. & Astrophys.*, **152**, 130.
101. d'Hendecourt, L.B., Allamandola, L.J., Grim R.J.A., & Greenberg J.M., 1986. *Astron. & Astrophys.*, **158**, 119.
102. d'Hendecourt, L.B., Allamandola, L.J., 1986. *Astron. & Astrophys. Supl. Ser.*, **64**, 453.
103. d'Hendecourt, L.B., & Jourdain de Muizon M., 1990. *Astron. & Astrophys.*, **223**, L5.
104. Henkel C., Mauersberger R., Wilson T.L., Snyder L.E., Menten K., & Wouterloot J.G.A., 1987. *Astron. & Astrophys.*, **182**, 299.
105. Hodapp, K.W., Sellgren K., and Nagata T. 1988. *Astrophys. J.*, **326**, L61.
106. Hollenbach, D., & Salpeter E.E., 1970. *J. Chem. Phys.*, **53**, 79.
107. Hollenbach, D., & Salpeter E.E., 1971. *Astrophys. J.*, **163**, 155.
108. Hudson, L., & Donn B. 1991. *Icarus*, **94**, 326.
109. Hudson, L., & Moore M., 1993. *Astrophys. J.*, **404**, L29.
110. van de Hulst, H.C., 1949. *Rech. Astr. Obs. Utrecht*, **11**, part 2.
111. van de Hulst, H.C., 1957. *Light Scattering by Small Particles*, (John Wiley, New-York).
112. Hunt, A.L., Taylor C.E., & Omohundro J.E., 1962. *Adv. Cryo. Eng.*, **8**, 100.
113. Iglesias, E., 1977. *Astrophys. J.*, **218**, 697.
114. van Ijzendoorn, L.J., 1985. *Ph.D. thesis*, Leiden University, The Netherlands.
115. Jenniskens, P., Baratta, G.A., Kouchi A., de Groot M.S., Greenberg J.M., & Strazzulla G. 1993. *Astron. & Astrophys.*, **273**, 583.
116. Johnson R.E., & Brown, W.L., 1983. *Nucl. Instrum. Methods B*, **209/210**, 469.
117. Jones, A.P., & Williams D.A., 1984. *Mon. Not. R. Astron. Soc.*, **209**, 955.
118. Jones, A.P., & Williams D.A., 1985. *Mon. Not. R. Astron. Soc.*, **217**, 413.
119. de Jong, T., Kamijo, F. 1973. *Astron. & Astrophys.*, **25**, 363.
120. Jura, M., 1975. *Astrophys. J.*, **197**, 575.
121. Kerr, T.H., Adamson A.J., & Whittet, D.C.B., 1991. *Mon. Not. R. Astron. Soc.*, **251**, 60p.
122. Kerr, T.H., Adamson A.J., & Whittet, D.C.B., 1993. *Mon. Not. R. Astron. Soc.*, **262**, 1047.
123. Knacke, R.F., & Larson H.P. 1991. *Astrophys. J.*, **367**, 162.
124. Knaap, H.F.P., van den Meijdenberg C.J.N., Beenakker J.J.M., & van de Hulst H.C., 1966. *Bull. Astron. Insts. Neth.*, **18**, 256.
125. Kouchi, A., 1990. *J. Cryst. Growth*, **99**, 1220.
126. Lacy, J.H., Baas F., Allamandola L.J., Persson S.E., McGregor P.J., Lonsdale C.J., Geballe T.R., & vande Bult C.E.P., 1984. *Astrophys. J.*, **276**, 533.
127. Lacy, J.H., Carr J.S., Evans N.J. II, Baas F., Achtermann J.M., & Arens J.F., 1991. *Astrophys. J.*, **376**, 556.
128. Lancaster, G.M., Honda F., Fukuda Y., & Rabalais J.W. 1979. *J. Am. Chem. Soc.*, **101**, 8.

129. Laufer D., Kochavi E., & Bar-Nun, A., 1987. *Phys. Rev.*, **B 36**, 9219.
130. Lee, T.J. 1972. *Nature Phys. Sci.*, **237**, 99.
131. Léger, A., 1983. *Astron. & Astrophys.*, **123**, 271.
132. Léger, A., Gauthier S., Défourneau D., & Rouan D., 1983. *Astron. & Astrophys.*, **117**, 164.
133. Léger, A., Jura M. & Omont A., 1985. *Astron. & Astrophys.*, **144**, 147.
134. Leitch-Devlin M.A., & Williams D.A. 1984. *Mon. Not. R. Astron. Soc.*, **210**, 577.
134b. Leitch-Devlin, M.A., & Williams D.A., 1985. *Mon. Not. R. Astron. Soc.*, **213**, 295.
135. Madey, T.E., & Yates J.T Jr., 1977. *Chem. Phys. Lett.*, **51**, 77.
136. Marenco, G., Schutte A., Scoles G., & Tommasini F., 1972. *J. Vac. Sci. Technol.*, **9**, 824.
137. Mattera, L., 1978. *Ph.D. Thesis,* University of Waterloo, Ontario.
138. Mayer, E., & Pletzer R., 1986. *Nature,* **319**, 298.
139. McFadzean A.D., Whittet, D.C.B., Longmore A.J., Bode M.F., & Adamson A.J., 1989. *Mon. Not. R. Astron. Soc.*, **241**, 873.
140. Mendoza-Gomez, C.X., 1992. *Ph.D. thesis,* Leiden University, The Netherlands.
141. Menten, K.M., Walmsley C.M., Henkel C., & Wilson T.L., 1986a. *Astron. & Astrophys.*, **157**, 318.
142. Menten, K.M., Walmsley C.M., Henkel C., Wilson T.L., Snyder L.E., Hollis J.M., & Lovas F.J., 1986b. *Astron. & Astrophys.*, **169**, 271.
143. Menten, K.M., Walmsley C.M., Henkel C., & Wilson T.L., 1988. *Astron. & Astrophys.*, **198**, 253.
144. Menzel, D., 1986. *Nucl. Instr. & Meth. Phys. Res.*, **B13**, 507.
145. Meyer, D.M., & Roth K.C., 1991. *Astrophys. J.*, **376**, L49.
146. Millar, T.J., & Nejad L.A.M. 1985. *Mon. Not. R. Astron. Soc.*, **217**, 507.
147. Millar, T.J., & Williams D.A., eds., 1993. *Dust and Chemistry in Astronomy,* (IOP Publ. Ltd., Bristol).
148. Moore, M., & Donn B., 1982. *Astrophys. J.*, **257**, L47.
149. Moore, M., Donn B., Khanna R., & A'Hearn M.F., 1983. *Icarus,* **54**, 388.
150. Moore, M., & Hudson L., 1992. *Astrophys. J.*, **401**, 353.
151. Nair, N.K., Adamson A.W., 1970. *J. Phys. Chem.*, **74**, 2229.
152. Nebeling, B., Roessler K., & Stoecklin G., 1985. *Radiochim. Acta,* **38**, 15.
153. Nishi, N., Shinohara H., & Okuyama T., 1984. *J. Chem. Phys.*, **80**, 3898.
154. Olofsson 1984. *Astron. & Astrophys.*, **134**, 36.
155. Omont, A., Moseley S.H., Forveille T., Glaccum W.J., Harvey P.M., Likkel L., Loewenstein R.F., & Lisse C.M. 1990. *Astrophys. J.*, **355**, L27.
156. Pauls, T.A., Wilson T.L., Bieging, J.H., & Martin R.N., 1983. *Astron. & Astrophys.*, **124**, 123.
157. Palumbo, M.E., & Strazzulla G., 1993. *Astron. & Astrophys.*, **269**, 568.
158. Pearl, J., Ngoh M., Ospina M., & Khanna R., 1991. *J. Geophys. Res.*, **96**, 17477.
159. Pickles, J.B., & Williams D.A., 1977a. *Astrophys. Space Sci.*, **52**, 443.
160. Pickles, J.B., & Williams D.A., 1977b. *Astrophys. Space Sci.*, **52**, 453.
161. Pirronello, V., 1987. *Nucl. Instrum. Methods B,* **19/20**, 959.

162. Pirronello, V., Brown W.L., Lanzerotti L.J., Marcantonio K.J., & Simmons E.H., 1982. *Astrophys. J.*, **262,** 636.
163. Pirronello, V., Brown W.L., Lanzerotti L.J., Lanzafame G., & Averna D., 1988a. In: *Dust in the Universe,* eds. M.E. Bailey & D.A. Williams (Cambridge Univ. Press, Cambridge), p. 281.
164. Pirronello, V., Brown W.L., Lanzerotti L.J., & Maclennan C.G., 1988b. In: *Experiments on Cosmic Dust Analogues.* Eds. E. Bussoletti, C. Fusco & G. Longo (Kluver, Dordrecht) p. 287.
165. Pirronello, V., 1993. Millar, T.J., & Williams D.A., eds., In: *Dust and Chemistry in Astronomy,* (IOP Publ. Ltd., Bristol), p. 297.
166. Plambeck, R.L., & Wright M.C.H., 1987. *Astrophys. J.*, **317,** L101.
167. Prince, R.H., & Floyd G.R., 1976. *Chem. Phys. Lett.*, **43,** 326.
168. Rocard, F., 1986. *Ph.D. Thesis,* Université Paris-Sud, Orsay, France.
169. Rocard, F., Bénit, J., Bibring J.-P., Ledu D., Meunier R., 1986. *Rad. Effects,* **99,** 97.
170. Roessler K., 1986. *Rad. Effects,* **99,** 21.
171. Roessler K., 1987. *Radiochim. Acta,* **42,** 123.
172. Roessler K., 1988. *Nucl. Instrum. Methods B,* **32,** 519.
173. Rosenberg, R.A., Rehn V., Jones V.O., Green A.K., Parks C.C., & Stulen R.H., 1981. *Chem. Phys. Lett.*, **80,** 488.
174. Rosenberg, R.A., Rehn V., Green A.K., LaRoe P.R., & Parks C.C., 1983. In: *DIET I, Springer Series in Chemical Physics,* **24,** N.H. Tolk, M.M. Traum, J.C. Tully, & T.E. Madey, Eds., (Springer-Verlag, Berlin Heidelberg).
175. Rouan, D., Omont A., Lacombe F., & Forveille T. 1988. *Astron. & Astrophys.*, **189,** L3.
176. Sandford, S.A., Allamandola L.J., Tielens A.G.G.M., & Valero G.J., 1988. *Astrophys. J.*, **329,** 498.
177. Sandford, S.A., & Allamandola L.J., 1988. *Icarus,* **76,** 201.
178. Sandford, S.A., & Allamandola L.J., 1990a. *Icarus,* **87,** 188.
179. Sandford, S.A., & Allamandola L.J., 1990b. *Astrophys. J.*, **355,** 357.
180. Sandford, S.A., & Allamandola L.J., 1993a. *Astrophys. J.*, **409,** L65.
181. Sandford, S.A., & Allamandola L.J., 1993b. *Astrophys. J.*, in press.
182. Sato, S., Nagata T., Tanaka M., & Yamamoto T., 1990. *Astrophys. J.*, **359,** 192.
183. Schmitt, B., 1986. *La surface de la Glace: Structure, Dynamique et Interactions - Implications Astrophysiques.* Thesis, University of Grenoble, France.
184. Schmitt, B., & Klinger J., 1987. In: *Symposium on the Diversity and Similarity of Comets. ESA Spec. Publ.*, **278,** 613.
185. Schmitt, B., Ocampo J., & Klinger J., 1987. *Suppl. J. de Physique C1,* **48,** 519.
187. Schmitt, B., Grim R.J.A., & Greenberg J.M., 1988a. In: *Experiments on Cosmic Dust Analogues.* Eds. E. Bussoletti, C. Fusco & G. Longo (Kluver, Dordrecht), *Astrophys. Space Sci. Lib.*, **149,** 259..
188. Schmitt, B., Grim R.J.A., & Greenberg J.M., 1988b. In: *Dust in the Universe.* M.E. Bailey & D.A. Williams, Eds (Cambridge Univ. Press, Cambridge), p. 291.
189. Schmitt, B., Greenberg J.M., & Grim R.J.A., 1989a. *Astrophys. J.*, **340,** L33.

190. Schmitt, B., Grim R.J.A., & Greenberg J.M., 1989b. In: *Infrared Spectroscopy in Astronomy. ESA Spec. Publ.*, **290**, 213.
191. Schmitt, B., Espinasse S., Grim R.J.A., Greenberg J.M., & Klinger J., 1989c. In: *Physics and Mechanics of cometary Materials, ESA Spec. Publ.*, **302**, 185.
192. Schutte, W.A., 1988. *The Evolution of Interstellar Organic Grain Mantles*, Thesis, Leiden University, The Netherlands.
193. Schutte, W.A., & Geenberg J.M., 1986. In: *Light on Dark Matter*, ed. F.F. Israel, (Reidel, Dordrecht), p. 229.
194. Schutte, W.A., & Geenberg J.M., 1991. *Astron. & Astrophys.*, **244**, 190.
195. Schutte, W.A., Geballe T.R., van Dishoeck E.F., & Geenberg J.M., 1994. In: *Physical Chemistry of Molecules and Grains in Space*, AIP Conf. Proceeding Series, This volume.
196. Schutte, W.A., Allamandolla L.J., & Sandford S.A., 1993a. *Science*, **259**, 1143.
197. Seki, J., & Hasegawa H. 1983. *Astrophys. Space Sci.*, **94**, 177.
198. Shalabiea, O.M., & Greenberg J.M., 1994. In: *Physical Chemistry of Molecules and Grains in Space*, AIP Conf. Proceeding Series, This volume.
199. Smith, R.G., Sellgren K., & Tokunaga A.T., 1988. *Astrophys. J.*, **334**, 209.
200. Smith, R.G., Sellgren K., & Tokunaga A.T., 1989. *Astrophys. J.*, **344**, 413.
201. Smith, R.G., Sellgren K., & Brooke T.Y., 1993. *Mon. Not. R. Astron. Soc.*, **263**, 749.
202. Smoluchowski, R., 1981. *Astrophys. Space Sci.*, **75**, 353.
203. Smoluchowski, R., 1983. *J. Phys. Chem*, **87**, 4229.
204. Soifer, B., Willner S.P., Capps R.W., & Rudy R.J., 1981. *Astrophys. J.*, **250**, 631.
205. Spitzer, L., 1948. *Astrophys. J.*, **107**, 6.
206. Strazulla, G., Calgagno L., & Foti G., 1983a. *Mon. Not. R. Astron. Soc.*, **204**, 59p.
207. Strazulla, G., Pirronello V., & Foti G., 1983b. *Astrophys. J.*, **271**, 255.
208. Sweitzer, J.S., 1978. *Astrophys. J.*, **225**, 116.
209. Tanaka M., Sato S., Nagata T., & Yamamoto T., 1990. *Astrophys. J.*, **352**, 724.
209. Tielens, A.G.G.M., 1983. *Astron. & Astrophys.*, **119**, 177.
210. Tielens, A.G.G.M., & Hagen W. 1982. *Astron. & Astrophys.*, **114**, 245.
211. Tielens, A.G.G.M., Allamandolla L.J., Bregman J., Goebel J., d'Hendecourt L.B., & witteborn F.C., 1984. *Astrophys. J.*, **287**, 697.
212. Tielens, A.G.G.M., & Allamandolla L.J. 1987a. In: *Physical Processes in Interstellar Clouds*, eds. Morfill G.E. & Scholer M. (Reidel, Dordrecht), 333.
213. Tielens, A.G.G.M., & Allamandolla L.J. 1987b. in *Interstellar Processes*, D.J. Hollenbach & H.A. Thronson Jr., eds., (Reidel, Dordrecht), p. 397.
214. Tielens, A.G.G.M., Tokunaga, A.T., Geballe T.R., & Baas F., 1991. *Astrophys. J.*, **381**, 181.
215. Turner, B.E., 1990. *Astrophys. J.*, **362**, L29.
216. de Vries, A.E. , Haring R.A., Haring A., Klein F.S., Kummel A.C., & Saris F.W., 1984. *J. Phys. Chem.*, **88**, 4510.
217. Walmsley, C.M., Hermsen W., Henkel C., Mauersberger R., & Wilson T.L., 1987. *Astron. & Astrophys.*, **172**, 311.

218. Watson, W.D., 1976. *Rev. Mod. Phys.*, **48**, 513.
219. Watson, W.D., & Salpeter E.E., 1972a. *Astrophys. J.*, **174**, 321.
220. Watson, W.D., & Salpeter E.E., 1972b. *Astrophys. J.*, **175**, 659.
221. Whittet, D.C.B., 1988. in *Dust in the Universe,* eds Bailey M.E.& williams D.A., Cambridge University Press, Cambridge, p. 25.
222. Whittet, D.C.B., 1993. eds. Millar, T.J., & Williams D.A., In: *Dust and Chemistry in Astronomy,* (IOP Publ. Ltd., Bristol), p. 9.
223. Whittet, D.C.B., Bode M.F., Longmore A.J., Baines D.W.T., & Evans A., 1983. *Nature,* **303**, 218.
224. Whittet, D.C.B., Longmore A.J., & McFadzean A.D., 1985. *Mon. Not. R. Astron. Soc.,* **216**, 45p.
225. Whittet, D.C.B., Bode M.F., Longmore A.J., Adamson A.J., McFadzean A.D., Aitken D.K., & Roche P.F., 1988. *Mon. Not. R. Astron. Soc.,* **233**, 321.
226. Whittet, D.C.B., Adamson A.J., Duley W.W., Geballe T.R., & McFadzean A.D., 1989. *Mon. Not. R. Astron. Soc.,* **241**, 707.
227. Whittet, D.C.B., & Duley W.W. 1991. *Astron. & Astrophys. Rev.,* **2**, 167.
228. Willacy, K., & D.A. Williams, 1993. *Mon. Not. R. Astron. Soc.,* **260**, 635.
229. Williams D.A., 1968. *Astrophys. J.,* **151**, 935.
230. Williams D.A., 1993. eds. Millar, T.J., & Williams D.A., In: *Dust and Chemistry in Astronomy,* (IOP Publ. Ltd., Bristol), p. 143.
231. Williams D.A., & Hartquist T.W., 1984. *Mon. Not. R. Astron. Soc.,* **210**, 141.
232. Willner, S.P., Gillett F.C., Herter T.L., Jones B., Krassner J., Merrill K.M., Pipher J.L., Puetter R.C., Rudy R.J., Russell R.W., & Soifer B.T., 1982. *Astrophys. J.,* **253**, 174.
233. Zhao, N.S., 1990. Ph.D. Thesis, Leiden University, The Netherlands.
234. Ziegler, J.F. 1980. *Stopping Cross Sections for Energetic Ions in All Elements* (Pergamon, New York).
235. Zinnecker, H., Webster A.S., & Geballe T.R., 1985. In: *Nearby Molecular Clouds,* ed. G. Serra, (Springer, Berlin), p. 81.
236. Averna, D., & Pirronello V., 1991. *Astron. & Astrophys.,* **245**, 239.
237. Dujardin, G., Hellner L., Philippe L., Azria R., & Besnard-Ramage M.J., 1991. *Phys. Rev. Letters,* **67**, 1844.
238. Hudgins, D.M., Sandford S.A., Allamandola L.J., & Tielens A.G.G.M., 1993. *Astrophys. J. Supl. Ser.,* **86**, 713.
239. Johnson R.E., Lanzerotti L.J., & Brown, W.L., 1982. *Nucl. Instrum. Methods,* **198**, 147.
240. Lanzerotti, L.J., Brown W.L., & Johnson R.E., 1985. In: Ices in the Solar System, eds J. Klinger, D. Benest, A. Dollfus & R. Schmoluchowski (Reidel, dordrecht), p. 317.
241. Lanzerotti, L.J., Brown W.L., & Marcantonio K.J., 1987. *Astrophys. J.,* **313**, 910.
242. Pirronello, V., & Averna D., 1988. *Astron. & Astrophys.,* **196**, 201.
243. Roux, J.A., Wood B.E., Smith A.M., & Plyler R.R., 1980. *Tech. Rep. AEDC-TR-79-81,* Arnold Eng. Dev. Cent., Arnold Air Force Station.
244. Schutte, W.A., Allamandolla L.J., & Sandford S.A., 1993b. *Icarus,* **104**, 118.
245. Trotta, F., & Schmitt B., 1994. In: *Physical Chemistry of Molecules and Grains in Space,* AIP Conf. Proceeding Series, This volume.

DISCUSSION

SCHUTTE — Comment on ice mantle desorption by chemical explosions: we have performed laboratory experiments on the chemical explosions of UV irradiated ices. Comparing the IR spectrum before and after the explosion, we find that only the most volatile species (CO, O_3) were desorbed while H_2O desorption was, at most, very small ($\leq 5\%$). Also a simple calculation comparing the energy produced by radical reactions with the H_2O binding energy shows that H_2O desorption by chemical explosions should be inefficient (see Schutte and Greenberg 1991).

LEGER — Do you consider that we understand the presence of NH_3 in gas phase in dense regions of molecular clouds, considering the short time needed for the condensation of the molecules and the difficulty to desorb them.

SCHMITT — This is not well understood yet, but it has been suggested that desorption of N_2 followed by gas-phase reactions may produce NH_3 in dense regions of molecular clouds.
The dense cores where NH_3 has been observed have higher dust temperatures than typical dense clouds. Average temperatures along the line of sight are found to range between 30 K and 60 K. Sublimation of NH_3 may then be expected from the hottest dust in dense cores, probably at $T > 100$ K.

WDOWIAK — I would like to make a comment about the topic of "chemical explosions" and request a response. In the laboratory the timescale for building up a concentration of free radicals is shorter than the diffusion time scale. This results in "bombs" that can be detonated by a modest increase in temperature. In the interstellar medium, the diffusion time scale is probably shorter than the time scale of production of a concentration of radicals. This condition would not yield "bombs"!

SCHMITT — The diffusion time scale for radicals in H_2O-rich ice mantles at about 10K has not been measured directly, but it must be very large. Theoretical extrapolations of measurements at higher temperature indicate that the diffusion time scale should be much larger than the timescale for building up a concentration of radicals of about 1% (see answer of Willem Schutte).

WDOWIAK — Are the frozen radicals stable over long timescales ?

SCHUTTE — Comment on the question by WDOWIAK : L. van Yzendoorn investigated this problem for his Ph. D. Thesis in the Leiden Laboratory. He tracked the diffusing and annealing of the radicals stored in a H_2O/CO ice at 10K by UV irradiations by monitoring the intensity of the chemtiluminescence upon slow warm-up. It was seen that chemiluminescence starts at ~ 23K and ends at ~ 50K. From theoretical considerations describing the diffusion of radicals as a function of temperature it was deduced that the radicals should, at 10K, be stable for very long timescales ($\sim 10^9$ years).

SHALABIEA — One of the interstellar grains contributions to the gas phase is the charge exchange reaction. My question is : Is such process already included i.e. taken in the consideration ? If yes, how much such process affects the ionization state of the interstellar gas phase ?

SCHMITT — The charge exchange reaction is taken into account in some models but I do not know how it affect the ionization state of the interstellar gas phase.

BENIT — It is quite inaccurate to compare the UV & CR effects on grains because it does not take into account the density of energy deposited parameter.
Why CR won't be efficient on icy grains ?
Why would they be only efficient on small grains ?
Why would the efficiency depend on the energy ?
(These questions are to be answered in the frame of heating processes and other if relevant).

SCHMITT — The comparison of the efficiency of molecule synthesis between UV photons and cosmic rays does take into account the density of energy deposited and the way it is deposited. The average efficiency of a CR is about 1000 times larger than the average efficiency of a UV photon, but this ratio may vary by at least an order of magnitude depending on several parameters such as the grain mantle compositions and thicknesses. Second, this grain efficiency ratio must be weighted by the relative fluxes of CR and UV photons in the

astrophysical environment considered. This ratio is certainly highly variable from the diffuse medium to dense molecular clouds. In places where grain mantles exist, there are numbers for these fluxes in the literature that, combined with laboratory efficiencies, may favor either of the processes, but with some advantage for UV photolysis in moderately dense clouds or in the external parts of clouds, and for CR induced synthesis inside dense clouds. The different type of chemistry induced by these two processes combined with the variation of composition of grain mantles from the internal to external parts of clouds should also be taken into account when comparing UV and CR effects. Cosmic rays are considered efficient for heating large grains (~ 0.1 - 0.2 μm). The limited efficiency of heating was about X-rays, not cosmic rays.
The efficiency of heating do not depend on the energy of the cosmic ray but rather on its atomic number (to the power of 2).

LEACH — Can you say something about clathrate hydrates, which have the useful property of trapping volatile species which can be released at phase transitions at interesting temperature ranges ?

SCHMITT — Clathrate hydrate is a non-stoechiometric H_2O crystalline structure that can trap as much as 17% of other molecules. Its stability depends on its dissociation pressure. At low temperature, it is expected to be more stable than the solid formed with the molecules trapped in the clathrates (i.e. the CO clathrate releases CO molecules more slowly than solid CO). However, a temperature of at least 100°K is required to form clathrate either by direct interaction of gas phase molecules with H_2O ice, or by crystallization of amorphous H_2O-rich ice mixtures. Therefore the probability to found clathrate in the interstellar medium is quite low. This is not the case at the surface or in the interior of icy planets and satellites of the solar system.

DETERMINATION OF THE OPTICAL CONSTANTS OF SOLIDS IN THE MID AND FAR INFRARED

F. Trotta and B. Schmitt

Laboratoire de Glaciologie et Géophysique de l'Environnement-CNRS,BP 96, 38402 Saint Martin d'Hères, France.

ABSTRACT

We have elaborated a numerical code in order to calculate the complex index of refraction of interstellar candidate solids (pure or mixed) in the mid and far infrared ranges. We present the numerical code, the experiment and preliminary results for ammonia and methane.

INTRODUCTION

The mid and far infrared ranges are very important for the identification of molecules condensed in interstellar grain mantles.
But for a good interpretation of the observed spectra we need to record infrared spectra in laboratory. for a large variety of possible situations (composition, temperature...).
In preparation to the ISO mission, which will cover the mid and far infrared ranges (2-200µ), we have to elaborate a laboratory data bank for the interpretation of the future interstellar spectra.
The essential parameters of the spectra to be determinated are:
- The complex index of refraction (n-ik), which allow us to compute the radiative transfert through grains.
- The characteristics of the various absorption bands (frequencies, integrated intensities, width at half maximum height-FWHM).
- The evolution of these characteristics with temperature.
For the determination of these parameters we perform laboratory transmission spectra of solids (pure or mixed), supposed to be present in interstellar grain mantles. We have developed a numerical code in order to compute the optical constants.

EXPERIMENTAL PROCEDURES

Thin films are deposited from the gas phase on a CsI

window cooled at about 10K by a He cryostat.

A Lake Shore DRC 93CA allows us to control the temperature with two diode thermometers.

The thicknesses are mesured by a laser interference technique with a He-Ne laser in reflection and in transmission.

We take transmission spectra for a serie of about 15 increasing thicknesses with a Nicolet 800 FTIR spectrometer at a resolution of 1 cm^{-1}.

THE NUMERICAL CODE

This code allows us to obtain iteratively the complex refractive index, as a function of wavenumber, from transmission spectra recorded on thin films of various thicknesses.

First, the program computes the theoretical spectrum corresponding to a film of thickness "d" deposited on a CsI window[1],[2] :

$$T_{bsl} = \frac{8 n_e^2 n_s(\upsilon)}{(n_e^2+1)(n_s(\upsilon)^2+1) + 4 n_e^2 n_s(\upsilon) + (1-n_e^2)(n_e^2 - n_s(\upsilon)^2)\cos(2\delta_1)} \quad (1)$$

where n_e is a constant reference index for the sample, and $n_s(\upsilon)$, the substract index. δ_1 is the difference in optical path introduced by the sample.

$$\delta_1 = 2\pi \upsilon n_e d + \delta' \quad (2)$$

with

$$\begin{array}{llll} \delta'=0 & \text{if} & n_e < n_s \\ \delta'=\pi/2 & \text{if} & n_e > n_s \end{array} \quad (3)$$

The experimental spectra, Φ_{spect}, is divided by the background flux, Φ_{bckg}, (with the window but without the sample), and normalized :

$$T_{norm} = T_{exp} T_{subs} \quad (4)$$

with

$$T_{exp} = \frac{\Phi_{spect}}{\Phi_{bckg}} \quad (5)$$

and,

$$T_{subs} = 2 \frac{n_s}{n_s^2 + 1} \quad (6)$$

T_{subs} is the theoretical transmission spectrum through the CsI window, without sample, if we assume there is no absorption and no interference.

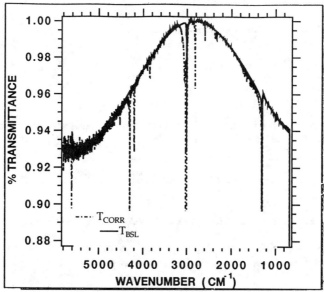

Figure 1: Example of the fit of the theoretical spectrum by the corrected spectrum.

Now we fit each normalized spectrum with its theoretical baseline, T_{bsl} (figure 1) using a correction fonction, f_{corr}, of the first order, proposed by Pearl et al.[1] :

$$T_{corr} = T_{norm} f_{corr} \qquad (7)$$

with

$$f_{corr}(\upsilon) = a_1 + a_2 \frac{\upsilon}{d\upsilon} + (a_3 + a_4 \frac{\upsilon}{d\upsilon}) \cos(2\delta_1) \qquad (8)$$

The coefficients a_1, a_2, a_3 and a_4 are computed using a least square method on the points supposed to be in nonabsorbing zones.

The result is a baseline corrected of all diffusive effects and others phenomena. We compute the absorbance for the serie of spectra recorded with increasing thicknesses:

$$A(d,v) = \ln\frac{T_{bsl}(d,v)}{T_{corr}(d,v)} \qquad (9)$$

We compute the extinction coefficient (imaginary part of the refractive index) using a least square method coupled with the Beer-Lambert law.:

$$k(v) = \frac{A(v,d)}{4\pi v d} \qquad (10)$$

Then we use the substractive Kramers kronig relation to obtain the real part:

$$n(v) = n(v_0) + \frac{2}{\pi} P \int_0^{+\infty} \left[\frac{k(v')v' - k(v)v}{v^2 - v^2} - \frac{k(v')v' - k(v_0)v_0}{v^2 - v_0^2} \right] dv' \qquad (11)$$

which is used in the next iterative loop.

The numerical code converges in about four iterations. We illustrate the convergence rate by the parameter τ (figure 2):

$$\tau = \frac{n_{iter} - n_{iter-1}}{n_{iter}} \qquad (12)$$

where iter is the iteration's number

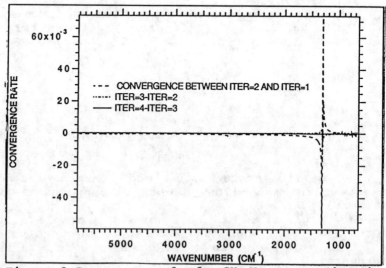

Figure 2 Convergence of n for CH_4. We can see that the numerical code converges in 4 iterations.

RESULTS

For CH$_4$ we use n(ν_0)=1.302 at ν_0=15802 cm^{-1} and a temperature T=33K.[3] (figure 3)

For NH$_3$ we use n(ν_0)=1.37 at ν_0=15802 cm^{-1} at a temperature T=20K [4] (figure 4).

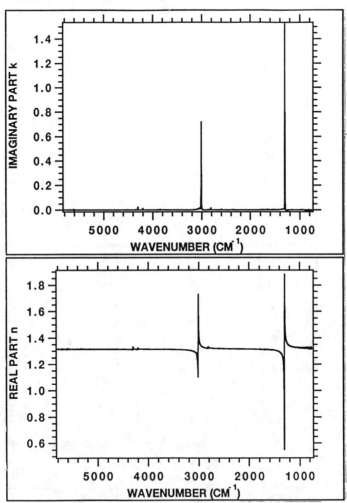

Figure 3 Real part and imaginary part of the refractive index for CH$_4$.

SOME RESTRICTIONS

The result on the real part depends on the choice of

the reference index $n(\nu_0)$. An error on this parameter propagate all along the wavenumber axis.

The value of this parameter is not always available in the literature. In this case we take the refractive index in the liquid-state, the densities of the liquid and solid phases, and we use the Lorentz-Lorenz relation to estimate the refractive index in the solid phase.

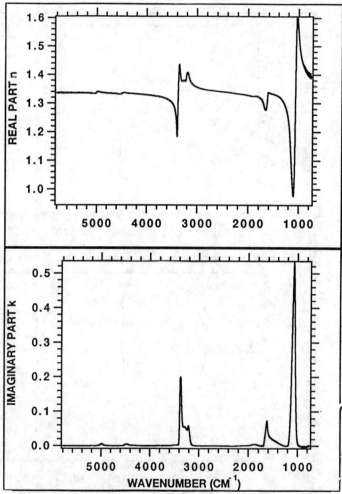

Figure 4 *Real part and imaginary part of the refractive index of ammoniac.*

Another critical point is the determination of non absorbing zones. For example we must take care about the bands wings effects.

Some solids, such as SO_2 (figure 5), are very sensitive to temperature and deposition conditions and show some evolution in their spectra as the deposition progress. This phenomena may be due to a temperature gradient in the film. We must choose with carefully the spectra used in the numerical code. Effectively, if we take a serie of spectra, with only small changes in the position of the bands, a distorsion and a broadening of the band calculated from the least square method will occurs.

REFERENCES

1- A. Vasicek, Optics of thin films, North-Holland publishing company, 1960.
2- J. Pearl, M. Ngoh, M. Ospina, and R. Khanna, Journal of Geophysical Research, 96, 477, 1991.
3- B. N. Khare, W. R. Thompson, C. Sagan, E. T. Arakawa, C. Bruel, J. P. Judish, R. k. Khanna, and J. B. Pollack, in First International Conference on Laboratory Research for Planetary Atmospheres, 1990
4- B. E. Wood, J. A. Roux, J. Opt. Soc. Am., 72, 720, 1982.

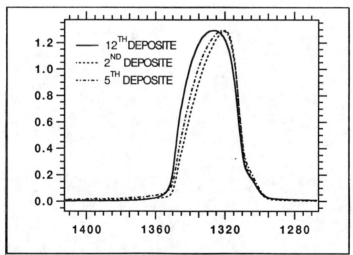

Figure 5 The 1320 cm^{-1} band of SO$_2$ shows a shift of about 10 cm^{-1} during the deposition. This could be due to a temperature gradient in the film.

MOLECULAR PHOTOPRODUCTION RATE COEFFICIENTS IN ICY GRAIN MANTLES AS APPLIED TO DUST/GAS CHEMICAL MODELING

Osama M. Shalabiea and J. Mayo Greenberg

Laboratory Astrophysics, Sterrewacht Leiden,
P.O.-Box 9504, 2300 RA Leiden, The Netherlands.

ABSTRACT

Rate coefficients for the photoproduction of molecules in interstellar icy grain mantles are essential input parameters for dust/gas chemical models. We present a general procedure for the deduction of these rate coefficients involving a combination of experimental, observational and theoretical data. Improved estimates of the rate coefficients for the photoproduction of some important molecules in the icy grain mantles are given. In addition to the external ultraviolet (UV) photons which penetrate the cloud, we include also the internal sources of UV photons - mainly the UV photons induced by cosmic rays -, which clearly have an important effect on the photoproduction rate especially in dense interstellar clouds where the external UV photons are greatly depleted.

1. INTRODUCTION

For any dust/gas chemical model there are two main problems concerning the grain chemistry, i.e, that of grain surface reactions and that of photochemical reactions. In this work we will focus on the production of some key molecules in the icy grain mantles via photochemical processes, as was first suggested by Greenberg & Yencha [1]. This has now become well established as one of the main molecular production sources in dust/gas chemical models. Laboratory simulation experiments have shown that UV irradiation produces new molecular species in the icy mantles [2,3,4,5]. The photoproduction of CO_2 and H_2CO were studied by d'Hendecourt et al. [2]. Based on their experimental results, tentative rate coefficients for these molecules were derived [5,6]. Recently CH_3OH photoproduction from H_2CO in the icy mantles was studied experimentally [4] and included in the dust/gas chemical models [7]. The gas phase and icy grain mantle abundances of dust/gas chemical models are very sensitive to variations of the molecular photoproduction rate coefficients.

The main objective of the present work is to obtain more precise estimations for these rate coefficients. We include the external as well as the internal sources of UV photons, the latter not having been specifically considered previously in dense clouds. The photoproduction of the important astrophysical molecules CO_2, H_2CO and CH_3OH in the icy grain mantles will be given as examples.

2. THEORETICAL FORMULATION OF THE RATE COEFFICIENTS

The number of molecular species (i) produced in icy grain mantles via photochemical processes, $(\frac{dn_i}{dt})_{pp}$ may be expressed by

$$\left(\frac{dn_i}{dt}\right)_{pp} = \kappa^i_{jk} \times n_j \times n_k \quad \text{cm}^{-3}\text{s}^{-1} \tag{1}$$

where:-
κ^i_{jk} is the molecular photoproduction rate coefficient of a molecular species (i) due to the photoreaction between two other species (j) and (k) in the grain mantles.
n_j and n_k are the number density of the parent molecular species (j) and (k).

The rate coefficient κ^i_{jk} is what is needed for any theoretical dust/gas chemical model. The question is, how can we derive more precisely the rate coefficients for photochemical processes in icy interstellar grain mantles from the laboratory results ?

The rate of photoproduction of molecules (i) can also be written as

$$\left(\frac{dn_i}{dt}\right)_{pp} = \alpha_i \times \Phi_{tot} \times 4\pi r_d^2 n_d \quad \text{cm}^{-3}\text{s}^{-1} \tag{2}$$

where:-
α_i is the net number of molecular species (i) photoproduced per incident photon,
Φ_{tot} is the total available UV flux within the interstellar cloud in photon s^{-1} cm^{-2},
$4\pi r_d^2 n_d$ is the average grain surface area per cm^3, where r_d and n_d are the radius and the number density of the interstellar grains respectively.

From Equations (1) and (2) it follows that :

$$\kappa^i_{jk} = \frac{\alpha_i \times \Phi_{tot} \times 4\pi r_d^2 n_d}{n_j \times n_k} \quad \text{cm}^3\text{s}^{-1} \tag{3}$$

The volume densities n_j and n_k of the parent molecules (j) and (k) in the grain mantles may be converted into the experimental column densities N$_j$ (cm^{-2}) and N$_k$ (cm^{-2}) in the ice layer. The rate coefficient for the photoproduction of a molecule (i) in the icy grain mantles is then given by

$$\kappa^i_{jk} = \left(\frac{\alpha_i \times \Phi_{tot}}{N_j \times N_k}\right) \times \left(\frac{1}{4\pi r_d^2 n_d}\right) \quad \text{cm}^3\text{s}^{-1} \tag{4}$$

With $r_d = 1.0 \times 10^{-5}$ cm and $n_d = 10^{-12} \times n_o$ cm^{-3}, n_o being the total hydrogen number density ($n_o = n_H + 2\, n_{H_2}$), we finally obtain

$$\kappa^i_{jk} = 7.96 \times 10^{20} \times \left(\frac{\alpha_i}{N_j \times N_k}\right)_{Lab} \times \left(\frac{\Phi_{tot}}{n_o}\right)_{ISM} \text{ cm}^3\text{s}^{-1} \quad (5)$$

It is clear from the general Equation (5) that κ^i_{jk} depends on two main parts. The *first* "experimental" part contains the parameters α_i (photon^{-1}), N_j (cm^{-2}) and N_k (cm^{-2}). These may be obtained from the results of UV irradiation experiments for different ice mixtures designed to be consistent with interstellar grain mantle composition:

$$\alpha_i = N_i / \Delta t \, \Phi_{UV} \quad photon^{-1} \quad (6)$$

where Δt is the irradiation time and Φ_{UV} is the photon flux in the laboratory experiment. The *second* part contains the interstellar cloud parameters Φ_{tot} (photon cm^{-2} s^{-1}) and n_o (cm^{-3}).

3. THE PHOTOPRODUCTION RATE COEFFICIENTS FOR CO_2, H_2CO and CH_3OH IN THE ICY GRAIN MANTLES

Based on equation (5) we will derive in this section the photoproduction rate coefficients for CO_2, H_2CO and CH_3OH molecules in the icy grain mantles. Since H_2O and CO are often the most dominant components of the grain mantles and, assuming, that each newly formed molecule is produced by a photoreaction scheme involving two parent molecules, we may write :

$$H_2O:gr + CO:gr + h\nu \longrightarrow CO_2:gr + 2H$$
$$H_2O:gr + CO:gr + h\nu \longrightarrow H_2CO:gr + O:gr$$
$$H_2O:gr + H_2CO:gr + h\nu \longrightarrow CH_3OH:gr + O:gr$$

The experimental parameters (α_i , N_j and N_k) needed to derive the production rate coefficients of CO_2, H_2CO and CH_3OH are given in table (I). The uncertainties in α_i are estimated to be $\sim 20\,\%$.

Table (I) The experimental parameters needed to derive the rate coefficients.

	α_i (ph^{-1})	N_j (cm^{-2})	N_k (cm^{-2})
CO_2 *	4.5 (-3)	N_{H_2O} = 1.1(18)	N_{CO}= 5.5(17)
H_2CO *	1.7 (-3)	N_{H_2O} = 1.1(18)	N_{CO}= 5.5(17)
CH_3OH **	1.5 (-3)	N_{H_2O} = 4.3(17)	N_{H_2CO}= 4.2(16)

* d'Hendecourt et al. (1986) [2]
** derived from experiments by Zhao (1990) [4]

In the interstellar cloud the total UV flux 912-2000 Å consists of two main sources, the external radiation field $\Phi_{ext} = 10^8 \times e^{-2A_v}$ photon s^{-1} cm^{-2} and the internal UV photons which are mainly produced by cosmic rays. Estimation by various authors of this source vary between 1.0×10^3 - 1.5×10^4 photons cm^{-2} s^{-1} [8,9,10]. Here we will use 5.0×10^3 photons cm^{-2} s^{-1} as an average value. It is possible that the internal flux could even be considerably higher as a result of star formation processes [11,12]. We do not take any wavelength dependence of the process into account. The photoproduction rate coefficients depend on both the interstellar cloud densities and geometries. As an example, the rate coefficients for CO_2 with (*) and without (**) inclusion of cosmic ray induced internal UV photons are given Table (II) for a uniform spherical cloud of 1 pc radius. For such a cloud A_v and n_o are related by

$$A_v = 1.63 \times 10^{-3} \times n_o \tag{7}$$

Table(II) The photoproduction rate coefficients for CO_2 for different values of n_o and A_v with (*) and without (**) inclusion of cosmic ray induced internal UV photons.

n_o cm^{-3}	A_v mag.	κ^i_{jk} (*)	κ^i_{jk} (**)
500	0.8	2.4(-13)	2.4(-13)
2×10^3	3.3	4.2(-16)	4.0(-16)
4×10^3	6.5	7.7(-18)	3.3(-19)
1×10^4	16.3	3.0(-18)	4.1(-28)

From Table (II) it is clear that the photoproduction rate coefficients in the interior of a cloud with $n_o \geq 2 \times 10^3$ cm^{-3} are strongly increased by the inclusion of the internal UV photons.

Using the experimental parameters in Table (I), the rate coefficients for CO_2, H_2CO and CH_3OH with (*) and without (**) internal UV photons are shown in Table (III) for our standard dense cloud model ($A_v = 4$ mag., $n_o = 2\times 10^4$ cm^{-3}, i.e size = 0.1 pc).

Table (III) The photoproduction rate coefficients for CO_2, H_2CO and CH_3OH with (*) and without (**) internal UV photons for the standard dense cloud model ($A_v = 4$ mag., $n_o = 2\times 10^4$ cm^{-3}).

	κ^i_{jk} (*)	κ^i_{jk} (**)
CO_2	1.2 (-17)	1.0 (-17)
H_2CO	4.4 (-18)	3.8 (-18)
CH_3OH	1.3 (-16)	1.2 (-16)

It is clear from Table (III) that the inclusion of the internal UV photons has very little effect on the rates for this standard cloud model.

Using these photoproduction rate coefficients gas phase abundances were calculated with our dust/gas chemical model [7]. We found that for; e.g, H_2CO, including some realistic desorption process from the grains gives about one order of magnitude higher gas abundances than the pure gas-phase model.

4. CONCLUSIONS

(*i*) The photoproduction rate coefficients depend not only on the UV photons but also on the number density of dust grains which, in turn, depends on the total hydrogen number density of the interstellar cloud.

(*ii*) The inclusion of internal sources of UV photons clearly gives higher molecular photoproduction rate coefficients in the interior of the clouds. This, of course, could lead not only to higher solid molecular abundances but also to higher gas phase abundances in the presence of some desorption process.

(*iii*) For any dust/gas chemical models, it is not only necessary to have reliable gas phase rate coefficients but also to have reliable solid state rate coefficients based on further accurate experimental studies. This will greatly enhance the accuracy of chemical models of interstellar clouds.

ACKNOWLEDGEMENT

We gratefully acknowledge partial support by NASA grant #NGR33-018-148. We thank Dr.Ewine Van Dishoeck and Dr. Willem Schutte for many helpful discussions. One of us (OMS) wishes to thank Cairo University and the World Laboratory for a fellowship to study in Leiden University.

REFERENCES

1- Greenberg, J.M. and Yencha, A.J. (1973) in Interstellar Dust and Related Topics ,eds.J.M. Greenberg and H.C. Van de Hulst, IAU symp. no. 52, (Reidel , Dordecht), p. 309.

2- d'Hendecourt, L.B., Allamandola, L.J., Grim, R.J.A. and Greenberg, J.M. (1986) A & A 158, 119.

3- Allamandola, L.J., Sandford, S.A. and Valero, G.J. (1988) Icarus 76, 225.

4- Zhao, N.S. (1990) Ph.D. thesis, Univ. of Leiden.

5- Greenberg ,J.M., Mendoza - Gómez C.X., de Groot, M.S. and Breukers, R. (1993) in Dust and Chemistry in Astronomy, eds. Millar, T.J. and Williams, D.A. (IOP publ. Ltd.), p. 265.

6- Breukers, R., d'Hendecourt, L.B. and Greenberg,J.M. (1992) in Chemistry and Spectroscopy of Interstellar Molecules, eds. Bohme,D.K.; Herbest,E.; Kaifu,N. and Saito,S. (Un. of Tokyo presss), p.273 .

7- Shalabiea, O.M., Greenberg, J.M. and Van Dishoeck, E.F. (1993) in preparation.

8- Prasad, S.S. and Tarafdar, S.P. (1983) ApJ 267, 603.

9- Gredel, R., Lepp, S. and Dalgarno, A. (1989) ApJ, 347, 289.

10- Cecchi-Pestellini, C. and Aiello, S. (1992) MNRAS 258, 125.

11- Norman, C. and Silk, J. (1980) ApJ 238, 158.

12- Greenberg, J.M. (1982) in Submillimeter Wave Astronomy, eds. Beckman, J.E. and Phillips, J.P. (Cambridge Univ. Press), p. 261.

PHOTODESORPTION FROM CO ICES

L. Hellner[1][2], G. Dujardin[1][2], T. Hirayama[1][3], L. Philippe[1][2],
M.J. Ramage[1][2], G. Comtet[2], M. Rose[1][4]

(1) Laboratoire de Photophysique Moléculaire, CNRS, Bât 213
(2) Laboratoire pour l'Utilisation du Rayonnement Electromagnétique
(LURE), CNRS, CEA, MEN, Bât 209D
Université Paris Sud, 91405 Orsay Cedex, FRANCE
(3) Department of Physics, Gakashuin University
Toshimaka 171 Tokyo, JAPAN
(4) Physik Department E20, Technical University, Munich
85747 Garching, FRG

ABSTRACT

The photon excitation of multilayers of condensed CO leads to the desorption of a great number of different ions, C^+, O^+, CO^+, C_2O^+, C_3O^+, C_4O^+, $(CO)_2^+$ and $C_3O_2^+$. The threshold energy for desorption of the most abondant ions (C^+, O^+) is around 30 eV. Such a high threshold is believed to result from a two electron excitation in the solid. The two main resonances observed at lower energies (15.35 eV and 17 eV) in the CO^+ ion yield curve of which the threshold energy is 14.3 eV are assigned to two excitonic levels of solid CO.

INTRODUCTION

Molecules such as H_2O, CH_3OH and CO have been clearly identified as constituents of ice mantles of cold interstellar grains[1]. In order to understand the exchange between the solid phase and the gas phase surrounding these grains, it is important to consider the desorption[2] induced by impact of photon or cosmic rays. Molecules condensed as multilayers on a cooled metallic substrate (10 K) can be considered as a faithful model for interstellar grains covered with molecular ices. Photodesorption experiments in laboratories are needed to i) identify the electronic excitations and determine the photon energies which can induce the desorption, ii) understand the dynamics of desorption at the surface which determines the desorption yield of various atomic or molecular species. In this communication, data on Photon Stimulated Ion Desorption (PSID) from solid carbon monoxide in the 10 - 60 eV energy range are presented and compared to previous results obtained for valence excitation (20 - 35 eV)[3] and for high energy [4] and K-shell excitation of oxygen (540 eV) and carbon (296 eV) [5].

EXPERIMENTAL

The experimental setup is composed of a UHV chamber (pressure below 10^{-10} mbar) equipped with a liquid-helium flow cryostat. The carbon monoxide multilayers are prepared by condensing pure CO on a polycrystalline platinum substrate cooled at 10 K. Synchrotron radiation from SuperACO at Orsay, dispersed by a grazing incidence monochromator, is used as a photon source of variable energy in the 14-60 eV energy range. Desorbed positive and negative ions are mass selected by a

Riber SQX 156 quadrupole filter and counted as a function of the photon energy. Ion yields are normalized to the photon flux monitored by the current from a gold mesh. An hemispherical electrostatic electron analyser allows photoemission measurements.

RESULTS

1. Ion Desorption.

By photoexcitation of CO multilayers in the 14 - 60 eV energy range, the following positive ions are observed to desorb: CO^+, C^+, O^+, C_2O^+, $(CO)_2^+$, C_3O^+, C_4O^+ and $C_3O_2^+$. Only the most abundant ions (C^+, O^+, CO^+, C_2O^+ and $(CO)_2^+$) were observed by Rosenberg et al [3], probably due to a lower sensitivity in this latter experiment. For hv = 600 eV, R. Scheuerer et al[5] observed ions up to the mass 252 corresponding to $(CO)_9^+$. Higher masses observed in this high energy experiment may be due to the formation, after Auger decay, of multiply charged ions. Two negative ions O^- and C^- are also observed in the present experiment, however with a very weak yield. Except for CO^+ the energy threshold for the desorption of those different ions is around 30 eV. Above this threshold the main ion is always C^+ whatever the photon energy is. The relative yield of the desorbed ions is given for a photon energy of 40 eV.

Table I   Relative yield (%) of the different ions for hv = 40 eV

C^+	O^+	CO^+	C_2O^+	C_3O^+	$(CO)_2^+$	C_4O^+	$C_3O_2^+$
77.9	13.2	1.6	3.4	0.8	1.7	0.1	1.3

The energy dependence of the ion desorption yield of these different ions is quite comparable (except for CO^+). A typical excitation spectrum of one of these ions (O^+) is depicted on figure 1 which shows a threshold around 30 eV. Quite surprisingly CO^+ ion desorption is observed at excitation energy lower than 25 eV. In figure 2 the excitation spectrum of CO^+ at low energy is displayed; this spectrum shows up two main resonances with maxima at 15.35 and 17 eV respectively. At these two excitation energies the ion yield of CO^+ is approximatively identical to the yield at hv = 40 eV.

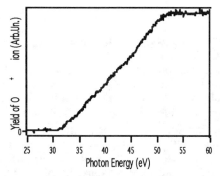

Figure 1. O$^+$ yield as a function of photon energy from 10 multilayers of CO at 10 K

Figure 2. CO$^+$ relative yield as a function of photon energy from 10 multilayers of CO at 10 K

2. Photoemission Spectra.

Photoemission measurements are performed in order to identify the ionic states which can initiate the desorption processes. The figure 3 displays a photoemission spectrum obtained for a photon energy of 100 eV. On the low energy side of the binding energy, the three main peaks at 12.85, 15.75 and 18.65 eV correspond respectively to the formation of the first three ionic states of CO$^+$, X($^2\Sigma^+$), A($^2\Pi$) and B($^2\Sigma^+$) [6]; the last feature on the high energy side at 38 eV should correspond to one hole in the 3σ shell of the molecule while the features at intermediate energies correspond to the formation of satellite states [7] for which ionization and excitation of outer valence shell electrons occur simultaneously.

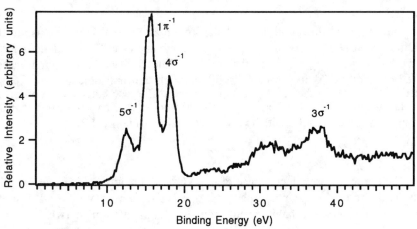

Figure 3. Photoemission spectra of condensed CO at 10 K for hv = 100 eV
(E_B = 0, vacuum level) CO ground state configuration: $1\sigma^2\ 2\sigma^2\ 3\sigma^2\ 4\sigma^2\ 1\pi^4\ 5\sigma^2$

DISCUSSION

1. Identification of precursor states of desorption processes.

The threshold for desorption of most of the ions (except CO$^+$) around 30 eV lies in the energy range for formation of satellite states (see figure 3). This energy also exactly coïncides with the onset of electron pair formation in solid CO, as previously measured in an electron-electron coincidence experiment [8]. This latter electron pair process has been assigned to the formation of satellite states [8]. It follows that ion desorption in this energy range most probably originates from the formation of two hole-one electron states [9]. Nevertheless it is still difficult at the present time to identify which of those satellite states are responsable of the ion desorption.

For the desorption of CO$^+$, the situation is markedly different since, according to the photoemission spectrum (figure 3) only single hole ionic state can be formed below 20 eV. The excited states leading to the desorption of CO$^+$ in this range should be neutral states since they give rise to resonances in the excitation spectrum (fig. 2). Rydberg series converging to CO$^+$ A$^2\Pi$ and B$^2\Sigma^+$ states are known in the gas phase [10]. In solid state excitonic levels can certainly be derived from these Rydberg states. In the absence of any photoabsorption data on condensed CO to support this statement, we speculate that the resonances at 15.35 and 17 eV should correspond to the excitation of such excitonic levels.

2. Formation of the different desorbed ions.

CO$^+$

The observed threshold for CO$^+$ desorption at 14.3 eV is very close to the ionization potential of carbon monoxide in gas phase at 14 eV [10b]. This latter energy plus the binding energy of CO in the solid ($\cong$ 3meV) [11] is the minimun energy required to eject a CO$^+$ ion from the solid although the first ionic state of CO$^+$ is situated at 12.85 eV in the solid phase. The small difference between the observed threshold (14.3 eV) and the thermodynamic threshold (14.0 eV) indicates that the CO$^+$ ions will leave the solid with a very reduced kinetic energy.

C$^+$ and O$^+$

The photon energy dependence of these C$^+$ and O$^+$ ion yields is very similar to what is observed for gas-phase photodissociative ionization of CO[12]. It follows that O$^+$ and C$^+$ most probably result from the fragmentation of two hole -one electron valence states which where assigned respectively to the G$^2\Sigma$ (5σ^{-2} 6σ) and C$^2\Sigma$ (5σ^{-1} 1π^{-1} 2π) states of CO$^+$ [13].

(C$_2$O)$^+$, (CO)$_2^+$.......

As mentioned above such complex ions have been already observed by PSID[3,5]. By SIMS[14] of condensed CO with 1 KeV Ar^+, He^+, Ne^+, Kr^+, Xe^+ beams similar ions have been also detected. In these different experiments the relative yield of the ions is different. The general formula of these species can be written as follows, $C_n(CO)^+$, $(CO)_n^+$, $C_n(CO)_2^+$. From SIMS experiments [14] it was stipulated that most of these ions are coming from ion-molecule and association reactions. In an ion cyclotron resonance spectrometer [15] the formation of C_2O^+ was observed from the following reaction :

$$CO^+ + CO \rightarrow C_2O^+ + O \qquad (1)$$

But according to the heat of formation of this ion, $\Delta H_f (C_2O^+) = 392 \pm 10$ Kcal/mol [15], this reaction is strongly endothermic and is observable only when CO^+ is excited. In our experiment where C_2O^+ is the major ion it is then more probable that C_2O^+ is coming from the association reaction :

$$C^+ + CO \rightarrow C_2O^+ \qquad (2)$$

which is slightly exothermic and which can then occur in a condensed medium. An other association reaction[3], highly exothermic,

$$CO + C_2O^+ \rightarrow C_3O_2^+ \qquad (3) \quad \Delta H = -142.8 \text{ kcal/mole}$$

can be responsible for the formation of $C_3O_2^+$. The enthalpy variation of this reaction is calculated with $\Delta H_f (C_3O_2^+) = 222$ kcal/mole [16]. $(CO)_2^+$ is more probably produced by the photoionization of $(CO)_2$ rather than from the reaction $CO^+ + CO$ since CO^+ and $(CO)_2^+$ don't have the same photon energy dependence.

CONCLUSION

It is shown in this paper that Photon Stimulated Ion Desorption from condensed CO occurs mainly, in agreement with previous papers, for photon energy higher than 30 eV through the formation of satellite states. However PSID of CO^+ is also observed at lower energies after one electron excitation in the 14-22 eV energy range. We propose that the excitation of excitonic levels of condensed CO induces this CO^+ ion desorption. The origin of C_2O^+, C_3O^+, C_4O^+, $C_3O_2^+$ desorbed ions is considered to be the result of ion-molecule reaction at the surface of the CO ice whereas $(CO)_2^+$ is expected be formed by the photoionization of dimers.

REFERENCES

1. Dust and Chemistry in Astronomy Ed. by T.J. Millar and D.A. Williams, Institute of Physics Publishing Bristol and Philadelphia p. 9 (1993).
2. A. Léger, M. Jura, and A. Omont, Astron. Astrophys. 144, 147 (1985).
3. R.A. Rosenberg, V. Rehn, A.K. Green, P.R. LaRoe and C.C. Parks, DIET I (Eds N.H. Tolk, M.M.Traum, J.C.Tully, T.E. Madey), Springer Series in Chemical Physics 24, 247, 1983.
4. R. Scheuerer, P. Feulner, G. Rodker, Zhu Lin and D. Menzel, DIET IV, Springer Series in Surface Sciences , 19, Eds G.Betz and P. Varga (Springer-Verlag, Berlin, Heidelberg) 1990 p.235.
5. R.A.Rosenberg, P.J.Love, P.R. LaRoe, V. Rehn and C.C. Parks, Physical ReviewB, 31, 2634, 1985.
6. P.R. Norton, R.L. Tapping, H.P. Broïda, J.W. Gadzuk and B.J. Waclawski, Chem. Phys. Lett. 53, 465 (1978).
7. W. Eberhardt and H.J. Freund, J. Chem. Phys. 78, 700 (1983).
8. G. Dujardin, L. Hellner and M.J. Besnard-Ramage, Bull. Soc. Roy. Sc. Liège 58, 269 (1989).
9. R.A. Rosenberg, V. Rehn, A.K. Green and P.R. LaRoe, Phys. Rev. Lett. 51, 915 (1983).
10a. M. Ogawa and S. Ogawa, J. Mol. Spect., 41, 393 (1972).
10b. J.H. Fock, P. Gürtler and E.E. Koch, Chem. Phys., 47, 87 (1980).
11. I. Amdur and L.M. Shuler, J. Chem. Phys. 38, 188 (1963).
12. G.R. Wright, M.J. Van der Wiel and C.E. Brion, J. Phys B: Atom. Molec. Phys.,9, 675 (1976).
13. D.E. Ramaker, J. Chem. Phys., 78, 2998 (1983).
14. H.T. Jonkman and J. Michl, J. Am. Chem. Soc., 103, 733 (1981).
15. M.T. Bowers, M. Chau and P.R. Kemper, J. Chem. Phys. 63, 3656 (1975).
16. Gas phase Ion and Neutral Thermochemistry, J. Phys. Chem. Ref. Data 17 (1988).

DISCUSSION

LEGER — Are you planning to go to lower energy photon as only photons with hv ≤ 13.6 eV are present in the ISM ?

HELLNER — We can plan to work at lower energy (hv < 12 eV) when we will be able with one experimental set up to detect neutral desorbed species.

DUTUIT — In the 14 eV photon energy range, where you observe CO^+ ions, could you have some second order light problems ?

HELLNER — No I dont think our CO^+ ions signal, in the 14 eV photon energy range, is due to second order light problems. I can give two arguments against it. If it was the case:
  1. We should observe 2 resonances at two times the energy of our observation (30.7 and 34 eV) and we dont
  2. We would observe also a signal for the others ions.

SCHMITT — Do you have an idea about the ion photodesorption yields (ion/photon) in your experiments ?

HELLNER — There is no obvious answer to such question and we can only give a rough estimation. If we suppose that the ion detection efficiency is between 10^{-2} and 10^{-3} we can say that :

$$10^{-6} < Y_{CO+} \ (hv = 17 \ eV) < 10^{-7}$$
$$10^{-3} < Y_{C+} \ (hv = 40 \ eV) < 10^{-4}$$

CAN Mg/Fe SULPHIDES SOLVE THE PROBLEM OF THE 30 μm BAND OF CARBON STARS?

B. Begemann, H. Mutschke, J. Dorschner and Th. Henning
MPG-Arbeitsgruppe "Staub in Sternentstehungsgebieten"
Schillergäßchen 2-3, D-07745 Jena, Germany

ABSTRACT

Since the middle of the last decade a growing number of carbon-rich sources with a prominent emission band at 30 μm has been detected. Based on laboratory transmission spectra, this feature was tentatively attributed to circumstellar dust grains consisting of MgS.
As in the outflow in which the grains condense the quantities of Mg and S are comparable with that of Fe, a series of sulphides of the bulk composition $Mg_xFe_{1-x}S$ (x=0...0.9) was prepared in our working group laboratories. Up to now, optical data of such sulphides are lacking. It is the main purpose of this paper to make them available in to get a more reliable base for the theoretical modelling of the observed 30 μm profiles.
The synthesis of the Mg/Fe sulphides in question has been done by the evacuated silica tube method followed by arc-melting and quenching. The chemical and mineralogical state of the samples has been determined by X-ray diffraction, scanning electron microscopy and energy dispersive micro-probe analysis.
In connection with the possible occurrence of a sulphide component in the interstellar dust, especially in molecular clouds, the chemical stability of the sulphides against H_2O has been examined. Thermochemical considerations yielded arguments that only the Fe-rich sulphides have a chance to survive in the interstellar environment.
Optical data of the prepared sulphides have been determined by reflectance spectroscopy on polished surfaces in the wavenumber range of 5000-20 cm^{-1}. For the sake of comparison, transmittance spectra of small particles have also been investigated.
A comparison of the calculated profile of the 30 μm band for a continuous distribution of ellipsoids using the data of the sample $Mg_{0.5}Fe_{0.5}S$ showed a satisfactory agreement with the observed spectrum of IRC +10 216.

INTRODUCTION

A steadily growing number of carbon-rich sources in advanced stages of stellar evolution, e.g. AGB stars, protoplanetary nebulae (PPN) and planetary nebulae (PN), shows a prominent solid-state emission band centred at about 30 μm [1]. Based on transmission spectra of commercial MgS powder, the band was identified with vibrations of solid MgS condensed in the outflow of the sources [2,3]. Because of the lack of optical data for suitable sulphides, up to now, a more accurate examination of this identification was not possible. Therefore, it has been a great challenge for a laboratory astrophysics group to make available such data, to characterize the synthesized sulphide dust analogues by analytical methods, and to compare the spectra of small particles calculated by using the measured data with the spectra of the observed sources. In this paper, these results will be presented. Furthermore, some important astrophysical questions will be discussed:
Are there other elements involved in the sulphide formation in stellar outflows?
Can solid sulphides formed this way survive in interstellar space and, thus, contribute to the refractory dust components present in the interstellar space, especially, in molecular clouds?
If they exist are these interstellar sulphides observable?

PREPARATIONS AND MINERALOGICAL ANALYTICS

The basic assumption for our laboratory approach was that the condensation of sulphides is not restricted to Mg. From the abundance table of the elements [4,5], Fe seems to be a similar candidate. Its importance is also underlined by the occurrence of FeS in primitive solar system solids [6,7]. For this reason, a series of sulphides of the bulk composition $Mg_xFe_{1-x}S$ with x=0.0, 0.1, 0.25, 0.4, 0.5, 0.6, 0.75, 0.9 was prepared in our laboratory. The sulphides were synthesized from the native elements by the evacuated silica tube method with an annealing at 1273 K for 24 h. The resulting sulphide powders were portioned, pressed to pellets, and melted in an arc furnace to obtain compact samples. The molten drop-like samples were quenched on a water-cooled copper plate. All preparational steps had to be performed in an Ar atmosphere because the Mg-rich sulphides are easily attacked by atmospheric moisture (cf. next section). This quenched drop-like samples were used for the determination of the chemical and mineralogical properties and of the optical constants. For their densities, values between 3.02 g cm^{-3} for $Mg_{0.9}Fe_{0.1}S$ and 4.6 g cm^{-3} for FeS were obtained. For electron-microscopic studies (SEM and EDX) and for the measurements of IR reflectivity in the wavenumber range from 5000 to 20 cm^{-1}, polished surfaces had to be prepared. For X-ray diffractometry and IR transmission spectroscopy, a part of the compact samples was ground to small particles.

In the samples, three mineral phases have been detected: troilite (FeS), niningerite ((Mg,Fe)S) and small amounts of metallic iron. Troilite and niningerite (solid solutions of MgS and FeS) are typically meteoritic minerals not common in terrestrial mineralogy [6,8]. On the SEM pictures, the sample with the bulk composition $Mg_{0.1}Fe_{0.9}S$ shows niningerite dendrites with typical lengths of 30 to 100 μm and widths of 10 μm near the centre of the sample. Near the surface, the dendrites are much smaller. These niningerite dendrites have grown into a troilite matrix containing very small Fe segregations (Fig.1). The sample with the complementary bulk composition $Mg_{0.9}Fe_{0.1}S$, shows the dominant niningerite phase with brain-like appearance containing less than 5 % troilite and traces of metallic Fe only as interstitial fillings (Fig.1). The Mg/Fe ratio of the niningerites for these final members of the bulk composition and two more samples can be found in Table I. For the characterization of the stoichiometry of the niningerites, the subscript $\bar{y}$ has been used. $\bar{y}$ has to be considered as a mean value because the niningerites show a systematically varying composition from the edge to the centre of the grain (zoning), an effect which is wellknown from the mineralogy of meteorites [8]. The niningerite grains with x=0.4 have average diameters of 10 μm. With growing x the grains become larger and more spherical.

Table I Bulk composition of the samples (index: x) compared with the composition of the niningerites (mean stoichiometric index: $\bar{y}$) and percentage of FeS in the sample.

Bulk composition $Mg_xFe_{1-x}S$ x	Niningerite composition $Mg_{\bar{y}}Fe_{1-\bar{y}}S$ $\bar{y}$	Percentage of FeS z
0.1	0.2	0.79
0.4	0.33	0.50
0.75	0.79	0.07
0.9	0.94	$\leq$0.05

X-ray diffraction patterns of the pulverized samples showed peaks of troilite, niningerite and small amounts of metallic iron. With increasing Mg content, the cubic unit cell volume of the niningerite solid solution grows. Because of the wide compositional range (zoning), the niningerite peaks are broadened.

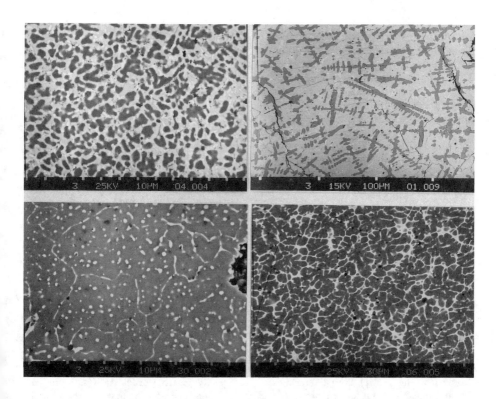

Figure 1: BSE-images of mixed Mg/Fe sulphides.

Top right picture: The sample with the bulk composition $Mg_{0.1}Fe_{0.9}S$ consists of niningerite dendrites (dark-grey) of the mean composition $Mg_{0.2}Fe_{0.8}S$ in a troilite matrix (light-grey). The percentage of FeS is 0.79. In the troilite matrix, very small white grains of metallic iron are embedded.

Top left picture: In the sample $Mg_{0.4}Fe_{0.6}S$ (bulk composition), the dark-grey niningerites (mean composition $Mg_{0.33}Fe_{0.67}S$) becomes more spherical in the troilite matrix (light-grey). The percentage of FeS is 0.5. The small white grains are metallic iron segregations. (Notice the change of magnification).

Bottom right picture: In the sample $Mg_{0.75}Fe_{0.25}S$ (bulk composition), the ratio of niningerite (dark-grey, mean composition $Mg_{0.79}Fe_{0.21}S$) to troilite (white) is 0.07. Metallic iron grains can not be seen in this magnification.

Bottom left picture: The sample with the highest Mg-content (bulk composition $Mg_{0.9}Fe_{0.1}S$), consist of a niningerite matrix (mean composition $Mg_{0.94}Fe_{0.06}S$) with troilite (white) only as an interstitial filling (percentage smaller than 0.05).

ABUNDANCE AND CHEMICAL STABILITY OF Mg/Fe SULPHIDES

Sulphide grains can only be formed in the cooling outflows of sources showing chemical conditions which exclude the effective oxidation of sulphur. Carbon-dominated AGB stars, PPN, and PN offer such a reducing chemical environment. If the strong 30 μm band observed in the spectra of those sources is indeed due to Mg-S vibrations as suggested [2], then this explanation presupposes that a sufficient quantity of sulphide dust is available. This again presupposes that enough sulphur beeing the limiting factor for the formation of MgS is available for the sulphidization of Mg. Up to now, reliable S abundances of AGB stars, PPN, and PN are still lacking. Estimates based on solar system ratios appearently allow the formation of a sufficiently large quantity of MgS in order to account for the strength of the observed band. However, little is known on the details of the condensation process.

Under the reducing conditions in the environment of their formation, MgS grains should be chemically stable. This situation, however, should change drastically when the grains leave the circumstellar space and are incorporated to molecular clouds. It is a well-known fact that pure solid MgS is a hygroscopic substance which is easily attacked by H_2O and transformed to MgO and $Mg(OH)_2$ according to the equations

$$MgS(s)+H_2O(g) \rightleftharpoons MgO(s)+H_2S(g) \qquad MgS(s)+2H_2O(g) \rightleftharpoons Mg(OH)_2(s)+H_2S(g) \quad (1)$$

In contrast to MgS, FeS, however, is chemically much more stable if it comes into contact with H_2O although the reaction equations are perfectly analogous to (1):

$$FeS(s) + H_2O(g) \rightleftharpoons FeO(s) + H_2S(g) \qquad FeS(s) + 2H_2O(g) \rightleftharpoons Fe(OH)_2(s) + H_2S(g) \quad (2)$$

The different behaviour of the reactions (1) and (2) becomes strikingly visible if one determines the free enthalpy (Gibbs' function) G, for the reaction products on the right-hand side and the same value for the parent substances on the left-hand side and subtracts the values from each other. For (1), the differences ΔG are strongly negativ (for 298.15 K(s,g) and 0.1 MPa(s), ideal gas(g) it amounts to -32.3 and -68.4 kJ per mole [9] for the first and the second reaction equation, respectively) whereas for the equations (2), the sign of ΔG is positive (for 298.15 K(s,g) and 0.1MPa(s), ideal gas(g), the corresponding values amount to +45.8 and +33.8 kJ per mole). Negative ΔG means that the reaction proceeds spontaneously from the left to the right because the reaction products are more stable than the parent substances. The positive values of ΔG for the reactions (2) confirm the great stability of FeS against H_2O.

In agreement with this result, experiments with our sulphide samples showed that solid solutions of MgS and FeS become the more stable against H_2O the higher the percentage of FeS is. An exact determination of ΔG for the reactions of mixed Fe/Mg sulphides with H_2O is not possible because the necessary thermochemical data are not available. When the mixed Mg/Fe sulphides get to a more oxidizing and colder environment present in molecular clouds, the Mg-rich sulphides will be decomposed by water-ice particles whereas FeS should remain stable. That FeS bands in this environment have not been detected up to now can be explained by two reasons: Generally, only a few FIR observations of molecular cloud sources are available at all and the vibrational bands of FeS are relatively weak.

DETERMINATION OF OPTICAL DATA

For studying optical properties of the prepared sulphide samples, specular reflectance measurements on polished surfaces (near-normal incidence) in the wavenumber range from 5000 - 20 cm^{-1} were carried out [10]. The reflectance spectrum of the semiconductor FeS is characterized by high reflectivity in the MIR, a group of weak bands in the 30 μm region, and a strong rise of the reflectivity in the submm range. With increasing Mg-content, the reflectivity of the sulphides in the MIR decreases because the refractive index of the insulator

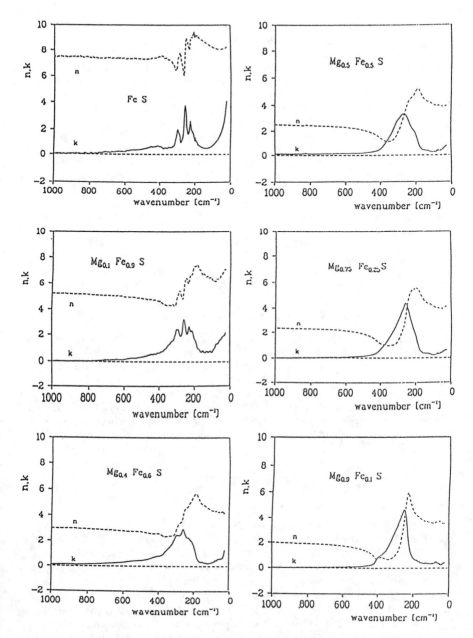

Fig. 2 Optical data in the wavenumber range from 1000 to 20 cm^{-1} derived from the reflectance values of the prepared sulphide samples of the bulk composition $Mg_xFe_{1-x}S$ with $x = 0, 0.1, 0.4, 0.5, 0.75, 0.9$. The bulk composition is given in each diagram. The curves of the magnesium-richest member of the series show some influence of MgO at about 400 cm^{-1}.

MgS is considerably lower than the one of FeS. This decrease of the MIR reflectivity is accompanied by a strong increase of the strength of the Mg-S vibration band at 300 cm^{-1} which becomes the dominating spectral feature at all. Simultaneously, the steep slope of the submm reflectivity typical of Fe-rich sulphides vanishes for the Mg-rich samples.

From the reflectance spectra, the optical constants n and k in the wavelength range of 1000 - 20 cm^{-1} were derived by the Kramers-Kronig analysis. In Fig. 2, the n and k values of the samples with the bulk compositions characterized by x= 0.0, 0.1, 0.4, 0.5, 0.75 and 0.9 are represented. In the plot for x=0.9, a slight influence of oxidation on the data is indicated.

Additionally to the reflection spectroscopy of the compact samples, the transmission spectra of powdered sulphides with grain diameters smaller than 2 μm embedded in polyethylene were measured in the spectral range from 680 to 20 cm^{-1}. The comparison of these spectra with those calculated for spheres in the Rayleigh limit and for a continuous distribution of ellipsoids (CDE) shows satisfactory agreement only for the CDE calculations of the sample with the bulk composition x=0.5. The reasons for these discrepancies are that the particles were too large in order to fulfill the condition of the Rayleigh limit (this is valid only for the Fe-rich samples because of the large n values) and that strong shape effects caused by surface plasmons play an important role. For the Mg-rich samples, obvious traces of the oxidation effect mentioned above can be seen. The influence on the transmission spectra is much stronger than that found for the reflectance data. At about 540 cm^{-1}, an additional absorption band appears which is related to MgO, a decomposition product of MgS.

COMPARISON WITH THE EMISSION BAND OF IRC +10 216

Radiation transfer calculations basing on laboratory data allow the most exactly comparison with astronomical spectra. For the 30 μm band, these calculations with sulphide data have not been performed yet. For a more qualitativ comparison, it is possible to compare the emission excess of C-rich sources with MgS absorbance spectra [1,11]. In Figure 3, we compare the emission excess of IRC +10 216 with the absorbance calculated from the optical data of $Mg_{0.5}Fe_{0.5}S$ for small spheres and CDE in vacuum and MgS using Cox [1]. The CDE calculation of the Mg/Fe sulphide fits very well the long wavelength wing of the band. The short wavelength wing could be fitted better if a higher amount of spherical particles would be assumed.

REFERENCES

1. P. Cox, Astron. Infrared Spectroscopy Conf. (1993), in press.
2. J. H. Goebel and S. H. Moseley, Astrophys. J. (Lett.) 290, L 35 (1985).
3. J. A. Nuth, S. H. Moseley, R. F. Silverberg, J. H. Goebel and W. J. Moore, Astrophys. J. (Lett.) 290, L 41 (1985).
4. H. Palme, H. E. Suess and H. D. Zeh, Landolt-Boernstein VI,2a, 267 (1981).
5. E. Anders and N. Grevesse, Geochim. Cosmochim. Acta 53, 197 (1989).
6. R. T. Dodd, Meteorites. Cambridge Univ. Press 1981.
7. H. Schulze, Untersuchungen zur stofflichen Charakterisierung des Kometen Halley auf der Grundlage von Staubimpakt-Massenspektrometern, Ph.D., Berlin (1993).
8. K. Ehlers and A. El. Goresy, Geochim. Cosmochim. Acta 52, 877 (1988).
9. M. W. Chase, Jr., C. A. Davies, J. R. Downey, Jr., D. J. Frurip, R. A. McDonald and A. N. Syverud, JANAF Thermochemical Tables, 3rd Ed.,Part II, (1985).
10. H. Mutschke, B. Begemann, J. Dorschner and Th. Henning, Infrared Physics, Proceedings of CIRP 5 (1993), in press.
11. R. F. Silverberg, S. H. Moseley and W. Glaccum, Interstellar Processes: Contributed Papers, ed. D. J. Hollenbach and H. A. Thronson (NASA TM-88342), p.110 (1987).

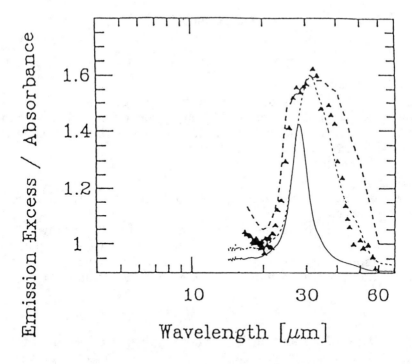

Fig. 3 A qualitative comparison of the 30 μm band profile of the carbon star IRC +10 216 (triangels) with the mass absorption coefficient derived from the optical data of the sample $Mg_{0.5}Fe_{0.5}S$. The solid line is based on spheres of the Rayleigh case, the dashed line on CDE calculations. The heavy dashed line has been taken from Cox [1] who used transmission data of MgS powder embedded in polyethylene [2,3].

DISCUSSION

BUSSOLETI — I have two questions :
1. Have you observed any other band in the IR range ?
2. How does the spectrum around 30μm appear when you use submicron grains rather than "theoretical" particles which are built by using bulk n and k values ?

BEGEMANN —
1. In the wavenumber range from 5000 to 20 cm^{-1} we have observed only this 30 μm band.
2. The best comparison of the particles embedded in PE and the calculated data gives the sample $Mg_{0.5} Fe_{0.5} S$. In this case the COE calculations and our measurements have only small differences. With increasing Mg-content our measured samples show our additional band of MgO.

SHUTTE — Spheroidal grains of $Fe_{0.5} Mg_{0.5} S$ give an excellent fit of the 30μm emission band, while spherical grains do not. How elongated do the grains have to be to obtain a good fit ?

BEGEMANN — For our calculations we used particles with diameter ratios of about 1:2 or 1:3 and this is in good agreement with our TEM-studies.

COX — As you know, the 30 μm emission band can contain up to 20% of the total bolometric luminosity in some sources (IC 418, IRAS 22272 + 5435). Did you check how much magnesium and sulfur is needed if MgS is indeed the carrier of the 30 μm band ?

BEGEMANN — We have estimated the ratio between the mass of sulphide dust for the 30 μm band and the mass of amorphous carbon dust for the FIR continuum of IRC + 10216. When we assume same temperatures of carbon and sulphide, grains are calculated with a sulphur-to-carbon ratio of 0.05. This value is of the same magnitude as the ratio in the solar gas (0.05), cometary dust (Halley, 0.09) and carbonaceous chondrites (0.62).

EXPERIMENTAL STUDY OF LABORATORY-SYNTHESIZED CARBONACEOUS GRAINS AND ASTROPHYSICAL IMPLICATIONS

L. Colangeli and V. Mennella
Osservatorio Astronomico di Capodimonte, Napoli

E. Bussoletti and G. Monaco
Istituto Universitario Navale, Napoli

P. Merluzzi, P. Palumbo, and A. Rotundi
Universitá Federico II, Napoli

ABSTRACT

We present here the most recent results obtained in the laboratory from various kinds of carbon-based submicron grains. The joint use of VUV, IR and Raman spectroscopy allows to characterize them and to determine their structural and spectral properties. We compare our results with similar data obtained by various authors on other kinds of carbonacous materials produced by UV irradiation as well as by ion bombardment. We are driven to conclude that carbon-based solid particles tend to evolve towards a common final structure, as a consequence of various kinds of processing. The merging of all these experimental evidences may bring to some interesting astrophysical implications about the actual composition of cosmic dust.

INTRODUCTION

Solid particles represent an important constituent of the interstellar and circumstellar medium. In fact, grains are directly involved in cooling and heating mechanisms and they play an important role in determining the gas phase composition of clouds; moreover, as metallic components carry the charge inside dense clouds, they play also a key role in regulating the ion-molecule chemistry in these regions. When we consider the solar system it is well known the interest that solid particles have in relation to many different items related to planets, planetary rings, asteroid and comet evolution. Also these considerations explain the blossoming of space missions which are expected to start in the coming years: MARS 94, Cassini and the third ESA cornerstone "ROSETTA".

In this contest a great impulse to solid material analysis has been certainly given by the findings related to Halley's comet as obtained expecially by means of the instruments flown aboard Giotto spacecraft. Different unexpected kinds of particles (CHON grains, for example) were found to be ejected by the nucleus which, actually, appeared unexpectedly black because of the large amount of carbonaceous materials covering the surface. IR spectra recently obtained on a variety of comets have presented emission features which indicate the presence of several different constituents for the grains.

So far, but in the case of meteorites and the two Giotto encounters with P/Halley and P/Grigg-Skjellerup, any information on dust grains is derived by interpreting the observed spectra on the basis of data obtained experimentally in the laboratory. In

Table I Laboratory production of carbon-based materials

Product	Production technique
Amorphous carbon grains (AC)	Arc discharge in Ar
Hydrogenated amorphous carbon grains (HAC)	Arc discharge in H_2
Ion produced hydrogenated amorphous carbon grains (IPHAC)[3,4]	Ion irradiation of refractory molecular solids

this contest, laboratory analysis of "analogue" cosmic materials is crucial to obtain physical and chemical parameters which allow a correct interpretation of astronomical data. Since many years the Laboratory of Cosmic Physics in Neaples has undertaken a systematic work aimed to study different kinds of materials (both grains and molecules) which may represent good "analogs" of cosmic matter.

In this paper we present an updating of results obtained on laboratory-sinthesized carbonaceous grains by using different techniques. The interest in this class of compounds resides in the fact that all solid carbon forms represent about half of the volume of interstellar dust[1] and are present also in planetary environments and in meteorites.

LABORATORY ANALYSIS

The philosophy of the general laboratory approach consists of three main steps: a) dust production by using selected materials, b) dust characterization, c) comparison of laboratory results with astronomical observations.

Phase (a) is conceived in such a way that we produce in the laboratory physical and chemical conditions which simulate, as far as possible, the environments in which the grain birth takes place. This allows to obtain various kinds of particles with different nature and structure. Details on the methods that we commonly use are largely described in previous papers to which we refer for further information[2]. Table I summarizes the grains that have been analysed in the laboratory as well as their production techniques. It is worth mentioning that our amorphous carbon (AC) and hydrogenated amorphous carbon (HAC) grains have average radii of about 50 - 100 Å. We also report in Table I information about materials which can be compared with our samples.

Transmission (T.E.M.) and scanning (S.E.M.) electron microscopy coupled with Energy Dispersive X-ray (E.D.X.) analysis allow to get information about size, shape, nature and chemical composition of the samples. Spectroscopy from VUV to FIR is used to obtain a deep analysis of the particles. Specific items concern the presence of IR bands, the possible existence of an UV peak at 220 nm, the rise in the VUV, and the value of the optical gap. All these features can be compared with spectroscopic observations in space. In addition, we have obtained further information about the structure of grains via Raman spectroscopy.

RESULTS AND DISCUSSION

The VUV spectra of both AC and HAC grains (Fig. 1) are characterized by a peak falling at about 80 - 90 nm which we interprete as due to $\sigma + \pi$ electronic transitions. Our data show also a peak falling at around 240 nm for AC grains while this is lacking in HAC particles. We note that an UV peak appears when HAC samples are annealed (Fig. 2). Consistently, the optical gap shifts from 1.23 eV, to 1.03 eV and to 0.45 eV respectively for HAC, annealed HAC and for AC.

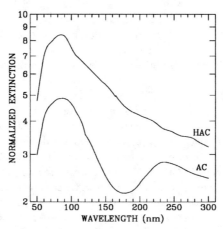

Fig. 1. The far UV extinction spectra of our AC and HAC samples.

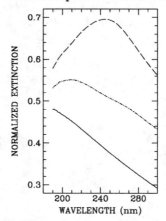

Fig. 2. UV extinction spectra of AC grains (dashed line), HAC grains (full line) and annealed (14 hours at 400 °C) HAC grains (dotted-dashed line).

Finally, C-H stretching resonances at 3.3 - 3.4 μm are present in the IR spectra of HAC samples while they are lacking both for annealed HAC and for AC (Fig. 3). All the previous experimental evidences suggest that the three kinds of considered carbon grains are different in terms of structure and hydrogen content. It seems likely that hydrogenated (HAC) samples tend to loose most of their H content under annealing and that they tend to change their structure towards a state in which the amount of graphitic bonds increases. The final product is quite similar to AC grains. This interpretation is also corroborated by the results of Raman spectroscopy, as the relative intensity of "G" and "D" bands (Fig. 4) varies according to the structure of the grains.

At this stage, it is interesting to compare our results with those reported for other carbonaceous materials (Table I). Strazzulla and Baratta[3] have studied the evolution driven by ion irradiation of frozen gases towards a refractory carbonaceous solid. Frozen benzene and butane tend to evolve towards a sort of Ion Produced Hydrogenated Amorphous Carbon (IPHAC) whose main characteristics appear similar to those observed for our samples. Actually, as the ice evolves under irradiation, Raman spectroscopy shows the birth of a band falling around 1600 cm^{-1} quite similar to what appears in our samples. In addition, their IR spectra evidence that bands typical of the parent-molecules tend to disappear completely. Moreover, Jenniskens et al.[4] have observed a similar evolution, under ion bombardment, for the organic residue (yellow stuff ?) produced by UV irradiation of frozen gases. We note also that the optical gap for these residues tends towards values typical of HAC and AC grains that we produce.

In conclusion, according to laboratory results, it can be derived that all carbon-based materials tend to evolve towards a sort of a more or less hydrogenated amorphous carbon when they are exposed to UV and/or ion bombardment. This evidence is of great importance for its astrophysical implications as it implies that any kind of carbon-based grain is somehow forced, during its life in space, to evolve (or degrade) towards an amorphous/disordered structure, regardless of how its structure and nature were at its birth.

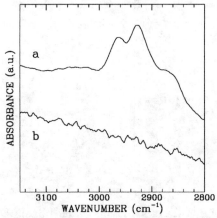

Fig. 3. IR spectra of HAC (a) and annealed (14 hours at 400 °C) HAC grains (b).

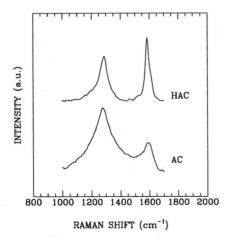

Fig. 4. Raman spectra of AC and HAC grains.

ASTROPHYSICAL IMPLICATIONS

Our experimental results can be directly applied to some astrophysical observations in order to derive indications on the chemical-physical status of the solid matter. In particular, the VUV rise often observed in the interstellar and circumstellar extinction curves can be explained simply as due to the low energy tail of a VUV peak produced by the presence of carbon grains having a structure similar to both AC and HAC. Similarly, the bands which appear in both interstellar and circumstellar extinction, respectively peaking at 220 nm and 240 nm, can be interpreted with the presence in the sources of amorphous carbon grains whose hydrogen content and structure varies according to the actual environment status. In fact, according to our experimental results, a lower hydrogenation of the grains implies the shift of the UV peak towards higher wavelengths. Therefore, a hump at longer wavelengths can be an indication that the particles are dehydrogenated. Finally, some circumstellar extinction curves are characterized by a lack of the UV hump: this would imply that the particles, are highly hyrogenated in these environments, as it appears to be reasonable.

Concerning the identification of the 3.4 μm emission band in comets, we note that, according to the results obtained by Mumma[5] for P/Swift-Tuttle, free methanol molecules are able to explain the long wavelength wing of the observed band profile quite well. On the other hand, in the case of P/Halley's spectrum our laboratory data suggest that hydrogenated amorphous carbon grains, similar to our HAC, can contribute to the

3.4 μm band[6]. A reasonable conclusion could be that a mixture of both molecular and grain materials is able to match quite well the whole shape of the observed feature.

ACKNOWLEDGEMENTS

This work was supported by MURST, ASI and CNR grants.

REFERENCES

1. A.G.G.M. Tielens and L.J Allamandola, Physical Processes in Interstellar Clouds (Reidel, Dordrecht, 1986), p. 160.

2. E. Bussoletti, L. Colangeli, A. Borghesi and V. Orofino, Astron. Astrophys. Suppl. 70, 257 (1987).

3. G. Strazzulla and G.A. Baratta, Astron. Astrophys. 266, 434 (1992).

4. P. Jenniskens, G.A. Baratta, A. Kouchi, M.S. deGroot, J.M. Greenberg and G. Strazzulla, Astron. Astrophys. 273, 583 (1993).

5. M.J. Mumma, Planet. Space Sci., in press (1993).

6. L. Colangeli, G. Schwehm, E. Bussoletti, S. Fonti, A. Blanco and V. Orofino, Astrophys. J. 348, 718 (1990).

AMORPHOUS CARBON GRAINS AND THE CIRCUMSTELLAR EXTINCTION AROUND C-RICH OBJECTS

A. M. Muci, A. Blanco, S. Fonti and V. Orofino

Department of Physics, University of Lecce
C.P. 193, 73100 LECCE (Italy)

ABSTRACT

We present here fits of the anomalous circumstellar extinction observed toward some carbon-rich objects. The calculated spectra are obtained using for the first time esperimental data of amorphous carbon grains with different hydrogen content produced in our laboratory.

INTRODUCTION

While the well known interstellar bump around 220 nm is fairly constant in position[1], some observations have evidenced that circumstellar extinction toward carbon-rich sources exhibits a broad absorption peak at longer wavelengths spreading from 230 to 250 nm [2-5]. In this work we present fits of such sources suggesting that the differences in spectra could be attributed to the hydrogen content of the amorphous carbon (AC) grains responsible for such spectra.

ANNEALED DEHYDROGENATED AC GRAINS

Submicron AC particles in this case have been produced by striking an arc discharge between two carbon electrodes in an inert argon atmosphere, according to the standard technique already described elsewhere [6, 7]. The extinction properties in the wavelength region 0.19 - 30 µm of such grains have been obtained for samples:
a) held at fixed temperatures in the range 10 - 700 K;
b) annealed at 700 K for 6 hours.

The results of the measurements show no major changes in the infrared extinction properties of the analysed samples [8]. In the ultraviolet (UV) region, instead, where the unprocessed AC grains exhibit an absorption bump falling on the average at 240 nm, we find that annealing shifts the peak position of about 8 nm toward longer wavelengths (Fig. 1).

ANNEALED HYDROGENATED AC GRAINS

With the same technique (arc discharge) but in hydrogen atmosphere we have also produced hydrogenated AC grains with the same size and morphological properties as those produced in argon. As already pointed out by Blanco and coworkers [9, 10], while freshly formed grains show a featureless UV spectrum rising monotonically with decreasing wavelength, the same annealed particles exhibit an UV bump with a peak position ranging from 205 to 250 nm depending on the temperature and the duration of the annealing process (Fig. 2).

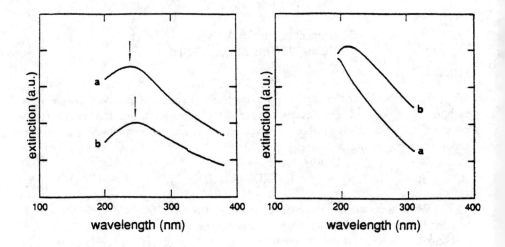

Fig. 1 - Laboratory extinction spectra of AC grains produced in Ar atmosphere before (curve a) and after (curve b) annealing for 6 hr at 700 K. Arrows indicate the position of the peak in the two cases.

Fig. 2 - Laboratory extinction spectra of AC grains produced in H_2 atmosphere before (curve a) and after (curve b) annealing for 14 hr at 700 K.

RESULTS AND DISCUSSION

By using the laboratory UV spectra of the various AC grains previously described, we fit (see Figs. 3-6) the extinction curves observed in some carbon rich circumstellar environments (see Table I).

As it can be seen from Figs. 3-6 and Table I, the extinction toward carbon-rich sources can be interpreted by the presence of circumstellar carbon grains similar to those studied in our laboratory. Freshly formed hydrogenated AC grains can explain the featureless spectrum of the H-rich source HD 89353. On the other hand H-poor sources, exhibiting extinction spectra with the peak falling in the 230 - 250 nm region, could be interpreted by the presence of AC grains with different hydrogen content. Dehydrogenation can be obtained either by annealing particles produced in H_2 atmosphere or by producing the same grains directly in H-poor atmosphere (argon). We want to stress, however, that even the latter type of particles contain a low amount of hydrogen which can be removed by subsequent annealing.

Our findings can be regarded as the experimental confermation of the theoretical scenario outlined by Hecht [12] and Sorrell [13].

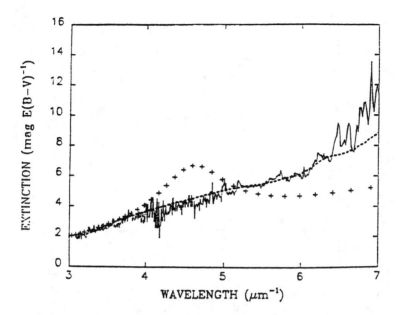

Fig. 3 - Comparison between the circumstellar extinction observed [4] toward HD 89353 (solid line) and the laboratory spectrum of freshly formed AC grains produced in H_2 (dashed line). The mean interstellar extinction derived by Seaton [11] is also shown (crosses).

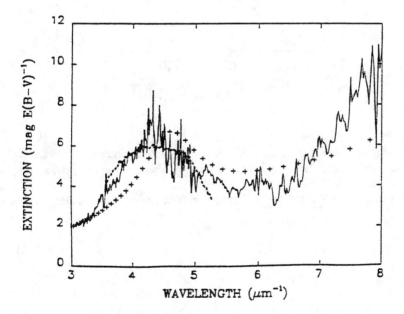

Fig. 4 - Same as Fig. 3 for the HD 213985 source [4] and AC grains produced in H_2 and annealed for 20 hr at 700 K.

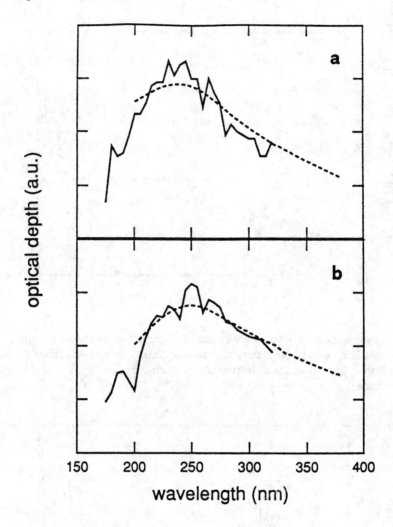

Fig. 5 - (a) Comparison between the circumstellar extinction observed [3] toward R CrB (solid line) and the laboratory spectrum of freshly formed AC grains produced in Ar (dashed line).
(b) Same as (a) for RY Sgr [3] and AC grains produced in Ar and annealed for 6 hr at 700 K.

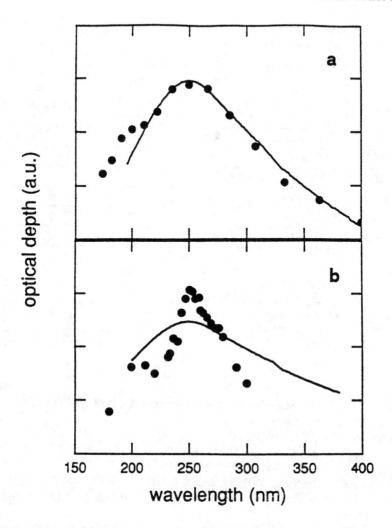

Fig. 6 - (a) Comparison between the circumstellar extinction observed [2] toward Abell 30 (dots) and the laboratory spectrum of AC grains produced in Ar and annealed for 6 hr at 700 K (solid curve).
(b) Same as (a) for V348 Sgr [5]. The different width of the bump in this peculiar object could be explained by clustering and/or size distribution effects of laboratory AC grains.

Table I - Circumstellar extinction observed in some carbon rich objects and laboratory carbon grains used to obtain the best fit

Source name	Source type	H content	Peak wavelength λ_p(nm)	Laboratory grain type	Ref.
HD 89353	Post-AGB star	H-Rich	-	a	4
HD 213985	Post-AGB star	H-Poor	230	b	4
R CrB	Carbon star	H-Poor	230 ÷ 250	c	3
Ry Sgr	Carbon star	H-Poor	230 ÷ 250	d	3
Abell 30	Planetary nebula	H-Poor	247	d	2
V348 Sgr	Peculiar	H Poor	250	d	5

NOTES
a - AC grains in H_2 freshly formed
b - AC grains in H_2 annealed for 20 hr a 700 K
c - AC grains in Ar freshly formed
d - AC grains in Ar annealed for 6 hr at 700 K

REFERENCES

1. Mathis J.S. 1987, in Scientific Accomplishments of the IUE, Y. Kondo ed., Reidel Publishing Comp. (Dordrecht)
2. Greenstein J.L. 1981, ApJ 245, 124
3. Hecht J., Holm A.V., Donn B., Wu C.C. 1984, ApJ 280, 228
4. Buss R.H., Lamers H., Snow T.P. 1989, ApJ 347, 977
5. Drilling J.S., Schönberner D. 1989, ApJ 343, L45
6. Borghesi A., Bussoletti E., Colangeli L., Minafra A., Rubini F. 1983, Infrared Phys. 23, 85
7. Bussoletti E., Colangeli L., Borghesi A., Orofino V. 1987, A&AS 70, 257
8. Muci A.M. 1992, Thesis, University of Lecce
9. Blanco A., Bussoletti E., Colangeli L., Fonti S., Stephens J.R. 1991, ApJ 382, L97
10. Blanco A., Bussoletti E., Colangeli L., Fonti S., Mennella V., Stephens J.R. 1992, ApJ 406, 739
11. Seaton M.J. 1979, MNRAS 187, 73p
12. Hecht J. 1986, ApJ 305, 817
13. Sorrell W.H. 1990, MNRAS 243, 570

VIBRATIONAL EXCITATION OF HYDROGEN DESORBED FROM A CARBON SURFACE

C. Schermann, S.F. Gough, F. Pichou, M. Landau, R.I. Hall,
L. D. M. A., CNRS et Université P. et M. CURIE
4 place Jussieu T12-B75, 75252 Paris Cedex 05, France.

and I. Čadež[†]
[†]Institute of Physics, Pregrevica 118, 11080 Zemun, Yugoslavia.

ABSTRACT

We have studied the recombination of H atoms adsorbed on a carbon surface as a function of the surface temperature between 90 and 300K. The molecules formed by this mechanism are excited to high vibrational states and this excitation shows a strong temperature dependence.

INTRODUCTION

H_2 is the most abundant molecule in the diffuse interstellar medium and is inferred to be the major form of hydrogen in dense clouds. Since H_2 is at the origin of most chemical reactions[1], its abundance and internal energy are of primary interest. The formation of H_2 by binary collision is ineffective, and, at low densities, the three body reaction is totally inefficient. Also, it is now accepted that the only way to obtain H_2 molecules in space is via the recombination of H atoms on the surface of interstellar grains. Theoretically, of the possible reaction surfaces, silicates[2], ice[3] or carbon (crystalline or amorphous), graphite seems to be the most effective[4]. Using our recently developed technique[5] for determination of relative vibrational populations of hydrogen molecules obtained by recombination on metal surfaces, we have investigated the recombinative desorption of atomic hydrogen from a carbon surface.

RECOMBINATIVE DESORPTION MECHANISMS

The studied reaction is:
  H + H + Surface → H_2 + Surface
This reaction can occur in two ways. In the first, called the Langmuir-Hinshelwood mechanism, two adsorbed atoms diffuse on the surface and recombine with each other before leaving the surface, taking the necessary energy from the surface: in this case the desorbed molecule will have an energy (internal and translational) corresponding to the surface temperature. In the

second process, called the Eley-Rideal mechanism, an atom comes from the gas phase, enters in collision with an adsorbed atom, recombines immediately with it and the two atoms leave the surface as a molecule: the molecule will then have an internal energy, as the incident atom is not thermalised with the surface. To study these processes it is necessary to know the binding energy of H and H_2 with carbon.

Experimentally, only a few studies have been made on the adsorption of atomic or molecular hydrogen by carbon or graphite, giving a large range of values for the recombination coefficient[6] (0.005 to 0.05), and a sticking probability[7] of 0.02 to 0.04 for H and zero for H_2, if no additional activation energy is given. To our knowledge, no adsorption energy has yet been determined. However, many quantum chemical studies[8,9,10] have recently been made and their results will be used as a basis for the following discussion.

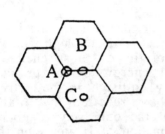

Fig. 1. Positions of H above graphite surface

An H atom has three possible positions for chemisorption on a graphite surface (fig. 1). Site A, which corresponds to the H atom directly above a C atom, is the most favoured, with a binding energy E_{ads} of 0.3 to 1.14eV according to the above authors. Site B posesses a slightly smaller binding energy (0.3 to 1.05eV) while site C, center of the hexagonal lattice cells, is not a stable binding site as the binding energy is zero or weakly negative (positive energies refer to binding).

The easiest way to represent the interaction between H or H_2 and a surface is to draw a one dimensional Lennard-Jones potential energy diagram, where z is the distance between the particle and the surface. On figure 2, we show the results of Ross and Olivier[11] for the H_2 physisorbed state, and of Aronowitz and Chang[9] for the chemisorbed state (on site A). Only half the energy of the H_2-Metal system is taken into account in order to compare two equivalent systems.

As has been observed experimentally, H_2 cannot dissociate into two chemisorbed atoms, as is the case for most metals but requires an activation energy, E_{act}, to be supplied. On the other hand, two chemisorbed atoms only have to pass through the barrier E_b in order to yield an H_2 molecule. Following the results of Aronowitz and Chang this barrier is of the order of 0.4eV.

The probability of obtaining H_2 by the Langmuir-Hinshelwood mechanism depends on the ability of the atoms to move on the surface. Calculations performed by Aronowitz and Chang[9], and confirmed by Klose[10], show that an atom does not stay more than 10^{-9} seconds in a particular site. This gives a great mobility to an H atom.

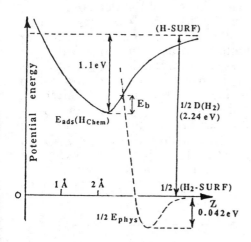

Fig.2. Schematic potential energy diagram for adsorption of H and H_2 on graphite.

However, to obtain an H_2 molecule it is necessary that the two atoms approach within approximately 1.5 Angstroms of each other and penetrate by tunnel effect the potential barrier E_b of the order of 0.4eV. The calculation[10] performed at 10K indicates that, at this temperature, the probabilty of two atoms leaving as H_2 is only 1 in 2000 approaches. This sets a lower limit, since at higher temperatures we expect the mobility of the atoms to be higher, and in any case our surface is not a perfect crystal and thus the depth of the well is variable, both of which favour recombination.

All these calculations have been performed for a weak rate of surface coverage.

EXPERIMENTAL SET-UP AND DIAGNOSTIC METHOD

The apparatus used consists of a cylindrical cell machined from stainless steel and surrounded by an outer jacket through which a cooling fluid could be circulated. With liquid nitrogen as circulating fluid a cell wall temperature as low as 80K was attainable. Carbon was evaporated onto the walls of the cell by resistive heating of a graphite filament. In this way all kinds of carbon can form on the walls, including carbon black, amorphous and polycrystalline carbon. The evaporation is carried out after heating the cell in good vacuum (about 10^{-8} Torr) to avoid contamination of the surface by residual gases. Atomic hydrogen was generated by dissociation of molecular hydrogen on a hot tungsten filament contained in the cell. The hydrogen beam exiting the cell contains those vibrationally excited hydrogen molecules formed by recombinative desorption of atomic hydrogen on the cell wall. The molecular beam is then crossed by a narrow, well-defined, low energy electron beam, and vibrational state determination is achieved by an orginal method based on the dissociative attachment process:

$$e^- + H_2(v) \rightarrow H_2^- \,(^2\Sigma_u^+) \rightarrow H^- + H$$

The negative molecular ion has a very short lifetime and dissociates to give H⁻ and H. The H⁻ signal is detected. Our interest in this method comes from the fact that the cross section for the process rises dramatically as one moves to higher vibrational states of the molecule (more than one order of magnitude for each of the first five vibrational states). As the internal energy of the molecule increases, i.e. one moves to higher vibrational excitation, less electron energy is required to produce dissociation, and the threshold thus moves to lower impact energies. In addition, the cross sections for the vibrational levels are peaked at threshold, allowing each vibrational level to be clearly resolved, as can been seen in figure 3. Scaling the experimental spectra with the known cross sections for the process enables the vibrational state populations to be deduced.

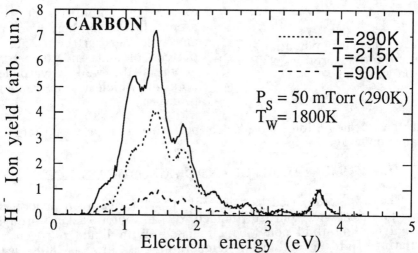

Fig. 3. Spectrum of the H⁻ yield as a function of incident electron energy for three temperatures of the recombinating carbon surface.

This method is simple and efficient: populations as low as 10^{-7} of the v=0 population can be observed, corresponding to a detection sensibility of about 10^5 particles cm⁻³ for the v=5. However, one drawback is that the translational energy is not known, which limits the study to rovibrational excitation. The particles are studied after they have left the cell, and since the calculations show that after approximately 5 collisions the molecules have relaxed to the v=0, it would seem that those molecules observed to be in high lying vibrational states are those which have left the cell after undergoing very few collisions. We believe that the populations of the high levels detected in our experiment are not far from the nascent distribution.

RESULTS AND DISCUSSION

On figure 3 we can compare the results obtained at three temperatures: ambient temperature of about 290K, the temperature corresponding to the maximum excitation of the molecules, 215K, and the lowest temperature attainable in the present experiment, 90K.

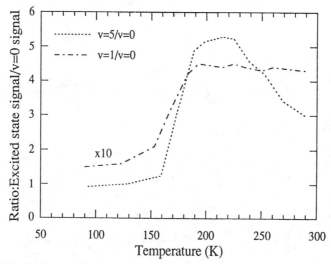

Fig. 4. Variation of the excited state signal ratio v=5/v=0 and v=1/v=0 as a function of temperature.

As we can see, the vibrational states are excited up to v=7 and rotational excitation is nonexistant. Comparing these spectra with those obtained under identical conditions, but with a tungsten surface, it is seen that the excitation is stronger for the high levels in carbon than in tungsten. The excitation of the v=7 is only possible if we retain the theoretical binding energy of the H atoms on site B of 0.3eV. If we take the maximum value of 1.14eV, only an excitation to the v=4 is possible and, as in the previous experiment on metal surfaces[12], we must invoke physisorbed states of H. In fact, when all the sites are filled up the surface becomes more inert, and a second layer of physisorbed H atoms must exist, with a weak binding energy. In this case, the atoms are more mobile and recombine more easily. Molecules formed in this way have a maximum attainable internal energy close to the dissociation limit of 4.48eV. The temperature effect is quite strong, as can been seen in figure 3. As the temperature decreases, the vibrational excitation of the first few levels remains constant down to about 200K, while at the the same time the excitation of the higher levels increases strongly. At temperatures lower than 200K the population of all the levels decreases in the same manner.

On figure 4, we have drawn the ratios of the peak heights $v=1/v=0$ and $v=5/v=0$ as a function of temperature. Clearly, there is a threshold at around 200K where the ratio for each falls by a factor of 5.

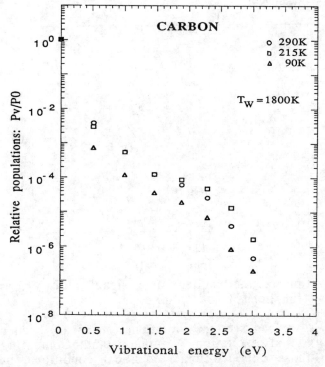

Fig. 5. Variation of the relative populations P_v/P_o as a function of the vibrational energy.

The difference in behaviour, at high temperature, between the first 4 vibrational states and the higher ones seems to indicate that the formation proceeds in two different ways, which gives support to the H_2 formation from the physisorbed states of H.

We presently explain the shape of the $v=5/v=0$ curve, i.e. the increase in excitation of the higher levels followed by a rapid decrease, as arising from the competition between two different mechanisms. Firstly, as the temperature descends the rate of covering of the surface by H atoms increases, but at the same time the mobility of the atoms decreases until this second effect becomes dominant.

Figure 5 shows the relative populations P_v/P_0 versus vibrational energy. We can see that even if the population of all the levels decreases at 90K in relation to those at higher temperatures, there is, none the less, still a strong excitation of the molecules. The Boltzmann temperature deduced from the slopes of this curve is still superior to 2300K.

CONCLUSION

In conclusion, we have shown that hydrogen molecule formation does indeed occur on a carbon surface in the temperature range between 90 and 300K. The molecules thus formed are vibrationally excited, and this excitation shows a strong dependence on the temperature of the carbon surface. However, even at 90K the population of the vibrational states is high. We are currently working on the further development of the experiment to study the formation of molecular hydrogen on other surfaces of astrophysical importance, for example amorphous hydrogenated carbon surfaces with and without the addition of oxygen.

REFERENCES

1. D.R. Flower and G. Pineau des Forêts, Mon. Not. ast. Soc. 247,500 (1990).
2. W.W. Duley and D.A. Williams, Mon. Not. ast. Soc. 223, 177 (1986).
3. R. Smoluchovwski, J. Phys. Chem. 87, 4229 (1982).
4. M.J. Barlow and J. Silk, Ast. J. 207, 131 (1976).
5. D. Popovic, I Cadez, M. Landau, F. Pichou, C. Schermann and R.I. Hall, Meas. Sci. Technol. 1, 1041 (1990).
6. A.B. King and H. Wise, J. Phys. Chem. 67, 1163 (1963).
7. G.A. Beitel, J. Vac. Sci. Technol. 8, 647 (1971).
8. A. J. Bennet, B. McCarroll, R.P. Messmer, Surf. Sci. 24, 191 (1971).
9. S. Aronowitz, S. Chang, Ast. J. 293, 243 (1985).
10. S. Klose, Astron. nachr. 310, 409 (1989).
11. S. Ross, J.P. Oliver, On Physical Adsorption (New York: Interscience) (1964).
12. I. Cadez, C. Schermann, M. Landau, F. Pichou, D. Popovic and R.I. Hall, Z. Phys. D: At., Mol. Clusters 26, 328 (1993).

DISCUSSION

<u>BUSSOLETTI</u> — You are presently observing a carbone surface which is some average between graphite, amorphous carbon and polycrystalline carbon. Therefore your results are an "average" somehow. Do you expect strong changes by using a single kind of surface ?

<u>SCHERMANN</u> — I dont know. Obviously the surface is not well defined and we would like to move towards a system such as yours which would allow us to select the nature of the carbon surface to be studied.

<u>SCHMITT</u> — These are really interesting results ! From the geometry of your experimental system the H_2 molecules have to collide one or more times the cell walls before escaping to the electron beam analyzer. Do you expect relaxation of H_2 upon collision, especially for high vibrational energies (n = 7 or larger). I will be very excited to see similar results on ice surfaces !

<u>SCHERMANN</u> — The mean number of collisions undergone by a molecule on the walls during its stay in the cell can be changed by a factor of 10 (between 100 and 1000). The spectra obtained are unchanged also, we believe that those molecules observed to be in high vibrational states have undergone only a few collisions, but that these are sufficient to relax the high levels (higher than 7, say). In any case, due to the dissociation limit of H_2^-, we can only observe H_2 excitation to the V = 9. Since the intensity of the electron beam below 0.3eV is very small this naturally perturbs the detection of the high levels. We are presently changing the geometry of our cell to try to collect molecules after only 1 collision in order to improve these measurements. F. Rostas has planned some such experiments on icy surfaces using optical detection techniques.

<u>MENZEL</u> — As a general comment, I should like to call attention to the fast amounts of information existing on sticking, desorption, surface reactions, and surface photochemistry in Surface Physics. While the systems investigated there are usually well-defined single crystal surfaces, many general properties can be carried over to more complicated surfaces. In connection to the present paper, I would suggest to link up to this type of information by taking measurements for H_2 desorption from Cu simple crystal surfaces. The dynamics of these systems is very well understood and could serve as a point of reference before going to more complicated systems. In this way the joining-up of the

information from these two fields —surface physics, astrochemistry— could be utilized, wich is certainly worthwhile.

SCHERMANN — I agree with the first remark. As for the second remark, it would be very interesting but would pose many technical problems.

ROUEFF — Can you comment on the rotational distribution obtained for H_2 in your experiments ?

SCHERMANN — We have not observed rotational excitation in the case of a carbon surface. However, the resolution of the apparatus is sufficient to discern rotational excitation, since we have previously observed rotation in the case of a copper surface.

RECENT MEASUREMENTS ON COAL IN THE NEAR AND MID-IR AS A MODEL OF INTERSTELLAR DUST

K. Ellis[1], O. Guillois[1], L. Nenner[1], R. Papoular[1,2], C. Reynaud[1]

1-Service des Photons, Atomes et Molécules, CEA, Centre d'Etudes de Saclay, 91191, Gif sur Yvette cedex, France
2-Service d'Astrophysique, CEA, Centre d'Etudes de Saclay, 91191, Gif sur Yvette cedex, France

ABSTRACT

High resolution Infrared spectra of three coals of different rank (age) are reported in the range 4000-500cm^{-1} using Diffuse Reflectance Infrared Fourier Transform (DRIFT) spectroscopy. High ranking coals provide the best fit to the interstellar IR spectra, including UIBs (Unidentified Infrared Bands) and continuum. The medium rank coals, when suitably heated, also give a good agreement with the observations.

Modulated Emission Spectroscopy (MES) is used in the mid-ir at low resolution on one of the high ranking coals to measure the absorbtivity of the material as a function of its temperature. IR bands are found not to be affected by temperature rise up to 700K but are progressively destroyed above (reminescent of the loss of UIBs in strong IS radiation fields). We show also that the underlying continuum in coal absorbtivity increases with the temperature and demonstrates a changing slope, which is directly connected to the semi-conductor properties of this material and has strong implications in the heating mechanism in the coal model.

1- THE IR SPECTRA OF COAL

Due to the high opacity of coal, measurements are not easily feasable by transmission so that DRIFT spectroscopy was used. The IR spectra of three coals, from different seams, namely Mericourt, Escarpelle and La Mûre in order of increasing rank are reported in figure 1. These results reinforce the initial statement[1] that coal exhibits all of the strongest unidentified bands (UIBs)[2], namely the 3040, 1615, ~1300, 1150, 890cm^{-1} (3.3, 6.2, ~"7.7", 8.6, 11.3μm) features as well as a sloping underlying continuum. The bands can be interpreted on the basis of the known and well studied composition of coals, mainly carbon, hydrogen and oxygen, and their complex and inhomogeneous structure[3] (disordered stacks of graphitic planes, linked by oxygen bridges and various hydrocarbon functional groups). Within the coal model, the UIBs are attributed to the functional groups attached to the carbon skeleton which demonstrate the aromatic C-H stretch near 3.3μm, C=0 and the C=C skeletal vibration around 6.2μm, mainly ether bridging groups within the carbon skeleton near 7.7-8.6μm and a single, isolated, aromatic, out-of-plane C-H bending vibration at 11.3μm.

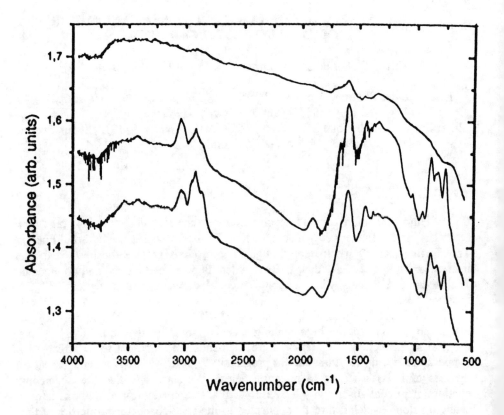

Fig. 1: DRIFT spectra from coals of increasing rank from the bottom to the top: Mericourt, Escarpelle and La Mûre respectively.

Different coals demonstrate different numbers and intensities of bands depending on the relative concentrations of C, H and O which depend on the age and stage of coalification of the coal. In the less evolved coals (as Mericourt), many more bands are observed which are not seen ubiquitously in the ISM for example. As the rank increases, we observe an evolution towards a more aromatic spectrum, as evidenced by the increase of the 3.3 μm feature compared to the 3.4μm aliphatic bands in the Escarpelle spectrum with respect to the Mericourt spectrum. This dependence of the ratio [3.3]/[3.4] with rank may account for the variation of this ratio from one astronomical source to another.

Going to the highest ranking coal, La Mûre, we observe a considerable reduction of the bands intensity compared to the underlying continuum. It corresponds to a highly graphitised structure. Unfortunately, the near-IR region is perturbed by scattering and electronic absorption, swamping the absorption bands due to functional groups so that the expected 3.3μm band is not observed. However the mid-ir part of the La Mûre spectrum is strongly simplified and demonstrates exclusively bands observed in the ISM.

2- THE EFFECT OF HEAT-TREATMENT

Heat is the main accelerating factor in the evolution of coal towards graphite. During this process, the physical and chemical properties of coal evolve and the relative quantities of C, H and O change, with a subsequent effect on the presence and intensity of the bands in the near and mid-ir. We studied the effect of heat-treatment at temperatures up to 1100K on the three coal samples. No changes occur below 700K in the shape and intensiy of the bands. Further treatments lead to a progressive aromatisation of the sample with the loss of the aliphatic 3.4, 6.9 and 7.3µm bands and the increase of the 3.3µm one (FIG 2). We note also the loss of hydrogen (decreasing of the 12.3 and 13.4 CH out of plane bending bands), and loss of oxygen with the reduction in the 6.2µm band. Finally, the only remaining features are coincident with the most prominent UIBs and that other features corresponding to functional groups not seen in the ISM are lost.

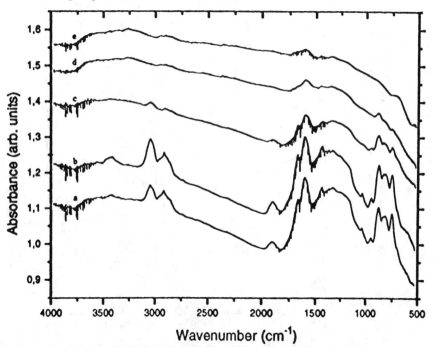

Fig.2 : Effect of heat-treatment on the IR spectra of Escarpelle coal. (a) untreated ; (b) treated at T=790K;(c)T=920K;(d)T=1000K and (e)T=1100K

All the Mericourt and Escarpelle spectra demonstrate the evolution to a flatter slope of the continuum absorption with increasing temperature. It is consistent with a smaller electronic band gap as a result of further graphitisation of the sample. La Mûre shows little spectral evolution on heating even up to 1100K. This is expected since this high ranking coal is already in a high graphitised form.

3- COMPARISON WITH NEBULAR OBSERVATIONS

The coal samples with the more aromatised molecular structure appear to be the closest to the interstellar observations. It is the case for the La Mûre seam sample and also for the Escarpelle seam sample heat-treated at T>920K. We attempt to compare directly the spectral bands of these samples to two interstellar observations, namely from the NGC7027 nebulae (FIG 3). We observe that agreement is not limited to the presence and position of the bands, but also widths and relative intensities within the mid-ir broad underlying feature very close to the observations.

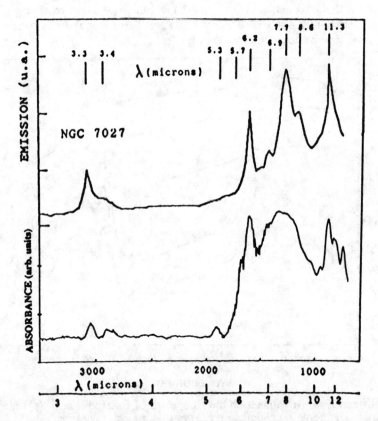

Fig 3: Comparison of the emission spectrum of NGC7027 (top) to the absorbance spectrum of heat-treated Escarpelle coal after substraction of underlying continuum (bottom).

4- THE EFFECT OF TEMPERATURE

Since the temperature of interstellar dust may vary significantly from one region of the ISM to another, the quantitative dependance of absorbtivity of coal on temperature is required. For this purpose, we use Modulated Emission Spectroscopy[4], an experimental method based on the direct measurement of the IR emission of a sample submitted to modulated UV/vis light. The sample temperature depends on the intensity of the exciting light and the method allows the determination of its surface temperature. Results obtained in the mid-ir are presented in FIG 4 in the case of a middle rank coal.

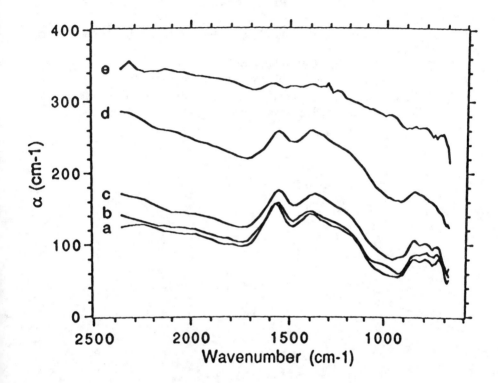

Fig 4: Absorbtivity spectra for the same coal sample, at different surface temperatures. (a) 550K; (b) 640K; (c) 715K; (d) 870K; (e) 1050K.

The underlying continuum is strong, presents a slope increasing to lower wavelength and is an increasing function of the temperature, a behaviour that is directly related to the semi-conductor character of the material. These characteristics are the basis of the heating mechanism we proposed for the grains in the ISM[5].

In spite of a lower resolution than in the DRIFT measurements, we observe that the bands are unchanged up to 715 K but we observe the structural changes induced by heating at higher temperatures. Thus, vanishing of the UIBs in strong IS radiation will occur at dust temperature of ~1000K in the coal model. The heating mechanism we propose[5] corresponds to dust temperature far below this range.

ACKNOWLEDGEMENTS

One of us (K.E.) is indebted to the Commission des Communautés Européennes for a fellowship.

REFERENCES

1- R. Papoular, J. Conard, M. Giuliano, J. Kister and G. Mille, Astron. Astrophys. 217 (1989) 204
2- M. Cohen et al. Astrophys. J. 302 (1986) 737
3- J.N. Rouzaud and A. Oberlin, Thin solid fibers 105 (1983) 75
4- O. Guillois, I. Nenner, R. Papoular, C. Reynaud, Appl. Spec. (submitted)
5- O. Guillois, I. Nenner, R. Papoular, C. Reynaud, in these conference

COAL MODEL FOR THE UV-VISIBLE INTERSTELLAR EXTINCTION CURVE

O. Guillois[1], R. Papoular[1,2], C. Reynaud[1] and I. Nenner[1,3]

1-Service des Photons, Atomes et Molécules, CEA, Centre d'Etudes de Saclay, 91191, Gif sur Yvette cedex, France
2-Service d'Astrophysique, CEA, Centre d'Etudes de Saclay, 91191, Gif sur Yvette cedex, France
3-LURE, Laboratoire mixte CNRS, CEA et MRES, Centre Universitaire de Paris-sud, Bât. 209D, 91405 Orsay cedex, France

ABSTRACT

Photoreflectance measurements followed by Kramers-Kronig analysis have yielded the dielectric constants and refractive indices of polycrystalline graphite and five coal samples ranging from high to low rank. The optical efficiencies Q_{abs} are calculated. All species show the $\sigma + \pi$ electronic resonance at ~800 Å but only two of the most graphitized samples display a strong π resonance around 2200 Å. This π feature weakens, broadens and fades into a structureless continuum when graphitization diminishes. We confirm that the most graphitized coal (anthracite) is the closest fit to the λ2175 bump in the interstellar extinction curve. In addition, we suggest that poorly graphitized coals are adequate model materials to account for the underlying continuum of the extinction curve from the UV onto the visible range.

1-INTRODUCTION

The λ2175 bump in the interstellar extinction curve (ISEC) is very well documented[1]. This band is known to lie on a broad continuum extending in the whole visible range and rising towards short wavelengths up to 912 Å, the limit of observations due to the absorption of hydrogen. Several attempts to interpret and assign this spectral band have been made and reviewed[2,3]. Models[4,5] as well as free molecules (PAH's)[6] have been proposed without offering a satisfactory agreement with observations. Recently, we have performed photoreflectance and electron energy loss measurements[7] on high ranking coals, (highly graphitized) which show that "La Mure" coal is a good model for the ISEC λ2175 bump, assuming this is sitting upon a continuum in $\lambda^{-0.85}$. Interestingly, this coal is also a good candidate for UIBs spectra, as reported elsewhere[8].

In this paper, we present new photoreflectance measurements, in the visible and UV region, performed on coals of very different ranks, including poorly graphitized ones, to offer an interpretation of the whole ISEC, i.e. both the bump and continuum.

2-STRUCTURE AND PROPERTIES OF COAL

Coal, as extracted from earth, is made up mainly of an organic part, minerals and free molecules trapped in pores. The organic part resembles kerogen, that is an insoluble, three-dimentional organic, macro-molecular skeleton which is also found in primitive meteorites[9]. This organic part, which is of interest for the dust model contains the most abundant elements in the universe, C, H and O with atomic concentrations proportional to 1 : 0.5 : 0.02 respectively. The structure and properties of coals is a well studied and documented subject in chemistry[10,11,12]. These materials are classified in data banks, according to their age or mining depth (their rank) because their properties do not depend much on their birthplace. Coal evolves with time and temperature. The higher the rank the higher is the graphitization of the coal, the lower is the abundance of hydrogen and oxygen compared to carbon. Coals are labelled by the name of the mine from which they are extracted, each of them being characterized by the C/H and C/O abundance and other characteristics like the microstructure. In the present work, we have studied in decreasing order of graphitization, samples of several french mines : "La Mure", "Escarpelle", "Méricourt", "Vouters" and "Gardanne". The microstructure of coal is composed of graphite-like "bricks" oriented at random and lying in a quasi-fluid amorphous aliphatic phase.

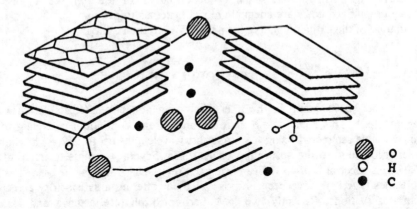

Fig.1: Schematic of the microstructure of coal

Each of these bricks, called "basic structural units" (BSUs) is made out of a few layers of aromatic carbon rings, of 10 to 20 Å in size ; the layers are in turbostratic disorder. As the rank increases, the mutual orientation of the BSUs change towards a more ordered arrangement. Various functional groups are attached to the BSUs, in the form of CH, CH_2, COH etc.... Oxygen atoms bridge between the bricks. These functional groups are responsible for the vibrational bands seen in the infra-red (IR) and those mimick UIBs. The aromatic carbon skeleton of BSUs provides the π-π^* electronic resonance in the UV, which is under study in the present paper.

Coal is a semiconductor, unlike graphite. The subsequent spectral properties are presented in a separate paper. Coal provides a family of earthly, natural material, with properties all correlated with the single dependance on rank (or carbon content). It gives a good model for dust in various environments.

3-EXPERIMENTAL

The experimental set-up used for photoreflectance spectroscopy has been described previously[7]. Briefly, monochromatized synchrotron radiation from Super ACO storage ring at the LURE facility in Orsay is used as the light source in the 400 to 4000 Å range. The beam was directed by a rotatable gold-plated mirror and focussed onto the sample placed into vacuum chamber. The nearly perpendicular reflected beam is detected by a salicylate window in front of a photomultiplier. The coal samples were in the form of fine, demineralized powder pressed into pellets. For each sample, the reflectivity spectrum was deduced by comparison with a reference beam directed from the mirror right into the detector.

Because of the drastic reduction of synchrotron luminosity at wavelengths longer than 5000 Å, the reflectivity spectra were completed by diffuse reflectance measurements in the 2000-9000 Å range, using a Cary spectrophotometer in the laboratory.

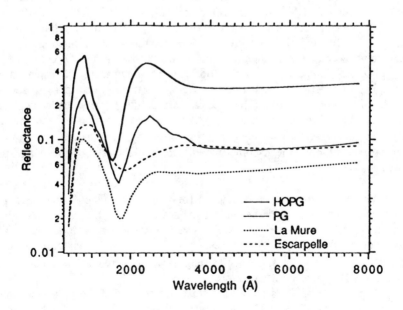

Fig.2a

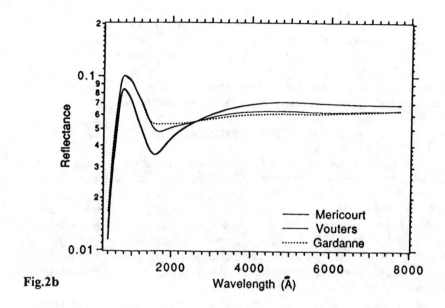

Fig.2b

4-RESULTS AND COMPARISON WITH THE EXTINCTION CURVE

The correctness of the experimental technique was tested with a highly oriented pyrolytic graphite (HOPG), as well as with a polycrystalline graphite (PG). Both corresponding spectra are presented on fig2a. Note the good agreement in abscissae of the two UV bumps in both cases with Taft and Phillip[13]. Differences in intensity between both spectra are due to the polycrystalline form of the latter (which is closer to IS dust form): The microcrystallites are randomly oriented, as a result the peaks are weaker, and the trough is shallower than those of the HOPG sample.

The reflectance spectra of the too highest ranking coals are displayed on the same figure in order to show the effect of graphitization in disordered material like coals. We note the similarity of the spectral reflectivities, with a tendency of the dip (near 1800 Å) and the peak (near 2500 Å) to weaken, broaden and red-shift as the coal rank decreases. This regular progression is terminated with the total disappearence of both the π-bump and dip for the lowest ranking coals such as Gardanne (fig.2b).

We have completed a Kramers-Kronig analysis (adapted from Stern[14]) of these results, which yielded the optical indices, n and k for each materials. The correctness of our procedure was proven numerically with an ideal double Lorentzian oscillator: That is to say, two Lorentzian oscillators, respectively centered on σ and π-bumps, and with strengths which were chosen so as to simulate each of the reflectance spectra measured.

Application of Mie's theory for small grains yielded the optical efficiencies. Normalized curves of $Q_{ext}(\lambda)$ are shown in figures 3a and 3b, for both graphite

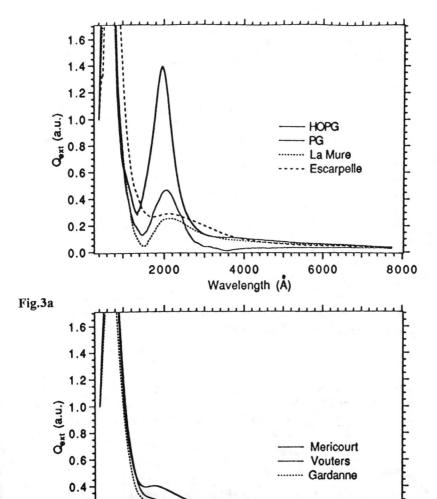

Fig.3a

Fig.3b

samples and coals. Note the regular evolution of the π-bump from the highest ranking coal La Mure, to the lowest evolved Gardanne coal which, finally, presents only a decreasing continuum. So, coals provide a continuous range of material. La Mure coal is the nearest fit to the IS λ2175 band, but provide too strong a contrast of the π-band, which indicates that this material can not be responsible for the underlying continuum of the IS band. Hence, we must try to fit the continuum with one or more coals of lower ranks (fig.4).

Thus, a fit of the whole IS extinction curve should require a combination of coals of different degrees of graphitization (probably in grains of various sizes). Such a mixture may account for the different bump widths and strengths observed in different directions of the sky. It could be the outcome of carbon grain processing in different radiation fields for different lengths of time.

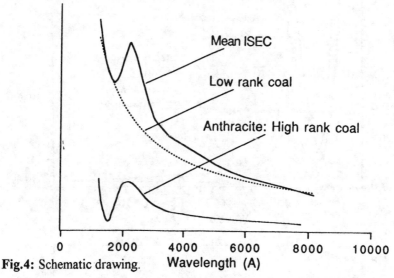

Fig.4: Schematic drawing.

Acknowledgements: We thank the staff of the Laboratoire pour l'Utilisation du Rayonnement Electromagnétique for the synchrotron radiation facilities, as well as Drs J. Breton and P. Martin for their assistance during the measurements.

REFERENCES

1. E. Fitzpatrick and D. Massa, ApJS, 72, 163 (1990) ; ibid. 307, 286 (1986)
2. B. Draine, Interstellar dust, IAU Symp. N°135, ed. L. Allamandola and A. Tielens (Dordrecht, Kluwer) (1989)
3. J. Mathis, ARA&A, ed. D. Burbidge, D. Layzer and A. Sandage (Annual Reviews, Palo Alto, USA) 22, 37 (1990)
4. J. Hecht, ApJ, 305, 817 (1986)
5. J. Mathis and G. Wiffen, ApJ, 341, 808 (1989)
6. C. Joblin, A. Leger and P. Martin, ApJ, 393, L79 (1992)
7. R. Papoular, J. Breton, G. Gensterblum, I. Nenner, R.J. Papoular and J.-J Pireaux, A&A, 270, L5 (1993)
8. O. Guillois, I. Nenner, R. Papoular, C. Reynaud, A&A, Submitted
9. J. Kerridge in *Carbon in the Galaxy*, NASA Conf. Publ. 3061, NASA, Moffet Field, CA, p.10. (1990)
10. E. Stach, M-Th Mackowsky, M. Teichmüller, G. Taylor, D. Chandra and R. Teichmüller, in *Stach's Textbook of Coal Petrology*, Gebruder Borntragen, Berlin (1986).
11. D. Van Krevelen, in *Coal*, Elsevier, Amsterdam (1961)
12. *Kerogen*, ed. B. Durand, Technip, Paris (1980)
13. E. Taft and H. Philipp, Phys. Rev., 138A, 197 (1965)
14. F. Stern, *Solid State Physics*, 15, 300 (1963)

THE DUST HEATING MECHANISM IN THE COAL MODEL

O. Guillois[1], I. Nenner[1], R. Papoular[1,2], C. Reynaud[1]

1-Service des Photons, Atomes et Molécules, CEA, Centre d'Etudes de Saclay,
91191, Gif sur Yvette cedex, France
2-Service d'Astrophysique, CEA, Centre d'Etudes de Saclay, 91191, Gif sur Yvette cedex, France

ABSTRACT

Recent laboratory developments and measurements have led to a better knowledge of the optical absorptivity of coals from ~ 400 Å to ~ 50 µm. This allows us to take up the problem of the heating mechanism of coal grains on a quantitative basis. It is shown that highly graphitized coal grains at a steady-state temperature of ~ 250-400 K adequately reproduce the general spectral characteristics of UIB emission. These temperatures can be reached and maintained if the grains are small (10-1000 Å) and subjected to radiative fluxes of ~$10^{-6} - 10^{-4}$ W/cm^2, which is the range of fluxes estimated in UIB emitting nebulae. The heating efficiency of the exciting source does not decrease dramatically with photon energy until the colour temperature of the source falls below ~2000 K: thus, hard UV photons are not necessary to excite the UIBs. These results are derived from measured properties of coal and can all be traced back to the semiconducting properties of this material, particularly to the dramatic decrease of absorptivity between the visible and the near-ir.

1- THE OPTICAL ABSORPTIVITY SPECTRUM

It is not possible to discuss the radiative heating mechanism of dust without a fair knowledge of the <u>continuum and band</u> absorptivity α (cm^{-1}) of the material over a very wide spectral range. Our recent efforts in this direction have led to a better knowledge of $\alpha(\lambda)$ between ~ 400 Å and ~ 50 µm, for the more graphitized coals. Figure 1 shows α for coal from the Escarpelle seam which was most thoroughly studied in our laboratory. This curve is a synthesis of measurements in the Vis/UV[1], in the near-ir[2], in the mid-ir[3] and in the far-ir[4]. All measurements, except the latter, yielded absolute values. The far-ir measurements showed that the opacity goes like ~ $\lambda^{-1.6}$. In fig.1, the mid-ir measurements were continued beyond 14.5 µm using this power-law.

In the following, we have used Urbach's rule to predict the changes of α in the near-ir and mid-ir, as a function of material temperature.

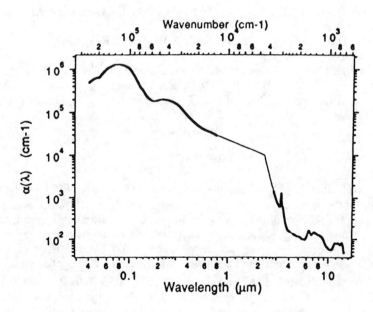

Fig.1: The absorptivity spectrum $\alpha(\lambda)$ of the semi-anthracite from Escarpelle.

2- EQUILIBRIUM TEMPERATURE OF A GRAIN

Consider a grain of radius a (cm), in a radiation field F (W/cm^2/μm) such that its temperature settles at T_d (K). Then, assuming the grain to be optically thin,

$$\pi a^2 \int_0^\infty d\lambda\, Q_{abs}(\lambda, a)\, F(\lambda) = 4\pi a^2 \int_0^\infty d\lambda\, Q_{abs}(\lambda, a)\, P(\lambda, T_d), \qquad (1)$$

where Q_{abs} is the optical efficiency and P, Planck's law. In the Rayleigh limit of small grains, $Q_{abs}/a \cong \alpha(\lambda)$, independent of a. Now, if the heating source is a black body of temperature T_* (K), we can defined for it an average absorptivity

$$\bar{\beta} = \frac{\int_{\lambda_0}^\infty d\lambda\, \alpha(\lambda) P(\lambda, T_*)}{\int_{\lambda_0}^\infty d\lambda\, P(\lambda, T_*)}, \qquad (2)$$

(where $\lambda_0 = 912$ Å, which corresponds to the threshold of the absorption of hydrogen) and an intensity I (W/cm^2) such that (1) becomes

$$\bar{\beta} I = 4 \int_0^\infty d\lambda\, \alpha(\lambda) P(\lambda, T_d), \qquad (3)$$

which is represented in fig.2 (solid line). The right-hand ordinates represent $I(T_d)$ for $\bar{\beta} = 5 \; 10^5$ cm^{-1} (corresponding to UV heating sources). In the range of estimated I's in reflection nebulae, $250 \leq T_d \leq 400$ K.

The dashed line in fig.2 represents $I(T_d)$ for amorphous carbon grains having $\alpha = \text{cst}.\lambda^{-1}$. Note the milder slope and the lower temperature for the same intensity I.

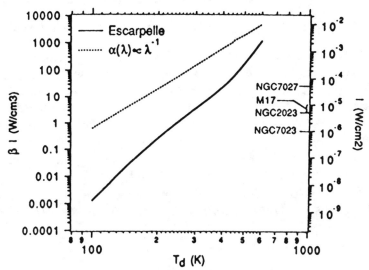

Fig.2

3. THE CORRESPONDING EMISSION SPECTRA

The emission spectrum of an optically thin grain is proportional to $\alpha(\lambda)P(\lambda, T_d)$. Three spectra are shown in fig.3 for the range of <u>thermodynamic</u>

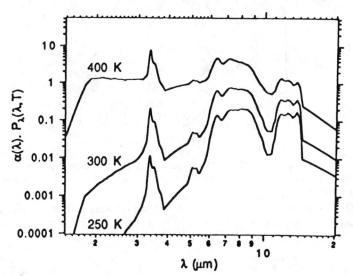

Fig.3

temperatures determined in §2. The corresponding colour temperatures, T_c, are quite high in the near-ir: 500 to 1000 K between 2.2 and 4.8 μm. This is because the absorptivity is very high below 1 μm, and orders of magnitude lower beyond 3 μm, ensuring a relatively high T_d and a still higher T_c. All this, in turn, is due to coal being an amorphous semiconductor with a gap energy of ~ 0.7 eV ($\lambda \sim$ 1.6 μm). The solid state nature of coal is also responsible for the spectral continuum necessary to absorb radiant energy and to fit observations. No free molecule has these properties.

4- COMPARISON WITH NEBULAR OBSERVATIONS

The feature-to-continuum ratio in fig.3 is higher than in UIB spectra. Also the ratio of the 3.3-μm-band flux to total flux ranges between 0.2 and 5 % while observations suggest a range 0.1-1 %[5]. This points to the conclusion that Escarpelle is still too rich in H, O functional groups. More graphitized coals, with weaker vibrational features are preferable, such as the anthracite from the La Mure seam. Moreover, the $\pi - \pi^*$ UV band of La Mure better fits the 2175 Å band of the IS extinction curve[1], and its mid-ir spectrum is closer to the UIB spectrum: no aliphatic 6.9 μm feature, only weak duo and trio aromatic vibrations between 12 and 14 μm. In figure 4, this mid-ir spectrum has been drawn together with the corresponding part of the observed spectrum of the Red Rectangle[6], for comparison.

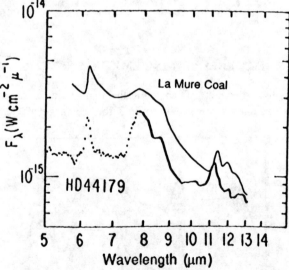

Fig.4: The mid-ir spectrum of La Mure coal (linear absorbances). For purposes of comparison with observation, the emission spectrum of the Red Rectangle is also shown (logarithmic scale).

Based on these considerations, we modeled observed nebular spectra by adding to the continuum an adjustable fraction of the feature strengths measured on

Escarpelle. Figure 5 shows the fit to NGC 7027: only ~ 1/6 of the measured strength is necessary; T_d=300 K.

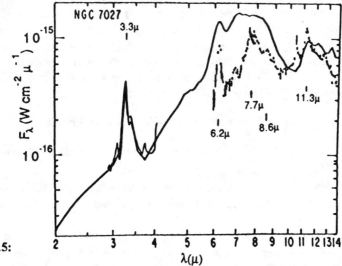

Fig.5:

5- HEATING EFFICIENCY OF PHOTONS

Using IRAS data on a number of reflection nebulae, Sellgren et al.[7] computed the ratio of the radiative flux of the 12-μm photometric band to I_{bol}(FIR), a quantity roughly equal to the total luminosity in the far ir wavelength range. We modeled this ratio with a mixture of optically thin coal grains and optically thick

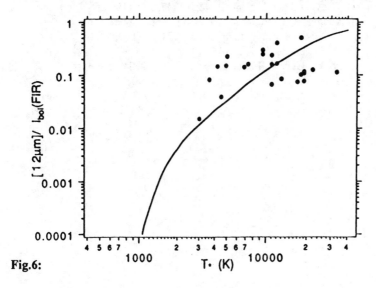

Fig.6:

silicate dust, which are supposed to account for the near/mid ir and the far ir, respectively.

With this simplified picture, we can compare in fig.6 the predictions of our model (solid line) with the observations (dots). The values of the adjustable parameters of the model are: average grain size of coal ~4 10^{-6} cm; relative abundance of C and Si estimated to be 1/4; average field intensity I=5 10^{-8} W/cm^2 (These low values result from the fact that these observations are confined to distance of 3 to 6 arcmin from the illuminating stars). Fig.6 shows that UIBs can be excited by photons of energy much less than 10 eV.

6- COMPARISON WITH OTHER OBSERVABLES

a) *Ratio of fluxes in the aromatic 3.3-μm band and in the IRAS 12-μm band.* Giard et al.[8] estimated this to be ~ 2%, on average, in the inner galaxy. In our model, this ratio depends only on T_d, and 2% corresponds to $T_d \cong 300$ K, quite in the right range of temperature needed to yield emission spectra similar to those that are observed.

b) *Ratio of stellar energy absorbed by "small" grains to the total absorbed by dust.* Sellgren et al.[6] estimated this to be of order 1/3. In our model, this ratio varies between ~0.1 and ~0.5 for 3000 ≤ T_* ≤ 30000 K, thus correctly bracketing the observed average.

REFERENCES

1. R. Papoular, J. Breton, G. Gensterblum, I. Nenner, R.J. Papoular, J.-J Pireaux, A&A 270, L5 (1993)
2. R. Papoular, C. Reynaud and I. Nenner, A&A, 247, 215 (1991)
3. O. Guillois, I. Nenner, R. Papoular, C. Reynaud, Appl. Spec. (submitted)
4. C. Sourisseau, G. Godens, R. Papoular, A&A, 254,L1 (1992)
5. J.L. Puget and A. Leger, ARA&A 27, 161 (1989)
6. R.W. Russell, Thesis, Univ. Cal., San Diego; fig.7b (1978)
7. K. Sellgren, L. Luan, M.W. Werner, ApJ 359, 384 (1990)
8. M. Giard, F. Pajot, J.M. Lamarre, G. Serra, E. Caux, A&A 215, 92.(1989)

DISCUSSION

GIARD — Since a significant fraction, 10 to 30%, of the total energy is absorbed and re-emitted by the dust particles emitting the near-infrared photons, should your model imply a step between 0.5 and 2μm in the extinction curve? Such a step is not observed.

REYNAUD — The IS extinction curve we usually refer to is a mean deduced from different measurements along different lines of sight. As described by Johnson et al (1965, ApJ., 141), the IS extinction curves display a great variety of behaviours, with a more or less rapid decrease in the near IR. Along a given line of sight, we must encounter different columns of dust, with different compositions and structures such as more or less amorphous silicates and carbon particles. These will have different extinction spectra. For example, the absorbance of amorphous carbon grains goes like λ^{-1}. Such a mixture may provide a less pronounced step (which is a characteristic of only the more evolved coals like anthracites). Moreover, according to Mie's theory, the step in Q_{ext} should also be milder for the larger grains. In conclusion, we can not invoke such a step in the extinction curve before having studied the effect of a mixture of grains of various compositions and sizes.

GIARD — When you compare the predictions of your model with the 3.3μm/Total IR and 12μm/Total IR ratios measured towards the inner Galaxy you should use a much smaller $\bar{\beta}I$ and here Td: as much as 90% of the inner Galaxy IR flux comes from diffuse molecular and HI clouds (see Sodrosky et al. 1989 and 1993) with $I \cong 2\ 10^{-8}$Wcm (~10 times the solar ISRF). $\bar{\beta}$ for ISRF spectra characteristic of the diffuse medium should be of the order of 10^5 (your Fig. 1). Thus $\bar{\beta}I \cong 2\ 10^{-3}$ W cm^{-3} and Td < 100 K (your Fig. 2).

REYNAUD — It is not obvious to us that the major source of UIBs in the inner Galaxy is composed of diffuse molecular and HI clouds with ISRFs as described by Giard.
1. As far as we know, there is no direct spectroscopic evidence of the 3.3 μm feature emission from such clouds. Even the IR emission of the other UIBs is only a surmise based on the <u>wide band</u> IRAS photometry.
2. The inner Galaxy is known to include dense and hot regions, and HII regions in particular. The question of the relative spectral contributions of various dense/hot and diffuse/cold regions is still open. The model of Sodroski et al. (1989), only seeks to fit the 60 and 100 μm IRAS maps, and does not even allude to UIB emission. Other models (Mezger, Cox et al., 1978 to 83; Puget, Ryter et al., 1977 to 85) take into account near and mid-

IR data and are therefore led to assign a much stronger FIR contribution to matter heated by near-by O and B stars (30 to 60 %).

3. Moreover, the study of the 3.3 µm emission by external galaxies reveals a strong correlation of its intensity with starburst and HII regions (Moorwood, 1986, A&A 166, 4) and with the Brγ hydrogen recombination line associated with HII regions (Mizutani et al., 1989, ApJ. 336, 762). This seems to tell us that the main source of 3.3 emission in the inner regions of galaxies lies in star formation regions, which are characterized by strong UV radiation. One is therefore entitled to assume for them a value of βI at least as high as for reflection nebulae (~ 1 W/cm^3), enough to sustain steady-state emission of the 3.3 feature.

Of course, if new data established the emission of this feature by cold, diffuse clouds, we would have to consider transient heating. In this event, coal grains of ~ 10 Å would be in a better position than PAHs, thanks to their strong absorptivity continuum in the vis/UV and low absorptivity in the IR.

Panel discussion on
POSSIBLE CARRIERS OF UNIDENTIFIED SPECTRA

*Edited by James Lequeux, Paris-Meudon Observatory
and Ecole Normale Supérieure, Paris*

Panel members: *Luigi Colangeli (Napoli), Yves Ellinger (Paris), Christine Joblin (NASA-Ames), James Lequeux (Paris, moderator), Cecile Reynaud (Saclay), Guy Serra (Toulouse), Thomas J. Wdowiak (University of Alabama), David Williams (Manchester: conclusions)*

James Lequeux: Introduction.
The aim of this panel discussion is to confront ideas on the nature of the carriers of both the Unidentified Infra-red Bands (UIBs) seen in emission from interstellar and circumstellar matter and the Diffuse Interstellar Bands (DIBs) seen in absorption in the spectra of stars.

Let me summarize first the observed properties of the **UIBs**:
- they are ubiquitous in the InterStellar Medium (ISM) whenever there is ultraviolet or even visible radiation (cf. reflection nebulae);
- they are strong thus their carrier(s) must contain abundant atoms, in particular carbon;
- their carrier(s) survives for some time a strong UV field, then is destroyed (cf. NGC 7027);
- there are important, small scale variations in their abundance (cf. variations in the 12 μm/100 μm flux ratio observed by IRAS), that cannot be due to variations in the radiation field, suggesting condensation-evaporation or quenching-dequenching processes;
- there are variations in the 3.3 μm/11.3 μm band intensity ratio (and in other ratios) connected with the hardness of the radiation field but perhaps with other parameters;
- the 3.3 μm band exhibits resolved fine structure when observed with sufficient wavelength resolution;
- the red fluorescence now seen in many objects (cf. P.J. Sarre's contribution) may well be produced by the carriers of the UIBs.

Now for the **DIBs**:
- they have been discovered as early as 1921 but are still a mystery; a good recent review is by G. Herbig (1993, ApJ, 407 142);
- there are about 200 known DIBs, mostly in the visible (P. Jenniskens & F.-X. Desert 1993, A&A,S, in press): in the UV one finds only the broad band at 2175 A, probably due to graphite, and in the IR only 2 bands at 1.18 and 1.32 μm; the DIBs have various widths, but none is as narrow as atomic IS absorption lines;
- there is some correlation between the equivalent width of DIBs and interstellar extinction, but it is rather poor; DIBs have been seen in circumstellar material;
- there have been claims for "families" of DIBs exhibiting a parallel behaviour; I personally do not find these claims very convincing;
- the carriers of the DIBs must be abundant given their strength;

- like the UIBs at 3.3 μm, some DIBs exhibit fine structure that might correspond to a C=C stretch (Jenniskens & Desert, 1993, A&A 274, 465);
- recent studies of *circumstellar* D"I"Bs by Le Bertre & myself (1992, A&A 255, 288; 1993, A&A 274, 288) show a conspicuous lack of correlation with UIBs (cf. NGC 7027 or IRAS 21282+5050); these diffuse bands exist in oxygen-rich and carbon-rich environments, but the strongest ones are found in *nitrogen-rich* environments (WN stars, HR Car, AG Car, Hen 1379), suggesting that nitrogen plays a key role in their carriers.

P.J. Sarre. Concerning the 3 μm/11μm distributions shown by the chairman for the Red Rectangle, I would like to draw attention to a comment by Tokunaga et al. (1991, ApJ 380, 452): mention is been made there of "an apparent evolution from a type-2 spectrum near the star to a spectrum more like Type 1 further from the star" (see definition of these types in the paper). In considering the intensity ratio of these bands this should be kept in mind.

My second point concerns the diffuse absorption bands recently seen in circumstellar shells. I think these results are of particular interest because the bands seen strongly (4430, 5780, 6284) are the very ones which are weak relative to 5797 and 6614, the latter being strong at high galactic latitudes.

Christine Joblin. The PAH model for UIBs and DIBs

I want to discuss several points concerning this model, bearing in particular on work at NASA-Ames.

1. Spectral agreement UBIs/vibrational bands of PAHs.
Neutral molecules can account for the positions of the bands but not for their relative intensities. Thus if PAHs are responsible they must be de-hydrogenated and/or ionized. Their theoretical spectra (Ellinger and collaborators) and laboratory spectra in Ar matrices (Szczepanski & Vala 1993, Nature; d'Hendecourt and Leger, poster, this conference) give a better match. The theoretical work of Pauzat and collaborators (this conference) yields positions for the neutral bands and intensities of cations (species with no hole in the π electronic system). The most stable ionized molecules are not necessarily the direct cations of PAHs but product species: ion-trap experiments by Boissel (poster, this conference) show for example that if $C_{14}D_{10}^+$ ions are submitted to radiation from a Xe lamp with $\lambda>330$ nm) the parent ions disappear and the very stable ion $C_{12}D_8^+$ is produced.

2. Position of the 3.3 μm features
Flickinger et al. (1991) consider that condensed PAHs reproduce well the position of this band. Joblin et al. (poster, this conference) obtained IR spectra of gaseous molecules at high temperatures, thus thermally excited, and show that the positions of the vibration bands depend on the vibrational temperature: they consider that the position and width of the 3.3 μm band is compatible with the carriers being free molecules at high temperatures.

3. Satellite bands of the 3.3 μm band
These may be the contribution of subgroups (-CH_3, -CH_2· for example) as shown by spectra of methyl-coronene and ethyl-coronene. The hot band if any should be broad and contribute only to the plateau.

4. DIBs
I report on several matrix experiments on PAH cations.
- Naphtalene cation ($C_{10}H_8^+$): Salama & Allamandola (1991). There is a good correspondance between the band positions and 7 DIBs but there is a problem with the relative intensities: is it due to matrix effect, or is the coincidence in position fortuitous?

- Pyrene cation ($C_{16}H_{10}^+$): Salama & Allamandola (1992). The main band of this ion is close to the strong DIB at 4430 Å. It is worth studying pyrene-like molecules. Leger & d'Hendecourt (poster, this conference) suggest that the methylene pyrene cation ($C_{16}H_9^+$-$CH_2\cdot$) is the carrier of the 4430 Å DIB. It might also be responsible for 4 other DIBs (4180, 4428, 4780 and 7564 Å).
- Phenantrene cation ($C_{14}H_{10}^+$): Salama, Joblin & Allamandola (1993). It could account for some DIBs but the main band of this molecule is in the near-IR (around 9000 Å) where no DIBs have been searched for yet.

I. Nenner, to C. Joblin. Have you analyzed the profile of structures around 11.3 μm in the spectra of PAHs you have studied at 3.3 μm? I raise this question because the mid-IR features are better "fingerprints" of the molecular structure than those at 3.3 μm.

C. Joblin: It is usually assumed that the 11.3 μm band is associated to solo hydrogens, i.e. hydrogens isolated in an aromatic cycle. This is confirmed by laboratory spectra. The fact that the 11.3 μm band is the dominant one in this spectral range could indicate that the PAHs are partly dehydrogenated. However, I do not exclude possible contribution of some small carbonaceous grains (HACs, coals, ...) to the 11.3 μm band.

Thomas J. Wdowiak: present and future experiments

UIRs. There is a general agreement, except for Thaddeus, that aromatic structures are involved, but there is a general disagreement as to what they are. There are several cults: PAHs, hudrogenated amorphous carbon, coal, fullerenes, etc. Experiments and computations are coming "hot and fast" to clear up the situation, but as usual the most important experiments are yet to come: they will take into account the problem of collisions when dealing with long time-scales that are involved in the emission process. For example:
- for neutrals the Barker (Michigan) and Saykally (UCB) experiments: shooting fast in the UV and measuring fast..
- for ions, one could use ion traps and "secret" single-photon counting detectors, shooting continuously and taking time for measuring. Who will do that?

It is very important to understand that a "UIR" experiment in the laboratory must extract photons from an excited molecule or ion *before* it collides with another molecule or ion or with the wall of the device. As a consequence of the long time scale for emission (≈0.1 second) very few photons are emitted at a given time and the detector must be very sensitive (a photon counter). In an ion trap only few ions are trapped and even though the isolation time is long with respect to these 0.1 second the flux is low. Again a very sensitive detector and very low background are required. On the other hand, if isolation for a long time can be achieved one has only to integrate over a long time while the ions are continuously illuminated with the excitation radiation.

Again, the new detectors mentioned by Dr. Brechignac are an exciting development not only for observations but for laboratory experiments as well.

DIBs. Here there is no agreement as to the carrier(s) identity. Theories abound: linear carbon chains and rings, ionized PAHs, fullerenes, porphyrine, etc. Experiments that have been performed specifically to address the mystery (not data from the literature!) fall into two categories:
- those that transform a molecule into a candidate for being a DIB carrier: ex. PAH cations;
- those that make a "soup" from a simple precursor: ex. CH_4 or C.

We are "carbon soup" makers (see our poster at this conference); in our soup we taste the flavor of the strong DIBs at 4472, 5778, 6269+6283, 6613 and 7240 Å, provided of course you are willing to accept matrix shifts for argon.

Luigi Colangeli: *contribution not communicated.* It was mainly summarizing laboratory work of the author's group on spectra of PAHs, HACs, etc.: see in particular the following papers: 1992, ApJ 385, 577; ApJ 391, L45; ApJ 392, 284; ApJ 396, 369; 1993, ApJ 406, 739; 1994, A&A, in press.

Cecile Reynaud: about the coal model for UIRs

We propose that UIRs are produced by grans of coal (coal in its demineralized form=kerogen). Coal is mainly composed of carbon with small amounts of H and O. It is a semi-conductor. The properties of coal depend only on its rank (=age). The oldest ones have more aromatic structure, i.e. a graphitic carbon skeleton with attached functional groups. The mid-IR spectrum of a highly-graphitized anthracite (La Mure) is strikingly similar to that of the Red Rectangle (see posters by Guillois et al. and Ellis et al., this conference) and the absorption in the UV resembles the IS extinction curve except at the shortest wavelengths. Due to the semi-conductor nature of the coal, its emission color temperature in the mid-IR is considerably higher than the physical temperature, hence the need for very small, qunatum-heated grains is alleviated: for a grain smaller than 1000 Å, color temperatures of 500-1000 K correspond to physical temperatures of 250-400 K. In this model, contrary to the PAH model, the heating efficiency of the exciting source does not decrease dramatically with photon energy: hard UV photons are no longer necessary to excite the UIBs.

Yves Ellinger: *contribution not communicated.* Its contents which summarized some of the work of the author's group can be found in several recent papers, in particular Parisel et al. 1992, A&A 266, L1, Talbi et al. 1993, A&A 268, 805, De Frees et al. 1993, ApJ 408, 530, discussing the origin of the UIBs, of some DIBs and of the structure of the 3.3 μm UIB on the basis of *ab initio* calculations.

Guy Serra: could organo-metallic chemistry contribute to the formation of UIR carriers?

It is quite possible that organometallic chemistry exists in the ISM or CSM. For example, iron is very abundant in the ISM (Fe/H=$3\ 10^{-5}$). It is also very depleted, and it is difficult to understand why it should be only depleted in silicates (Klotz et al., poster, this conference). On the other hand aromatic molecules are very abundant in the ISM (Ristorcelli et al. poster, this conference) and a Fe-PAH chemistry is very favorable: Fe and Fe^+ react and bind with aromatic cycles (experiments by Boissel, poster, this conference, and *ab initio* calculations by Klotz et al.). The binding energy Fe-aromatic is >1.5 eV and the activation barrier, if any, is <0.15 eV. The formation and photodissociation of these complexes has been studied by Marty et al. (poster, this conference) and they have been shown to be stable inside molecular clouds beyond 4-5 visual magnitudes of extinction: they might contribute to the depletion of PAHs inside molecular clouds.

Consequences of organo-metallic chemistry are:
- the formation of such complexes can yield new materials and contribute to the depletion of PAHs and to the formation of grains;
- they might trigger catalytic reactions, e.g. help the formation of aromatic cycles from acetylene (poster by Klotz et al, this conference);
- new transitions in emission and absorption can be produced; could they be further candidates for UIBs?

At present there is no clear evidence for such complexes in the ISM, but studies are just starting in this field. We should not ignore in any case the possible contribution of organo-metallic compounds to interstellar chemistry.

DISCUSSION

There was a lively but somewhat erratic general discussion after the talks; here are some edited excerpts of this discussion, attempting to give its flavor by scanning the range of opinions between the participants.

L. Ziurys. I happen to be the "other" astrophysicist in this room which agrees with Pat Thaddeus! The experimental evidence from the laboratory supporting the PAH theory is strongly lacking. Other possibilities such as the long carbon chains have to be seriously examined before the "truth" will be arrived at.

M. Guelin. We have seen (C. Reynaud's presentation in this session) infrared spectra of coals. These fit beautifuly the mid-IR interstellar bands. Since carbon grains, we know, are present in the circumstellar and interstellar media, do we really still need PAHs?

I. Nenner. Indeed, in our coal model, the agreement with UIBs (see e.g. the Red Rectangle) is such that there is no need to add mixtures of PAHs to the coal. This is because the micro structure of this material includes aromatic carbon skeletons on which are attached functional groups responsible for both the continuum and the bands seen in astrophysical objects.

E. Bussoletti. The presence of amorphous carbon grains is clearly proven by the presence of an IR continuum in many sources.

L. d'Hendecourt. There are however two evidences for molecular PAHs in the ISM:
1) the presence of the 3.3 and 6.2 mm emission bands in the cirrus clouds: the coal model will not be able to produce those two emission lines for a lack of sufficiently high equilibrium temperature in diffuse clouds;
2) at this meeting, very impressive results on the fit of a few DIBs by PAH^+ have been presented (Allamandola & Salama, Leger & d'Hendecourt). It is the first time that a very close fit to a few bands within two electronic states is obtained and although such results have to be confirmed in the gas phase a link between DIBs and molecular species (PAHs?) is very encouraging for the PAH hypothesis.

C. Reynaud. Concerning the cirrus medium, it is perhaps preferable to wait for future astronomical observations before making definitive statements: in particular, an entire spectrum would be necessary to discuss the UIBs in this medium.

M. Guelin. There is some (growing?) evidence that many small circumstellar and interstellar molecules are desorbed from the grains, for example by photo-desorption at the edge of circumstellar envelopes or interstellar clouds. In this case, many of these molecules are still attached to the grain surfaces, only a small fraction (10^{-1} - 10^{-2} ?) getting into the gas. The same may apply to the PAH-carbon grain couple. The IR domain is certainly the worse one for discriminating between the detached (i.e. gas phase) and attached aromatic rings. The observed UIBs are likely to arise from the most abundant of these two forms, presumably the aromatic rings attached to the grains, if we follow the small circumstellar molecule parallel.

C. Joblin. In my opinion, small carbon grains (such as the coal model grains) and PAHs as gas phase molecules co-exist in some objects.

J. Lequeux. I agree. If this is the case, the emission of UIBs will be dominated by the grains if the interstellar radiation field is strong, or by the PAHs if it is weak.

A. Zavagno. We have pointed out the importance of the underlying continuum in planetary nebulae relative to UIBs. If one looks at the emission maps there are strong evidences for the presence of a shell near the star which emits the 10 μm underlying continuum and of "PAH" emission at much larger distance. This kind of mapping in different bands could be a good way for testing the idea of a distribution of populations (grains and PAHs). This should be done in different environments (pre-main sequence stars, planetary nebulae, reflection nebulae...) in order to try to see an evolution of the dust and large molecules.

CONCLUDING REMARKS, by D. Williams

I have probably been chosen to make an attempt at summarizing this panel discussion because the chairman thinks I am the only person in this room who doesn't have a strong opinion about the origin of DIBs. I confess to have written two papers on this topic more than 20 years ago, and I am pleased that they have been forgotten! Does that disqualify one? Old papers are sometimes embarrassing in their crude simplicity...or sheer error. (A further confession: I once made an error by a factor 10^{21} in an accepted paper, then withdrew it. The referee stole the idea and published it in a well known journal. I hope the standards have improved a little since then!).

The idea of a possible link between UIBs and DIBs (and near-IR luminescence, not much discussed here) is interesting. However it is not easy to check observationally as DIBs are seen in absorption in the diffuse medim while UIBs and luminescence are seen in emission from regions of (mainly) higher excitation. I am not sure we have gone very far, and the chairman himself produced a counter-example of a lack of correlation between UIBs and DIBs. Christine Joblin claims that ionized PAHs could perhaps do both, but Thomas Wdowiak did not say anything about the ability of his "soup" to carry UIBs. Thus I will rather discuss UIBs and DIBs separately.

UIBs. It is useful to ask: what constitutes an identification? We may be able to decide if small ionized, partly dehydrogenated PAHs, or others, can do it, but will we be able to be more specific? Probably not. In any case, a variety of carriers seems to be implied by the large variations from one band to another. These variations from site to site have been strongly stressed by Jean-Loup Puget two days ago. Thomas Wdowiak gave the impression that the best experiments are yet to come; or perhaps we should hold back and retain our scepticism: the interpretations of existing experiments can be significantly different. Of course, the major controversy is: solids vs. isolated molecules. Alain Leger suggested a simple experiment: irradiate a solid with UV and try to see the 3.3 μm emission. Has anyone done it, or do we feel we know the answer? I heard rumors of such experiments that lack repeatability. Luigi Colangeli stresses the amorphous carbon link to PAHs...

I am pleased that someone had something good to say about coal. As you may know the UK coal industry is being dismantled and sold off, and has received bad publicity. The demand of French scientists for deep-mined coal may preserve jobs of miners!

There seems to be a fashion for small PAHs, now. We should pay credit to Duley and Jones who first noted good agreement with anthracene and naphtalene in their paper "Mothballs in the ISM". Smaller PAHs are likely to take part in chemistry, ionization, etc. The role of these molecules in gas is not clear: this is a topic that the speakers did not have time to address, but which is of importance. On the other hand, who would have imagined a few years ago that a hundred of scientists would sit till after 10 o'clock at night for a discussion of quantum-mechanical calculations of large molecules, of their accuracy and reliability (although Yves Ellinger claimed that "there is no error"). The advance has been spectacular. It is now said to be faster and more reliable to make these calculations than experiments (or is this a plea for supercomputers?). Anyway, these beautiful numerical experiments test the importance of dehydrogenation and ionization in a way not yet available in the laboratory. Theoreticians say they can give strong constraints and support to the PAH hypothesis -but there is no information yet for DIBs.

DIBs. Will DIBs tell us something new (e.g. long chains) or are they a manifestation of the content of the ISM as we now understand it? What about the list of observational constraints set by Herbig? Do we pay enough attention to it, or are we merely playing a game in spectroscopy? Now, the major effort seems to be made not by astronomers (who however have done a lot recently), but by chemists and spectroscopists. May be the problem is too hard for astronomers, or may be they feel that there is unlikely to be much new information in DIBs? However there is still new information arriving, even after more than half a century. The new work on circumstellar D"I"Bs seem particularly important, with the idea of connection to nitrogen. It would be interesting to hear spectroscopists speculate about this connection. Did they? No one responded to this. Could theory do something (more easily?). This raises also the question of other abundant "impurities", e.g. Fe as discussed by Guy Serra. This is reminiscent of some old ideas....The solid-state model for DIBs is less favored that it used to be (the coal people did not say anything about DIBs). Is this fair? We do not have total agreement on grains, and solids can show sharp absorptions. An argument against is the presence of variable wavelength shifts in lattices, but this might apply also to gas phase in some circumstances.

To conclude, most of the discussion today concentrated on UIBs. It is not clear that we have made much progress on DIBs, and I look forward to discussions on these continuing for many more years. Conversely, real progress is being made on UIRs bands both through theory and through laboratory experiments. Carbon grains and PAHs seem to be closer, but the emission mechanisms for solids deserve to be looked at more closely. As some contributions have emphasized, we need to pay more attention to the variability between lines of sight. This has been a good and lively discussion, and I congratulate the participants for the constructive spirit of their interventions.

CONCLUDING OVERVIEW

MOLECULES AND GRAINS IN SPACE: AN OVERVIEW

A. Dalgarno
Harvard-Smithsonian Center for Astrophysics, Cambridge, MA 02138

ABSTRACT

A summary overview of the presentations at the meeting on the Physical Chemistry of Molecules and Grains in Space is given.

INTRODUCTION

Molecules and dust are ubiquitous in the Universe, existing in extraordinary abundance and variety of forms in a diverse range of environments ranging from protostars in the process of gravitational collapse to the ejecta of exploding supernovae. The observations of molecules and dust provide major challenges to chemistry, challenges of identification and challenges of interpretation. The chemistry, dynamics and the evolution of astronomical objects are intimately linked. Of special interest is the chemistry, physics and astronomy of those species, very small grains or very large molecules, that occupy the interface between the gas phase and the solid phase.

OVERVIEW

My introductory comments were made either explicitly or implicitly in the authoritative survey with which David Williams initiated our discussions. I was impressed by the amount of material he covered and by the amount of material to which he could make only passing reference, like molecular and dust formation in such exotic environments as the early universe, novae and supernovae. They are discussed in the written version. David Williams did comment on the chemistry of diffuse molecular clouds where the environment is relatively simple and well-characterized so that chemical ideas can be directly tested. It is generally agreed that hydrogen molecules are made on the surfaces of dust grains (on which some important experiments have been carried out by C. Schermann et al.), and ejected into the interstellar medium (though L. P. Cassam-Chenai et al. raised the possibility that reactions with polycyclic aromatic hydrocarbons (PAHs) may be a significant source). The interstellar gas is ionized by ultraviolet photons and by cosmic rays and the ionization drives a gas phase ion-molecule chemistry. David Williams pointed out that the resulting chemistry, simple though it is and well-determined as it appears to be, does not yield an entirely satisfactory explanation of the measured molecular abundances. He argued, as he and others have for many years, that gas phase chemistry must be supplemented by a grain chemistry, most evidently as a source of the molecule NH recently found in diffuse clouds but which apparently cannot be produced in significant abundance by reactions occurring in cold interstellar clouds. David Williams is probably correct, but the question will remain open until the chemistry leading to CH^+ is understood. It seems certain that the efficient production of CH^+ requires either a warm environment or a cold environment, in which ions are accelerated above thermal velocities. In a warm environment, NH may be produced by the endothermic reaction of N with H_2.

The introductory talk proceeded on with a discussion of dark clouds from the interiors of which the interstellar radiation field is excluded by absorption by the dust grains and in which the cold surfaces of the grains act as sinks on to which atoms and molecules condense. There must be mechanisms by which molecules are released from the grains and put back into the gas phase where they are observed, so that grain chemistry is an unavoidable component of the chemistry of dark clouds. Models of dark clouds including gas and surface chemistry have been constructed for many years but the treatments of grain chemistry are highly uncertain, and much chemistry, in the laboratory and in theoretical calculations, needs to be done before realistic models can be constructed and molecular composition can become a secure diagnostic probe of astronomical events.

Observations are a crucial ingredient of progress. W. A. Schutte reported on the detection of molecules on grain surfaces by observations in the infrared. Water ice is the dominant constituent, but CO, CH_3OH and probably H_2CO have been identified in some directions by the presence of infrared absorption features. The relative abundance of H_2O and CO depends, I assume, on the relative abundance of oxygen and carbon monoxide in the interstellar gas. I wonder if we understand the chemistry well enough to derive the gas phase O/CO ratio from the surface absorption data. It is a number of considerable interest.

In contrast to the small number of molecules detected in the solid phase, nearly one hundred distinct species have been identified in the gas phase, almost all in the radio or millimetre region of the spectrum. To find them requires a combination of astronomy and chemistry. We heard from Jean-Luc Destombes (see J.-L. Destombes et al.) about the challenging laboratory experiments on the millimetre and sub-millimetre spectroscopy of transient species like free radicals and molecular ions and a method for measuring radiative lifetimes of vibrationally excited molecular ions was described in a paper by S. Jullien et al. Millimetre spectra of deuterated isopotomers of CH_3CN and CH_3CCH were measured by G. Wlodarczak et al. and CH_2DCN and CH_2DCCH but not CH_3CHCH_2 were detected in the interstellar medium. The molecules CH_3NH_2 and CH_3CHCH_2 were also investigated by Wlodarczak et al. Experimental studies of an electronic transition in C_3 were reported by J. Baker et al, in CO by J. Baker, J. L. Lemaire et al. and D. Maimasson et al. and in CS^+ by D. Cossart et al. A valuable compilation of spectroscopic data was described by J. Crovisier. We heard from Peter Botschwina about the remarkable progress that has been achieved in ab initio quantum-mechanical calculations of molecular geometries, transition frequencies and dipole moments and we look forward to much more. We learnt also about the difficulties and achievements in laboratory and interstellar measurements on refractory material containing metals from Lucy Ziurys and in a paper by J-M. Denis and J-L. Ripoli, and valuable spectroscopic data were presented by M. A. Anderson et al. and J. Chevaleyre et al.

Excitation also occurs by optical pumping and the process for water by near infrared photons was investigated by A. Dheeger and M. Giard.

The possible role of co-ordination reactions involving metals in space was discussed in papers by A. Klotz et al., P. Marty et al. and N. Mauron and Ch. Cuilain. A laboratory study of co-ordination reactions between Fe^+ and napthaline was reported by P. Boissel and of hydrogen and nitrogen atoms with $Fe(CO)^+_n$ by H. Mestdagh et al. It is suggested that metal aggregates and complexes may participate in the catalytic formation of molecules.

The interpretation of molecular emission and absorption lines involves a knowledge of collisional excitation cross sections and Malcolm Walmsley chaired a panel which examined as an example the excitation of ammonia in collisions with helium and with ortho and parahydrogen. Poster papers were presented by C. Rist and P. Valiron on calculations on collisions of NH_3. Despite its complexity, the collision problem seems to be well in hand—the greater challenge lies in the determination of the potential energy surfaces. Several calculations of potential energy surfaces were reported at the meeting in poster papers by H. Lavendy et al., M. Hanus et al., A. Spielfiedel et al., and D. Talbi and Y. Ellinger, though most with the intent of exploring chemical reactivity.

Reliable potential energy surfaces are also a crucial element in calculations of photodissociation which were described by Reinhard Schinke in an elegant presentation of an elegant theory. Photodissociation is important in circumstellar shells, diffuse molecular clouds, regions (PDR's) subject to intense radiation fields and in dense clouds in which photons are generated internally by cosmic rays. Branching ratios in dissociation are difficult to obtain by experiment and theory is a valuable input. There is much to be done. Important advances have occurred in our understanding of the photodissociation of CO as evidenced by papers by S. Warin et al. and F. Rostas et al. and there was a poster paper by A. Spielfiedel et al. on the mechanisms of photodissociation of SiO but no rates were given. Nevertheless, reliable data on photodissociation are rare. An interesting experimental study of the photodynamics of acetylene by J. H. Fillion et al., was summarized by Dolores Gauyacq, who made the bold claim that the measurements had no relevance to astronomy. In fact, we need to understand as best we can the chemistry of such an astrophysical important molecule.

Having found the molecules in space and determined their abundances comes the problem of constructing plausible chemical models. For gas phase reactions, few rate coefficients are available at the temperatures characteristic of molecular clouds and none in a physical environment where the rotational level populations are not in thermal equilibrium. These unusual conditions raise questions of deep chemical significance which have stimulated much theoretical activity represented here by David Clary, by M. Ramillon et al., A. Orel et al., I. F. Schneider et al., and P. Honvault et al. and much experimental activity represented here by Dieter Gerlich, Bertrand Rowe et al., Michel Costas et al., A. Sorgenfrei and D. Gerlich and H. Abouelaziz et al. These are pioneering experiments driven by astronomy. They may have dramatic consequences. The teaching of chemistry at Bristol University may be transformed, if I understand David Field correctly, and possibly of still more cosmic importance the newly measured rate coefficients have had a devastating impact, if I understand Tom Millar correctly (see E. Herbst et al.), on models of interstellar chemistry. Whereas before, the models worked, more or less, at early times in the evolution of molecular clouds, now they fail at all times in making sufficient abundances of complex molecules. Perhaps the situation can be saved by turning to interstellar grains as a source as well as a sink. In fact, in regions of star formation there is ample evidence for the injection into the gas phase from grain surfaces of water and methyl alcohol. There have been several theoretical investigations of increasing sophistication in the details of grain chemistry— two reported here by S. Charnley and by O.M. Shalabiea et al.—but most of the basic data are lacking. It was encouraging to see here a study of the photodesorption from CO ices by L. Hellner et al. With photons of sufficient energy, CO^+ is released. A useful theoretical analysis of interstellar rates of photodesorption by the interstellar radiation field and by cosmic ray induced photons was given in a poster paper by O. M. Shalabiea and J. M. Greenberg, and an instructive exploration of the transfer of ultraviolet radiation in dark clouds was given by S. Aiello et al. and a study of the effects of varying the ionization rates by P.R.A. Farquhar et al. Purely gas phase models may need more careful attention. Jacques Le Bourlot et al. have discovered the existence of two distinct solutions for the chemical composition at astrophysically interesting values of density, ionization rate and depletion. With one solution, the C/CO ratio is higher (which may be an explanation of the observations in some dense clouds). If the neutral carbon density is high, the formation rates of complex molecules are enhanced but also their destruction rates. The discovery of LeBourlot et al. raises the question of multiple solutions and the possible occurrence of chaotic regimes.

There were a few papers describing molecular line observations (A. Howe et al., D. Field et al., R. Gredel et al., M. Womack et al., Kopp et al., Scappini et al. and G. Wlodarczak et al.) which is where the subject begins.

We should be able to learn something useful about molecular processes in interstellar clouds from the study of comets and in any case the interstellar-comet connection is an exciting issue. Presumably comets retain information about the interstellar medium as it was about five billion years ago when the solar system was formed but that memory requires chemistry to unlock it. We heard from the experts. J. Crovisier (and D. Bockelée-Morvan) reported on the infrared and radio spectroscopy of comets and C. Arpigny on the ultraviolet and visible spectroscopy, who also gave a brief account of the chemistry. I wish there had been more time to examine the relationships between cometary and interstellar atomic and molecular abundances. At a simple level, photodissociation is a fundamental aspect of cometary chemistry that determines the spatial distribution of the primary and the daughter products. Cometary data may provide information about photodissociation processes useful also for interstellar models. Molecules are also released from the solid particles of the cometary nuclei by the action of solar radiation so providing information on photodesorption and the effects of bombardment by the solar wind. The gas-grain chemistry of comets and the protosolar nebula was the topic of a poster paper by Z. Moravec and V. Vanýsek.

Dr. Arpigny mentioned the presence in cometary spectra of the Cameron bands of carbon monoxide and surmised they arise from photodissociation of carbon dioxide. They of course appear also in the airglow of the planets Venus and Mars where photodissociation of CO_2 is surely one of the sources. The unexpected discovery of S_2 in one comet was interpreted by V. Vanýsek, and the unexpected emission of the Swan bands of C_2 in the spectrum of Halley was analyzed by Rousselot et al.

Observations of complex molecules in the atmospheres of the giant planets and the satellite Titan were presented by Thérèse Encrenaz. I gather that when it rains on Titan, it rains methane and ethane. On another occasion, it would be instructive to compare and contrast the chemistries of the interstellar gas, comets and planetary atmospheres. The chemistries are intimately related, each informing the others. Perhaps the subject is too large.

J. L. Puget presented an absorbing survey of future space missions and describing the exciting prospects that we hope lie ahead, including the detection of molecules forming in galactic haloes and in the early Universe.

In the last third of his talk, David Williams mentioned a number of important topics encountered in the chemistry of near-stellar regions, also discussed in papers by J. M. C. Rawlings et al, and he raised the question of the dust extinction curve and its variability. He made a long list of carbonaceous materials as possible candidates. The case for the coal model was argued by I. Nenner (see O. Guillois et al.). At this point David Williams uttered the dreaded word PAH or polycyclic aromatic hydrocarbon.

The meeting was rich in contributions on the chemical and spectroscopic properties of grains and of large molecules including PAH's, fullerenes and clusters. Whatever else PAH's and fullerenes have done for us, they have stimulated a fascinating chemistry. Philippe Brechignac (see P. Hermine and P. Brechignac) described some extraordinary experiments in which jet-cooled PAH's were investigated, and they noted that an emission feature of the comet Halley could be attributed to phenanthrene (see G. Moreels et al.). Wei Lee and T. J. Wdowiak reported on an experiment that suggested meteoritic material could originate as PAH's, formed in stars. Françoise Pauzot (F. Pauzot et al.) described theoretical calculations on ionized and dehydrogenated PAH's and suggested that PAH's in reflection nebulae and in the diffuse interstellar gas may have different spectral characteristics. Photofragmentation of PAH ions was investigated experimentally by P. Boissel et al., and by H. W. Jochims et al., basic theoretical work using Hücker theory was described by F. Torrens et al and laboratory infrared spectroscopy by C. Joblin et al. and L. d'Hendecourt and A. Léger.

Much research was reported on surface interactions and clusters in laboratory and astronomical studies. Poster papers were presented by B. Begemann et al., J. P. Baluteau and A. Zavagno, A. Zavagno et al., L. Colangeli et al., J. Leclercq et al., M. Comeau et al., K. Ellis et al., O. Guillois et al., S. D. Taylor and D. A. Williams, F. Trotta and B. Schmitt, B. van der Hoek and A. Omont et al.

Martin Jarrold gave a fascinating account of the chemistry of carbon, silicon and germanium cluster ions and of fullerenes and polycyclic ring isomers. Louis d'Hendecourt and Bernard Schmitt gave stimulating presentations on the physics of grains and ice mantles which led effectively into a discussion of the unidentified infrared emission bands and the diffuse interstellar bands. Peter Sarre spoke on the diffuse interstellar bands raising the possibility of vibrationally excited molecules as sources of absorption. T. J. Wdowiak considered absorption by PAH ions. There were several significant papers on the diffuse interstellar bands by S. Edwards and S. Leach, A. Léger and L. d'Hendecourt, O. Parisel and Y. Ellinger and T. J. Wdowiack et al.

A notable contribution to the debate about the role of PAH's was made by Pat Thaddeus, who advanced an original and comprehensive alternative. His theory has much to commend it. In addition to the elegance of its expression, his explanation of the unidentified emission bands and the diffuse interstellar bands as due to carbon chain molecules has the decided advantage over PAH's that there are no experimental data to refute it or confirm it, and acquiring the needed data will be difficult. But a similar comment was true at the time J-L. Puget and A. Léger suggested PAH's.

James Lequeux chaired a panel discussion on the emission and absorption bands which demonstrated well that there are conflicting views and much basic chemistry has to be done before a secure explanation will be found. It was perhaps the major issue of the meeting. We would like to adopt the economical hypothesis that the emission bands and the absorption bands have a common origin. That belief, as I understand it, places the PAH hypothesis in grave danger, not because of the emission bands for which the hypothesis was first advanced, but because they seem to be failing the absorption band test. The reverse suggestion that whatever is responsible for the diffuse absorption bands is also responsible for the emission bands is more secure if only because there is no identification yet of the absorption bands. A combination of laboratory measurements, astronomical observations and theoretical calculations in a continuing dialogue between chemistry and astronomers of the kind that this meeting has encouraged is needed to answer the question.

ACKNOWLEDGMENT

This work was supported by the National Science Foundation, Division of Astronomical Sciences, Grant AST-93-01099.

LIST OF PARTICIPANTS

ABOUAF R.
LCAM/Univ. Paris Sud
Bât. 351
Univ. Paris Sud
F-91405 ORSAY Cedex

AIELLO S.
Dipartimento di Astronomia
e Scienza dello Spazio
Univ. degli studi di Firenze
Via Leone Pancaldo,2/45
I-50127 FIRENZE

ALLAIN T.
DAMAP
Observatoire de Paris-Meudon
etInst. f. Astronomie et Astrophysik
T.U. Berlin
Hardenbergstrasse 36
D-10623 BERLIN

ANDERSON M.
Dept. of Chemistry
Arizona State Univ.
TEMPE AZ 85287-1604
U.S.A.

ARPIGNY C.
Inst. d'astrophysique
5, av. de Cointe
B-4000 LIEGE

BARTHES M.G. Mme
C.E.C.M./CNRS
15, rue Georges Urbain
F-94407 VITRY sur Seine

BAUMGARTEL H.
Freie Univ. Berlin
Inst. f. Phys. Chemie
TAKUSTRASSE 3
D-14 195 BERLIN

BEEGLE L.
Physics Depart.
Univ. of Alabama at Birmingham
310, Campbell Hall, Univ. Bld
BIRMINGHAM AL 35294-1170
U.S.A.

BEGEMANN B. Mme
M.P.G./Working Group
"Dust in star forming regions"
Schillergasschen 3
D-0-6900 JENA

BENIT J.
IAS ORSAY
Bât. 121
F-91405 ORSAY Cedex

BOGEY M.
Univ. de Lille
Lab. de spectroscopie hertzienne
Bât. P5/ U.S.T.L.
F-59655 VILLENEUVE D'ASCQ Cedex

BOISSEL P.
Photophysique moléculaire
Bât. 213
Univ. Paris Sud
F-91405 ORSAY Cedex

BOTSCHWINA P.
Inst. f. Physikalische Chemie
der Univ. Gottingen
Tammannstr. 6
D-3400 GOTTINGEN

BRECHIGNAC P.
Univ. Paris Sud
Photophysique moléculaire
Bât. 213
F-91405 ORSAY Cedex

BUSSOLETTI E.
Ist. di Fisica Sperimentale
Ist. Univ. Navale
Via de Gasperi 5
I-80133 NAPLES

CABARET L.
Lab. Aimé Cotton
Bât. 505
F-91405 ORSAY Cedex

CACCIANI P.
Lab. Aime Cotton
Bât 505
F-91405 ORSAY Cedex

List of Participants

CANOSA A.
Physique Atomique et Moléculaire
Univ. Rennes I
Campus de Beaulieu
F-35042 RENNES Cedex

CASSAM-CHENAI P.
Astrochimie quantique
ENS Radioastronomie
24, rue Lhomond
F-75005 PARIS

CASTETS A.
Observatoire de Grenoble
Lab. d'Astrophysique
414 rue de la Piscine
BP 53 X
F-38041 GRENOBLE Cedex

CHARNLEY S.
NASA Ames Research Center
Space Science Division, MS 245-3
Moffett Field/CA 94035
U.S.A.

CHEVALEYRE J.
Spectrometrie Ionique
et Moléculaire
Univ. C. Bernard (Lyon I)
Bât. 205
43, Bd du 11 Novembre 1918
69622 VILLEURBANNE Cedex

CLARY D.
Cambridge University
Depart. of Chemistry
Lensfield Rd,
CAMBRIDGE CB2 1EW-U.K.

COLANGELI, L.
Oss. Astronomico di Capodimonte
Via Moiariello 16
I-80131 NAPOLI

COSTES M.
Photophysique photochimie moléculaire
Univ. Bordeaux I
F-33405 TALENCE Cedex

COX P.
Observatoire de Marseille
2, place Leverrier
F-13248 MARSEILLE Cedex 04

CROVISIER J.
Observatoire de Paris-Meudon
5, place J. Janssen
F-92190 MEUDON

DALGARNO A.
Harvard-Smithsonian Center
for Astrophysics
60, Garden Street
CAMBRIDGE, MA 02138
U.S.A.

DE BOISANGER C. Mlle
Service PTN
Centre d'études de Bruyères le Chatel
F-91680 BRUYÈRES LE CHATEL

DESPOIS D.
Observatoire de Bordeaux
BP 89
F-33270 FLOIRAC

DEROUAULT J.
Spectroscopie moléculaire et cristalline
Univ. Bordeaux I
351, cours de la libération
F-33405 TALENCE Cedex

DESTOMBES J.L.
Spectroscopie Hertzienne
Univ. Lille I
F-59655 VILLENEUVE D'ASCQ Cedex

D'HENDECOURT L.
I.A.S./CNRS
Univ. Paris XI
Bât 121
91405 ORSAY Cedex

DORTHE G.
Photophysique et photochimie moléculaire
Univ. Bordeaux I
F-33405 TALENCE Cedex

DUTUIT O. Mme
Physico chimie des rayonnements
LPCR/Bât. 350
Univ. Paris Sud
F-91405 ORSAY Cedex

DUVERT G.
Observatoire de Grenoble
BP 53X
414, rue de la Piscine
F-38041 GRENOBLE Cedex

List of Participants

ELHANINE M.
Physique moléculaire et application
Univ. Paris Sud
LPMA/Bât. 350
F-91405 ORSAY Cedex

ELLINGER Y.
Astrochimie quantique
ENS/Radioastronomie
24, rue Lhomond
F-75005 PARIS

ENCRENAZ M.T. Mme
Observatoire de Paris-Meudon
DESPA
Place J. Janssen
F-92195 MEUDON Cedex

FARQUHAR P.
UMIST/Depart. of mathematics
PO BOX 88
MANCHESTER MGO IQD- U.K.

FEAUTRIER N. Mme
DAMAP
Observatoire de Paris Meudon
5, place Janssen
F-92195 MEUDON Cedex

FEMINELLA F.
C.E.S.R.
9, av. du Colonel Roche
BP 4346
F-31029 TOULOUSE Cedex

FESTOU M.
Observatoire Midi Pyrenées
14, rue E. Belin
F-31400 TOULOUSE

FIELD D.
School of chemistry
Univ. of Bristol
CANTOCK'S CLOSE
BRISTOL BS8 ITS-U.K.

FILLION J.H.
Univ. Paris Sud
LPPM/ Bât. 213
F-91405 ORSAY Cedex

FLOWER D.
Univ. de Durham
Depart. of Physics
Science lab./South Road
DURHAM DH1 3 LE-U.K.

GAUYACQ D. Mme
Photophysique moléculaire
Bât. 213
Univ. Paris Sud
F-91405 ORSAY Cedex

GARGAUD M. Mme
Observatoire de Bordeaux
BP 89
F-33279 FLOIRAC

GERLICH D.
Fakultät fur Physik
Univ. Freiburg
Hermann-Herder Strasse 3
D-7800 FREIBURG

GERAKINES P.
Leiden Observatory
Sterrewacht Leiden
Postbus 9513
NL-2300 RA LEIDEN

GIARD M.
CESR/CNRS
9, av. du Colonel Roche
BP 4346
F-31029 TOULOUSE Cedex

GOUGH S.
Dynamique moléculaire et atomique
4, place Jussieu
Tour 12 E5 B75
F-75252 PARIS Cedex 05

GREDEL R.
European Southern Observatory,
Casilla 19001
Santiago 19
CHILI

GUELIN M.
IRAM
300, rue de la Piscine
Domaine universitaire
F-38406 ST MARTIN D'HERES Cedex

GUILAIN C. Mlle
CESR/CNRS
9, av. du Colonel Roche
BP 4346
F-31029 TOULOUSE Cedex

GUILLOIS O.
SPAM/Bât 522
CE SACLAY
F-91191 GIF SUR YVETTE Cedex

List of Participants

HANUS M. Mme
ENS/Radioastronomie
24, rue Lhomond
F-75005 PARIS

HELLNER L. Mme
Photophysique moléculaire
Univ. Paris Sud
Bât 213
F-91405 ORSAY Cedex

HENINGER M.
Physico chimie des rayonnements
Univ. Paris Sud/Bât. 350
F-91405 ORSAY

HERMINE P.
Photophysique moléculaire
Univ. Paris Sud
Bât. 213
F-91405 ORSAY Cedex

HONVAULT P.
LDMA
Tour 12 /5ème étage
4, place Jussieu
F-75252 PARIS Cedex 05

JARROLD M.F.
Depart. of Chemistry
Northwestern Univ.
2145 Sheridan Road
EVANSTON,IL 60208
U.S.A.

JOBLIN C. Mlle
NASA AMES RESEARCH CENTER
MS-245-6
MOFFETT Field-CA 94035-1000
U.S.A.

KLOTZ A.
C.E.S.R.
9, av. du Colonel Roche
F- 31029 TOULOUSE Cedex

LAVENDY H.
Dynamique moléculaire et photonique
Bât. P5
F-59655 VILLENEUVE D'ASCQ Cedex

LEACH S.
DAMAP
Observatoire de Paris Meudon
5, place J. Janssen
F-92195 MEUDON

LEE WEI
Univ. of Alabama
Depart; of Physics
310, Campbell Hall
BIRMINGHAM/AL 35294-1170
U.S.A.

LEGER A.
Inst. d'astrophysique spatiale
Univ. Paris XI-Bât. 121
F-91405 ORSAY Cedex

LELEYTER M. Mme
Groupe de simulation numérique
en physique des agrégats
Faculté des sciences
33, rue St Leu
F-80039 AMIENS Cedex

LEQUEUX J.
DEMIRM
Observatoire de Paris-Meudon
5, place J. Janssen
F-92195 MEUDON Cedex

LEMAIRE J.L.
DAMAP
Observatoire de Paris-Meudon
5, av. J. Janssen
F-92195 MEUDON Cedex

MALMASSON D.
Observatoire de Paris Meudon
DAMAP-EPSA
5, place J. Janssen
F-92195 MEUDON Cedex

Mc CARROLL R.
Dynamique moléculaire
et atomique/UPMC
Tour 12/Boite 75
4, place Jussieu
F-75252 PARIS Cedex 05

MARTY Ph.
C.E.S.R.
9, av. du Colonel Roche
BP 4346
F-31029 TOULOUSE Cedex

MAURON N.
C.E.S.R.
9, av. du Colonel Roche
BP 4346
F-31029 TOULOUSE Cedex

List of Participants

MENNELLA V.
Observatorio astronomico
di Capodimonte
via Moiariello 16
I-80131 NAPOLI

MENZEL D.
Physik-dept. E. 20
Techn. Univ. Munchen
D-85748 GARCHING

MILLAR T.
UMIST
Depart. of Mathematics
PO Box 88
U.K.-MANCHESTER M60 1QD

MOUTOU C. Mlle
Inst. d'astrophysique spatiale
Bât. 121
Univ. Paris XI
F-91405 ORSAY Cedex

MOREELS G.
Observatoire de Besançon
BP 1615
25010 BESANCON Cedex

NENNER I. Mme
LURE/ Bât. 209 D
Univ. Paris Sud
91405 ORSAY Cedex

OREL A. E.Mme
Livermore National Lab.
Depart. of Applied Science
PO BOX 808 L.794
LIVERMORE CA 94550
U.S.A.

OROFINO V.
Univ. Degli Studi di Lecce
Dipartimento di Fisica
Via Arnesano
I-73100 LECCE

PARISEL O.
Astrochimie quantique
ENS/ Radioastronomie
24, rue Lhomond
F-75005 PARIS

PASCOLI G.
GSINPA
Depart. de Physique
33, rue Saint Leu
F-80039 AMIENS Cedex

PAUZAT F.Mme
ENS/Radioastronomie
ENS
24, rue Lhomond
F-75005 PARIS

PHILIPPE L.
Photophysique moléculaire
Univ. Paris Sud
Bât 213
F-91405 ORSAY

PUGET J.L
Inst. d'astrophysique Spatiale
Univ. Paris XI
Bât. 121
F-91405 ORSAY Cedex

PINEAU DES FORETS G.
DAEC/Observatoire de Paris-Meudon
5, place J. Janssen
F-92195 MEUDON cedex

QUEFFELEC J.L.
Depart. Physique atomique
Univ. Rennes I
Campus de Beaulieu
F-35042 RENNES Cedex

RAMILLON M.
Univ. P. et M. Curie
LDMA/Tour 12/Boite 75
4, place Jussieu
F-75252 PARIS Cedex 05

RAYEZ M.T. Mme
Physicochimie théorique
Univ. Bordeaux I
33405 TALENCE Cedex

REYNAUD C. Mme
SPAM-Bât. 522
CE SACLAY
F-91191 GIF SUR YVETTE Cedex

RIPOLL J.L.
ISMRA-Composés thio-organiques
6, Bd Marechal Juin
F-14050 CAEN

RIST C. Mme
Observatoire de Grenoble
Lab. d'astrophysique
BP 53 X
F-38041 GRENOBLE

List of Participants

RISTORCELLI I.
C.E.S.R.
9, av. du Colonel Roche
BP4346
F-31029 TOULOUSE Cedex

ROBINSON M.
Depart. of Physics
Univ. of Alabama at Birmingham
310, Campbell Hall, University Blvd
BIRMINGHAM, Alabama 35294-1170
U.S.A.

ROSTAS F.
DAMAP
Observatoire de Paris-Meudon
5, place J. Janssen
F-92195 MEUDON Cedex

ROUEFF E. Mme
Observatoire de Meudon
DAEC
5, place J. Janssen
F-92195 MEUDON

ROWE B.
DPAM/URA 1203 CNRS
Univ. Rennes I
Campus de Beaulieu
F-35042 RENNES Cedex

SABLIER M.
Lab. de chimie
24, rue Lhomond
F-75231 PARIS Cedex

SARRE P.J.
Depart. of Chemistry
Univ. of Nottingham
Univ. Park
NOTTINGHAM NE7 2RD-U.K.

SCAPPINI F.
Spettroscopia Molecolare, C.N.R.
Via de Castagnoli, 1
I-40126 BOLOGNE

SHALABIEA O.M.
Leiden Observatory
Lab. astrophysics
PO Box 9513
NL-2300 RA Leiden

SCHERMANN C.
CNRS/Univ. P. et M. Curie
Lab. dynamique moléculaire et atomique
Tour 12-E5 B 75
4, place Jussieu
B-75252 PARIS Cedex 05

SCHINKE R.
MPI fur Stromungsforschung
Bunsenstrasse 10
D-3400 GOTTINGEN

SCHREEL K.
Molecular and Laserphysics
Fac. of Science
Univ. of Nijmegen
Toernooiveld 1
NL 6525ED NIJMEGEN

SCHMITT B.
LGGE-CNRS
54, rue Molière
BP 96
F-38402 ST MARTIN D'HERES Cedex

SCHUTTE W.
Leiden Observatory Lab.
Sterrewacht Leiden
Postbus 9513
NL-2300 R.A. LEIDEN

SERRA G.
CESR-CNRS
9, av. du colonel Roche
BP 4346
31029 TOULOUSE Cedex

SORGENFREI A. Mlle
Fakultat fur Physik
Universitat Freiburg
Hermann-Herder-Str. 3
D-7800 FREIBURG

SPIELFIEDEL A. Mlle
DAMAP
Observatoire de Meudon
5, place J. Janssen
92195 MEUDON Cedex

TCHANG-BRILLET W.U.L.
DAMAP
Observatoire de Meudon
F-92195 MEUDON

THADDEUS P.
Harvard center for astrophysics
60, Garden street
CAMBRIDGE MA 02138
U.S.A.

TALBI D.
ENS/Radioastronomie
24, rue Lhomond
F-75005 PARIS

TAYLOR S.
UMIST
Depart. of Mathematics
PO BOX 88
U.K.-MANCHESTER M60 1QD

TROTTA F.
LGGE-CNRS
BP 96
F-38402 ST MARTIN D'HERES

TROYANOWSKY C.
SFC/DIVISION DE CHIMIE PHYSIQUE
Lab. de chimie physique
11, rue P. et M. Curie
F-75005 PARIS

VALIRON P.
Observatoire de Grenoble
Lab. d'astrophysique
B.P. 53 X
F-38041 GRENOBLE

VAN DEN HOEK B.
Sterrenkundig Inst. Anton Pannekoek
Kruislaan 403
NL-1098 S J AMSTERDAM

VANYSEK V.
Astronomical Inst.
Charles University
Svedska 8
150000 PRAHA 5
CZECH REP.

VERVLOET M.
Photophysique moléculaire
Univ. Paris Sud
Bât. 213
F-91405 ORSAY Cedex

VERSTRAETE L.
CEA/PTN
B.P.12
F-91680 BRUYERES LE CHATEL

VETTER R.
Arbeitsgruppe physikalisch-chemische
dynamik im WIP
Rudower Chaussee 5,
Geb 2.14
D-12484 BERLIN

VIENT A.
DAMAP
Observatoire de Meudon
5, place J. Janssen
F-92195 MEUDON Cedex

WALMSLEY C.M.
M.P.I. fur Radioastronomie
Auf dem Hugel 69
53 BONN
R.F.A.

WARIN S. Mlle
Lab. d'astrophysique
Observatoire de Grenoble
414, rue de la piscine
BP 53X
F-38041 GRENOBLE Cedex

WDOWIAK T.
Depart. of Physics
Univ. of Alabama at Birmingham
310, Campbell Hall, University Blvd
BIRMINGHAM, Alabama 35294-1170
U.S.A.

WILLIAMS D.A.
UMIST
Depart. of Mathematics
P.O. BOX 88
MANCHESTER M60 1 QD-U.K.

WLODARCZAK G.
Spectroscopie Hertzienne
Univ. de Lille 1
Bât. P5
F-59655 VILLENEUVE D'ASCQ Cedex

WOMACK M. Mme.
Depart. of Physics and Astronomy
Northern Arizona Univ.
FLAGSTAFF, AZ 86011-6010
U.S.A

List of Participants

ZAVAGNO A. Mme
Observatoire de Marseille
2, place Leverrier
F-13248 MARSEILLE Cedex 4

ZIURYS L.M.Mme
Depart. of Chemistry
Arizona State Univ.
TEMPE, AZ 85287-1604
U.S.A

Author Index

A

Abouelaziz, H., 537
Aiello, S., 149
Allen, M. D., 311
Anderson, M. A., 297, 311
Apponi, A. J., 305, 311
Arpigny, C., 205

B

Baker, J., 355
Baluteau, J. P., 59, 195
Barlow, M. J., 437
Barsella, B., 149
Barthelat, J.-C., 705
Baumgärtel, H., 659
Beegle, L. W., 687
Begemann, B., 781
Benayoun, J. J., 387
Bergman, P., 39
Billy, N., 529
Blanco, A., 795
Bocherel, P., 445
Bockelée-Morvan, D., 107
Bogey, M., 269
Boissel, P., 667, 699
Bordas, C., 373
Botschwina, P., 321
Boutou, V., 373
Brechignac, P., 261, 613
Bruni, G., 39
Burie, J., 289
Bussoletti, E., 789

C

Čadež, I., 801
Canosa, A., 537
Capron, L., 529
Casey, S., 53
Cassam-Chenaï, P., 543
Caubet, P., 429
Cecchi-Pestellini, C., 149
Chambaud, G., 337
Charnley, S. B., 155
Chaudret, B., 183, 705
Chevaleyre, J., 373
Clairemidi, J., 255, 261
Clary, D. C., 405
Colangeli, L., 789
Comeau, M., 595, 605
Comtet, G., 773
Cordonnier, M., 269
Cossart, D., 367
Costes, M., 423
Couris, S., 355
Cox, P., 53, 59
Crovisier, J., 107, 393

D

Dalgarno, A., 841
Daudey, J.-P., 705
Daugey, N., 429
deCaro, D., 705
Défourneau, D., 629
Demaison, J., 289
Demuynck, C., 269
Denis, J.-M., 293
Destombes, J.-L., 269
D'Heeger, A., 199
d'Hendecourt, L., 629
d'Incan, J., 373
Dorschner, J., 781
Dorthe, G., 423, 429
Drew, J. E., 437
Drira, I., 337
Duflot, D., 343
Dujardin, G., 773

E

Edwards, S. A., 589
Ellinger, Y., 515, 543, 635, 649
Ellis, K., 811
Encrenaz, T., 117

Erba, B., 373
Evans, N. J., 189

F

Farquhar, P. R. A., 135
Feautrier, N., 337
Fenistein, S., 381
Field, D., 29
Fillion, J. H., 395
Flament, J. P., 343
Flower, D. R., 167, 477
Flügge, J., 321
Fonti, S., 795
Forveille, T., 53
Fye, J. L., 571

G

Gargaud, M., 519
Gauyacq, D., 395
Geballe, T. R., 73
Gerakines, P. A., 73
Gerin, M., 29, 33
Gerlich, D., 489, 505
Ghanem, N., 423
Giard, M., 63, 199, 705
Glaccum, W. J., 53
Goidet-Devel, B., 255
Gomet, J. C., 537
Gouédard, G., 529
Gough, S. F., 801
Gredel, R., 47
Greenberg, J. M., 73, 767
Guilain, Ch., 99
Guillois, O., 811, 817, 823

H

Hall, R. I., 801
Halvick, P., 429
Hanus, M., 515
Hellner, L., 773
Heninger, M., 381
Henning, Th., 781

Herbst, E., 135, 141
Hermine, P., 261, 613
Hibbins, R. E., 87
Hirayama, T., 773
Höper, U., 321
Horani, M., 367
Horn, M., 321
Howe, D. A., 141
Hunter, J. M., 571

J

Jarrold, M. F., 571
Joblin, C., 629
Jochims, H. W., 659
Julienne, P. S., 361
Jullien, S., 381

K

Klotz, A., 705
Kopp, M., 33
Kulander, K. C., 549

L

Lamarre, J. M., 63
Landau, M., 801
Lavendy, H., 343
Leach, S., 29, 589, 659
Le Bourlot, J., 161, 167
Leclercq, J., 595, 605
Lee, H. H., 141
Lee, W., 675, 687, 693
Lefèvre, G., 667
Le Floch, A., 349
Léger, A., 629
Le Guennec, M., 289
Leleyter, M., 595, 605
Lemaire, J., 381
Lemaire, J. L., 29, 349, 355
Le Naour, C., 63
Lengsfield, B. H., III, 549
Léotin, J., 63
Lequeux, J., 831

Levy, B., 429
Lukac, P., 537

M

Malmasson, D., 349, 355
Marty, P., 183, 705
Marx, R., 381
Mauclaire, G., 381
Maurelli, J., 373
Mauron, N., 99
McCarroll, R., 519
Mennella, V., 789
Mény, C., 63
Merluzzi, P., 789
Mestdagh, H., 529
Miles, J. R., 87
Millar, T. J., 135, 141
Millié, P., 429
Mladenowić, M., 321
Monaco, G., 789
Moravec, Z., 239
Moreels, G., 255, 261
Moseley, H., 53
Muci, A. M., 795
Mutschke, H., 781

N

Naulin, C., 423
Nedelec, L., 537
Nenner, I., 811, 817, 823

O

Offer, A., 477
Omont, A., 53
Orel, A. E., 549
Orofino, V., 795
Oswald, M., 321
Oswald, R., 321

P

Pajot, F., 63
Palumbo, G. G. C., 39
Palumbo, P., 789
Papoular, R., 811, 817, 823
Parisel, O., 515, 649
Pascoli, G., 595, 605
Pasquerault, D., 537
Pauzat, F., 543, 635
Philippe, L., 773
Pichou, F., 801
Pineau des Forêts, G., 29, 161, 167

R

Ramage, M. J., 773
Ramillion, M., 519
Rawlings, J. M. C., 189, 437
Rayez, J. C., 429
Rayez, M. T., 429
Rebrion, C., 537
Reynaud, C., 811, 817, 823
Ripoll, J.-L., 293
Ristorcelli, I., 63, 183, 705
Robbe, J. M., 343
Rolando, C., 529
Rose, M., 773
Roskamp, E. J., 571
Rosmus, P., 337
Rostas, F., 29, 349, 355
Rotundi, A., 789
Rouan, D., 29
Roueff, E., 33, 161, 167
Rousselot, P., 255, 261
Rowe, B. R., 445, 537
Rühl, E., 659

S

Sablier, M., 529
Sarre, P. J., 87
Scappini, F., 39
Schermann, C., 801
Schick, E., 321
Schinke, R., 557

Schmitt, B., 735, 759
Schutte, W. A., 73
Seeger, S., 321
Serra, G., 63, 183, 705
Shafizadeh, N., 395
Shalabiea, O. M., 767
Simons, D., 29
Sims, I. R., 445
Smith, I. W. M., 445
Sorgenfrei, A., 505
Spielfiedel, A., 337
Steimle, T. C., 297

T

Talbi, D., 635
Taylor, S. D., 81
Tchang-Brillet, W.-Ü L., 361
Thaddeus, P., 711
Thiébot, Ph., 667
Tobita, S., 659
Trotta, F., 759

V

Valente, A. M., 373
van den Hoek, B., 173

van Dishoeck, E. F., 73
Vanysek, V., 239, 247
Vervloet, M., 367
Viala, Y. P., 337, 387
Vient, A., 349, 355
Vigué, J., 529

W

Walmsley, C. M., 463
Walters, A., 269
Warin, S., 387
Wdowiak, T. J., 675, 687, 693
Williams, D. A., 3
Wlodarczak, G., 289
Womack, M., 305

Y

Yoder, J. T., 305

Z

Zavagno, A., 59, 195
Zhou, S., 189
Ziurys, L. M., 297, 305, 311

AIP Conference Proceedings

		L.C. Number	ISBN
No. 170	Nuclear Spectroscopy of Astrophysical Sources (Washington, DC, 1987)	88-71625	0-88318-370-6
No. 171	Vacuum Design of Advanced and Compact Synchrotron Light Sources (Upton, NY, 1988)	88-71824	0-88318-371-4
No. 172	Advances in Laser Science—III: Proceedings of the International Laser Science Conference (Atlantic City, NJ, 1987)	88-71879	0-88318-372-2
No. 173	Cooperative Networks in Physics Education (Oaxtepec, Mexico, 1987)	88-72091	0-88318-373-0
No. 174	Radio Wave Scattering in the Interstellar Medium (San Diego, CA, 1988)	88-72092	0-88318-374-9
No. 175	Non-neutral Plasma Physics (Washington, DC, 1988)	88-72275	0-88318-375-7
No. 176	Intersections Between Particle and Nuclear Physics (Third International Conference) (Rockport, ME, 1988)	88-62535	0-88318-376-5
No. 177	Linear Accelerator and Beam Optics Codes (La Jolla, CA, 1988)	88-46074	0-88318-377-3
No. 178	Nuclear Arms Technologies in the 1990s (Washington, DC, 1988)	88-83262	0-88318-378-1
No. 179	The Michelson Era in American Science: 1870–1930 (Cleveland, OH, 1987)	88-83369	0-88318-379-X
No. 180	Frontiers in Science: International Symposium (Urbana, IL, 1987)	88-83526	0-88318-380-3
No. 181	Muon-Catalyzed Fusion (Sanibel Island, FL, 1988)	88-83636	0-88318-381-1
No. 182	High T_c Superconducting Thin Films, Devices, and Applications (Atlanta, GA, 1988)	88-03947	0-88318-382-X
No. 183	Cosmic Abundances of Matter (Minneapolis, MN, 1988)	89-80147	0-88318-383-8
No. 184	Physics of Particle Accelerators (Ithaca, NY, 1988)	89-83575	0-88318-384-6
No. 185	Glueballs, Hybrids, and Exotic Hadrons (Upton, NY, 1988)	89-83513	0-88318-385-4
No. 186	High-Energy Radiation Background in Space (Sanibel Island, FL, 1987)	89-83833	0-88318-386-2
No. 187	High-Energy Spin Physics (Minneapolis, MN, 1988)	89-83948	0-88318-387-0

No. 188	International Symposium on Electron Beam Ion Sources and their Applications (Upton, NY, 1988)	89-84343	0-88318-388-9
No. 189	Relativistic, Quantum Electrodynamic, and Weak Interaction Effects in Atoms (Santa Barbara, CA, 1988)	89-84431	0-88318-389-7
No. 190	Radio-frequency Power in Plasmas (Irvine, CA, 1989)	89-45805	0-88318-397-8
No. 191	Advances in Laser Science—IV (Atlanta, GA, 1988)	89-85595	0-88318-391-9
No. 192	Vacuum Mechatronics (First International Workshop) (Santa Barbara, CA, 1989)	89-45905	0-88318-394-3
No. 193	Advanced Accelerator Concepts (Lake Arrowhead, CA, 1989)	89-45914	0-88318-393-5
No. 194	Quantum Fluids and Solids—1989 (Gainesville, FL, 1989)	89-81079	0-88318-395-1
No. 195	Dense Z-Pinches (Laguna Beach, CA, 1989)	89-46212	0-88318-396-X
No. 196	Heavy Quark Physics (Ithaca, NY, 1989)	89-81583	0-88318-644-6
No. 197	Drops and Bubbles (Monterey, CA, 1988)	89-46360	0-88318-392-7
No. 198	Astrophysics in Antarctica (Newark, DE, 1989)	89-46421	0-88318-398-6
No. 199	Surface Conditioning of Vacuum Systems (Los Angeles, CA, 1989)	89-82542	0-88318-756-6
No. 200	High T_c Superconducting Thin Films: Processing, Characterization, and Applications (Boston, MA, 1989)	90-80006	0-88318-759-0
No. 201	QED Structure Functions (Ann Arbor, MI, 1989)	90-80229	0-88318-671-3
No. 202	NASA Workshop on Physics From a Lunar Base (Stanford, CA, 1989)	90-55073	0-88318-646-2
No. 203	Particle Astrophysics: The NASA Cosmic Ray Program for the 1990s and Beyond (Greenbelt, MD, 1989)	90-55077	0-88318-763-9
No. 204	Aspects of Electron-Molecule Scattering and Photoionization (New Haven, CT, 1989)	90-55175	0-88318-764-7
No. 205	The Physics of Electronic and Atomic Collisions (XVI International Conference) (New York, NY, 1989)	90-53183	0-88318-390-0
No. 206	Atomic Processes in Plasmas (Gaithersburg, MD, 1989)	90-55265	0-88318-769-8
No. 207	Astrophysics from the Moon (Annapolis, MD, 1990)	90-55582	0-88318-770-1

No. 208	Current Topics in Shock Waves (Bethlehem, PA, 1989)	90-55617	0-88318-776-0
No. 209	Computing for High Luminosity and High Intensity Facilities (Santa Fe, NM, 1990)	90-55634	0-88318-786-8
No. 210	Production and Neutralization of Negative Ions and Beams (Brookhaven, NY, 1990)	90-55316	0-88318-786-8
No. 211	High-Energy Astrophysics in the 21st Century (Taos, NM, 1989)	90-55644	0-88318-803-1
No. 212	Accelerator Instrumentation (Brookhaven, NY, 1989)	90-55838	0-88318-645-4
No. 213	Frontiers in Condensed Matter Theory (New York, NY, 1989)	90-6421	0-88318-771-X 0-88318-772-8 (pbk.)
No. 214	Beam Dynamics Issues of High-Luminosity Asymmetric Collider Rings (Berkeley, CA, 1990)	90-55857	0-88318-767-1
No. 215	X-Ray and Inner-Shell Processes (Knoxville, TN, 1990)	90-84700	0-88318-790-6
No. 216	Spectral Line Shapes, Vol. 6 (Austin, TX, 1990)	90-06278	0-88318-791-4
No. 217	Space Nuclear Power Systems (Albuquerque, NM, 1991)	90-56220	0-88318-838-4
No. 218	Positron Beams for Solids and Surfaces (London, Canada, 1990)	90-56407	0-88318-842-2
No. 219	Superconductivity and Its Applications (Buffalo, NY, 1990)	91-55020	0-88318-835-X
No. 220	High Energy Gamma-Ray Astronomy (Ann Arbor, MI, 1990)	91-70876	0-88318-812-0
No. 221	Particle Production Near Threshold (Nashville, IN, 1990)	91-55134	0-88318-829-5
No. 222	After the First Three Minutes (College Park, MD, 1990)	91-55214	0-88318-828-7
No. 223	Polarized Collider Workshop (University Park, PA, 1990)	91-71303	0-88318-826-0
No. 224	LAMPF Workshop on (π, K) Physics (Los Alamos, NM, 1990)	91-71304	0-88318-825-2
No. 225	Half Collision Resonance Phenomena in Molecules (Caracas, Venezuela, 1990)	91-55210	0-88318-840-6
No. 226	The Living Cell in Four Dimensions (Gif sur Yvette, France, 1990)	91-55209	0-88318-794-9
No. 227	Advanced Processing and Characterization Technologies (Clearwater, FL, 1991)	91-55194	0-88318-910-0

No. 228	Anomalous Nuclear Effects in Deuterium/Solid Systems (Provo, UT, 1990)	91-55245	0-88318-833-3
No. 229	Accelerator Instrumentation (Batavia, IL, 1990)	91-55347	0-88318-832-1
No. 230	Nonlinear Dynamics and Particle Acceleration (Tsukuba, Japan, 1990)	91-55348	0-88318-824-4
No. 231	Boron-Rich Solids (Albuquerque, NM, 1990)	91-53024	0-88318-793-4
No. 232	Gamma-Ray Line Astrophysics (Paris-Saclay, France, 1990)	91-55492	0-88318-875-9
No. 233	Atomic Physics 12 (Ann Arbor, MI, 1990)	91-55595	088318-811-2
No. 234	Amorphous Silicon Materials and Solar Cells (Denver, CO, 1991)	91-55575	088318-831-7
No. 235	Physics and Chemistry of MCT and Novel IR Detector Materials (San Francisco, CA, 1990)	91-55493	0-88318-931-3
No. 236	Vacuum Design of Synchrotron Light Sources (Argonne, IL, 1990)	91-55527	0-88318-873-2
No. 237	Kent M. Terwilliger Memorial Symposium (Ann Arbor, MI, 1989)	91-55576	0-88318-788-4
No. 238	Capture Gamma-Ray Spectroscopy (Pacific Grove, CA, 1990)	91-57923	0-88318-830-9
No. 239	Advances in Biomolecular Simulations (Obernai, France, 1991)	91-58106	0-88318-940-2
No. 240	Joint Soviet-American Workshop on the Physics of Semiconductor Lasers (Leningrad, USSR, 1991)	91-58537	0-88318-936-4
No. 241	Scanned Probe Microscopy (Santa Barbara, CA, 1991)	91-76758	0-88318-816-3
No. 242	Strong, Weak, and Electromagnetic Interactions in Nuclei, Atoms, and Astrophysics: A Workshop in Honor of Stewart D. Bloom's Retirement (Livermore, CA, 1991)	91-76876	0-88318-943-7
No. 243	Intersections Between Particle and Nuclear Physics (Tucson, AZ, 1991)	91-77580	0-88318-950-X
No. 244	Radio Frequency Power in Plasmas (Charleston, SC, 1991)	91-77853	0-88318-937-2
No. 245	Basic Space Science (Bangalore, India, 1991)	91-78379	0-88318-951-8

No. 246	Space Nuclear Power Systems (Albuquerque, NM, 1992)	91-58793	1-56396-027-3 1-56396-026-5 (pbk.)
No. 247	Global Warming: Physics and Facts (Washington, DC, 1991)	91-78423	0-88318-932-1
No. 248	Computer-Aided Statistical Physics (Taipei, Taiwan, 1991)	91-78378	0-88318-942-9
No. 249	The Physics of Particle Accelerators (Upton, NY, 1989, 1990)	92-52843	0-88318-789-2
No. 250	Towards a Unified Picture of Nuclear Dynamics (Nikko, Japan, 1991)	92-70143	0-88318-951-8
No. 251	Superconductivity and its Applications (Buffalo, NY, 1991)	92-52726	1-56396-016-8
No. 252	Accelerator Instrumentation (Newport News, VA, 1991)	92-70356	0-88318-934-8
No. 253	High-Brightness Beams for Advanced Accelerator Applications (College Park, MD, 1991)	92-52705	0-88318-947-X
No. 254	Testing the AGN Paradigm (College Park, MD, 1991)	92-52780	1-56396-009-5
No. 255	Advanced Beam Dynamics Workshop on Effects of Errors in Accelerators, Their Diagnosis and Corrections (Corpus Christi, TX, 1991)	92-52842	1-56396-006-0
No. 256	Slow Dynamics in Condensed Matter (Fukuoka, Japan, 1991)	92-53120	0-88318-938-0
No. 257	Atomic Processes in Plasmas (Portland, ME, 1991)	91-08105	0-88318-939-9
No. 258	Synchrotron Radiation and Dynamic Phenomena (Grenoble, France, 1991)	92-53790	1-56396-008-7
No. 259	Future Directions in Nuclear Physics with 4π Gamma Detection Systems of the New Generation (Strasbourg, France, 1991)	92-53222	0-88318-952-6
No. 260	Computational Quantum Physics (Nashville, TN, 1991)	92-71777	0-88318-933-X
No. 261	Rare and Exclusive B&K Decays and Novel Flavor Factories (Santa Monica, CA, 1991)	92-71873	1-56396-055-9
No. 262	Molecular Electronics—Science and Technology (St. Thomas, Virgin Islands, 1991)	92-72210	1-56396-041-9
No. 263	Stress-Induced Phenomena in Metallization: First International Workshop (Ithaca, NY, 1991)	92-72292	1-56396-082-6

No. 264	Particle Acceleration in Cosmic Plasmas (Newark, DE, 1991)	92-73316	0-88318-948-8
No. 265	Gamma-Ray Bursts (Huntsville, AL, 1991)	92-73456	1-56396-018-4
No. 266	Group Theory in Physics (Cocoyoc, Morelos, Mexico, 1991)	92-73457	1-56396-101-6
No. 267	Electromechanical Coupling of the Solar Atmosphere (Capri, Italy, 1991)	92-82717	1-56396-110-5
No. 268	Photovoltaic Advanced Research & Development Project (Denver, CO, 1992)	92-74159	1-56396-056-7
No. 269	CEBAF 1992 Summer Workshop (Newport News, VA, 1992)	92-75403	1-56396-067-2
No. 270	Time Reversal—The Arthur Rich Memorial Symposium (Ann Arbor, MI, 1991)	92-83852	1-56396-105-9
No. 271	Tenth Symposium Space Nuclear Power and Propulsion (Vols. I–III) (Albuquerque, NM, 1993)	92-75162	1-56396-137-7 (set)
No. 272	Proceedings of the XXVI International Conference on High Energy Physics (Vols. I and II) (Dallas, TX, 1992)	93-70412	1-56396-127-X (set)
No. 273	Superconductivity and Its Applications (Buffalo, NY, 1992)	93-70502	1-56396-189-X
No. 274	VIth International Conference on the Physics of Highly Charged Ions (Manhattan, KS, 1992)	93-70577	1-56396-102-4
No. 275	Atomic Physics 13 (Munich, Germany, 1992)	93-70826	1-56396-057-5
No. 276	Very High Energy Cosmic-Ray Interactions: VIIth International Symposium (Ann Arbor, MI, 1992)	93-71342	1-56396-038-9
No. 277	The World at Risk: Natural Hazards and Climate Change (Cambridge, MA, 1992)	93-71333	1-56396-066-4
No. 278	Back to the Galaxy (College Park, MD, 1992)	93-71543	1-56396-227-6
No. 279	Advanced Accelerator Concepts (Port Jefferson, NY, 1992)	93-71773	1-56396-191-1

No. 280	Compton Gamma-Ray Observatory (St. Louis, MO, 1992)	93-71830	1-56396-104-0
No. 281	Accelerator Instrumentation Fourth Annual Workshop (Berkeley, CA, 1992)	93-072110	1-56396-190-3
No. 282	Quantum 1/f Noise & Other Low Frequency Fluctuations in Electronic Devices (St. Louis, MO, 1992)	93-072366	1-56396-252-7
No. 283	Earth and Space Science Information Systems (Pasadena, CA, 1992)	93-072360	1-56396-094-X
No. 284	US-Japan Workshop on Ion Temperature Gradient-Driven Turbulent Transport (Austin, TX, 1993)	93-72460	1-56396-221-7
No. 285	Noise in Physical Systems and 1/f Fluctuations (St. Louis, MO, 1993)	93-72575	1-56396-270-5
No. 286	Ordering Disorder: Prospect and Retrospect in Condensed Matter Physics: Proceedings of the Indo-U.S. Workshop (Hyderabad, India, 1993)	93-072549	1-56396-255-1
No. 287	Production and Neutralization of Negative Ions and Beams: Sixth International Symposium (Upton, NY, 1992)	93-72821	1-56396-103-2
No. 288	Laser Ablation: Mechanismas and Applications-II: Second International Conference (Knoxville, TN, 1993)	93-73040	1-56396-226-8
No. 289	Radio Frequency Power in Plasmas: Tenth Topical Conference (Boston, MA, 1993)	93-72964	1-56396-264-0
No. 290	Laser Spectroscopy: XIth International Conference (Hot Springs, VA, 1993)	93-73050	1-56396-262-4
No. 291	Prairie View Summer Science Academy (Prairie View, TX, 1992)	93-73081	1-56396-133-4
No. 292	Stability of Particle Motion in Storage Rings (Upton, NY, 1992)	93-73534	1-56396-225-X
No. 293	Polarized Ion Sources and Polarized Gas Targets (Madison, WI, 1993)	93-74102	1-56396-220-9
No. 294	High-Energy Solar Phenomena A New Era of Spacecraft Measurements (Waterville Valley, NH, 1993)	93-74147	1-56396-291-8
No. 295	The Physics of Electronic and Atomic Collisions: XVIII International Conference (Aarhus, Denmark, 1993)	93-74103	1-56396-290-X

No.	Title	Report No.	ISBN
No. 296	The Chaos Paradigm: Developments an Applications in Engineering and Science (Mystic, CT, 1993)	93-74146	1-56396-254-3
No. 297	Computational Accelerator Physics (Los Alamos, NM, 1993)	93-74205	1-56396-222-5
No. 298	Ultrafast Reaction Dynamics and Solvent Effects (Royaumont, France, 1993)	93-074354	1-56396-280-2
No. 299	Dense Z-Pinches: Third International Conference (London, 1993)	93-074569	1-56396-297-7
No. 300	Discovery of Weak Neutral Currents: The Weak Interaction Before and After (Santa Monica, CA, 1993)	94-70515	1-56396-306-X
No. 301	Eleventh Symposium Space Nuclear Power and Propulsion (3 Vols.) (Albuquerque, NM, 1994)	92-75162	1-56396-305-1 (Set) 156396-301-9 (pbk. set)
No. 302	Lepton and Photon Interactions/ XVI International Symposium (Ithaca, NY, 1993)	94-70079	1-56396-106-7
No. 303	Slow Positron Beam Techniques for Solids and Surfaces Fifth International Workshop (Jackson Hole, WY 1992)	94-71036	1-56396-267-5
No. 304	The Second Compton Symposium (College Park, MD, 1993)	94-70742	1-56396-261-6
No. 305	Stress-Induced Phenomena in Metallization Second International Workshop (Austin, TX, 1993)	94-70650	1-56396-251-9
No. 306	12th NREL Photovoltaic Program Review (Denver, CO, 1993)	94-70748	1-56396-315-9
No. 307	Gamma-Ray Bursts Second Workshop (Huntsville, AL 1993)	94-71317	1-56396-336-1
No. 308	The Evolution of X-Ray Binaries (College Park, MD 1993)	94-76853	1-56396-329-9
No. 309	High-Pressure Science and Technology—1993 (Colorado Springs, CO 1993)	93-72821	1-56396-219-5 (Set)
No. 310	Analysis of Interplanetary Dust (Houston, TX 1993)	94-71292	1-56396-341-8